T0140078

Advances in Intelligent Systems and Computing

Volume 835

Series editor

Janusz Kacprzyk, Polish Academy of Sciences, Warsaw, Poland
e-mail: kacprzyk@ibspan.waw.pl

The series "Advances in Intelligent Systems and Computing" contains publications on theory, applications, and design methods of Intelligent Systems and Intelligent Computing. Virtually all disciplines such as engineering, natural sciences, computer and information science, ICT, economics, business, e-commerce, environment, healthcare, life science are covered. The list of topics spans all the areas of modern intelligent systems and computing such as: computational intelligence, soft computing including neural networks, fuzzy systems, evolutionary computing and the fusion of these paradigms, social intelligence, ambient intelligence, computational neuroscience, artificial life, virtual worlds and society, cognitive science and systems, Perception and Vision, DNA and immune based systems, self-organizing and adaptive systems, e-Learning and teaching, human-centered and human-centric computing, recommender systems, intelligent control, robotics and mechatronics including human-machine teaming, knowledge-based paradigms, learning paradigms, machine ethics, intelligent data analysis, knowledge management, intelligent agents, intelligent decision making and support, intelligent network security, trust management, interactive entertainment, Web intelligence and multimedia.

The publications within "Advances in Intelligent Systems and Computing" are primarily proceedings of important conferences, symposia and congresses. They cover significant recent developments in the field, both of a foundational and applicable character. An important characteristic feature of the series is the short publication time and world-wide distribution. This permits a rapid and broad dissemination of research results.

Advisory Board

Chairman

Nikhil R. Pal, Indian Statistical Institute, Kolkata, India
e-mail: nikhil@isical.ac.in

Members

Rafael Bello Perez, Universidad Central "Marta Abreu" de Las Villas, Santa Clara, Cuba
e-mail: rbellop@uclv.edu.cu

Emilio S. Corchado, University of Salamanca, Salamanca, Spain
e-mail: escorchado@usal.es

Hani Hagras, University of Essex, Colchester, UK
e-mail: hani@essex.ac.uk

László T. Kóczy, Széchenyi István University, Győr, Hungary
e-mail: koczy@sze.hu

Vladik Kreinovich, University of Texas at El Paso, El Paso, USA
e-mail: vladik@utep.edu

Chin-Teng Lin, National Chiao Tung University, Hsinchu, Taiwan
e-mail: ctlin@mail.nctu.edu.tw

Jie Lu, University of Technology, Sydney, Australia
e-mail: Jie.Lu@uts.edu.au

Patricia Melin, Tijuana Institute of Technology, Tijuana, Mexico
e-mail: epmelin@hafsamx.org

Nadia Nedjah, State University of Rio de Janeiro, Rio de Janeiro, Brazil
e-mail: nadia@eng.uerj.br

Ngoc Thanh Nguyen, Wroclaw University of Technology, Wroclaw, Poland
e-mail: Ngoc-Thanh.Nguyen@pwr.edu.pl

Jun Wang, The Chinese University of Hong Kong, Shatin, Hong Kong
e-mail: jwang@mae.cuhk.edu.hk

More information about this series at http://www.springer.com/series/11156

Anna Burduk · Edward Chlebus
Tomasz Nowakowski · Agnieszka Tubis
Editors

Intelligent Systems in Production Engineering and Maintenance

Editors
Anna Burduk
Faculty of Mechanical Engineering
Wrocław University of Science and Technology
Wrocław, Poland

Tomasz Nowakowski
Faculty of Mechanical Engineering
Wrocław University of Science and Technology
Wrocław, Poland

Edward Chlebus
Faculty of Mechanical Engineering
Wrocław University of Science and Technology
Wrocław, Poland

Agnieszka Tubis
Faculty of Mechanical Engineering
Wrocław University of Science and Technology
Wrocław, Poland

ISSN 2194-5357 ISSN 2194-5365 (electronic)
Advances in Intelligent Systems and Computing
ISBN 978-3-319-97489-7 ISBN 978-3-319-97490-3 (eBook)
https://doi.org/10.1007/978-3-319-97490-3

Library of Congress Control Number: 2018950093

© Springer Nature Switzerland AG 2019
This work is subject to copyright. All rights are reserved by the Publisher, whether the whole or part of the material is concerned, specifically the rights of translation, reprinting, reuse of illustrations, recitation, broadcasting, reproduction on microfilms or in any other physical way, and transmission or information storage and retrieval, electronic adaptation, computer software, or by similar or dissimilar methodology now known or hereafter developed.
The use of general descriptive names, registered names, trademarks, service marks, etc. in this publication does not imply, even in the absence of a specific statement, that such names are exempt from the relevant protective laws and regulations and therefore free for general use.
The publisher, the authors, and the editors are safe to assume that the advice and information in this book are believed to be true and accurate at the date of publication. Neither the publisher nor the authors or the editors give a warranty, express or implied, with respect to the material contained herein or for any errors or omissions that may have been made. The publisher remains neutral with regard to jurisdictional claims in published maps and institutional affiliations.

This Springer imprint is published by the registered company Springer Nature Switzerland AG
The registered company address is: Gewerbestrasse 11, 6330 Cham, Switzerland

Preface

The Second International Conference on Intelligent Systems in Production Engineering and Maintenance ISPEM 2018 was organized by the Faculty of Mechanical Engineering, Wrocław University of Science and Technology and the Committee on Production Engineering of the Polish Academy of Sciences. The high prestige of the conference is confirmed by the honorary patronage of the Rector of the Wroclaw University of Science and Technology. The conference was held in Wrocław (Poland) in September 2018.

The ISPEM conference was established as an interdisciplinary forum in the areas of intelligent systems, methods and techniques, and their applications in production systems. The importance of these issues attracted both academic and industrial participants. The exchange of ideas and opinions on research efforts and industry needs related to intelligent systems in industrial applications was very fruitful. The primary concerns of the conference attendees were new solutions for innovative plants, research results, and case studies taking into account advances in production and maintenance from the point of view of Industry 4.0.

Manuscripts submitted for ISPEM 2018 were accepted for publication in the Conference Proceedings on the basis of two independent reviews by members of the Program Committee appointed by the Conference Chairs. The reviewers were selected according to their competences and the principle of avoiding a conflict of interest—a personal relationship between a reviewer and an author, occupational subordination, direct scientific cooperation over the last two years prior to the review. Each review was concluded with an explicit recommendation to accept or reject. Minor or major revisions with additional review cycles were also possible—about 40% of the finally accepted manuscripts were subjected to a revision procedure. If a negative review was given, the submission was rejected. This situation happened in the case of 30% of the submitted manuscripts. The names of the referees for publications were not revealed. On receiving the required number of reviews and finishing the evaluation phase, the Conference Chairs passed them to authors. Before the final decision, the authors' replies to reviewers' comments were required. The entire evaluation and acceptance process was carried out electronically with the use of EasyChair Conference Service.

We would like to thank all the authors for presentations and the representatives of both industry and academia for participating in lively discussions. Special thanks go to the members of the Program Committee for the reliable process of reviewing the papers. We are also grateful to the Organizing Committee for the hard work at the conference preparatory stage.

September 2018

Anna Burduk
Edward Chlebus
Tomasz Nowakowski
Agnieszka Tubis

Organization

ISPEM'2018 is organized by the Faculty of Mechanical Engineering, Wrocław University of Science and Technology and Production Engineering Committee of the Polish Academy of Sciences.

Executive Committee

Honorary Chair

Cezary Madryas	Rector of Wrocław University of Science and Technology, Poland

General Chairs

Edward Chlebus	Wrocław University of Science and Technology, Poland
Tomasz Nowakowski	Wrocław University of Science and Technology, Poland

Co-chairs

Anna Burduk	Wrocław University of Science and Technology, Poland
Agnieszka Tubis	Wrocław University of Science and Technology, Poland

Program Committee

Ali Aidy	National Defence University of Malaysia, Malaysia
Akimov Oleg	National Technical University, Ukraine
Antosz Katarzyna	Rzeszow University of Technology, Poland
Awasthi Anjali	Concordia University, Canada
Avila Paulo	Polytechnic Institute of Porto, Portugal
Banaszak Zbigniew	Koszalin University of Technology, Poland
Baraldi Piero	Politecnico di Milano, Italy
Barni Andrea Francesco	University of Applied Sciences of Southern, Switzerland
Basl Josef	University of West Bohemia, Czech Republic
Berenguer Christophe	Grenoble Institute of Technology, France
Bernard Alain	Ecole Centrale de Nantes, France
Bożejko Wojciech	Wrocław University of Science and Technology, Poland
Bracke Stefan	University of Wuppertal, Germany
Bučinskas Vytautas	Vilnius Gediminas Technical University, Lithuania
Burduk Anna	Wrocław University of Science and Technology, Poland
Bureika Gintautas	Vilnius Gediminas Technical University, Lithuania
Buscher Udo	Technical University of Dresden, Germany
Cambal Milos	Slovak University of Technology in Bratislava, Slovakia
Capaldo Guido	University of Naples Federico II, Italy
Cariow Aleksandr	West Pomeranian University of Technology in Szczecin, Poland
Chlebus Edward	Wrocław University of Science and Technology, Poland
Chlebus Tomasz	Wrocław University of Science and Technology, Poland
Cholewa Mariusz	Wroclaw University of Science and Technology, Poland
Chromjaková Felicita	Tomas Bata University in Zlín, Czech Republic
Ćwikła Grzegorz	Silesian University of Technology, Poland
Davidrajuh Reggie	University of Stavanger, Norway
Demichela Micaela	Politecnico de Torino, Italy
Despotis Dimitris	University of Piraeus, Greece
Deuse Jochen	Technical University of Dortmund, Germany
Diakun Jacek	Poznan University of Technology, Poland
Donatelli Gustavo	Federal University of Santa Catarina, Brazil

Dossou Paul-Eric	ICAM University, France
Dostatni Ewa	Poznan University of Technology, Poland
Drevetskyi Volodymyr	National University of Water and Environmental Engineering, Ukraine
Duda Jan	Cracow University of Technology, Poland
Dulina Ľuboslav	University of Zilina, Slovakia
Dybała Bogdan	Wrocław University of Science and Technology, Poland
Edl Milan	University of West Bohemia, Czech Republic
Ferreira Luis	University of Porto, Portugal
Filipenko Oleksandr	Kharkiv National University of Radio Electronics, Ukraine
Fomichov Serhii	National Technical University of Ukraine, Ukraine
Fridgeirsson Thordur Vikingur	Reykjavik University, Iceland
Fumagalli Luca	Politecnico di Milano, Italy
Furch Jan	University of Defence, Czech Republic
Furman Joanna	Silesian University of Technology, Poland
Gawlik Józef	Cracow University of Technology, Poland
Gierulski Wacław	Kielce University of Technology, Poland
Gola Arkadiusz	Lublin University of Technology, Poland
Górski Filip	Poznan University of Technology, Poland
Grabowik Cezary	Silesian University of Technology, Poland
Grajewski Damian	Poznan University of Technology, Poland
Grall Antoine	University of Technology of Troyes, France
Gregor Milan	University of Žilina, Slovakia
Greenwood Allen G.	Mississippi State University, USA
Grozav Sorin	Technical University of Cluj-Napoca, Romania
Grzybowska Katarzyna	Poznan University of Technology, Poland
Gwiazda Aleksander	Silesian University of Technology, Poland
Gyula Mester	Óbuda University, Hungary
Hamrol Adam	Poznan University of Technology, Poland
Jodejko-Pietruczuk Anna	Wrocław University of Science and Technology, Poland
Jurdziak Leszek	Wrocław University of Science and Technology, Poland
Jurko Jozef	Technical University of Kosice, Slovakia
Kacprzyk Janusz	Polish Academy of Sciences, Poland
Kalinowski Krzysztof	Silesian University of Technology, Poland
Kantola Jussi	University of Vaasa, Finland
Kasprzyk Rafał	Military University of Technology, Poland
Kawałek Anna	Czestochowa University of Technology, Poland
Kaźmierczak Jan	Silesian University of Technology, Poland
Kłos Sławomir	University of Zielona Góra, Poland

Knapiński Marcin	Czestochowa University of Technology, Poland
Knosala Ryszard	Opole University of Technology, Poland
Korytkowski Przemysław	West Pomeranian University of Technology in Szczecin, Poland
Krause-Jüttler Grit	Technische Universität Dresden, Germany
Krenczyk Damian	Silesian University of Technology, Poland
Kristal Mark	Volgograd State Technical University, Rosja
Król Robert	Wrocław University of Science and Technology, Poland
Krykavskyi Yevhen	Lviv Technical University, Ukraine
Kuczmaszewski Józef	Lublin University of Technology, Poland
Kuric Ivan	University of Zilina, Słowacja
Lewandowski Jerzy	Lodz University of Technology, Poland
Lis Teresa	Silesian University of Technology, Poland
Loska Andrzej	Silesian University of Technology, Poland
Ładysz Rafał	George Mason University, USA
Łebkowski Piotr	AGH University of Science and Technology, Poland
Majstorovic Vidosav	University of Belgrade, Serbia
Malec Małgorzata	Instytut Techniki Górniczej (KOMAG), Poland
Matuszek Józef	University of Bielsko-Biala, Poland
Michalak Dariusz	Solaris Bus & Coach S.A., Poland
Mikler Jerzy	KTH Royal Institute of Technology, Sweden
Milazzo Maria Francesca	University of Messina, Italy
Milecki Andrzej	Poznan University of Technology, Poland
Modrak Vladimir	Technical University of Košice, Slovakia
Nazarko Joanicjusz	Bialystok University of Technology, Poland
Noel Frederic	Grenoble Institute of Technology, France
Nowacki Krzysztof	Silesian University of Technology, Poland
Nowakowski Tomasz	Wrocław University of Science and Technology, Poland
Pandilov Zoran	Ss. Cyril and Methodius University, Republic of Macedonia
Patalas-Maliszewska Justyna	University of Zielona Góra, Poland
Pasichnyk Vitalii	National Technical University of Ukraine, Ukraine
Pasquier Frederique	ICAM University, France
Pawlewski Paweł	Poznan University of Technology, Poland
Peres Francois	University of Toulouse, France
Perez Pereales David	Universidad Polytechnica de Valencia, Spain
Plinta Dariusz	University of Bielsko-Biala, Poland
Putnik Goran D.	University of Minho, Portugal
Reiner Jacek	Wrocław University of Science and Technology, Poland
Rojek Izabela	Kazimierz Wielki University, Poland

Roumpos Christos	Technical University of Crete, Greece
Saniuk Anna	University of Zielona Góra, Poland
Saniuk Sebastian	University of Zielona Góra, Poland
Santarek Krzysztof	Warsaw University of Technology, Poland
Sęp Jarosław	Rzeszow University of Technology, Poland
Simon Silvio	Brandenburg University of Technology Cottbus-Senftenberg, Germany
Skoczypiec Sebastian	Cracow University of Technology, Poland
Skołud Bożena	Silesian University of Technology, Poland
Stadnicka Dorota	Rzeszow University of Technology, Poland
Świć Antoni	Lublin University of Technology, Poland
Terzi Sergio	Politecnico di Milano, Italy
Tormos Bernardo	Universitat Politècnica de Valencia, Spain
Trebuna Peter	Technical University of Kosice, Slovakia
Tuokko Reijo	Tampere University of Technology, Finland
Turmanidze Raul	Georgian Technical University, Georgia
Ucal Sari Irem	Istanbul Technical University, Turkey
Ungureanu Nicolae	Technical University of Cluj-Napoca, Romania
Valis David	Brno University of Defence, Czech Republic
Werbińska-Wojciechowska Sylwia	Wrocław University of Science and Technology, Poland
Więcek Dorota	University of Bielsko-Biala, Poland
Wirkus Marek	Gdańsk University of Technology, Poland
Wyczółkowski Ryszard	Silesian University of Technology, Poland
Xiao-Guang Yue	Wuhan University, China
Xie Min	City University of Hong Kong, Hong Kong
Zio Enrico	Ecole Centrale Paris LGI-Supelec, France

Organizing Committee

Będza Tomasz	Wrocław University of Science and Technology, Poland
Chlebus Tomasz	Wrocław University of Science and Technology, Poland
Górnicka Dagmara	Wrocław University of Science and Technology, Poland
Kotowska Joanna	Wrocław University of Science and Technology, Poland
Krot Kamil	Wrocław University of Science and Technology, Poland
Krowicki Paweł	Wrocław University of Science and Technology, Poland

Musiał Kamil	Wrocław University of Science and Technology, Poland
Rusińska Małgorzata	Wrocław University of Science and Technology, Poland
Werbińska-Wojciechowska Sylwia	Wrocław University of Science and Technology, Poland
Woźna Anna	Wrocław University of Science and Technology, Poland

Organized by

Contents

Modelling and Simulation of Production Processes

Product Design and Product Manufacturing in Industry 4.0

Intelligent Systems in Production Engineering and Maintenance

3D Geometry Recognition for a PMI-Based Mixed Reality Assistant System in Prototype Construction

Matthias Neges, Stefan Adwernat, Mario Wolf(✉), and Michael Abramovici

Chair for IT in Mechanical Engineering (ITM), Ruhr-Universität Bochum, Universitätsstraße 150, 44780 Bochum, Germany
mario.wolf@itm.rub.de

Abstract. The purpose of prototypes in an industrial mass production context is the assessment and validation of desired product characteristics, relating to functional, geometrical or aesthetical aspects. Therefore, a prototype is an abstracted model with a subset of selected properties. Before single prototype parts are assembled, a quality inspection on receiving is mandatory.

Based on a project conducted with the construction facility of an automotive supplier, the authors propose a concept for the automated quality inspection of received prototype parts. This inspection is performed by using reference points and characteristics, defined in a 3D CAD document. As part of the concept, this paper elaborates a module for the automated object recognition by means of computer vision, utilizing 3D CAD geometry and images as reference objects. The authors implemented and validated the module under laboratory conditions.

Keywords: Object recognition · Visual inspection · Prototyping

1 Introduction

In the context of industrial mass production, e.g. the automotive industry, the term "prototype" describes an abstracted physical or digital model or a mock-up depicting selected properties of the specific product. The purpose of prototypes comprises the assessment and validation of the respective product design in terms of functional, geometrical and aesthetical aspects. The feedback information from testing with prototypes helps to adjust the initial product design at an early stage to minimize the risk of costly design changes after start of mass production. As the properties of the prototype may differ widely from the sellable product the manufacturing processes and the utilized materials may vary as well in order to be cost-efficient and rapidly available [1, 2].

Since the emergence of terms like "digital mockup" and "virtual prototyping" [3] the digitalization of prototype construction became even more important due to recent technical improvements of virtual or mixed reality technologies (VR, MR) and their increasing acceptance in both private and professional context. These technologies visualize digital contents to different extents and allow the user to interact with the displayed objects [4]. Since the construction and adaption of physical prototypes is costly and time-consuming numerous applications in academia and industry aim at reducing the number of physical prototypes by creating virtual prototypes utilizing VR

© Springer Nature Switzerland AG 2019
A. Burduk et al. (Eds.): ISPEM 2018, AISC 835, pp. 3–11, 2019.
https://doi.org/10.1007/978-3-319-97490-3_1

and MR technologies. Examples for the proven feasibility for these approaches are described by Rademacher [5] and Schreiber [6]. However, physical prototypes are still essential in today's production facilities and are being regarded as indispensable even beyond 2040 by Winkelhake [7].

Prototypes are built at different stages of the product development process and for different purposes, but the prototype construction is usually located before the start of serial production. It commonly incorporates a small batch production of multiple versions of the same digital 3D component to build several slightly different pre-production parts or assemblies for assessment and validation. The prototype parts are usually manufactured either in-house or by an external supplier. In both cases, the parts are spot-checked in a randomized quality inspection on receiving before further processing.

A critical factor for the quality inspection is the process speed, as the single parts must be rapidly available for further operations. Based on a project conducted with an automotive supplier, this paper addresses problems for the quality inspection of received parts in prototype construction facilities.

An initial step for the quality assessment involves the comparison of the target geometry (as designed) and the part's actual geometry (as manufactured), to detect potential deviations, e.g. a missing drill hole. In the considered scenario, inspection speed matters more than recognition accuracy. Therefore, an automated visual inspection regarding previously defined characteristic points is preferred over a reverse engineering approach, i.e. scanning and reconstruction of the entirety of the part's surface, or using mechanical probes [8].

An essential prerequisite for the visual inspection is the detection of defined characteristics on the physical prototype. In this context the research field of Computer Vision (CV) addresses techniques and theoretical approaches including image acquisition, processing and analysis, with one goal being able to compare a given image with a reference image. This is achieved by the recognition of elementary features, such as edges, corners, circles or straight lines in the given image [9]. Following the detection of the features within an image, the interest locations are described through a feature descriptor, for the subsequent comparison, to find a relationship between the given and the reference image. The detection and description of these salient points rests upon different mathematical methods, which can be computed by different algorithms, e.g. Features from Accelerated Segment Test (FAST), Speeded-up Robust Features (SURF), Scale-Invariant Feature Transform (SIFT) or Oriented FAST and Rotated BRIEF (ORB). The actual comparison of features in the given image and the reference is performed by feature matching algorithms, such as Fast Library for Approximate Nearest Neighbors (FLANN) or Brute Force (BF) [10]. Each of these algorithms has various strengths and weaknesses, which must be considered for the respective application.

With a clear focus on automated visual inspection in prototype construction facilities, this paper elaborates a concept for the detection of previously defined characteristics by means of CV. These characteristics may refer to product and manufacturing information (PMI), stored in the prototype's 3D CAD document. In general, PMI are attached to a CAD model, documenting a product in regards to design, manufacturing and inspection, including data about e.g. dimensions, tolerances, manufacturing information or generic annotations [11]. As mixed reality is a proven facilitator for the context sensitive visualization and interaction with the displayed information [12], this technology is being considered as part of the proposed overall concept.

2 Aims and Requirements

The related use case is derived from a project conducted with the prototype construction facility at Adient Ltd. Co. KG (Burscheid, Germany). Adient's IT infrastructure supports and accelerates business processes especially in production, as this is imperative as an automotive supplier. The project entails investigations to digitalize the prototype part receiving in the corresponding facility, with the purpose to reduce errors and speed-up the prototype construction process.

As mentioned above, the received prototype parts require a quality inspection before further processing, e.g. assembly and testing. In assistance with common image processing and analysis methods, an automated visual inspection process is able to recognize objects and their positioning, to check the completeness, shape, geometry and dimensions, as well as the surface or to detect defects [8].

The visual inspection process within the proposed overall concept is focused on the recognition of actual received prototype parts. Therefore, the aim of the paper at hand is to find a way for the automatic identification of the actual part to be inspected to provide the corresponding quality and inspection information of the specific prototype part.

Since the use case is derived from an industrial project, the general requirement is the real-world usability. This implies a setup for the prototypical object recognition comparable to the productive one. Therefore, an installation with cameras mounted over a conveyer belt is ideal. In addition to this, specific requirements for the part recognition arise. The first requirement concerns the clear identification of a prototype part by reference to a previously defined set of potential parts. The recognition should ideally be carried out based on 3D CAD geometry or alternatively with reference camera images. Following a successful identification, the third requirement includes the automatic retrieval of a part's respective data, as this is necessary for the subsequent visual quality inspection. Finally, high detection precision, repeating accuracy and a sufficient level of efficiency are mandatory, as the detection should offer real-time results.

3 Concept

The overall concept for the PMI-based mixed reality assistant system consists of several distinct modules. The main goal of the presented approach is an accelerated visual quality inspection of received parts in a prototype construction facility by means of computer vision (CV). To enable a quick analysis of the actual shape towards the target shape, the inspection is carried out based on special previously defined characteristics, regarding the part's geometry, instead of scanning and analyzing the entire surface of a part. As described in the introductory section, existing CV algorithms are able to detect salient elements within an image to compare them towards a reference image. Therefore, the geometrical characteristics must be visible, when transferred to the reference image, or, to the contrary, covered and hidden features are not suitable for a visual inspection.

The concluding step of the overall concept is the inspection results visualization to the prototyping worker for a qualitative assessment.

For a better overview, Fig. 1 illustrates only the major components of the overall concept. The previously described use case consists of four major process steps or modules, which, in turn, comprise several activities. This paper focuses on the partial aspects for the part recognition, labelled as module 1 and module 2 (Fig. 1).

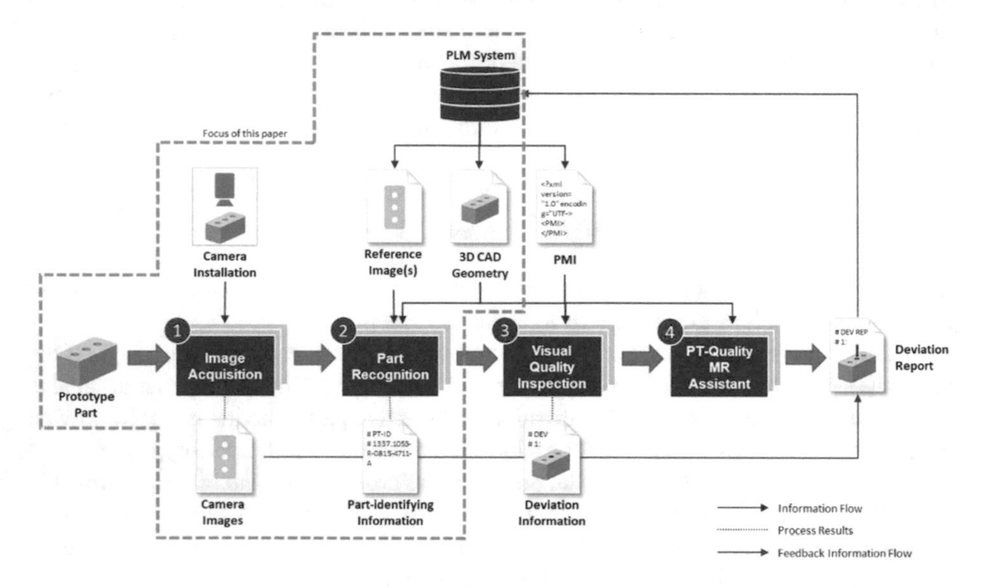

Fig. 1. Concept overview

The first step (Fig. 1) begins with the image acquisition of the specific prototype part. Therefore, a physical input for this module is the respective part itself. As mentioned in the requirements section, the part recognition works on an installation, equipped with cameras and mounted over a conveyer belt for an in-line supply of the prototype parts. The output of this step is an image of the actual part.

Whereas step 1 can be considered as a preparation, the second module entails the main activities for the CV-based part recognition. This step requires comparative data, which means different images, for the evaluation of similarities within these pictures. On the one hand, there are the previously taken pictures of the actual prototype part from module 1 and on the other hand, 3D geometry models or reference images as an alternative serve as potential "objects" to be identified (Fig. 1). The desired result of this module is the successful identification of the actual prototype part, including an exact mapping to the part-identifying information, e.g. a prototype part identification number (ID).

Based on the identification (step 2), the correlating input documents for the third module are provided. This novel approach for a visual quality inspection utilizes both geometric data as well as product and manufacturing information (PMI), derived from 3D CAD documents (Fig. 1). These characteristics serve as reference points, which need to be detected in the previously acquired images and checked for the purpose of an automated visual inspection. An example for a geometric reference in this context may be a drill hole feature in the respective 3D CAD model, which must exist in the physical part as well. Additional reference points can be extracted from the annotations and inspection information in the model's PMI data. A precondition for this approach is the prior accumulation of the 3D CAD models with the required information in a suitable way, e.g. through a PLM system. A result of this module is the information about deviations between the physical part and the 3D CAD model, e.g. a missing drill hole.

The fourth module (Fig. 1) addresses an assistant system using mixed reality (MR) visualization, to support the prototyping worker in the quality assessment and decision-making in an intuitive and interactive way. The deviation information resulting from step 3 is therefore used to visualize the delta between the part and CAD model. The worker can then generate a Deviation Report that bundles all gathered information.

As one requirement for this approach concerns the real-world usability, the integration into an existing IT infrastructure is mandatory. Therefore, the necessary data for the steps 2 to 4 are provided by a central data source, e.g. a PLM system.

4 Prototype Implementation and Validation

4.1 Module 1: Image Acquisition

As mentioned before, the goal of the paper at hand is the creation of a module for the recognition of three-dimensional objects as part of a holistic concept for automated visual inspection and measurement data management in prototype construction.

For this purpose, the authors created the following demonstrator hardware configuration in the laboratory, visible in Fig. 2. The demonstrator consists of four Raspberry Pi 3, three of those with the Raspberry Camera Modules v2 as autonomous camera systems and the fourth as host for the calculation and desktop application. The authors mounted the three camera systems in a tunnel spanning over the conveyer belt, which transports the prototype parts into the cameras' field of view. The camera systems

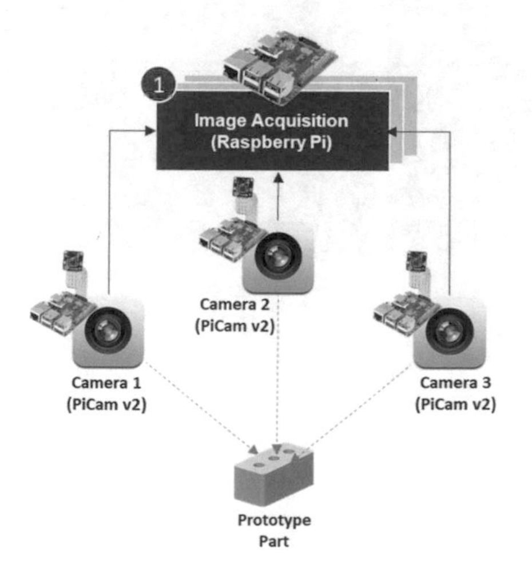

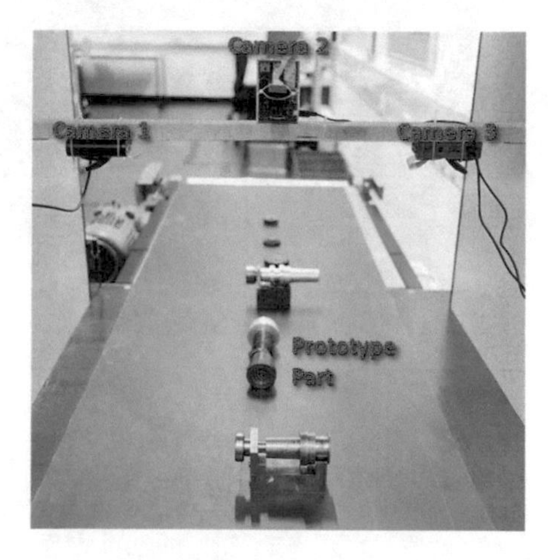

Fig. 2. Demonstrator schematic (left) and construction (right)

provide a live video stream to the calculation unit. The prototype parts in this demonstrator scenario are randomly picked items from the universities workshops, consisting of cogwheels, shafts and other mechanical parts or assemblies.

4.2 Module 2: Part Recognition

The physical demonstrator forms the basis for four following use cases, which validate the concept through implementation and evaluation of the measurements.

1. Recognition by 3D CAD geometry: In each recognition phase, another prototype part is placed under the camera system. The software compares the found features in the camera pictures with features found in the stored 3D CAD geometry.
2. Recognition by reference image: In each recognition phase, another prototype part is placed under the camera system. The software compares the found features in the camera pictures with features found in the stored reference image.
3. Repetition accuracy: For ten recognition phases, the same prototype part is placed under the camera system. This procedure repeats for each prototype part with both reference image and 3D geometry basis.
4. Precision/exactitude: Same procedure as mentioned before but with geometrically very similar prototype parts, that only differ in size (cogwheels).

The process for the optical part or assembly recognition starts with placing the concerning prototype parts on the conveyer belt. Then the image capturing process initializes and the conveyer belt transports the part towards the camera systems. Once the cameras register a part, the computing unit analyzes the still images of the camera live streams to identify the part. The GUI then informs the user about the results of the identification process, providing additional identifying information to the detected part.

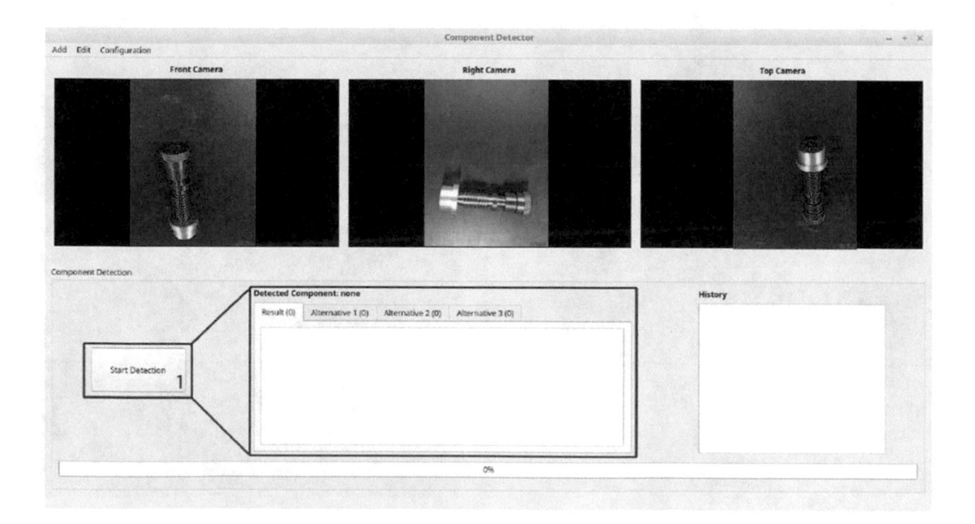

Fig. 3. Graphical user interface of the assembly detector

The described process was implemented in the demonstrator's software. As the demonstrator consists of four Raspberry Pi computers with their Raspbian operating system, the authors used the following software development tools and libraries. Python Version 3 as scripting language, PyQt5 for GUI programming, PyAssimp for 3D assembly importing, PyOpenGL for rendering 3D models, FFmpeg for video streaming and OpenCV 3 with contributed modules for the computer vision aspects of the demonstrator.

Figure 3 shows the graphical user interface of the assembly detector that offers the option to view the live camera streams to ensure a valid part placement and start or stop the optical recognition process.

4.3 Evaluation

Following to the hardware installation and the prototypical software implementation, the authors tested the entire recognition process, i.e. to detect the actual part or assembly. Beside an accurate part recognition the run time of the recognition process is a significant value, that should be minimal, to offer real-time results. As mentioned above in the introductory section, the different detection, description and matching algorithms should be evaluated against each other within the given use case considering recognition accuracy and speed in particular. Therefore, the recognition of one out of five different prototype parts with ten reference images each was tested with the following feature detection and description algorithms regarding their run time:

- Features from Accelerated Segment Test (FAST),
- Speeded-up Robust Features (SURF),
- Scale-Invariant Feature Transform (SIFT),
- Oriented FAST and Rotated BRIEF (ORB).

Based on these algorithms, the authors also tested the following matching algorithms regarding their run time and the overall detection accuracy:

- Fast Library for Approximate Nearest Neighbors (FLANN),
- Brute Force (BF)

Table 1 summarizes the results of the performed tests including measured run times and the accuracy of the used feature matching algorithms. As indicated in the table as well, the FAST algorithm does not support the usage as a feature descriptor. Regarding the run times for feature detection and description, ORB and SURF are prominent, though the feature matching and overall accuracy must be considered as well. The lowest run time of the feature matching algorithms is achieved with ORB, thus the recognition accuracy declines under 50%, which is inacceptable for this use case. Therefore, the authors used a combination of both algorithms SURF and SIFT in the prototypical implementation for the detection and description.

The subsequent step is the evaluation of the four above mentioned use cases based on the same algorithm combination. Scenario 1 and 2 (Recognition by 3D CAD geometry and by reference image) do not show significant differences and deliver nearly the same detection accuracy. In Scenario 3 (Repetition accuracy), the same part was placed under the camera ten times. Each time, the part was successfully detected. In the last scenario regarding precision and exactitude, cogwheels of different sizes were placed under the camera. Due to the utilized edge detection, the particular cogwheel was detected correctly.

Table 1. Feature detection, description, matching run times and accuracy

	SURF	SIFT	ORB	FAST
Feature detection and description run time [s]	5,31	23,30	1,84	9,58
Run time BF [s]	9,59	3,92	2,23	n/a
Run time FLANN [s]	35,53	11,12	2,27	n/a
Overall accuracy [%]	60	89	40	n/a

5 Conclusion

The authors have shown that the approach at hand enables the three-dimensional object recognition by means of existing computer vision algorithms, utilizing both 3D CAD geometry and camera images as reference objects. The presented approach is only the foundational part of a holistic concept for automated visual inspection and the measurement data management in prototype construction. Future work will consider the extraction of product and manufacturing information (PMI) and geometrical characteristics from a 3D CAD model as reference points for an automated visual quality inspection within the prototype construction. Requirements and interfaces for the viability of the concept, such as a suitable information accumulation, will be considered in the future work.

Furthermore, the authors seek to implement a mixed reality based assistant system for the interactive visualization of the inspection results, such as deviations, to enable the prototyping shop worker a qualitative assessment of the prototype parts.

References

1. Feldhusen, J., Grote, K.-H.: Pahl/Beitz Konstruktionslehre. Springer, Heidelberg (2013). https://doi.org/10.1007/978-3-642-29569-0
2. Pahl, G., Beitz, W., Blessing, L., et al. (eds.): Engineering Design: A Systematic Approach, 3rd edn. Springer-Verlag London Limited, London (2007). https://doi.org/10.1007/978-1-84628-319-2
3. Spur, G., Krause, F.-L.: Das virtuelle Produkt: Management der CAD-Technik. Hanser, München (1997). ISBN 3-446-19176-3
4. Dörner, R., Broll, W., Grimm, P., et al. (eds.): Virtual und Augmented Reality (VR/AR) Grundlagen und Methoden der Virtuellen und Augmentierten Realität. eXamen.press. Imprint: Springer, Heidelberg (2013). https://doi.org/10.1007/978-3-642-28903-3
5. Rademacher, M.H.: Virtual Reality in der Produktentwicklung. Springer Fachmedien Wiesbaden, Wiesbaden (2014). https://doi.org/10.1007/978-3-658-07013-7
6. Schreiber, W., Zimmermann, P.: Virtuelle Techniken im industriellen Umfeld. Springer, Heidelberg (2011). https://doi.org/10.1007/978-3-642-20636-8
7. Winkelhake, U.: Die digitale Transformation der Automobilindustrie. Springer, Heidelberg (2017). https://doi.org/10.1007/978-3-662-54935-3

8. Keferstein, C.P., Marxer, M., Bach, C.: Fertigungsmesstechnik. Springer Fachmedien Wiesbaden, Wiesbaden (2018). https://doi.org/10.1007/978-3-658-17756-0
9. Priese, L.: Computer Vision. Springer, Heidelberg (2015). https://doi.org/10.1007/978-3-662-45129-8
10. Awad, A.I., Hassaballah, M. (eds.): Image Feature Detectors and Descriptors. Studies in Computational Intelligence. Springer International Publishing, Cham (2016). https://doi.org/10.1007/978-3-319-28854-3
11. International Organization for Standardization. ISO 14306:2017(E): Industrial automation systems and integration - JT file format specification for 3D visualization (2017)
12. Abramovici, M., Wolf, M., Adwernat, S., et al.: Context-aware maintenance support for augmented reality assistance and synchronous multi-user collaboration. Procedia CIRP **59**, 18–22 (2017). https://doi.org/10.1016/j.procir.2016.09.042

A Computer Application for Drone Parametrization: Developing Solution for Drone Manufacturing

Christopher Nikulin[1(✉)], Marcos Zuñiga[3], Constanza Cespedes[1], Cristopher Rozas[2], Sebastian Koziolek[5], Tomás Grubessich[4], Pablo Viveros[4], and Eduardo Piñones[1]

[1] Departamento de Ingeniería en Diseño, Universidad Técnica Federico Santa Maria, Valparaíso, Chile
christopher.nikulin@usm.cl

[2] Departamento de Ingenieria Mecanica, Universidad Técnica Federico Santa Maria, Valparaíso, Chile

[3] Departamento de Electronica, Universidad Técnica Federico Santa Maria, Valparaíso, Chile

[4] Departamento de Industrias, Universidad Técnica Federico Santa Maria, Valparaíso, Chile

[5] Mechanical Department, Wroclaw University, Wrocław, Poland

Abstract. In this article, a solution for parametrization of drone parts by following a structured approach is proposed. Through this research, the authors attempt to contribute with a solution for those users that can have a 3D printer, but not necessarily have the specific knowledge to create appropriate parts for their drone or related modifications. A master 3D model has been created, which can be modified through a simple user interface, allowing to modify the general 3D model. The solution aims to manufacture the master model according to different drone sizes.

Keywords: Parametrization · 3D printer · Manufacturing process

1 Introduction

From an industrial point of view, the Additive Manufacturing application has been increasing during the last years. For instance, Birtchnell and Hoyle [1] highlighted that additive manufacturing technology has been expanding in the last years. In fact, Birtchnell and Hoyle [1] forecast more than a million of users for 2018. Nevertheless, many of these new users will have several limitations to exploit the real benefits of this type of technology. In this context, the combination of additive manufacturing technology, along with the development of a 3D parameterized model is relevant for the creation of favourable solutions in the development of more customized products [2]. On one hand, additive manufacturing is characterized for its high level of personalization and capacity to produce complex models, which can be suitable to be modified in many different cases [3, 4]. On the other hand, the parameterization provides a set of solutions for a certain problem, combined with a previous definition of the design, which can be used as link between lacks of knowledge and technology [5].

© Springer Nature Switzerland AG 2019
A. Burduk et al. (Eds.): ISPEM 2018, AISC 835, pp. 12–21, 2019.
https://doi.org/10.1007/978-3-319-97490-3_2

In this scenario, this research proposes an integrated solution, for instance, design, parametrization, and manufacturing processes capable of generating drone modifications. With this proposal, the authors attempt to contribute with a solution for those users that can reach the technology, but don't have the specific knowledge to create an appropriate modification (i.e. different motors, length, diameters, etc., for customization). Indeed, the authors contribution aims to update a "master model" to be easily and quickly made, producing therefore variations of the initial model for a specific drone, combining this solution with an interface that enables the users to access the parameters, which allows the generation of a 3D model by a user that does not necessarily have 3D modelling knowledge, and with this the later production with additive manufacturing.

2 A Brief Framework of Drone

Due to the several types of drones that have emerged in the latest years, and also due to the large amount of parts, this research has developed a study that allows to define the boundaries of analysis to be solved. In this scenario, it is necessary to identify the different kinds of systems, parts and components in order to find which might be a

Table 1. Main variables involved in parametrization of a drone-design.

Requirements			Dimension	Symbol	CAD design
Drone base	Control/receiver subsystem	Motors control unit	Max. length motors control unit	$[L_B]$	Max. length of controllers to install
		Motors control unit	Max. width motors control unit	$[A_B]$	Max. width of controllers to install
		System screw	Screw diameter	$[D_{SC}]$	Screw drilling diameter to electronical devices anchorage
		Anchorage system	Screw configuration	–	Screw drilling distribution
Drone wings	Work subsystem	Motor	Motor diameter	$[D_M]$	Anchorage diameter motor-structure
		Motor screw	Screw diameter	$[D_P]$	Drilling screw diameter to anchorage on the structure
		Motor anchorage	Screw configuration	–	Drilling screw distribution
		Propeller	Propeller length	$[L_H]$	Wings minimal length
	Transmitter subsystem	ESC	ESC width	–	Groove separation in wings to ESC anchorage

possible solution to be developed with rapid prototyping technology. Frequently, drones are characterized, generally, by the motor system; that refers to the propulsion of the drone in the air; the control and transmission system; the structure system, and others.

According to authors experience, a number of relevant variables has been organized by using a simple benchmarking methodology, which is a tool used by industries as a comparative data study to improve the decision-making process [6, 7]. On one hand, a benchmarking analysis will help to determine the range of solutions in which additive manufacturing can be a contribution to the stated problem. On the other hand, the benchmarking tool applied to product design, allows to make decisions and also to propose solutions by considering users' viewpoint, because this comparative analysis enables to quantify and organize data gathered from the different variables in the problem that authors want to address [8]. Table 1 shows an abstract of the comparative analysis performed in this research. Figure 1 presents drone types and general shape. The next classification corresponds to the main fields of application of the drones.

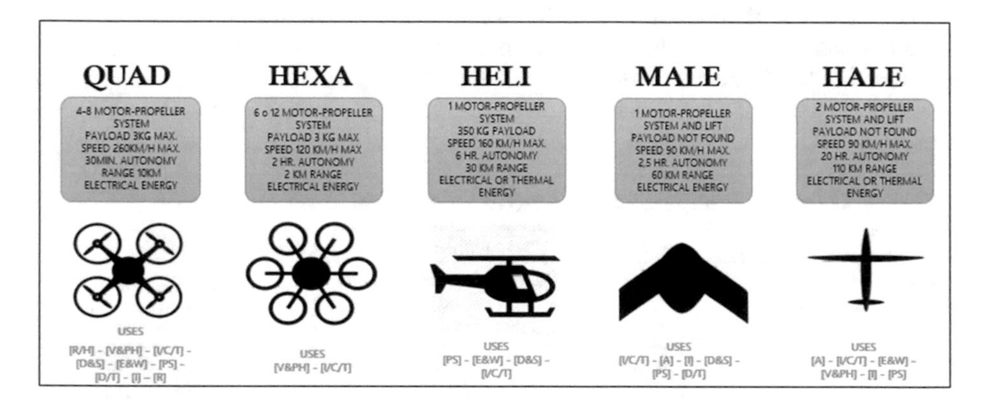

Fig. 1. Drone comparison analysis using a simple benchmarking approach. Parameters: quantity of motor-propeller systems, maximum payload, maximum speed, autonomy, range, and typical uses.

In this research, the QUAD Drone model has been selected as the first step of parametrization. This decision allows to state the boundaries for this research and also for the technology to be used.

3 Methodological Approach

In this section, the methodology used for the creation of the proposed solution is presented. The next diagram (Fig. 2) shows the complete process for the proposed solution, starting from ideas generation until prototype (Fig. 2).

Step 1: In the initial step, gathering data is recommended to collect information about the problem, more than one observation method is possible to be used [9]. Indeed, during this step is essential to firstly identify all the system and related elements, which allows to understand the drone functionality and related performance.

Step 1.1: It is important to mention that an engineering solution is based upon the identification of main functions of the analysed system [10]. With this perspective, it is important to obtain all the systems that constituted the drone when is flying, and to identify all systems required to enable this activity.
Step 1.2: The capability to identify and analyse design systems allows to identify the necessary stages to perform a function in the system. This is required, because this process may be different for each drone application. Once identified and analysed the systems that enable the drone flight, designers have to decide about which elements have to be simplified in order to replicate it in a parameterized model.
Step 2: In this step, a sketching which simplifies the results of step 1 is proposed, in order to start the parametrization process. This step is made either in a sequential manner or in parallel, depending on how the system analysis was performed. This step complements step 1, because it allows to generate the first sketches for iterations within computer assisted drawing models (CAD) [11].
Step 2.1: Once simplified the systems that allow the development of the drone activity to be validated, it is necessary to create a conceptual design, where all the systems and related parts are identified as well. In this step is where the parameterization and the dimension' relations begin.
Step 2.2: Once the dimensions that will command the CAD model are known, a master 3D model is generated using as a link for parameters analysed in step 2.1. It is important to use previous step dimension during the linking process, because it enables to have a more consistent solution during the parameterization process [12].
Step 3: In this step, the conceptual model of the created solution is simplified, through the link of a CAD model that can be parameterized according to step 2 specifications. The proposal of a graphic interface allows users without 3D modelling knowledge to enter (i.e. access input data) the dimensions of the drone in order to generate a proper 3D model.
Step 3.1: In this step, a graphic and generic model of the drone is created. The graphic model is presented to the user as an image with dimensions and instructions about how to use the general model to create an appropriate drone. Registering the values associated to each parameter, the generation of the 3D model begins.
Step 3.2: The need of creating a link between the CAD model and a simple interface, is for minimizing the human errors by users which do not have the tools to run a 3D modelling software [12]. Therefore, the graphic interface should be considered as a solution to reduce the lack of knowledge for those users that have 3D printers but not necessarily the knowledge to create an appropriate solution [14].
Step 4: This step corresponds to the manufacture of parts and links of the components developed by the 3D printer.
Step 4.1: Additive manufacturing allows the creation of components in an immediate and personalized approach [12, 13]. Nevertheless, some considerations are necessary to be taken into account when additive manufacturing is used, such as type of material, density, type of deposit, displacement speed, among other relevant factors for the manufacture of components. These factors influence directly the quality of production of each part, and also the performance parameters.
Step 4.2: The assembly stage definitely aims to link all the parts that were created independently by using rapid manufacturing process. This process allows to obtain

the final product, ending up ready for the next stage, which consists in testing it. In this stage, it is important to consider terminations (detailed design) of each one of the manufactured parts (i.e. product details).

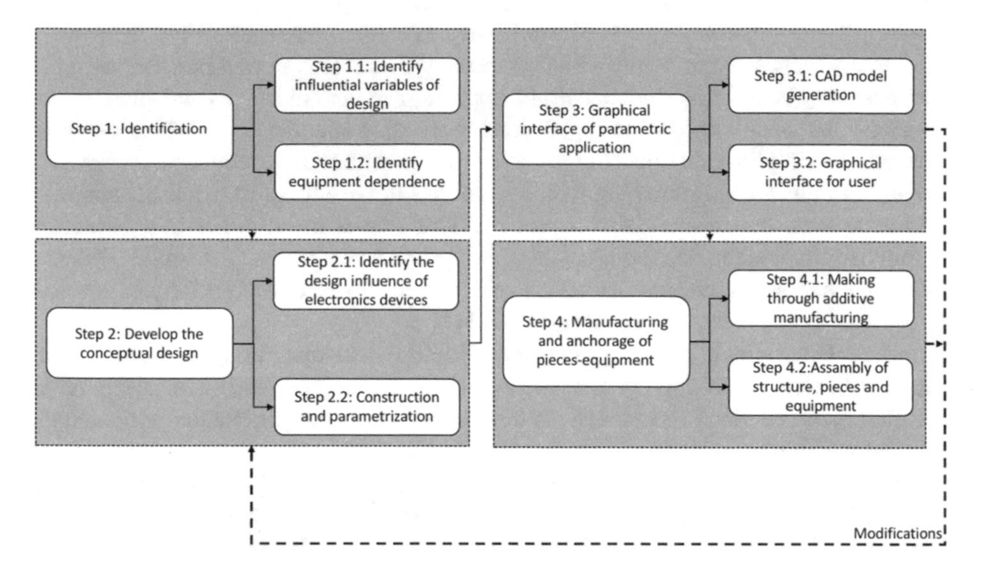

Fig. 2. Methodological steps followed in this research.

4 Case Study

Step 1: In this step, the conditions of the solution are defined, created with basis on the state of the art. In this step, the QUAD Drone model was defined as a solution, that will be used only for entertainment activities.

Step 2: In this step, the dynamic conceptual models of the solution were created, according to the system and related parts involved in the problem, where the minimum systems required to accomplish the drone design, and also a simple diagram was applied to identify the main variables involved in the master model (Fig. 3).

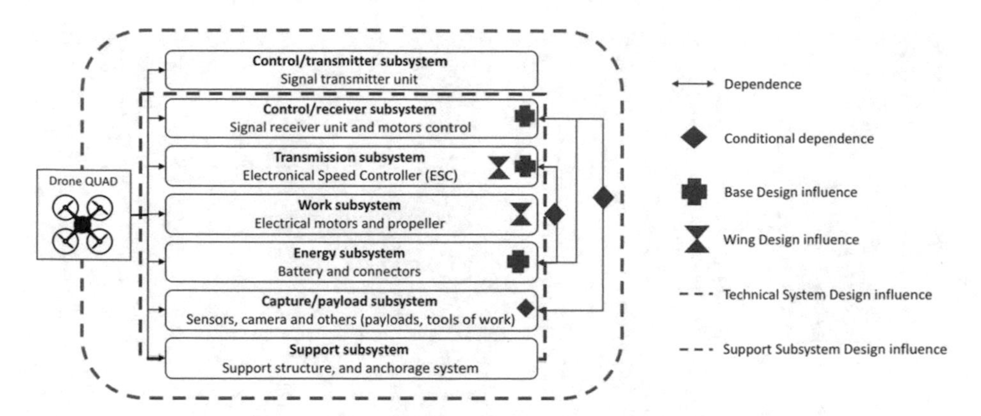

Fig. 3. Drone model to be analysed

Besides, in this step was created a graphic model that shows relations of the dimensions required to link a CAD model (Fig. 3). This representation allows to visualize the total of parameters to be used and the interrelations among them (Figs. 4 and 5).

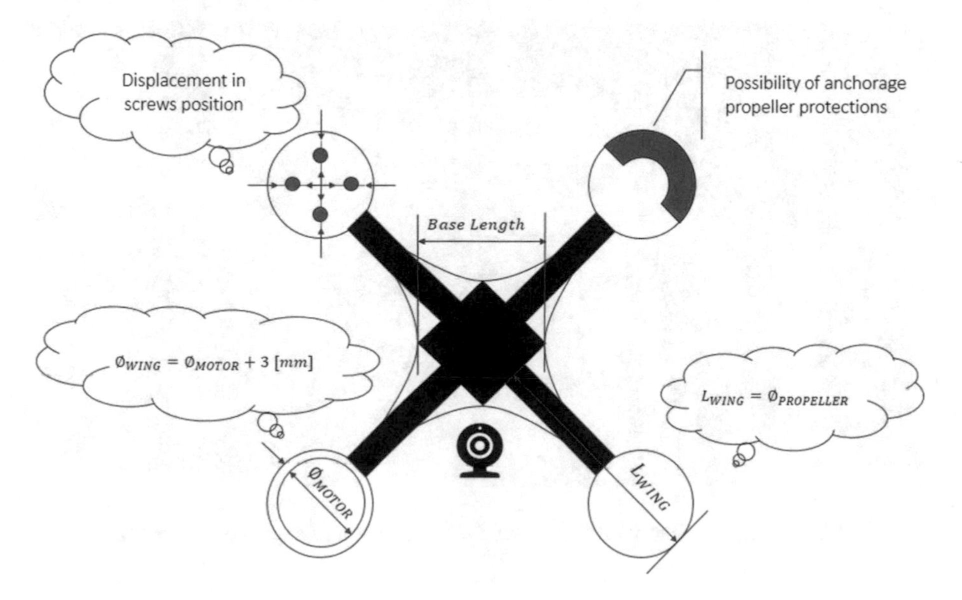

Fig. 4. Setting main variables for parametrization.

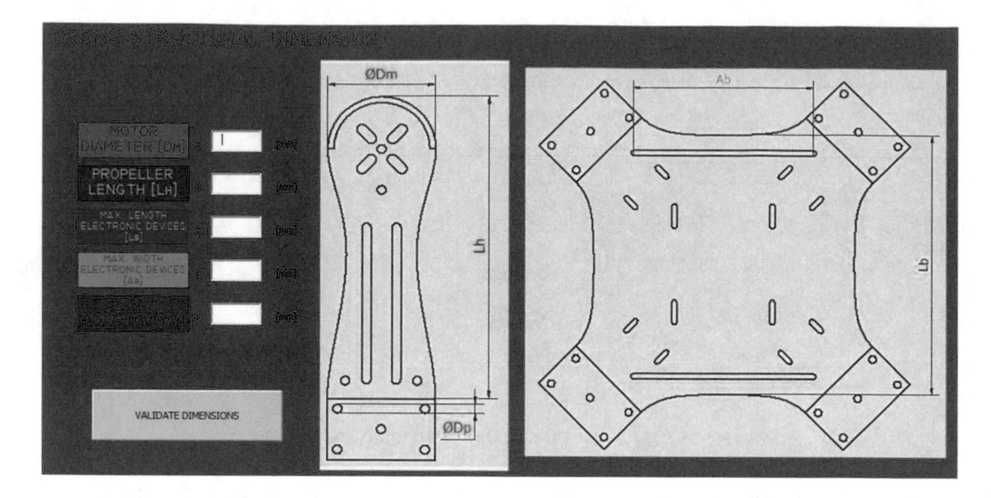

Fig. 5. Evolution of computer model.

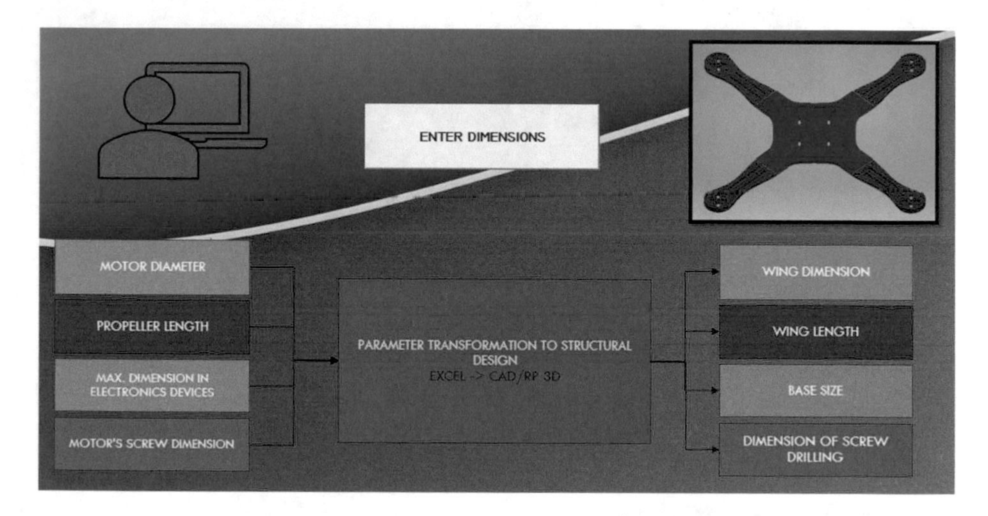

Fig. 6. Graphic interface of the user for the creation of a parametric model.

It is important to mention that the resistance of the parts is related to the resolution and density of the 3D prototype. Nonetheless, this research has not taken into account specific parameters of each 3D printer, therefore users need to pay attention when selecting parameters in the manufacture process. Figure 5 shows 3D designs of initial prototypes based on different parameters.

Step 3: Once the parameterized CAD model was finished, the interface was created with a reference image of the Drone's parts to visualize where and how the dimensions will be applied, taken as reference with the parametrized master 3D model (Fig. 6).

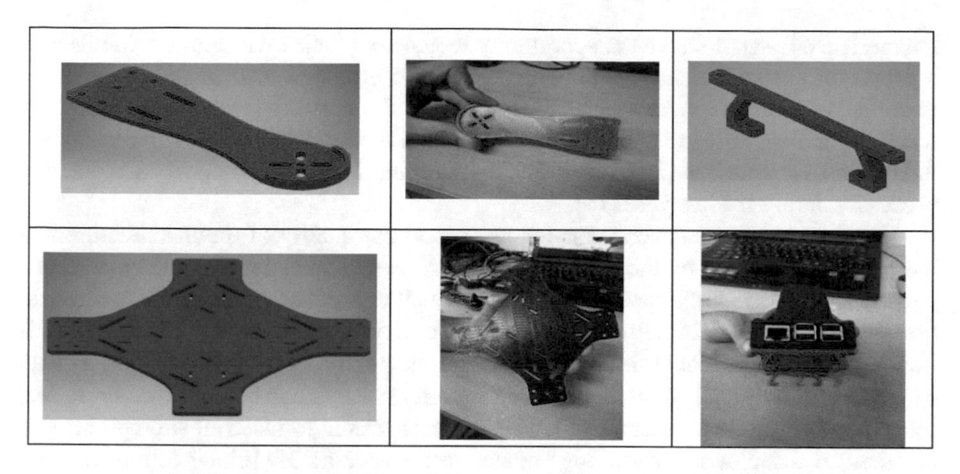

Fig. 7. CAD parts and relation with the prototype in reality.

In Fig. 5, the basic aspects of the graphic interface are shown: a reference image was created to understand the relation among variables and drone dimension. Entering the specific dimension of the drone is possible for modifying the 3D model directly.
Step 4: In this step, the final model of the 3D prototype is presented (Fig. 6). In the following picture, a general 3D model is shown, where personalized dimension of the prototype will emerge by considering the user parameters.
Figure 8 presents a parameterized model as an example to compare two variations of the master 3D model.

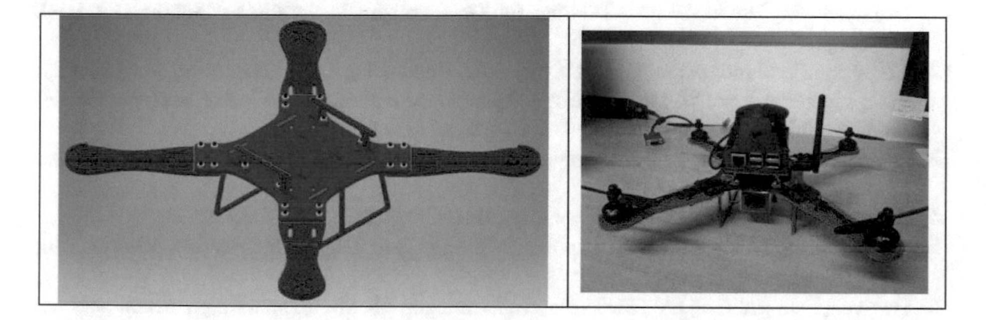

Fig. 8. Drone prototype developed with the master model.

5 Discussion and Conclusions

This research has proposed a step by step method for the creation of parametrization QUAD Drones. The research proposal allows users with lack of CAD skills to use a parametric model for drone customization, according to different motors, and dimension of the drone. The development of a tool to create drones with parameterization, allows inexperienced users to have the opportunity to generate their own solution, in this case a drone for entertainment. On the other hand, the production of parts with additive manufacturing seems to be a plausible alternative for those that have rapid manufacturing technology, given its quick massification on the market. Indeed, rapid manufacturing has become an alternative system-solution to more complex manufacture processes during the last year [15].

However, there are several constraints around the quality of the manufacture that have to be considered by the final user in this research, such as: (i) a complete drone parametrization might be too complex for every alternative of elements for customization; consequently, multiplatform solutions can be a envisioned as an alternative for this issue; (ii) the parts created through 3D printing lack the adequate surface quality and, in some cases, material strength, therefore material strength can be improved by adding resin (ex: epoxy); other alternative is that subjects related to material strength have to be assessed based on the knowledge related to the specific 3D printer or through trial and error.

Further research will be focused on improving the aspects of detail design and trying to explore a multifunctional parametrization system for drones.

Acknowledgement. The authors would like to acknowledge the support of CONICYT through the project FONDECYT-Iniciación (ID 11170227) and, FONDEF-ID16110114 and Centro Científico Tecnológico de Valparaíso (CCTVal FB-0821).

References

1. Birtchnell, T., Hoyle, W.: 3D Printing for Development in the Global South: The 3D4D Challenge. Springer, London (2014)
2. Conner, B.P., Manogharan, G.P., Martof, A.N., Rodomsky, L.M., Rodomsky, C.M., Jordan, D.C., Limperos, J.W.: Making sense of 3-D printing: creating a map of additive manufacturing products and services. Addit. Manuf. **1**, 64–76 (2014)
3. González Santos, R.: Las tecnologías de prototipado rápido en la cirugía. Rev. Cuba. Estomatol. **50**(3), 331–338 (2013)
4. Berman, B.: The new industrial revolution. Bus. Horiz. **55**(2), 155–162 (2012)
5. Yanagawa, K.: Confluence of Parametric Design and Digital Fabrication Restructuring Manufacturing Industries (2015)
6. Boxwell, R.J., Rubiera, I.V., McShane, B., Zaratiegui, J.R.: Benchmarking para competir con ventaja. McGraw-Hill, New York (1995)
7. Camp, R.C.: Benchmarking: The Search for Industry Best Practices that Lead to Superior Performance (1989)
8. Shyam, R.: Benchmarking in Product Design (2015)
9. Huntington, H., Callaghan, T., Fox, S., Krupnik, I.: Matching traditional and scientific observations to detect environmental change: a discussion on Arctic terrestrial ecosystems. Ambio **33**, 18–23 (2014)
10. Pahl, G., Beitz, W.: Engineering Design a Systematic Approach. Springer Science & Business Media, London (2013)

11. Feyen, R., Liu, Y., Chaffin, D., Jimmerson, G., Joseph, B.: Computer-aided ergonomics: a case study of incorporating ergonomics analyses into workplace design. Appl. Ergon. **31**(3), 291–300 (2000)
12. Roller, D.: An approach to computer-aided parametric design. Comput. Aided Des. **23**(5), 385–391 (1991)
13. Chu, C.H., Song, M.C., Luo, V.C.: Computer aided parametric design for 3D tire mold production. Comput. Ind. **57**(1), 11–25 (2006)
14. Nikulin, C., Ulloa, A., Carmona, C., Creixell, W.A.: Computer-aided application for modeling and monitoring operational and maintenance information in mining trucks. Arch. Min. Sci. **61**(3), 695–708 (2016)
15. Campbell, T., Williams, C., Ivanova, O., Garrett, B.: Could 3D Printing Change the World. Technologies, Potential and Implications of Additive Manufacturing. Atlantic Council, Washington, DC (2011)

A New Approach to Design of a Cyberphysical System Exemplified by Its Use in the Electro-Hydraulic Hybrid Drive

Lech Knap, Wiesław Grzesikiewicz, and Michał Makowski(✉)

Institute of Vehicles, Warsaw Univeristy of Technology, Narbutta 84, 02-524 Warsaw, Poland
{l.knap,wgr,mmakowski}@simr.pw.edu.pl

Abstract. Cyber-Physical Systems (CPS) are the systems that link cyberspace with the physical world by means of a conglomerate of interrelated elements (sensors and actuators) and computational capabilities. The paper is focused on a new and uniform method of CPS systems design. The task of designing the complex CPS system requires the solution of multi disciplinary problems which arise from the use of many different technologies. In this paper we use a new, task-oriented, method of designing the CPS systems which is illustrated by its application in electro-hydraulic hybrid drive of a city vehicle. In the considered hybrid drive, an electric drive is cyclically supported by a hydrostatic drive during acceleration or regenerative braking of the vehicle. The results of experimental studies presented in this paper were obtained from a designed and built laboratory model of a lightweight delivery van for city traffic and equipped with the studied hybrid drive. The obtained results suggest the possibility to considerably increase the effectiveness of energy conversion in the electric drive of the vehicle by means of the hydrostatic support. By applying the proposed task-oriented design method it was possible to design and build an efficient hybrid drive in which the load on the electric battery can be significantially decreased by use of hydrostatic support.

Keywords: Cyber-physical systems · Electro-hydraulic drive · Hybrid drive Vehicle · Experimental studies

1 Introduction

In the world surrounding us, for the last decade or two, an increasingly growing use of mechatronic systems consisting of mechanical systems cooperating with electronic systems, has been observed. In the human environment, many such systems can be found, supporting people in their duties or just providing a better quality of life. Even though many of these systems fulfill vital functions, such as controlling drivetrains, vehicle active safety systems, robots, vehicle traffic, etc., they still remain mostly invisible and we often do not realize their existence [6, 8, 9]. Each of the mentioned mechatronic systems is characterized by typical features, i.e. has an electronic control unit and physical interfaces (sensors and actuators) allowing for the mechatronic system's influence on surrounding physical environment (physical processes). The electronic unit works under the control of the software which provides a certain level of so-called

© Springer Nature Switzerland AG 2019
A. Burduk et al. (Eds.): ISPEM 2018, AISC 835, pp. 22–31, 2019.
https://doi.org/10.1007/978-3-319-97490-3_3

"integrated intelligence" enabling interaction and coordination of many physical processes at the same time. Such a systems are customarily named cyber-physical systems or CPS [2, 4].

One of the important problems the CPS designers have to face is lack of universal and widely acceptable methods of description of these systems' design and their components. This can be seen as resulting from a particularly high degree of complexity of their design and the necessity of applying various forms of requirement description among designers from different domains. This is why a task-oriented method for designing the CPSs [3] has been used in this work. It is based on the process approach and the constant improving of the project. This work illustrates on the example of electro-hydrostatic hybrid drive how this new method can be successfully used during a design stage of the project.

Moreover in the article we also discuss in details mentioned hybrid drive which is built in the form of an electric drive with hydrostatic support. Such a drive is intended for a city car characterized by cyclic movement. The aim of our study is also to evaluate the influence of the hydrostatic support on the effectiveness of energy conversion and on the relief of the electric drive. It will also be shown that such a hydrostatic support results in decreasing traction energy consumption of a vehicle.

The obtained results of numerical studies confirm the thesis that it is possible to improve the effectiveness of energy conversion process in the electric drive of a city car by means of the hydrostatic drive support. The experimental research has also shown that employing the hydrostatic support reduces the load on the electric battery, which results in smaller energy losses in the electric

2 Description of Process-Oriented Approach to a Design of a Hybrid Drive CPS

The process-oriented approach in organization management has undergone changes and transformations over the years, recently, however, it has been subject to development and revival again. It means employing the system of processes in the organization together with identification and interaction, as well as managing these processes. According to the assumptions of the theory of process management, their use should lead to achieving goals efficiently and effectively. In order to provide for the correct operation of the process, it is necessary to identify the numerous connections and interactions that can take place within the process or among processes. Fulfilling the tasks in turn, as part of the process, is possible due to controlling the necessary resources.

Taking into consideration the possibility of ensuring efficient and effective realization of the processes, it is also necessary to determine the methods of monitoring the process, thus of achieving the goals. Monitoring the processes is most frequently based on coefficients, which are determined and measured on an ongoing basis in the course of the process. Thanks to the carefully selected process indicators, taking decisions regarding correctness of the process functioning or necessity of its modification is possible - which is understood as perfecting of the process functioning.

An example of application of the process approach is discussed using the case of designing the system controlling the mentioned laboratory model of the electro-hydrostatic hybrid drive which is shown conceptually in Fig. 1. The mathematical description of such a system was described in detail in [1].

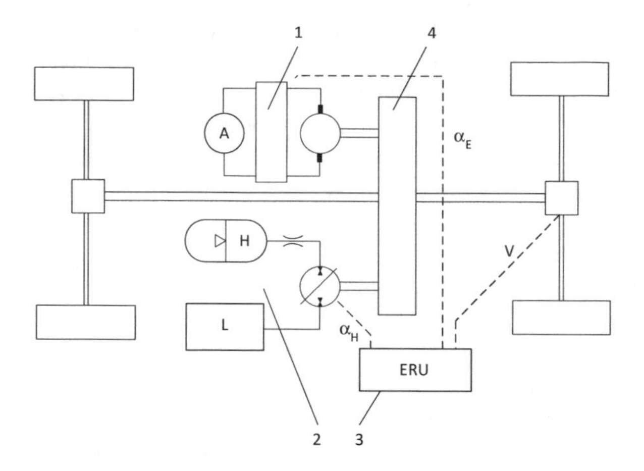

Fig. 1. Scheme of the hybrid electro-hydrostatic drive: 1 – electric drive, 2 – hydrostatic drive, 3 – electronic regulation unit (ERU), 4 – power transmission mechanism.

The first step in a successful design of a CPS is to understand all processes which occur inside the cybernetic and physical layer. For example, in the presented system, a process of energy flow occurs among three reservoirs, in which energy is stored in three forms: electric (in the electric drive) and mechanical kinetic (kinetic energy of a vehicle) as well as thermodynamic, i.e. hydraulic potential and thermal (in the hydrostatic drive). While the vehicle accelerates, the energy is collected from the electric battery and partly from the hydro-pneumatic battery (when it is possible), and then is converted into kinetic energy of the vehicle. During vehicle braking, kinetic energy, having been converted, returns to the hydro-pneumatic battery and, in part, to the electric battery. The above-mentioned processes of energy flow and conversion are accompanied by the energy dissipation process. The energy conversion occurs in the electric machine operating as a motor or generator and in the hydraulic machine operating as a motor or a pump.

In the second step, it is necessary to identify all main activities - which occur inside the mentioned processes - and combine them with physical assets and components (e.g. actuators, sensors, electronic units and other interfaces) needed for the proper functioning of those activities. Such a relationship, in the case of the presented hybrid drive CPS, is shown in Fig. 2. Based on the sensors mounted on the CPS, it is possible to assess and analyze the motion of the vehicle as well as the values of current parameters of electric and hydrostatic drives defining physical quantities characterizing energetic state of both drive systems. Further on, this information can be used to compute in the ERU appropriate values of two functions α_E - the voltage of electric motor powering a vehicle and α_H - describing the unit capacity of the hydrostatic pump-motor. Although the operation of both machines is controlled by the mentioned electronic control system (ERU – cf. Fig. 1) the proper operation also requires an appropriate algorithm of control. The optimal signal values can be established using various methods. In the case of an error information is given to the driver through human-machine interface (HMI). Similar issues are considered in traditional hybrid drives, in which the combustion engine is supported by an electric drive (HEV) [7] or in hydrostatic drives (HHV) [5], as well as in electric drives with hybrid energy storage system (HESS) [10].

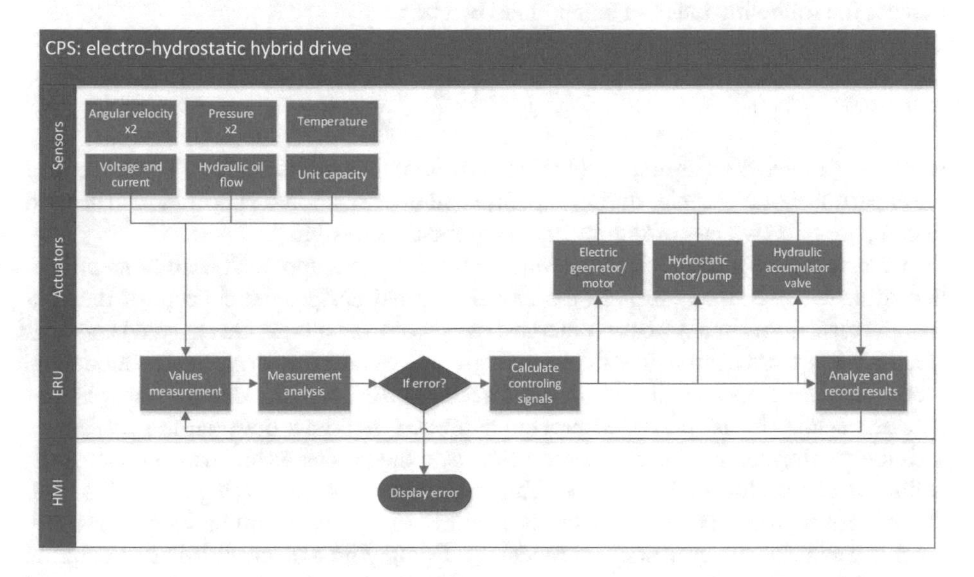

Fig. 2. Process diagram of energy conversion in the electro-hydrostatic hybrid drive CPS

The diagram presented in Fig. 2 has yet another advantage because it shows relationships between activities and assets. In this way it is possible to follow a flow of signals between different components of the CPS and identify potential drawbacks or bottlenecks. But this can be also used as an input to a risk analyses allowing to asses related threats and vulnerabilities. Certainly, at the initial stage of designing, it is difficult to define precisely all those relationships. But, the first stage of designing, based on the processes, enables selection of the majority of main (or all) physical interfaces necessary to complete the process. On the basis of the specified sensors, it is possible to select and design an electronic system ensuring the supply for the sensors, measurement of physical quantities, and generating the signals controlling the proportional valves. This is what makes possible performing the subsequent stages of designing and improving, allowing for increasingly precise selection of all components of the system. Together with the development of the CPS project description, it is necessary to modify the process diagram, which will thus become growingly complete, but also more and more detailed. It is clear that from the beginning of the cyclic designing the process indicates also, how the algorithm will have to work in the controller. The algorithm needs to complete the actions in the successive steps, which were defined in the process diagram. It is thus apparent that the process diagram can be used by the engineers specializing in designing the physical as well as the cybernetic layers. The process approach is therefore a designing method that can be employed by engineers from various domains and allows for finding a common ground while building a system at the level of integration of different domains.

The last stage of the process-oriented approach to the CPS design is to propose and measure effectiveness of the process in achieving assumed goals. For example, in the mentioned case of the hybrid drive, to assess effectiveness of the energy conversion process the following index of battery load can be used:

$$I^2 = \frac{1}{N}\int_0^T i^2(t)dt \tag{1}$$

where i – denotes discharging or charging current of the electric battery, T – the time of the experiment of N – cycles during experimental investigations. This index can be used to compare effectiveness of improvements in the CPS design.

Comparing the model-oriented design with the process approach enables identification of several common areas. In the model-oriented design, the concept of the CPS design consisting of many phases emerges; the alternating phases are analyzed and realized in the process approach with every single process. In the user-oriented modeling, a considerable emphasis is laid on the director, actor, and the defining of relations between actors. In the presented process approach, relations between the successive activities performed by different resources within the processes are also defined. If the definition of an actor is adopted describing them as the resource acting within the CPS, the process approach becomes an approach similar very close, and in many cases very much the same as the user-oriented modeling. Employing a deterministic or stochastic approach to describe the CPS model is possible, too, as well as the connection in serial or in parallel. The process approach – as can be seen on the diagram of the exemplary process – also allows for the locating in time of the events realized by particular resources.

3 Experimental Investigation

On the basis of the method described above, a model of the CPS hybrid drive was designed and built. In order to verify the thesis on the possibility of increasing the effectiveness of energy conversion in the electric drive due to the application of the hydrostatic support, also a laboratory station was built. The station allows for experimental investigations of the CPS hybrid drive. The structure of the CPS is to some extent analogous to the system described previously, which is presented in Fig. 1. The scheme of the structure of the experimentally investigated CPS shown in Fig. 3 consists of:

- hydrostatic drive consisting of the hydro-pneumatic battery and the pump-motor,
- electric drive consisting of the electric battery and the electric motor-generator,
- system mapping a vehicle, consisting of the flywheel and the hydraulic pump.

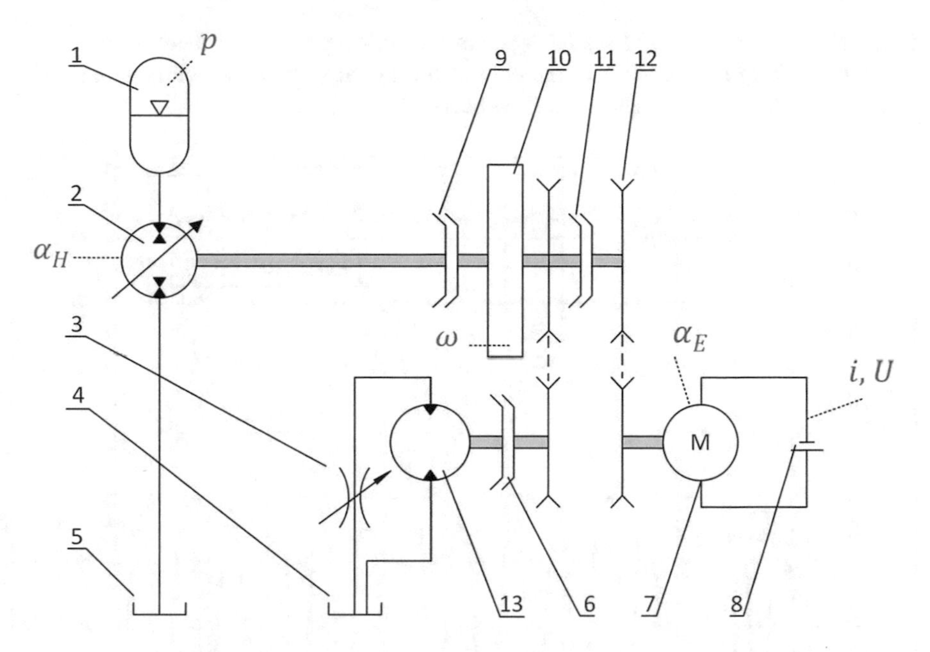

Fig. 3. Scheme of the laboratory station to study hybrid electro-hydrostatic drive together with measurement-control signals: 1 – hydro-pneumatic battery, 2 – pump-motor with control system, 3 – throttling elements, 4 – tank, 5 – tank, 6 – clutch, 7 – electric motor with control system, 8 – electric battery, 9 – clutch, 10 – flywheel, 11 – clutch, 12 – transmission belt, 13 – safety valve, 14 – hydraulic pump.

The subsystems listed above are connected by electromagnetic clutches, thanks to which the flywheel may be powered in an electric or hybrid way. Additionally, there is a measure-control system in the station, consisting of: a control computer, data filing system, conditioner of measurement-control signals and metioned previously sensors - cf. Fig. 2. As it was mentioned above due to features of measure-control system it is possible to calculate signals α_H, α_E.

In the laboratory station, inertia of the vehicle is rendered by the flywheel placed on the shaft of the pump-motor. The resultant moment of inertia of the spinning elements of the station amounts to $J = 0.84\ \text{kgm}^2$. With the assumed transmission ratio between the pump-motor/electric motor and the wheels of the vehicle $j_H = j_E = 8$, and the wheel radius amounting to $r = 0.3$ m, the above-mentioned moment of inertia renders the inertia of the vehicle with the mass $m = \left(\frac{i}{r}\right)^2 J \approx 600$ kg. Besides, rotational speed of the flywheel 1250 rev/min corresponds to the vehicle speed of 5 m/s.

Figure 4 shows the graphs illustrating the results of measuring U, i, ω; obtained in the course of electric (hydraulic drive is disengaged) accelerating and braking of the flywheel. At the beginning of the each cycle and in the first stage of the cycle $t \in [5, 10]$s the flywheel is accelerated to the speed of $\omega = 1250\,\text{rev/min}$ and the current i flowing from the battery increases to the value of about 70 A; then, when $t \in [10, 15]$s, the flywheel rotates at the constant speed and the current value drops to approximately 12 A;

at the last stage, when t ∈ [15, 20]s regenerative braking of the flywheel occurs, during which the battery is charged with the current the initial value of which is about –37 A. In all 16 cycles the same changes can be observed.

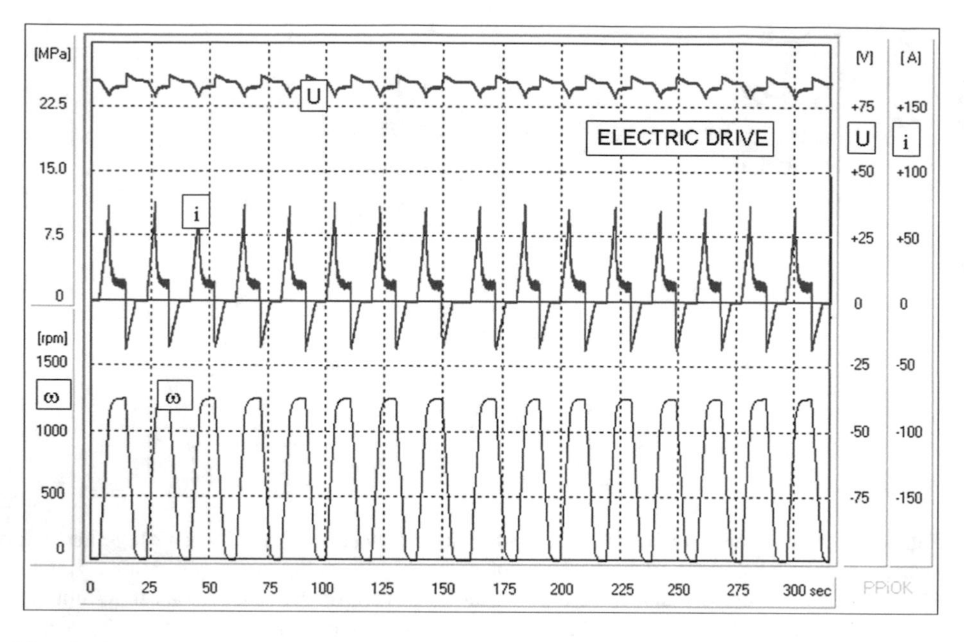

Fig. 4. Results of measurements during 16 cycles of electric accelerating and braking of the flywheel

Analogous measurements, however, for the electric drive with hydrostatic drive support, were performed also for 16 cycles of the flywheel accelerating and braking. The results of measurement for the hybrid drive are shown in Fig. 5. While accelerating (the first cycle and t ∈ [5, 10]s), the highest value of the current decreased (compared to the electric drive) to approximately 33 A in the first cycle and about 50 A in the last. At the second stage (t ∈ [10, 15]s) the flywheel is accelerated only electrically, and the value of the collected current amounts to 12 A, while braking (t ∈ [15, 20]s) the electric drive is disconnected, i.e. the current flowing through the battery amounts to zero (i = 0).

Figure 5 also illustrates the curve of gas pressure p in the hydro-pneumatic battery. While accelerating (t ∈ [5, 10]s, $\alpha_H = 0.5$) energy is collected from the electric battery and hydro-pneumatic battery, and thus the gas pressure decreases from 29 MPa to 18.3 MPa. At the second stage (t ∈ [10, 15]s) the hydrostatic drive is disconnected ($\alpha_H = 0$), and the visible slight increase in gas pressure results from heating of the gas during isochoric process. During hydrostatic braking of the flywheel (t ∈ [15, 20]s, $\alpha_H = 1$) the hydro-pneumatic battery is charged, i.e. compressing gas to the value of 24 MPa. It has to be borne in mind that in the first cycle of the hydrostatic drive operation (t ∈ [5, 20]s) described above, the pressure of gas decreased by 29–24 = 5 MPa; it means that the amount of energy accumulated in the hydro-pneumatic battery also decreased.

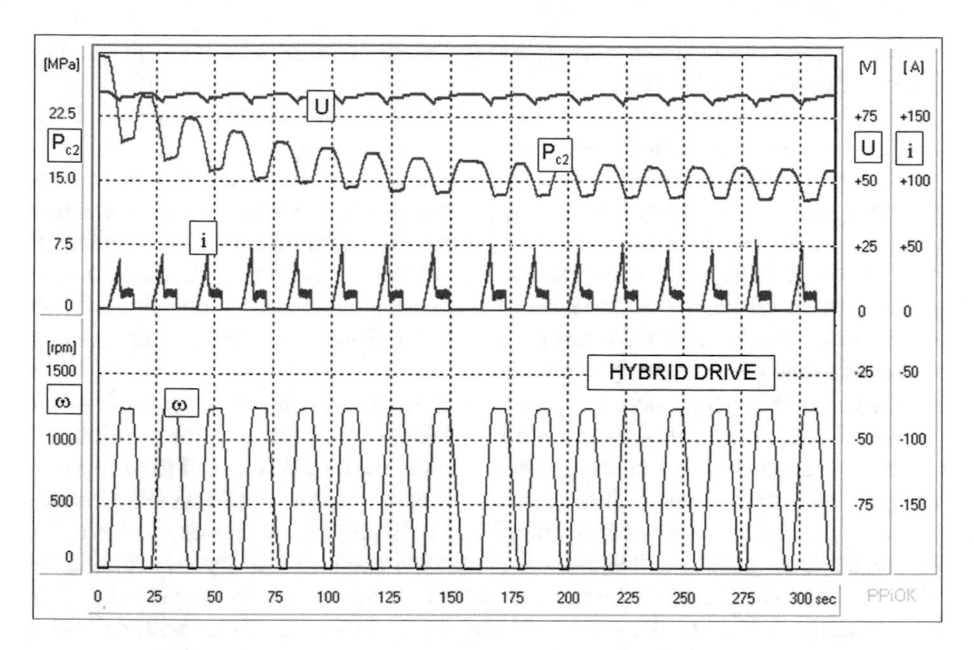

Fig. 5. Results of measurements during 16 cycles of hybrid accelerating and braking of the flywheel

During the second cycle (cf. Fig. 16) ($t \in [25, 40]$s) gas pressure decreased by 2 MPa to the value of 22 MPa. Having performed 8 cycles ($t \approx 160$s), the stabilization of gas pressure occurs, which means that in the following tests the values of pressure pulsate within 13–16 MPa.

In order to compare the studied drives, the diagrams in Figs. 4 and 5, illustrating the currents flowing through the electric batteries in both drives, are analyzed. According to a visual evaluation of the diagrams, the values of current intensity in the hybrid drive are clearly lower than in the electric drive. To compare them, the index of battery load mentioned in Eq. (1) was calculated and the following the following values were obtained:

- for the electric drive $I_E^2 = 10.15 \cdot 10^3 [A^2 s]$,
- for the hybrid drive $I_H^2 = 4.05 \cdot 10^3 [A^2 s]$.

By comparing these values it may be concluded, that in the hybrid drive, the average value of squared current intensity is 2.5 times lower. Hence, the electric energy losses (e.g. on the electric battery, wires) in the installation of hybrid drive are significantly lower.

4 Conclusions

The presented approach to designing and building the CPSs allows for creation of the universal description of the system's project, which can be seen as a kind of language, the so-called "meta-description". Applying this description allows for interchangeability of components within the CPSs – assuming that the processes within would realise the same goals. In the custom software, this corresponds to the possibility of using different libraries, which despite being different still realise the same goal.

The suggested method of the CPS description is based on the assumption that its task is nothing short of the goal of one or many processes. The processes are, in turn, sets of activities realised by individual CPS components. The activities can be performed both in serial or in parallel, and the proposed record allows for envisaging this fact. The key advantage of the discussed method is drawing special attention to identification not only of the connections between resources but also determination of precise requirements of all the interested parties as early as possible.

The presented example describing the CPS of the electro-hydrosttic hybrid drive and the results of its experimental studies confirmed the assumed thesis on the possibility of increasing the effectiveness of energy conversion in the electric drive of the vehicle by using the hydrostatic support. The hydrostatic support consists in alternating processes of charging and discharging the hydro-pneumatic battery, occurring during the cyclic load of the drive of the vehicle, resulting from the conditions of the city traffic. The increased effectiveness of conversion means that the amount of energy collected from the electric battery is smaller and so are the energy losses connected with collecting energy for accelerating and returning the energy during regenerative braking.

Acknowledgments. This project was funded by the Polish National Science Center allocated on the basis of the decision number DEC-2011/01/B/ST8/06822.

References

1. Grzesikiewicz, W., Knap, L., Makowski, M., Pokorski, J.: Study of the energy conversion process in the electro-hydrostatic drive of a vehicle. Energies **11**, 348 (2018)
2. Khaitaan, S.K., i McCalley, J.D.: Design techniques and applications of cyber physical systems: a survey. IEEE Syst. J. **9**(2) (2015)
3. Knap L. Projektowanie i budowa zadaniowo zorientowanego system CPS (in English: Design and Construction of the task-oriented Cyber-Physical System). Wydawnictwo Naukowe Instytutu Technologii Eksploatacji – PIB, Radom (2017)
4. Lee, E.A., i Seshia, S.A.: Introduction to Embedded Systems, A Cyber-Physical Systems Approach. MIT Press, Los Angeles (2011)
5. Liu, T., Jiang, J., Sun, H.: Investigation to Simulation of Regenerative Braking for Parallel Hydraulic Hybrid Vehicles, Measuring Technology and Mechatronics Automation (2009)
6. Loos, S., Platzer, A., i Nistor, L.: Adaptive cruise control: Hybrid, distributed, and now formally verified. In: Formal Methods, FM 2011. Lecture Notes in Computer Science, vol. 6664, pp. 42–56. Springer (2011)
7. Tribioli, L.: Energy-based design of powertrain for a re-engineered post-transmission hybrid electric vehicle. Energies **10**, 918 (2017)
8. Wei, L.: Introduction to Hybrid Vehicle System Modeling and Control. Wiley. Published by John Wiley & Sons Inc., Hoboken, New Jersey (2013)

9. Williams, B., Martin, E., Lipman, T., Kammen, D.: Plug-in-hybrid vehicle use, energy consumption, and greenhouse emissions: an analysis of household vehicle placements in Northern California. Energies **4**, 435–457 (2011). https://doi.org/10.3390/en4030435
10. Wieczorek, M., Lewandowski, M.: A mathematical representation of an energy management strategy for hybrid energy storage system in electric vehicle and real time optimization using a genetic algorithm. Appl. Energy **192**, 222–233 (2017)

An Analysis of the Efficiency of a Parallel-Serial Manufacturing System Using Simulation

Sławomir Kłos(✉) and Justyna Patalas-Maliszewska

University of Zielona Góra, Licealna 9, 65-417 Zielona Góra, Poland
{s.klos, j.patalas}@iizp.uz.zgora.pl

Abstract. The efficiency of discrete manufacturing systems and the level of work-in-progress are most important topics, especially for producers of automotive parts. In the present paper, an analysis of the throughput and average product lifespan of a parallel, serial manufacturing system, with varied buffer allocations and operating times, is presented. A model of the manufacturing system has been prepared using Tecnomatix Plant Simulation Software. This study was conducted using theoretical data sets and various statistical distributions of processing times. The main goal of the research was to analyse the impact of buffer allocation on the behaviour of a general, parallel manufacturing system. The methodology for preparing simulation experiments is here proposed.

Keywords: Parallel serial manufacturing system · Computer simulation
Buffer allocation · Throughput · Product lifespan

1 Introduction

Parallel production lines are typical topology used in the manufacturing systems of many sectors. The structure of the manufacturing system plays an important role in the effectiveness, maintenance and work-in-progress of production processes. In this paper, the impact of the allocation of buffer capacity and the batch size, of a discrete manufacturing system, on the throughput and average lifespan of products, has been analysed using the simulation method. Computer simulation is a research method often used for analysing and developing discrete manufacturing systems because of its flexibility and availability in the modelling of complex structures within systems. Computer simulation techniques are often used for the study of manufacturing systems [1] such as:

- general system design and facility design/layout,
- material handling system design,
- cellular manufacturing system design,
- flexible manufacturing system design,
- manufacturing operations, planning and scheduling,
- maintenance operations, planning and scheduling,
- real-time control,
- operating policies,
- performance analysis.

© Springer Nature Switzerland AG 2019
A. Burduk et al. (Eds.): ISPEM 2018, AISC 835, pp. 32–43, 2019.
https://doi.org/10.1007/978-3-319-97490-3_4

The study presented in this paper belongs to this last category, viz., analysis of the performance of manufacturing systems. The buffer-allocation problem is an NP-hard, combinatorial, optimisation issue and is well known in industrial engineering research. On the one hand, the correct allocation of buffer capacities on production lines can result in an increase in the overall efficiency of a production system. Many scientific papers address the buffer-allocation problem with the use of computer simulation, in the general design of discrete-manufacturing systems and in the analysis of the operation, production planning and scheduling of such systems. The problem of maximising the throughput of production lines, by changing buffer sizes or locations, using simulation methods, was studied by Vidalis et al. [2]. A critical literature overview of buffer allocation and production-line performance was carried out by Battini et al. (2009). Demir et al. proposed a classification scheme in order to review the literature on the subject and presented a comprehensive survey of the buffer-allocation problem in production systems [3]. Staley and Kim presented the results of simulation experiments carried out for buffer allocations, in closed-series production lines [4]. Kłos and Patalas-Maliszewska analysed the impact of buffer allocation on the effectiveness of manufacturing systems [5]. Jagstam and Klingstam used 'discrete event simulation' as an aid for the conceptual design and preliminary study of manufacturing systems, through the development of a virtual factory. They proposed a simulation handbook in order to fully integrate simulation as a tool in engineering processes. They also identified problems associated with the integration of 'discrete event simulation' in the design of manufacturing systems. Simulation methods are often used to design and analyse the effectiveness of automated, guided vehicles in manufacturing plants [6]. Varela et al. analysed two, alternative, manufacturing-scheduling configurations, in a two-stage, product-oriented, manufacturing system, exploring the environments of a hybrid flow shop (HFS) and a parallel flow shop (PFS). They compared the results of the research on production scheduling in the hybrid and parallel flow, taking into account the 'Makespan Minimisation Criterion' [7]. Kochańska and Burduk presented a concept for the improvement of production support processes, in the ordering of tools, parts and consumables for machines. In this concept, some selected Lean Manufacturing (LM) tools and simulation models of primary and secondary processes, created in the iGrafx Process for Six Sigma software, were applied. Based on the results of the simulation studies, implementation of the scheduling method, along the stream of values, was proposed [8]. Bocewicz et al. proposed the effective sequencing and scheduling of material handling systems (MHSs) based on fuzzy processing time constraints in transportation operations, which is a problem in production flow scheduling resulting from an assumed set of constraints, imposed by admissible production routes and schedules of automated, guided vehicles [9]. The study of improving of the efficiency of production processes using computer simulation methods is proposed in publication cycle by Bartkowiak et al. [10–12].

In the present paper, the impact of buffer allocation on the throughput of the parallel, serial manufacturing system is analysed. The general research problem can be formulated as follows: We are given a parallel, serial, discrete manufacturing system that includes machines and buffers. The buffer capacity allocation in the system and determination of operating times and batch sizes have a significant impact on the rate of the performance of the system.

The model of a parallel, serial manufacturing system includes the structure of machines and buffers, processing and setup times, batch sizes and dispatching rules. It has been assumed that the system is fully automated, that is, having no human input. A detailed description of the model is presented in the next chapter.

2 The Model of the Manufacturing System

The structure of a parallel, serial manufacturing system is presented in Fig. 1. The system includes N technological operations and M production lines.

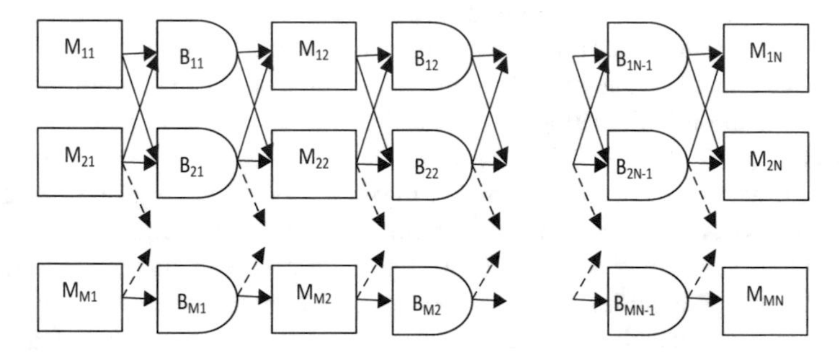

Fig. 1. The structure of a parallel serial manufacturing system

Each production line includes N machines and N-1 intermediate buffers. Each machine is connected to neighbouring buffers and each buffer is connected to neighbouring machines in 2 or 3 connections. The elements are transferred between machines and buffers, using the Round-Robin dispatching rule. If a buffer or machine is busy, then the next, empty buffer or machine is chosen. The model of the system has been prepared using Tecnomatix Plant Simulation Software (Fig. 2). The model of the system includes 9 CNC machines and 6 buffers. Three technological operations are realised in the system, firstly by machines CNC11, CNC21, CNC31, secondly by machines CNC12, CNC22, CNC32 with the final operation being realised by machines CNC13, CNC23 and CNC33. This means that production in the system can be realised via several, alternative routes. For example, if an element is located, initially, on machine CNC11, 6 alternative routes can be taken into consideration: (1) CNC11->CNC12->CNC13; (2) CNC11->CNC22->CNC23; (3) CNC11->CNC22->CNC33; (4) CNC11->CNC32->CNC23; (5) CNC11->CNC32->CNC33; (6) CNC11->CNC22->CNC13. It has been assumed that the machines, located in each column, complete the same, technological operations, therefore, each route enables the same production process to be completed. The level of machine availability has been defined as 95%. The operation and setup times are defined using Uniform Distribution. Uniform Distribution can be applied when little is known about random numbers, in such as processing times.

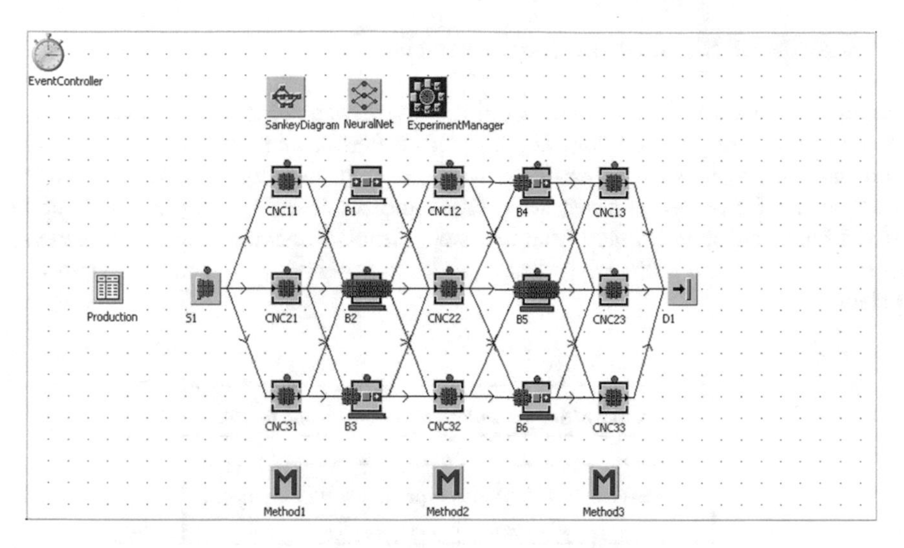

Fig. 2. The structure of a parallel, serial manufacturing system

The probable density function of the Uniform Distribution for start <x<stop is:

$$f(x) = \frac{1}{stop - start} \tag{1}$$

where the average value of the distribution is

$$\mu = \frac{start + stop}{2} \tag{2}$$

and the variance takes on the value of

$$\sigma^2 = \frac{(stop - start)^2}{2} \tag{3}$$

After the technological operation has finished, the elements are located in the buffers, using the same Round Robin dispatching rule. The machines can only send products to neighbouring buffers. Initially, the processing times for all CNC machines are defined in seconds, as follows: Start = 440 and Stop = 480 s. Simulation experiments were conducted for several variants of the operation and setup times. Initially, the setup times for the machines were defined as: Start = 60 and Stop = 600 s. The system produces four batches of products A, B, C and D, with the initial batch size, defined repetitively, as 100, 300, 80, or 120 units. The research methodology and results of the simulation experiments are presented and discussed in the next chapter. The impact of the allocation of buffer capacity on the total performance of the system was analysed.

3 Results of Simulation Experiments

Analysis of a parallel, serial manufacturing system was completed in seven steps; the simulation research methodology is presented in Fig. 3. The system throughput, along with the average lifespan of products, was analysed, initially, in order to obtain the initial values for operation times, setup times and batch sizes, using randomly defined sets of buffer capacity levels. In the first step, simulation experiments were generated for randomly defined buffer capacities, in order to observe the general behaviour of the system.

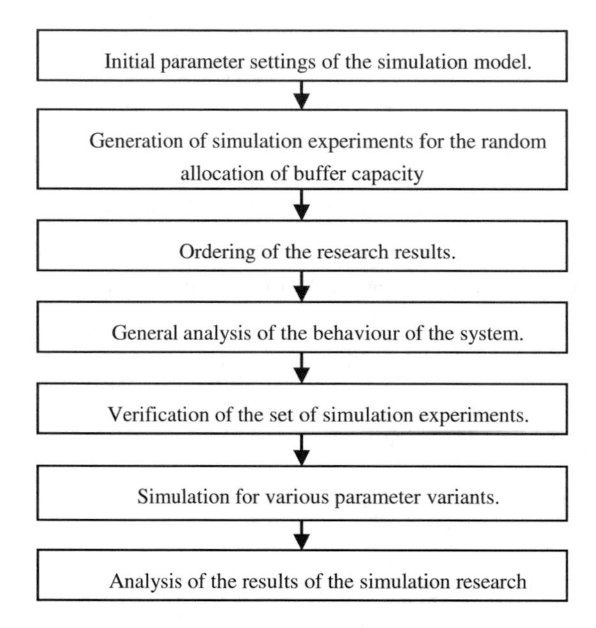

Fig. 3. Methodology for research into the simulation of a parallel serial manufacturing system

For the initial values of operation times, presented in the last chapter, 50 simulation experiments were generated with random buffer capacities from 1 to 20 units.

The input values of the experiments were the allocated buffer capacities and the output was throughput per hour. For each simulation experiment, 3 observations were performed. Figure 5 presents the results of the experiments. The throughput values are in ascending order. The throughput range was from 20,80 to 21,78 products per hour. In Table 1, the allocation of buffer capacity for the lowest throughput value is presented, that is, for the first three experiments. Analysis of buffer capacity shows that if buffers B1 and B4 are significantly larger than the rest of the buffers, then system throughput is low. To check the above conclusion, a new allocation of buffer capacity was proposed; this is presented in Table 2, q.v. new experiments 1–5 (Fig. 4).

The lowest throughput value was obtained for the buffer allocation presented in experiment Exp. 10.

Table 1. Results of simulation experiments – the lowest throughput of the system

Experiment	B1	B2	B3	B4	B5	B6	Throughput
1	19	2	3	19	1	1	20,59
2	19	8	2	16	5	1	20,80
3	17	1	16	15	19	7	20,82

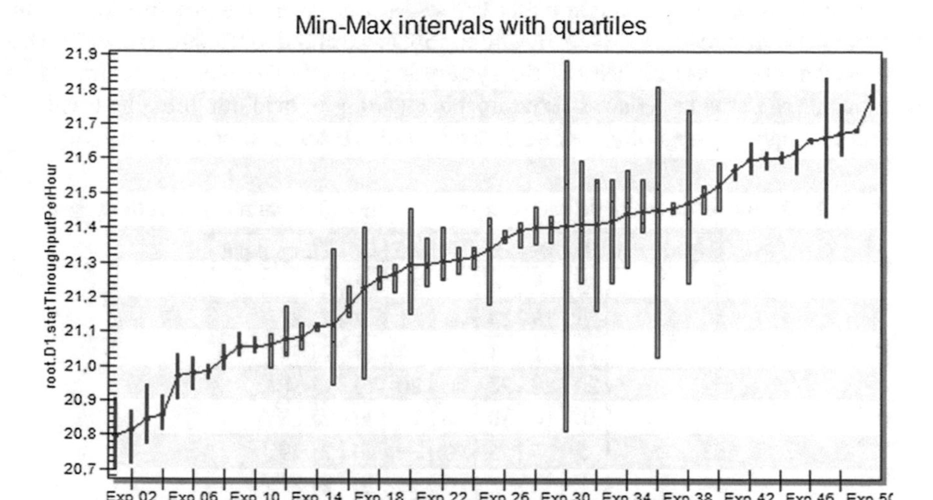

Fig. 4. Throughput values per hour ordered

Table 2. Results of new simulation experiments – the lowest throughput of the system

Experiment	B1	B2	B3	B4	B5	B6	Throughput
1	20	1	1	20	1	1	20.56
2	20	2	2	20	2	2	20.54
3	20	3	3	20	3	3	20.58
4	20	2	2	20	1	1	20.53
5	20	5	5	20	1	1	20.82
6	19	2	3	19	1	1	20.59
7	19	8	2	16	5	1	20.80
8	17	1	16	15	19	7	20.82

The highest throughput value was obtained for the allocation of buffer capacities shown in Table 3. The highest throughput value was obtained where buffers B4, B5 and B6 had higher capacities. Five, new, simulation experiments were prepared in order to confirm the thesis.

The results are presented in Table 4. The highest throughput value was obtained for experiment 54. The results presented in the table show that similar throughput values

Table 3. Results of the simulation experiment – the highest throughput of the system

Experiment	B1	B2	B3	B4	B5	B6	Throughput
48	8	5	6	7	12	3	21.67
49	15	13	14	9	11	7	21.68
50	6	1	1	11	18	19	21.78

can be obtained for quite different buffer capacities, q.v., compare experiments 48 and 52 where the difference between the throughputs is about 0.02 and the difference between the total buffer capacity of the system is 53 units. The number of simulation experiments can now be reduced. Experiments resulting in medium throughput values, or with the highest variability, can be deleted from the set of simulation experiments.

Table 4. Results of new simulation experiments – the lowest throughput of the system

Experiment	B1	B2	B3	B4	B5	B6	Throughput
48	8	5	6	7	12	3	21.67
49	15	13	14	9	11	7	21.8
50	6	1	1	11	18	19	21.78
51	20	20	20	20	20	20	21.56
52	10	10	10	20	20	20	21.65
53	5	5	5	20	20	20	21.79
54	2	2	2	20	20	20	21.81
55	1	1	1	20	20	20	21.76

The goal of this step is to prepare a new set of experiments, in order to analyse the impact of other system parameters on throughput and the average lifespan of a product. The set of verified simulation experiments is presented in Table 5.

Table 5. A new set of simulation experiments

Experiment	B1	B2	B3	B4	B5	B6
1	20	2	2	20	1	1
2	20	2	2	20	2	2
3	20	1	1	20	1	1
4	20	3	3	20	3	3
5	17	1	16	15	19	7
6	19	18	1	16	8	17
7	10	18	1	7	2	14
8	10	20	7	17	6	15
9	13	20	8	11	9	3
10	11	2	14	12	5	19
11	15	2	8	17	18	19
12	12	12	20	2	19	7
13	10	20	12	9	6	11

(continued)

Table 5. (*continued*)

Experiment	B1	B2	B3	B4	B5	B6
14	18	9	12	2	16	4
15	14	17	20	17	20	10
16	2	5	13	15	5	1
17	14	11	6	9	16	17
18	12	18	18	16	4	1
19	15	12	16	6	19	9
20	9	14	8	7	11	11
21	18	12	17	11	7	16
22	20	20	20	20	20	20
23	16	11	15	4	10	8
24	16	15	14	3	8	16
25	18	14	15	10	9	5
26	10	10	10	20	20	20
27	15	13	14	9	11	7
28	6	1	1	11	18	19
29	5	5	5	20	20	20
30	2	2	2	20	20	20

Results of the 30 simulation experiments are presented in Figs. 5 and 6. The experiments are ordered according to the increasing throughput value. The throughput value ranges from $P_{Exp01} = 20.53$ to $P_{Exp30} = 21.81$ elements *per* hour, this being the highest throughput value. The average product lifespan ranges from 56 min to 3 h and 27 min. The lowest average lifespan values were obtained for experiments Exp28 and Exp30 (respectively $\omega_{Exp28} = 56{:}41$ and $\omega_{Exp30} = 59{:}20$).

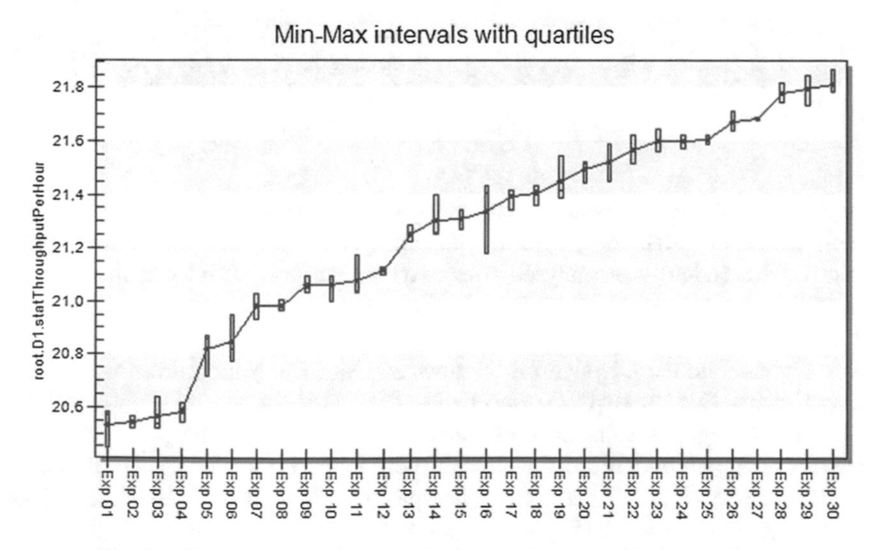

Fig. 5. Throughput *per* hour, ranked according to verified experiments

The effectiveness of a system includes not only its throughput but also *work-in-progress*, the lower, the better. To evaluate the total effectiveness of a parallel, serial manufacturing system, the following system performance rate θ is proposed:

$$\theta = \frac{P}{\omega} \tag{4}$$

where P – throughput of the system, ω – average product lifespan. The values for the performance rate of the system are presented in Fig. 7. Values for the performance rate of the system are presented in Fig. 8. The results presented in the chart show that the highest efficiency, within the system, was obtained for the buffer capacity allocations in Exp28 and Exp 30 (θ > 500). The lowest, system performance rate value was reached in experiments Exp5, Exp6, Exp8, Exp11, Exp15, Exp 23 (θ < 200).

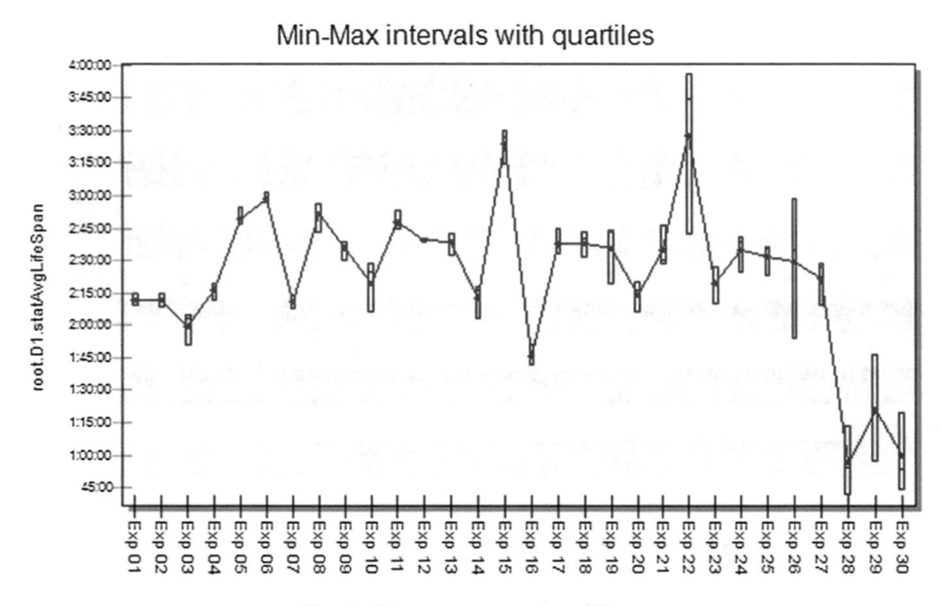

Fig. 6. The average product lifespan

In the next step of the research methodology, the impact of various parameters on a parallel, serial manufacturing system was analysed. The throughput and average lifespan of the system was analysed over a wider range of CNC machine operation times, that is, with new, Uniform Distribution with the following parameters Start = 300 and Stop = 600).

The highest throughput value for the new variant of operation times was obtained for experiments Exp16, Exp26 (respectively, $P'_{Exp16} = 22.34$ and $P'_{Exp26} = 22.36$). The lowest throughput value was obtained for Exp1, where $P'_{Exp01} = 22.87$. The values of average product lifespan were relatively lower for a wider range of operation times. The lowest values of the average lifespan were obtained for Exp28 and Exp30 (respectively, $\omega'_{Exp28} = 45{:}35$ and $\omega_{Exp30} = 43{:}13$). The values of the system

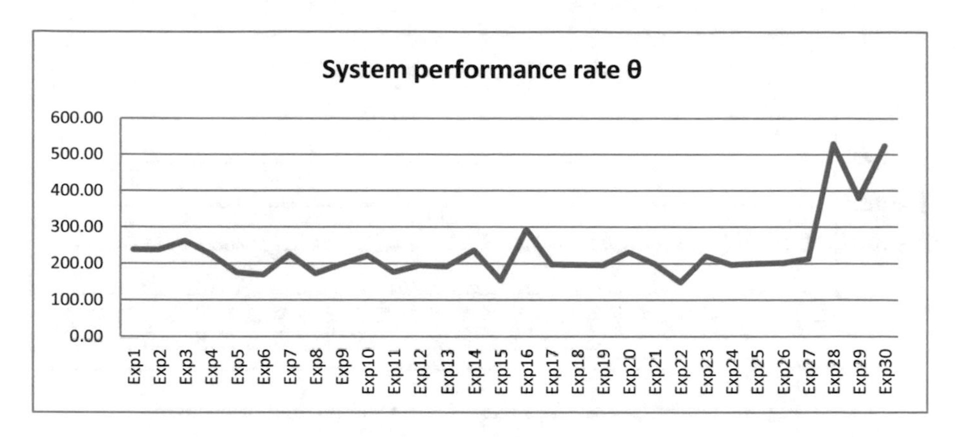

Fig. 7. System performance rate θ for the initial operation times

performance index are presented in Fig. 8. Generally speaking, system effectiveness was higher, *q.v., higher average throughput and lower average product lifespan.* The best performance rate values were obtained for the buffer allocations in experiments Exp28 and Exp30 (θ' > 700). The lowest values of the system performance rate were obtained, similarly, as in the first variant, in Exp22. Generally speaking, analysis of the simulation results show that a change of the range of operation times had no impact on the total performance rate of the parallel, serial manufacturing system.

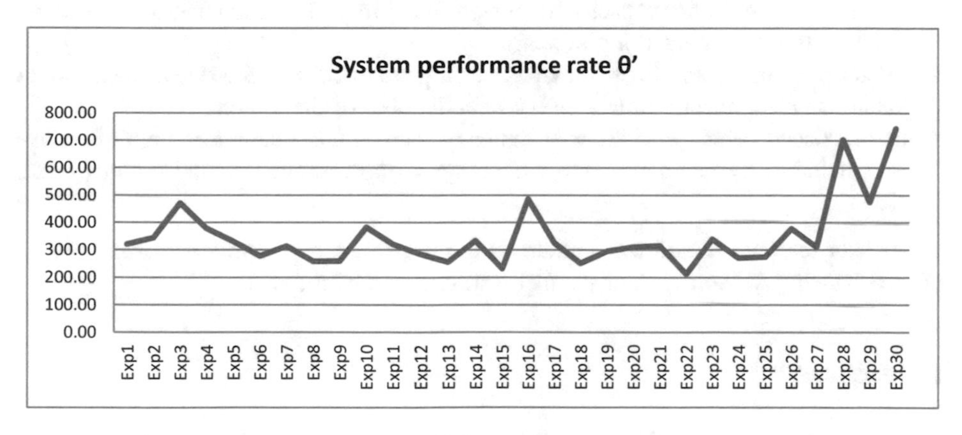

Fig. 8. System performance rate θ' for a wider range of operation times

The last part of the study was conducted for initial operation times and equal lot sizes for all products (A = B = C = D = 200). Even though the variability of the results was higher, the behaviour of the system was, in general, similar to the initial parameters. The average product lifespan for equal and higher batch sizes was relatively lower than was the case with the initial parameters. The system performance rate for equal product batch sizes is presented in Fig. 9. The highest value of the system performance rate was obtained in Exp28 (θ" = 680.91).

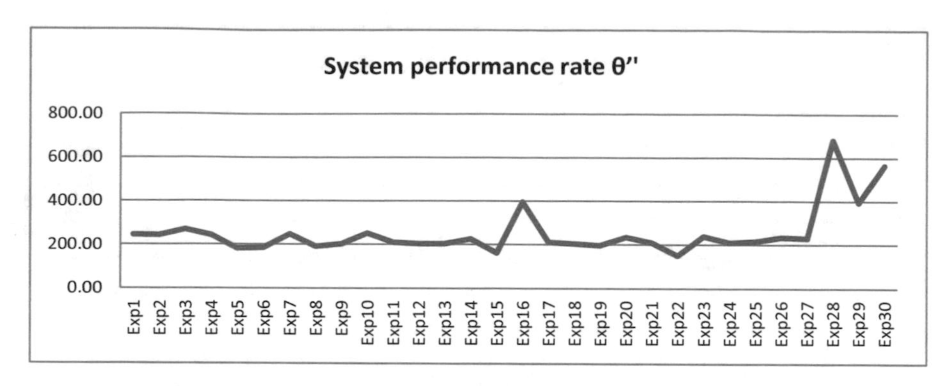

Fig. 9. System performance rate θ" for equal product batch sizes

4 Conclusions

A model of a parallel, serial manufacturing system has been proposed. On the basis of the assumptions defined, the simulation model of a parallel, serial manufacturing system was created using Tecnomatix Plant Simulation Software. The methodology of the simulation research was proposed. On the basis of this methodology, the behaviour of a parallel, serial manufacturing system, with 9 CNC machines and 6 buffers, was analysed. The following conclusions can be formulated:

- the allocation of buffer capacity has a significant impact on the effectiveness of the parallel, serial manufacturing system,
- the range of operation times of manufacturing resources and the size of the product batch has a significant impact on the effectiveness of the system,
- there exists a certain allocation of buffer capacities that guarantees high effectiveness, within a system and over a wide range of operation times and product batch sizes.

Further research should concentrate on the impact on a system's performance rate of the changing of structures of parallel, serial manufacturing.

References

1. Negahban, A., Smith, J.S.: Simulation for manufacturing system design and operation: literature review and analysis. J. Manuf. Syst. **33**, 241–261 (2014). https://doi.org/10.1016/j.jmsy.2013.12.007
2. Vidalis, M.I., Papadopoulos, C.T., Heavey, C.: On the workload and 'phase load' allocation problems of short reliable production lines with finite buffers. Comput. Ind. Eng. **48**, 825–837 (2005). https://doi.org/10.1016/j.cie.2004.12.011
3. Demir, L., Tunali, S., Eliiyi, D.T.: The state of the art on the buffer allocation problem: a comprehensive survey. J. Intell. Manuf. **25**, 371–392 (2014). https://doi.org/10.1007/s10845-012-0687-9

4. Staley, D.R., Kim, D.S.: Experimental results for the allocation of buffers in closed serial production lines. Int. J. Prod. Econ. **137**, 284–291 (2012). https://doi.org/10.1016/j.ijpe.2012.02.011
5. Kłos, S., Patalas-Maliszewska, J.: The topological impact of discrete manufacturing systems on the effectiveness of production processes. In: Recent Advances in Information Systems and Technologies, Advances in Intelligent Systems and Computing, vol. 571, pp. 441–452. Springer International Publishing (2017). https://doi.org/10.1007/978-3-319-56541-5_45
6. Jagstam, M., Klingstam, P.: A handbook for integrating discrete event simulation as an aid in conceptual design of manufacturing systems. In: Proceedings of the 2002 Winter Simulation Conference, vol. 2, pp. 1940–1944 (2002)
7. Varela, M.R.L., Trojanowska, J., Carmo-Silva, S., Costa, N.M.L., Machado, J.: Comparative simulation study of production scheduling in the hybrid and the parallel flow. Manag. Prod. Eng. Rev. **8**(2), 69–80 (2017). https://doi.org/10.24425/119404
8. Kochańska, J., Burduk, A.: Optimization of production support processes with the use of simulation tools. In: Information Systems Architecture and Technology: Proceedings of 38th International Conference on Information Systems Architecture and Technology, ISAT 2017, pp. 275–284. Springer (2017). https://doi.org/10.1007/978-3-319-67223-6_26
9. Bocewicz, G., Nielsen, I.E., Banaszak, Z.A.: Production flows scheduling subject to fuzzy processing time constraints. Int. J. Comput. Integr. Manuf. **29**(10), 1105–1127 (2016). https://doi.org/10.1080/0951192X.2016.1145739
10. Bartkowiak, T., Pawlewski, P.: Reducing the negative impact on the production and filling process of the simulative study. In: Proceedings - Winter Simulation Conference, pp. 2912–2923 (2017). Art. Well. 7822326. https://doi.org/10.1109/wsc.2016.7822326
11. Bartkowiak, T., Ciszak, O., Jablonski, P., Myszkowski, A., Wisniewski, M.A.: A simulative study approach for improving the efficiency of production process of floorboard middle layer. In: Lecture Notes in Mechanical Engineering, pp. 13–22 (2018). https://doi.org/10.1007/978-3-319-68619-6_2
12. Bartkowiak, T., Gessner, A.: Modeling performance of a production line and optimizing its efficiency by means of genetic algorithm. In: ASME 2014 12th Biennial Conference on Engineering Systems Design and Analysis, ESDA 2014 (2014). https://doi.org/10.1115/esda2014-20141
13. Battini, D., Persona, A., Regattieri, A.: Buffer size design linked to reliability performance: A simulative study. Comput. Ind. Eng. **56**, 1633–1641 (2009)

Aperiodic Surface Topographies Based on High Precision Grinding Processes: Analysis of Cutting Fluid and Cleaning Process Influences Using Non-parametric Statistics

Stefan Bracke(✉) and Max Radetzky

University of Wuppertal, Chair of Reliability Engineering and Risk Analytics, Gaussstrasse 20, 42119 Wuppertal, Germany
{bracke,radetzky}@uni-wuppertal.de

Abstract. High precision manufacturing processes of technical product surfaces have to fulfil demanding requirements regarding functional characteristics like roughness, gloss and colour. Especially the reproducibility and control of aperiodic surface topographies based on grinding processes are influenced by many manufacturing process factors and cleaning methods. This paper outlines a concept for the multivariate analysis of measurement data regarding aperiodic surface topographies (characteristics: roughness and gloss) based on small sample sizes, varied process parameter and different cleaning methods. The concept and analysis are based on non-parametric statistics, due to the particular challenge which is the missing knowledge of the distribution models regarding to the characteristics roughness and gloss. The application of the worked out measurement strategy and the multivariate non-parametric statistical method is shown within a case study "Grinding surfaces of cutlery" comparing four different processes. The compared cleaning compounds are cold degreaser (standard method) and acetone (optional method) in combination with different technical grinding processes (cutting fluid fat respectively water based).

Keywords: Multivariate analysis · Non-parametric statistics · Grinding process Cleaning influence · Surface topography

1 Introduction

For the manufacture of cutlery, where in addition to sharpness and durability, the optical perception of the blades play a major role, a special fine grinding process has been developed. It is characterised by a high amount of process parameters, which strongly influence the result of the blades surface topography. The background of the research work is the development of a new grinding machine generation, using water based cutting fluid. The goal is to detect grinding parameter sets, which ensure the reproducibility and repeatability of the grinding results regarding a small scattering of surface topography characteristics (roughness and gloss). Out of two different production processes a variety of blades is manufactured and cleaned with two different methods.

© Springer Nature Switzerland AG 2019
A. Burduk et al. (Eds.): ISPEM 2018, AISC 835, pp. 44–55, 2019.
https://doi.org/10.1007/978-3-319-97490-3_5

It is difficult, to distinguish the blades visually and subjectively (qualitative) regarding the surface topography. To solve the challenge of receiving an uniform production output, multivariate surface analyses with the use of non-parametric statistics can be set up in order to evaluate the aperiodic surface topographies based on different manufacturing processes. This statistical analytics allows the user, to distinguish the blades in a quantitative way. The application of the research work is shown within the case study "Grinding surfaces of cutlery".

2 Goal of Research Study

The goal of the research work is the comparison of aperiodic surface topographies with non-parametric statistics applied at the example of four different grinding processes. The analyses consider surface topographies, which are manufactured based on varied process parameters and cleaning methods. The combination of varied process parameters are as follows:

- Comparison of two grinding processes utilising cutting fluid based on fat without additional cleaning respectively with the use of acetone (cleaning).
- Comparison of two grinding processes utilising water based cutting fluid without additional cleaning respectively with the use of acetone (cleaning).

The goal of the research work is to provide an approach based on statistical methods, how to analyse surface topographies with aperiodic profiles in a quantitative way, considering small sample sizes. The key questions with regard to the analysis are as follows:

1. Which of the manufacturing processes shows the highest cleaning effect?
2. What is the difference between the cleaning effects of the four possible combinations of processes?
3. Which of the analysed surface characteristics (R_a, R_z and gloss value) is most affected by the cleaning influence?
4. How can the cleaning effect be quantitatively determined?

3 Base of Operations

3.1 Manufacturing and Cleaning of Cutlery Samples

Figure 1 shows an excerpt of the most important cutlery manufacturing steps. Concerning to [1] the production of a knife can contain up to 55 working-steps. The main focus is the fine grinding process of the blade, being classified to the group of side grinding [2, 3]: The rotating grinding tool is in contact with one side of the work piece predominantly by linear feed. The utilised material is knife steel 1.4116 (X45CrMoV15) with a Rockwell hardness of 56 HRC.

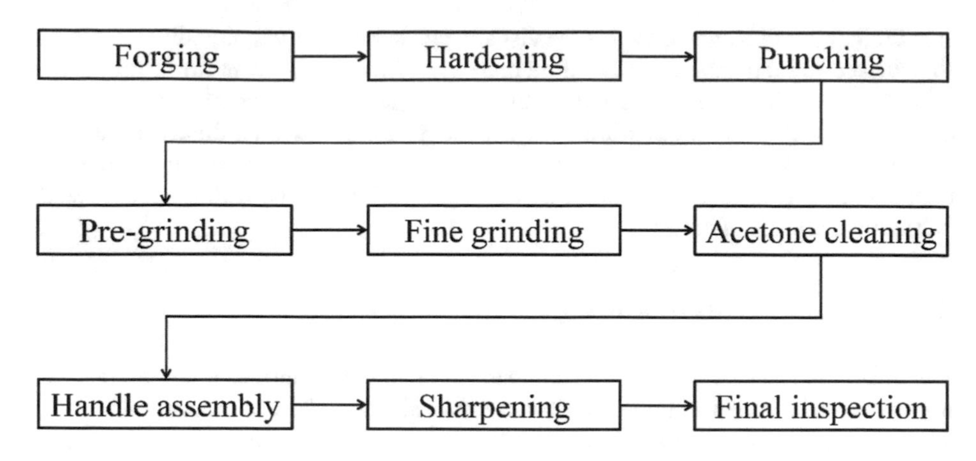

Fig. 1. Excerpt of cutlery manufacturing process (Franz Güde GmbH).

Two cleaning methods can be distinguished after the grinding process. The standard method of the manufacturer, using cold degreaser (UN number 3295), in the following described as "no additional cleaning" and the optional cleaning of the blade using acetone (CAS number 67-64-1, UN number 1090). The organic compound with the formula C_3H_6O is a colourless, volatile liquid and is utilised as a fat solvent for cleaning and degreasing metal surfaces.

Table 1 sums up important influencing factors of the grinding process regarding the surface topographies besides the influence of cleaning methods.

Table 1. Influencing process parameters regarding surface topography [5].

Factor	Range	Unit
Grinding disc compound	1–2 types	–
Cutting fluid mixture	1–2 types	–
Flow rate of cutting fluid	10–90	l/min
Cutting fluid temperature	20–100	°C
Feed rate	100–500	mm/min
Cutting speed	5–35	m/s
Revolutions per minute	100–1000	rpm
Motor current	5–100	A
Contact force	10–1000	N
Blade temperature	50–500	°C

Summarised in Table 2, the four types of analysed samples can be found. The analysed knives of the same type have a blade length of 21 cm and differ regarding the grinding process and the cleaning method. The cutlery of process A are grinded with a cutting fluid based on fat, without additional cleaning (cold degreaser only). Cutlery of process B differ regarding the cleaning method utilising acetone. The varied grinding method used in process C is based on water with additive compounds without additional cleaning (cold degreaser only) and process D differs concerning the cleaning with acetone.

Table 2. Summary of investigated samples.

Description	Cutlery type	Grinding	Cleaning method	Sample size
Process A	1765/21 (8" Slicer)	Fat based	No additional cleaning	32
Process B	1765/21 (8" Slicer)	Fat based	Acetone	32
Process C	1765/21 (8" Slicer)	Water based	No additional cleaning	20
Process D	1765/21 (8" Slicer)	Water based	Acetone	20

Detailed surface topography recordings of process A, B, C and D (shown in Fig. 2) are generated with a Carl Zeiss Axio Imager.A1 m light microscope, under constant exposure, with magnification factor of 50 and a EC Epiplan-Neofluar 50x/0,80 Pol lens.

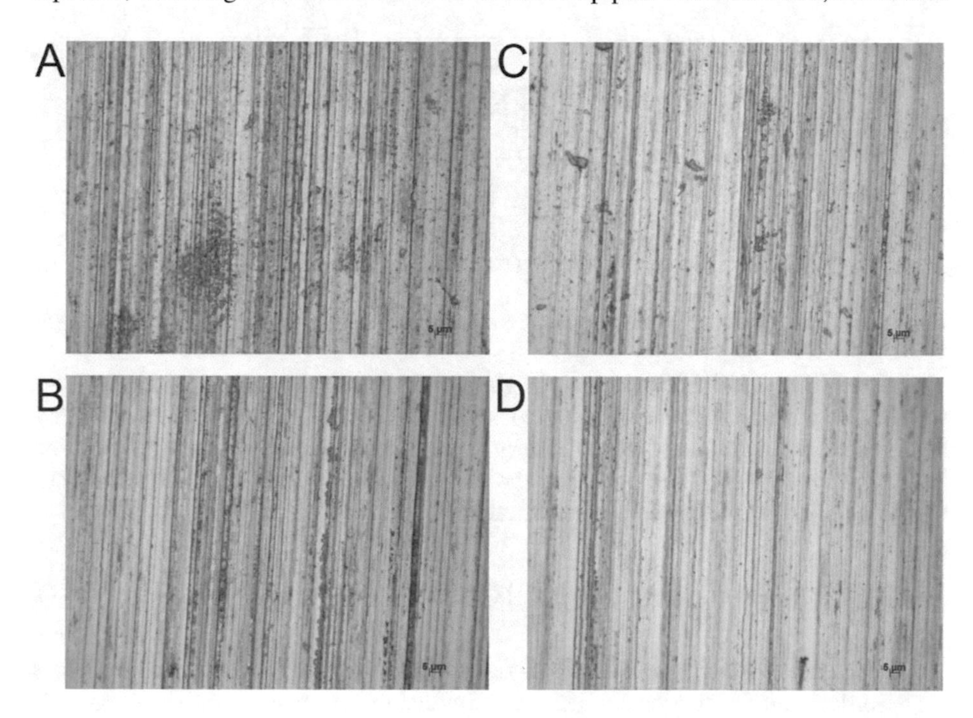

Fig. 2. Aperiodic surface profiles based on grinding processes, using different cutting fluids and cleaning methods (Photo: TH Köln, Germany).

3.2 Measurement Systems and Key Characteristics Describing Surface Topology

The most relevant characteristics regarding the analysis and comparison of surface topographies are the arithmetic average roughness R_a and the average surface roughness R_z, recorded with PCE-RT11, having a tactile diamond tip radius of 10 μm (± 1 μm).

Furthermore, the gloss value measured at 60° angle with PCE-PGM-60 is important. The definition of the characteristics R_a, R_z and gloss are simplified shown in Fig. 3.

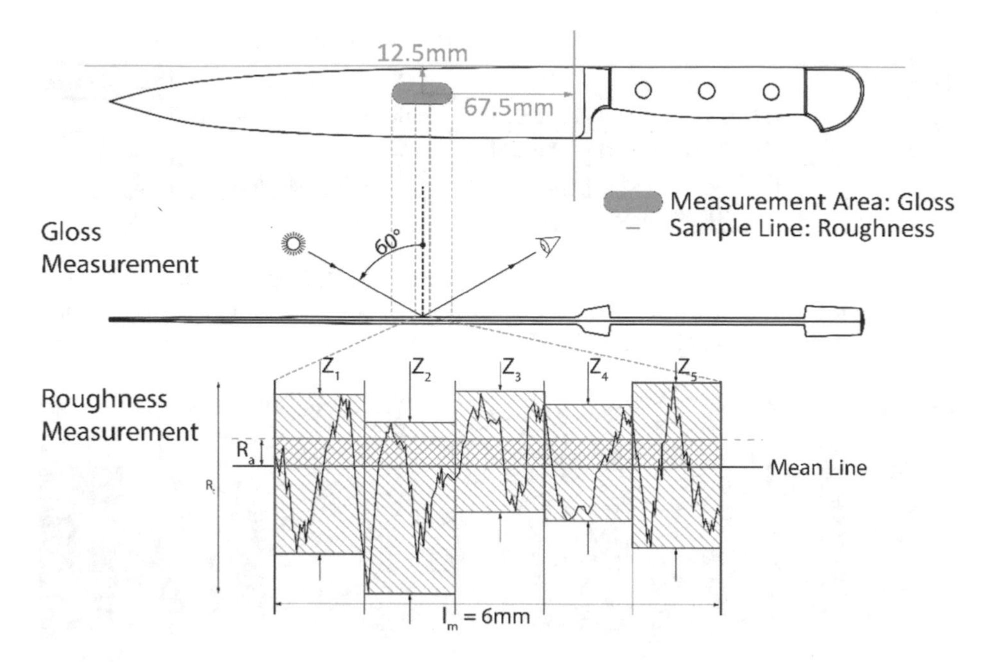

Fig. 3. Schematic diagram of gloss/roughness measurement position at knife type 1765/21 [5].

It is very important to ensure that the gloss and roughness measurements take place at the same position on the different blades. For this reason, a spacer made of ABS is designed to fix the different types of cutlery and ensure the same measurement position.

According to [4, 5], R_a is defined as the arithmetic mean of the ordinate values Z(x) within a single measuring section indicated in Eq. 1.

$$R_a = \frac{1}{l} \int_0^l |Z(x)| dx \tag{1}$$

The average surface roughness R_z is defined as the arithmetic mean of the largest single-order values of several contiguous individual measurement sections Z_i exemplified in Eq. 2, illustrated in Fig. 3 [4].

$$R_z = \frac{1}{n}(Z_1 + Z_2 + \ldots Z_n) \tag{2}$$

In the particular use case, the roughness measuring device is specified by a total sampling length of 6 mm.

According to [6], gloss is described as the optical property of a surface to reflect directed light (cf. also [5]). The gloss value indicated in GU (gloss units) is the ratio multiplied by 100 of reflected light fluxes by a sample and a glass surface with the refractive index of 1.567 at a wavelength of 587.6 nm, where the reflection angle and the aperture angles of the light source and receiver are fixed. Due to [7], the specular gloss G is defined as the ratio between the measured irradiance of the sample and the gloss reference, as defined in Eq. 3.

$$G = \frac{I_{sample}}{I_{reference}} \cdot 100 \tag{3}$$

I_{sample} is the irradiance of the knife surface and $I_{reference}$ the irradiance of the gloss reference. The reference is a black glass plate with the refractive index of 1.567 and its gloss reading is defined to be equal 100 GU [8]. Furthermore, gloss at a specified angle (20°, 60° or 85°) can be calculated from the ratio of the reflected light of the sample and the reflected light of the glass standard indicated in Eq. 4 using the Fresnel's equation [8].

$$G_\alpha = \frac{\left(\frac{cos\alpha - \sqrt{b^2 - sin^2\alpha}}{cos\alpha + \sqrt{b^2 - sin^2\alpha}}\right)^2 + \left(\frac{b^2 cos\alpha - \sqrt{b^2 - sin^2\alpha}}{b^2 cos\alpha + \sqrt{b^2 - sin^2\alpha}}\right)^2}{\left(\frac{cos\alpha - \sqrt{1.567^2 - sin^2\alpha}}{cos\alpha + \sqrt{1.567^2 - sin^2\alpha}}\right)^2 + \left(\frac{1.567^2 cos\alpha - \sqrt{1.567^2 - sin^2\alpha}}{1.567^2 cos\alpha + \sqrt{1.567^2 - sin^2\alpha}}\right)^2} \cdot 100 \tag{4}$$

Here, the 60° selection for the α value is the universal measuring angle for all gloss levels and has a measuring range from 0 to 1000 GU. The refraction index of the knife surface is represented by b.

3.3 Non-parametric Statistics

Non-parametric models differ from parametric models in the model structure, which is not determined in advance, but is determined from the data. The term non-parametric does not mean that such models have no parameters at all. The type and number of parameters is not fixed from the outset. E.g. non-parametric methods are mathematical procedures for testing statistical hypotheses. Unlike parametric statistical tests, non-parametric methods are not based on the assumption of a certain probability distribution regarding the analysed variables. Therefore, they are also applicable independent from distribution model regarding the measured data [5].

For the comparison of two independent samples (e.g. location and dispersion) the most common approaches of non-parametric statistic significance tests can be used according to [9], as follows:

Wilcoxon-Mann-Whitney-Test: Two samples F and G have the same location "μ" (null hypothesis H_0), or differ significantly regarding the location (alternative hypothesis H_1), as shown in Eq. 5.

$$H_0{:}\mu_F = \mu_G; \quad H_1{:}\mu_F \neq \mu_G \tag{5}$$

Levene's test: Two samples F and G are equivalent (null hypothesis H_0), or differ regarding to variance "σ^2" significantly (alternative hypothesis H_1), seen in Eq. 6.

$$H_0{:}\sigma_F^2 = \sigma_G^2;\ \ H_1{:}\sigma_F^2 \neq \sigma_G^2 \tag{6}$$

Kruskal-Wallis test: A certain number of samples (i) have the same location "μ" (null hypothesis H_0), or one or more samples differ significantly regarding the location (alternative hypothesis H_1), cf. Eq. 7.

$$H_0{:}\mu_F = \mu_G = \ldots = \mu_i;\ \ H_1{:}\mu_F \neq \mu_G \neq \ldots \neq \mu_i \tag{7}$$

Bartlett test (generalised Levene's test): A certain number of samples (i) is equal (null hypothesis H_0) or the dispersion "σ^2" of one or more samples differs significantly to other samples (alternative hypothesis H_1), cf. Eq. 8.

$$H_0{:}\sigma_F^2 = \sigma_G^2 = \ldots = \sigma_i^2;\ \ H_1{:}\sigma_F^2 \neq \sigma_G^2 \neq \ldots \neq \sigma_i^2 \tag{8}$$

Precondition is the independency of the analysed samples, which can be evaluated by using the Spearman correlation coefficient (rank based analysis; independent of characteristic distribution model), cf. Eq. 9 [10].

$$r_s = \frac{\sum_{i=1}^{n}\left(R(x_i) - \overline{R(x)}\right)\left(R(y_i) - \overline{R(y)}\right)}{\sqrt{\sum_{i=1}^{n}\left(R(x_i) - \overline{R(x)}\right)^2 \sum_{i=1}^{n}\left(R(y_i) - \overline{R(y)}\right)^2}} \tag{9}$$

3.4 Measurement Strategy

In order to maintain homogeneous conditions, the measurements are performed in a laboratory with constant air and light conditions to eliminate interference factors. The measurements are carried out based on the described technical measurement systems in Sect. 3.2. The goal of the measurement strategy is the detection of dependencies between surface topography characteristics and the cleaning effect by bivariate and multivariate comparison of the R_a, R_z and gloss values (cf. [5]; adapted approach). The measuring position centrally arranged at the blade surface (cf. Fig. 3) is chosen to compare the cutlery type 1765/21.

Step 1 Calibration of measuring devices with setting gauges (glass standard plate of 99.1 GU, R_a roughness standard of 1.64 µm).

Step 2 Measurement of gloss and roughness without knife fixation change at the spacer, with removal of the measuring devices.

Step 3 Cleaning process of the blade, using a defined quantity of acetone and cleaning tissue free of residues.

Step 4 Second measurement of gloss and roughness without knife fixation change at the spacer, with removal of the measuring devices.

Step 5 Analysis using statistical tests, described in Sect. 3.3, with the goal to detect surface heterogeneity and failures regarding the specifications.

Step 6 Analysis by means of pairwise sample topography comparison with the goal to find differences between different sets of cutlery manufactured with different process technologies.

Step 7 Analysis based on comparison of several sample topographies with multivariate methods

4 Case Study "Grinding Surfaces": Application and Results

Based on explained measurement strategy (cf. Sect. 3), a summary of the analysing results is outlined in this section. It can be stated, that the surface topographies grinded with the parameters of processes A and B (cf. Table 2) contain low gloss values and comparable high values of the arithmetic average roughness R_a and the average surface roughness R_z (cf. Fig. 4). In contrast, surface topographies grinded with parameters of processes C and D (cf. Table 2) show a comparatively high gloss and lower R_a and R_z values. Furthermore, Fig. 4 shows the regression models regarding R_a and gloss respectively R_z and gloss.

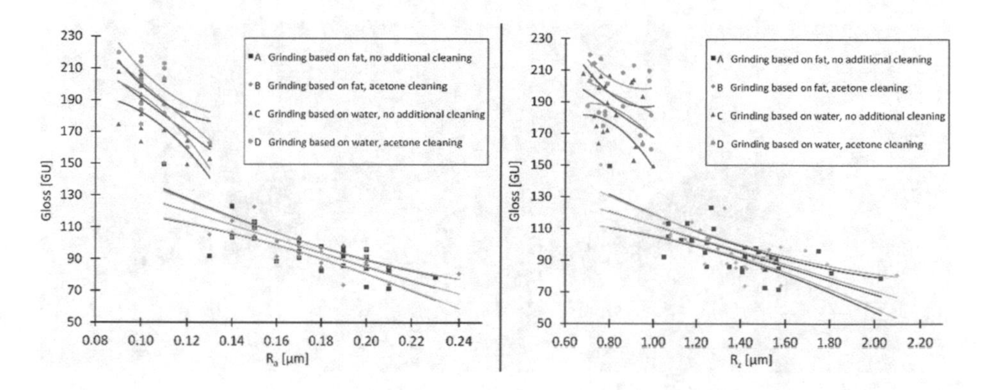

Fig. 4. Comparison of grinding processes A-D (left side: R_a and gloss; right side: R_z and gloss).

Implemented confidence bands (95% level) [11] regarding the regression models show the similarities of the processes: The overlapping of the confidence bands is a strong indicator for the non-differentiability of the regression models.

Based on carried out non-parametric significance tests with a significance level of 5% (cf. Sect. 3.3), regarding the measurement values, the relevant statistic values and results are listed in Table 3.

For the multivariate analysis of the processes A, B, C and D (cf. Table 2), a multiple linear regression model is calculated based on [9] including the variables arithmetic average roughness R_a, the average surface roughness R_z and the gloss value. The definition of the target value function is conducted by considering the measured minimum and maximum values of the surface parameters. R_a and R_z are selected as influencing variables and the gloss is set to be the target value. With this approach the cleaning effects can be compared taking in account the three-dimensional surface correlation functions. The results of the determined multiple linear regressions are shown in Fig. 5 (process A and B) and Fig. 6 (process C and D).

Table 3. Statistical results of surface topography parameters.

Value	Process A	Process B	Process C	Process D
n	32	32	20	20
R_a average [μm]	0.176	0.173	0.106	0.106
R_a median [μm]	0.180	0.170	0.100	0.100
R_a standard deviation [μm]	0.028	0.028	0.011	0.010
R_z average [μm]	1.382	1.368	0.822	0.833
R_z median [μm]	1.415	1.370	0.790	0.785
R_z standard deviation [μm]	0.247	0.248	0.093	0.108
Gloss average [GU]	95.544	97.052	184.403	193.788
Gloss median [GU]	92.200	95.355	185.550	194.465
Gloss standard deviation [GU]	15.254	14.885	19.582	18.876

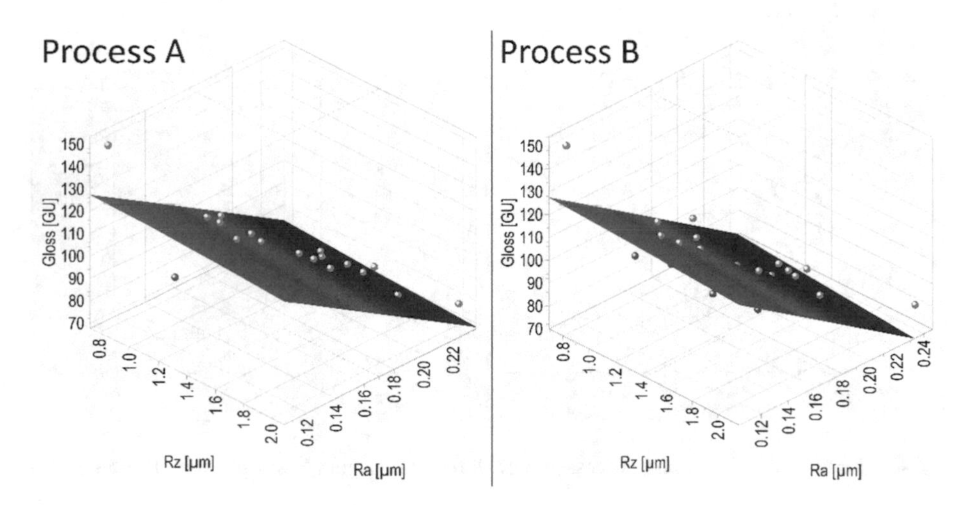

Fig. 5. 3D graphs of the regression models based on parameters R_a, R_z and gloss showing the surface topography dependency (left side: process A; right side: process B).

The perspective setting of the 3D graphs is equivalent, therefore a visual comparison of the processes is possible. The resulting surface equations and correlation coefficients are listed in Table 4. Processes A and B show the highest correlation and are very similar with regard to the surface equation. The correlation coefficients and surface equations of processes C and D – especially process C - differ and can be distinguished visually. The parameters of the regression models A and B in comparison to C and D are different, the reason is the influence of the cutting fluid (water based versus fat based, cf. Table 2).

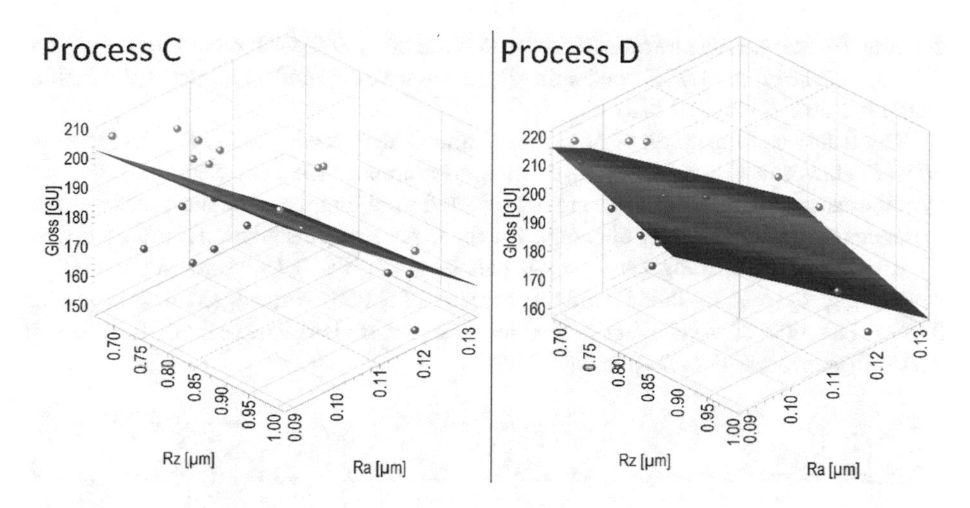

Fig. 6. 3D graphs of the regression models based on parameters R_a, R_z and gloss showing the surface topography dependency (left side: process C; right side: process D).

Table 4. Results of multiple linear regression (process A-D).

Process number	Mathematical surface equation	Correlation coefficient
Process A	$y_A = 172.667 - 3.51387 \cdot X_1 - 410.751 \cdot X_2$	$r_A = 0.7984$
Process B	$y_B = 171.800 - 3.35782 \cdot X_1 - 405.961 \cdot X_2$	$r_B = 0.7980$
Process C	$y_C = 294.806 + 16.2372 \cdot X_1 - 1167.37 \cdot X_2$	$r_C = 0.6250$
Process D	$y_D = 330.298 + 100.799 \cdot X_1 - 2079.95 \cdot X_2$	$r_D = 0.7487$

Figure 7 shows a detailed comparison of process A and B considering R_a, R_z and the gloss value. Based on conducted non-parametric tests (cf. Sect. 3.3) it can be stated that the null hypotheses of equal averages, medians and variances are not rejected, which can be interpreted as no significant difference regarding the surface topographies of process A and B. The mean percentage deviations before (process A) and after the

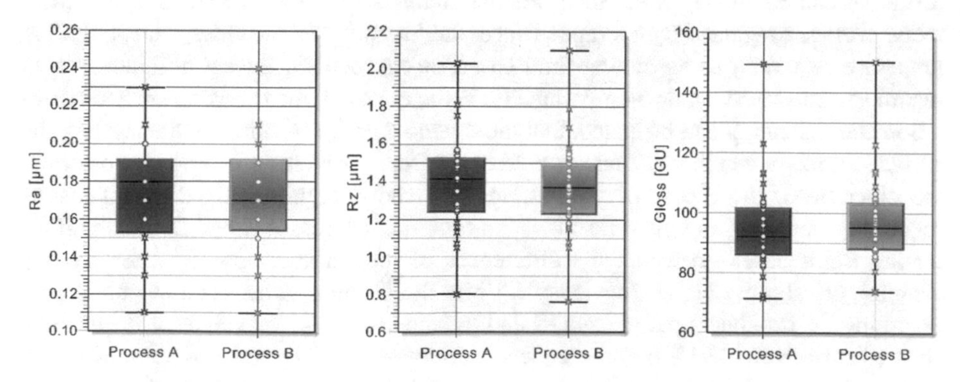

Fig. 7. Boxplot - Comparison of grinding processes A and B considering R_a, R_z and gloss.

cleaning (process B) can be quantified to a decline of −1.84% with regard to R_a (p-value 0.570), a decrease of −1.22% concerning R_z (p-value 0.773) and an increase of the gloss value of 1.60% (p-value 0.528).

The detailed comparison of processes C and D with regard to R_a, R_z and the gloss value is shown in Fig. 8. Resulting from non-parametric tests (cf. Sect. 3.3), the null hypotheses of equal distributions are not rejected implying that averages, medians and variances are not differing significantly. The mean percentage deviations before (process C) and after the cleaning (process D) can be quantified to a decrease of −0.02% concerning R_a (p-value 0.863) and an increase of 1.06% with regard to R_z (p-value 0.892). The most obvious effect of the cleaning process is the increase of the gloss of 4.91% from process C to D (p-value 0.105).

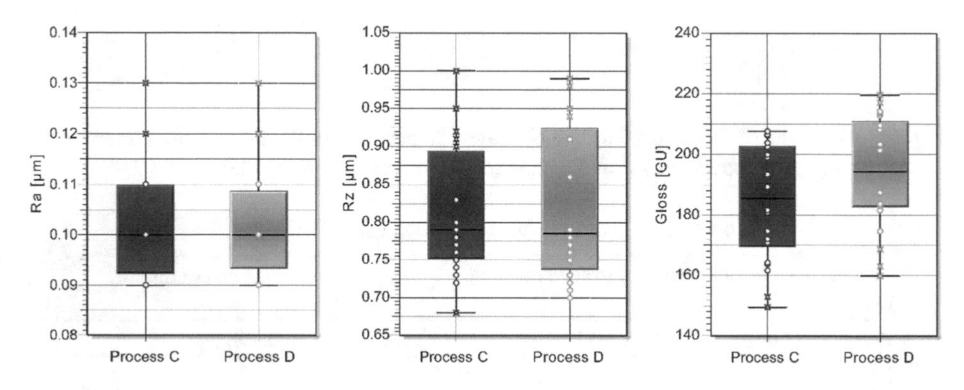

Fig. 8. Boxplot - Comparison of grinding processes C and D considering R_a, R_z and gloss.

5 Summary

This paper outlines an approach for the multivariate analysis of aperiodic surface topographies within high precision grinding processes based on small sample sizes and shows the influence of different cutting fluids and acetone cleaning processes. Four different processes are analysed varying in grinding method (cutting fluid) and cleaning procedure (different combinations). Main focus are the analyses of the surface parameters arithmetic average roughness R_a, average surface roughness R_z and the gloss value. Grinding processes A and B, using cutting fluid based on fat, contain lowest differences with regard to confidence bands, non-parametric tests, multiple linear regression and correlation coefficients. It can be stated, that there is no significant difference regarding the surface topology whether acetone or cold degreaser is used for cleaning purposes. On the other hand, the choice of the cutting fluid shows significantly different results regarding roughness and gloss results. With regard to processes C and D, which utilise cutting fluids based on water, the differences of the surface topology after acetone cleaning are slightly higher (tendency). While the cleaning process shows no impact regarding the roughness parameters R_a and R_z, the average gloss values of the acetone cleaned surface are 4.91% higher (tendency, from statistical hypothesis point of view). It may be assumed that the analysis of large sample sizes will lead to significant results or differences in gloss. Nevertheless a significant difference of the surface topology parameters regarding roughness reasoned by acetone usage cannot be found.

Acknowledgements. The authors thank Peter Born (Franz Güde GmbH, Solingen Germany) for providing reference samples and thank Pit Fiur (BUW, Wuppertal Germany) supporting the analysing activities. Furthermore the authors thank the project sponsor (Federal Ministry of Education and Research (BMBF, Germany); project organisation by Karsruher Institut für Technologie (KIT), Produktion und Fertigungstechnologien (PTKA-PFT) for supporting the project MuPro2 (multivariate process optimisation 2), which is the base for the shown research work.

References

1. Franz Güde GmbH Homepage. http://www.guede-solingen.de. Accessed 20 Apr 2018
2. Klocke, F.: Manufacturing Processes 2 – Grinding, Honing, Lapping. 1st edn. Springer, Heidelberg (2009). https://doi.org/10.1007/978-3-540-92259-9
3. DIN 8589-11:2003-09, Manufacturing processes chip removal – Part 11: Grinding with rotating tool – Classification, subdivision, terms and definitions
4. DIN EN ISO 4287:2010-06, Geometrical Product Specifications (GPS) – Surface texture: Profile method – Terms, definitions and surface texture parameters (ISO 4287:1997 + Cor 1:1998 + Cor 2:2005 + Amd 1:2009)
5. Bracke, S., Radetzky, M., Born, P.: Multivariate analysis of aperiodic surface topography within high precision grinding processes. In: 12th CIRP Conference on Intelligent Computation in Manufacturing Engineering, 18–20 July 2018, Gulf of Naples, Italy (2018). (submitted)
6. DIN EN ISO 2813:2014-02, Paints and varnishes – Determination of gloss value at 20°, 60° and 85° (ISO 2813:2014)
7. Kuivalainen, K.: Glossmeters for the measurement of gloss from flat and curved objects. In: Publications of the University of Eastern Finland, Dissertations in Forestry and Natural Sciences, No 44. University of Eastern Finland, Joensuu (2011)
8. Szajna, G., Szewczul, J.: Calibration of Glossmeter. In: Lighting Conference of the Visegrad Countries (Lumen), Karpacz, Poland; 13–16 September 2016. IEEE (2016)
9. Hartung, J., Elpelt, B.: Multivariate Statistik. Oldenbourg, Munich (1999). (in German language)
10. Backhaus, K., Erichson, B., Plinke, W., Weiber, R.: Multivariate Analysemethoden. Springer, Heidelberg (2011). (in German language)
11. Sachs, L.: Applied Statistics – A Handbook of Techniques. 2nd edn. Springer, New York (1984). https://doi.org/10.1007/978-1-4612-5246-7

Application of a Multidimensional Scaling Method to Identify the Factors Influencing on Reliability of Deep Wells

Edward Kozłowski[1(✉)], Dariusz Mazurkiewicz[2], Beata Kowalska[3], and Dariusz Kowalski[3]

[1] Department of Quantitative Methods in Management, Faculty of Management, Lublin University of Technology, Nadbystrzycka 38D, 20-618 Lublin, Poland
e.kozlovski@pollub.pl

[2] Department of Production Engineering, Faculty of Mechanical Engineering, Lublin University of Technology, Nadbystrzycka 36, 20-618 Lublin, Poland
d.mazurkiewicz@pollub.pl

[3] Department of Water Supply and Wastewater Disposal, Faculty of Environmental Engineering, Lublin University of Technology, Nadbystrzycka 40B, 20-618 Lublin, Poland
{b.kowalska, d.kowalski}@pollub.pl

Abstract. Elements of the water distribution systems (WDS) as removable objects can be repaired or replaced, however their failures are difficult for analysis. The impact of WDS failures can be reduced if preventive actions are taken based on their potential of occurrences or if a failure occurs and is detected within a minimum period of time after its occurrence. This requires the development of a forensic system for WDS failures. This is why in this paper we will present an analysis of reliability data from a water supply sources consisting of deep wells taking into consideration additional, potential failure reasons. The aim of the work was also application possibility analysis of a multidimensional scaling method to identify the factors influencing on reliability of deep wells.

Keywords: Multidimensional scaling · Reliability · Water Distribution System

1 Introduction

In accordance with the reliability theory, water supply systems are classified as renewable objects and devices [1]. Thus, they can be repaired or replaced. In this case, the time of recovering initial functional properties, which constitutes one of the basic and key elements of exploitation and conditions, is essential. In turn, it constitutes the basis for selecting and correctly determining the indicators of reliable operation of water supply objects, resulting in their reliability models.

Despite significant research it has not yet been possible to identify the key factors which influence on WDS (Water Distribution System) failures. As a result, WDS experience considerable water loss and water quality problems each year, which also have a significant influence on the economical aspects of their functioning. The reasons for WDS failures include the complexity of the WDS itself and uncertainties in their

© Springer Nature Switzerland AG 2019
A. Burduk et al. (Eds.): ISPEM 2018, AISC 835, pp. 56–65, 2019.
https://doi.org/10.1007/978-3-319-97490-3_6

modelling. These uncertainties are caused also by a lack of knowledge of different parameters (epistemic uncertainty) or the randomness of parameter values (aleatory uncertainty) or both [2, 3]. Therefore, water supply systems and their elements are the subject of multi-aspect reliability studies, which aim, i.e. at identification of the factors causing damage and elimination of their negative impact.

The main purpose of a water distribution system is to deliver safe water of desirable quality, quantity and continuity to consumers. However, in many cases, a WDS fails to fulfill its goal owing to structural and associated hydraulic failures and/or water quality failures. The impact of these failures can be reduced significantly if preventive actions are taken based on their potential of occurrences or if a failure occurs and is detected within a minimum period of time after its occurrence, which requires the development of a forensic system for WDS failures according to [2]. However, every decision concerning the future behavior of technical objects is charged with some uncertainty, which in this approach is reflected in the volume and variety of possible paths of exploitation procedures to choose (operating and maintenance). This uncertainty is further increased by the long-time horizon of the intended effects or consequences of decision making with the use of intelligent data analysis tools and methods [3, 4]. On the other hand, the consequences of a failure can be reduced to an acceptable limit if an intelligent prognostic analysis is carried out and necessary preventive measures are taken on time. In the event of a failure, the consequences can be reduced significantly if the failure is detected and necessary action is taken within a minimum amount of time after its occurrence. Therefore, to reduce the likelihood and consequences of a failure, water utility managers must intervene accordingly. The interventions could be in day-to-day operations and maintenance activities such as failure detection, location and repair, or long-term improvement activities such as assets, or as part of an asset rehabilitation or replacement [1].

There are few papers on reliability of groundwater intakes and they usually concern only some of the issues. According to [4], there are several reasons for this status quo; there are numerous types of groundwater intakes, as well as plethora of construction solutions within wells. Their damages are of stochastic nature and the distribution of damages may be exponential; alternatively, if damages are related to the aging process (corrosion, colmatation, incrustation, etc.), Weibull distribution can be employed.

All of these factors considered, it is still extremely important to conduct analyses aimed at efficient and repeatable identification of the factors influencing the reliability of groundwater intake systems. Especially when the problem becomes quite complicated, because the system consists of multivariable dynamic constrained subsystems that are interconnected collection of wells, valves, tanks, regulators, intersection nodes, and sources. These subsystems are subject to continuously constraints and varying conditions which makes, that some of subsystems are susceptible to failures. The knowledge obtained in this way constitutes the basis for maintaining efficient performance of objects, as well as control over their operation in order to utilize the available equipment in the most optimal way. This aspect is highly important and previously was also pointed by [6], according to whom there is a need an application of control strategies that takes into account the system and components reliability. This is why, the aim of the work was to analyze possibility of a multidimensional scaling method application to identify the factors influencing on reliability of deep wells.

2 Survival and Hazard Functions

Usually, the survival and hazard functions are determined to analyze the reliability of engineering objects. The survival function is used to designate the probability that the system will survive beyond specified time. The hazard function is utilized to calculate the instantaneous rate of occurrence of the event. With terminology survival and hazard is connected the aging intensity term. Below we describe and explain this terminology.

Let the non-negative random variable $T : \Omega \rightarrow \Re_{+}$ represents the waiting time to failure (death of plant). In literature the variable T is called a survival time. The random variable T has a probability density function $f(t), t > 0$. The cumulative distribution function:

$$F(t) = P(T < t) \tag{1}$$

designates the probability that the failure (defect) will occur by duration t. On the other hand, the value of survival function at time t designates the probability that the failure (defect) will not occur until time t. The survival function is calculated as:

$$S(t) = P(T \geq t) = 1 - F(t) = \int_{t}^{\infty} f(s)ds \tag{2}$$

and presents the probability of correct work of device just before time t (the probability of surviving to duration t). The hazard function denotes the instantaneous rate of occurrence of failure and can be presented by:

$$h(t) = \lim_{\varepsilon \to 0} \frac{P(t \leq T < t + \varepsilon | T \geq t)}{\varepsilon} = \lim_{\varepsilon \to 0} \frac{\int_{t}^{t+\varepsilon} f(s)ds}{\varepsilon P(T \geq t)} = \frac{f(t)}{S(t)} = -\frac{d}{dt} \ln S(t) \tag{3}$$

The hazard function presets changing the survival characteristic of device (plant) over time. From (3) we obtain the dependence between survival function and cumulative hazard function $H(t) = \int_{0}^{t} h(s)ds$ as follows:

$$S(t) = \exp(-H(t)). \tag{4}$$

The cumulative hazard function represents the sum of risks occurring in duration from 0 to t.

When the random variable T has an exponential distribution with a scale $\alpha > 0$, then from (2)–(4) for any $t \geq 0$ the survival, hazard and cumulative hazard functions are:

$$S(t) = \exp\left(-\frac{t}{\alpha}\right), h(t) = \frac{1}{\alpha}, H(t) = \frac{t}{\alpha} \tag{5}$$

respectively. Additionally the expected time to first failure (time of reliable operation) is eual to:

$$\tau = \min\{t \geq 0 : H(t) \geq 1\} = \alpha.$$

With the hazard, cumulative hazard and survival functions are connected the aging intensity function (see e.g. [9]). The aging intensity is defined as the ratio of the instantaneous failure rate (hazard function) to a baseline failure rate, which is defined as:

$$\frac{H(t)}{t} = \frac{1}{t}\int_0^t f(s)ds.$$

Thus, at the moment $t > 0$ the aging intensity function $L(t)$ is a ratio of the density function to the baseline failure rate and is given by:

$$L(t) = \frac{h(t)}{\frac{1}{t}H(t)}. \tag{6}$$

The aging property of plant (system) can be represented as follows:

- if a failure rate is increasing ($L(t) > 1$) then the system is called aging (positive aging);
- if a failure rate is constant ($L(t) = 1$) then the system is called non-aging;
- if a failure rate is decreasing ($0 < L(t) < 1$) then the system is called anti-aging (negative aging).

In various aspects of survival, failure and reliability analysis we often use the Weibull distribution [7]. The different modification of Weibull distribution was presented in [8]. The special case of Weibull distribution is called an exponential distribution. This distribution will be used to describe the time to failure for special type of deep well pumps.

Let (Ω, Σ, P) be a probability space and the random variable $T : \Omega \to \Re_+$ has an exponential distribution. Thus, the density function and distribution functions are given respectively:

$$f(x) = \frac{1}{\alpha}\exp\left(-\frac{x}{\alpha}\right) \quad \text{for } x \geq 0, \tag{7}$$

$$F(t) = \begin{cases} 1 - \exp\left(-\frac{x}{\alpha}\right), & \text{for } x > 0, \\ 0, & \text{for } x \leq 0, \end{cases} \tag{8}$$

where $\alpha > 0$ represents a scale parameter. When the random variable T has an exponential distribution with a scale $\alpha > 0$, then the expected value and variance are $EX = \alpha$, $VarX = \alpha^2$.

From (7) we see that if the random variable T has an exponential distribution, then the aging intensity is constant and $L(t) = 1$. Thus when the times to failure for different objects have the exponential distribution with different scale parameters, then it is not sufficient to compare the reliability between plants (objects). Below we will present a better tool to differentiate the plants with respect distributions of time to failure.

3 Comparison of Reliability of Objects

Let us consider the situation, where the similar technical objects are working under different conditions; thus, the various factors influence them. Some of the factors are constant and do not change over time (e.g. identical pumps are working in different wells, where the parameters of water in these wells are different and do not change over time). Using the classic methods, for example Least Square Regression, Logistic Regression, it is not always possible to determine factors that significantly influence the reliability of objects (engineering systems). Our approach is based on defining the relationship between the latent variables and external factors. The values of latent variables we determine by applying the multidimensional scaling method, where the differences between reliability of engineering objects is calculated on the basis of Kolmogorov-Smirnov statistic.

On the probability space (Ω, Σ, P) we define the sequence of random variables $\{X_i\}_{1 \le i \le n}$, where for any $1 \le i \le n$ the random variable $X_i : \Omega \to \Re$ represents the time to failure of i-th system, engineering object. We assume that the random variable X_i has a cumulative distribution function $F_i(t), t \ge 0$ for $1 \le i \le n$.

Definition 1. The difference d_{ij} of reliability between i and j objects with respect to survival is greatest difference between survival functions of these objects.

From Definition 1 the difference d_{ij} of reliability between objects i and j is equal:

$$d_{ij} = \sup_{t \ge 0} \left| S_i(t) - S_j(t) \right| \tag{9}$$

From (4) we have:

$$d_{ij} = \sup_{t \ge 0} \left| F_i(t) - F_j(t) \right| \tag{10}$$

The distance d_{ij} is a Kolmogorov-Smirnov statistic between random variables X_i and X_j for $1 \le i, j \le n$.

In order to determine the similarities and differences of reliability between systems (objects), we employ the multidimensional scaling (MDS). Using the formula (10) we determine the differences between of all analyzed objects and create a matrix, which contains these differences. Applying the technique of multidimensional scaling we can define the sequence of points $z_i \in \Re^2$, $1 \le i \le n$, where the elements of this sequence correspond to objects. These points are determined by solution the task:

$$\min_{z_1, z_2, \ldots, z_n} D(z_1, z_2, \ldots, z_n) \tag{11}$$

where:

$$D(z_1, z_2, \ldots, z_n) = \sum_{1 \leq i,j \leq n} \left(d_{ij} - \|z_i - z_j\|\right)^2 \tag{12}$$

is called a stress function, $\| \ \|$ is an Euclidean norm.

MDS is a statistical technique, which allows us to visualize the locations of individual objects as points in $\Re^2$ space. The objects which have a similar reliability are located close together (see e.g. [10]), but the objects with different reliability are away from each other.

4 Analysis of Well Reliability

The methodology of the statistical analysis described above was applied in relation to an existing water supply station supplying a large city. The station is equipped with groundwater intake, comprising 8 deep wells operating in rare siphon system, i.e. without pumps. The water from the well is supplied to the tank and then to the water distribution network by means of a set of pumps. The reliability analysis was based on operational data from 1995 to 2012. The technological scheme of the intake was presented in Fig. 1. In the further analysis, each well was considered as an integral object.

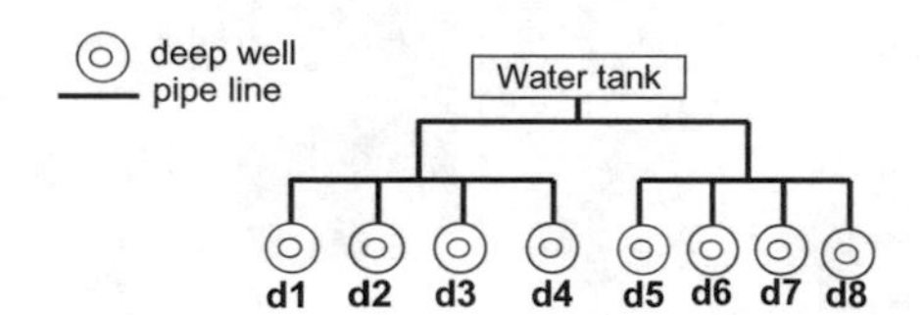

Fig. 1. Technological scheme of the water intake in the considered water supply station

For each well, it is assumed that the time of its reliable operation till failure is subordinated to exponential distribution (working hypothesis). These hypotheses were verified by means of chi-squared test χ^2. The values of χ^2 statistic are presented in Table 1. On the basis of the obtained results at the significance level of 0.05, there is no basis for rejecting the working hypotheses. Therefore, it is assumed that the time of correct operation of the considered wells is subordinated to exponential distribution. Table 1 also contains the values of parameters $1/\alpha$, mean operation time of each device, as well as standard deviations of the reliable operation time.

The density functions of reliable operation of each well are presented in Fig. 2, which visualizes also the density function of reliable operation of each well. In order to compare the reliable operation of intakes, the Kolmogorov-Smirnov distances between times of reliable operation of water supply stations were determined by using formula (10).

From Eq. (11) the Kolmogorov-Smirnov distance is the greatest differences between survival functions of these objects. Next, we define the distance matrix $\{d_{ij}\}_{1 \leq i,j \leq 8}$ for water supply stations and determine the positions for these objects in

Table 1. Inverse scale coefficient of the exponential distribution for each well, mean value standard deviation and the value of chi-squared (χ^2)

Deep well	$1/\alpha$	Mean	St.dev	χ^2 stat
d1	0.0015	649.7619	5846.751	0.6622
d2	0.0016	643.0838	5846.751	0.6697
d3	0.0016	642.9581	5846.751	0.6698
d4	0.0016	642.9581	5846.751	0.6698
d5	0.0016	639.5911	5846.751	0.6740
d6	0.0016	639.5911	5846.751	0.6740
d7	0.0016	634.4568	5846.751	0.6810
d8	0.0016	632.3948	5846.751	0.6839

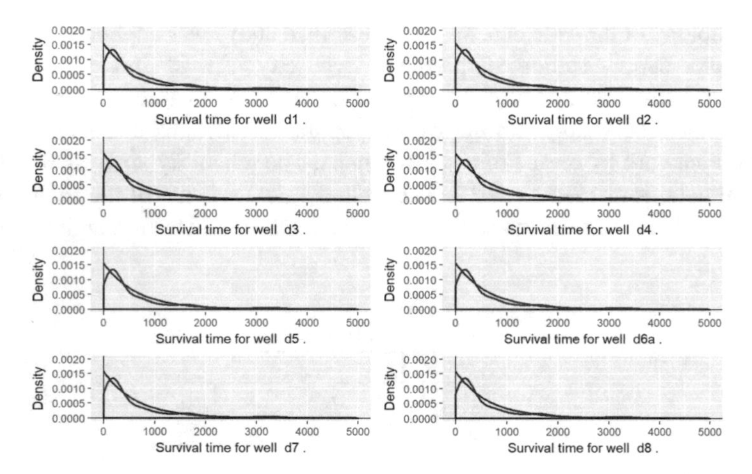

Fig. 2. Kernel estimators of density function (blue curves) for each intake and density functions of exponential distribution (red curves)

$\Re^2$ space. For this purpose we apply the MDS (multidimensional) method. Figure 3 presents the position map of object reliability determined in the set of latent variables X and Y. The behaviour of density functions for various intakes (see Fig. 2) is similar; however, the MDS method enables to differentiate between particular wells.

Figure 3 shows that the reliabilities of d1, d7, and d8 wells differ from the rest; nevertheless, the greatest difference between d1 and d8 approximates 0.01. This means that the largest differences between survival functions for d1 and d8 wells amount to roughly 0.01.

Afterwards, the analysis pertaining to the impact of well operation efficiency, iron content, hardness and turbidity of water on the reliable well operation was conducted. Table 2 presents the values of coefficients of correlation with latent variables *X* and *Y*, which were determined from position maps of objects. It can be seen that variable *X* is poorly correlated with Efficiency, Iron content, Hardness, Turbidity, whereas the correlation of *Y* with Efficiency and Turbidity is strong.

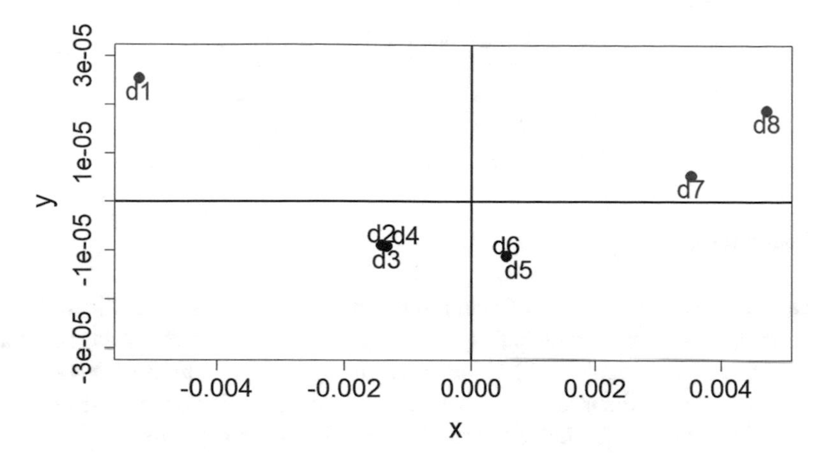

Fig. 3. Reliability position map of intakes prepared with Kolmogorov-Smirnov distances

Afterwards, we analyzed the relation between latent variable Y and Efficiency, Iron content, Hardness and Turbidity. On the basis of Akaike Information Criterion, a linear model which describes the dependence of variable Y on pump efficiency Q and water hardness H was selected:

$$Y = \alpha_1 \cdot Q + \alpha_2 \cdot H + \varepsilon \tag{13}$$

The basic statistics (values of estimators, their standard deviations, values of t and probability) for the model (13) are presented in Table 3. Therefore, at the significance level of 0.05, the working hypotheses ($H_0 : \alpha_j = 0$ for $j \in \{1, 2\}$) pertaining to the non-existing impact of parameters in model α_1 and α_2 (13) should be rejected. Thus, the efficiency of wells and hardness of water have a significant impact on the variable Y.

Table 2. Coefficients of correlation with latent variables X and Y

	Efficiency	Iron	Hardness	Turbidity
X	−0.0448	0.1077	0.1437	0.1536
Y	0.8313	−0.3085	−0.1470	−0.4188

The coefficient of multiple correlation R2 equals 0.6529. Its significance was investigated as well (working hypothesis) $H_0 : R = 0$ (the coefficient of multiple correlation differs insignificantly from 0). The value of test statistic F_{val} equals 5.643 and $P(F > F_{val}) = 0.0418$. Therefore, at the significance level of 0.05, the working hypothesis has to be rejected in favor of the alternative hypothesis. This means that the value of variable Y depends on the well efficiency and water hardness to a significant degree. No significant impact of efficiency, iron content, hardness and turbidity on variable X was observed. On the basis of the obtained data it can be concluded that the efficiency of wells and water hardness have a significant influence of the reliable operation of these objects.

Table 3. Statistics for the linear model

	α_j	S_{α_j}	t	$P(>\vert t\vert)$
$j = 1$	$9.66 * 10^{-7}$	$2.87 * 10^{-7}$	3.3595	0.0152
$j = 2$	$-1.13 * 10^{-5}$	$3.39 * 10^{-6}$	−3.3234	0.0159

5 Conclusions

This paper presents the method of determining the factors having significant influence on reliable operation of water supply stations. The use of the MDS method enabled us to determine latent variables which influence on reliable operation of devices. However, determining the linear dependences between latent variables and factors allowed us to identify factors affecting the reliable operation of the pumps. Out of the four analyzed factors, efficiency and hardness of water had a significant influence on the reliability of well operation. The conducted analysis indicates the need of expanding typical well operation monitoring, usually involving efficiency and pressure. In the considered case, it is also recommended to monitor water hardness as an indicator of increased susceptibility to malfunctions. The conducted analysis of risk factors will enable to take actions aimed at improving the reliability of the entire intake by devising new procedures of well exploitation.

References

1. Romaniuk, M.: Optimization of maintenance costs of a pipeline for a V-shaped hazard rate of malfunction intensities. Eksploatacja i Niezawodnosc – Mainten. Reliab. **20**(1), pp. 46–56 (2018). http://dx.doi.org/10.17531/ein.2018.1.7
2. Islam, M.S., Sadiq, R., Rodriguez, M.J., et al.: Water distribution system failure: a framework for forensic analysis. Environ. Syst. Decis. **34**, 168–179 (2014). https://doi.org/10.1007/s10669-013-9464-3
3. Pietrucha-Urbanik, K., Studziński, A.: Case study of failure simulation of pipelines conducted in chosen water supply system. Eksploatacja i Niezawodnosc – Mainten. Reliab. **19**(3), pp. 317–323 (2017). http://dx.doi.org/10.17531/ein.2017.3.1
4. Kozłowski, E., Mazurkiewicz, D., Kowalska, B., Kowalski, D.: Binary linear programming as a decision-making aid for water intake operators. In: Burduk, A., Mazurkiewicz, D. (eds.) Intelligent Systems in Production Engineering and Maintenance – ISPEM 2017. Advances in Intelligent Systems and Computing, vol. 637, pp. 199–208. Springer International Publishing. https://doi.org/10.1007/978-3-319-64465-3_20
5. Loska, A.: Scenario modeling exploitation decision-making process in technical network systems. Eksploatacja i Niezawodnosc – Mainten. Reliab. **19**(2), pp. 268–278 (2017). http://dx.doi.org/10.17531/ein.2017.2.15
6. Salazar, J.C., Weber, P., Sarrate, R., Theilliol, D., Nejjari, F.: MPC design based on a DBN reliability model: application to drinking water networks. IFAC- Papers On Line **48–21**, 688–693 (2015)
7. Weibull, W.: A statistical distribution function of wide applicability. J. Appl. Mech. **18**(3), 293–297 (1951)

8. Almalki, S.J., Nadarajah, S.: Modification of the Weibull distribution: a review. Reliab. Eng. Syst. Saf. **124**, 32–55 (2014)
9. Jiang, R., Ji, P., Xiao, X.: Aging property of unimodal failure rate models. Reliab. Eng. Syst. Saf. **79**, 113–116 (2003)
10. Hastie, T., Tibshirani, R., Friedman, J.: The Elements of Statistical Learning: Data Mining, Inference and Prediction. Springer, New York (2009)

Case Study of Production Planning Optimization with Use of the Greedy and Tabu Search Algorithms

Łukasz Łampika(✉), Kamil Musiał(✉), and Anna Burduk(✉)

Faculty of Mechanical Engineering, Wroclaw University of Science and Technology, Wybrzeze Wyspianskiego 27, 50-370 Wroclaw, Poland
{lukasz.lampika,kamil.musial,anna.burduk}@pwr.edu.pl

Abstract. In the paper, optimization of production order execution in a production company is suggested. Until now, decisions regarding prioritization of orders were made by people. Therefore, in this research, comparison of a human-based method and two algorithm methods was performed, in order to find optimal set of production orders. Tabu Search and Greedy algorithms were selected. The aim of the described case study was maximization of profit and production of selected orders. Costs of turning tools were taken under consideration, which made the case more complex. In order to choose the best method of solving this case, real historical data from the company were collected concerning orders from the considered period of time. Moreover, basing on the same data, Greedy and Tabu Search algorithms were prepared. Finally, all the two methods were compared and, in conclusion, Tabu Search solutions acquired highest profit for the whole tested period.

Keywords: Production planning · Greedy algorithm · Tabu search
Optimization methods · Production systems · Decision-making processes

1 Introduction

Current global economy forces companies to maximize profits by reducing costs and executing customers' orders in proper time, amount and quality. In this paper, execution of production orders, as one of the most important processes for gaining profit, are considered [6–8]. Reaching highest profits by correct management means choosing most suitable orders to be executed with possibly smallest usage of resources such as financial, human, material or temporal ones. Nowadays, many algorithms help with decision-making process by analyzing significant numbers of data and taking under consideration many constraints [9–11]. Decisions made by humans have their limitations regarding choosing the best solution especially within complex systems. Algorithmic applications reduce numbers of errors, accelerate labour and reduce time of conducting production decisions.

In this paper, two types of algorithms are tested in order to optimize real production planning by choosing best set of orders and reaching highest profits. In Sect. 2 differences between algorithms are compared, Sect. 3 describes case study, Sect. 4 presents algorithms codes and compare results, Sect. 5 summaries results of presented case study.

© Springer Nature Switzerland AG 2019
A. Burduk et al. (Eds.): ISPEM 2018, AISC 835, pp. 66–75, 2019.
https://doi.org/10.1007/978-3-319-97490-3_7

The Tabu Search and the Greedy algorithms are compared one to another and also to real company data, based on human decisions. The case concerns prioritization of orders, simultaneously focusing on maximisation of profit, costs and customer's requirements regarding obligatory orders. The Greedy algorithm method uses a locally optimal decision with no assessment of its effect on next steps [4, 12]. It is a very popular method in production optimization tasks close to human way of thinking. On the other hand, the Tabu Search algorithm looks for the solution space by means of specific sequence of movements [1–3] and check all neighbouring solutions. Tabu search is a smart and fully deterministic method to move forward better and better solution.

Production companies face complex planning problems that, without artificial intelligence methods, would be impossible to be solved in the most optimum way. None of the mentioned algorithms guarantees finding the optimum solution, but they produce the solution sufficiently acceptable from the company point of view. The goal of this research is to choose a fast and most accurate method of finding optimal solution maximising profits.

2 Differences Between Algorithms

2.1 Methodology

In the paper 2 different methodologies will be tested. As mentioned above, the greedy algorithm chooses a locally best solution without next steps analysis. It works close to the natural human approach in case of decision making process. In the paper the Greedy algorithm result has been compared with real company data. Tabu Search algorithm scrolls the solution space and using the fact that good solution very often occur next to another good solution always checks all nonboring solutions.

2.2 Greedy Algorithm

The way of finding a solution by the Greedy algorithm method is making a selection in each step [4]. Therefore, the chosen solution is the one that seems to be the best at the moment. The Greedy algorithm finds the locally optimal rating, but it does not analyse, whether the data in the following steps are correct. [4, 12]. This method chooses the target function that locally minimises or maximises the target value, so it is mostly used for maximisation or minimisation of the considered value. Key advantages of this type of algorithm are fast operation, reasonable memory requirements and compliant implementation [4, 5].

2.3 Tabu Search

Tabu Search is an algorithm which searches the space created by all possible solutions with a particular sequence of movements [1–3]. Between them, there are forbidden (taboo) movements. This algorithm avoids spinning around the local optimum by archiving information about proven solutions in a Taboo List (TL). The Tabu Search algorithm is deterministic, therefore processing the same data with the same Tabu Search algorithm will give identical results.

3 Case Study

3.1 Description of the Company's Technological Process

The problem described in this article was inspired by a real company from automotive industry. The core business of this company is production of wheel rims of a metal alloy. The process is divided to melting row material, casting, machining and painting.

Production starts from melting one of two types of row material: AlSi11 or AlSi7. Then the melted material is cleaned by purging with argon. Clean liquid metal is transported to one of 48 low-pressure casting machines. Each of them has a different mould to cast proper wheels for customers. The molten alloy is conveyed by gas pressurization from the holding furnace to a steel mould via a riser pipe. This is followed by visual inspection for quality control purposes. Casting burrs are removed and the wheel is measured. Every wheel is 100-% inspected by automatic and semi- automatic X-ray machines. A special program identifies the design and inspects the wheel quality.

After that, wheels cast of AlSi7 pass through heat treatment process lasting around five hours and are then stress-relieved at lower temperatures. This makes the structure of the material more homogeneous and ensures good mechanical properties.

Machining of the casting takes place on automatic CNC production systems. One line consists of three different machines arranged in a row. Each side of the wheel rim is machined by one CNC lathe. At the end of each line, a drilling machine drills holes for screws and valve. Keeping dimensions in tolerance is crucial at this step of the process. After machining, the wheels are additionally tested for any imbalance.

All wheels are then washed, dried and tested for leaks, using helium. Machining of the wheels is followed by mechanical brushing and final deburring by hand. This stage ensures optimum preparation for the coating process.

Each wheel is subjected to surface finishing and a complex coating process, which is characterized by multilayer structure. Application of primer is followed by a sprayed base coat, before the wheel is finally covered with a transparent finish. This final coat protects the rim from corrosion. After this process, final products are stored in the warehouse and shipped to a client.

3.2 Problem Description

The case study was selected within a large manufacturing, automotive company to examine different methods of decision-making process on a complex and ordinary situation. Production enterprises are often challenged with this type of optimization problems. Despite that, in this company, the planning department bases only on human factor and does not use any artificial intelligence methods.

The analysed case study regards machining area, which consists of 18 CNC lines. Production plan is made separately for this department. Clients prepare orders in weekly periods and send them to the planning department. Some orders have high priority due to previous delay or penalties contained in a sale contract, therefore they are mandatory to be produced. There are different product types with different selling prices. Each week, customers order different numbers of products. Turning knives are the main cost in the machining process, being the main constraint in the decision-making process. There are 6 knife types (A, B, C, D, E, F), each of them having its lifetime, measured in numbers of manufactured products. After full deterioration of a knife, a new tool is needed to be bought. The historical lifetime of a tool, taken under consideration in the research, was ten weeks. Each week was treated as a separate unit and variations within it were not considered by the research.

Regarding the above, the aim of this case study was to gain maximum benefit by choosing orders to be executed. There are several constraints, such as mandatory orders, production time, duration and costs of knives.

3.3 Chosen Parameters

Within elaboration of the algorithm, the following parameters were accepted for each order:

- Production time [pt] - [pt_1, pt_1, …, pt_n] - required for the order execution;
- Profit [p] - [p_1, p_2, …, p_n] - that the company can gain thanks to its execution;
- Costs [kc] – [kc_A, kc_B, …, kc_N] - cost of all knives used in the following calculation.

The purpose of the algorithm was selection of orders in the way guaranteeing possibly highest profit, including obligatory orders, profit and additional criteria like additional cost for disposal of worn knives. The issue was solved with use of the Tabu Search and the Greedy algorithms.

Effectiveness of the selected algorithms was examined by simulations of subsequent working periods. For the examined case, the Tabu Search and Greedy algorithms were considered. The aim was to find the calculation method giving a solution in a relatively short time using a mid-range PC computer. This would allow the company to modify orders and parameters easily, as well as to obtain calculated solutions in nearly real-time.

4 Case Study Algorithms

4.1 Solutions of the Greedy Algorithm

In the first step, the data are sorted in descending order. In this step, the algorithm divides the profit of each non-mandatory order by its production time and sorts the products in descending sequence:

$$\frac{p_{1'}}{pt_{1'}} \geq \frac{p_{2'}}{pt_{2'}} \geq \frac{p_{3'}}{pt_{3'}} \geq \cdots \geq \frac{p_{n'}}{pt_{n'}} \tag{1}$$

p_1, p_2, …, p_n are the profits of specific orders, pt_1, pt_1, …, pt_n are their production times.

The second step adds the highest orders. Then, the algorithm adds all times of mandatory production times – the orders that must be executed. Time that is left can be spent for the remaining, non-mandatory orders. The algorithm adds all times of the highest sorted values until it reaches the maximum value, not exceeding the maximum available time:

$$pt_{1'} + pt_{2'} \ldots + pt_{m'} \leq T_{max}. \tag{2}$$

The third step adds the next orders. When adding another order results in exceeding the maximum volume (maximum available time), this solution is skipped, the next order is added and the criterion is checked again.

If:

$$pt_1 + pt_2 + pt_3 pt_4 > T_{max}, \tag{3}$$

then:

$$pt_1 + pt_2 + pt_3 + pt_5 \leq T_{max}. \tag{4}$$

The algorithm will terminate, when the available time is utilized:

$$pt + pt_{2'} + \ldots + pt_{m'} = T_{max} \tag{5}$$

or when all the orders are checked:

$$pt_{1'} + \ldots + pt_{m'} + pt_{m+2'} + pt_{n-1'} + pt_{n'} \leq T_{max}. \tag{6}$$

A result of the algorithm action is total profit from execution of the orders, i.e. sum of profits from all the accepted orders minus cost of worn knives:

$$p_{sum} = (p_{1'} + p_{2'} + \ldots + p_{m'} + p_{m+2'} + p_{n-1'} + p_{n'}) - (x \times KC_A + y \times KC_B + z \times KC_D) \tag{7}$$

where x, y, z are numbers of knives used for production.

4.2 Solutions of the Tabu Search Algorithm

Algorithm 1. Tabu Search pseudo-code for the considered problem

```
Tabu Search algorithm (S0, var S, max_m, max_it)
Set S = S0 and n_iteration = 0
Repeat:
        m = 0
        best = 0
        it = it + 1
            Repeat
                  m = m+1
                  Execute Check_the_neighboring_solution (S,Sm)

                  Execute Check_Tabu_list (S,Sn)
                        if (f(Sm) > best then (best = f(Sm) and (m2= m))
            until (m = max_m)

      Execute Add_to_Tabu_list (S,Sm2)

      S= Sm2

      If best > solution then solution = best
   Until n_iteration = max_it
```

The analysed problem may be presented as a binary sequence:

1 1 0 0 0 …,

in which each bit corresponds to the order assigned to the working week. The value "1" means that the order is taken and proceeded, the value "0" means that the order is rejected. Implementation of five orders can be described as a bit sequence, e.g.:

1 1 1 0 0 .

Check the Neighbouring Solution:
In this step, the algorithm checks all of the neighbouring solutions (differing in one bit):

1 1 1 0 **1**
1 1 1 **1** 0
1 1 **0** 0 0
1 **0** 1 0 0
0 1 1 0 0.

Check Tabu List:
The algorithm checks, whether the movement between previous and actual solutions were already made. If so, this solution is omitted and then the best solution is chosen, according to the specified criterion.

Add to Tabu List:
The algorithm adds it to the Tabu List, i.e. the list of prohibited movements (TL). The algorithm remembers, which solution is the best one and repeats the steps the specified number of times.

In the case when each bit represents an order [o] - [o_1, o_2.....o_n], its parameters are:

- production time [pt] - [pt_1, pt_2, ..., pt_n] - required for the order execution;
- profit [p] - [p_1, p_2, ..., p_n] - that the company can gain thanks to its execution;
- cost [kc] – cost of all knives used in the following calculation.

According to the assumptions, total production time is strictly defined. The algorithm has a limit: total time of the orders multiplied by the bit value representing the order can not exceed the maximum production time T_{max}.

$$1pt_1 + 1pt_2 + 0pt_3 + \cdots + 0pt_n \leq T_{max} \tag{8}$$

According to the adapted objective function, the criterion for choosing the best solution is profit of the orders including disposal of knives.

$$F(c) = (0p_1 + 1p_2 + 1p_3 + \cdots + 0p_n) - \mathrm{kc} \rightarrow MAX \tag{9}$$

The algorithm was set to 100 iterations.

4.3 Comparison of Algorithm Results

Analysis of the results indicates that, in most weeks, results were equal or very close. The greatest difference occurred in weeks 9 and 10 (Fig. 1). These were the weeks with the largest order numbers and product types, which leads to a conclusion that the Greedy algorithms and human-based decision making have a problem in finding the optimum solution when variables increase. In summary, the Tabu Search generates the highest income. In the week 9., the company scheduler made a mistake omitting the most profitable order and the Greedy algorithm found a local maximum. The Tabu Search maximised the profit nearly twice comparing to that of the Greedy algorithm.

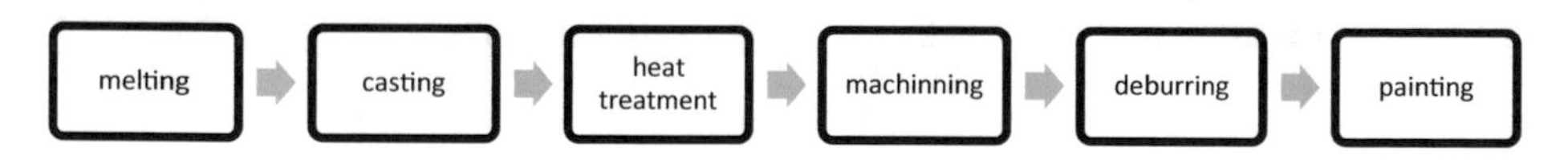

Fig. 1. Production process

Another difference between the compared methods occurred in the week 10. As it is presented in Fig. 2, in the week 10. the Tabu Search had the lowest deterioration percentage of knives on the level of 92%. Both methods and the Greedy algorithm reached 96%. Moreover, the Tabu Search used 12 knives comparing to 10 knives used by two other methods. In spite of that, the Tabu Search reached the highest profit in the week 10. This example shows a major difference in optimisation methods. The Greedy algorithm and human decision making in the described company were more focused on cost reduction and less willing to increase costs, even if it would bring higher profits.

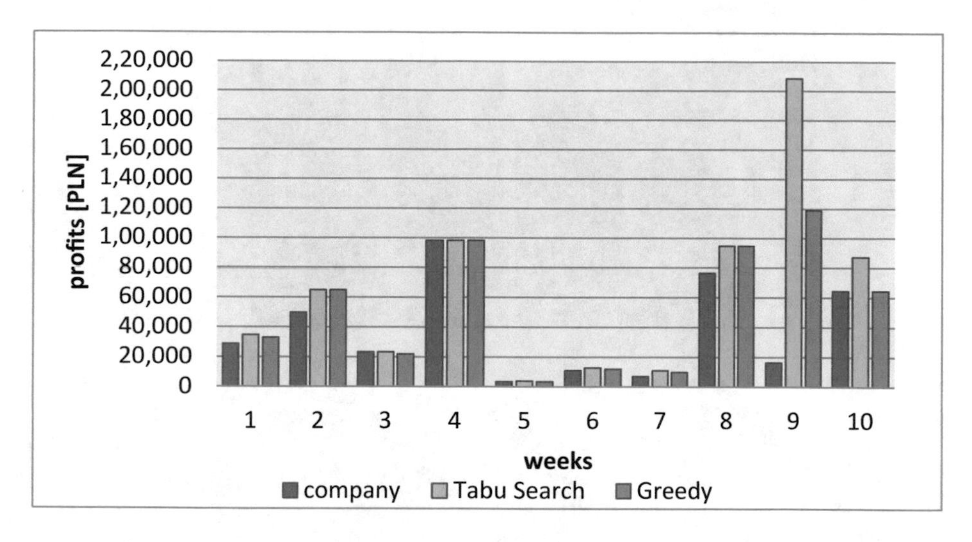

Fig. 2. Profits earned by the company and estimated by algorithms during 10 weeks

In was not easy to see differences in each week, but overview through the whole observed period of time shows repeatable proceeding way (Fig. 3.). People from the company are more willing to produce more and to fulfill client's needs, but therefore number of knives and their types increases. On the other hand, the Greedy algorithm executed only 152 orders, being the best at reducing costs. The Tabu Search is in the

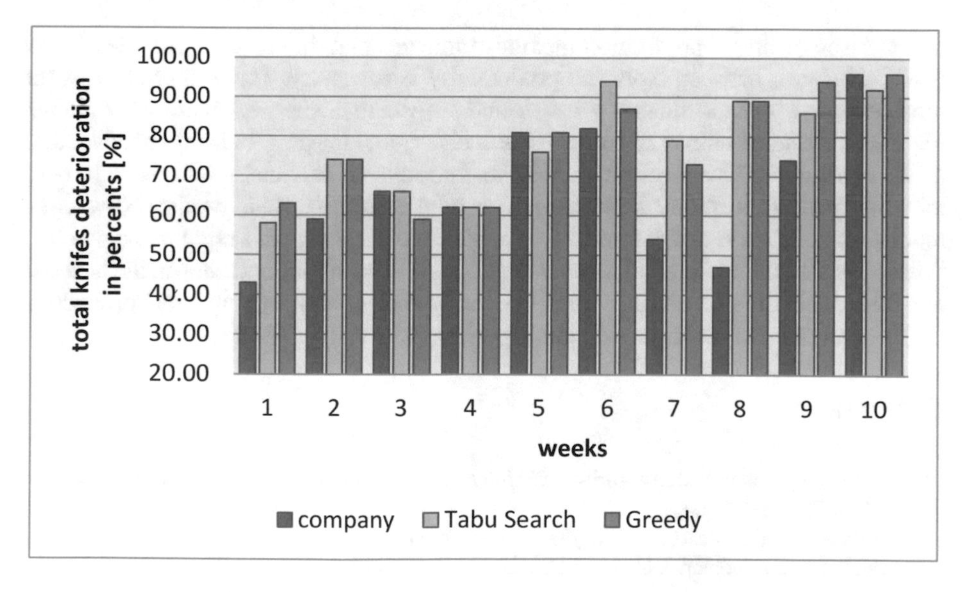

Fig. 3. Total percentage deterioration of all knives used in each week

middle between philosophy of low costs and philosophy of high production, which led it to best solutions in maximization of profit (Fig. 4).

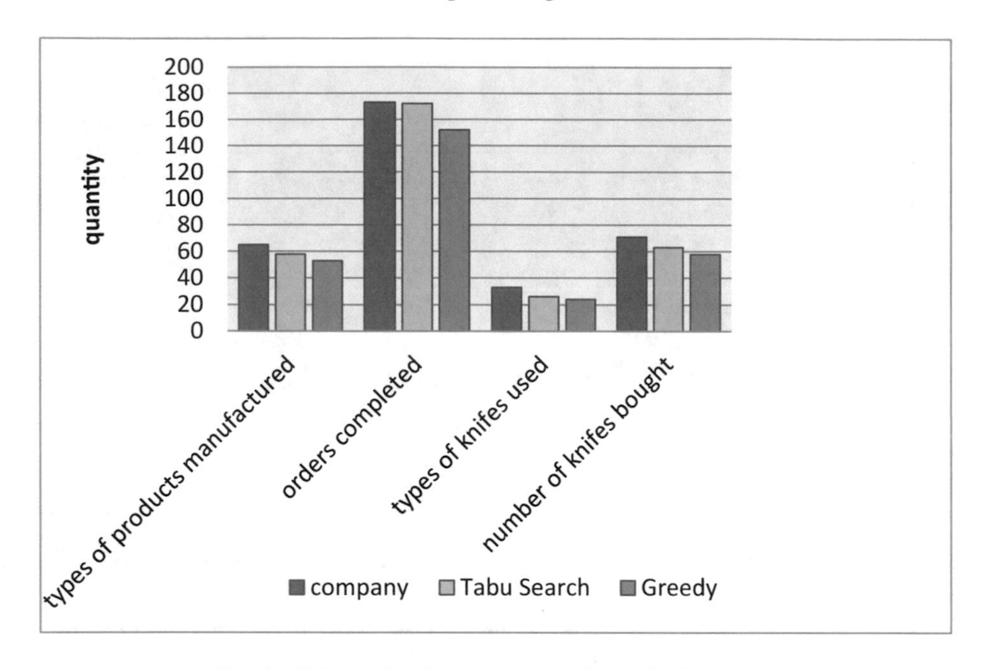

Fig. 4. Values of various parameters for total of all weeks

5 Summary

Comparison of three optimisation methods indicates that, in order to gain the highest profit, a balance between costs and productivity is necessary. The Tabu Search is the most effective method among those tested, especially when numbers of variables increase. Excluding human errors from the decision-making process is a great advantage of the algorithms. This research showed that absence of artificial intelligence in optimization process in production companies might decrease their profits. Nowadays, number of constraints and complexity of production planning rises rapidly and, in close future, people will not be able to compare all possible solutions and choose the best one in a reasonable time. The Tabu Search showed its potential in solving a real production problem and further researches on this algorithm will be carried-out.

References

1. Brandão, J.: A tabu search algorithm for the open vehicle routing problem. Eu. J. Oper. Res. **157**(3), 552–564 (2004)
2. Cordeau, J.F., Gendreau, M., Laporte, G.: A tabu search heuristic for periodic and multi-depot vehicle routing problems. Networks **30**(2), 105–119 (1997)

3. Grabowski, J., Wodecki, M.: A very fast tabu search algorithm for the permutation flow shop problem with makespan criterion. Comput. Oper. Res. **31**(11), 1891–1909 (2004)
4. Kahraman, C., Engin, O., Kaya, I., Öztürk, R.E.: Multiprocessor task scheduling in multistage hybrid flow-shops: a parallel greedy algorithm approach. Appl. Soft Comput. **10**, 1293–1300 (2010)
5. Zhang, Z., Schwartz, S., Wagner, L., Miller, W.: A Greedy algorithm for aligning DNA sequences. J. Comput. Biol. **7**, 203–214 (2000)
6. Betz, F.: Strategic business models. Eng. Manage. J. **14**(1), 21–28 (2002)
7. Guillén, G., Badell, M., Espuña, A., Puigjaner, L.: Simultaneous optimization of process operations and financial decisions to enhance the integrated planning/scheduling of chemical supply chains. Comput. Chem. Eng. **30**(3), 421–436 (2006)
8. Papageorgiou, L.G.: Supply chain optimisation for the process industries: advances and opportunities. Comput. Chem. Eng. **33**(12), 1931–1938 (2009)
9. Jones, D.F., Mirrazavi, S.K., Tamiz, M.: Multi-objective meta-heuristics: an overview of the current state-of-the-art. Eur. J. Oper. Res. **137**(1), 1–9 (2002)
10. Pohekar, S.D., Ramachandran, M.: Application of multi-criteria decision making to sustainable energy planning - a review. Renew. Sustain. Energy Rev. **8**, 365–381 (2004)
11. Kotowska, J., Markowski, M., Burduk, A.: Optimization of the supply of components for mass production with the use of the ant colony algorithm. In: Intelligent Systems in Production Engineering and Maintenance, ISPEM 2017. Advances in Intelligent Systems and Computing, pp. 347–357. Springer (2018)
12. DeVore, R.A., Temlyakov, V.N.: Some remarks on greedy algorithms. Adv. Comput. Math. **5**, 173–187 (1996)

Collaborative Network Planning Using Auction Parallel to Pool-Based Energy Market with Shifting Bids

Izabela Zoltowska(✉)

Warsaw University of Technology, Nowowiejska 15/19, Warsaw, Poland
i.zoltowska@ia.pw.edu.pl

Abstract. In this paper producers and distributors in supply chain need to agree on orders quantities and schedules. Instead of classical integrated production-distribution planning model we propose a reformulation in form of centralized auction, oriented on maximization of individual benefits. Production costs and inventory costs are handled as parts of competitive offers/bids. We reformulate the inventory costs, so they can be interpreted as costs of reduction or increase in ordered quantities. Ordered distribution quantities are balanced in each period with offered production. We point to the similarity of the proposed model to the centralized double energy market for demands with shifting capabilities. This allows us to use pricing model based on uplifts minimization, originally developed for pool-based auction. As a result we obtain prices complemented with minimum uplifts that support competitive producers and distributors to follow schedules efficient for entire supply chain. This novel coordination mechanism is formulated as a set of two clearing and pricing mixed integer linear programming problems (MILP). Simple example illustrates the approach.

Keywords: Production · Distribution · Supply chain integration
Double auction · Market clearing · Pricing non-convexities
Dynamic programming

1 Introduction

There is a growing need for coordination mechanisms in networks with collaborative enterprises. Supply chains require efficient cooperation between supplies, producers, distributors and retailers. Simultaneously, each local participant of the chain is focused on its own goals, that may be in opposite to the whole system operation [5].

Cooperation within the company's supply chain involves coordinated, or centralized flow of orders, according to information sharing within an enterprise. The company can achieve highest benefits when coordinating activities such as purchases and logistics. The decisions involve agreement on volumes, frequencies, and sources of interchange. Naturally, these shared decisions impact greatly both sides of the chain: for example producers may prefer to agree on high-volume, less frequent orders, while distributors may insist on every-day orders to avoid excess inventory. Satisfactory results should require significant effort from collaborating partners, in bilateral negotiations it is hard to consider more than only a handful of copartners.

© Springer Nature Switzerland AG 2019
A. Burduk et al. (Eds.): ISPEM 2018, AISC 835, pp. 76–86, 2019.
https://doi.org/10.1007/978-3-319-97490-3_8

Research effort is driven toward decentralized planning in order to respond to needs of multi-enterprises supply chain. It is unrealistic to call for the existence of a central decision-maker who has a global information and authority to decide for each local entity. One research stream is concentrated on agents-based planning models – interesting solution for one retailer and multiple suppliers is provided in [2], together with literature review. Others focus on designing market-based mechanisms to bridge the system welfare objectives with those of the profit-seeking market participants [1].

This paper uses concept of a centralized auction that is a reminiscent of the integrated production-distribution minimum cost problem stated in centrally planned economies. However, a fundamental difference is that auction participants are competitive, self-interested decision-makers. Auction provides coordination tools, namely the exchange prices to support centrally determined purchasing schedules. We draw analogies between production-distribution auction and energy generation-demand exchange, which enables us to use mechanism designed to successfully coordinate multiple participants in highly constraint, non-convex auction settings.

This paper makes following contribution in a field of collaborative supply chain mechanisms. We show how to reformulate a classical integrated production-distribution planning model to fit market-based requirements. Moreover, we develop individually rational pricing model based on findings in area of non-convex electricity market pricing. As a result we provide auction that optimizes and supports schedules efficient for entire supply chain and for individual, competitive participants.

The paper is structured as follows: Sect. 2 redefines the classical production-distribution planning model as the centralized auction model. In Sect. 3 we address the issue of adequate pricing and formulate MILP pricing model based on non-convex electricity market pricing model. The approach is illustrated with a simple numerical example in Sect. 4. Finally, Sect. 5 discusses and concludes the paper.

2 Production-Distribution Planning Model

The following Table 1 gives a summary notation.

Table 1. Notation.

Sets	
$l = 1..L$	Producers
$d = 1..D$	Distributors
$t = 1..T$	Planning periods, plus 0 as initial period
Parameters	
D_{dt}	Customer demand of distributor *d*, period *t*
c_d^{inv}	Unit inventory cost of distributor *d*
e_d	Unit customer demand price of distributor *d*

(*continued*)

Table 1. (*continued*)

Sets	
I_d^{max}	Maximum inventory level of distributor d
P_l^{max}	Maximum capacity of producer l
P_l^{min}	Minimum capacity of producer l
s_l	Unit production cost of producer l
S_l^{fix}	Fixed production cost of producer l
Variables	
d_{dt}	Quantities ordered by distributor d, period t
f_{ldt}	Quantity ordered from producer l by distributor d in period t
I_{dt}	Inventory level of distributor d, at beginning of period t
p_{lt}	Production level (order size) of producer l, period t
v_{lt}	Production state (0 or 1) of producer l, period t

2.1 Traditional Integrated Production-Distribution Planning Model

The ideal solution to production-distribution problem in a collective network would be to optimize order quantities to minimize costs within the whole system:

$$\min \sum\nolimits_t \left(\sum\nolimits_l \left(s_l p_{lt} + S_l^{fix} v_{lt} \right) + \sum\nolimits_d c_d^{inv} I_{dt} \right), \quad (1)$$

subject to:

$$P_l^{min} v_{lt} \leq p_{lt} \leq P_l^{max} v_{lt}, \quad \forall l, t \quad (2)$$

$$\sum\nolimits_d f_{ldt} = p_{lt}, \quad \forall l, t \quad (3)$$

$$\sum\nolimits_l f_{ldt} = d_{dt}, \quad \forall d, t \quad (4)$$

$$I_{dt} = I_{d,t-1} + d_{dt} - D_{dt}, \quad \forall d, t \geq 2 \quad (5)$$

$$0 \leq I_{dt} \leq I_d^{max}. \quad \forall d, t \quad (6)$$

Constraint (2) restricts produced quantities. Equations (3) and (4) are flow constraints connecting variables p_{lt} (producers) and d_{dt} (distributors), used here to facilitate connection with further auction-based formulation.. In (5) inventory balance is imposed. In (6) we limit maximum size of inventories. Such a basic model may be further enhanced for, i.e. delivery cost, size of ordered quantity, etc. [5].

2.2 Auction-Based Production-Distribution Planning Model

The ideal, integrated production-distribution problem may be hard to incorporate in a real collective network, as was previously discussed. To optimize order quantities we propose centralized auction formulation, with objective to maximize participants surpluses, i.e. sum of differences between benefits of distributors and costs of producers, determined from their multipart offers/bids, that constitute of trading constraints and valuations. In such a framework the participants are evaluated (priced) based on their characteristics, which have impact on the overall system surplus. This introduces competitiveness conditions into supply chain operations.

Distribution (Demand) Side of the Auction. Revenue of a single distributor d is the product of demanded quantities and prices: $\sum_t e_d D_{dt}$. As a result of cooperation the distributor receives ordered quantities that may deviate from demanded quantities, i.e. $d_{dt} = D_{dt} + d_{dt}^+ - d_{dt}^-, d_{dt}^+, d_{dt}^- \geq 0$. Surplus or deficit quantity may be interpreted as a shift of demand. Distributor speeds up, or postpones the order, with a cost of holding inventory for each day.

Surplus quantity will be utilized in a future to cover the demand, each period with a surplus is associated with cost involving also previous surpluses: $c_d^{inv} \sum_{h=1}^{t} d_{dh}^+$. Maximum valuation of surplus order in a period t is then equal to: $e_d d_{dt}^+ - c_d^{inv} \sum_{h=1}^{t} d_{dh}^+$.

On the other hand, reduction (deficit) quantity may be interpreted as a voluntary resale of the ordered quantity. Such a reduction is profitable for the distributor at the price recovering at least the inventory cost incurred so far. The increase of valuation resulting from resale is then equal to $c_d^{inv} \sum_{h=1}^{t} d_{dh}^-$. Simultaneously, resale quantity decreases benefit by the reduced volume, so the overall valuation in period t is equal: $c_d^{inv} \sum_{h=1}^{t} d_{dh}^- - e_d d_{dt}^-$. Notice, that deficit valuations account for potential deficits in previous periods, since distributor may be interested in not using the whole available inventory. I.e. consider a situation, when distributor buys large quantity in a period with low price, to use it partially in consecutive periods with high prices.

Summarizing, total benefit of distributors is the sum of all their partial valuations:

$$B = \sum_d \sum_t \left(e_d D_{dt} + e_d d_{dt}^+ - e_d d_{dt}^- - c_d^{inv} \sum_{h=1}^{t} (d_{dh}^+ - d_{dh}^-) \right). \tag{7}$$

Last part in (7) may be rewritten as follows:

$$\sum_d \sum_t c_d^{inv} \sum_{h=1}^{t} (d_{dh}^+ - d_{dh}^-) = \sum_d c_d^{inv} \sum_t (T + 1 - t)(d_{dt}^+ - d_{dt}^-), \tag{8}$$

which gives proper calculation of total inventory cost – each surplus or deficit quantity should be accounted from a period under consideration until last period in planning horizon. It is clear that surplus/reduction quantities have different meaning from inventory state, but they contribute to the transitory level of inventory, since also:

$$I_{dt} = \sum\nolimits_{h=1}^{t} (d_{dh}^{+} - d_{dh}^{-}), \quad \bigvee d, t \tag{9}$$

and further, after we substitute $d_{dt}^{+} - d_{dt}^{-} = d_{dt} - D_{dt}$ into (9), we obtain traditional inventory balance constraint, identical to (5):

$$I_{dt} = \sum\nolimits_{h=1}^{t-1} (d_{dh}^{+} - d_{dh}^{-}) + d_{dt} - D_{dt} = I_{d,t-1} + d_{dt} - D_{dt}. \quad \bigvee d, t \geq 2 \tag{10}$$

Naturally, surpluses may not exceed maximum inventory capacity:

$$\sum\nolimits_{h=1}^{t} (d_{dh}^{+} - d_{dh}^{-}) \leq I_{d}^{max}. \quad \bigvee d, t \tag{11}$$

Production (Supply) Side of the Auction. Cost of a single producer l is simply the offered cost of produced quantity, including fixed costs, etc. Total production cost is equal $C = \sum_{l} \sum_{t} \left(s_{l} p_{lt} + S_{l}^{fix} v_{lt} \right)$, where production schedules must meet the constraints declared by each producer:

$$P_{l}^{min} \cdot v_{lt} \leq p_{lt} \leq P_{l}^{max} \cdot v_{lt}. \quad \bigvee l, t \tag{12}$$

Production-distribution Auction Objective and Balance Constraint. Demand and supply are matched in the trade balance constraint ensured in each period:

$$\sum\nolimits_{l} p_{lt} - \sum\nolimits_{d} d_{dt} = 0. \quad \bigvee t \tag{13}$$

Besides (13) constraints (10)–(12) of auction participants must be met.

The goal of the trade lies in maximization of the overall profits of auction participants, i.e. $max\, Q = B - C$. The auction objective is then stated as:

$$\begin{aligned} Q = & \sum\nolimits_{d} \left(\sum\nolimits_{t} e_{d} (D_{dt} + d_{dt}^{+} - d_{dt}^{-}) - c_{d}^{inv} \sum\nolimits_{t} (T + 1 - t)(d_{dt}^{+} - d_{dt}^{-}) \right) \\ & - \sum\nolimits_{lt} \left(s_{l} p_{lt} + s_{l}^{fix} v_{lt} \right) = \sum\nolimits_{d} \left(\sum\nolimits_{t} e_{d} (D_{dt} + d_{dt}^{+} - d_{dt}^{-}) - c_{d}^{inv} \sum\nolimits_{t} I_{dt} \right) \\ & - \sum\nolimits_{lt} \left(s_{l} p_{lt} + S_{l}^{fix} v_{lt} \right) = \sum\nolimits_{t} \left(\sum\nolimits_{d} e_{d} d_{dt} - c_{d}^{inv} I_{dt}) - \sum\nolimits_{l} (s_{l} p_{lt} + S_{l}^{fix} v_{lt}) \right). \end{aligned} \tag{14}$$

Optimization criteria (14) is a reformulation of traditional system objective (1), what we shown in the last line of (14), after some simple rearrangements. Since all the ordered quantities should satisfy demanded quantities, we can exclude term $\sum_{t} \sum_{d} e_{d} d_{dt}$ from (15) and obtain exactly the same objective as in traditional, centralized planning model focused on minimization of costs.

Indeed, optimization problem (10)–(14) uses concept of a centralized auction that is a reminiscent of the minimum cost problems stated in centrally planned economies. However, a fundamental difference is that auction participants are competitive, self-interested decision-makers driven by the need to achieve high individual profits. Auction provides coordination tools, so the independent entities follow most efficient schedules. The main tool of the auction are the exchange prices.

3 Pricing Production-Distribution Auction

Following the economic theory [3] market-clearing prices for each period can be explicitly obtained from the solution of auction models, as shadow prices to the balance constraint (14). However, in a case of a centralized auction using complex structure of bids/offers such an approach may be inadequate. Non-convexities derived from binary variables (production states) and time-coupling constraints (inventories states) diminish auction clearing model from its simple properties.

Main challenge of the centralized auction is in providing fair market-clearing prices, that support centrally imposed schedules. The individual goals may be in conflict with the overall system objective. Moreover, fair prices may not even exist in the non-convex environment. For example, the least expensive producer may be excluded because of too restrictive minimum production level. This fact, known as the lack of the competitive equilibrium was observed by many authors, especially in the context of a pool-based electricity auctions – see for example [6] for a comprehensive review.

Closer observation of our auction problem (10)–(14) reveals analogies between production-distribution formulation and energy generation-demand exchange. Production within supply chain relates very closely to energy generation: production limits, start-up times and costs, quantity-dependent marginal costs – all these technical characteristics are shared between participants representing supply. In fact, it is the supply that draws most attention of researchers, since it is challenging, yet really needed to consider competitiveness conditions in highly constraint technical environment [4].

Distribution, on the other hand, may be viewed as a price-responsive energy customer with capabilities to shift its demand – that is exactly what distributor is doing ordering less, or more of demanded quantity. However, there is a negligible number of literature that deal with shifting bids of demands and generation non-convexities. To author's best knowledge the only attempt to pricing pool-based auction with shifting bids is provided in previous paper by Zoltowska [9].

Here we build on previous findings to derive new pricing model adapted to the needs of distributors. It uses uplift minimizing model for pricing, as described in [8].

3.1 Idea of Uplift Minimizing Pricing Scheme

The general idea for pricing non-convex auction is as follows. The clearing model allows us to maximize social welfare. Uplifts (compensations) are paid separately from market clearing payments to unfairly priced buyers and sellers. Uplifts cover the difference between the optimal and actual benefits/profits. They are directly minimized in the model that sets out relation between uplifts (variables denoted with R), prices (variables denoted with π), market optimal schedules (parameters denoted with $\hat{p}, \hat{v}, \hat{d}$)

and individual bids/offers (parameters denoted as in previous paragraph). Main challenge is in deriving formulation for optimal individual benefits/profits.

Let π_t variable denote the trade price in period t. This price should be used to settle the orders. To investigate fairness condition it is useful to define auxiliary surplus and deficit prices λ_{dt}^{+}, λ_{dt}^{-}, λ_{lt}^{+}, λ_{lt}^{-}, for each distributor d and each producer l, as functions of prices. If the surplus price is positive, the corresponding participant is competitive – should make profit. If the bid/offer is non-competitive, it's surplus price is zero. This concept was described for the first time in [7].

However, when bids/offers involve the time-coupled constraints, the adequate pricing may not be addressed independently in particular periods, rather the entire planning horizon must be examined. One method, developed in the author's previous papers [8]–[10] uses forward dynamic programming model to investigate individual profits conditions in multi-period setting. Pricing model formulated in this paper is based on the same technique, while new profit relations are stated.

3.2 Uplifts Required for Individual Producers

As already discussed, production offers within supply chain under consideration match directly typical energy generation offers. Thus we use formulation for uplifts of individual producers directly from previous work. Only the relevant constraints are included, for more details the interested reader is referred to the previous papers.

$$\lambda_{lt}^{+} - \lambda_{lt}^{-} = \pi_t - s_l, \quad \forall l, t \tag{15}$$

$$\tilde{P}_{lt} = \lambda_{lt}^{+} P_l^{max} - \lambda_{lt}^{-} P_l^{min} - S_l^{fix}, \quad \forall l, t \tag{16}$$

$$P_{lt} = (\pi_t - s_l)\hat{p}_{lt} - S_l^{fix}\hat{v}_{lt}, \quad \forall l, t \tag{17}$$

$$R_{lt} \geq \tilde{P}_{lt} - P_{lt}, \quad \forall l, t \tag{18}$$

$$R_{lt} \geq -P_{lt},\ R_{lt} \geq 0 \quad \forall l, t \tag{19}$$

In (15) relation between auxiliary individual surplus/deficit prices, offered prices and market prices is stated. Uplift R_{lt} for producer l in period t is defined in (18) as a difference between optimal profit $\tilde{P}_{lt}$ (16) and actual profit P_{lt} defined in (17). Optimal profit as given in (16) may be negative when price does not cover fixed cost – lost is minimal either when producing at full capacity (when price exceeds unit production price, so λ_{lt}^{+} is positive) or production should be minimal (λ_{lt}^{-} is positive in other case). Additionally, in (19) we assure uplift if production makes negative profit.

Notice, that in case of no production actual profit is equal zero, so uplift will be assigned in (18) only if optimal profit is positive.

3.3 Uplifts Required for Individual Distributors

Even though there is analogy between operation of distributors and flexible demands, still the most important difference is that dependencies between temporal benefits are time-coupled. In [9] it was sufficient to investigate part of sum of highest benefits within whole trading period. To properly formulate uplift conditions for distributors we need to address multi-period dependencies between surpluses and deficits. In fact, we need to make use of dynamic programming, similarly to method introduced in [8].

First, to consider surplus we check if valuation price exceeds market price:

$$\lambda_{dt}^{\mathrm{sur},+} - \lambda_{dt}^{\mathrm{sur},-} = e_{dt} - c_d^{inv}(T+1-t) - \pi_t. \quad \forall d,t \tag{20}$$

While to assess competitiveness conditions of reducing the order we write:

$$\lambda_{dt}^{\mathrm{def},+} - \lambda_{dt}^{\mathrm{def},-} = \pi_t - e_{dt} + c_d^{inv}(T+1-t). \quad \forall d,t \tag{21}$$

We can observe that $\lambda_{dt}^{\mathrm{sur},+} - \lambda_{dt}^{\mathrm{sur},-} = \lambda_{dt}^{\mathrm{def},-} - \lambda_{dt}^{\mathrm{def},+}$, so it is enough to consider single pair of prices, i.e. $\lambda_{dt}^{+} - \lambda_{dt}^{-}$, also in a specific period distributor may either buy or sell.

Surplus from buying is inseparable from some future deficit from selling. That is why we define maximum benefit ω_{dt}^{i} obtained from period 1 to *t* for distributor *d* when in state *i*, where state denotes how many surplus orders are in inventory. As Toczyłowski and Zoltowska have shown in [8], ω_{dt}^{i} functions can be determined with the use of the Bellman inductive functions:

$$\omega_{dt}^{i} = \max\left\{\omega_{d,t-1}^{i}; \omega_{d,t-1}^{i-1} + \lambda_{d,t-1}^{+}; \omega_{d,t-1}^{i+1} - \lambda_{d,t-1}^{+}\right\}, \quad \forall d,t,\mathrm{i} \geq 1 \tag{22}$$

where ω_{dt}^{0} should be considered similarly to (22), but without a middle term. Since we end up with empty inventory, in starting period 0 and in last period T only 0 state should be considered – see Fig. 1 illustrating how the particular states *i* may be reached in consecutive periods. Transition between different states is related with increment (transition to higher state) or decrement (transition to lower state) of benefit, in amount given by surplus price in particular state. Transition between the same states has no effect on particular maximum benefit ω_{dt}^{i} obtained in period *t*.

To guarantee proper realization of the max{ } functions, binary variables x_{dt}^{ik} must be included, each for one term inside max{ } function ($k = 1..3$). They are used along with sufficiently large parameters Υ as follows:

$$0 \leq \omega_{dt-1}^{i} \leq \omega_{dt}^{i} \leq \omega_{dt-1}^{i} + \left(1 - x_{dt}^{i1}\right)\Upsilon, \quad \forall d,t,\mathrm{i} \tag{23}$$

$$0 \leq \omega_{dt-1}^{i-1} + \lambda_{dt-1}^{+} \leq \omega_{dt}^{i} \leq \omega_{dt-1}^{i-1} + \lambda_{dt-1}^{+} + \left(1 - x_{dt}^{i2}\right)\Upsilon, \quad \forall d,t,\mathrm{i} \geq 1 \tag{24}$$

$$0 \leq \omega_{dt-1}^{i+1} - \lambda_{dt-1}^{+} \leq \omega_{dt}^{i} \leq \omega_{dt-1}^{i+1} - \lambda_{dt-1}^{+} + \left(1 - x_{dt}^{i3}\right)\Upsilon, \quad \forall d,t,\mathrm{i} \tag{25}$$

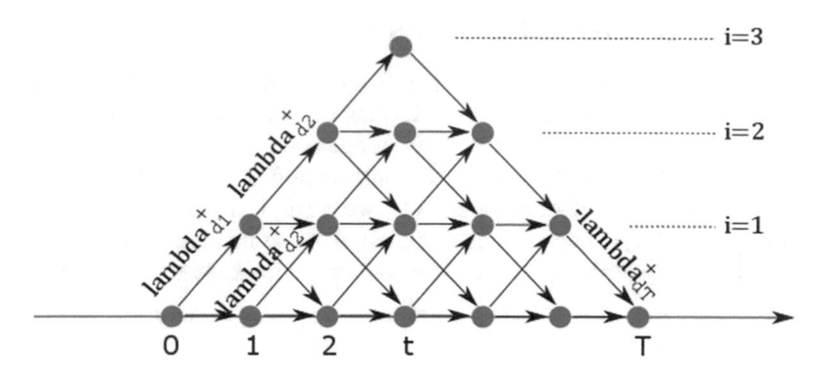

Fig. 1. Illustration of method for determining maximum benefit, according to (23)–(25)

$$\sum\nolimits_{k} x_{dt}^{ik} = 1. \quad \bigvee d, t, \mathrm{i} \tag{26}$$

In summary, actual benefit of distributor is equal:

$$B_d = \sum\nolimits_{t} ((e_d - \pi_t)\hat{d}_{dt} - c_d^{inv}\hat{I}_{dt}), \quad \bigvee d \tag{27}$$

$$R_d \geq \omega_{dT}^{0} - B_d, \; R_d \geq 0. \quad \bigvee d \tag{28}$$

Then, the uplift (non-negative) for distributor d is defined in (28) as a difference between optimal, maximum profit obtained in last period and actual profit from (27).

The objective of pricing problem is to minimize uplifts of all participants:

$$\min \text{Uplifts} = \sum\nolimits_{d} R_d + \sum\nolimits_{lt} R_{lt}, \tag{29}$$

with constraints stated in (15)–(19), (23)–(28).

Such a direct minimum-uplift pricing model (DMU) is formulated as a highly complex multi-period mixed-integer programming problem. However, as was discussed in previous works, market prices minimizing uplifts do not differ significantly from dual prices to (13), that can be derived from the relaxed solution to auction clearing problem (10)–(14) with fixed commitment variables. Effective procedure is to solve DMU pricing model with additional constraints imposing fixed values of prices, equal to the dual prices. The resulting restricted model is easily solvable and gives integer feasible solution, that improves the computations of original model.

4 Illustrative Example

We show the applicability of the proposed auction on a very simple example. Here we consider 2 producers and 1 distributor collaborating to agree on orders quantities in the next 7 periods. Details of offered production capabilities are shown in Table 2.

Table 2. Offered production capabilities.

Production characteristics	P_l^{min}	P_l^{max}	s_l	S_l^{fix}
Producer $l = 1$	10	90	40	190
Producer $l = 2$	20	200	45	100

Table 3 provides quantities demanded by the customers of distributor in each period. Inventory holding cost equals 3$, while unit customer demand price is 60$.

Table 3. Quantities demanded and auction results: orders together with prices.

Period	$t = 1$	2	3	4	5	6	7
Demanded quantity D_t	98	104	74	127	103	58	108
Ordered quantity d_t	112	90	90	124	90	76	90
Surplus quantity d_t^+	14	0	16	0	0	18	0
Deficit quantity d_t^-	0	14	0	3	13	0	18
Inventory level I_t	14	0	16	13	0	18	0
Production of prod. 1	90	90	90	90	90	76	90
Production of prod. 2	22			34			
Unit auction price π_t	45	48	42	45	48	40	43

Table 3 shows also results of auction clearing and pricing. It can be observed, that distributor is assigned orders exceeding demanded quantities in periods with lower prices: periods 1, 3 and 6. On the other hand, in periods with higher prices distributor is able to reduce order size and use quantities stored in inventories. It should be noticed, that prices rise after periods with nonempty inventory, exactly by the cost of inventory, so benefit of distributor is maximized. It is equal to 10,187$ and there is no other schedule that could yield higher benefit. What is interesting, realization of orders following exactly the demanded quantities yields the same benefit for distributor, however production cost would have to raise in such a case from 28,690$ to 29,296$.

5 Conclusions

In this paper we propose a centralized auction that may be used as coordination mechanism for agreement of orders quantities and schedules, in a competitive environment of producers and distributors collaborating within supply chain setting. As a result we obtain schedules efficient for entire supply chain, that are supported with auction prices. Such a market-based approach is a promising alternative to traditional, centralized methods. It enhances organizational communication and coordination.

We explain analogy of our auction with non-convex energy auction, which allowed us to base pricing model on the concept known in the electricity market as uplifts minimization. The resulting model derives fair, computationally tractable prices.

This research could be further extended into several directions. One would be to extend the supply chain participants technical characteristics to include for example delivery costs, expandable capacities, costs depending on the size of orders, different types of products in orders etc. Future work should also focus on verification of the approach on a larger setting, ideally already published in other research in this area.

Other natural line of research would be to analyze if this newly proposed mechanism could be sufficient to provide right incentives for strategic investments and improvements that should result in progressive increase in local efficiency.

Finally, the next step in would be to investigate how to design an information system that could be implemented to support collaboration within a supply chain.

References

1. Danloup, N., Allaoui, H., Goncalves, G.: Literature review on OR tools and methods for collaboration in supply chain. In: Proceedings of 2013 IEEE International Conference on Industrial Engineering and Systems Management (IESM), pp. 1–7 (2013)
2. Jung, H., Jeong, B.: Decentralised production-distribution planning system using collaborative agents in supply chain network. Int. J. Adv. Manuf. Technol. **25**(1–2), 167–173 (2005). https://doi.org/10.1007/s00170-003-1792-x
3. Kalagnanam, J., Parkes, D.C.: Auctions, bidding and exchange design. In: Simchi-Levi, D., Wu, S.D., Shen, Z. (eds.), Handbook of Quantitative Supply Chain Analysis: Modeling in the E-Business Era, pp. 143–212. Kluwer, Boston (2004). https://doi.org/10.1007/978-1-4020-7953-5_5
4. Kaleta, M.: Price of fairness on networked auctions. J. Appl. Math. **2014**, 1–7 (2014). https://doi.org/10.1155/2014/860747
5. Kempf, K.G., Keskinocak, P., Uzsoy, R. (eds.): Planning production and inventories in the extended enterprise: a state of the art handbook, vol. 1. Springer (2011). https://doi.org/10.1007/978-1-4419-6485-4
6. Madani, M., Van Vyve, M.: Computationally efficient MIP formulation and algorithms for European day-ahead electricity market auctions. Eur. J. Oper. Res. **242**(2), 580–593 (2015). https://doi.org/10.1016/j.ejor.2014.09.060
7. Toczyłowski, E.: Optimization of Market Processes under Constraints. EXIT Publishing Company, Warsaw (2002)
8. Toczylowski, E., Zoltowska, I.: A new pricing scheme for a multiperiod pool-based electricity auction. Eur. J. Oper. Res. **197**(3), 1051–1062 (2009). https://doi.org/10.1016/j.ejor.2007.12.048
9. Zoltowska, I.: Demand shifting bids in energy auction with non-convexities and transmission constraints. Energy Econ. **53**, 17–27 (2016). https://doi.org/10.1016/j.eneco.2015.05.016
10. Zoltowska, I.: Direct minimum-uplift model for pricing pool-based auction with network constraints. IEEE Trans. Power Syst. **31**(4), 2538–2545 (2016). https://doi.org/10.1109/TPWRS.2015.2468083

Combination of the Earned Value Method and the Agile Approach – A Case Study of a Production System Implementation

Dorota Kuchta(✉)

Wroclaw University of Science and Technology, Wybrzeze Wyspianskiego 27, 50-370 Wroclaw, Poland
Dorota.Kuchta@pwr.edu.pl

Abstract. A combination of the Earned Value Method and the Agile approach is proposed by means of a case study of new production system implementation. The project was managed in the traditional way. Advantages of introducing Agile elements (mainly concerning the communication among all the project stakeholders) into the traditional Earned Value Method are shown.

Keywords: Earned Value · Agile approach
Production system implementation

1 Introduction

The Agile approach to project management is a fairly recent development, and it is in fashion in many circles today. Unfortunately, it cannot be used for each project type ([16]). Some projects have to be managed in a traditional way. However, it seems certain that the traditional project management will benefit from the introduction of some Agile elements. This has already been shown in the literature [4, 9].

The Earned Value Method belongs to the traditional project management. It is a method to be applied at intervals during project implementation in order to try to capture all the currently available information which might provide an insight about the final project outcome in the future. Here we propose to combine it with some Agile elements.

The objective of this paper is to show the advantage of combining the Earned Value Method (which belongs to the traditional project management) with elements of the Agile approach.

In Sect. 2 the Earned Value Method is presented, in Sect. 3 the Agile approach and its proposed combination with the Earned Value Method, and finally in Sect. 4 the case study is analysed. The paper terminates with some conclusions.

2 Earned Value Method

The Earned Value Method [3] is a method of project implementation control whose main aim is to provide a warning system, which should allow to foresee problems with project implementation as soon as possible in order to be able to react in time before the

© Springer Nature Switzerland AG 2019
A. Burduk et al. (Eds.): ISPEM 2018, AISC 835, pp. 87–96, 2019.
https://doi.org/10.1007/978-3-319-97490-3_9

problems influence project success (which is more and more often understood today as the satisfaction of principal stakeholders [5]) negatively and substantially. Once the project is started, at regular intervals a control procedure has to be put into practice, where the following questions should be answered (below they are presented as referring to the whole project, but they are also referred to individual activities or groups of activities if needed):

1. Is the project implementation so far acceptable from the point of view of cost?
2. Is the project implementation so far acceptable from the point of view of scope and time?
3. Is the project implementation so far acceptable from the point of view of due payments?
4. What is the current prediction of the project completion time and is it acceptable?
5. What is the current prediction of the total project cost and is it acceptable?

The project characteristics allowing to answer the above questions are named as follows ($n = 1,2,\ldots,N$ are the numbers of the subsequent control points, in which the control procedure of the Earned Value Method is implemented):

1. Budget at Completion(n), BAC(n): the current value of the total budget of the project. BAC(0) denotes the original budget of the project.
2. Budgeted Cost of Work Scheduled(n), BCWS(n): the value of the project work scheduled up to the moment of the n-th control point, expressed in terms of originally planned cost;
3. Budgeted Cost of Work Performed(n), BCWP(n): the value of the project work actually performed up to the moment of the n-th control point, expressed in terms of originally planned cost;
4. Actual Cost of Work Performed(n), ACWP(n): the actual cost of the project work actually performed up to the moment of the n-th control point;
5. Estimate of Cost at Completion(n), ECAC(n): the estimated cost of the implementation of the whole project according to the knowledge in the moment of the n-th control point. In the basic form of the Earned Value Method it is estimated according to the formula:

$$\mathrm{ECAC}(n) = \mathrm{BAC}(0) \cdot (\mathrm{ACWP}(n)/\mathrm{BCWP}(n)) \quad (1)$$

   However, formula (1) is a very rough one, as it takes into account only the planned and actual cost of the already performed project work. Other formulae have to be considered in many practical cases;
6. Variance of Cost at Completion(n), VCAC(n) = BAC(n)-ECAC(n). IF VCAC(n) is negative and has a substantially high absolute value, it means that there is a danger that the project will not be able to keep to the available budget. This is a piece of information concerning the future, thus there is time to undertake necessary measures before the project is finished and important project stakeholders are aware of the problem.
7. Estimate of Duration at Completion(n) - EDAC(n), Variance of Duration at Completion(n) - VDAC(n) are respective characteristics for project duration,

expressed in time units. Their underlying idea is the same as for ECAC(n) and VCAC(n). The role of BAC(n) plays here the maximal admissible deadline of the project in the moment of the n-th control point or simply the first estimate of project duration. Formulae for EDAC(n) are discussed e.g. in [11, 14, 15] and will of course be based, among others, on BCWP(n) and BCWS(n), analogously to (1), but will be more complicated, due to the fact that, contrary to the project cost, project duration is not a simple sum of individual activity duration.

The values ECAC(n), VCAC(n), EDAC(n), VDAC(n) are the most important ones. They deliver the warnings for the future: already in the n-th control point we can see what the estimates of the total project cost and duration are and whether they are acceptable. If they are not, there is time to implement corrective measures before the important project stakeholders are unsatisfied with the project outcome.

It is important to use as many pieces and sources of information as possible while determining those values. Formula (1) for ECAC(n) is a very simplistic one. It is used in the most basic version of the Earned Value Method. It shows that obviously the information about the planned and actual past performance of the project and their relationships (expressed by means of BCWS(n), BCWP(n), ACWP(n), CV(n), SV(n)) should be taken into account. But already in the existing extensions of the Earned Value Method [2, 6, 8] it is advised not to rely on simplistic formulae, but to gather all the relevant pieces of information, e.g. the emails and phone calls with the suppliers (any signaled problem with the delivery has to be considered), customers (any sign of dissatisfaction or of changing requirements has to be noted) etc. Also, various scenarios have to be taken into account and any uncertainty noticed and considered (e.g. in [7] interval or fuzzy numbers are used to express the uncertainty). The better the quality of the values ECAC(n), VCAC(n), EDAC(n), VDAC(n) for smaller values of n, the more stakeholders are asked and talked to also for small values of n, the better the warning system and the highest the probability of project success in the eyes of the principal stakeholders. We propose here to use in this step the agile philosophy, which will be shortly presented in the next section.

3 Agile Approach and Its Combination with the Earned Value Method

The Agile approach [16] was first formally presented in a so called Agile Manifesto [1] in 2001. Its first destination were IT projects, but nowadays it is used also for other project types [9]. Its main idea is openness to changes (instead of keeping to the initial plan at all cost) and a close and frequent communication within the project team and with the outside: with the project environment in the general sense of the word. It is assumed that the main goal of the project implementation is the satisfaction of principal project stakeholders, that is why the continuous communication with them plays a very important role.

A short exempt from the Agile principles introduced by the authors of the Agile Manifesto will convey its basic sense: “Business people and developers must work together daily throughout the project. Build projects around motivated individuals. Give them the environment and support they need, and trust them to get the job done. The most efficient and effective method of conveying information to and within a development team is face-to-face conversation" [1].

The Agile approach is the counterpart of the called traditional approach [16], where the changes with respect to the initial project plan are not so often, so deep and especially, so easily accepted (the Agile Manifesto says that in the Agile approach they are even "welcome").

In the literature combinations of the Agile approach with Earned Value have been proposed, mainly within the Agile approach [10, 12]. But there have also been attempts to introduce elements of the Agile approach into the traditional project management [4, 9]. The Agile approach in its fullness is not appropriate for all project types (e.g. in many construction projects deep changes at later stages are simply impossible), but selected Agile elements can be useful in most cases.

The Agile approach combined with the Earned Value Method, which we propose here, is a logical extension of the Earned Value Method modifications mentioned in the previous section. At each Earned Value control point agile meetings, thorough analyses of the environment (also from the psychological point of view), reception of even weak signals from the entire project environment should take place. The case study below presents an idea of how this could be put into practice.

4 Case Study

The case study presented here is a project discussed in [13]. It concerns the implementation of a new production system. It is based on a real world case, but here it will be presented in a simplified form due to the size limitation of the paper.

The objective of the project was to implement a new production system. Table 1 presents project activities (on a high abstraction level), together with their planned durations and the quantity of man-months required for 3 types of human resources: HR1, HR2, HR3.

Table 1. Project activities, their planned duration (in months) and the required usage of human resources HR1, HR2 and HR3 (man-months) [13].

Activity name	Activity acronym	Planned duration	Required resources
Elaboration of the system concept	SC	1	1 RH1
Elaboration of the requirements for Subsystem 1	RS1	1	1 RH1
Elaboration of the requirements for Subsystem 2	RS2	0,5	1 RH1
Development of Subsystem 1	DS1	1,5	1 RH2
Development of Subsystem 2	DS2	1	1 RH3
Integration of both subsystems	I	1	1 RH1 and 1 RH2 and 1 RH3

Figure 1 presents the network of the project, where planned durations of the activities and the dependencies between them are shown. There are two paths in the project. The longest one, regarding Subsystem 1, determines the minimal planned project duration: 4,5 months.

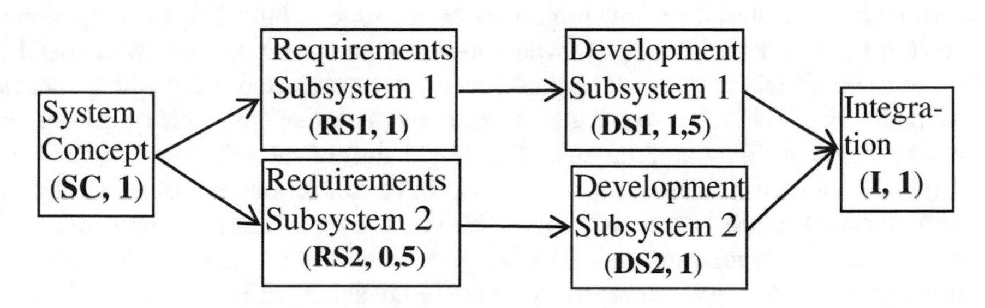

Fig. 1. Network of the project [13].

The planned project schedule (elaborated before the project start) was as follows:

The schedule shows that it was indeed possible to plan the project for 4,5 months. If we calculate the total number of man-months needed for the project, we obtain the planned project budget in man-months (BAC(0)): 8 mm.

Let us now consider three control points of the Earned Value method.

Control point 1: at the end of the 1th month

The situation is as follows: Resource RH1 started to work on activity SC in the middle of the 1. month, he worked for 0,5 month, the progress of activity SC is 50%, no other activity has been started.

We have the following values of the Earned Value Method (in mm. – man-months, except for EDAC and VDAC, which are expressed in months):

BCWS(1) is equal to 2 mm., because this is the value of the work planned before the end of the 1. month in mm. according to Table 2. BCWP(1) is the value in mm. of 50% of task SC according to the plan in Table 2. ACWP(1) has the same value, because we are told that resource RH1 has actually worked on the 50% of activity SC for 0,5 month. CV(1) and SV(1) are calculated according to formula (1) (CV) or an analogous one (SV). In SV(1) it was assumed that the schedule from Table 2 will simply be shifted to the right by half a month.

Table 2. The planned project schedule ("m." stands for "month", "h." for "half) [13].

	1. m.		2. m.		3. m.		4. m.		5. m.
	1.h.	2.h.	1.h.	2.h.	1.h.	2.h.	1.h.	2.h.	1.h.
RH1	SC	SC	RS1	RS1	RS2			I	I
RH2					DS1	DS1	DS1	I	I
RH3						DS2	DS2	I	I

The four last columns of Table 3 are the most important values for the project control and here the agile approach should be used. Values from Table 3 are only proposals. In reality we might have there different values, even expressed by means of fuzzy numbers, intervals, probability distributions or sets of scenarios. The first question that should be asked the project team and the important stakeholders is about the estimation of cost of the whole project according to their knowledge and opinion at the end of the 1. month. Formula (1) would indicate there is no problem with cost, but we have to verify this statement taking into account the projected delay with respect to the duration from Table 2. It seems that duration estimation of the whole projects in the 1. control moment might be 5 months, if a simple shift of the schedule from Table 2 (0,5 months forward) can be assumed. But we have to ask the project team why the project start took place later than planned. Maybe the resources are overloaded with other tasks in the organisation and EDAC(1) = 5 months is unrealistic? Or maybe the resources are able and ready to make up for the delay and a smaller value of EDAC(1) can be assumed? Before EDAC(1) is chosen to be equal to 5, the question has to be asked whether this delay is acceptable by the principal stakeholders and if so, whether this delay will not have any influence on ECAC(1) (fixed cost of the project, penalties for the delay, etc.). And we have to ask the principal stakeholders, also the project team, but above all the customer for whom the new production system is implemented, whether in this first month of the project anything has changed in their environment, in their expectations, availabilities, plans etc., which might influence the total cost and time of the project as well as the expected quality of the product, i.e. of the production system. If any essential change has to be introduced based on these conversations, it should be done so in an agile way, even changing the whole future schedule.

Let is assume that values in Table 3 were accepted and let us now consider the next control point.

Table 3. Earned Value characteristics for the 1. control point.

BAC	BCWS	BCWP	ACWP	CV	SV	ECAC	EDAC	VCAC	VDAC
8	1	0,5	0,5	0	−0,5	8	5 months	0	−0,5 month

Control point 2: at the end of the 2nd month

The situation is as follows: since the previous control point resource RH1 has worked for one month on activity SC, which has been finished, no other activity has been started.

We may have here the following values of the Earned Value Method:

The values in the first 6 columns of Table 4 were calculated analogously to the respective columns of Table 3. The future relating values (the last four columns) are here expressed in interval values and such values may be the basis for the exchange between principal project stakeholders. Two extreme scenarios gave rise to the interval values:

Table 4. Earned Value characteristics for the 2. control point.

BAC	BCWS	BCWP	ACWP	CV	SV	ECAC	EDAC	VCAC	VDAC
8	2	1	1,5	−0,5	−1	[8,5, 12,5]	[5,5, 7,5] months	[−0,5, −4,5]	[−1, −3] months

- Optimistic scenario: resource RH1 needed 50% more time to accomplish task SC than planned, but we assume that this will not influence future productivities of the resources in the tasks remaining to accomplish – in other words, the other tasks will take as much time as planned, which is shown in Table 5:

Table 5. The project schedule updated in the 2. control point, optimistic scenario.

	3. m		4. m		5. m		6.m.
	1.h.	2.h.	1.h.	2.h.	1.h.	2.h.	1.h
RH1	RS1	RS1	RS2			I	I
RH2			DS1	DS1	DS1	I	I
RH3				DS2	DS2	I	I

- Pessimistic scenario: resource RH1 has to be asked why it took him more time than it was planned to elaborate the concept of the production system. It has to be analysed whether the estimate of other tasks to be performed by RH1 should not be increased – or maybe even of all the remaining tasks. In Table 6 it is assumed that all the other tasks will take more time than planned – the actual information about task SC was assumed to force a correction in the future productivity in all the tasks. The respective values were found in an agile way – in a full openness to changes, received pieces of information about a lower productivity or about the overload of the resources etc.

Table 6. The project schedule updated in the 2. control point, pessimistic scenario.

	3. m		4. m		5. m		6.m.		7.m.		8 m.
	1.h.	2.h.	1.h.	2.h.	1.h.	2.h.	1.h.	2.h.	1.h.	2.h.	1.h
RH1	RS1	RS1	RS1	RS2	RS2				I	I	I
RH2				DS1	DS1	DS1	DS1	DS1	I	I	I
RH3					DS2	DS2	DS2		I	I	I

The cost estimates (ECAC, VCAC) were done on the basis of the assumption that there will be no penalties or other financial repercussions of the delay. But this has to be decided in the body of stakeholders and a wider spectrum of representatives of the project environment.

The agile approach would induce the necessity to make difficult decisions. Is the pessimistic scenario acceptable? If not, by whom not and to which extent? What can be done to prevent the pessimistic scenario from happening? How much this would cost additionally? Is it necessary to introduce deeper changes in the project?

For the moment we assume that it was decided to keep the interval-formed information, because both extremes were found to be possible and there was too much uncertainty to make crisp decisions. The next month has passed.

Control point 3: at the end of the 3rd month

The situation is as follows: since the previous control point resource RH1 has worked for half a month on activity RS1 and half a month on activity RS2. Both activities have been finished, no other activity has been started.

We can see that a sudden increase in the productivity of resource RH1 has taken place. An agile meeting should decide what the reason for this difference was. There may be a problem with the quality of the work accomplished by the resource RH1 in the 3rd month, if he were pressed to hurry up because of the delay he had caused in the first two months or the nature of the two tasks accomplished in the 3rd month (elaboration of production subsystems requirements), being different from the nature of the first, delayed task (which was Elaboration of the production system concept), was responsible for the difference. This would be a sign that resource competences and experience have to be taken into account to a higher extent in the estimations.

Also, the agile approach requires gathering all the signals, even weak or uncertain ones, about any changes in the project environment that might influence its future course. Here such a signal was received: resource RH3 will be needed in another project. Resource RH1 will have to overtake task DS2, but his productivity in this task will be equal to 50% of the respective productivity of resource RH3. Also, because of the lack of resource RH3, task Integration of the subsystems will take 50% more time than planned.

We may have here the following values of the Earned Value Method:

The first 6 columns of Table 7 are calculated as in the previous case. They refer to the past, thus to the first 3 months of the project. They show that there is a delay (negative SV, as tasks DS1 and DS2, planned partially for the 3rd month, were not started) and that CV = 0 might be misleading and has to be analysed deeply in the agile way. CV = 0 means that the actual cost and the planned cost of the work performed so far are equal. But this is true only on the level of the project. In order to perform a sufficiently deep analysis, we have to lower ourselves to the level of activities (Table 8).

So, we can see that CV(3) is negative for one activity. The total CV(3) is zero, because another activity has a positive value here, but the reasons of both non-zero

Table 7. Earned Value characteristics for the 3. control point.

BAC	BCWS	BCWP	ACWP	CV	SV	ECAC	EDAC	VCAC	VDAC
8	4	2,5	2,5	0	−1,5	8,5	7 months	−0,5	2,5 months

Table 8. Earned Value characteristics for the 3. control point, for activity SC, DS1, DS2.

Activity	BAC	BCWS	BCWP	ACWP	CV	SV
SC	1	1	1	1,5	−0,5	0
DS1	1	1	1	0,5	0,5	0
DS2	0,5	0,5	0,5	0,5	0	0

values of CV(3) should be analysed, with the underlying question whether they may have any consequences for the remaining project course.

In the analysed moment deep changes in the project have happened (changes in productivity, withdrawal of one of the resources). All other changes, also those in the moods of the persons involved (e.g. is RH1 happy to take over the duties of RH3 and how this might influence his motivation and consequently, his productivity?) have to be discussed in the agile meeting. Here the following remaining schedule was assumed (one scenario has been agreed upon), which was the basis for the values included in the last column of Table 7.

The project continued and, apart from minor changes, it followed the schedule from Table 9, which is very different for the initial one (Table 2). An agile approach, openness to discussions, complaints and signs of negative motivation, receptivity of even weak signals from the environment, combined with the classical Earned Value Methods, made it possible to be aware already since the end of the 2nd month that project duration 4,5 months from Table 2 was unrealistic and to prepare all the stakeholders for this delay and achieve their acceptance. If they had not accepted the schedules updated in each control point, thanks to the agile approach it would have been possible to introduce other changes, ask for additional resources etc. in order to satisfy the principal stakeholders (here e.g. the management board of the plant where the production system was being implemented, its future users, the project team, the suppliers etc.), which should be the main goal of the project.

Table 9. The project schedule updated in the 3. control point.

	4. m.		5. m.		6. m.	
	1.h.	2.h.	1.h.	2.h.	1.h	2.h.
RH1	DS2	DS2	DS2	I	I	I
RH2	DS1	DS1	DS1	I	I	I

5 Conclusions

The case study, based on a real world project of a new production system implementation [13], presents an initial, rough proposal of how the Earned Value Method, which has proven itself to be very useful in the traditional project control, can be improved (with respect to its project management support possibilities) thanks to the introduction of some elements of the Agile approach.

The case study shows that completing the Earned Value Method with enhanced communication and receptivity of signals from the project environment makes it more efficient in supporting the project manager in achieving the project success, which today is more and more often understood as the satisfaction of the principal project stakeholders.

The main conclusion is that the Earned Value Method, which per se belongs to traditional project management, may benefit strongly from the application of several agile elements. Further case studies in the area of production and logistic projects and an attempt to formalize the proposed approach are necessary in order to arrive at a mature Agile Earned Value Method.

References

1. Agile Manifesto homepage. http://agilemanifesto.org/. Accessed 22 Apr 2017
2. Dałkowski, T.B., Kuchta, D.: Classical and modified earned value method in project management. Badania Operacyjne i Decyzje **10**, 58–65 (2000). (in Polish)
3. Fleming, Q.W., Hoppelman, J.M.: Earned Value Management. Project Management Institute (1996)
4. Hughes, R.: Starting and scaling agile data warehousing. In: Hughes, R. (ed.) Agile Data Warehousing Project Management, pp. 303–344. Morgan Kaufmann, Boston (2013). Chap. 9
5. Johansen, A., Eik-Andresen, P., Ekambaram, A.: Stakeholder benefit assessment – project success through management of stakeholders. Procedia - Soc. Behav. Sci. **119**, 581–590 (2014)
6. Joly, M., Le Bissonnais, J., Muller, J.-L.G.: Maîtriser le coût de vos projets. AFNOR, Paris (1995)
7. Kuchta, D.: Fuzzyfication of the earned value method. WSEAS Trans. Syst. **4**(12), 2222–2229 (2005)
8. Kuchta, D.: On a certain extension of the earned value method. In: Problemi Teorìï Ta Metodologìï Buhgalters'Kogo Obliku, Kontrolû i Analìzu - Mìžnarodnij Zbìrnik Naukovih Prac'. Seria Buhgalters'Kij Oblìk, Kontrol' i Analìz, vol. 1, pp. 305–311 (2006)
9. Kuchta, D., L'Ebraly, P., Marchwicka, E.D.: Agile-similar approach based on project crashing to manage research projects. In: 18th International Conference on Enterprise information systems, ICEIS 2016, pp. 225–241. Springer (2017)
10. Kuchta, D., Skowron, D.: From time-driven activity-based costing model to earned value method in organisations managed by projects by means of scrum. In: Nauka i praktyka w zarządzaniu projektami, Project Management Excellence Forum Wroclaw 2012, pp. 61–85 (2012). (in Polish)
11. Solomon, P.J.: Performance-based earned value®. In: 15th Annual International Symposium of the International Council on Systems Engineering, INCOSE, pp. 180–197 (2005)
12. Sulaiman, T., Barton, B., Blackburn, T.: AgileEVM - earned value management in scrum projects. In: Proceedings - AGILE Conference, pp. 7–16 (2006)
13. Vallet, G.: Techniques de suivi de projets. Dunod, Paris (1997)
14. Vanhoucke, M.: Measuring Time: Improving Project Performance Using Earned Value Management. Springer, Cham (2009)
15. Vanhoucke, M.: Integrated Project Management and Control. Springer, Cham (2014)
16. Wysocki, R.K.: Effective Project Management: Traditional, Agile, Extreme. Wiley Publishing, Indianopolis (2009)

Computer Simulation of the Operation of a Longwall Complex Using the "Process Flow" Concept of FlexSim Software

Marek Kęsek[1(✉)], Agnieszka Adamczyk[1], and Monika Klaś[2]

[1] Faculty of Mining and Geoengineering, AGH University of Science and Technology, 30-059 Cracow, Poland
{kesek,aadamczyk}@agh.edu.pl

[2] Faculty of Management, AGH University of Science and Technology, 30-059 Cracow, Poland
monika.klas.897@zarz.agh.edu.pl

Abstract. The article discusses the essence and legitimacy of computer simulations to plan, monitor and optimize the course of various types of systems, in order to increase their productivity, shorten working time and achieve the desired economic efficiency without having to introduce costly and risky changes in the real process. This paper demonstrates a method of simulation of technological processes, using the example of mining production system realized with longwall complex strategies, using the FlexSim software. The possibility of simulation is presented through two approaches, where the first is based on the standard model structure, while the second is based on new software functionality, which is modeling in the "Process Flow" concept that is particularly useful during the simulation of complex operations. An advantage of the second concept was demonstrated by presenting a faster and more accurate construction process of the aforementioned model. The developed model allows for the exact selection of the best parameters and the estimation of the main indicators characterizing the production process carried out by the longwall complex. Modeling results can be used in the production planning process in mining and at the same time can contribute to maximize its efficiency.

Keywords: Modeling · Computer simulation · FlexSim · Longwall complex

1 Introduction

Computer simulation is mainly used to analyze a process, activity or complex operation in order to improve its performance. A great advantage of this type of solutions is that there is no need to introduce changes to the actual model. Such behavior is subject to a considerable risk of failure as the operation of the physical model could deviate from previously set standards, which would result in the loss of significant amount of time and unnecessary financial expenses. Computer modeling and simulation allows excluding such a situation, helps to precisely estimate the efficiency of the analyzed process and to catch any irregularities in a short time. This allows performing multiple experiments on a single virtual model and selecting the most appropriate parameters for

© Springer Nature Switzerland AG 2019
A. Burduk et al. (Eds.): ISPEM 2018, AISC 835, pp. 97–106, 2019.
https://doi.org/10.1007/978-3-319-97490-3_10

individual devices, optimum settings of workplaces or the order of particular activities. It also enables to take into account the number of employees, modes of transport, storage capacities, any downtime caused by the need to perform inspection and maintenance of machines and many others, thus contributing to the efficiency of the entire process, achieving optimum economic efficiency, or shortening the total working time [1]. In particular, modeling and simulation is used to solve problems, but it is equally important for designing a new process from scratch, as it can contribute to a much faster achievement of high production capacity. This solution has many advantages, but it requires a thorough knowledge of the individual factors that make up the entire system, as well as experience in working with software designed for modeling and simulation [2, 3].

Creating a model using the simulation method consists of several steps. In order to start building a model, it is necessary to define the problem, formulate research hypotheses and preliminary calculations first, and then determine the assumptions for the model and acquire precise information regarding actual and input data. The next step is to actually build the simulation, taking into account the factors necessary for the process to work, and to plan and conduct experiments aimed at optimizing performance. The next step is to analyze the results and draw appropriate conclusions. If the results are not satisfactory, the experiments should be repeated until satisfactory results are obtained [1, 2].

2 Overview of Software for the Modeling and Simulation

Computer simulation began in the 1950s. First programming languages appeared in the next decade, the main purpose of which was to facilitate the implementation of all modeling activities. Over time, ready-made simulation programs were also created, which allowed for the mappinf of processes in the 3D dimension [4].

Currently, there are many programs available on the market for modeling and computer simulation. Due to increasing efforts of companies towards cost reduction and process optimization, manufacturers are competing in creating ever more perfect software, helpful in building virtual models of various processes, and thus facilitating making appropriate decisions. This makes it easier and faster for companies to improve their work efficiency, locate bottlenecks and weaknesses in their processes, increase employee safety, and reduce cost, working time and other equally important factors. Examples of modern simulation programs include: SIMUL8, FlexSim, SprutCAM, Plant Simulation, Show Flow, Witness Simulation Software, Enterprise Dynamics Simulation Software, and others [1, 4].

Simple statistical simulations can also be carried out using Excel spreadsheets. The capability of cooperation with databases plays an important role, especially in the era of "Industry 4.0", because a number of calculations can be made outside the spreadsheet by the database engine itself [5].

In addition to the basic available functions, the spreadsheets also provide the capability to create macros and to work with a VBA application, however, despite the relatively simple operation when creating models, they cannot compare to modern, professional computer simulation programs [4].

3 Modeling and Simulation in Mining

Computer simulation can be successfully used in both underground and opencast mining. May include a wide range of problems of investment finance [6] by planning [7] to simulation.

The capability of building a simulation model of mining processes with the availability of such a wide range of modern software seems to be almost essential nowadays. The capabilities to simulate and adapt many variants without incurring unnecessary investments and risks can be effective in mines as well as in industrial companies, and their application can contribute to increasing production while optimizing costs, minimizing downtime and delays, or improving employee safety.

3.1 Mining of Hard Coal with a Longwall Complex

An example of the use of computer simulation in underground mining is the modeling of the coal mining process with the use of a longwall complex. Currently coal is mined most often using this solution. The entire complex consists mainly of a longwall shearer, which directly mines coal in longwall excavations whose lengths range from 60 to even 300 m. The shearer moves on a longwall conveyor that discharges output in the direction of the tunnel, where the ore goes to the longwall conveyor and is transported towards the mining shaft. After completing the full passage of the shearer, the longwall conveyor is moved to the wall and the whole cycle is repeated. The longwall complex also includes mechanized housings, ensuring the safety of workers in the vicinity [8–10].

The longwall complex is characterized by the interdependence of all the machines comprising it. This means that if one device fails, the operations of the entire system stop. Modern mining machines achieve high power and technical parameters, ensuring their durability and stability. Therefore, under ideal conditions, the entire complex could be operated continuously without interruptions or downtimes. However, in practice, the capabilities of the devices cannot be fully utilized due to many changing circumstances [8]. Therefore, the process of ore mining with the use of longwall complexes requires a simulation taking into account many factors, and its results may provide ample room for improvement and optimization of the operation of the simulated process.

In 2015, Dalin Cai (University of Wollongong, Australia) has built a simulation model of a production process carried out with a longwall complex, using FlexSim software, one of the best evaluated programs for modeling and simulation. The model was based on data from underground mines located in Australia. The author took into account the frequency of any delays in the operation of the complex, which were caused both by organizational issues and downtime related to possible equipment failures. By utilizing a model made in a way typical for FlexSim, he created a simulation which allowed for the analysis of various variants of the system operation, each time making changes to the assumed parameters, which included mainly the speed of the shearer and the depth of its cutting, the capacity of the conveyors and the width of the longwall. Thus, it was possible to estimate the production efficiency with the selection of the most optimal parameter settings. Research carried out by Dalin Cai has been extensively described in his work [11].

3.2 FlexSim Software

FlexSim enables to carry out simulations in virtually any field, from production to logistics, material handling, warehouse planning to mining, healthcare, customer service and more, with access to the library containing a large number of objects. The program enables simulation and visualization of processes in 3D, which greatly facilitates understanding and interpretation of the model built. FlexSim is also user-friendly, namely, objects selected from the library can be easily and quickly dragged into the dimension area with a single mouse click. In addition, the available library can be extended with new proprietary objects in order to best represent the modeled system. In order to interpret the conclusions and obtained results, the program enables to generate appropriate reports, allowing for more accurate estimation of the places requiring optimization and the necessity of introducing changes. FlexSim also allows you to create models that are freely expandable, which means that the space is unlimited [12].

3.3 "Process Flow" Concept

Initially, FlexSim only allowed for building models in a traditional way, which means use of drag&drop library, connecting elements and writing code to make some special functions. This is how the simulation of the operation of a longwall complex was created, that was made by Dalin Cai and described in his original work cited above. However, since 2016, the software manufacturer has been offering a next generation of modeling and simulation called "Process Flow". It is particularly useful when building large and complex models and allows to significantly reduce the time of their creation from a few days to several dozen minutes. This tool enables to build a logical model using block diagrams, which greatly simplify the organization of work compared to traditional computer codes. It also makes it possible to place the main logic of the designed model in the very center, which will make it easier to find and determine certain activities which are not progressing in an adequate way in the process [12]. Being inspired by the work of Dalin Cai, the authors decided to simulate a similar model, but using the concept described above in order to compare both methods and draw the appropriate conclusions.

4 Simulation Model of a Longwall Complex

The simulation model of the longwall complex was built in FlexSim 18.0.3 software, and the logic of the longwall shearer was mapped using ProcessFlow technology built into the software. The model has been built in such a way that it can be further extended with subsequent modules and variants of the analysis. Therefore, the logic of ProcessFlow is fully independent of the size of the shearer and the assumed operating parameters. The simulation work carried out in this analysis can be divided into three main areas:

- The construction of the 3D model of a longwall shearer
- Development of the operating logic of a longwall shearer
- Preparation of the panel for entering model parameters and analyzing of results

4.1 3D Model of a Longwall Shearer

TaskExecuter was the basic element of the standard object library in FlexSim used to build the model. Both the mining head itself as well as the individual conveyor modules and the housing are reproduced in the model using TaskExecutors. In addition, a queue was used to generate output.

The basic element of the model TaskExecuter is shown in Fig. 1, to which the bi-directional mining head (first central port) and subsequent modules of the conveyor are connected using central ports (in ascending order). Each conveyor module is a combination of two elements: TaskExecuter (responsible for the movement of the module during the mining operation) and a standard conveyor (responsible for the movement of output during the shearer operation). Each of the conveyor modules is connected to a queue (via an input port), which automatically sends the material to individual conveyors according to the movement of the shearer. In addition, the individual modules are connected to the housing via a central port. In the case of this model, it was necessary to distinguish between the conveyor and the housing modules, as each of these elements performs an independent movement and is characterized by different operating parameters. The longwall shearer moves along the path (NetworkNode) along individual modules. The model is designed to work regardless of the number of modules adopted. The route of the mining head is therefore shown in Table 1, which contains information on the location of the individual conveyor modules. The base longwall complex consists of 120 conveyor modules, each conveyor module is 1.7 m long, and the width of a single housing is 1.5 m. The model also takes into account the speed of the conveyor and the speed with which the single housing and the conveyor module move.

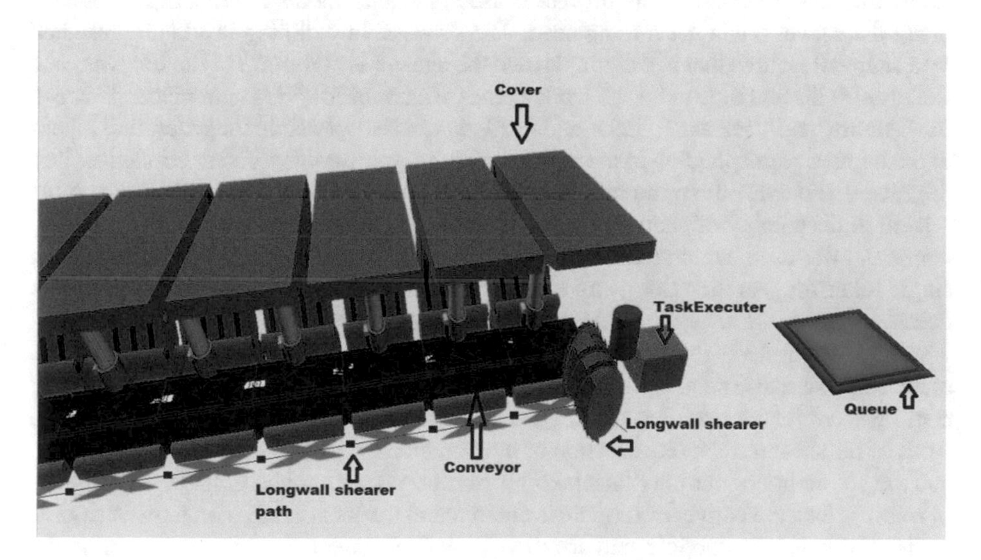

Fig. 1. View of the longwall shearer model in the FlexSim software (Source: own study)

Table 1. A fragment of the route of the mining head

	Module	Localization
Row 1	/Kombajn/TaskExecuter2	/Kombajn/TaskExecuter2/NN1
Row 2	/Kombajn/TaskExecuter13	/Kombajn/TaskExecuter13/NN2
Row 3	/Kombajn/TaskExecuter14	/Kombajn/TaskExecuter14/NN3
Row 4	/Kombajn/TaskExecuter15	/Kombajn/TaskExecuter15/NN4
Row 5	/Kombajn/TaskExecuter9	/Kombajn/TaskExecuter9/NN3
Row 6	/Kombajn/TaskExecuter10	/Kombajn/TaskExecuter10/NN3
Row 7	/Kombajn/TaskExecuter11	/Kombajn/TaskExecuter11/NN3
Row 8	/Kombajn/TaskExecuter12	/Kombajn/TaskExecuter12/NN3
Row 9	/Kombajn/TaskExecuter13~2	/Kombajn/TaskExecuter13–2/NN3
Row 10	/Kombajn/TaskExecuter9~2	/Kombajn/TaskExecuter9–2/NN3

Source: own study

4.2 Model Operation Logic – Process Flow

The model logic built using ProcessFlow consists of several interrelated blocks – subprocesses – as shown in Fig. 2. Model control begins in the "**Depth of extraction**" area, where the parameters such as wall thickness and backhoe depression of the longwall shearer are used to calculate the number of steps involved in moving the longwall complex into the wall. Each movement related to the movement of the longwall complex is a separate token, which starts the "**Specify the movement**" subprocess through the "Run SubFlow" block. This subprocess is used to determine the number of movements along the conveyor via the mining head. The ProcessFlow mining head is represented by a shared Resource that is Acquired when the shearer is at the end of the longwall and Released at the end of the shearer work cycle. "Run SubFlow" is again placed between the "Aquire" and "Release" blocks, and in this case it is responsible for generating tokens in the number corresponding to the number of conveyor modules. As mentioned earlier, the model is universal, so the number of conveyor modules is not fixed. It is a value related to the number of central ports connected to the main TaskExecuter (the "nrcp" command was used, which returns the number of central ports of a given object). Each single token triggers the "**Longwall shearer movement**" logic, in which it is precisely determined how the shearer should move along the conveyor.

The model reflects the standard movement of a longwall shearer, however, certain maneuvers were distinguished. At the beginning it is checked whether the shearer is just starting work (block deciding "Is this the first ride?"), if it is not the beginning of the work of the shearer, then the direction of movement is verified ("Is this the even ride?") and the possibility of mining ("Is it possible to mine?"). If the shearer has the possibility of mining, then the conveyor module is entered and the location to which the shearer is to drive ("Enter the module and location"). A SQL query is used to determine the conveyor and its location. The query searches the table with the shearer route and selects the applicable location based on the currently considered central port indicating the conveyor module. When the shearer makes the first pass or moves in a certain direction and mining is possible, output is created in a queue ("Create Object") which is sent

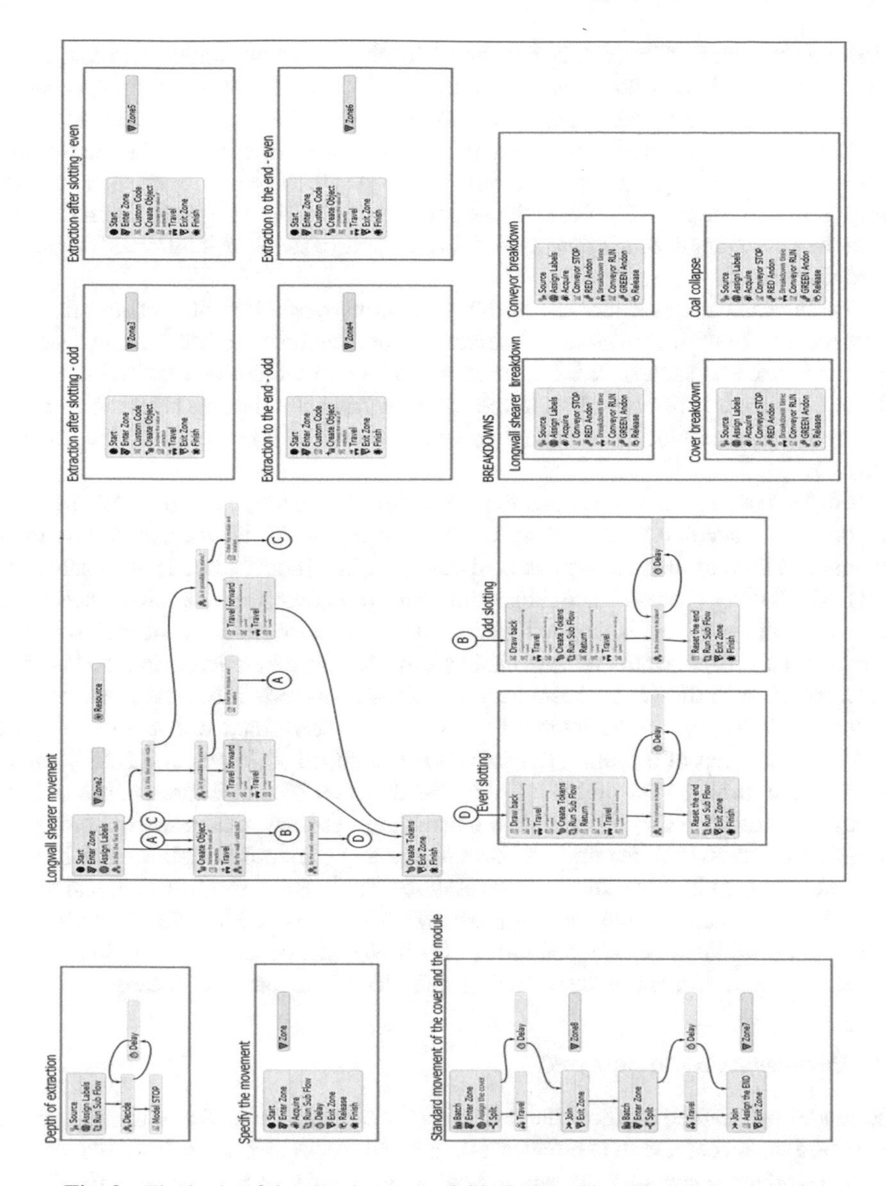

Fig. 2. The logic of the simulation model in ProcessFlow (Source: own study)

straight to the applicable conveyor module. The value of the extraction is then increased ("Increase the value of extraction") and the mining head moves to the indicated location ("Travel"). In the next step, the position of the mining head is checked ("By the wall - odd ride?" and "By the wall - even ride?"). If it is not next to the longwall, then the token responsible for the module movement is created in the "**Standard movement of the cover and the module**" area and the logic starts anew.

If the shearer is next to the wall, the logic responsible for withdrawing the mining head and changing its speed to the maneuvering speed ("Draw back", "Longwall shearer maneuvering speed", "Travel") is activated. The shearer reverses at the length specified in the "Data" table and does not mine during this time. At the same time, conveyor modules and housing are moved to the wall. In turn, the shearer cuts itself into a longwall and mines at a certain length (the "**Extraction after slotting odd/even**" logic, which is responsible for making a mining the coal stone after the shearer reverses is activated using the "Run SubFlow"). Then, the shearer moves to the longwall, extracting coal along the entire length of the conveyor ("**Extraction to the end -odd/even**" logic) and the logic restarts.

The standard movement of the housing and conveyor takes into account the delay of the movement of these elements in relation to the movement of the mining head. The "Batch" block was used in order to force this delay, in which the required number of tokens is collected corresponding to the distance from the mining head which allows moving the elements of the longwall complex. First, the housing is moved, then the conveyor.

In order to reflect the actual operating conditions of the longwall complex, the model also takes into account various types of failures: failure of the shearer, failure of the conveyor, failure of the housing and coal collapse. The failure logic is presented in the "**BREAKDOWNS**" area. It should be emphasized that each time a failure occurs, the entire longwall complex is stopped. Each failure has been modeled in the same way, only the related time parameters are changed. A token is generated using the "InterArrival Source" according to the assumed probability distribution for a given failure, then the longwall shearer should interrupt the currently performed operation ("Acquire" block). In the event of a failure, the conveyor is stopped ("Conveyor STOP") and the andon of the mining head lights up red. The duration of the failure is shown in the "Breakdown time" block. To simulate mean time between breakdowns authors used normal distribution with parameters like average and standard deviation, which can be changed by user via "Breakdowns parameters" Table from Control Panel. After the failure has been rectified, the conveyor restarts ("Conveyor RUN"), the andon lights up green and the mining head can continue ("Release"). The occurrence of individual failures is taken into account in the operating statistics of the longwall shearer.

4.3 Parameters to Be Analyzed

The model is designed to allow the analysis of the operation of the longwall complex depending on different wall parameters (such as thickness and height) and the longwall shearer (working speed, maneuvering speed, haulage and backhoe depression). The user can freely change these parameters depending on the needs. In addition, due to the fact that the longwall complex is prone to failures, the possibility of entering the parameters of particular failures has been added (average time between failures and average time for repair). After data is entered in the prepared control panel and the model is restarted, the results for the current assumptions are obtained. The result of the model is information on the effectiveness of the operation of the longwall shearer and on the impact of individual failures on the operation of the shearer. In addition, the coal output in m^3 and tons is calculated. Figure 3 shows a view of the control panel where one can enter data and analyze the simulation results. This simulation uses parameters of the KOPEX longwall shearer [13].

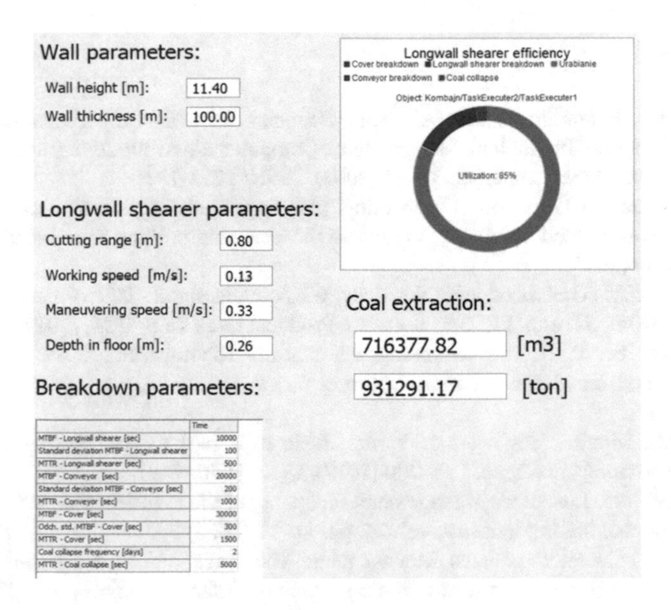

	Time
MTBF - Longwall shearer [sec]	10000
Standard deviation MTBF - Longwall shearer	100
MTTR - Longwall shearer [sec]	500
MTBF - Conveyor [sec]	20000
Standard deviation MTBF - Conveyor [sec]	200
MTTR - Conveyor [sec]	1000
MTBF - Cover [sec]	30000
Odch. std. MTBF - Cover [sec]	300
MTTR - Cover [sec]	1500
Coal collapse frequency [days]	2
MTTR - Coal collapse [sec]	5000

Fig. 3. Sample results of simulation analysis (Source: own study)

5 Conclusions

Simulation modeling allows for the analysis and comparison of different variants and the selection of the best considerations. As the article shows, it is possible to successfully create models also for mining (underground and opencast) and thus to build a model of a longwall complex in a simple way. Thanks to the use of ProcessFlow technology, the model is versatile and has the potential to expand further, both in terms of the size of the complex and in terms of additional analysis. However, it should be remembered that only a properly validated and verified model can be the basis for any analyzes and conclusions. At this stage, the model can be a tool to assist decision makers in determining the effectiveness of a longwall complex. In addition, it enables to test different technical parameters of the shearer and thus to compare solutions at no cost.

It is worth emphasizing that when building a simulation model, special attention should be paid to the correct representation of the system logic. ProcessFlow enables to show dependencies in a clear and user-friendly way by means of block diagrams.

The approach presented by the authors of this work is the first attempt of such a simulation made in the ProcessFlow concept and can be further developed and improved. At the current stage, it was found that using ProcessFlow to build a coal production model is more transparent, effective, allows for a significant reduction of time of simulation and easier implementation of possible changes in the model.

Acknowledgments. This paper was supported by AGH University of Science and Technology [no 11.11.100.693].

References

1. Ciszak, O.: Scientific Notebooks from Poznan University of Technology. Machine Construction and Production Management: Computer aided modeling and simulation of production processes, no. 6, pp. 39–45 (2007). ISSN: 1733-1919
2. Kłos, S., Kuc, P.: Production Engineering, planning, modeling, simulation: Modeling and simulation of production processes based on the Tecnomatix Plant Simulation software, no. 9, pp. 19–30 (2015)
3. Beaverstock, M., Greenwood, A., Nordgren, W.: Applied Simulation: Modeling and Analysis Using FlexSim, 5th edn. FlexSim Software Products Inc, Orem, USA (2017)
4. Zdanowicz, R.: PAR, measurements, automation, robotics: Selection of software for modeling and simulation of manufacturing processes, no. 1, pp. 10–17 (2006). ISSN: 1427-9126
5. Kęsek, M.: Mineral Engineering: Visual Basic as a tool for monitoring and analyzing machines, Cracow, no. 2, pp. 195–200 (2017). ISSN 1640-4920
6. Bąk, P.: Mineral Resources Management: Financing of the investment activity based on the example of coal mining industry, vol. 24, pp. 11–17 (2008)
7. Fuksa, D.: Mineral Resources Management: The ways of solving non-linear decision problems through application of optimal productions plans for mines, vol. 23, pp. 97–108 (2007)
8. Brodny, J., Stecuła, K.: Support Systems in Production Engineering, Technical Systems Engineering: Determining the efficiency of using a longwall shearer, vol. 2(14), pp. 77–85 (2016). ISSN: 2391-9361
9. Brodny, J., Stecuła, K.: Polish Society of Production Management, Innovation in Management and Production Engineering: Analysis of the effectiveness of using a set of mining machines, vol. 1(3), pp. 413–423 (2016)
10. Snopkowski, R., Sukiennik, M.: Archives Of Mining Sciences: Selection Of The Longwall Face Crew With Respect To Stochastic Character Of The Production Process - Part 1-Procedural Description, vol. 57 (2012)
11. Cai, D.: University of Wollongong Thesis Collection: Using Flexsim® to simulate the complex strategies of longwall mining production systems, Australia (2015)
12. Main page FlexSim Software Products, Inc. https://www.flexsim.com/
13. Longwall shearer parameters. http://www.kopex.com.pl/upload/user/file/OFERTA/kompleksy-scianowe/kopex_pniowek_folder_ok.pdf

Concept of Power Grid Resiliency to Severe Space Weather

Olga Sokolova(✉) and Victor Popov

Peter the Great St. Petersburg Polytechnic University, St. Petersburg, Russia
olga.sokolova@gmx.ch
https://ch.linkedin.com/in/onsokolova

Abstract. Space weather (SW) as a type of natural hazard can trigger disasters resulting in large number of fatalities and economic losses. In the recent history, this has happened twice: in, 1989, in North America, and in, 2003, in multiple mid- and high-latitude regions. Population growth and wider usage of technologies sensitive to reliable and high quality electricity supply increase the economic loss in case of such an event recurrence. Contrary to other natural hazards, industry has little realwork operational experience. SW's echo on Earth affects large geographic areas and may result in simultaneous loss of multiple network elements. The used $N-1$ principle for ensuring power grid resiliency is not adequate in this case. Hence, enhancing the grid's resiliency to such an event is of high interest. In this paper, the concept of boosting power grid resiliency to SW is given. First, the idea of SW impact on power grid and the role of critical factors of different nature are described. The list of actions for enhancing power grid resiliency to SW is given in the second part of the paper.

Keywords: Blackout · Critical factors · Natural hazard
Network elements · Mitigation · Power grid · Space weather
Resiliency · Risk assessment

1 Introduction

While power grid planners and operators know how to deal with the more common natural hazards, bringing space weather (SW) as another natural hazard into the picture raises new questions. Severe SW events may have a substantial impact on modern power grids. There are number of documented evidences of severe SW impact on power grids. In terms of magnitude, the strongest ever registered GMD was the so-called Carrington storm in 1859 [1]. Despite the failure of telegraph systems in parts of Europe and North America, this GMD did not cause any substantial technical or economic losses, due to the relatively undeveloped infrastructure in the mid-19th century. The Easter Sunday Storm in 1940 was the first evidence of SW impact on power grids. Afterwards, other examples of destructive SW impact on power grids were registered.

© Springer Nature Switzerland AG 2019
A. Burduk et al. (Eds.): ISPEM 2018, AISC 835, pp. 107–117, 2019.
https://doi.org/10.1007/978-3-319-97490-3_11

The storm in 1989 resulted in Hydro-Quebec blackout which caused a nine-hour outage in most of the Quebec region of Canada and the north-eastern parts of the United States. In total, more than 200 anomalies were registered in the US power grid during the 1989 storm [2]. Less intense activity in October 2003 provoked widespread power outages in Scandinavia, North America, Russia and post-poned power transformer outage in South Africa [3,4]. In November 2004, a moderate SW activity resulted in the abnormal overheating of the power transformers on several stations in the Chinese power grid [5]. The measures taken for achieving more economic benefits in power transfer by developing high voltage power grids, decreases their robustness to severe SW effects. The voltage levels expansion cause the average circuit resistance to decrease, which in turn results in smaller power losses. Further development of the power grids increases their vulnerability to SW. It consequently leads to the growth of "high-risk" zones i.e. the area where the high risk of negative SW impact on the power grid is considered. Strength, impact of SW and its duration are subject to uncertainty. Moreover, SW impact on power grid operation and society well-being is much larger than the hazard stricken region itself. Hence, even countries located in the regions with medium and low risk to SW can profit from developing mitigation strategies beforehand. The ability to respond to severe events depends strongly on the level of preparedness. The motivation of this paper is to classify the list of actions for enhancing the power grid resiliency to SW through the identification of critical factors.

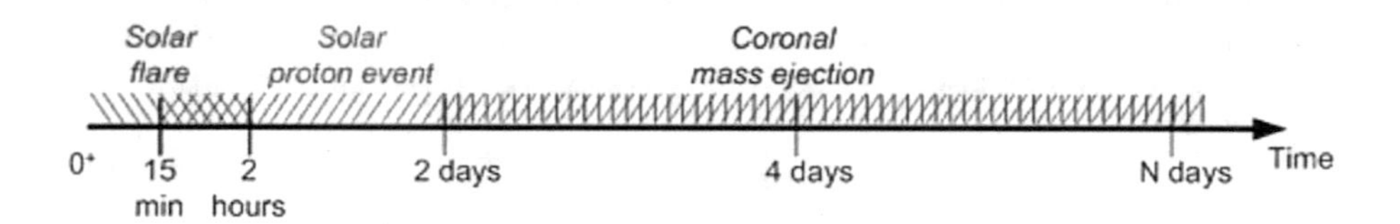

Fig. 1. Time sequence of solar storm events

2 Principles of Space Weather Impact on Power Grids

The term SW is used to describe the changing conditions on the Sun and in the solar wind, magnetosphere, ionosphere and thermosphere. If severe enough, SW can influence the performance and reliability of space-born and ground-based infrastructure and can endanger human life or health. The Sun emits electromagnetic radiation over a spectrum from gamma, X-ray and UV radiation to solar radio radiation. Emissions of the Sun's surface are classified in three types (Fig. 1):

- Solar flare - magnetically driven explosions on the Sun's surface. The Sun also produces high-energy cosmic rays. The particles have energies in the range of 10 MeV to 100 MeV.

- Solar proton events - high energy solar cosmic rays. Very energetic solar proton events are capable of producing protons in the order of 20 GeV.
- Coronal mass ejections (CME) - vast clouds of seething gas, charged plasma with embedded magnetic fields which blasts from the Sun into interplanetary space.

Power grid risk to SW activity is associated with CME. When CME strikes the Earth, the compressed magnetic field in the leading edge smashes into Earth's magnetic field like a battering ram. This produces disturbance in the Earth's magnetic field. Due to Faraday's law, the change of geomagnetic field causes in turn the change of electric field and induces currents called geomagnetically induced currents (GIC). These currents are distributed over any conductor system which has grounding points e.g. power grids, pipelines, railways, etc. CME vary in size and in terms of their impact on critical infrastructure. Level of GIC impact on power grids can be the basis of SW activity scaling. The National Oceanic and Atmospheric Administration (NOAA) in USA, the European Incoherent Scatter Scientific Association (EISCAT) and other agencies provide SW activity forecasting products in an index-style classification. The most familiar one is the K_p index, which shows the deviation of the Earth's magnetic field over a 3-hour period from 13 observatories. K_p classifies the severity of SW activity in a similar way to Richter scale for earthquakes (from 0 to 9). Severe SW can be severe with respect to one parameter, but moderate or weak with respect to others. In [6], the list of parameters the high amplitude of which allows us to classify the conditions as severe ones is given. These parameters are: massive soft and hard X-rays, CME traveling rapidly away from the Sun, the creation of large fluxes of accelerated particles, substational variations in a range of geomagnetic parameters, strong GIC. It is believed that CME which result in high GIC are more hazardous.

The real picture is more complicated. Three levels of GIC can be distinguished:

- an intense GIC that flow for a short duration can lead to cascade failures as during Hydro-Quebec blackout
- long-lasting high GIC can cause severe transformer overheating, leading to failure in days (as it was the case for the loss of power transformer at Salem nuclear power plant)
- moderate GIC can initiate localized degradation of power transformer isolation that continues even after GIC ends (as it was the case for the power transformer loss in South Africa)

The severity of SW impact on power grids is determined by a set of critical factors, which are convenient to subdivide into four groups: SW parameters (activity type, frequency, direction, polarity, ground conductivity), power grid parameters (architecture, voltage level, power system state, geographic location), power system equipment parameters (equipment type, construction scheme, isolation characteristics) and awareness (social awareness, forecast). In other words, power grid robustness to SW is primarily determined by power system equipment characteristics, its location within the grid and developed response plan.

The most prone equipment to SW effects is power transformer, which is the key element for a reliable power system operation. They also differ from other power system equipment as they cannot be replaced quickly and their replacements involves high cost. The power transformer unit can be replaced in the period of 12–18 months and an associated cost can reach USD 20 million [7]. The power transformers capability to withstand GIC depends on DC flux path [8]. The relative power transformer susceptibility to GIC can be scaled as follows: single phase transformer - 1.0; three phase five limb transformer - 0.4; three phase three limb transformer - 0.05.

Power transformer's saturation due to GIC flow over its core endangers not only transformers operation by itself but power grid operation as a whole. Another power system equipment which is crucial for its reliable operation is synchronous machine. High harmonics distortion caused by the power transformer saturation limits synchronous machine maximum admissible power due to windings degradation. Estimation done for the turbo generator TBB-500 shows that in case of single phase transformer saturation, admissible power has to be reduced by 50%, and in case of three phase five limb power transformer saturation - by 25%. As a result, it limits the dispatching capacity for correcting power imbalance provoked by the GIC. The GIC impact on other power system equipment is less significant. Table 1 summarizes the SW effects on power system equipment. The robustness of each equipment type to GIC impact is represented in the first column. The level of impact on system operation in case of unit loss and the repair cost are given in the columns two and three respectively. The repair cost includes also replacement cost in case an equipment unit cannot be repaired on site.

Table 1. Comparison of power system equipment susceptibility to GIC

System equipment	Equipment robustness	System effect	Repair cost
Power transformers	Low	High	High
Instrument transformers	Low	High	Low
Synchronous machines	Medium	High	High
Shunt reactors	High	High	High
Circuit breakers	High	Medium	Medium
Capacitors	Medium	High	High
DC substations	High	High	High
Transmission lines	High	High	Medium

The analysis showed that high GIC levels are obtained in the nodes with the high number of connections which are longer than 100 km ($l > 100$ km) [9]. In other words, the higher the density of long high voltage transmission lines in the grid, the higher is the vulnerability. This dependence is more expressed than the dependence on geomagnetic latitude in the regions with constant ground conductivity, e.g. Scandinavia. Scandinavian grid is located in the high latitude region i.e. aurora region and consists of three synchronously working power grids of Finland, Sweden and Norway. Scandinavian grid has a strong historical record on technical failures caused by GIC. CME parameters that affect the grid are similar or almost similar. Nevertheless, Swedish grid experiences the strongest negative impact from GIC. It is also shown in [10] that ends and corners of a grid are prone to large GIC. It means that low-latitude systems may be affected by the GIC as well. The power grid connection scheme is dictated by its operation state. The increase in the interregional bulk power transfers leads power grids to be operated closer to their limits. The minimum load states are determined by heavy transfer patterns via backbone long-distance interconnections. Thereby, node loss may result in heavier voltage avalanche with minimal capacity for mitigation.

Industry awareness of SW depends on the sector and their geographical location. [11] states that even countries with a perceived low domestic SW risk can benefit from a global approach to increase resiliency. Three following examples show the importance of this work:

1. Eruption of Eyjafjallajökull, 2010, when volcanic ash clouds restricted air transportation over 70% of Europe. The level of preparedness is one of the escalating factors in the crisis and, despite the existence of well-known precursors throughout the world, volcanic ash clouds were not included in the risk registers of many countries [12].
2. Tsunami in Japan, 2011, the assessment of which was limited by pre-existing mitigation procedures of such event, since tsunami risk was not identified as required. The study scenario did not take into account the maximum historically recorded wave in the region, which resulted in underestimating the risk [13].
3. Hurricane Sandy, 2012, which affected power grid equipment and utilities. In consequence, electricity undersupply became a driver of another crisis, which lasted up to 2 weeks and required White House to take mitigation actions.

The concept of power grid resiliency and the actions for its implication are described in the next chapter.

3 Properties of Resiliency and Its Implication

According to United Nations Office for Disaster Risk Reduction, resiliency is the ability of a system, community or society exposed to hazards to resist, absorb, accommodate, adapt to, transform and recover from the effects of a hazard in a timely and efficient manner, including through the preservation and restoration of its essential basic structures and functions through risk management [14]. [15] introduced four properties of a resilient system:

1. Robustness: strength, or the ability of elements, systems, and other measures of analysis to withstand a given level of stress or demand, without suffering degradation or loss of function.
2. Redundancy: capacity of satisfying functional requirements in the event of disruption, degradation or loss of functionality.
3. Rapidity: the capacity to meet priorities and achieve goals in a timely manner in order to contain losses, recover functionality and avoid future disruption.
4. Resourcefulness: the capacity to identify problems, establish priorities, and mobilize alternative external resources when conditions exist that threaten to disrupt some element, system, or other measure.

Quantifying resiliency is required for evaluating the implemented and planned mitigation strategies. It is not a straightforward process, since it is a multidimensional and dynamic process. Numerous resiliency metrics exist such as the frequency and duration of electricity undersupply, the number of customers disconnected, affected area, average load restoration time, etc. Both short-term and long-term resiliency metrics are needed to be boosted.

Severe SW event is categorized as a typical high impact low frequency event. This kind of events has the potential to cause widespread or catastrophic impact to the sector and reach maximum impact with little indications. On the other hand, industry has little real work operational experience which makes forecast and impacts quantification difficult. Each CME is unique. Modern geophysics cannot precisely describe the morphological characteristics of different CME and their impact. Because the scientific data describing the hazard goes back a few decades only and technology for SW observation improves immensely, any correlation of the data registered in the different solar cycles is limited. The lack of data poses an additional difficulty. Two approaches exist to forecast extreme SW events: event-based prediction and probabilistic forecast. In the first case, the CME properties and its echo on the Earth's surface - z-component of the magnetic field B_z- should be predicted before they reach the Earth.

CME are forecasted with the help of coronagraphs installed on the aged SOHO and STEREO missions. Existing capabilities limits our prediction window to 10–20 h for severe conditions. This time interval is too little to implement any meaningful mitigation actions. Probabilistic forecast is based on the likelihood analysis of $\frac{dB}{dt}$ value using historical records. Limitations for this approach are the following: choice of distribution law and the data accuracy. It keeps the question, "how extreme an extreme event can be", open. We have limited time series for the period more than 40 years which brings uncertainties in modeling 1 - in 50 years and 1 - in 100 years events. [16,17] estimated that Carrington like event can occur approximately 1,13 times per century.

If required the resiliency can be boosted with the combination of technical and operational actions supported by enhanced SW forecast. All kinds of actions for resilience implication have a total cost of their implementation. The resiliency

Table 2. Overview of actions for boosting power grid resiliency

Accept	Mitigate	Avoid
Implementation of HV power transformers with higher resiliency against GIC	Obligatory Buchholz relay usage	Relay protection schemes redesign
Zigzag power transformer connections	Change the maximum error limit for instrumental transformers	Dispatching algorithms adaptation
Implementation of power transformers with a clearance gap in the core	Implementation of a special protective relay containing no microelectronic components and based on discrete high voltage elements resistant to electromagnetic interference and surge overvoltage [18]	
Implementation of power transformers with special compensative windings	Installing GIC blocking devices in transformer neutrals	
Construction of compensated transmission lines or redesign of the cross-cut compensated lines to lengthwise compensated ones		

enhancement measures are identified and prioritized depending on their criticality and contribution. In general, the actions can be subdivided into three groups: accept, mitigate, avoid. The first group "accept" includes actions that provide the highest possible power grid robustness to SW effects. It is important to mention that a congestion point exists after which further investments do not improve the system's resiliency. The active and passive actions which target the GIC blocking devices installation refer to the second group. The implementation of the actions from the third group involves the minimum cost. The overview of actions is given in the Table 2.

The time after an event can be considered as a window of opportunities for implementing the actions. The relevant stakeholders in order to avoid the same loss in the future are eager to implement actions with higher cost and boost long-term resiliency. For example, the Hydro-Quebec blackout in 1989 gave a kick to redesign the entire high voltage grid by installing lengthwise compensation devices.

4 Discussion and Conclusions

The growth of population and economic development shifts the energy consumption towards a higher share of electric power. New technology and process development result in higher cost of electricity undersupply. Moreover, the society expects an increased reliability of power supply and a reduced restoration time. The resistance to the SW effects and redundancy of critical infrastructures are the key elements of resiliency. The SW impact on power grids is a complex and technical issue but many of the potential consequences are common to other risks. In this paper, authors identify and analyze the extended set of critical factors that determine power grid vulnerability to SW. Based on the outcome, a set of solutions for boosting power grid resiliency to SW is proposed. These solutions correspond to different enhancement strategies described in Table 2. Plans to boost resiliency to severe SW can be dealt under existing strategies for other events, but in some parts the existing capability is insufficient. There is no "end-to-end" product which covers the whole spectrum of activities needed for fully operational SW services. Nevertheless, the set of space-born and ground-based mitigation actions exist. In general, the following principles, as shown in Fig. 2, should be applied: **preparedness** - all relevant stakeholders should be properly prepared, including clarity of the roles and responsibilities; **continuity** - the actions should be grounded within existing functions and be familiar; **forecast** - stakeholders should forecast the risks and model direct and indirect impacts in advance; **integration** - appropriate guidance should be developed and effective training exercises should be performed; **coordination** - engagement based on mutual trust facilitates the information sharing; **communication** - two-way communication is crucial.

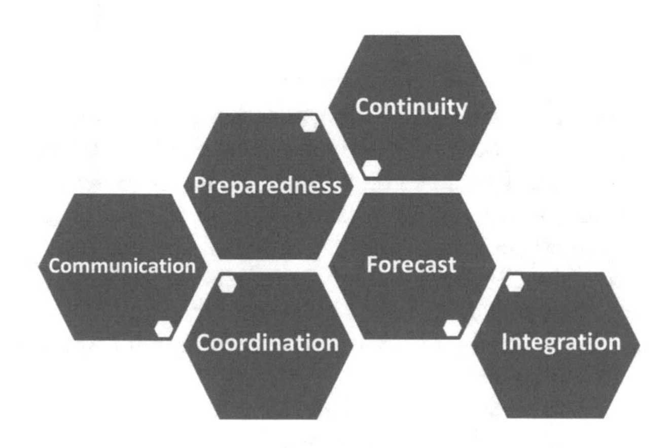

Fig. 2. The combination of principles for boosting power grid resiliency

Natural hazards can have devastating impacts that may be impossible to prevent or control. The consequences of the severe SW event can be estimated as "Losses = $\sum_{i=N}$ Cost of each event × Probability of an event" (N is the number of events). Modern infrastructure systems are highly interconnected. Moreover, GIC affect any grounded system of conductors, not only power grids. The change of space radiation impacts simultaneously space-born infrastructure.

Nevertheless, the primary avenue of catastrophic damage caused by solar storm is through power system infrastructure. This, in turn, can impair the operation of other critical infrastructures. The impact is both direct and indirect. Direct impact refers to system outages that are directly caused by GMD. The indirect impact refers to infrastructures that are affected by primary industry outage. The indirect impact can be both immediate and delayed. Modern society is strongly dependent on reliable electricity provision. The critical infrastructure systems which are vital for societal and economic well-being and safety are described and listed in [19]. The potential scenario of multiple critical infrastructure outage caused by SW is given by the authors in Fig. 3. The vulnerability of other sectors e.g. banking, navigation, government functions, internet, etc. - are produced by the responses to electric and communication systems outage.

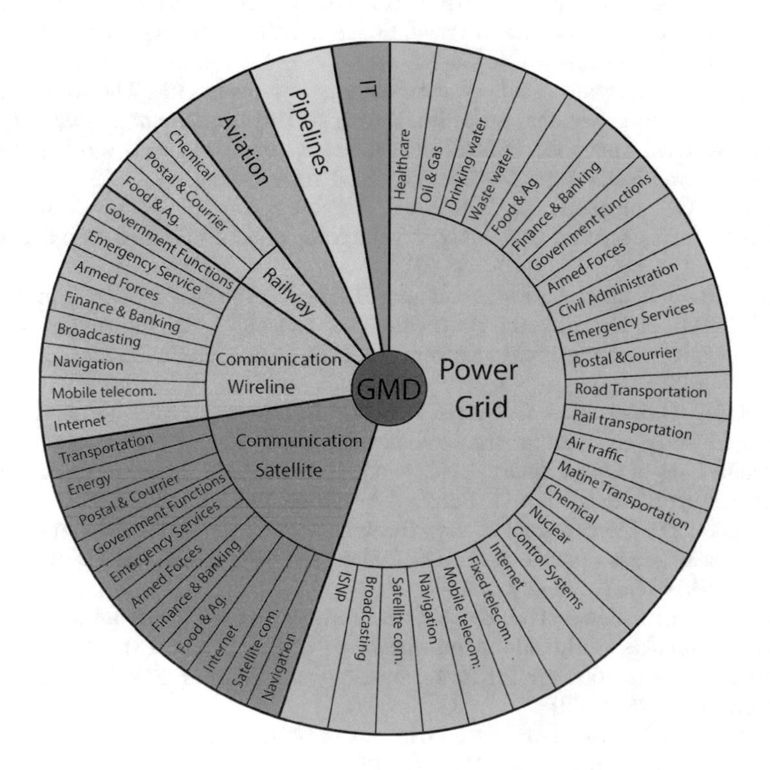

Fig. 3. Primary and secondary critical infrastructure disruptions

The GMD research and science are relatively young fields. The links between science and resilience planning are not mature. In closing, there are still a variety of scientific challenges that must be addresses in order to create power grid resilient to severe SW.

References

1. Kappenman, J.: Geomagnetic Storms and their Impacts on the US Power Grid. Metatech, US (2010)
2. Bolduc, L.: GIC observations and studies in the Hydro-Québec power system. J. Atmos. Solar Terr. Phys. **65**(16), 1793–1802 (2002)
3. Lundstedt, H.: The sun, space weather and GIC effects in Sweden. Adv. Space Res. **37**, 1182–1191 (2006)
4. Gaunt, C.T.: Reducing uncertainty-responses for electricity utilities to severe solar storms. SWSC **4**, A01 (2014). https://doi.org/10.1051/swsc/2013058
5. Liu, C., Liu, L., Pirjola, R.: Geomagnetically induced currents in the high-voltage power grid in China. IEEE Trans. Power Deliv. **24**, 2368–2374 (2009). https://doi.org/10.1029/2012GL051431
6. Riley, P., Baker, D., Liu, Y., Verronen, P., Singer, H., Güedel, M.: Extreme space weather events: from cradle to grave. Space Sci. Rev. **214**(1), 21 (2018). https://doi.org/10.1007/s11214-017-0456-3
7. Sokolova, O., Burgherr, P., Schwarzman, A., Collenberg, W.: The Impact of Solar Storms on Power Systems. Swiss Re, Zurich (2014). *Order no: 1505890_14_en*
8. Sokolova, O., Popov, V.: Critical infrastructure exposure to severe solar storms. In: Safety and Reliability: Methodology and Applications, pp. 1327–1340 (2017)
9. Sokolova, O., Burgherr, P., Collenberg, W.: Solar storm impact on critical infrastructure. In: Safety and Reliability: Methodology and Applications, pp. 1515–1521 (2015)
10. Pirjola, R., Viljanen, A., Pulkkinen, A.: Research of geomagnetically induced currents (GIC) in Finland. In: Proceedings of the 2007 7th International Symposium on Electromagnetic Compatibility and Electromagnetic Ecology, pp. 269–272. IEEE (2007)
11. UNOOSA: United Nations/United Arab Emirates High Level Forum: Space as a driver for socio-economic sustainable development. United Nations Committee on Peaceful Uses of Outer Space (UNOOSA), Dubai (2017). http://www.unoosa.org/documents/pdf/hlf/HLF2017/Book_of_Abstracts.pdf
12. Alexander, D.: Volcanic ash in the atmosphere and risks for civil aviation: a study in European crisis management. Int. J. Disaster Risk Sci. **4**, 9–19 (2013). https://doi.org/10.1007/s13753-013-0003-0
13. Mohrbach, I.L., Power Tech eV, V.G.B.: Fukushima two years after the tsunami-the consequences worldwide. Atomwirtschaft **58**(3), 152 (2013)
14. United Nations: 2009 UNISDR Terminology on disaster risk reduction. United Nations, Geneva (2009)
15. Cimellaro, G.P., Reinhorn, A., Bruneau, M.: Seismic resilience of a hospital system. Struct. Infrastruct. Eng. **6**, 127–144 (2009). https://doi.org/10.1080/15732470802663847
16. Love, J.J.: Credible occurrence probabilities for extreme geophysical events: Earthquakes, volcanic eruptions, magnetic storms. Geophys. Res. Lett. **39**(10), (2012). https://doi.org/10.1029/2012GL051431
17. Love, J.J., Rigler, E.J., Pulkkinen, A., Riley, P.: On the lognormality of historical magnetic storm intensity statistics: implications for extreme? event probabilities. Geophys. Res. Lett. **42**, 6544–6553 (2015). https://doi.org/10.1002/2015GL064842

18. Amuanyena, L.A.: Effects of geomagnetically induced currents on power transformers and reactors Doctoral dissertation, University of Cape Town, Cape Town (2003)
19. The Council Of The European Union: Council Directive 2008/114/EC of 8 December 2008 on the identification and designation of European critical infrastructures and the assessment of the need to improve their protection. Official Journal of the European Union, Brussels (2008)

Promoting Cross-Border Cooperation Between Science and Small Businesses as a Source of Innovation

Grit Krause-Juettler(✉)

CIMTT Centre for Production Engineering and Management,
Technische Universtität Dresden, Helmholtzstr. 7a, 01069 Dresden, Germany
grit.juettler@tu-dresden.de

Abstract. In the course of European integration, cross-border cooperation between science and enterprises becomes important. Following paper presents results of two sub-studies. 1st study aimed to identify factors promoting the occurrence of cross-border cooperation of small enterprises with research organisations. Empirical results from a sample of SMEs in Saxon-Czech borderland (SMEs; N = 263) – mainly belonging to the manufacturing sector – show, that i.e. willingness to collaborate, corporate innovation culture, enterprise's absorptive capacity, positively perceived regional cooperation climate, adequate funding opportunities for collaboration and regional availability of qualified staff are factors contributing significantly to emergence of cross-border collaboration between firms and research organisations. 2nd sub-study investigated how collaboration between SMEs and research organisations proceeds. Main result of a qualitative study (N = 89) is a model of social roles describing tasks which are necessary for successful collaboration processes. Based on empirical results, strategic recommendations are deduced.

Keywords: Collaboration · Innovation · Knowledge and technology transfer Networking

1 Introduction

SMEs – which account for the largest share of companies in Europe (99,8%) – and their ability to innovate, have a high impact on the European economic capacity [13]. However, mostly this kind of enterprises are not able to develop innovations on their own terms. They depend on assistance of research organisations (ROs) within collaborative innovation projects for improving their innovative power [5]. However, as for example own investigations show, there exist many barriers hindering especially SMEs from cooperating with universities and other ROs [18]. In the course of European integration, collaborative projects between actors from science and businesses on cross-border and transnational level become important. Because, by combining different national competencies innovation capabilities are released [20]. However, currently in many cross-border regions, framework conditions and infrastructures mostly fit only to national respectively regional requirements. They do not match to the needs of cross-border collaboration [20]. Differences in administrative structures and

© Springer Nature Switzerland AG 2019
A. Burduk et al. (Eds.): ISPEM 2018, AISC 835, pp. 118–127, 2019.
https://doi.org/10.1007/978-3-319-97490-3_12

procedures, still existing stereotypes at personal level that seem to be unchangeable as well as a lack of knowledge about neighboring regions and countries still prevent the development of synergies and encumber the emergence of cross-border collaboration between science and economy [15, 16]. Currently, it is not fully clear, which factors could have a positive influence on the emergence of cross-border collaboration. Knowledge of these factors would be a precondition for developing a systematic approach for enhancement of collaboration. Therefore, following paper aims to elaborate factors that facilitate cross-border cooperation between science and companies by involving all relevant actors of a regional (cross-border) innovation system.

2 Theoretical Framework of the Study

For presented study, regional innovation system approach was applied. Core of the concept is the assumption that there is a subsystem generating the knowledge (science) as well as a subsystem exploiting the knowledge economically (economy) in order to contribute to the competitiveness of a whole region [10–12] – in this case a cross-border region. Cooke and colleagues find reasons for regional disparities in the different quality of regional research infrastructure, in the different skills of companies to generate or absorb innovations and innovative knowledge. Furthermore, different qualities of interconnections of both subsystems are responsible for regional differences. In this context, aspect of trust between regional actors plays an important role. Thus, concept of the regional innovation system considers the innovation process as a social phenomenon in which various regional actors are involved. Actors including representatives e.g. from companies, ROs as well as various intermediary institutions supporting the implementation of joint projects (e.g. innovation and technology parks, chambers of commerce, public administrations) [17]. Consequently, joint innovation projects between science and economy are a regional task, which has necessarily to involve all relevant regional actors. Present paper focuses on cooperation between universities and companies in collaborative research and development projects. Former studies have already shown that such projects have a significant impact on the economic success of companies [5, 23].

3 Influencing Factors, Research Subjects and Course of the Study

Paper summarises results of two sub-studies. **1st sub-study** highlights necessary preconditions in companies facilitating cooperation with research organisations as well as regional framework conditions that could have positive impact on occurrence of joint projects between these two spheres. Occurring collaborative projects between science and economy are assumed as dependent variable measured as 'collaborative projects within the last three years'. Following variables are defined as independent variables explaining the occurrence of joint projects between science and economy. Therefore, empirical part of 1st sub-study refers to following intra-company aspects:

- company's willingness to cooperate (own hypothesis)
- corporate innovation culture [7]
- company's absorptive capacity [9, 27]
- existence of intra-company barriers for collaboration (e.g. lack of finances, missing capacity for marketing of new products) [25].

Additionally, 1st sub-study also included listed regional factors, appraised by representatives of responding companies:

- perceived regional 'social climate' in terms of cooperation [14]
- regional barriers for innovation and cooperation (e.g. bureaucracy regards application for funding, language barriers, limited applicability of research results) [1]
- existence of public funding for collaboration between science and companies
- availability of qualified staff for research and development.

Data was obtained by a quantitative questionnaire survey in enterprises, which was finished by 263 respondents (year 2014). Questionnaire included mainly self-developed scales that were pretested in former surveys.[1] Selection of enterprises was based on a stratified random sample. Respondents received questionnaire by post and could return it in a stamped envelope. Response rate was 20%. Geographical area of investigation was Saxon-Czech border region covering part of North-West Bohemia (Czech Republic) and South-Eastern part of Free State of Saxony (Germany). Area can be regarded as a common cross-border region [4]. Investigated region is characterised by a strong dominance of small enterprises (Ústí Region, 99.8%; Saxony, 99.7% [24]) accompanied by an absence of large, research-intensive company headquarters. At the same time, region is distinguished by a large number of universities and an above-average density of non-university ROs [22].

2nd sub-study concentrated on different intermediary actors supporting ROs and SMEs in overcoming cooperation barriers and ensuring their successful collaboration [19]. Sub-study aimed to identify roles and tasks of intermediary actors, which can positively contribute to implementation of joint research and development projects between science and especially SMEs. Study targeted on answering the following research questions:

- Which social roles of intermediary actors can be identified within collaborative projects between science and businesses?
- What are the contents of these roles and how do they contribute to implementation of joint projects between science and businesses?

For answering these questions, an interview study based on a field manual was conducted. Considered regional areas for the empirical investigation were firstly the Free State of Saxony in the Northeast of Germany as well as three federal states in Northern Germany. Both areas are characterised by a similar economic structure

---

[1] All questionnaires, field manuals etc. published in: Krause-Juettler, G., Lohse, K., Jandova, A., Jerabek, M., Berrova, E. & Lauterbach, P.: Region und Innovation am Beispiel des sächsisch-böhmischen Grenzraums [Region and Innovation. Example of Saxon-Czech border area], Dresden/Prag, (2014).

dominated by SMEs drawing their essential power of innovation from cooperative relations with universities and other ROs. For qualitative study, 89 interviews with representatives of scientific, economic and intermediate organisations (see Table 1; e.g. administrations, technology centres etc.) were carried out. Interviewees were selected based on their expertise. While realising the interviews, a third person prepared an analogous transcript that formed the basis for the later content analysis [21]. Table 1 documents composition of interview sample.

Table 1. Composition of the interview sample (2nd sub-study) describing organisational affiliation of interviewees (*e.g. chambers of commerce, technology parks and centres, clusters)

SMEs		ROs		intermediary organisations*	
Saxony	N-Germany	Saxony	N-Germany	Saxony	N-Germany
8	16	16	20	20	9
24		36		29	
		N=89			

4 Empirical Findings

4.1 1st Sub-study: Companies

Descriptive Results

Only about five percent of surveyed companies in Saxony and almost 15% of respondents from Czech area of investigation employ more than 249 employees. This corresponds roughly to the described economic structure of the region. Most of the companies belong to the manufacturing sector (see Fig. 1).

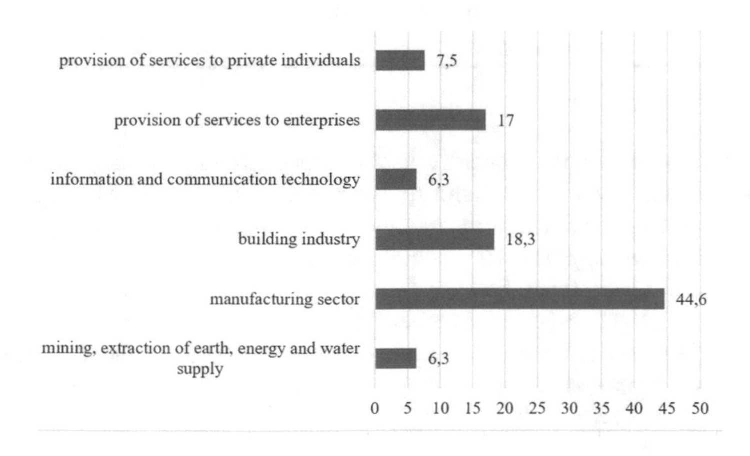

Fig. 1. Industry sector affiliation of surveyed enterprises in percent (N = 224)

More than half of the surveyed enterprises in Saxony (53%) implemented research and development projects together with a research organisation in the last three years. In observed Czech region, a quarter of surveyed companies (27%) have done so over the last three years. Examples for cross-border as well as international research and development collaboration between science and economy are very rare in the existing sample (11 in Saxony vs. 31 in Czech region). Nevertheless, nearly half of the surveyed enterprises (49,2%) report that they are willing to cooperate with research organisations. Most collaborative projects of enterprises and ROs are implemented on regional or national level.

Results of Correlation Analysis
First of all the results show that research collaboration between science and SMEs contribute significantly to the development of innovations in the surveyed enterprises ($r = .51$; $p < .01$). Intra-company factors *willingness to cooperate* ($r = .58$, $p < .01$), high pronounced *innovation culture* ($r = .28$, $p < .01$) and *absorptive capacity* ($r = .30$, $p < .01$) are aspects fostering significantly the occurrence of collaboration between enterprises and ROs. *Internal barriers* like a lack of financial resources for research and development or missing capacities for marketing of new products reduce significantly the emergence of research cooperation ($r = -.16$, $p < .01$) (Table 2).

Table 2. Correlation analysis of internal aspects perceived by responding SMEs (N = 263) and occurrence of collaboration with ROs (*p < .05; **p < .01)

		1
1	collaboration with science in last 3 years	-
	enterprise´s willingness to cooperate	.58**
	corporate innovation culture	.28**
	corporate absorptive capacity	.30**
	internal barriers	-.16**

Investigation of regional framework conditions shows, that a positive regional *cooperation climate* defined as company's perception of trustful and open relations between relevant regional actors regards potential cooperation ($r = .41$, $p < .01$), *adequate public funding opportunities* for collaboration ($r = .32$, $p < .01$) and *availability of qualified staff* for research projects ($r = .22$, $p < .01$) are facilitators for research cooperation. Perceived *regional barriers* like bureaucracy regards public funding, language barriers between potential project partners and limited practical applicability of research results are significantly negative related to the appearance of research collaboration ($r = -.17$, $p < .01$) (Table 3).

Table 3. Correlation analysis of regional framework conditions perceived by responding SMEs and occurrence of collaboration with ROs (*p < .05; **p < .01)

		1
1	collaboration with science in last 3 years	-
	perceived cooperation climate	.41**
	public funding opportunities	.32**
	availability of qualified staff	.22**
	regional barriers	-.17**

4.2 2nd Sub-study: Intermediary Actors and Organisations

Concerning research questions raised for the 2nd sub-study, role model described in Table 4 derived. Interview study proved, that defined roles are of high importance for implementing joint projects between science and businesses [19]. Elaborated model provides an overview about roles that have to be assumed by different actors for successful implementation of collaborative innovation projects (answer for research question 1). Furthermore, table also describes contents of these identified roles and how they contribute to implementation of joint projects (answer for research question 2).

Table 4. Identified social roles in collaborative innovation projects between science and businesses and description of their contents [19]

role	contents
moderator	• providing a platform for meeting potential partners • supporting the networking between actors from scientific and economic sphere • constant up-to-date overview about potential partners
translator	• mediation between and translation of different 'professional languages' and objectives in scientific and economic sphere • promoting the overcome of barriers between the two spheres
driver	• development of visions for future regards specific fields of technology and topics of research • political networking (lobbying)
risk-seeking driver	• willingness for risky investments in new technologies
executor	• complete professional/ scientific handling of collaborative research projects based on expertise
supporter	• technology scouting • legal and economic consulting

5 Summary

Based on results of 1st sub-study, it is to conclude that in the observed period of time, there were hardly any cross-border cooperation between science and economy observable in the monitored region. Nevertheless, study delivers many starting points for promoting this kind of cross-border cooperation. Influencing factors are to find on enterprise (e.g. shaping innovative culture, improving absorptive capacity) as well as regional level (enhancing cooperation climate by providing platforms for meetings, simplifying funding procedures). 2nd sub-study pointed out that due to already discussed barriers between science and economy a complex scenario of social roles is required for successful implementation of collaborative innovation processes [19]. Not only roles assumed by persons who work in those projects mainly based on professional expertise are essential for successful collaboration, but also people overtaking mediating roles. They serve as *boundary-spanners* [2, 3, 8] (e.g. moderator, translator, supporter) between different actors and participating organisations. These so-called 'boundary-spanning roles' help negotiating barriers between science and enterprises and thereby promote the establishment of trustful basis for cooperation.

Starting from regional innovation system approach considering the innovation process as a social phenomenon in which various regional actors are involved, results show, that for shaping conditions and fostering joint development of innovative products and processes, it is necessary to include all relevant regional actors – especially in a (cross-border) region that is dominated by a small-scale economic structure.

6 Conclusions and Strategic Recommendations

Results of 1st sub-study showed, that factors influencing the emergence of (cross-border) collaboration between science and economy are to find inside as well as outside of enterprises. In terms of 'inside factors', it has been shown that enterprises must be able to implement such collaborative projects especially in terms of own innovative capacities (e.g. innovation culture) and the ability to absorb and incorporate external knowledge and transform it into new products and processes (e.g. absorptive capacity).

Furthermore, numerous internal factors exist that hinder collaborative projects. But also 'outside factors' like the social climate for collaboration (e.g. availability of regional platforms for finding partners) or framework conditions (e.g. funding, bureaucracy) have an impact on development and implementation of projects between science and industry. Additionally, results of 2nd sub-study indicate, that especially in an environment dominated by SMEs, boundary spanners like intermediary organisations play an important role and have to be taken actively into account when shaping conditions for collaboration. Hence, measures for promoting establishment of cooperation have to be introduced simultaneously in different fields of action, considering internal as well as external factors influencing enterprises and their innovation activities. Therefore, six different fields of action for fostering cross-border cooperation between science and businesses are identified [6, 18]. These fields are:

1. fostering corporate innovation culture,
2. supporting especially SMEs in managing research collaboration,
3. providing collaboration platforms for improving regional cooperation climate,
4. enhancing regional conditions for research collaboration between science and SMES (e.g. legal arrangements, funding procedures, political importance),
5. providing adequate public funding opportunities,
6. implementing measures for facilitating the recruitment of qualified staff for research projects.

According to these foreseen fields of action, INTERREG Central Europe project TRANS^{3}Net is implemented, focusing on actors called 'transfer promotors' and aiming to improve their transnational cooperation by establishing a network in the border triangle of Czech Republic, Germany and Poland. So-called 'transfer promotors' are individuals or organisations initiating, implementing and supporting joint innovation projects between science and economy. They serve as 'boundary spanners' [8], by arranging contacts, moderating misunderstandings and by keeping the whole transfer process running [26]. 'Transfer promotors' are located in universities and research organisations, e.g. transfer units, in associations close to economy, e.g. chambers of commerce, in regional administrations as well as various intermediary institutions, e.g. technology parks and innovation centres [22]. Main objective of the transnational network of 'transfer promotors' will be the reduction of barriers that currently prevent transnational collaboration between science and economy.

7 Limitations of the Studies and Future Research

Limitations of the 1st sub-study are very apparent. Theoretical foundations of the study are very selective. Only a limited set of variables was considered. 1st sub-study is therefore to be understood as a pilot study. Further investigations are urgently needed to gain more reliable insights. Due to the small sample, accompanied by a high self-selectivity of respondents a generalisability of the results is not possible. They are only valid for the surveyed population. Moreover, it should be noted that the quantitative study is merely a cross-sectional survey, so the presented results are only a snapshot of a current situation. Due to the large sample of 2nd sub-study it can be assumed that the content representativeness is given. Though, because of some specifics in the German economic and scientific system (e.g. public funding structures, importance of industrial sector resp. a still low transfer orientation of universities, educational sovereignty of the Federal states, and organisation of the transfer-system etc.) the results can not be fully generalised on an international level.

Presented results are only considering cooperation between science and industry in Saxon-Czech border area respectively in different regions of Germany. A further investigation is necessary gaining more reliable insights about influencing factors responsible for development, implementation and possible results of cooperation on cross-border and transnational level [6, 18]. Furthermore, a substantiated theoretical embedding of findings is necessary. Already available theoretical work from other European border regions can be made fruitful for this (e.g. [20]). This should also be done for transferring positive experiences from other border regions. During implementation of interview study, another research field became obvious. Currently, knowledge and technology transfer carried out in collaboration between science and enterprises finds only low reputation in the scientific community. Here, it is essential to find approaches for increasing incentives for transfer, otherwise transfer continues to be a result of coincidences without a systematic procedure for transforming inventions into marketable products.

References

1. Albors-Garrigós, J., Rincon-Diaz, C.A., Igartua-Lopez, J.I.: Research technology organisations as leaders of R&D collaboration with SMEs: role, barriers and facilitators. Technol. Anal. Strateg. Manag. **26**(2), 37–53 (2014)
2. Aldrich, H., Herker, D.: Boundary spanning roles and organization structure. Acad. Manag. Rev. **2**(2), 217–230 (1977)
3. Ancona, D.G., Caldwell, D.F.: Bridging the boundary: external activity and performance in organizational teams. Adm. Sci. Q. **37**, 634–665 (1992)
4. Anděl, J., Jerabek, M., Orsulak, T.: Vývoj Sídelní strukturya obyvatelstva pohradnicních okresu Usteckeho kraje, Usti nad labem (2004)
5. Belderbos, R., Carree, M., Lokshin, B.: Cooperative R&D and firm performance. Res. Policy **33**, 1477–1491 (2004)
6. Berrová, E., Jerabek, M., Krause-Juettler, G.: Research and Practice: Partners and/or competitors? General findings and regional specifics in the cooperation of research and practice sphere on the example of Czech-Saxon borderland. GeoSCape **9**(2), 33–46 (2015)
7. Brettel, M., Cleven, N.J.: Innovation culture, collaboration with external partners and NPD performance. Creativity Innov. Manag. **20**(4), 253–272 (2011)
8. Comacchio, A., Bonesso, S., Pizzi, C.: Boundary spanning between industry and university: the role of technology transfer centres. J. Technol. Transf. **37**, 943–966 (2012)
9. Cohen, W.M., Levinthal, D.A.: Absorptive capacity: a new perspective on learning and innovation. Adm. Sci. Q. **35**, 128–152 (1990)
10. Cooke, P.: Regional innovation systems: competitive regulation in the new Europe. Geoforum **23**, 365–382 (1992)
11. Cooke, P.: Regional innovation systems – an evolutionary approach. In: Cooke, P., Heidenreich, M., Braczyk, H.-J. (eds.) Regional innovation systems, pp. 1–18. Routledge, London (2004)
12. Doloreux, D., Parto, S.: Regional innovation systems: current discourse and unresolved issues. Technol. Soc. **27**, 133–153 (2005)
13. Gagliardi, D., Muller, P., Glossop, E., Caliandro, C., Fritsch, M., Brtkova, G., Bohn, N.U., Klitou, D., Avigdor, G., Marzocchi, C., Ramlogan, R.: A recovery on the horizon? Annual report on European SMEs 2012/2013. https://www.escholar.manchester.ac.uk/api/datastream?publicationPid=uk-ac-man-scw:212438&datastreamId=FULL-TEXT.PDF. Accessed 27 Apr 2018
14. Hauser, C., Tappeiner, G., Walde, J.: The learning region: the impact of social capital and weak ties on innovation. Reg. Stud. **41**(1), 75–88 (2007)
15. Hoekman, J., Frenken, K., van Oort, F.: The geography of collaborative knowledge production in Europe. Ann. Reg. Stud. **43**(3), 721–738 (2009)
16. Koschatzky, K.: A river is a river – cross-border networking between Baden and Alsace. Eur. Plan. Stud. **8**, 429–449 (2000)

17. Koschatzky, K., Schnabl, E., Zenker, A., Stahlecker, T., Kroll, H.: The Role of Associations in Regional Innovation Systems. Working Papers Firms and Region No. R4/2014 (2014)
18. Krause-Juettler, G., Lohse, K., Jandová, A., Jerabek, M., Berrová, E., Lauterbach, P.: Region und Innovation am Beispiel des sächsisch-böhmischen Grenzraums [Region and Innovation. Example of Saxon-Czech border area]. Grada, Dresden/Prag (2014)
19. Krause-Juettler, G.: Scenarios of transfer: social roles within collaborative innovation processes. In: Academic Proceedings 2014 University-Industry Interaction Conference: Challenges and Solutions for Fostering Entrepreneurial Universities and Collaborative Innovation, pp. 451–462 (2014)
20. Lundquist, K.-J., Trippl, M.: Towards cross-border innovation spaces. A theoretical analysis and empirical comparison of the Öresund region and the Centrope area. SRE Discussion Papers, 2009/05, WU Vienna University of Economics and Business, Vienna (2009)
21. Mayring, P.: Qualitative content analysis. Forum qualitative Sozialforschung/Forum: Qualitative Social Research **1**(2). Article ID 20. http://nbn-resolving.de/urn:nbn:de:0114-fqs0002204. Accessed 25 Apr 2018
22. Mogiła, Z., Tiukało, A., Giebel, M., Krause-Juettler, G.: A strategy for a transnational network of transfer promotors. http://141.76.19.93/trans3net/wp-content/uploads/2018/03/Strategy_complete_online_small.pdf. Accessed 27 Apr 2018
23. Robin, S., Schubert, T.: Cooperation with public research institutions and success in innovation: evidence from France and Germany. Res. Policy **42**, 149–166 (2013)
24. Statistisches Bundesamt: Anteile kleiner und mittlerer Unternehmen an ausgewählten Anteilen [Ratios of SMEs]. https://www.destatis.de/DE/ZahlenFakten/GesamtwirtschaftUmwelt/UnternehmenHandwerk/KleineMittlereUnternehmenMittelstand/Tabellen/Insgesamt.html. Accessed 03 May 2018
25. Strobel, N., Kratzer, J.: Obstacles to innovation for SMEs: evidence from Germany. Int. J. Innov. Manag. **21**(3), 234–249 (2017)
26. TRANS^{3}Net: National understandings of transfer processes and reference model of transnational transfer (2016). http://141.76.19.93/trans3net/wp-content/uploads/2016/10/Report-on-national-understandings-of-transfer.pdf. Accessed 27 Apr 2018
27. Zahra, S.A., George, G.: Absorptive capacity: a review 'reconceptualization and extension'. Acad. Manag. Rev. **27**(2), 185 (2002)

Definition of Characteristic Values for the Efficient and Safe Implementation of Electronic Cam Gears

Armin Schleinitz(✉), Holger Schlegel, and Matthias Putz

Technische Universität Chemnitz, Straße der Nationen 62, 09111 Chemnitz, Germany
armin.schleinitz@mb.tu-chemnitz.de

Abstract. Automated and concatenated processes can provide a solution to market conditions, such as more powerful and efficient manufacturing systems. In this context, frequently occurring automation components are electronic drive systems linked by electronic cam gears. The focus of this work is the determination of characteristic values for the design of electric drives, which allow a safe and efficient implementation of electronic cam gears taking into account a given accuracy. The characteristic value determination is based on the connection of the time and the image domain. For this purpose, the movement task is transferred by means of Laplace transformation into an image domain necessary for the implementation. The examinations are based on a simulation model whose results are to be confirmed at a test bench. In addition to previously common design methods, the determined characteristic values are intended to stating more precisely the static and dynamic design of drive systems for the implementation of electronic cam gears, and to enable them at an earlier point in time.

Keywords: Automation · Cam gear · Electronic cam · Movement task
Laplace transformation · Control and feedback control technology
Cascade control · Simulation

1 Introduction

With increasing economic, technical, ecological and social requirements, companies are requested to provide customer-friendly and high-quality products and services in a time- and cost-efficient manner [1]. This framework resulting from the market can be countered by automating and chaining manufacturing processes. The linkage of these components can be done by electronic cam gears [2]. Here, the manufacturing process specifies a movement task. This is described by an electronic cam gear and implemented by electronic drives.

For mechanical cam gears, the movement task can be described as a mathematical function with respect to the rotation angle of the master shaft [3]. When using electronic cams, the production cycle is specified by a central controller and transmitted to the servo inverter of the master drive. This master drive transmits its angle information to the servo inverters for the auxiliary drives (slave). The corresponding electronic cam

© Springer Nature Switzerland AG 2019
A. Burduk et al. (Eds.): ISPEM 2018, AISC 835, pp. 128–138, 2019.
https://doi.org/10.1007/978-3-319-97490-3_13

is stored in the servo inverter of the respective auxiliary drive and converts the linear path information into a non-linear one. Consequently, higher jerk, acceleration or speed values can be derived for the auxiliary drives in comparison to the master drive.

For the implementation of electronic cam gears, therefore, jerk, acceleration or speed characteristics are often used. Moreover the design of the required drives is based, amongst other things, on maximum current values as well as static and dynamic torque characteristics [3–5]. Effective characteristic values are proposed especially for servomotors, which in turn relate to accelerations or moments [5]. These interpretations become more difficult by the high complexity and lack of holistic consideration of the mechatronic production system [6]. For this reason, in practice some iterations will be made consisting of the choice of modified components and their recalculation, which are accompanied by simulations to configure such a system.

In this work, characteristic values for the design of electrical drives are to be defined, which allow an efficient and safe implementation of electronic cam gears and take into account a given accuracy. In other words, based on an given cam and time requirements, it should recommendations for selecting suitable drives with regard to static and dynamic properties are made or vice versa, for an existing machine cams are to be checked for practicability. The importance of the achievable positioning accuracy in current systems becomes clear in [7–9].

As starting point for this investigation is a polynomial of the 5th order for the description of a dwell rise dwell movement [3]. In previous work, promising results for the control parameterization from the combination of time and frequency domain could be shown [10]. In analogy of this purpose, a transfer of the time domain into the image domain necessary for the examination is also carried out by means of Laplace transformation. The advantages arising from this design process are reflected in the savings of a complete simulation of the entire drive system or a hardware-in-the-loop simulation (see [11]). Thus it is possible, with the characteristic value of the cam from the frequency domain of the position control loop, to read out specific deviations in the time domain.

2 Parameter Identification

2.1 Modeling

By means of analytical functions, laws of motion of cam gears can be described. A systematization with regard to the determination of the laws of motion for cams is listed in the VDI guideline VDI 2143 [3]. This allows the description of relative movements of two gear members. As a result, a time dependence is introduced, which results from the fixing of a drive movement. As a basis for this characteristic value determination a normalized dwell rise dwell movement is used, which can be described by means of (1):

$$x(z) = 10z^3 - 15z^4 + 6z^5 \qquad (1)$$

It applies to $0 \leq z \leq 1$. In the case of constant rotational speed or constant speed of the master drive and by substitution of z = at, 'a' can be regarded as a scaling factor for the time. It turns out:

$$x(z) = 10(at)^3 - 15(at)^4 + 6(at)^5 \tag{2}$$

with $0 \leq t \leq 1$. This time course can be transferred into the image domain by means of Laplace transformation:

$$X(s) = \int_0^\infty e^{-st}(10(at)^3 - 15(at)^4 + 6(at)^5)dt \tag{3}$$

By using the sum rule, consideration of the corresponding limits and neglect of initial values it turns out:

$$X(s) = \frac{60s^2a^3 - 360sa^4 + 720a^5}{s^6} \tag{4}$$

The gained function has to be transferred into a suitable simulation model. For this purpose, the considered signal description is divided according to the transfer function in the time domain. For the necessary jump and the integrator 1/s is factored out respectively. The model is supplemented by two scaling factors fak_{MS} and fak_S. The scaling factor fak_{MS} refers to the master and the slave drive, fak_S only to the slave drive. They allow a freely selectable positional scaling. The described signal description is shown in Fig. 1.

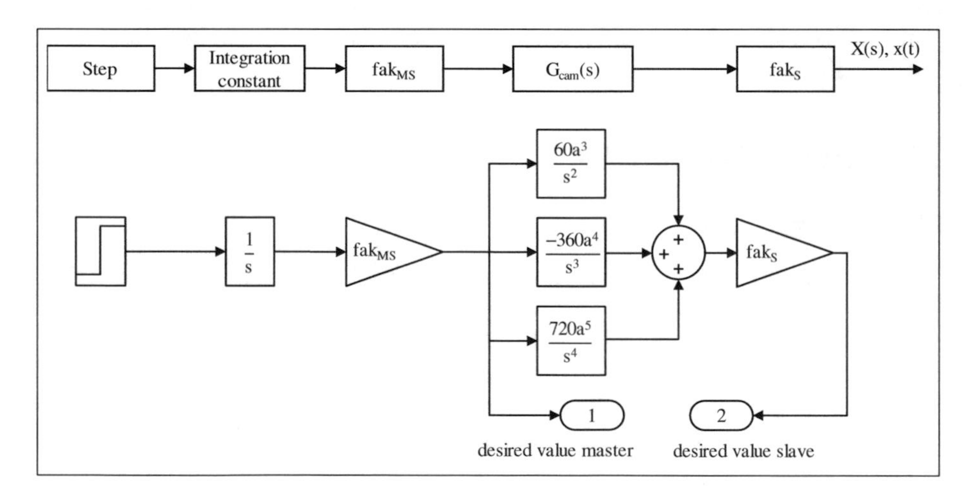

Fig. 1. Schematic signal description

This results the transfer functions listed for the master (5), the cam (6) and the slave (7).

$$G_M(s) = \frac{1}{s} * \frac{1}{s} * fak_{MS} \tag{5}$$

$$G_{cam}(s) = \left[\frac{60a^3}{s^2} - \frac{360a^4}{s^3} + \frac{720a^5}{s^4}\right] * fak_S \tag{6}$$

$$G_S(\mathrm{s}) = \frac{1}{s} * \frac{1}{s} * fak_{MS} * \left[\frac{60a^3}{s^2} - \frac{360a^4}{s^3} + \frac{720a^5}{s^4}\right] * fak_S \tag{7}$$

Following, the signal sequence can be determined using conventional simulation systems in the time domain while maintaining the limits $0 \leq t \leq 1$ and a = 1, and is shown in Fig. 2. Diagram 1 shows therein the position of the master drive, thus of the drive that processes the linear path information. The diagrams 2 to 4 show the position, speed and acceleration curves of the slave drive. The simulation results clearly show that the specifications from VDI 2143 for the maximum speed and acceleration values are complied with.

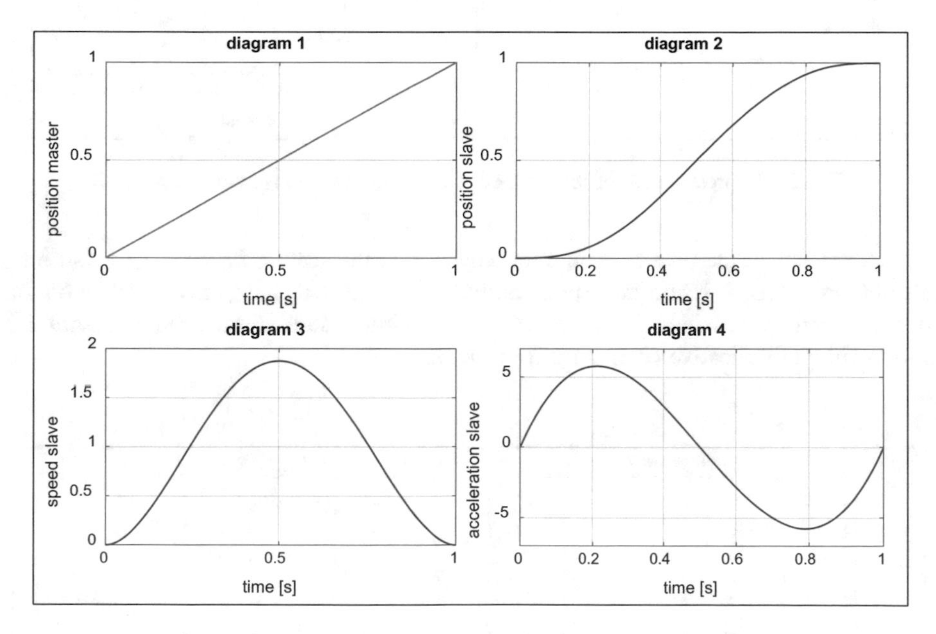

Fig. 2. Representation of normed dwell rise dwell movement

The effects of a change in 'a', that is the scaling of time, are shown in Fig. 3 for the time domain. An increase of 'a' is therefore equivalent to a shortened execution time of the movement task. However, the end position of the master and the slave remain unaffected by this change.

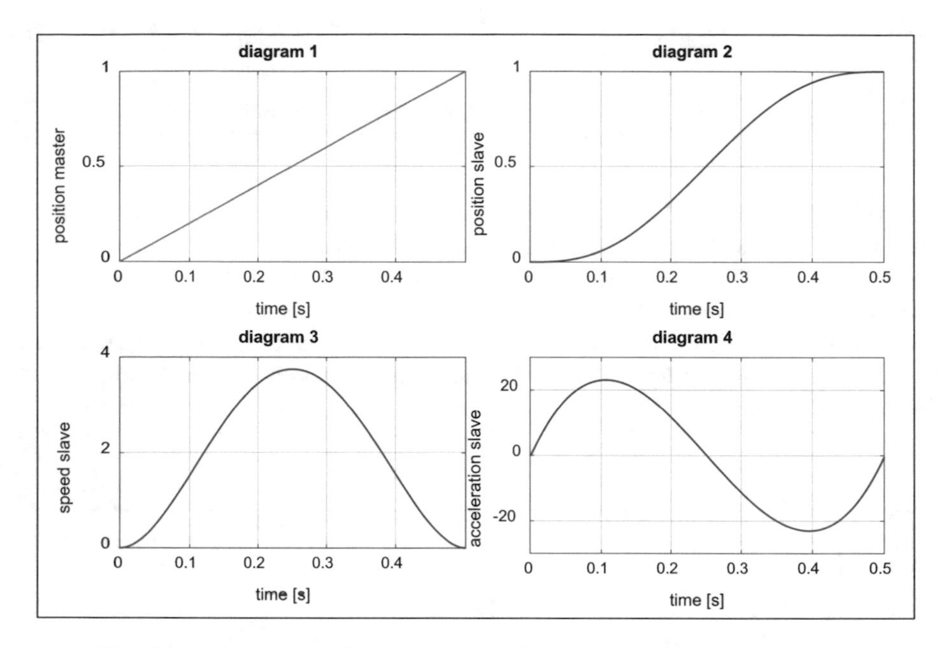

Fig. 3. Representation of the normed dwell rise dwell movement with a = 2

The effects in the time domain upon changes in the scaling factors fak_{MS} and fak_S are shown in Fig. 4. It becomes apparent that fak_{MS} and fak_S have only an effect on the position and the position increases with increasing values. Furthermore, diagram 2 shows the multiplicative chaining of the factors.

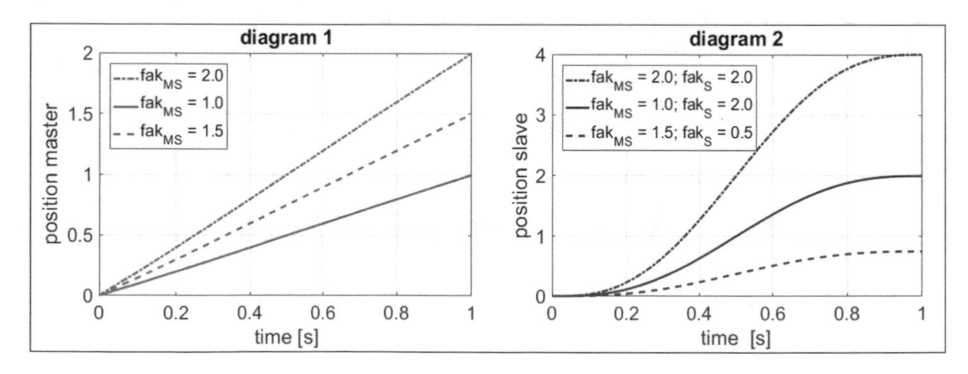

Fig. 4. Variation of the scaling factors fak_{MS} and fak_S with a = 1

To continue the examination in the image domain, an analysis of the frequency responses is necessary. By substitution of s = jω in the formulas (5) to (7), the analysis can be carried out in the Bode diagram. Figure 5 shows the corresponding Bode diagram which displays the behavior of the frequency response of the cam with variation of the parameter ‘a’. It can be seen that the frequency response with rising ‘a’ responds with an increasing frequency. The dimension of the amplitude and the phase response remain constant.

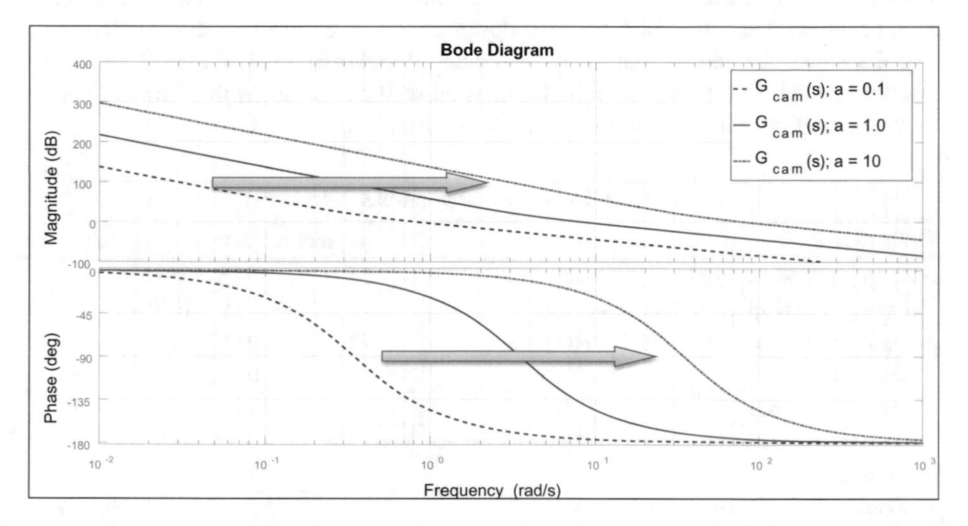

Fig. 5. Bode diagram of $G_{cam}(s)$ with variation of parameter a

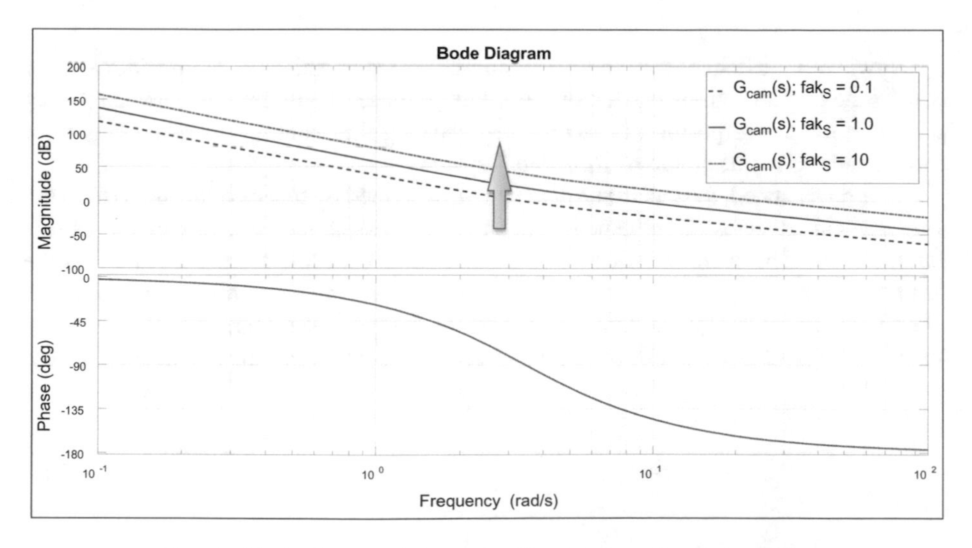

Fig. 6. Bode diagram of $G_{cam}(s)$ with variation of parameter fak_S

The effects of the scaling factor fak_S are shown in Fig. 6. This only affects the amplitude. If the value of said factor increases, the amplitude also increases. The graphs of the phase response are all congruent and show no reaction to the variation of the scaling factor fak_S.

2.2 Extension for the Feedback Control Technology

To consider not only individual components, but the entire system, it is necessary to include the drive, the control system and the already developed cam in the analysis. For this purpose, a cascade control is set up with a simplified speed controller and a higher-level position controller, which are both designed as a PI controller. The creation of the model follows the remarks of Groß, Hamann, Wiegärtner [12]. Figure 7 shows the resulting signal flow plan. The initial values as well as the determined model parameters for creating this simulation model can be found in Table 1.

Table 1. Initial values for controller calculation

Designation	Symbols	Value	Unit
Sampling period current loop	T_{Ai}	$62.5 * 10^{-6}$	s
Sampling period speed control loop	T_{An}	$125 * 10^{-6}$	s
Sampling period position control loop	T_{Ax}	$250 * 10^{-6}$	s
Motor constant	K_m	0.923	Nm/A
Total output inertia	J	$14 * 10^{-5}$	$kg*m^2$
Sum of the smallest delay times in the speed control loop	$T_{\sigma n}$	$4.3 * 10^{-4}$	s
Reset time	T_N	$17 * 10^{-4}$	s
Proportional coefficient in the speed controller	K_P	0.1764	Nm*s/rad
Speed set point delay	T_{Gn}	$17 * 10^{-4}$	s
Proportional coefficient in the position controller	K_V	195.86	s^{-1}

Now, with the help of the model, the simulative investigation of the positional error in the time domain can be carried out. For this purpose, the difference between the set point and the actual position is recorded by varying the scaling factor ‘a’. This error corresponds to the accuracy of the overall system.

As already stated, the electronic cam can be described by means of formula (6) in the time domain. A decisive influencing factor on the frequency behavior of the cam is parameter ‘a’. The characteristic size of the electronic cam can be the zero and thus the center frequency (MF), which can be determined by zeroing the numerator polynomial. The connection of the results from time and image domain is summarized in Table 2. The resulting constant factor points out a direct relationship between time and image domain.

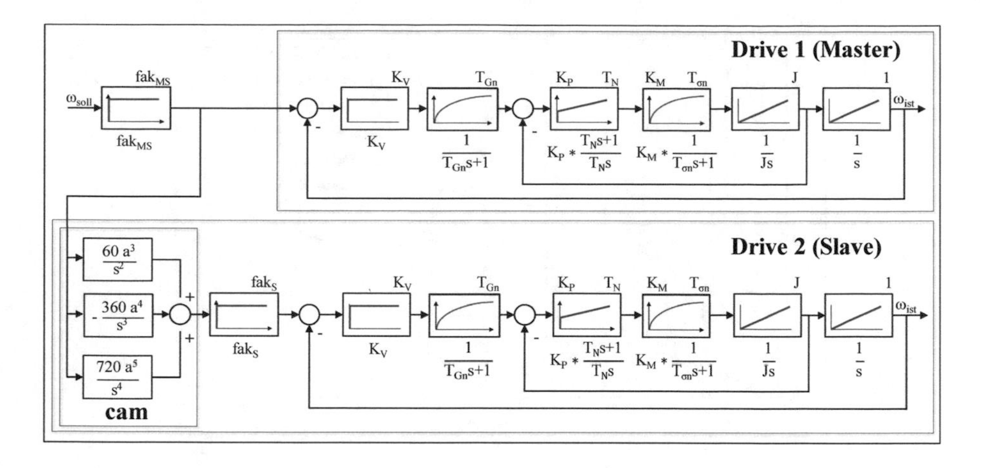

Fig. 7. Signal flow plan of the simulation model

Table 2. Connection of the results from time and image domain

Scaling factor	Image domain	Time domain	Factor
a	MF [rad/s]	Positional deviation Δ s	MF/Δ s
0.25	0.866	0.002393	361.888842
0.5	1.7321	0.004786	361.909737
1	3.4641	0.009572	361.899290
2	6.9282	0.019140	361.974922
4	13.8564	0.038250	362.258824

2.3 Transfer to a Test Bench

For the further investigation, the simulative results are transferred to a test bench. Here, too, a cascade control is used, which can be parameterized by the Simotion Engineering System Scout V4.3.1.3 at the test bench. The controller design is carried out analogous to that of the simulation model. Deviations exist only in the sampling times and the drives used, which are specified by the test bench.

Besides, nonlinearities in the system must be excluded for the further investigation, since the characteristic value determination refers to a linear system [13]. The limits of the linear system are determined, for example, by the standstill current of the motor. Likewise, the torque limit is such a limit. Especially the torque limitation limits the analysis in the phase response to relatively low frequencies. However, in this area, the influences from the superposition of unknown dead times due to the low phase value are stronger than in the higher frequency range. The reason for exceeding the limit can be attributed to the path to be traveled by the cam. The Simotion Engineering System Scout allows scaling of the cam which corresponds to the described factor fak_S. As a result, the frequency of the cam can be increased without exceeding the torque limit.

The deviations between the set point and the actual position in the time domain are also recorded at the test bench, with variation of the scaling factors ‘a’ and fak_S.

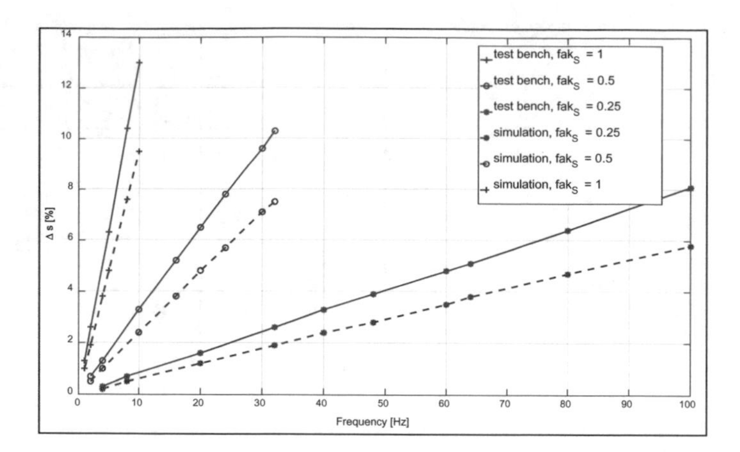

Fig. 8. Position deviation above the frequency of the cam

Figure 8 shows the results of this examination and compared with the corresponding values from the simulation model. The scaling factor ‘a’ is reflected in the frequency. It can be seen that the data series from the simulation and the test bench have comparable orders of magnitude. In particular, the ratio between the slopes of the graphs from the model and those of the test bench are identical. The differences between the graphs with the same scaling factor fak_S can be attributed to the different hardware.

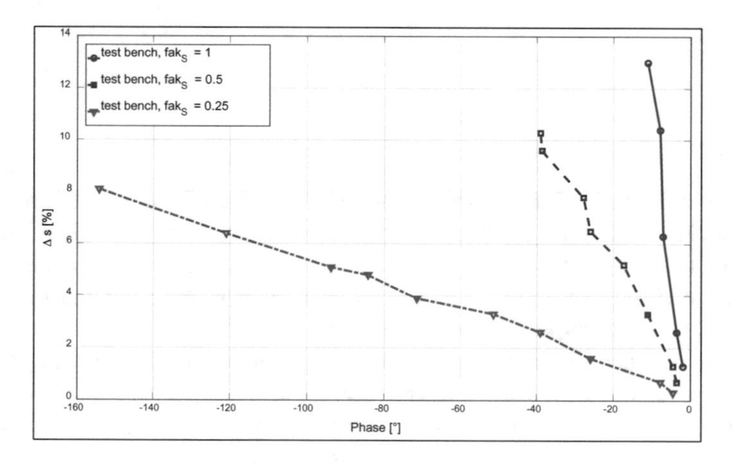

Fig. 9. Connection time and image domain

Subsequently, the time domain and the phase response from the image domain are merged. For this purpose, the phase shift of the individual frequencies from the Bode diagram is assigned to the associated position deviations for the test bench. This results in the diagram in Fig. 9. It assigns the position deviation from the time domain to the corresponding phase shift from the image domain and takes into account the scaling factor fak_S.

Due to the linear behavior of the graphs, in particular of the graph with $fak_S = 0.25$, it is possible to calculate the resulting positional deviation for every phase shift. If the position deviations between −20° and 0° are neglected due to the described restrictions, then this graph can be approximated by formula (8). The remaining two straight lines each have a fourfold steeper slope and can therefore also be determined from (8). With (8) it is therefore possible to show the effect on changes of the cam.

$$y = -0.05x + 0.56 \tag{8}$$

3 Conclusions

In the present work it could be shown that the dwell rise dwell movement listed in the VDI guideline VDI 2143, which can be described with a polynomial of the 5th order, is transferable by a Laplace transformation into an image domain necessary for the examination. This results in the possibility of integrating this movement task into simulation models and further analyzing them. Based on the model the direct connection between time and image domain could be demonstrated. The results of the simulation were able to transfer to a test bench and confirmed on it. This work thus provides the possibility already in the design process of electronic drives to define their accuracy or to show the performance limits for existing systems under specification of specified accuracies.

References

1. Schäppi, B., Andreasen, M.M., Kirchgeorg, M., Radermacher, F.-J.: Handbuch Produktentwicklung. Carl Hanser Verlag, München (2005)
2. Lohse, G.: Konstruktion von Kurvengetrieben. Expert verlag, Rennigen-Malmsheim (1994)
3. VDI 2143 Blatt 1, Bewegungsgesetze für Kurvengetriebe, Theoretische Grundlagen, Verein Deutscher Ingenieure e.V., Düsseldorf (1980)
4. Kiel, E. (Hrsg.): Antriebslösungen Mechatronik für Produktion und Logistik. Springer, Heidelberg (2007)
5. Heine, A.: Ein Beitrag zur kennwertorientierten Entwicklung kurvengesteuerter, ebener Schrittgetriebe. Universitätsverlag Chemnitz, Chemnitz (2015)
6. Isermann, R.: Mechatronische Systeme Grundlagen 2 Auflage. Springer, Heidelberg (2008)
7. Altintas, Y., Verl, A., Brecher, C., Uriarte, L., Pritschow, G.: Machine Tool Feed Drives. Ann. CIRP **60**(2), 779–796 (2011)
8. Neugebauer, R., Ihlenfeldt, S., Hellmich, A., Schlegel, H.: Modelling feed drives based on natural excitation—improving accuracy. CIRP Ann. Manuf. Technol. **66**(1), 369–372 (2017)
9. Schlegel, H., Hellmich, A., Hipp, K., Quellmalz, J., Neugebauer, R.: Improved controller performance for electromechanical axes. Solid State Phenom. **251**, 113–119 (2016)
10. Hipp, K., Uhrig, M., Hellmich, A., Schlegel, H., Neugebauer, R.: Combination of criteria for controller parameterisation in the time and frequency domain by simulation-based optimization. J. Mach. Eng. **16**(4), 70–81 (2016)
11. Abel, D., Bollig, A.: Rapid Control Prototyping: Methoden und Anwendungen. Springer, Heidelberg (2006)

12. Groß, H., Hamann, J., Wiegärtner, G.: Elektrische Vorschubantriebe. Publicis KommunikationsAgentur GmbH, GWA, Erlangen (2006)
13. Lunze, J.: Automatisierungstechnik Methoden für die Überwachung und Steuerung kontinuierlicher und ereignisdiskreter Systeme 4, überarbeitete Auflage. Walter de Gruyter GmbH, Berlin (2016)

Design of Performance Indicators Based on Effective Time and Throughput Variability. Case Study in Mining Industry

Tomás Grubessich[1], Raúl Stegmaier[1,2], Pablo Viveros[1], Mónica López-Campos[1], Fredy Kristjanpoller[1], Christopher Nikulin[3(✉)], and Sebastian Koziolek[4]

[1] Departamento de Industrias, Universidad Técnica Federico Santa Maria, Valparaíso, Chile
tomas.grubessich@usm.cl
[2] DICAR, University of Catania, Catania, Italy
[3] Departamento de Ingeniería en Diseño, Universidad Técnica Federico Santa Maria, Valparaíso, Chile
christopher.nikulin@usm.cl
[4] Mechanical Department, Wroclaw University, Wroclaw, Poland

Abstract. The following paper proposes a method on how to analyze productive systems to achieve performance indicators that allows to know the state of the system. In particular, the objective of the paper is to analyze performance indicators that allow to understand the state of the production line in systems that present variability conditions in the performance of their equipment, and in their operational conditions that will not allow direct calculation of the effective time. It is proposed to begin with the utilization of a methodology to increase the understanding of the system, which will generate a conceptual model that will concentrate the required knowledge through a logic structure that will ease the subsequent analysis. Then, a step by step process is proposed to define the system, its performance indicators of interest, and the most efficient and effective way to obtain those, considering the existing restrictions. Finally, system and subsystem level indicators will be obtained, which will be a representation of the real state of the process, by representing the effective times, and variable throughput. All of the above will be applied in a case study in the mining industry from Chile.

Keywords: Historical analysis · Increase the knowledge of the system
Performance indicators · Variable throughput

1 Introduction and Problem Statement

Historical analysis is fundamental to the realization of studies in varied areas within the organizations. The main benefit of this type of analysis is to increase the comprehension degree that the system possesses, and once taking this basis, varied possibilities are open to deepen the knowledge, so much to support the management of the system, as to support the process of the associated decision making [6].

© Springer Nature Switzerland AG 2019
A. Burduk et al. (Eds.): ISPEM 2018, AISC 835, pp. 139–150, 2019.
https://doi.org/10.1007/978-3-319-97490-3_14

Within the most concrete applications that can be done about the historical analysis, is the definition of the actual and past state of the process through the indicators of performance. Some of the indicators considered are related to reliability, maintenance, availability, etc. [1, 4, 7, 10], in order to describe both the functioning of the equipment, and of the system in a global manner. This analysis helps the detection of opportunities of improvement, and the information here obtained is fundamental for a second analysis, where a behaviour of the equipment according to the different perspectives of interest will be defined [9, 11]. Some of these perspectives of study are the time between failures, repairing time, production rates, among others.

Nevertheless, there are some conditions that difficult the performance indicators analysis in a complex system representation. A complex system will be known as a system where it is not possible to extract its behaviour directly from sources of information that are available. In the particular case of this paper, two types of these difficulties will be reviewed. The first one is about the difficulty to determine effective time given the operational characteristics of the system. The second one is about a variable throughput given the raw material. These conditions may make not possible to quantify the impact of the operational decisions on the system, such as, to raise the number of equipment, or to change operational conditions, in other words, by performing these changes, it is not possible to certainty know what will happen to the system in particular. This condition also restricts the applications of diverse types of models that set up methodologies to represent the system [2, 3, 5, 8, 12].

To resolve this, this document aims to analyze performance indicators that allow understanding the state of the production line in complex systems such as those described above. The structure of the document to address its objective, begins with the application of a methodology to increase the knowledge of the system [6], followed by a step by step process that guides the design of the performance indicators through which the productive system is modelled. The result of the application of the methodology noted above will be having performance indicators designed for each subsystem based on their own operational characteristics, throughput variability, and state of the historical information. With these performance indicators, the state of the system will be understood, complying with the objective of the paper. All of the above will be applied in a case study in mining industry.

2 Historical Analysis and Design of Performance Indicators

In order to address the raised problem, the following sequence is carried out. First, the methodology to increase the organizational learning based on experts' knowledge and information [6] is selected to increase the knowledge of the system. Then, the first step in this section corresponds to show the main steps and results when applying the methodology. Once this is done, it will be shown how the generated knowledge is used to define the bases for the design of the performance indicators.

2.1 Methodology to Increase the Understanding of the System

Project Definition

The productive process, comprehends copper ore production line from its extraction from the mine, transport by a truck fleet, primary crushing of the ore and stockpile accumulation. The system analyzed in this project includes from the truck fleet to the belts transporting the material to the stockpile. The first subsystem corresponds to a fleet composed of 54 trucks, which are assigned different routes to transport the ore from the mine to a crusher. The second subsystem corresponds to two independent primary crushers whose purpose is to reduce the size of the mineral. Finally, there is a conveyor belt system consisting of three belts in a series logic to transport the ore from the crushers to an accumulator.

The primary objective corresponds to a design of a performance indicator for the produced tons each day, and how these are related to the behaviour of the process of interest. It is known that when the system is analyzed, the following conditions must be acquainted: there are relevant differences in the operation of each part of the productive line; the characteristics of the copper ore generate variability in the throughput of specific equipment; and, each part of the system has its own historical information with specific characteristics.

Given the mentioned conditions, the design of indicators must respond to the possible difficulties. Following the suggestions of the methodology [6], a project is created, whose objective is to design performance indicators for the system that represent the behaviour, considering the limitations and characteristics of each subsystem. The system consists of the fleet of trucks, primary crushing and conveyor belts. The project team consists of:

- Team Responsible of the Project: It is composed of members of the organization responsible for the management and analysis of the system. They become the direct responsible and respond to the top management of the company for the results.
- Expert Panel: It's conformed by the person in charge of the management of the whole system, the person in charge of the operation and the experts members of this area. All these actors have years of experience.
- IT Department: In this particular case, it is decided that it is not necessary to have the exclusive participation of people from the IT Department, since the access to the sources of information required is available.

Step by Step Implementation Process

As proposed by the methodology, it is composed of a virtuous cycle in four areas, generating a spiral progression as the understanding of the system increases. As more iterations are performed, it is possible to increase the degree of development of the desired objective. Figure 1 presents the outline proposed by the methodology.

In this way, the interaction between the panel of experts and the data coming from the computer systems will be the basis for the iterative process that will increase the understanding of the system.

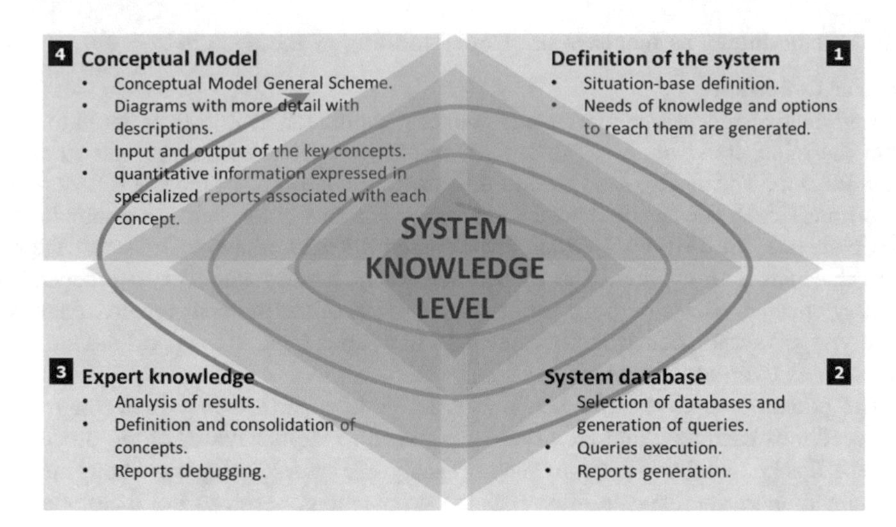

Fig. 1. Scheme of the development of conceptual model system. Source: own elaboration.

Conceptual Model

After implementing the methodology, a conceptual model is obtained with the formalization of the identified and relevant knowledge, which is found with a logical structure easy to understand and at different levels of detail to simplify its explanation.

As shown in the Fig. 2, the conceptual model consists of different elements that must be analyzed and from which it is possible to achieve the objective.

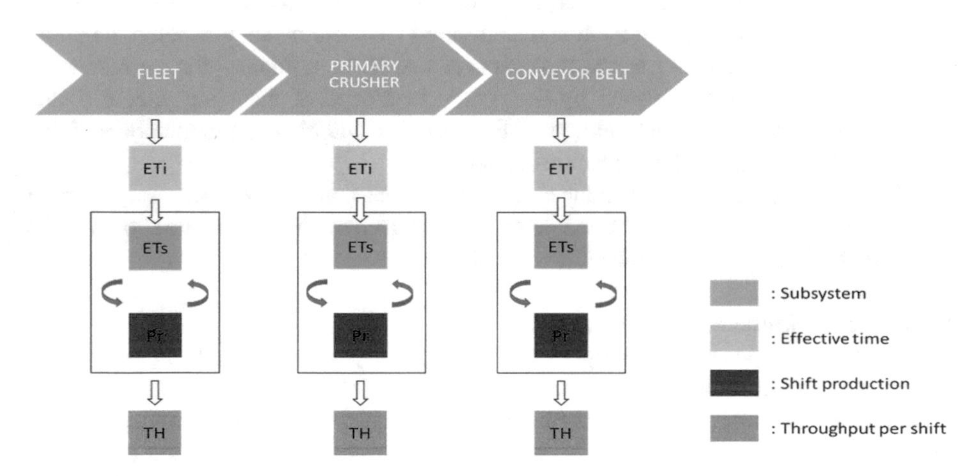

Fig. 2. General scheme of the conceptual model. Source: own elaboration.

First, three subsystems are defined corresponding to: the truck fleet, the primary crushing and the belt conveyor. The system has a continuous flow of ore, but there are small stockpiles that give autonomy to each subsystem in the production of each twelve-hour shift. This required an analysis that considers a certain level of independence between subsystems and between work shifts. Second, the effective time can be obtained in each subsystem from the historical information, but each subsystem has different characteristics, so the way to calculate this time must be customized. Third, the production per work shift for each subsystem is obtained from historical information. Finally, there is the throughput of each subsystem, whose calculation must be careful and customized for each subsystem. This is because the throughput is variable, mainly due to the characteristics of the copper ore.

It should be noted that the developed conceptual model includes the general scheme shown, and a set of elements such as reports, databases, queries, technical reports, among others.

2.2 Design of the Indicators

Step by Step Process Proposal

Now, with the knowledge of the interest aspects detected in the previous section, the proposed step-by-step process for the design of the performance indicators will be presented. To begin with, the Fig. 3 shows the proposed step-by-step process.

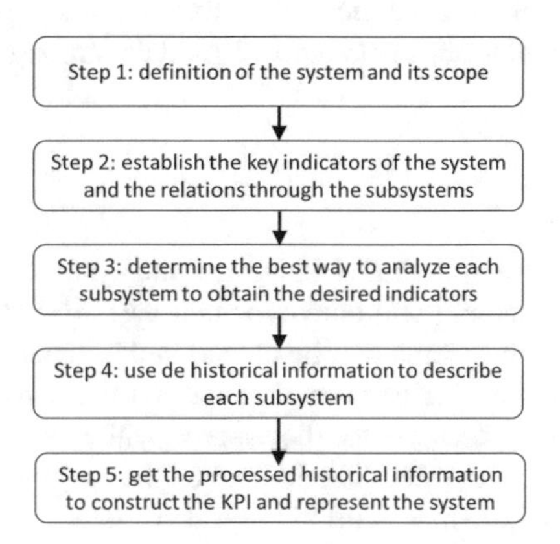

Fig. 3. Step by step process proposal. Source: own elaboration.

For a better understanding of this step by step process, a brief description of each of its stages will be given below:

- Step 1: definition of the system and its scope: the system is defined, with its purpose, scope and main characteristics. It also describes its subsystems and how they contribute to the global production.
- Step 2: establish the key indicators of the system and the relations through the subsystems: the required performance indicator is defined in detail, and at the same time how each subsystem, affects or influences its construction. The relationships and functions between subsystems are described, as well as the main characteristics of each subsystem in relation to the indicator of interest are established.

- Step 3: determine the best way to analyze each subsystem to obtain the desired indicators: knowing the requirements of the desired indicator, the characteristics of each subsystem and the experience of the experts, the best strategy to analyze each subsystem is determined. The strategy may be different for each subsystem, but it must respond coherently to the desired indicator.
- Step 4: use the historical information to describe each subsystem: a descriptive analysis of the different subsystems is carried out to verify that they behave as expected. To do this, it is recommended to start with an exploratory analysis, then, step by step, build each variable of interest until finally achieving the productive performance indicator of the subsystem that contributes to the global productive performance.
- Step 5: get the processed historical information to construct the KPI and represent the system: with the purified information of each subsystem, two fundamental types of information are generated. The first has to do with the results of the historical behaviour of the productive performance indicators of the system and of each subsystem, and the second, with the construction of the historical data of the variables to build the behaviour of the system. From the first category, it is possible to analyze the historical behaviour of the indicator, identifying atypical behaviour, seasonal factors, the impact of external factors, among others. With the second type of information, we will have the ideal input that will allow performing another type of study, such as simulation, mathematical models, among others.

3 Case Study

- Step 1: definition of the system and its scope: the system under study includes the truck fleet, primary crushing and conveyor belt system that unloads the product in the stockpile. These subsystems will be considered independent due to the presence of small stockpiles before the primary crusher and the conveyor belt. It is required to design performance indicators for the system, which allow a description from the subsystems to the whole system. The impact of the variability of copper ore on the throughput of the equipment should also be considered.
- Step 2: establish the key indicators of the system and the relations through the subsystems: given the independence of the subsystems in the calculation of the indicators, the performance indicators will be designed based on effective times and equipment throughput. For the calculation of throughput per subsystem, the historical production will be divided by the effective time.
- Step 3: determine the best way to analyze each subsystem to obtain the desired KPI: since each subsystem has characteristics that differentiate it from the rest, a brief description of how the analysis will be performed in each of the subsystems will be given below:

Fleet subsystem: The fleet is composed of a certain number of trucks, of which, the number of hours that each truck was in different states is recorded. The behaviour of each truck is very variable, in some cases it can be highly used, in others it can be waiting for work routes, it can be in a long preventive maintenance, and thus different states. Representing the behaviour of each truck in the future represents a highly demanding work. That is why the expert team decides to analyze the behaviour of the fleet as a whole, since its behaviour is more stable. Therefore, the effective time will be calculated for each shift, considering that the entire fleet and the throughput will be equal to the production of the subsystem, divided by the effective time. It should be noted that it is necessary to distinguish the production of the fleet, towards the crusher 1 of the production towards the crusher 2. In this way, the fleet would be represented as a total effective time, with a percentage of time dedicated to each crusher and a throughput per route that is equal to the production divided by the effective time.

Primary Crusher: This subsystem is composed of two primary crushers that receive the copper ore by independent routes from the mine. For each one of them, there is historical information about the detention events. For each of these events, the information system has the start time, the finish time and it can assign a category. The categories defined by the experts are: preventive maintenance, corrective maintenance, operational detention and waiting for ore. Each work shift has twelve hours, and by subtracting the sum of the detention times of the four categories mentioned, the effective time will be obtained. Finally, dividing the production of each crusher by its effective hours of work per shift, will have the real performance of that shift.

Conveyor belt: This subsystem is formed by a set of belts with a functional logic of serial system, that is, if one of them fails, the entire subsystem stops. The effective time can be built in the same way as in the crushers, identifying each detention event in any of the categories mentioned and subtracted to the shift hours. The throughput will be calculated as the production of the shift divided by the effective time.

- Step 4: use the historical information to describe each subsystem: As each subsystem has its own historical information, a brief description of how the information will be used in each of them will be given:

Fleet subsystem: the analysis will be made from the states presented by the entire fleet in each shift. The Fig. 4 shows the result when analyzing the information obtained from the historical database purified:

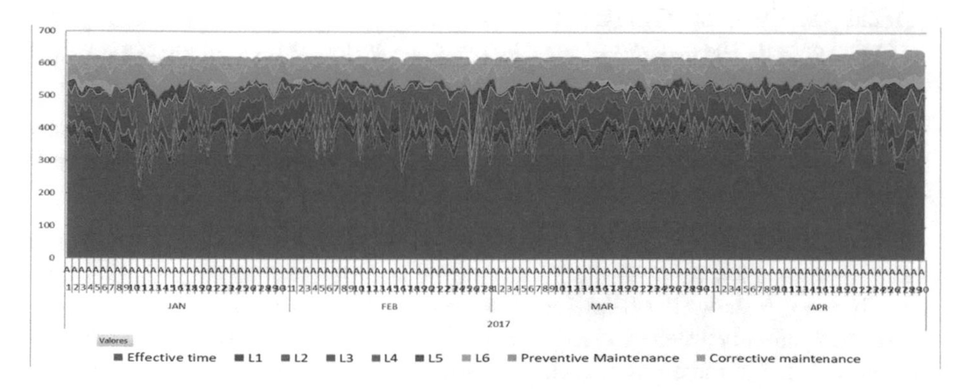

Fig. 4. Representation of the behaviour of the fleet. Source: own elaboration.

There are ten types of states for each truck and the behaviour of the fleet could be represented. It is decided to group these categories in the categories mentioned above and that correspond to preventive maintenance, corrective maintenance, operational detention and waiting for ore. In the next step, the behaviour of each of these states should be analyzed, and the effective time will be the subtraction of the time of the shift minus the sum of the times of the states of detention. Regarding the production, there is an exact record of production per shift that has a high precision. Then, and from the effective times and production, the throughput behaviour is built. It must be distinguish the production that goes to the Crusher one, of the production that goes to the Crusher two.

Primary Crusher: To begin the description of this subsystem, the effective time of a crusher is shown below as an example in the Fig. 5.

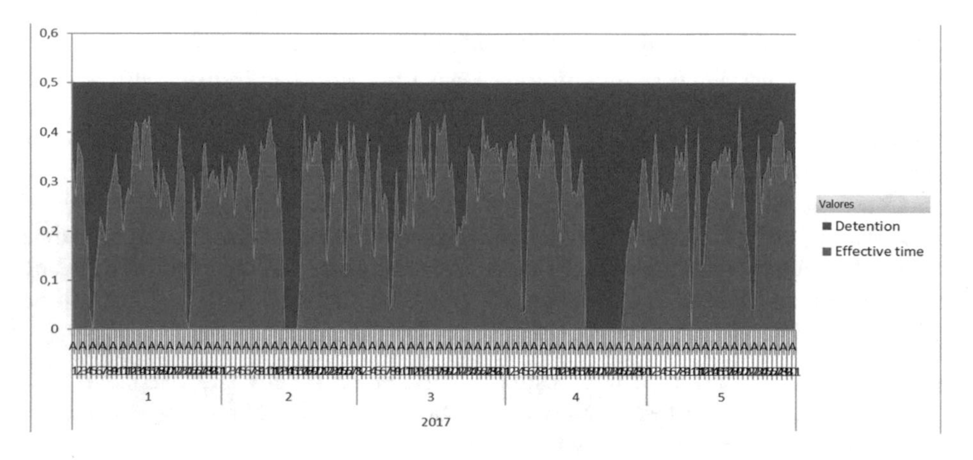

Fig. 5. Representation of the behaviour of the primary crusher. Source: own elaboration.

In the case of crushers, there is a great variety of detention times, which are grouped in the categories mentioned. Then, with the effective time of each crusher and the historical production, the throughput is calculated.

Conveyor belt: The analysis of the conveyor belt system is done in the same way as in the crusher system.

- Step 5: get the processed historical information to construct the KPI and represent the system: Once the sources of information have been refined, the representation of the indicator of interest guided by the defined process is carried out.

Primary Crusher: Before obtaining the performance indicator, the existence of an acceptable correlation between the effective time and the production per shift is analyzed. The Fig. 6 shows the behavior of these variables over time:

The correlation between production and effective time is over 80%. The analysis of the representation of the times of each detention categories is executed. Given the large number of behaviours that can be obtained, some examples will be given in the next figure (Fig. 7).

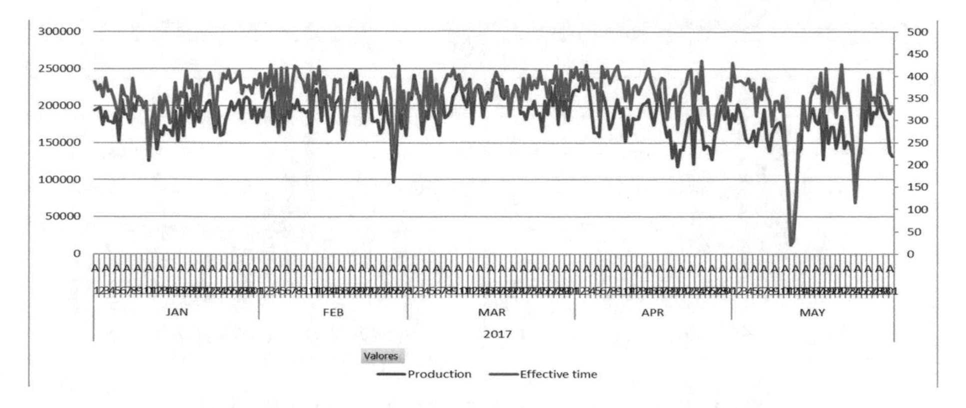

Fig. 6. Correlation between effective time and production. Source: own elaboration.

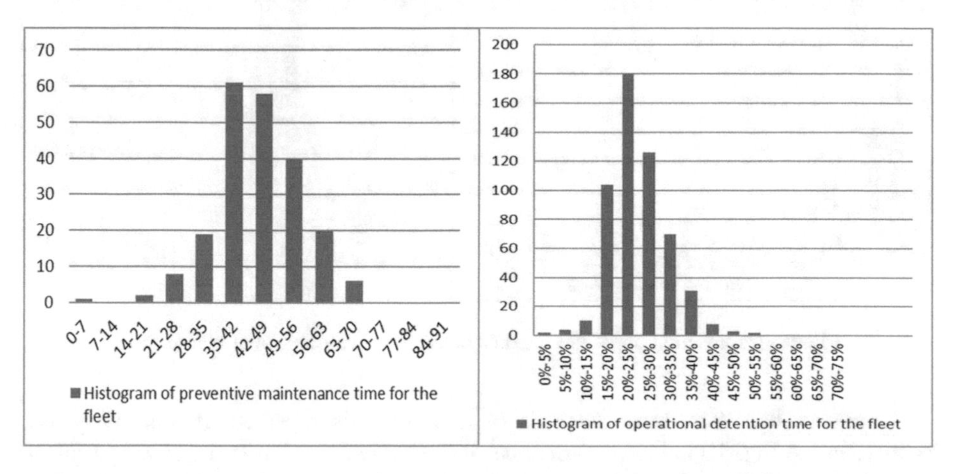

Fig. 7. Examples of behaviour of detention times of the fleet. Source: own elaboration.

The behaviour of detention by waiting for ore is not analyzed, since it was concluded that this type of detention is the result of what happens in the rest of the subsystems. Although the subsystems have some autonomy because of the intermediate stockpiles, the potential production they can achieve is restricted by the rest of the processes, which is reflected by the waiting for ore detention. Then, the fleet performance is calculated by the destination route what is shown in Fig. 8.

The final analysis for this subsystem indicates that, the potential production of the fleet will be equal to the potential effective time multiplied by the performance of per shift. The potential effective time is calculated as the time of the shift, minus the sum of the preventive and corrective maintenance and operational detentions.

Primary Crusher: Analyzing the correlation of effective time and production, this is over 80%, so the throughput can explain the rest of the relationship. Due to the size of the paper, only one example of the duration of a type of detention and the performance of one crusher is shown in the Fig. 9.

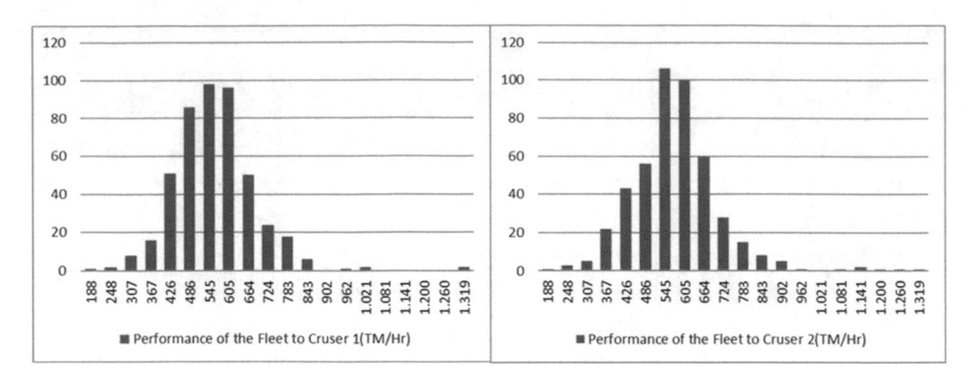

Fig. 8. Performance of the fleet. Source: own elaboration.

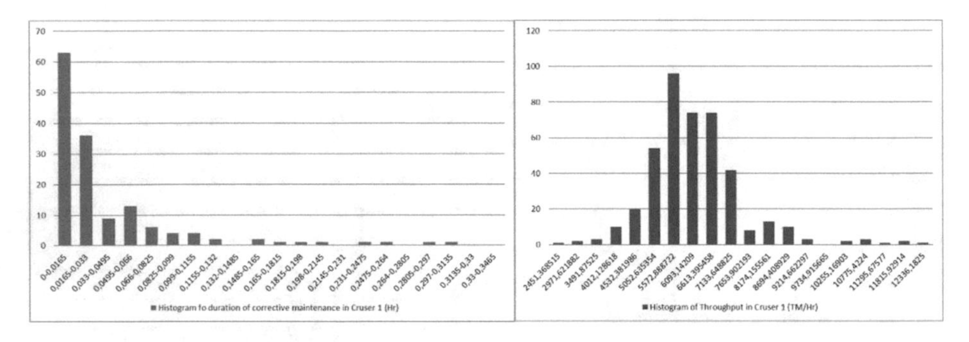

Fig. 9. Representation of the crusher behaviour. Source: own elaboration.

Conveyor belt: The same analysis of a crusher, is repeated in this subsystem. Finally, in the Fig. 10 is the performance of the system, which is the result of all the analysis.

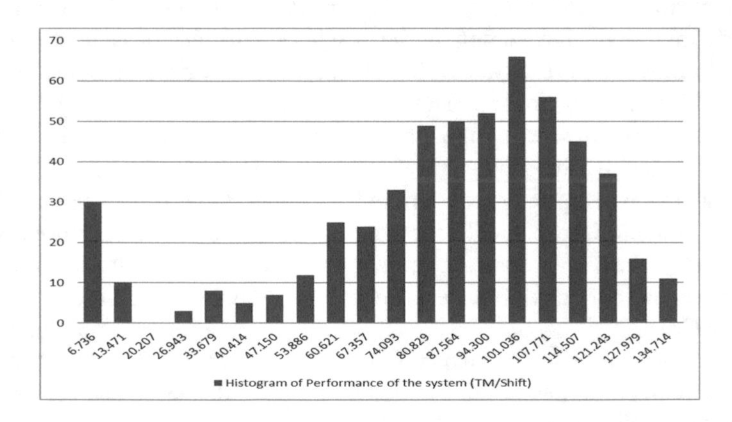

Fig. 10. Performance of the system. Source: own elaboration.

4 Discussion and Conclusions

With the development of this paper, it has been possible to design indicators that represent the key variables identified, achieving the objective set. This despite the difficulties presented by the process due to the particularities of its subsystems and the characteristics of copper ore. For this, it was necessary to increase the understanding of the system and to organize the knowledge with a logical structure through a conceptual model. Another problem that was addressed was having information systems with different data types and variability in the throughput of the equipment.

With the analysis carried out, it was possible to represent the system by identifying potential productions by subsystem and how this affects the overall performance of the system. This leads to the possibility of constructing customized indicators for each subsystem, which represent in a more detailed and efficient way its functioning.

In this way, this paper manages to describe the state of the complex system through performance indicators that are adapted to its particular characteristics.

It should be noted that the novelty of this work lies in how it is possible to obtain appropriate indicators that describe the system of interest, based on information with important limitations and characteristics of the operation that hinder the analysis. A more effective solution would be to improve the information systems related to the measurement of the load that passes through each equipment, and the information system related to the detention events. This is a solution that implies investment and execution time, but in the long term, it is the best alternative.

References

1. Ahumada, M.: Establishing and improving manufacturing performance measures. Robot. Comput. Integr. Manufact. **18**, 171–176 (2002)
2. Bause, F., Kritzinger, P.: Stochastic Petri Nets – An Introduction to the Theory, 2nd edn. Techniche Universität Dortmund, Dortmund (2002)
3. Buzacott, J.A., Shanthikumar, J.G.: Stochastic Models of Manufacturing Systems. Prentice-Hall, Englewood Cliffs, New Jersey (1993)
4. Calixto, E.: Reliability, availability, and maintainability analysis. In: Gas and Oil Reliability Engineering, Chapter 4, pp. 169–347 (2013)
5. Fuqua, B.: Markov Analysis. J. RAC, Third Quarter (2003)
6. Grubessich, T., Viveros, P.: Methodological proposal in order to increase the organizational learning based on experts' knowledge and information systems in the field of asset management and maintenance. DYNA Management **4**, 14 (2016)
7. Muchiri, P.: Development of maintenance function performance measurement framework and indicators. Int. J. Prod. Econ. **131**, 295–302 (2013)
8. Schryver, J.: Nutaro, M: Metrics for availability analysis using a discrete event simulation method. Simul. Model. Pract. Theor. **21**, 114 (2012)
9. Sharma, R.K.: Performance modeling in critical engineering systems using RAM analysis. Reliab. Eng. Syst. Saf. **93**, 891–897 (2008)
10. Van Horenbeek, A., Pintelon, L.: Development of a maintenance performance measurement framework—using the analytic network process (ANP) for maintenance performance indicator selection. Omega **42**, 33–46 (2014)

11. Viveros, P., Zio, E.: Integrated system reliability and productive capacity analysis of a production line. A case study for a Chilean mining process. Proc. Inst. Mech. Eng. Part O: J. Risk Reliab. **226**, 305–317 (2012)
12. Zio, E., Pedroni, N.: Reliability estimation by advanced Monte Carlo simulation. In: Simulation methods for reliability and availability of complex systems. Springer Series in Reliability Engineering, Part I, pp. 3–39 (2010)

Development of an Intelligent Drainage-Humidifying Control System Based on Neo-Fuzzy Neural Networks

Svitlana Matus[1], Anastasia Stetsenko[1], Viktor Krylovets[2], and Vitalii Kutia[1,3](✉)

[1] National University of Water and Environmental Engineering, Rivne, Ukraine
{s.k.matus,a.m.stetsenko}@nuwm.edu.ua
[2] Empeek, Lviv, Ukraine
viksoft@ukr.net
[3] Wroclaw University of Science and Technology, Wroclaw, Poland
vitalii.kutia@pwr.edu.pl

Abstract. In this work the mathematical models of non-saturated part of soil as the controlled object and the control methods of agricultural cultures' water well-being upon underground moistening on the bases of specialized artificial neural networks and structure of control loops were developed. The hardware and software components of the automated control system of water well-being are described.

Keywords: Automated control system · Neural network · Fuzzy logic Drained-humidifying system · Groundwater level

1 Introduction

Monitoring, remote control, control and automatic check out on hydro-land reclamation systems requires the availability of effective multilevel distributed control system. Because the land reclamation systems, as the complex objects of control, are determined by large time constants that characterize the dynamic properties (response time) of the object of the study, by presence of uncontrolled disturbances as a result of sharp change of meteorological conditions, distribution of regulated parameters in accordance with spatial coordinates and remoteness from the centralized control points. According to the results of research of Ukrainian scientists [1, 2], one can recommend that during development of projects of reconstruction and modernization of existing drainage-humidifying system (DHS), the use of modular systems (Fig. 1) constructions on the internal economical network. Development of methods for water regulation control on the DHS is carried out on the basis of the mathematical model of the object. During control of groundwater level, as the result of nonlinearity of the object and the hysteresis phenomena connection between humidity of the root layer and the groundwater level is ambiguous and it causes low quality of regulation. In dynamic modes, exactly the aeration zone is the most inertial link due to low moisture transfer coefficients that determines the mode of humidification.

© Springer Nature Switzerland AG 2019
A. Burduk et al. (Eds.): ISPEM 2018, AISC 835, pp. 151–160, 2019.
https://doi.org/10.1007/978-3-319-97490-3_15

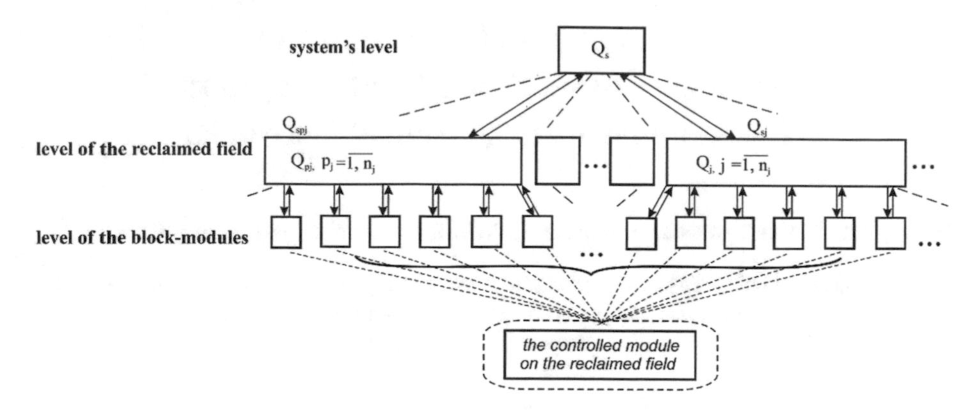

Fig. 1. Hierarchical structure of water-regulation on the modular type drainage-humidifying system

Therefore, the synthesis of regimes of water-regulation control on the DHS should be carried out on the basis of the analysis of the process of moisture transfer in the unsaturated zone, taking into account the influence of disturbances factors (total evaporation and precipitation). Today there are a number of methods of control soil moisture based on a multi-layered model of moisture transfer and combination of short-term and long-term meteorological forecasts. However, the questions of adaptation and learning of the computerized systems for controlling soil moisture in conditions of random weather factors, change of characteristics of the control object; increasing accuracy of control due to operational taking into account operation of disturbances on the object, ensuring receipt of planned crop capacity of agricultural crops at rational usage of energy and water resources are remain. Thus, development of the methods of automated control of moisture-supply of agricultural crops, taking into account disturbances, is an actual scientific and practical task.

2 The Mathematical Models of Moisture Transfer in Unsaturated Zone of the Module Soil Area

The drainage-humidifying system of Rivne experimental agricultural station with an area of 693 ha as a typical object of soil moisture control for agricultural crops was selected. The schematic structure of a cascade-combined ACS of water availability of a modular soil area by changing of the groundwater level (GWL) is shown in the Fig. 2, where W_z and W are the set and measured pressure head respectively; L_z and L are the set and measured water levels in the control well, respectively; $E_W = W_z - W$; $E_L = L_z - L$ is difference error; $L_{gr.w.}$ is the groundwater level; C_W is a controller of the pressure head; C_L is a water level controller in the control well; AM is an actuator mechanism; RB is the regulating action.

The internal control loop is the water level control circuit in the control well (control object Obj1), which is measured by the LE1 sensor and the external, setting loop is the control loop of the pressure head, which characterizes humidity.

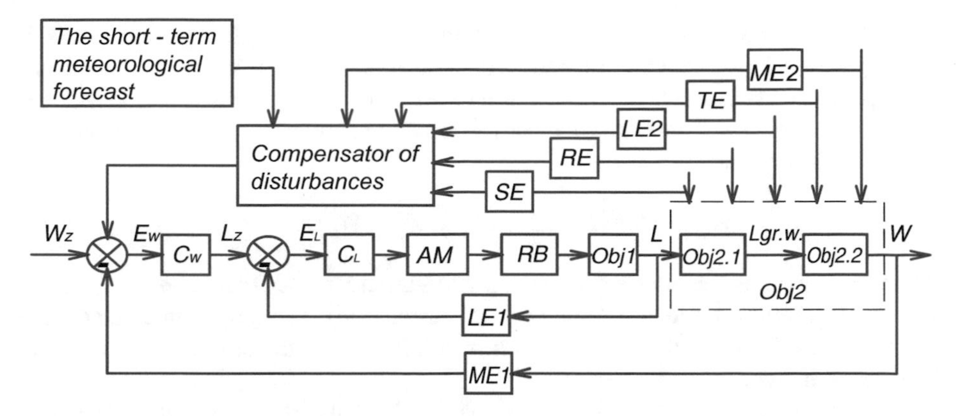

Fig. 2. Structure of the cascade-combined ACS of agricultural plants water availability within the modular field with the help of ground-water level's change

The object Obj2 is a soil area with a collector-drainage system. It is presented in the form of two links on the diagram. The first link Obj2.1 displays the collector-drainage system and the saturated soil zone and transforms the water level in the control well on the pressure in the collector-drainage system, and the pressure – in to the groundwater level Lgr.w. The second link Obj2.2 represents an unsaturated soil zone and transforms the GWL of the modular area into the pressure head of the root-contained soil layer W, which is measured by the sensor ME1. The following disturbances impact on the control object Obj2: 1 - ambient temperature (TE), 2 - precipitation (LE2), 3 - illumination intensity (RE), 4 - wind speed (SE), 5 - relative air humidity (ME2), 6 - structure of the soil, 7 - the phase of the plant development. Disturbances 1–5 constantly change over time, and disturbances 6 and 7 are constant for a quite long period of time. In addition, the values of disturbances 1–5 can be predicted with some accuracy on the basis of meteorological supervision data. In order to improve control accuracy and the level of technical operation of the DHS it is necessary to take into account influence of disturbances on the system. Since disturbances always change over time, the ACS must always calculate the regime of subsoil moisture and implement it on the controlled DHS modules using the technical means of water control.

The control object Obj2.2 is an unsaturated zone of soil and transforms the GWL of the modular area into the pressure head of the root-content layer of soil, the object is spatially distributed. There are some variable disturbances influence on the object except of the GWL that is the output value of the control object Obj2.1. Under these conditions, it is quite difficult to take out the object transfer functions for each of the channels. Therefore, it is proposed to develop a model of the object Obj2.2 in the form of a neural network (NN) formed on the basis of experimental data. For this purpose, it was formed a training sample of experimental data for 2 vegetation periods of two adjacent years, which includes 346 points for modelling of moisture supply of agricultural plants on the controlled module of the reclamation system. Data of one vegetation season will be used for training (173 units), and the data of other vegetation season will be used for testing (173 units). The modelling system was constructed for loamy and sandy soil mechanic composition of the Rivne experimental agricultural station. The output parameter of NN is the pressure head, which is related to the soil moisture with the help of the basic hydrophysical characteristics. Since the soil moisture in the next control period depends on weather conditions, the GWL and current moisture content in soil, NN for moisture prediction is presented as:

$$W_{k+1}^{h} = NN\left(P_{k+1}, D_{k+1}, L_{k+1}, W_{k-1}^{h}, W_{k}^{h}\right), \tag{1}$$

where the input parameters are amount of precipitation P (mm), the air humidity deficit D (mbar), groundwater level L from the light surface (m). The output parameter is the pressure head W^h (m) in the specified layer of soil h. NN is a transformation carried out by a neural network; k is the current step.

The conducted study of influence of the depth of the prehistory, the method of training, the type of activation function and the number of hidden layer neurons, the number of hidden layers on accuracy of the NN (1). The multilayered neural network of such architecture showed the best results: 3 neurons of the first layer with sigmoid activation functions, 1 neuron of the second layer with a linear activation function, learning method - Bayesian, the function of adjustment of weights and displacements - gradient with an inertial component, the error function is quadratic. The hybrid neo-fuzzy networks have been also used to model a non-saturated area of soil. The neo-fuzzy networks were initially introduced in [3] and there has been growing interest in their applications [4–6]. These networks combine fuzzy logic and neural networks. Instead of the usual synaptic weights the neo-fuzzy neuron contains nonlinear synapses NS_i, $i = 1, 2 \ldots n$, formed by a set of triangular symmetric functions uniformly distributed on the interval [0, 1], the membership function μ_{ji}, $j = 1, 2 \ldots m$, and own adjusted weight w_{ji} is connected with each of them. The output reaction of the neo-fuzzy neuron to the input data vector $x(k) = (x_i(k), x_2(k) \ldots x_n(k))^T,\ k = 1, 2 \ldots N$, is presented in the form:

$$y(k) = \sum_{i=1}^{n} f_i(x_i(k)) = \sum_{i=1}^{n} \sum_{j=1}^{m_i} \mu_{ji}(x_i(k)) \cdot w_{ji}(k), \tag{2}$$

where $w_{ji}(k)$ is a current value of synoptic weight, which is adjusted at the moment of time k at j-function of belonging of the i component of the input signal. The standard quadratic error is used as the criterion of the neo-fuzzy neuron training:

$$E(k) = \frac{1}{2}(d(k) - y(k))^2 = \frac{1}{2}e^2(k) = \frac{1}{2}(d(k) - \sum_{i=1}^{n} \sum_{j=1}^{m_i} \mu_{ji}(x_i(k)) \cdot w_{ji})^2, \tag{3}$$

minimization which with the help of the gradient procedure leads to the algorithm of training:

$$w_{ji}(k+1) = w_{ji}(k) + \eta \cdot e(k) \cdot \mu_{ji}(x_i(k)), \tag{4}$$

where $d(k)$ is the external training signal, η is the parameter of the search step that is selected from empirical consideration and determines the speed of convergence of the training process. Each network variable (1) is divided into 6 equal intervals with triangular membership functions [6], which satisfy Ruspini fragmentation (5). The value of the membership function of a variable x_i we can determine according to dependence (6). The results of work of the neural models of architecture 5-3-1, Logsig-Purelin and the neo-fuzzy networks for forecasting of pressure head in different layers of soil are given in the Table 1.

$$\sum_{j=1}^{m_i} \mu_{ji}(x_i(k)) = 1, i = 1, 2, \ldots, n, \tag{5}$$

$$\mu_{ji} = \begin{cases} \dfrac{x_i - c_{j-1,i}}{c_{ji} - c_{j-1,i}}, x \in [c_{j-1,}c_{ji}], \\ \dfrac{c_{j+1,i} - x_i}{c_{j+1,i} - c_{ji}}, x \in [c_{ji}, c_{j+1,i}], \\ 0 - \text{ in the other cases.} \end{cases} \tag{6}$$

Table 1. The results of neural models of architecture 5-3-1, Logsig-Purelin and neo-fuzzy networks for pressure head forecasting in different soil layers

Soil's layer	Mean square error, mH2O			
	Network Logsig-Purelin		Neo-fuzzy network	
	Training	Testing	Training	Testing
h = 0–10 cm	1.438	1.207	0.827	1.036
h = 10–20 cm	2.057	1.647	0.625	0.696
h = 30–40 cm	1.210	0.852	0.309	0.282

The maximum mean square error (MSE) of the neo-fuzzy networks is 1.04 mH2O. The results of the neural network's testing for pressure head forecasting in the soil's layer h = 30–40 cm are shown in Fig. 3. Increase of pressure head ±1.04 mH2O, in accordance with the basic hydrophysical characteristics, corresponds to an increase of volumetric soil moisture ±(2–4)% within the working range. It is enough to ensure the accuracy ±5% of volumetric moisture during pressure head's (moisture's) control. The verification of the adequacy of the developed neuromodels was carried out using Fisher's F-criterion.

The developed neo-fuzzy models for pressure head forecasting provide higher precision than multilayer direct propagation networks, have a simpler architecture that facilitates practical implementation and increases the training speed.

Thus, predictive mathematical models of the unsaturated zone of the modular soil area as a control object based on static multilayer artificial neural networks of direct propagation and neo-fuzzy neural networks have been developed to improve the accuracy of pressure head (moisture) forecasting. The developed neo-fuzzy models are used in the operator's automated work station (AWS) of the drainage-humidifying system and serve as a convenient tool for planning and controlling the humidification regimes of agricultural crops.

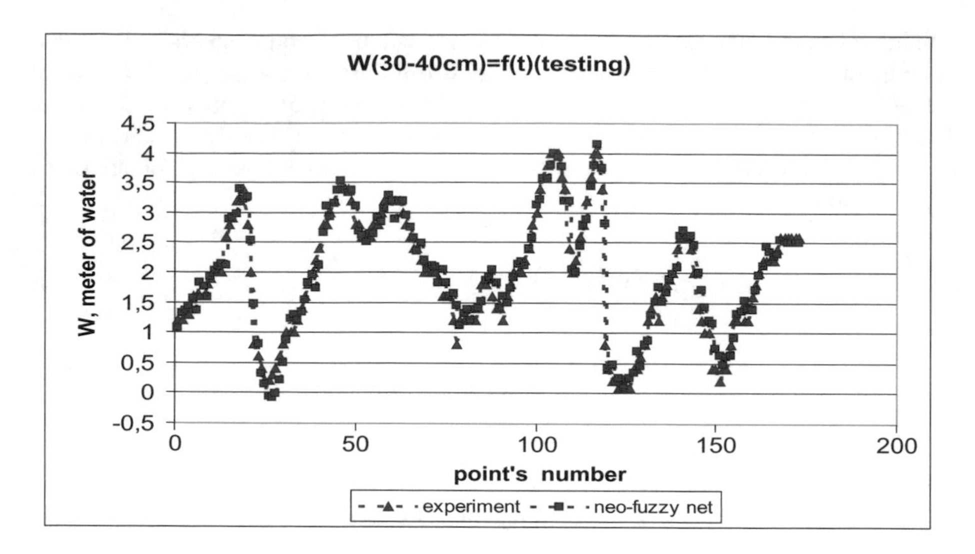

Fig. 3. Results of neo-fuzzy neural network's testing for pressure head forecasting in the soil's layer $h = 30$–40 cm

3 Control Methods of the Moisture Content of Agricultural Crops

The task of the external, setting, controller of water supply ACS (Fig. 2) is the definition of GWL, which must be kept on the modular section of the DHS during the calculation period to provide a given value of the pressure head, taking into account the variable meteorological parameters, therefore it is proposed to present it as a neural network type:

$$L_{k+1}^{h} = NN\left(P_{k+1}, D_{k+1}, L_{k-1}, L_{k}, W_{k+1}^{h}\right), \tag{7}$$

where the input parameters are the amount of precipitation P (mm), the air humidity deficit D (mbar), pressure head W^h (m) in a certain layer of soil h. The output parameter is the groundwater level from the light surface L (m). NN is the transformation, which is carried out by the neural network; k is the current step.

The study of the influence of the depth of prehistory, the teaching method, the type of activation function and the number of hidden layer neurons, the number of hidden layers on the accuracy of the neural controller (7) was carried out. The best results were given by the multilayer neural network of such architecture: 3 neurons in the first layer with sigmoid activation functions, 1 neuron of the second layer with a linear activation function, a teaching method that implements a variant of the error-reverse propagation algorithm in conjunction with the Polak-Ribiere optimization method, adjustment function of weights and displacements is a gradient with an inertial component, the error function is quadratic.

In the process of growth, the plants undergo several phases of development, during which the both ground and underground (root) parts are developing. In view of this, several controllers of the form (7) designed to control the soil moisture in a certain layer

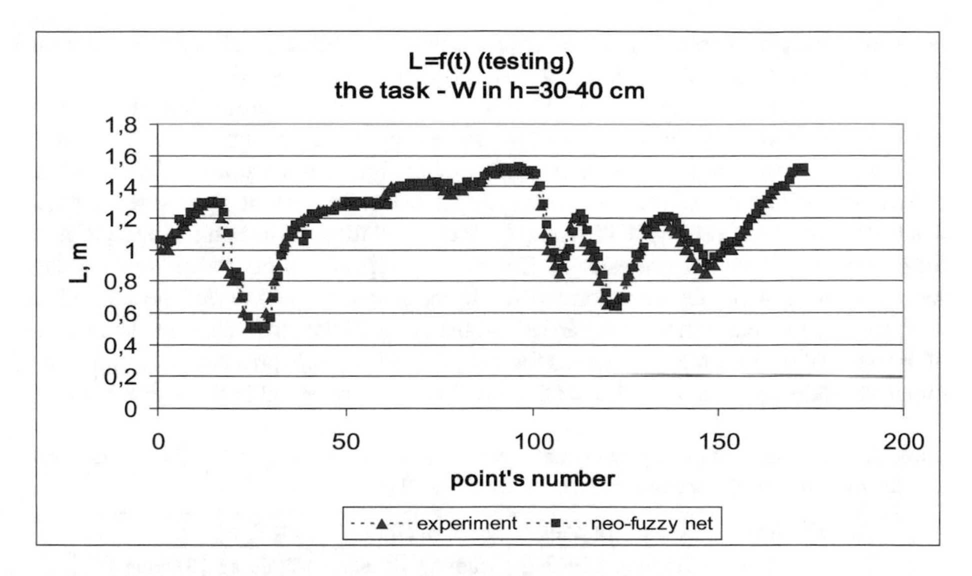

Fig. 4. Results of neo-fuzzy controller's testing for ground-water level's definition for ensuring the task of pressure head in the soil's layer h = 30–40 cm

(0–10, 10–20, 30–40 cm) have been developed. Switching between controllers is carried out with the development of plants root system.

Neurocontrollers of the form (7) for the determination of GWL on the basis of neo-fuzzy networks were also developed. The results of both types of controllers are given in Table 2. The results of testing the neo-fuzzy controller to determine the GWL to provide a given pressure head in the layer $h = 30\text{–}40$ cm are shown in Fig. 4. The calculation of adequacy of neurocontrollers is performed on the basis of Fisher's F-criterion at a level of significance $\alpha = 0.05$. According to the results of the work of the neo-fuzzy network, the maximum MSE value is 0.066 m. GWL increase or decrease at 0.07 m causes decrease or increase of the pressure head at (0.06–0.2) m water accordingly. Increase of the soil's pressure head ±0.2 m water station according to the basic hydrophysical characteristics corresponds to increase of volumetric soil moisture $\pm(0.5 - 1)\%$ within the working range.

Table 2. The results of the work of neurocontrollers of architecture 5-3-1, Logsig-Purelin and neo-fuzzy controllers for determination of the groundwater level

Setting soil's layer	Mean square error, m			
	Network Logsig-Purelin		Neo-fuzzy network	
	Training	Testing	Training	Testing
h = 0–10 cm	0.127	0.097	0.048	0.055
h = 10–20 cm	0.136	0.113	0.051	0.059
h = 30–40 cm	0.133	0.139	0.055	0.066

It is enough to ensure accuracy ±5% of the volumetric humidity during pressure head's (moisture's) control. The simulation of the ACS work with water content in three layers of soil and research of its dynamic properties was carried out. In this case, settling time is 8.33–12.5 days, delay time is 2.1–4.1 days, overshoot is 0%.

The possibility of adaptation of the developed models for prediction of the soil's pressure head and methods of control of moisture supply of agricultural crops to various DHS with subsoil moistening is shown. Implementation of the results of the thesis was carried out in Obukhiv DHS. The models for prediction of the soil's pressure head and determination of the setting GWL value for the calculating period on the basis of neo-fuzzy networks have been developed. The database from 428 points and the test data sample from 214 points were formed for training neural networks. The control of adequacy of the neural models is carried out using the Fisher's F-criterion. The results of the work of the neo-fuzzy networks for forecasting the soil's pressure head and determination of the necessary GWL for Obukhiv DHS are presented in the Table 3.

Table 3. The results of the neo-fuzzy networks' work for forecasting the soil's pressure head and determination of the necessary GWL for Obukhiv DHS

Soil's layer	Number of points		MSE, m H2O		MSE, m	
	Training	Testing	Training	Testing	Training	Testing
h = 0–10 cm	428	214	0.057	0.031	0.024	0.006
h = 10–20 cm			0.032	0.016	0.015	0.004
h = 30–40 cm			0.011	0.008	0.010	0.006

Thus, the developed intelligent methods for controlling the moisture content of agricultural crops with subsoil moistening, taking into account disturbances can improve accuracy of control, efficiency of the decision-making process, save water and energy resources while ensuring provision of planned harvest.

4 Software and Hardware of the Automated Control System

The hardware of the ACS of water supply is developed on the following structural units: (1) hydraulic controller equipped with a microprocessor control unit based on two Microchip microcontrollers (PIC18F4620 and PIC16F690) for measuring the soil's pressure head, the water level in the control well, data exchange with the measuring station (MS) by means of radio communication on the basis of the Radiocrafts module according to the RC-232C protocol; supply of control signals on to the actuator mechanism for change the water level in the control well and, accordingly, the GWL; (2) the MS, equipped with a microprocessor unit based on two PIC16F690 microcontrollers for measurement of meteorological parameters, data exchange with the automated work station (AWS) of the controller using the GSM modem SIM300C and microprocessor units of hydraulic controllers using a radio-modem in accordance with the RC-102 protocol; (3) the AWS of DHS's operator on the basis of the PC, which performs the functions of calculating the GWL task, data exchange with the MS through GSM communication, maintaining the common database technological parameters, the process visualization and communication with a controller, network communication with Internet; (4) portable operator control panel; (5) a block of discrete power outputs for controlling the electromagnetic valves of the hydraulic controller on the basis of the relay elements HLS-14F2 and transistor switches; (6) a voltage control unit based on the PIC16F690 microcontroller, a seven-segment indicator for input of initial settings, the LED indicators for displaying the current state of work, the solar accumulation panel with a serial transducer and an output voltage stabilizer, a high-capacity nickel-cadmium accumulator; (7) the sensors of technological parameters: water level in the control well, soil's pressure head, air humidity, air temperature, rainfall. The structure of the ASC moisture content of agricultural crops within the modular field with subsoil moistening is shown in Fig. 5, where Cws is a water supply controller; Cdf is a drain flow controller; MS is a measuring station; AWS is an automated work station of the operator; ME1-1 is a soil's pressure head sensor, LE-2-1 – is a water level sensor in the control well, TE-3-1 is an air temperature sensor, ME-4-1 is a sensor of relative air humidity, LE-5-1 is a rain sensor.

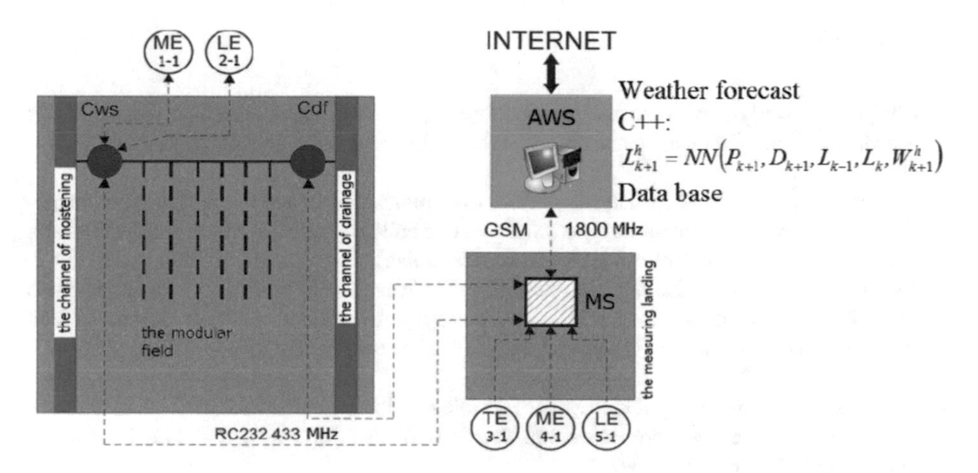

Fig. 5. Structure of the ACS of agricultural cultures water well being within the modular field with underground moistening of Rivne DHS

The software for the lower and higher levels of ACS was developed. The ACS can work in automatic, manual on-site and remote modes, and the software for all nodes of the control system was developed. The software of the AWS includes two DHS: Rivne and Obukhiv. The symbolic circuit of the corresponding station, the weather data, the measured GWL values and the soil's pressure head in three layers are displayed in this case. At the same time, according to the chosen mode of operation and the soil's layer, one can calculate the task of GWL value with the help of a neural network. In this case, the weight coefficients in accordance with the soil's layer and DHS take the place in to the neural network.

If the manual remote mode is selected, then the task of GWL value is entered by the operator. The recalculation of the task value of pressure head is carried out according to the basic hydrophysical characteristics and is displayed on the screen. The diagrams of change of current and setting GWL values and measured pressure head in three layers in real time mode; the value of temperature change and air relative humidity are also available. One can choose the numbers of the GWL and the pressure head sensors. The tab Events displays information about weather updates.

5 Summary

In this work an actual scientific and practical task in the field of agriculture production automation was solved. Its goal is development of intelligent models and control methods of water availability of agricultural plants on the DHS with underground moistening, software and hardware of ACS. The developed models, control methods and software are multipurpose for different DHS with underground moistening. The software can be easily adapted to another DHS, it is enough to retrain neural networks on the new data.

References

1. Chaly, B.: Design of drainage-moisturizing systems of block-modular type. Water Management of Ukraine: Scientific and Production Magazine (6), 55, Kyiv (2009). (in Ukrainian)
2. Kovalenko, P., Jacyk, M., Chaly, B., et al.: Recommendations for the design of drainage-moisturizing systems of modular type. Guide to state building codes. Reclamation systems and structures, 34, UkrNDIGIM, Kiev (2010). (in Ukrainian)
3. Yamakawa, T., Uchino, E., Miki, T., Kusanagi, H.: A neo-fuzzy neuron and its applications to system identification and prediction of the system behavior. In: Proceedings of 2nd International Conference on Fuzzy Logic and Neural Networks "IIZUKA – 92", pp. 477–483. Iizuka, Japan (1992)
4. Bodyanskiy, Y., Pliss, I., Vynokurova, O.: Flexible neo-fuzzy neuron and neuro-fuzzy network for monitoring time series properties. Inf. Technol. Manag. Sci. **16**(1), 47–52 (2014). https://doi.org/10.2478/itms-2013-0007
5. Bodyanskiy, Ye., Zaychenko, Yu., Pavlikovskaya, E., Samarina, M., Viktorov, Ye.: The neo-fuzzy neural network structure optimization using the GMDH for the solving forecasting and classification problems. In: Proceedings of 3rd International Workshop on Inductive Modelling, pp. 77–89 (2009)
6. Zurita, D., Delgado, M., Carino, J.A., Ortega, J.A., Clerc, G.: Industrial Time Series Modelling by Means of the Neo-Fuzzy Neuron. IEEE Access **4**, 6151–6160 (2016). https://doi.org/10.1109/ACCESS.2016.2611649

Evolution of Technical Systems Maintenance Approaches – Review and a Case Study

Tomasz Nowakowski, Agnieszka Tubis, and Sylwia Werbińska-Wojciechowska(✉)

Wroclaw University of Science and Technology, Wroclaw, Poland
{tomasz.nowakowski, sylwia.werbinska}@pwr.edu.pl

Abstract. The importance of technical objects maintenance issues takes on particular significance in the area of increasing competition an increasingly higher demands in the area of quality, reliability and productivity of performed system's functions and tasks. The main objectives of maintenance have evolved for the last fifty years. Thus, the article is aimed at the investigation of maintenance approaches evolution. The authors focus on the presentation of basic literature review covering the main maintenance approaches, from Maintenance 1.0 to Maintenance 4.0. First, the authors provide the reader with the main definitions connected with this research area and present few classifications of maintenance strategies with their historical background. The presented state of art was based on a review of available literature sources in the form of non-serial publications, scientific journals publications and conference proceedings. As a result, an overview of the literature includes the issues published in different times of the last forty years, and investigates the most well-known maintenance problems. Later, a simple case study on transportation company's maintenance management issues is provided.

Keywords: Maintenance · Technical system · Review

1 Introduction

One of the most important issue when ensuring high availability and reliability during assets life time is their maintenance performance [27]. Recently, maintenance is in a huge area of interest and research for engineers [19], because poorly maintained equipment may lead to more frequent equipment failures, poor utilization of equipment and delayed operational schedules. Wrongly selected or scheduled maintenance strategy of any equipment may result e.g. in scrap or products of questionable quality manufacturing. Following this, more and more companies are undertaking efforts to improve the effectiveness of maintenance functions [77].

Recently, a lot of researchers and publications in the field of maintenance decision models and techniques have been published to improve the effectiveness of maintenance process (see e.g. [57] for review). The known solutions have evolved in time, from Maintenance 1.0 level to at least Maintenance 4.0 level (that uses advances analyses and big data). On the other hand, organizations are aimed at improving their maintenance maturity. According to the authors of the report [65], which was aimed at

© Springer Nature Switzerland AG 2019
A. Burduk et al. (Eds.): ISPEM 2018, AISC 835, pp. 161–174, 2019.
https://doi.org/10.1007/978-3-319-97490-3_16

survey research on maintenance strategies implementation in companies in Belgium, Germany and the Netherlands, only 11% of the respondents (total of 280 respondents) have already achieved level 4.0. Following this, it is of utmost importance to investigate the possibilities and limitations of different maintenance approaches implementation in practice.

The article is aimed at the investigation of maintenance approaches evolution. The authors provide the reader with the main definitions connected with this research area and present few classifications of maintenance strategies with historical background.

Following this, the structure of the article is as follows: In the Sect. 2, the main definitions and objectives of maintenance are presented. Later, the comprehensive literature review on maintenance approaches is provided. The authors focus on the presentation of basic literature review covering the main maintenance approaches, from Maintenance 1.0 to Maintenance 4.0. This gives a possibility to present a case study of passenger transportation company and its maintenance management issues. The article ends up with a summary and conclusions for further research.

2 The Main Definitions and Scope of Maintenance

Maintenance theory has still been developing since 1960s of the XX century. Thus, in the literature there can be found many definitions of terms of maintenance, maintenance strategy, or maintenance policy. According to the European Standard PN-EN 13306:2010 [61], maintenance *is a combination of all technical, administrative and managerial actions during the lifecycle of an item intended to retain it, or restore it to a state, in which it can perform the required function.* The similar definition may be presented based on [25, 33] and is compliant with the PN-IEC 60300-3-10 standard [63], where maintenance is defined as *a combination of activities to retain a component in, or restore it to, a state (specified condition) in which it can perform its designated function.* These activities generally involve repairs and replacement of equipment items of a system and the maintenance decision is based on the system condition or on a definite time interval [25].

Based on these definitions, the main objective of maintenance, which is linked to the overall organizational objectives, should be to maximize the profitability of the organization by performing activities which retain working equipment in an acceptable condition, or return the equipment to an acceptable working condition [73]. Thus, following [mono] the principal objectives of maintenance are connected with (Fig. 1):

- ensuring system basic functions (availability, efficiency and reliability),
- ensuring system life through proper connections between its components (asset management),
- ensuring safety for human operators, environment and system itself,
- ensuring cost effectiveness in maintenance, and
- enabling effective use of resources, energy and raw materials.

The acquisition of these goals is possible taking into account opportunities and constraints that are connected with the main maintenance research areas, like

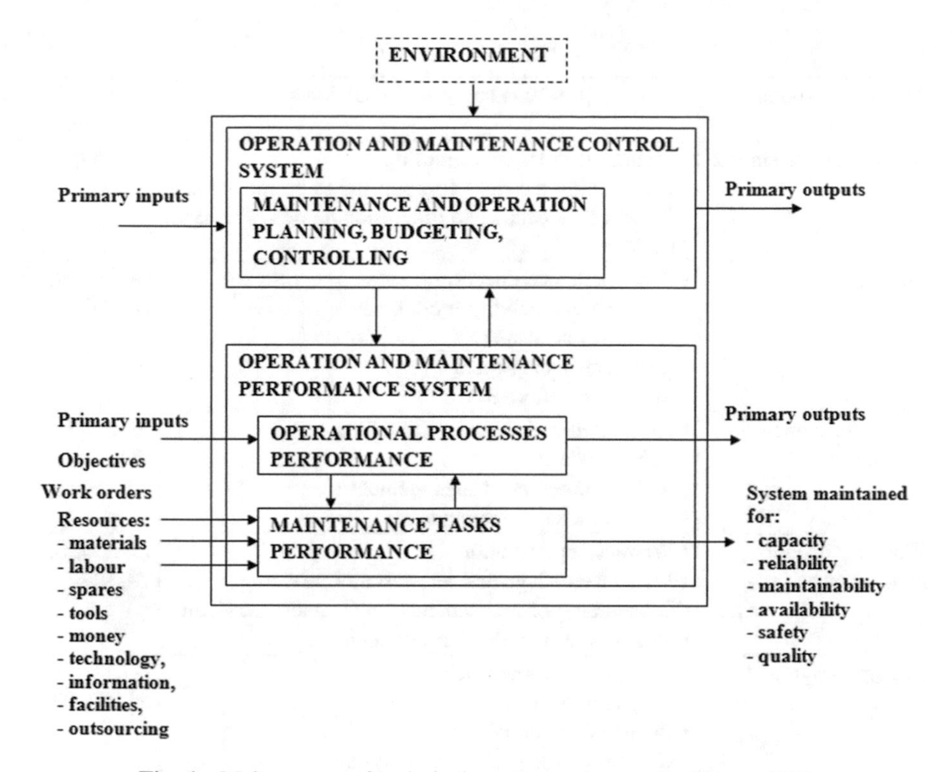

Fig. 1. Maintenance of technical systems – the scope, Source: [88]

maintenance strategy selection, maintenance planning, spare parts provisioning, or risk management. The short summary is given in Table 1.

The authors focus on the first subdomain connected with maintenance strategy selection. Thus, maintenance strategy is *a systematic approach to upkeep the technical objects* [29]. The maintenance strategy involves identification, researching and execution of many repair, replacements, and inspect decisions and may vary from facility to facility [29] (Fig. 2).

Table 1. The short summary of maintenance studies, Source: Based on [88]

The main maintenance subdomains	The main problems analysed in subdomains	Basic references
Maintenance strategy selection	• Selection of the maintenance policy for an element/system (CM, PM, PdM, CBM, RCM, …) • Maintenance optimization modelling • Maintenance integration	[2, 4, 69, 79]
Failure prediction/degradation modelling	• Aging management • RUL estimation • Uncertainty analysis • Accident analysis	[10, 31, 68, 75, 86]

(*continued*)

Table 1. (*continued*)

The main maintenance subdomains	The main problems analysed in subdomains	Basic references
Maintenance planning	• Maintenance tasks scheduling • Determining the right components to be maintained • Resource allocation and dimensioning of maintenance resources	[13, 60]
Spare parts provisioning	• Spare parts classification • Spare parts reliability modelling • Demand forecasting • Inventory management • Spare parts allocation	[9, 11, 39, 80, 84]
Risk management in maintenance	• Risk-based maintenance modelling • Safety indicators • Risk informed asset management • Human factor in maintenance	[6–8, 20, 40, 53]
Warranty and maintenance	• Warranty optimization, • Maintenance logistics for warranty servicing • Outsourcing of maintenance for warranty servicing • Warranty data collection and analysis	[54, 55, 66, 72, 91]
System design	• Design for maintenance • LCC approach • Redundancy modelling • Components dependence analysis • Dynamic reliability • Human factor in the design phase • Impact on health and environment • Logistic support planning	[21, 22, 44, 52, 89, 92]
Maintenance performance measurement	• Benchmarking analysis • Performance indicators assessment • Best practices identification • Customer satisfaction surveys • Maintenance process diagnosis and audits • Quality in maintenance • Maintenance reengineering	[42, 59, 67, 76]

Selecting the best maintenance strategy always depends on several factors such as the goals of maintenance, the nature of the technical object to be maintained, operational process patterns, and the work environment [56]. Following this, in the next Subsection the authors focus on the main maintenance strategies and investigates the four maturity levels of maintenance evolution.

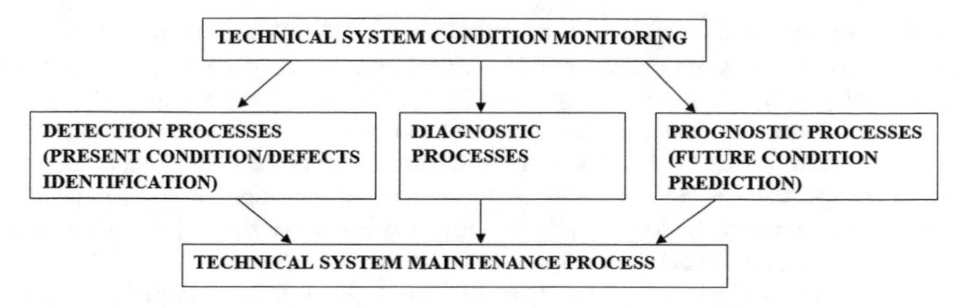

Fig. 2. The main problems in maintenance of technical systems, Source: [88]

3 Maintenance Approaches Evolution

The evolution of maintenance approaches in the last fifty years may be presented by a simple graph, given in the Fig. 3.

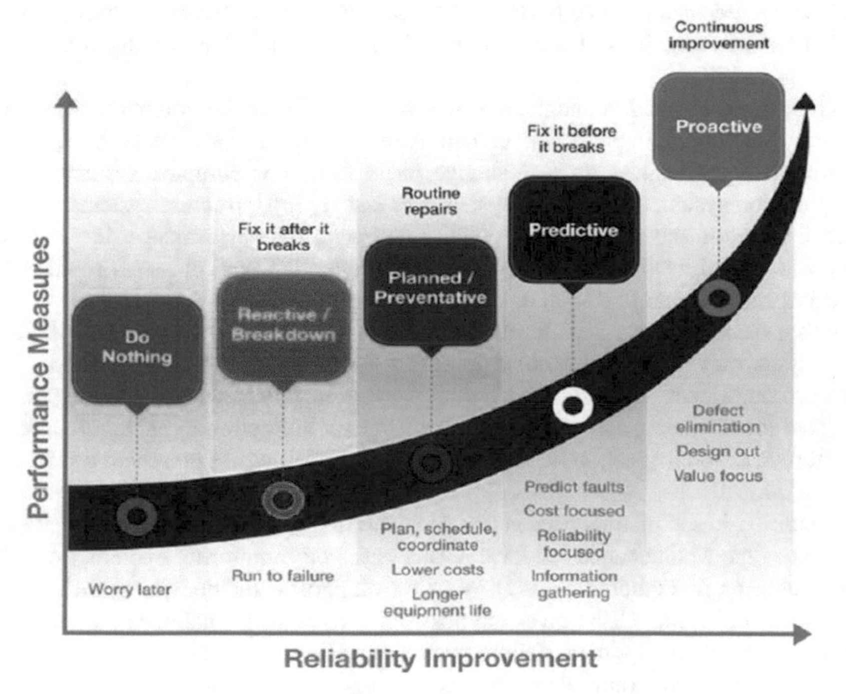

Fig. 3. The main maintenance approaches, Source: [87]

The first approach to maintenance (the Maintenance 1.0), often called as "run to failure" or corrective maintenance (CM) strategy, was very popular in the time period 1940–1960 [1]. CM is reactive and regards to any maintenance action that occurs when a system has been already failed, so there is no possibility to optimize its performance with respect to a given economic or reliability criteria [57]. While, a failure is defined *as an event, or inoperable state, in which any item or part of an item does not, or would not, perform as previously specified* [47]. This type of maintenance cannot be planned and has the associated consequences connected with system unavailability being the result of the failure. Therefore, using this type of technical system maintenance policy, there is no possibility to make any optimization of operational and maintenance parameters (see e.g. [30, 35, 36]). On the other hand, this maintenance strategy is still popular due to the low cost of its implementation.

In the situation, when it is necessary to avoid system failures during operation, especially when such an event is costly or/and dangerous, it is important to perform planned maintenance actions. Thus, we may choose the Maintenance 2.0 level connected with preventive maintenance (PM) performance. PM, according to MIL-STD-721C [47], means *all actions performed in an attempt to retain an item in a specified condition by providing systematic inspection, detection and prevention of incipient failures*. Basically, this approach tries to forecast or predict the wear and tear of life of equipment by using different approaches and recommends a corrective action. In this area the most commonly referred strategies in the literature are time-based PM and condition-based maintenance (CBM) [29]. Moreover, the difference between CM and PM is illustrated e.g. in [45] and the comparison of the main maintenance strategies is given e.g. in [48].

Time-based inspection and maintenance are still ones of the dominant maintenance policies used in an industry for certain types of assets that cannot be condition-monitored or maintained on a predictive basis [85]. For complex systems such as transportation systems, production systems, or critical infrastructure systems, the time-based inspection and maintenance policies can improve performance, increase reliability and capability of assets concerned, and reduce the cost of assets running [85]. More information can be found e.g. in [2, 18].

At the second maintenance level usually is classified inspection maintenance. Many components may become defective prior to failure and still remain operable. These types of components may benefit from an inspection policy whereby a component is inspected for the defect and consequently replaced at inspection to prevent failure [17]. Recent reviews on inspection maintenance modelling issues are presented e.g. in works [15, 16, 34, 78].

Condition-based maintenance is treated as the first maintenance strategy that can be included to the Maintenance 3.0 level. CBM bases on monitoring operating condition of a system or its components [41] by using diagnostic methods/measures [12, 28]. When it is applicable, CBM gives the possibility to perform maintenance actions just before the system/components failure occurrence. Hence, unlike CM and PM, CBM focuses not only on fault detection and diagnostics of components but also on degradation monitoring and failure prediction. Thus, CBM can be treated as the method used to reduce the uncertainty of maintenance activities [74]. The literature review on CBM policy is presented e.g. in works [2, 3, 18, 32, 64, 74]. A framework for condition monitoring and classification of decisions about appropriate maintenance actions performance based on two decision criteria (average downtime per failure and frequency of failure) are presented e.g. in [70].

Another maintenance policy, which usually is treated as a synonymous to CBM or is named as risk-based maintenance, is predictive maintenance (PdM) [71, 82]. This maintenance policy is used in these sectors where reliability is paramount, like nuclear power plants, transportation systems or emergency systems. Its main scope is to foresee faults or failures in a deteriorating system in order to optimize maintenance efforts by monitoring of equipment operating conditions to detect any signs of wear that are leading to a failure of a component [71]. The goal of the PdM program is to track component wear with a methodology that insures that any impending failure is detected [51]. The most commonly used monitoring and diagnostic techniques include, among others, vibration monitoring, thermography, tribology, or visual inspection [51]. The advantages of predictive or online maintenance techniques in identifying the onset of equipment failure are discussed e.g. in [37]. For more information, the author recommends reading e.g. [24, 51]. The advantages of this maintenance policy implementation are presented e.g. in [14].

The last maintenance approach is Maintenance 4.0. One of the first maintenance strategy being classified at this level is RCM (Reliability Centered Maintenance). According to the MIL-STD-3034 [46], RCM is a *method for determining maintenance requirements based on the analysis of the likely functional failures of systems/ equipment having a significant impact on safety, operations, and lifecycle cost*. RCM supports the failure-management strategy for any system based on its inherent reliability and operating context. RCM uses different tools (e.g. FMECA) to determine the relationships between the system elements and the level of its operation and then develops the effective maintenance management strategy (RCM Task Selection) [17, 49]. A comprehensive overview of this concept is presented e.g. in [deAlme'15, 24, 38, 43]. The main principles are, however, given in the PN-EN 60300-3-11 standard [62]. Moreover, the authors in their work [23] discuss the optimal maintenance policies for manufacturing companies introducing two other approaches aimed at improvement, namely autonomous maintenance and design out maintenance. The authors investigate the maintenance models and classify them based on the certainty theory.

Recently, based on the relevant research studies, the next level in predictive maintenance is connected with Proactive maintenance, often called as Predictive Maintenance 4.0 and incorporates the principals of continuous improvements method. One of the used concept here is Internet of Things that takes machine-to-machine technology to the next level by including a third element: data. According to the [90], all the machine data are to be available in one virtual network, which gives the producers the ability to aggregate and analyze the data to generate better predictive analytic models. For more information the author recommends reading e.g. [65, 81].

The main classification of maintenance strategies is given in the standard PN-EN 13306 [61]. The overview of maintenance approaches may be found e.g. in [26, 50, 58, 83] and analysis of maintenance philosophies development is given in [5]. General classification of maintenance strategies may be found in [88]. Following this, the next step is to investigate how these maintenance approaches may be implemented to the real-life technical systems performance.

4 Case Study

Many Polish enterprises still apply basic maintenance strategies in their everyday performance activities. Among the arguments in favor of maintaining the current status quo are usually:

- no costs for spare parts warehouse management,
- lack of knowledge and good practices in the area of other maintenance strategies implementation,
- lack of competence to prepare quantitative analyzes that improve the decision-making process in higher-level maintenance strategies,
- no data available in the electronic version that could be the basis for the preparation of quantitative analyzes.

Many maintenance managers are not aware that a strategy that seems to them to be simple and cost-effective (costs of repair and new spare parts purchasing are incurred only when the failure occurs), in fact generates numerous additional and hypothetical costs. Lack of this awareness often results from the lack of physical registration of these cost elements in the accounting system and their actual attribution to a given hazard event. A good example here is the company, in which the authors conducted research in the area of vehicle fleet maintenance.

The surveyed passenger transport company is a leading carrier on the Polish public transport market. Currently 156 vehicles are used to carry out transport tasks. These buses from the point of view of the service process can be divided into the following groups according to the criterion of the service life (Table 2).

Table 2. The number of vehicles with the defined service life

Vehicle's service life	Number of vehicles
Less than 10 years	16
10–20 years	101
More than 20 years	39

The transport company for the implementation of basic tasks (regional transport provided within the framework of collective public transport and regular employee and school transport) uses its transport fleet in 95%. Any surplus transport capacity is sold in the course of occasional additional orders, e.g. in order to support mass events, trips, etc.

Such a high operating coefficient of transport base indicate that in the event of failure of any vehicle, the company seeks to minimize the time when the bus is out of service. This is particularly the case when the company does not have a spare vehicles that may substitute the failed one. The situation is significantly hampered by the fact that the current vehicle maintenance strategy on which the company is based is a reactive strategy. This means:

1. lack of spare parts necessary to remove the occurred failure - necessary elements are acquired at the moment of demand occurrence,

2. partial repairs carried out in the jump system - partial removal of the failure, in order to allow short-term operation, and re-repair at the time of receiving the necessary parts,
3. extended waiting period for materials being necessary to carry the repair, resulting in an extended time of shutting the vehicle out of use.

The analysis of maintenance data carried out by the employees of the repair workshop proved that the effect of the given maintenance strategy is:

1. longer holding times of buses in the repair shop - the maximum registered time is 15 weeks with the exclusion of use of the vehicle in relation to the spare part lead time,
2. multiple minimal repairs performance for keeping the buses in operation till the required parts delivery,
3. implementation of repairs at a time when further use of the vehicle is not-possible.

As part of the risk assessment related to the adopted vehicles' maintenance strategy, the following hazards and their consequences have been defined in the company (Table 3).

Table 3. The main hazards and their consequences identified in the transportation company

Hazards	Consequences
• Long-term exclusion from the use of the vehicle • Multiple repairs of the same item in the same vehicle • Multiple vehicle's returns to the repair shop for minimal repairs in the short term • Vehicle failure during transport services performance	• Partial repair costs, enabling the operation of the vehicle to the required delivery time of spare parts • Costs several returns to the workshop in order to make further major repairs • Losses resulting from canceled courses operated by a given vehicle • Loss of potential additional orders • Loss of passengers' confidence due to cancellations • Costs of external repairs in situations when the vehicle is failed during the course performance • Towing costs of the vehicle • Additional costs of emergency purchases carried out at the time of failure occurrence (selection criterion is the time of delivery and not the price) • The difference in the price of the spare part resulting from the maintenance of the safety margin for the carrier by the supplier • Costs related to the substitution of failed vehicles

The company for many years records all information regarding performed repairs and replacements and used for this purpose maintenance materials. This registration was carried out so far in the written form, which significantly hindered its analysis.

However, for many years, the company uses IT system, which enables the collection and storage of the same data in an electronic form only. Following this, the analysis of data from 2016–2017 has proved that it is possible to determine the statistical repeatability of the selected failures in vehicles belonging to the same brand and age group. This allows the manager to estimate the possibility of selected failures occurrence. The consequences of their occurrence are also known. This is the basis for considering the possibility of changing the current vehicles' maintenance strategy to a risk based maintenance strategy. This would reduce the waste currently occurring in the enterprise related to the repair of vehicles and increase the security of services.

5 Summary

The results presented in this paper are a short summary of research conducted by the authors in the area of maintenance management performance. The effects of this study clearly demonstrate the need for detailed quantitative analyses performance in order to properly choose the maintenance strategy. Every organization gathers and analyses specific data whose acquisition is both time consuming and capital intensive. Thus, in order to ensure the effectiveness of forthcoming analysis and its actuality, it is necessary to define the main requirements for performed maintenance and limitations of the known maintenance approaches implementation in the analyzed company.

References

1. Abramczyk, A.: Market research: maintenance strategies in small and medium enterprises. Inżynieria i Utrzymanie Ruchu, vol. 2 (2017). (in Polish)
2. Ahmad, R., Kamaruddin, S.: An overview of time-based and condition-based maintenance in industrial application. Comput. Ind. Eng. **63**, 135–149 (2012)
3. Alaswad, S., Xiang, Y.: A review on condition-based maintenance optimization models for stochastically deteriorating system. Reliab. Eng. Syst. Saf. **157**, 54–63 (2017)
4. Alrabghi, A., Tiwari, A.: State of the art in simulation-based optimisation for maintenance systems. Comput. Ind. Eng. **82**, 167–182 (2015)
5. Arunaj, N.S., Maiti, J.: Risk-based maintenance – techniques and applications. J. Hazard. Mater. **142**, 653–661 (2007)
6. Aven, T.: Risk assessment and risk management: review of recent advances on their foundation. Eur. J. Oper. Res. **253**(1), 1–13 (2016)
7. Aven, T.: Risk analysis and management. Basic concepts and principles. Reliab. Risk Anal. Theor. Appl. **2**, 57–73 (2009)
8. Aven, T., Zio, E.: Foundational issues in risk assessment and risk management. Risk Anal. Int. J. **34**(7), 1164–1172 (2014)
9. Bacchetti, A., Saccani, N.: Spare parts classification and demand forecasting for stock control: investigating the gap between research and practice. Omega **40**, 722–737 (2012)
10. Bhargava, C., Banga, V.R., Singh, Y.: Failure prediction and health prognosis of electronic components: a review. In: Proceedings of 2014 RAECS, UIET Panjab University Chandigarh, 06–08 March 2014
11. Bijvank, M., Vis, I.F.A.: Lost-sales inventory theory: a review. Eur. J. Oper. Res. **215**, 1–13 (2011)

12. Blischke, W.R., Prabhakar Murthy, D.N.: Reliability: Modelling, Prediction and Optimization. Willey, New York (2000)
13. Budai, G., Dekker, R., Nicolai, R.P.: A review of planning models for maintenance and production. Econometric Institute, Erasmus School of Economics, Erasmus University (2006)
14. Bukowski, L., Jaźwiński, J., Majewska, K.: Maintenance of technical systems oriented to reliability – a concept of modified RCM. In: Proceedings of XXXV Winter School on Reliability – Problems of Systems Dependability. Publishing House of Institute for Sustainable Technologies, Radom (2007). (in Polish)
15. Chelbi, A., Ait-Kadi, D.: Inspection strategies for randomly failing systems. In: Ben-Daya, M., Duffuaa, S.O., Raouf, A., Knezevic, J., Ait-Kadi, D. (eds.) Handbook of Maintenance Management and Engineering. Springer, London (2009)
16. Chelbi, A., Ait-Kadi, D., Aloui, H.: Optimal inspection and preventive maintenance policy for systems with self-announcing and non-self-announcing failures. J. Qual. Maint. Eng. **14** (1), 34–45 (2008)
17. Cho, I.D., Parlar, M.: A survey of maintenance models for multi-unit systems. Eur. J. Oper. Res. **51**(1), 1–23 (1991)
18. Chowdhury, C.: A systematic survey of the maintenance models. Period. Polytech. Mech. Eng. **32**(3–4), 253–274 (1988)
19. Cunningham, A., Wang, W., Zio, E., Allanson, D., Wall, A., Wang, J.: Application of delay-time analysis via Monte Carlo simulation. J. Mar. Eng. Technol. **10**(3), 57–72 (2011)
20. De Almeida, A.T., Cavalcante, C.A.V., Alencar, M.H., Ferreira, R.J.P., De Almeida-Filho, A.T., Garcez, T.V.: Multicriteria and Multiobjective Models for Risk, Reliability and Maintenance Decision Analysis. Springer, Cham (2015)
21. Desai, A., Mital, A.: Design for maintenance: basic concepts and review of literature. Int. J. Prod. Dev. **3**(1), 77–121 (2006)
22. Dhillon, B.: Life Cycle Costing: Techniques, Models and Applications. Routledge (2013)
23. Ding, S.-H., Kamaruddin, S.: Maintenance policy optimization – literature review and directions. Int. J. Adv. Manuf. Technol. **76**(5–8), 1263–1283 (2015)
24. Duffuaa, S., Raouf, A.: Planning and control of Maintenance Systems. Modelling and Analysis. Springer, Cham (2015)
25. Emovon, I., Norman, R.A., Murphy, A.J.: An integration of multi-criteria decision making techniques with a delay time model for determination of inspection intervals for marine machinery systems. Appl. Ocean Res. **59**, 65–82 (2016)
26. Endrenyi, J., Aboresheid, S., Allan, R.N., Anders, G.J., Asgarpoor, S., Billinton, R., Chowdhury, N., Dialynas, E.N., Fipper, M., Fletcher, R.H., Grigg, C., Mccalley, J., Meliopoulos, S., Mielnik, T.C., Nitu, P., Rau, N., Reppen, N.D., Salvaderi, L., Schneider, A., Singh, Ch.: The present status of maintenance strategies and the impact of maintenance on reliability. IEEE Trans. Power Syst. **16**(4), 638–646 (2001)
27. Eruguz, A.S., Tan, T., Van Houtum, G.-J.: A survey of maintenance and service logistics management: classification and research agenda from a maritime sector perspective. Comput. Oper. Res. (2017). https://doi.org/10.1016/j.cor.2017.03.003
28. Feliks, J., Majewska, K.: Proactive maintenance as an aid for logistic of exploitation. In: Proceedings of Total Logistics Management, TLM 2006 Conference, Zakopane (2006). (in Polish)
29. Gandhare, B.S., Akarte, M.: Maintenance strategy selection. In: Ninth AIMS International Conference on Management, pp. 1330–1336, 1–4 January 2012
30. Goel, G.D., Murari, K.: Two-unit cold-standby redundant system subject to random checking, corrective maintenance and system replacement with repairable and non-repairable types of failures. Microelectron. Reliab. **30**(4), 661–665 (1990)

31. Gorjian, N., Ma, L., Mittynty, M., Yarlagadda, P., Sun, Y.: A review on degradation models in reliability analysis. In: Proceedings of the 4th World Congress on Engineering Asset Management, Athens, 28–30 September 2009
32. Guizzi, G., Gallo, M., Zoppoli, P.: Condition based maintenance: simulation and optimization. In: Proceedings of ICOSSSE 2009: Proceedings of the 8th WSEAS International Conference on System Science and Simulation in Engineering, 17–19 October 2009, Genoa, pp. 319–325 (2009)
33. Gulati, R., Kahn, J., Baldwin, R.: The professional's guide to maintenance and reliability terminology. Reliabilityweb.com (2010)
34. Gupta, P.P., Kumar, A.: Cost analysis of a three-state parallel redundant complex system. Microelectron. Reliab. **25**(6), 1021–1027 (1985)
35. Guo, H., Szidarovszky, F., Gerokostopoulos, A., Niu, P.: On determining optimal inspection interval for minimizing maintenance cost. In: Proceedings of 2015 Annual Reliability and Maintainability Symposium (RAMS), pp. 1–7. IEEE (2015)
36. Gupta, P.P., Sharma, R.K.: Cost analysis of a three-state repairable redundant complex system under various modes of failures. Microelectron. Reliab. **26**(1), 69–73 (1986)
37. Hashemian, H.M., Bean, W.C.: State-of-the-art predictive maintenance techniques. IEEE Trans. Instrum. Meas. **60**(10), 3480–3492 (2011)
38. Jones, R.B.: Risk-based Management: A Reliability Centered Approach. Gulf Publishing, New York (1995)
39. Kennedy, W.J., Wayne, P.J., Fredendall, L.D.: An overview of recent literature on spare parts inventories. Int. J. Prod. Econ. **76**, 201–215 (2002)
40. Khan, F., Rathnayaka, S., Ahmed, S.: Methods and models in process safety and risk management: past, present and future. Process Saf. Environ. Prot. **98**, 116–147 (2015)
41. Koochaki, J., Bokhorst, J.A.C., Wortmann, H., Klingenberg, W.: Condition based maintenance in the context of opportunistic maintenance. Int. J. Prod. Res. **50**(23), 6918–6929 (2012)
42. Kumar, U., Galar, D., Parida, A., Stenstrom, Ch., Berges, L.: Maintenance performance metrics: a state-of-the-art review. J. Qual. Maint. Eng. **19**(3), 233–277 (2013)
43. Legutko, S.: Development trends in machines operation maintenance. Eksploatacja i Niezawodnosc – Maintenance and Reliability **2**(42), 8–16 (2009)
44. Ling, D.: Railway renewal and maintenance cost estimating. Ph.D. thesis, Cranfield University (2005)
45. Lu, L., Jiang, J.: Analysis of on-line maintenance strategies for k-out-of-n standby safety systems. Reliab. Eng. Syst. Saf. **92**, 144–155 (2007)
46. Mil-Std-3034: Military Standard: Reliability-Centered Maintenance (RCM) process. Department of Defense, Washington D.C. (2011)
47. Mil-Std-721c: Military Standard: definitions of terms for reliability and maintainability. Department of Defense, Washington D.C. (1981)
48. Mishra, R.P., Anand, G., Kodali, R.: Strengths, weaknesses, opportunities, and threats analysis for frameworks of world-class maintenance. Proc. Inst. Mech. Eng. Part B J. Eng. Manuf. **221**(7), 1193–1208 (2007)
49. Młyńczak, M.: Preventive maintenance supported by dependability. In: Proceedings of PIRE 2005 Conference. Problemy i innowacje w remontach energetycznych. OBR Gospodarki Remontowej Energetyki, Wroclaw (2005). (in Polish)
50. Młyńczak, M., Nowakowski, T., Werbińska-Wojciechowska, S.: Technical systems maintenance models classification. In: Siergiejczyk, M. (ed.) Maintenance Problems of Technical Systems, pp. 59–77. Warsaw University of Technology Publishing House, Warsaw (2014). (in Polish)

51. Mobley, R.K.: Predictive maintenance. In: Mobley, R.K. (ed.) Maintenance Engineering Handbook. McGraw-Hill Professional, New York (2014)
52. Mobley, R.K.: Maintenance engineer's toolbox. In: Mobley, R.K. (ed.) Maintenance Engineering Handbook. McGraw-Hill Professional, New York (2014)
53. Mobley, R.K.: Simplified failure modes and effects analysis. In: Mobley, R.K. (ed.) Maintenance Engineering Handbook. McGraw-Hill Professional, New York (2014)
54. Murthy, D.N.P.: Product reliability and warranty: an overview and future research. Produção **17**(3), 426–434 (2007)
55. Murthy, D.N.P.: Product warranty and reliability. Ann. Oper. Res. **143**, 133–146 (2006)
56. Nowakowski, T.: Methodology for reliability prediction of mechanical objects. Research Work of the Institute of Machine Designing and Operation. Wroclaw University of Technology, Wroclaw (1999). (in Polish)
57. Nowakowski, T., Werbińska, S.: On problems of multi-component system maintenance modelling. Int. J. Autom. Comput. **6**(4), 364–378 (2009)
58. Pariazar, M., Shahrabi, J., Zaeri, M.S., Parhizi, S.H.: A combined approach for maintenance strategy selection. J. Appl. Sci. **8**(23), 4321–4329 (2008)
59. Parida, A., Kumar, U., Diego, D., Stenstrom, C.H.: Performance measurement and management for maintenance: a literature review. J. Qual. Maint. Eng. **21**(1), 2–33 (2015)
60. Phanden, R.K., Jain, A., Verma, R.: Review on integration of process planning and scheduling. In: DAAAM International Scientific Book, Chapter 49, pp. 593–618 (2011)
61. PN-EN 13306: 2010 Maintenance – Maintenance Terminology. European Committee for Standardization, Brussels (2010)
62. PN-EN 60300-3-11: 2010 Dependability Management, Application Guide - Reliability Centred Maintenance. PKN, Warsaw (2010)
63. PN-IEC 60300-3-10: 2006 Dependability Management, Application Guide – Maintainability. PKN, Warsaw (2010)
64. Prajapati, A., Bechtel, J., Ganesan, S.: Condition based maintenance: a survey. J. Qual. Maint. Eng. **18**(4), 384–400 (2012)
65. Predictive Maintenance 4.0: Predict the unpredictable. Report of PwC Belgium (2017). https://www.pwc.be/en/news-publications/publications/2017/predictive-maintenance-4-0.html. Accessed 21 Feb 2018
66. Rahman, A., Chattopadhay, G.: Review of long-term warranty policies. Asia Pac. J. Oper. Res. **23**(4), 453–472 (2006)
67. Samat, H.A., Kamaruddin, S., Azid, I.A.: Maintenance performance measurement: a review. Pertanika J. Sci. Technol. **19**(2), 199–211 (2011)
68. Sankararaman, S.: Significance, interpretation, and quantification of uncertainty in prognostics and remaining useful life prediction. Mech. Syst. Signal Process. **52–53**, 228–247 (2015)
69. Sarkar, A., Behera, D.K., Kumar, S.: Maintenance policies of single and multi-unit systems in the past and present. Int. J. Curr. Eng. Technol. **2**(1), 196–205 (2012)
70. Scarf, P.A.: A framework for condition monitoring and condition based maintenance. Qual. Technol. Quant. Manag. **4**(2), 301–312 (2007)
71. Selcuk, S.: Predictive maintenance, its implementation and latest trends. Proc. Inst. Mech. Eng. Part B J. Eng. Manuf. (2016). https://doi.org/10.1177/0954405415601640
72. Selviaridis, K., Wynstra, F.: Performance-based service contracting: a literature review and future research directions. Int. J. Prod. Res. **53**(12), 3505–3540 (2015)
73. Shenoy, D., Bhadury, B.: Maintenance Resource Management: Adapting Materials Requirements Planning MRP. CRC Press, London (2003)
74. Shin, H.-H., Jun, H.-B.: On condition based maintenance policy. J. Comput. Des. Eng. **2**, 119–127 (2015)

75. Si, X.-S., Wang, W., Hu, C.-H., Zhou, D.-H.: Remaining useful life estimation – a review on the statistical data driven approaches. Eur. J. Oper. Res. **213**, 1–14 (2011)
76. Simon, R.M.: Cannibalization policies for multicomponent systems. SIAM J. Appl. Math. **19**(4), 700–711 (1970)
77. Swanson, L.: Linking maintenance strategies to performance. Int. J. Prod. Econ. **70**, 237–244 (2001)
78. Tang, T.: Failure finding interval optimization for periodically inspected repairable systems. Ph.D. thesis, University of Toronto (2012)
79. Valdez-Flores, C., Feldman, R.: A survey of preventive maintenance models for stochastically deteriorating single-unit systems. Naval Res. Logist. **36**, 419–446 (1989)
80. Van Horenbeek, A., Bure, J., Cattrysse, D., Pintelon, L., Vansteenwegen, P.: Joint maintenance and inventory optimization systems: a review. Int. J. Prod. Econ. **143**, 499–508 (2013)
81. Varoneckas, A., Mackute-Varoneckiene, A., Rilavicius, T.: A review of predictive maintenance systems in Industry 4.0. Int. J. Des. Anal. Tools Integr. Circuits Syst. **6**(1), 68 (2017)
82. Veldman, J., Wortmann, H., Klingenberg, W.: Typology of condition based maintenance. J. Qual. Maint. Eng. **17**(2), 183–202 (2011)
83. Wang, H.: A survey of maintenance policies of deteriorating systems. Eur. J. Oper. Res. **139** (3), 469–489 (2002)
84. Wang, W.: A joint spare part and maintenance inspection optimisation model using the delay-time concept. Reliab. Eng. Syst. Saf. **96**, 1535–1541 (2011)
85. Wang, W., Carr, J., Chow, T.W.S.: A two-level inspection model with technological insertions. IEEE Trans. Reliab. **61**(2), 479–490 (2012)
86. Weber, P., Medina-Oliva, G., Simon, C., Iung, B.: Overview on Bayesian networks applications for dependability, risk analysis and maintenance areas. Eng. Appl. Artif. Intell. **25**, 671–682 (2012)
87. www2.deloitte.com/insights/us/en/focus/industry-4-0/using-predictive-technologies-for-asset-maintenance.html. Accessed 01 June 2018
88. Werbińska-Wojciechowska, S.: Technical system maintenance. Delay-time-based modelling (in rev., Springer)
89. Wilson, R.M.: Design for maintenance. Electric Traction Systems, Course on IET Professional Development, 3–7 November 2008, pp. 363–385 (2008). https://doi.org/10.1049/ic:20080524
90. www.reliableplant.com/Read/29962/internet-of-things. Accessed 01 June 2018
91. Wu, S.: Warranty data analysis: a review. Qual. Reliab. Eng. Int. **28**, 795–805 (2012)
92. Zaitseva, E.N., Levashenko, V.G.: Design of dynamic reliability indices. In: Proceedings 32nd IEEE International Symposium on Multiple-Valued Logic, ISMVL 2002, 15–18 May 2002. https://doi.org/10.1109/ismvl.2002.1011082

Forecasting the Mountability Level of a Robotized Assembly Station

Rafał Kluz(✉), Katarzyna Antosz, and Tomasz Trzepiecinski

Faculty of Mechanical Engineering and Aeronautics, Rzeszow University of Technology, Rzeszów, Poland
{rkktmiop,katarzyna.antosz,tomtrz}@prz.edu.pl

Abstract. This article presents the problem of determining the mountability level of the assembly station using an artificial neural network (ANN). The results of ANN modelling were compared with the results of experimental research and classical mathematical modelling. It was found that the error in predicting the mountability level using the artificial neural network is about two-fold lower than in the case of the error determined by classical mathematical modelling. Although the neural network ensures a lower prediction error, to obtain a good prediction it is necessary to conduct many experiments in the whole workspace of the robots to build a training set. Despite the worst prediction, a mathematical model of the mountability level only requires an analytical description of the kinematic structure of the assembly robot, so in industrial applications this is preferred due to the lower labour requirement.

Keywords: Assembly · Assembly station · Mountability level
Neural networks

1 Introduction

The basic condition for achieving high reliability of a robotized assembly station is to meet the mountability condition for all connecting parts. In fact, these conditions can only be met with a specified probability, therefore, by the mountability of the product, one should understand the probability of assembling its parts while maintaining the quality requirements [1].

Assembly operations using robotized assembly stations have increased in both number and complexity over the years because of the increasing requirements for product quality and quantity [2]. Intelligent industrial robotic systems are widely used in assembly applications because they can perform assembly tasks with high autonomy and adaptability to the environment. The concept of using robots for assembly operations has already been discussed in numerous research studies such as in [3–5]. The adoption of robots in assembly lines allows one to reduce space and improve cost efficiency by eliminating clamping devices and fixtures [6]. The use of a dual arm robot enables operations that are carried out by humans to be performed, and is characterised by number of advantages [4]:

© Springer Nature Switzerland AG 2019
A. Burduk et al. (Eds.): ISPEM 2018, AISC 835, pp. 175–184, 2019.
https://doi.org/10.1007/978-3-319-97490-3_17

- an increase in the automation level in manual assembly stations,
- a decrease in the cost of setting up a cell by selecting a dual arm robot,
- an increase in robot workspace,
- elimination of the synchronisation of single arm robots by adopting control functions for bi-manual actions.

The design process of automatic assembly devices and their equipment should not focus only on ensuring the correctness of the assembly operation, but must also take into account the economic aspect. An increase in the probability of correct completion of the assembly operation reduces the cost of downtime, which naturally leads to a reduction in the operating costs of the station. On the other hand, it increases the requirements for accuracy of the equipment assembly station, and its cost. Therefore, the usage of too precise and complicated means of assembly robotization can be highly uneconomic and put a question mark over the profitability of robotization of the specific assembly process. The solution to the problem presented may be the system characterised by the mountability level which is acceptable to the company [1, 7].

The experimental determination of the mountability level of the assembly station is a difficult and labour-intensive task, and consists in the decomposition of the total error generated on the assembly station into error components, i.e. kinematic, static and dynamic. Based on these components, the probability of joining parts is determined. Due to the complexity of the above-mentioned process, in many cases the analytical models are burdened with a significant error, which can also influence the accuracy of predicting the level of mountability [8, 9]. An alternative solution to the problem presented may be the use of artificial neural networks (ANN), which do not require the modelling of component errors, because they operate on the basis of the results of experimental investigations. ANNs are modern computational systems inspired by networks of biological neurons, wherein the neurons compute output values from input ones [10–12]. ANNs are tools to build and analyse linear and nonlinear models of complex regression and classification problems. These tools are used in particular when the dependence between inputs and outputs is very complicated.

Due to the possibility of including many factors in the modelling process, neural networks are able to build a model that gives the possibility of forecasting the mountability level of an assembly station. However, this requires gathering data about the process and the selection of the vector of the input variables and the configuration of the neural network.

This paper attempts to develop a numerical model using an artificial neural network to forecast the mountability level of the assembly station. The results of ANN were compared with the results of experimental investigations and classical mathematical modelling.

2 Description of a Neural Model

One of the main tasks necessary to build an optimal model of the neural network is a selection of sufficient input variables that essentially influence the value of the output variable. Too large a number of variables may cause noisy data whereas not taking into account even a single variable that has a critical influence on the output variable may lead to wrong results. Furthermore, adding more input network results in excessive expansion of the network architecture and at the same time the value of training data is increased. In turn, omission of essential variables in the input can cause a decrease in the quality of the network. This indicates that there are no universal criteria to select the architecture of an ANN. The architecture of an ANN, except the input and output layers, contains one or more hidden layers.

ANNs are tools to build and analyze linear and nonlinear models of complex regression and classification problems [13]. These tools are used particularly when dependence between inputs and outputs are very complicated. The network consists of elements named neurons that are connected together and the processing data is supplied as input. In general, the work of neural networks is based on the parallel processing idea. Each input signal x_i, where $i = 1, \ldots$, n is loaded to the neuron by weighted connections w_i (Fig. 2). Every neuron has a threshold value specified as its activation level. Sum of input signal values x_n multiplied by weight factors is calculated at k^{th} neuron. This value is then increased by external signal value which is referred to as a bias term Θ_k. Calculated in this way e value is the neuron activation value which is converted by established activation function $f_k(e)$ of kth neuron. The value determined by activation function is output neuron value and specifies the nonlinear relationship between resultant input signal and output signal y neurons. In some cases, the output of a unit can be a nonlinear function or a stochastic function of the total input of the unit [13] (Fig. 1).

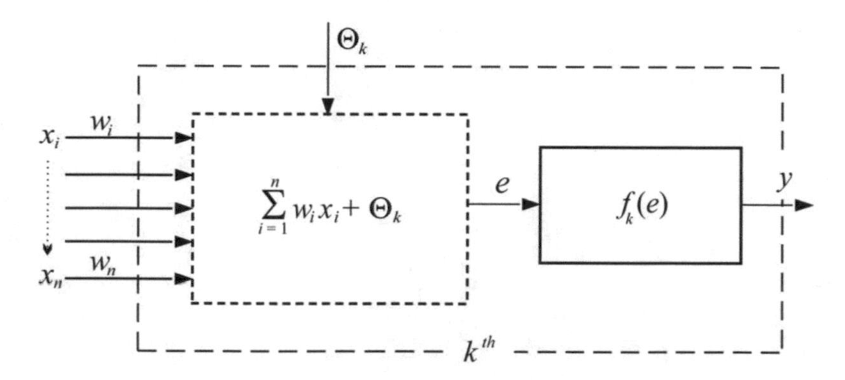

Fig. 1. Structure of nonlinear neuron k^{th}

A single layer neural network is characterized by the simplest structure. However, multilayer networks named multilayer perceptrons (MLP) are mostly utilized. A MLP with a suitable number of hidden layers and neurons is theoretically sufficient to approximate any nonlinear function [14]. In order to calculate the output value of neuron of MLP network the hyperbolic tangent function (Eq. (3)) is applied:

$$f(z) = tanh(z) = \frac{e^z - e^{-z}}{e^z + e^{-z}} \tag{1}$$

Data for the ANN analysis were obtained on the basis of experimental tests carried out on a robotized assembly station equipped with a Mitsubishi RV-M2 industrial robot (Fig. 2). The tests consist in assembly of cylindrical assembly parts with a clearance equal to: 22 μm, 46 μm, 62 μm, 78 μm, and 104 μm, in various places in the workspace of the station. The assembly of each unit was repeated 100 times, while the mountability levels were estimated as the ratio of the number of connections successfully carried out to the total number of attempts. The selection of variables that have a critical influence on the value of the mountability level is difficult because of the complex interactions of many factors which are additionally correlated with each other. Formula describing the output value of a neuron can be written mathematically as [13]:

$$y = f\left(\sum_{i=1}^{n} w_i u_i\right) \quad (2)$$

where $f(x) = 1$ when $x \geq 0$,
$f(x) = 0$ when $x < 0$, $w_0 = \delta$, $u_0 = -1$,
y – output of the neuron,
w_i – the weight of the connections that feed into neuron j,
u_i – input signal of the neuron i,
δ – threshold of the neuron.

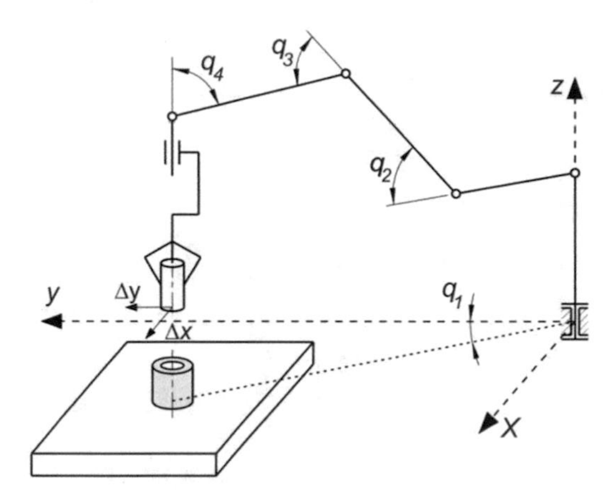

Fig. 2. The kinematic scheme of an industrial robot carrying out an assembly treatment of cylindrical parts

To determine the weighted sum and threshold activation value of separated neurons, it is necessary to prepare the training data set consisting of input signals and the corresponding values of output signals. The following input sets of variables were assigned as input signals:

- The values of joint coordinates (q_i) of the industrial robot which determine the place of realisation of the assembly operation in the workspace of the station,
- The value of joint clearance δ during assembly of cylindrical parts.

To check if all the variables selected essentially influence the value of mountability level the genetic algorithm (GA) module implemented in Statistica Neural Networks is used. Genetic algorithms are based on natural selection mechanisms as well as heredity; and operated on population of individuals that are potential solutions of the problem. Analogous to natural conditions individuals are subjected to reproduction. Mechanisms of natural selection depend on survival of the most adapted individuals in a specified environment. In other words, only the strongest individuals survive and are able to transmit genetic information to their offsprings [13].

The genetic encoding of a real or artificial organism is contained within their chromosomes [13]. A suitable representation of potential results should be able to be decoded in order to find solution for the input data structure. Fundamentals and details on the evolution process of chromosomes and the mechanisms of genetic operators is available in the open literature [15–18]. For optimization of a number of input variables, classical Holland's [19] genetic algorithm was used. The evaluation of the population was carried out with the help of a mechanism that, for each solution set, loads the best individuals so far found. The task of the genetic algorithm is thus to check the quality of the network that realizes the generalized regression for a given set of input variables to the ANN resulting from the reproduction mechanism of initial population [13]. The computations conducted confirm that the selected set of input variables essentially influence the value of the mountability level.

The expected signal at the output of the ANN was the value of probability of the correct completion of the assembly process. "The best" neural network architecture was determined in the Statistica Neural Networks program on the basis of the inputted experimental data. As a prediction model, a multilayer perceptron was used with five neurons in the input layer and one neuron in the output layer (Fig. 3). Two neural models of the best quality, differing in the number of neurons in the hidden layer, i.e. 7 neurons (ANN1) and 9 neurons (ANN2), were evaluated.

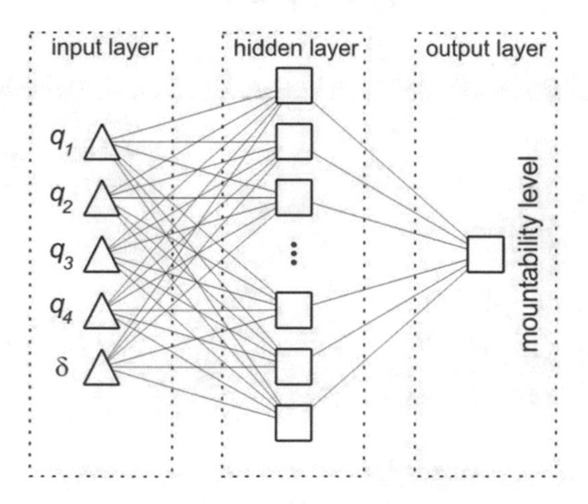

Fig. 3. Architecture of the multilayer perceptron

3 Training of the Artificial Neural Network

Among all the experimental sets of input data that correspond to the output signal, 20% were separated and assigned as a test set [13]. Data vectors from a test set did not participate in the training process and served only for the purpose of evaluating the ANN prognosis. From the remaining set of experimental data belonging to the training set, 10% was separated and assigned as a verification set. Data from this group were used for an independent check of the back propagation (BP) training algorithm [11, 12]. BP is an efficient technique for evaluating the gradient of an error function $e(w)$ for a neural network, which enables us to find good weight values using gradient descent methods. The term BP refers to the idea of propagating computed errors backwards through the network. The BP algorithm can be decomposed in the following four steps: feed-forward computation, back propagation to the output layer, back propagation to the hidden layer, and weight updates. The BP algorithm is stopped when the value of the error function has become sufficiently small [20]. To prevent over-learning, the learning process was stopped when the value of the verification root mean square error (Eq. 3) for the verification set had ceased declining (Fig. 4) [13].

$$RMS = \sqrt{\frac{\sum_{i=1}^{N} (c_i - y_i)^2}{N}} \tag{3}$$

where: N – number of vectors of training set,

c_i – expected signal of output neuron for ith standard,

y_i – signal of output signal for ith standard.

The value of the learning coefficient η, which is the parameter responsible for the stability and convergence rate of the training algorithm, was assumed to be equal to 0.01 [12].

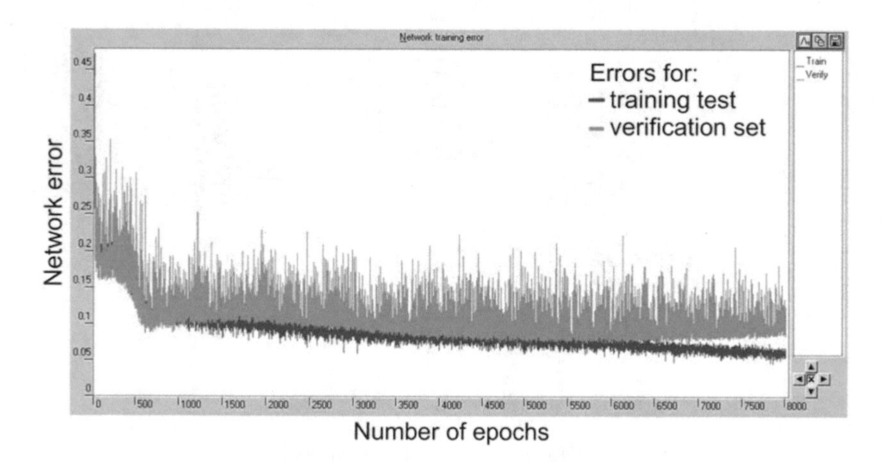

Fig. 4. Value of the RMS error during the training of network ANN1

4 Mathematical Modeling

The error of the assembly robot at any point in the workspace, defined by joint coordinates (q_i)) may be characterised by a random variable subjected to the normal probability distribution. The cross-sections of the probability density function of this distribution through planes perpendicular to the axis of the joined parts have an elliptic form. The ellipses of probability concentration are the geometric locations of the points at which the probability density of a robot error has a constant value. The equation of a family of ellipsoids, for a different probability value, makes it possible to determine the limits of the stopping area of the robot gripper in a position close to the nominal position. It permits the determination of the probability of joining parts with a specified clearance δ, based on the relationship:

$$\delta = \sqrt{\frac{8\chi_\alpha^2}{\Lambda_{xx} + \Lambda_{yy} - \sqrt{\Lambda_{xx}^2 - 2\Lambda_{xx}\Lambda_{yy} + \Lambda_{yy}^2 + 4\Lambda_{xy}^2}}} \quad (4)$$

where χ_α^2 is a quantile of the order of α of the Chi-square distribution with two degrees of freedom, Λ_{jk} - elements of the matrix inverse to the covariance matrix of the corresponding random variable of the error of the position of the joined part.

By making a simple transformation of the relationship (4), the quantiles of the order of α corresponded to the cylindrical assembly unit with a guaranteed clearance δ, which can be determined according to the relationship:

$$\chi_\alpha^2 = \frac{\delta^2}{8}\left(\Lambda_{xx} + \Lambda_{yy} - \sqrt{\Lambda_{xx}^2 - 2\Lambda_{xx}\Lambda_{yy} + \Lambda_{yy}^2 + 4\Lambda_{xy}^2}\right) \quad (5)$$

and then, using statistical mathematical tables, to read the mountability level of the assembly, i.e. the probability of correct completion of the assembly with a set clearance δ.

5 Results

To assess the network model, particular attention should be paid to the standard deviation ratio S.D. Ration and Pearson's correlation coefficient R [12, 21]. For a very good model the value of the standard deviation ratio is below 0.1. The high value of Pearson's R correlation coefficient in connection with the smallest value of SDR for the test set clearly demonstrates the very good quality of the network. Testing of predicting the neural network was carried out on the basis of four randomly selected sets of input data for which the probability of parts joining was determined experimentally. The comparison of the results of the neural network, the experimental investigations and the results obtained on the basis of mathematical modelling [1] is presented in Table 1.

Table 1. Comparison of the prediction of the mountability level using artificial neuronal networks with the results of mathematical modelling

Number of observation sets	Input variables					Mountability level*			
	q_1, rad	q_2, rad	q_3, rad	q_4, rad	δ, mm	Exp.	ANN1	ANN2	MM
10	0.5236	1.2217	1.3962	1.3962	0.021	0.24	0.23	0.23	0.26
31	0.6981	0.8726	0.8726	1.5707	0.078	0.96	0.95	0.94	0.98

* Exp. – experiment, ANN1 – network MLP 5:5-7-1:1, ANN2 – network MLP 5:5-9-1:1, MM – mathematical model

If the value of the verification error [12] is assumed as the criterion of the network assessment, the ANN1 network with 9 neurons in the hidden layer has the best prognostic properties, for which the RMS error value for the test set is equal to 0.0156. This error for the ANN2 network and the mathematical model was equal to 0.0189 and 0.0264, respectively. The ANN2 network shows the most similar error values of the prediction error (Fig. 5) for the four selected observation sets.

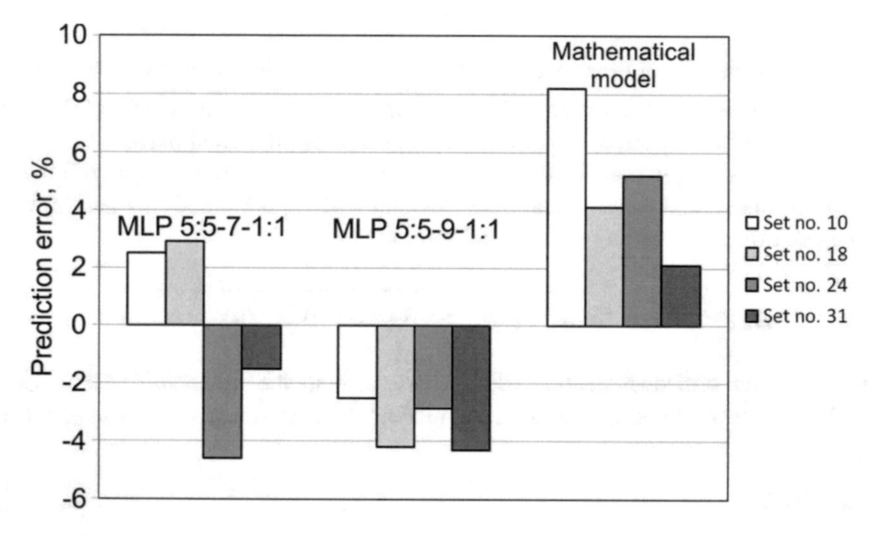

Fig. 5. Comparison of prediction errors of the mountability level by artificial neural networks and a mathematical model

All error values for ANN2 are negative, which means that the network underestimates the forecasted values. Furthermore, the average absolute error of the forecasting for the ANN1 network is the smallest and amounts to 2.87%. The values of the standard deviation ratio for the test set of the ANN1 and ANN2 networks were equal to 0.317 and 0.498, respectively. The value of Pearson's correlation coefficient R for test set for the ANN models analysed was equal to 0.827 (ANN1) and 0.916 (ANN2). The ANN2 network is characterised by the highest value of the standard deviation ratio, which excludes the model created.

6 Summary

The research conducted on the application of artificial neural networks to forecasting the mountability level of the assembly station gives optimistic premises to the possibility applying them in this type of task. The structure of the neural network was found in the Statistica Neural Network program based on the results of experiments. The average prediction error for the "best" network found did not exceed 2.87% and it was considerably smaller than in the case of the classical mathematical model (4.94%).

Despite the many advantages, the use of neural networks is limited. In the case analysed, it is necessary to perform very labour-intensive experimental tests aimed at determining the probability of joining parts in many points of the working space of the assembly station. These data are necessary for proper training of the neural network. Therefore, considering this aspect, it seems that in industrial applications, classical mathematical modelling can be used more frequently Despite the mathematical model having the worst prediction of the mountability level, it only requires an analytical description of the kinematic structure of the assembly robot, so in industrial applications this is preferred due to the lower amount of labour required.

References

1. Kluz, R.: Theoretical and experimental investigations of mountability of cylindrical parts. Technol. Autom. Assembly **1**, 6–9 (2008)
2. Chen, H., Zhang, G., Wang, J., Eakins, W., Fuhlbrigge, T.: Flexible assembly automation using industrial robots. In: Proceedings of IEEE International Conference on Technologies for Practical Robot Applications, pp. 46–51. IEEE, Woburn (2008)
3. Makris, S., Michalos, G., Eytan, A., Chryssolouris, G.: Cooperating robots for reconfigurable assembly operations: review and challenges. Procedia CIRP **3**, 346–351 (2012)
4. Tsarouchi, P., Makris, S., Michalos, G., Stefos, M., Fourtakas, K., Kalsoukalas, K., Kontrovrakis, D., Chryssolouris, G.: Robotized assembly process using dual arm robot. Procedia CIRP **23**, 47–52 (2014)
5. Wang, L., Mohammed, A., Onori, M.: Remote robotic assembly guided by 3D models linking to a real robot. CIRP Ann. **63**(1), 1–4 (2014)
6. Krüger, J., Schreck, G., Surdilovic, D.: Dual arm robot for flexible and cooperative assembly. CIRP Ann. **60**(1), 5–8 (2011)
7. Cho, N., Tu, J.F.: Quantitative circularity tolerance analysis and design for 2D precision assemblies. Int. J. Mach. Tools Manuf **42**(13), 1391–1401 (2002)
8. Kluz, R.: Mountability of sleeve joining realised by using assembly robots. Technol. Autom. Assembly **2**, 17–20 (2007)
9. Kluz, R.: Determination of the optimum configuration of robotized assembly station. Arch. Mech. Technol. Mater. **29**, 113–122 (2009)
10. Yegnanarayana, B.: Artificial Neural Networks. Prentice-Hall, New Delhi (2006)
11. Patterson, D.W.: Artificial Neural Networks—Theory and Applications. Prentice-Hall, Englewood Cliffs (1998)
12. StatSoft Inc.: Manual of STATISTICA Neural Networks Software. StatSoft Inc., Tulsa (1998)
13. Trzepieciński, T., Lemu, H.G.: Application of genetic algorithms to optimize neural networks for selected tribological tests. J. Mech. Eng. Autom. **2**(2), 69–76 (2012)

14. Hertz, J., Krogh, A., Palmer, R.G.: Introduction to the Theory of Neural Computation. Addison-Wesley, Reading (1991)
15. Sivanandam, S.N., Deepa, S.N.: Introduction to Genetic Algorithms. Springer Verlag, Berlin Heidelberg (2008)
16. Gen, M., Cheng, R.: Genetic Algorithms and Engineering Optimization. John Wiley & Sons Inc., New York (2000)
17. Davis, L.: Handbook of Genetic Algorithms. Van Nostrand Reinhold, New York (1991)
18. Gen, M., Cheng, R.: Genetic Algorithms and Engineering Design. John Wiley & Sons. Inc., New York (1997)
19. Holland, J.H.: Adaptation in Natural and Artificial Systems: An Introductory Analysis with Applications to Biology, Control and Artificial Intelligence. MIT Press, Cambridge (1992)
20. Cilimkovic, M.: Neural networks and back propagation algorithm. http://www.dataminingmasters.com/uploads/studentProjects/NeuralNetworks.pdf. Accessed 22 Feb 2018
21. Myers, R.H., Montgomery, D.C., Anderson, C.M.: Response Surface Methodology Process and Product Optimization using Designed Experiments. John Wiley and Sons Inc., New York (2009)

Identification of Challenges to be Overcome in the Process of Enhancing Innovativeness Based on Implementation of Central European Projects Funded from Interreg Programme

Mariusz Cholewa, Joanna Helman(✉), Mateusz Molasy, and Maria Rosienkiewicz

Faculty of Mechanical Engineering, Wroclaw University of Science and Technology, Wroclaw, Poland
joanna.helman@pwr.edu.pl

Abstract. The paper presents a feedback from implementation process of two projects focused on increasing innovation and enhancing technology transfer in advanced manufacturing-oriented organizations in Central Europe. The midterm review enabled to conduct analysis of results and experiences gathered during project realization. The comparative analysis is based on selected criteria. More complex image revealed during the qualitative analysis allowed to identify barriers and challenges related to implementation process. These findings allow to propose number of actions and recommendations that may help to avoid identified obstacles.

Keywords: Innovation · Technology transfer · Advanced manufacturing Knowledge management · Entrepreneurship · SMEs

1 Introduction

The growth of innovations is crucial for structural change towards sustainable development. Therefore the European Union (EU) innovation policy aims to create framework conditions that favors innovation and that allow bringing ideas to market. It will boost global competitiveness of the EU countries and enhance its economic position. Innovation has been placed at the heart of the Europe 2020 strategy for growth.

EU, as a society, is in a position to learn and innovate more than ever before. Despite large number of funding programs there are still some bottlenecks in terms of involvement of regional innovation actors into interregional cooperation or financial support of innovative activities [1]. This problem shows a need to develop new services and tools that need to be open, simple, international, easy-accessible and efficient. Another issue is to increase global innovation performance of EU (0,630) which is low in comparison to South Korea (0,74), US (0,73) or Japan (0,71) [2].

Significant trends may be observed in CE in manufacturing, especially focus of industrial policy on regions as well as re-industrialization of advanced countries (Germany, Italy, Austria) that are re-discovering innovative research-based manufacturing [3]. There is a need to fully exploit experience and achievements of different

© Springer Nature Switzerland AG 2019
A. Burduk et al. (Eds.): ISPEM 2018, AISC 835, pp. 185–194, 2019.
https://doi.org/10.1007/978-3-319-97490-3_18

innovation organizations to deliver new and innovative products, technologies and services or transferring them to different sectors and applications (e.g. transfer of lean manufacturing to mining industry) [4, 5]. Moreover, challenge to be solved by Interreg projects is the reduction of administrative barriers of innovation, public procurement of innovative products, services and social innovation [6, 7].

Development of common methods, action plans, tools and platforms will increase actors' innovation capacities through transnational networking and technology transfer to support advanced manufacturing capabilities in key enabling technologies which answers specific objective of improving sustainable linkages among actors of innovation systems for strengthening regional innovation capacity in central Europe.

Direct response for the Europe 2020 strategy is Interreg Central Europe (CE) programme. The main aim of the programme is to enhance innovativeness in European regions through strengthening linkages and beyond border cooperation to create synergy between the actors: SMEs, business, industry, research and educational organizations and public bodies [8]. The programme strategy takes into account not only potentials for territorial development, but also challenges to be tackled. Besides absence of common innovation strategies accompanied by lack of cooperation and knowledge transfer, one major limiting factor for exploiting existing innovation potential is the lack of highly skilled personnel in CE and Europe in general [9, 10].

The main aim of the paper is to present and compare the process of implementation of two selected EU's projects funded by Interreg Central Europe and focused on innovation and technology transfer in advanced manufacturing-oriented organizations. Additional aim of the presented research is to create a list of recommended actions on project and policymakers levels resulting from the analysis of the observations and identified problems during projects realization.

2 Methodology

To carry out the research inductive strategy was used [11]. It is a way of testing, in which on the basis of empirical data generalization is used, and the order of the research process is based on the model of induction including, among others, the selection of cases, conducting field research, data analysis and shaping and formulating generalizations.

The conducted study took the form of field research [12]. In it, the researcher, during his presence in the field, all the time interprets what he observes, experiences or what he can hear from others. The collected data he records in the form of notes from the site, which facilitates systematizing the material and re-confrontation with earlier ideas [13].

The comparative analysis will be based on selected criteria. The research has been made upon observations and periodic report analysis made by project's executors. On the basis of comparison challenges to overcome will be identified to improve the effectiveness of implementation of the projects.

3 The Introduction of Projects

Interreg Central European Programme with a budget of 246 million EUR from the European Regional Development Fund, supports transnational partnerships consisting of public and private organizations from nine Central European countries: Austria, Croatia, Czech Republic, Germany, Hungary, Italy, Poland, Slovakia and Slovenia [14]. "The minimum requirement for a project partnership is the participation of 3 financing partners from at least 3 countries. The typical project duration is around 36 months" [15]. The co-financing rates reach up to 85% of each partner's budget depending on the geographical location of an institution type. All the projects analysed in the paper are co-funded from this Programme. It results in their similar structure.

3.1 Introduction to NUCLEI Project

NUCLEI project "Network of Technology Transfer Nodes for Enhanced open Innovation in the Central Europe advanced manufacturing and processing industry" (http://www.interreg-central.eu/Content.Node/nuclei.html) is composed of six work packages including WP T1 Profiling existing advanced manufacturing technology transfer services and CE companies requirements, WP T2 Genetic modification of business services for innovation & tech-transfer in CE advanced manufacturing, WP T3 Fast-lane to CE advanced manufacturing innovation by one transnational collaborative environment, WP T4 Sustainability of i-service across CE advanced manufacturing industry & replication as well as WP Management and WP Communication. The consortium is comprised of ten organizations from 6 countries (Italy, Poland, Austria, Germany, Slovakia and Czech Republic) that represent Business Support Organizations, Higher Education and Research, Sectoral Agency and SMEs[1].

NUCLEI's main objective is to change the obsolete innovation management model from a "local-based" technology scouting approach to a transnational pool of knowledge supporting advanced manufacturing innovation beyond regional borders. This increases economic interdependences among seven manufacturing regions and encourages more effective transnational value chains in automotive, electrical industry, IT sector, robotic and mechanic automation. Such joint knowledge sourcing approach helps NUCLEI industrial clusters and its end-beneficiaries (corporations, SMEs, R&D performers) to foster process of emergence of new consortia/business deals for execution of technological, product, market projects. The creation of one broad and collaborative environment is expected to increase linkages with innovators outside own regions by 40–50%, accelerate the time-to-market of R&D concepts (from EU-funded research & CE laboratories to companies) by 15–20%, increase R&D expenditure and patent applications by 2–3% of mid-term turnover. The project's specific objectives include:

1. Spark a genetic modification of innovation management services for the advanced manufacturing and processing industries for a quicker commercial & exploitation

[1] Following description of the project is based on the application form and internal project's documents.

of the KET potentials. The enhancement of the technology intermediates roles federated in an organic system (clusters, brokers, technology excellence nodes) would help the Central European public and private labs as well as automation and mechatronics industries to cross the "valley of death" between research results and commercial exploitation and encourage the transnational exploitation of mature KET technologies concepts (TRL > 5–8) in new components and applications.

2. Increase linkages, commercial partnership and R&D-based expenditures through the transnational cooperation between CE excellence nodes and automation & mechatronics companies. Create a NUCLEI fast-lane to innovation for the advanced manufacturing companies by increasing the economic interdependencies among the seven regions involved in NUCLEI and encouraging more effective business linkages in automotive, electrical industry, IT sector, robotic & mechanic automation. Such collaborative and transnational open environment consists in 3 pilots (tech-brokerage; knowledge transformation; mechatronic standardization) to accelerate the time-to-market to transpose KET in industrial processing and the first move towards the standardization of mechatronics in the fields of environmental sustainability, internet of things & data security.
3. Consolidate, sustain & expand the NUCLEI innovation management approach within each Regional S3 and enhancement of the KET brokerage transnational value chain across less connected CE regions. Address Business Innovation Strategy to federate the excellence nodes (clusters, R&D, companies) in a stable organization to: (1) easily convey R&D over specific market segments to old and new members; (2) extend the sphere of transnational brokerage and tech-transfer to SMEs and organization of less connected regions wishing to be closely integrated in the excellence-nodes and cross-European value chains.

3.2 Introduction to TRANS3net Project

"Increased effectiveness of transnational knowledge and technology transfer through a trilateral cooperation network of transfer promotors" TRANT3net (http://www.interreg-central.eu/Content.Node/TRANS3Net.html) project is composed of six work packages including Transfer promotors, competences and reference model on transnational, Network's strategy, action plan and consultation of policymakers, Compiling and testing of network's portfolio, Implementation and sustainability of the transnational cooperation network of "transfer promotors"[2]. The consortium is comprised of nine organizations from three bordering countries Germany, Czech Republic, and Poland - three universities, three business support organizations and three representatives of local authorities. The border area between the countries is characterised by a low level of transnational cooperation between science and industry. Thus, the most important aim of TRANS3net is to shape conditions for building up a well working innovation system in this tri-national region. This objective is realised by establishing strong ties and a self-sustaining cooperation between "transfer promotors" and further actors of the scientific,

[2] Following description of the project is based on the application form and internal project's documents.

economic and public sphere. The activities of "transfer promotors" are aimed at initiating, implementing and supporting knowledge and technology transfer projects between science and economy. "Transfer promotors" work in research institutions (e.g. transfer offices), associations related to enterprises (e.g. chambers of commerce), regional administrations, various intermediary institutions (such as technology parks and centres) and further independent institutions.

1. Establishment of trustful relations between transfer promotors as a precondition for the emergence of contacts among representatives of the scientific and economic sphere on the transnational level.
   Knowledge about transfer promotors in all participating countries is in turn a precondition for initiating contacts and emergence of trustful relations. Finally, these personal relations are indispensable for the establishment of transnational collaborations because they guarantee sustainability of cooperation even after the project lifetime. In this regard social media are a tool to support sustainable relations.
2. Development, test, standardisation of event formats to overcome barriers between science and economy to support transnational research cooperation and dissemination of tested instruments in Central Europe.
   The standardization of event formats is a crucial condition in TRANS3net because it is assumed that economic applicable research results and approaches as well as problems and working methods of enterprises have to presented in an intelligible way. This serves as starting point for the creation of transnational research collaborations between academia and industry.
3. Establishment of a sustainable transnational innovation network of transfer promotors on the one hand and representatives of science and economy on the other hand. Apart from transfer promotors (being one-member group of the transnational innovation network) representatives from academia and economy are the second member group and - at the same time - the main target group of the network. They will use the network of transnational transfer promotors as mediators of contacts, as supporters during the implementation of transnational research cooperation (e.g. project management, legal and economic advice) as well as providers of additional consulting services.

4 Analysis of Project Implementation Effectiveness

The specificity of analysed projects is presented in the Table 1. It includes project partnership structure (different types of organizations are distinguished: Higher education and research (HE&R), Business support organisation (BSO), Sectoral agency (SA), Large enterprises (LE), Regional public authority (RPA), Local public authority (LPA), National public authority (NPA, SME, NGOs)), budget and duration.

In order to assess implementation process of abovementioned projects, a set of criteria was established based on the projects documentation. First a quantitative comparison between the outputs, tools and other types of project products and the values assumed in the application was made. Table 2 presents indicators defined in the application (proposal) and their measured values during the mid-term review (review) based on Joint Progress Reports from 3 previous reporting periods.

Table 1. Basic information about projects

Criteria vs. Project	NUCLEI	TRANS3net
No. of countries	6	3
No. of Partners	10	9
No. of HE&R/BSO/SA/SME/RPA/LPA	2/6/1/1/0/0	3/3/1/0/1/1
Total eligible budget (euro)	2405259,92	1821923,69
Duration (months)	30	36

Table 2. Outputs and indicators of the projects

Criteria vs. Project	NUCLEI (proposal)	NUCLEI (review)	TRANS3net (proposal)	TRANS3net (review)
No. of outputs	12	5	18	4
No. of tools & services	4	4	3	1
No. of innovation networks	1	0	1	1
No. of trainings	9	2	2	1
No. of pilot actions	3	2	10	3
No. of strategies & action plans	9	10	2	1
Unique visits to the project website	500	2153	1620	2198
Participants at project events	700	3595	515	331

As can be observed, NUCLEI project in two categories (tools & services, strategies & action plans) has already achieved its goals, even exceeding them. TRANS3net achieved its target score of just one category - Innovation networks.

Interreg Programme would like to force projects consortia to foresee capitalisation and communication activities (i.e. making the results available and transfer them to a wider audience) in order to roll-out and mainstream the achieved results. Also, two communication results indicators of analysed projects are presented in Table 2.

Both projects are going very well – most of the indicators are exceeded. So, it seems that projects are progressing as intended, but some possible risks can be seen, e.g. in the number of training sessions in NUCLEI or in the number of pilot actions in TRANS3net.

A more complex picture of the implementation process of abovementioned projects can be seen after qualitative analyse of feedbacks and project experiences gathered during project realisation.

5 Identified Problems in Projects Implementation

Quantity indicators do not give a full picture of the implementation status of a given project. Only the qualitative analysis allows to identify real barriers and problems that may affect the implementation of all project objectives. The main negative qualitative remarks that were observed during the implementation of both projects include:

1. Semantic barrier
   In both projects, a barrier among project's participants and beneficiaries was observed due to different levels of knowledge about the topics and issues discussed, and as a result a different level of understanding of the knowledge or expected results and benefits of the project implementation can be observed. Even if the participants of the same meeting had a similar level of knowledge, the definitions and expressions used in the projects, e.g. "transfer promotor", "e-service", "pool of excellence" and other introduced misunderstanding, which in turn often caused participants' resigning from active discussion.
2. Language barrier
   In above-mentioned projects the activities related to the international cooperation are carried out. International seminars, workshops and trainings are organized during which invited experts with well-established experience and a recognizable position in Europe provide the latest knowledge of selected thematic areas. Most of these events are carried out on the regional and local level. Participants, mainly from SMEs, during these meetings often have a problem in active participation, which is caused by poor knowledge of English. That is why they often passively spend time in these events and do not use the opportunity to ask questions or network with foreigners in order to gain contacts and establish cooperation. One of the examples could be TRANS3net project involving cooperation in a specific consortium consisting of representatives of neighbouring regions from 3 different countries, who are also representatives of 3 different groups: science, regional authorities and intermediary organizations. Poor knowledge of national languages of neighbours and English posed problems in free communication between representatives of target groups during events organized as part of the project. It was especially visible during visits and matchmaking meetings with entrepreneurs and regional authorities.
3. Organizational barriers
   The regulations applicable to projects financed by European funds have also appeared to be an obstacle for the organizations implementing the project. Restrictions related to public aid created a particularly troublesome element. These regulations imposed a state aid subsidy for all participants of organized events, and even for companies placed only in catalogues and databases. Any support in the form of: participation in training, financing travel for a matchmaking meeting, participation in such a meeting or demonstration of is treated as an aid increasing the competitive advantage of the company - the beneficiary of the program. However, not only the need to accept the amount of public aid, but the formalities required from the project partners and from the participants/beneficiaries of events caused the problem that many organizations did not want to and did not participate in the activities carried

out under the project. An additional observed complication is the system of defining public aid and its scope which differs from country to country. This situation makes it difficult to develop standards and procedures for the proceedings within the project in the process of granting that aid by individual project partners.

4. Resource barriers
   Another example of the observed problem characteristic for transnational projects, which may have a key impact on the implementation of projects, is the problem with access to data necessary for the implementation of selected project activities, e.g. the development of a strategy. The availability of data in the partner regions vary considerably in scope, detail and timeliness. Very often, at the stage of preparing a project application, the main applicant and partners are not aware of that differences between regions in terms of collecting and making available the data and information sought. Hence, it is very difficult to produce common analytical documents covering different regions or countries, and it is often necessary to carry out one's own research and analyses in order to collect the necessary and comparable data and information within a consortium.
5. Project specific barriers
   In addition, in each of the projects implemented, which differ in scope and activities carried out during their implementation, certain specific constraints have been identified. Despite the local character of these problems, it is worth listing them out, as they are characteristic for some of the measures, which in turn are often implemented in particular types or groups of projects. The NUCLEI project identified problems caused by insufficiently effective internal communication. It indirectly affected the international character of the activities. It was observed that the individual centres, involved in project activities, improved the quality and international character of their activities, but that they continued to be implemented at regional level and not at international level through an international network. This was caused by an inefficient flow of information between project partners about their activities carried out in particular regions, which resulted in the lack of their promotion in partner networks or clusters. This was evidenced by the small number of participants in each event from outside the region or foreign countries. An example of a problem encountered during the project implementation is the common definition of objectives and results of activities that would meet the needs of all the involved regions. A good example of this problem is the network's strategy and action plan developed under WP2 in TRANS3net project. The project aims to implement actions in three neighbouring regions and therefore its tools and strategies should respond to the specific needs of these regions. It turned out that the implementation of this assumption is very difficult because the regions have different needs resulting from their status, e.g. economic or development level. Therefore, the strategies or tools developed in the project must respond to the needs of future users in these regions and match their current situation, and at the same time build common standards, tools and procedures, so that cooperation in the area of e.g. technology transfer can be implemented easily and in the same way in all 3 regions.

6 Recommendations and Conclusions

The list of recommended actions resulting from the analysis of the observations and problems identified in the previous chapter is presented below. They can be divided on those on project level and those for the policymakers.

The first project level recommendation can be implemented already at the stage of preparing a project application. It is the assessment of particular tasks feasibility and products of the project by the members of the project consortium. However, this postulate may be difficult to achieve, as it requires a big volume of work for the pre-project activities. In addition, it is often the situation that the full membership of a project consortium is known shortly before the end of proposal preparation period, making it impossible to conduct detailed tasks analyses for all partners in a timely manner. Another solution seems to be to considered and plan activities related to obtaining necessary data, including predicting the necessity of conducting own research and analytical activities.

Another recommendation worth considering is to build a common repository of knowledge in order to build a common level of knowledge among all participants of the project. Common immersion in the project topic will allow to create the foundations for effective communication both inside the consortium and externally - with the beneficiaries. This involves the development of knowledge which should be effectively disseminated among the participants of the events.

One of the possible methods of solving the language and communication barrier is to actively support the invited beneficiaries during the events organized within the project. This assistance should be provided personally by the project partner during the project events. This could include assistance and language support during matchmaking meetings or interviews with potential business partners. In addition, ensuring the availability of a translator from local language to English and other way round during such meetings should also reduce this barrier.

On the policymakers level you can find organisational issues, one of which is the problem of state aid. An unambiguous consortium-wide procedure for defining and awarding aid, which would be developed at the level of the programme management organisation, would seem to eliminate this problem to a large extent.

Apart from the above recommendations, the following activities and issues can also be mentioned, which undoubtedly have a great impact on the effectiveness of the project implementation:

- Forming of a well-working project team
- Communication as basis for project success
  - integration of project partners
  - motivation of target groups
  - increase of bilateral talks
- Creating the interest, motivating for contributing to project activities
- Integrating various stakeholders in project progress–continuously, adequately

Presented in this paper number of actions and recommendations may help to avoid the above-mentioned problems and, as a result, to carry out the project implementation process without any difficulties.

References

1. Dibrov, A.: Innovation resistance: the main factors and ways to overcome them. Procedia–Soc. Behav. Sci. **166**, 92–96 (2015)
2. Commitment and Coherence – Ex-Post Evaluation of the 7th EU Framework Programme, High Level Expert Group, November 2015
3. Luby, Š., Lubyová, M.: Predictions of the success rate of EU new member states in receiving horizon 2020 funding. Informatol **49**(1–2), 41–46 (2016)
4. Urbancová, H.: Competitive advantage achievement through innovation and knowledge. J. Compet. **5**(1), 82–96 (2013)
5. Chlebus, E., Helman, J., Olejarczyk, M., Rosienkiewicz, M.: A new approach on implementing TPM in a mine – A case study. Arch. Civ. Mech. Eng. **15**(4), 873–884 (2015)
6. Lašáková, A., Bajzíková, L., Dedze, I.: Barriers and drivers of innovation in higher education: Case study-based evidence across ten European universities. Int. J. Educ. Dev. **55**, 69–79 (2017)
7. Sandberg, B., Aarikka-Stenroos, L.: What makes it so difficult? a systematic review on barriers to radical innovation. Ind. Mark. Manag. **43**(8), 1293–1305 (2014)
8. Interreg CENTRAL EUROPE Cooperation Programme, European Territorial Cooperation 2014–2020, June 2016
9. Schomaker, M.S., Zaheer, S.: The role of language in knowledge transfer to geographically dispersed manufacturing operations. J. Int. Manag. **20**(1), 55–72 (2014)
10. Lendel, V., Varmus, M.: Identification of the main problems of implementing the innovation strategy in Slovak businesses. Acta Univ. Agric. et Silvic. Mendel. Brun. **60**(4), 221–234 (2012)
11. Czakon, W.: Łabędzie Poppera – case studies w badaniach nauk o zarządzaniu [Swans of Popper - case studies in management sciences research]. Przegląd Organizacji **9**(2006), 9–12 (2006)
12. Kostera, M.: Kultura oraganizacji. Badania etnograficzne polskich firm [Ethnographic studies of Polish companies] (2005)
13. Rosen, M.: Coming to terms with the field: understanding and doing organizational ethnography. J. Manag. Stud. **28**, 2 (1991)
14. http://www.interreg-central.eu/Content.Node/apply/priorities/funding.html. Accessed 27 Mar 2018
15. http://www.interreg-central.eu/Content.Node/apply/home.html. Accessed 27 Mar 2018

Improvement of Production Process Scheduling with the Use of Heuristic Methods

Kamil Musiał, Dagmara Górnicka[(✉)], and Anna Burduk

Faculty of Mechanical Engineering, Wroclaw University of Science and Technology, Wybrzeze Wyspianskiego 27, 50-370 Wroclaw, Poland
{kamil.musial, dagmara.gornicka, anna.burduk}@pwr.edu.pl

Abstract. The paper deals with production scheduling optimization problem in the automotive company. Companies need to search for better solutions of production process scheduling, because of variety of data that need to be analyzed. Until now, company engineers used to schedule production basing on data about customers' orders and their know-how. The problem was that not all orders have the same priority and need different time of rearming machines, which makes large difference in total production time. In the paper he use of greedy algorithm and Tabu Search algorithm to verify current method of production process scheduling and improve the process has been proposed.

Keywords: Production process · Scheduling · Tabu Search · Greedy algorithm Heuristic methods · Intelligent methods in manufacturing

1 Introduction

Nowadays manufacturing companies are required to develop good quality products that are both at a competitive price and delivered on time. These requirements can be fulfilled only by companies that are able to compare various aspects while scheduling, so continuous improvement and searching for new solves became a standard in automotive industry [1, 2, 12, 13]. To keep up and be competitive with other producers, companies need to find new methods of organization, for example intelligent solutions [5, 6].

The paper examines the case where production schedules are being created by production engineers, based on customers' orders. This is very difficult task to for schedule using only the human resources because of various aspects that need to be analyzed, i.e. amount of shifts, different time of rearming, variety of products, differences between orders etc. Current scheduling plans are based on engineers knowledge, intuition and very simple calculations. Thus, schedules can be based on subjective beliefs instead of real profit possibility. The aim of the study was to verify if the methods of scheduling used in company can be improved by using intelligent methods also releasing human resources. There was a need to compare various aspects of production process and orders in different options. That is why the greedy algorithm and Tabu Search were used.

© Springer Nature Switzerland AG 2019
A. Burduk et al. (Eds.): ISPEM 2018, AISC 835, pp. 195–204, 2019.
https://doi.org/10.1007/978-3-319-97490-3_19

2 Case Study

Using the computer aided methods, i.e. heuristic algorithms, in production scheduling is getting more and more popular [14–16]. In this article, the example of improvement in scheduling process with the use of heuristic method was made in cooperation with large manufacturing company from automotive industry. The company is a supplier of brakes for global manufacturers from automotive industry. The production process is based mainly on assembly and CNC machines. The customers' orders need to be fulfilled at term, with good products quality.

First information about scheduling, very important in the light of using algorithms, is that customers make two different types of orders. There are *priority orders*, that need to be fulfilled regardless of any conditions, because without doing them the company would lose a customer. Therefore orders, named as priority, are being manufactured in the first place. The second type of orders are *optional orders*, that can be fulfilled to earn the profit, but these are not as important – company is not obliged to proceed them. Different orders need different time to be fulfilled. They also guarantee different profit.

The company works 8 h per day with 20 min break. Thus, full production time is 20 working days multiplied by 7 h and 40 min and 2 shifts each day, that gives 306 working hours monthly. The company is also able to perform third shift, what generates additional cost. The cost of production employee on first and second shift costs 30 euro per hour. On the third one, which is night shift, employee cost is 45 euro per hour. Additionally, organization, preparation and starting night shift costs 50.000 euro. There is also an option to make overtime work, with corresponding cost. In this case, with three shifts, total production time is 459 h.

In the issue being under investigation, there is a need to rearm the machines between different variants of manufactured product, which last between 2 and 130 min, depending on order type. In machine park there are 25 machines. Rearming time depends on two machines between whom rearming is necessary and is given as the matrix with 300 cells. Changing one order in the schedule two rearming times may change what change total time and cost and makes calculation much more complex. With the knowledge about products price, production cost and amount of orders, the planist make schedules every month. The way of scheduling, that is being used currently in the company is based on manual estimation -with no computer support - and does not include all of these conditions. The step-by-step algorithm of scheduling process was shown in Fig. 1.

As can be seen, company does not have the third shift option in their current scheduling. They analyze the orders by its priority, available time and by profit. They only start the night shift if it is obvious that it is profitable, without calculations of all options. They are afraid of planning more orders, because they do not want to risk mistakes in rearming time calculations. Thus, they very rarely use the possibility of starting third shift. That could be a reason of not considering third shift at all by some of the workers.

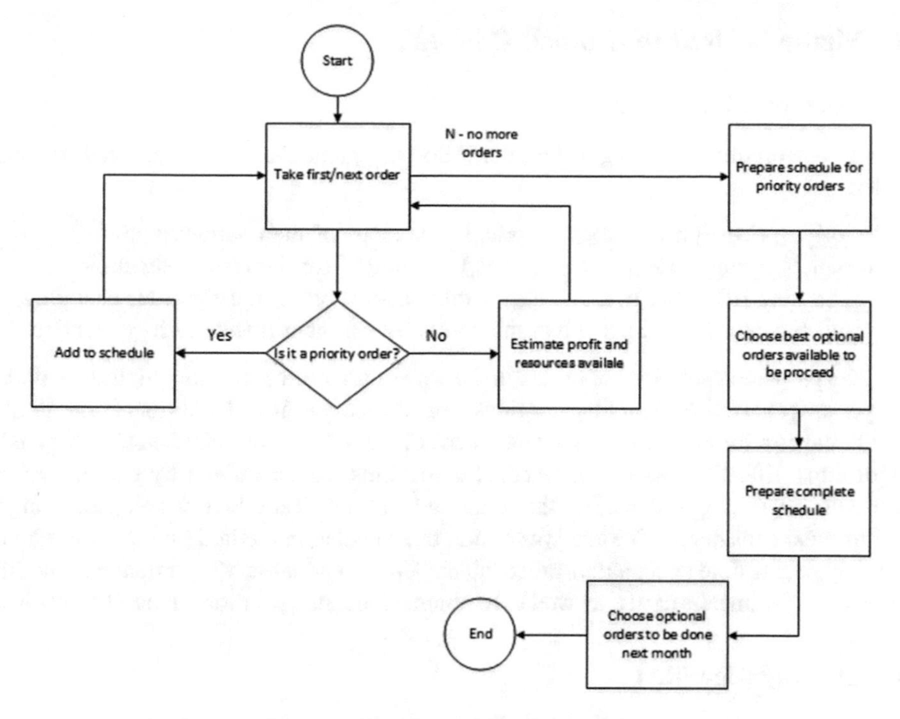

Fig. 1. Scheduling process in company

2.1 Example Used in Algorithms

Base for using greedy algorithm and Tabu Search methods in scheduling were data of 10 months of company functioning. Meantime the company got 246 different orders from customers, including 87 priority orders (Table 1).

Table 1. Amount of orders

Month	Amount of all orders	Amount of priority orders
1	18	9
2	23	12
3	12	5
4	17	6
5	26	7
6	33	14
7	22	8
8	26	13
9	36	3
10	33	10

3 Methods Used in Studied Company

3.1 Accepted Parameters

Within elaboration of the algorithm, the following parameters were accepted for each order:

- priority [prio]- [$prio_1$, $prio_2$, …, $prio_n$] – depends of the contract type;
- production time [pt]- [pt_1, pt_2, …, pt_n] – required for the order execution;
- rearm time [rt]- [rt_{12}, rt_{13}, …, rt_{nn}] – for machine rearm during order changing;
- profit [p]- [p_1, p_2, …, p_n] – that the company can gain thanks to its execution;

The purpose is selection of orders in the way guaranteeing possibly highest profit by proper order selection including priorities, rearm time, profit and additional criteria like additional cost for extra hours. Issue has been solved with use of Tabu Search and greedy algorithms. Effectiveness of the selected algorithms was examined by simulations of subsequent working months. For the examined case, the Tabu Search and greedy algorithms were considered. The aim was to find the calculation method giving a solution in relatively short time using a mid-range tablet. This would allow the company to modify orders and parameters easily, as well as to obtain calculated solutions in nearly real-time.

3.2 Greedy Algorithm

The first method used to verify scheduling is greedy algorithm. It determines a solution *"by making a selection in each step"* [1], which was shown in the next part of the article. Greedy algorithm is a heuristic method, what means that it does not guarantee that found solution is the optimal one, but it allows to get optimal in studied location [3]. Thus, it only rate the local optimum, without analyzing the data in next step [7, 8].

3.3 Solutions of the Greedy Algorithm

Sort in descending order: In this step, the algorithm multiplies profit of each order by its priority value, divides by production time, and sorts the products in descending sequence:

$$\frac{prio_{1'}p_{1'}}{pt_{1'}} \geq \frac{prio_{2'}p_{2'}}{pt_{2'}} \geq \frac{prio_{3'}p_{3'}}{pt_{3'}} \geq \ldots \geq \frac{prio_{n'}p_{n'}}{pt_{n'}} \tag{1}$$

Rearm times depend on to order that rearm is between and it's not taken under consideration in the equation above.

Add highest orders: Thereafter, the algorithm adds production times and rearm times from the highest sorted values until it reaches the maximum value, not exceeding the maximum available time:

$$pt_{1'} + rt_{12'} + pt_{2'} + rt_{23'} \ldots + pt_{m'} \leq T_{max}. \tag{2}$$

Add next: When adding another order results in exceeding the maximum volume (maximum available time), this solution is skipped, the next order is added and the criterion is checked again.

If:

$$pt_{1'} + rt_{12'} + pt_{2'} + \ldots + pt_{m'} + rt_{m,m+1'} + pt_{m+1'} > T_{max}, \tag{3}$$

then:

$$pt_{1'} + rt_{12'} + pt_{2'} + \ldots + pt_{m'} + rt_{m,m+2'} + pt_{m+2'} \leq T_{max}. \tag{4}$$

The algorithm will terminate, when the available time is utilized:

$$pt + rt_{12'} + pt_{2'} + \cdots + pt_{m'} = T_{max} \tag{5}$$

or when all the orders are checked:

$$pt_{1'} + rt_{12'} + \cdots + pt_{m'} + rt_{m,m+2} + pt_{m+2'} + \cdots + pt_{n-1'} + rt_{n-1,n'} + pt_{n'} \leq T_{max}. \tag{6}$$

A result of the algorithm action is total profit from execution of the orders, i.e. sum of profits from all the accepted orders:

$$p_{sum} = p_{1'} + p_{2'} + \cdots + p_{m'} + p_{m+2'} + p_{n-1'} + p_{n'} \tag{7}$$

3.4 Tabu Search

The second method that was used in scheduling verification is Tabu Search. It is also heuristic method, which is based on searching among all possibilities of solving the problem, with a sequence of movements [9–11]. Some of movements are temporarily forbidden, which is a reason of the algorithms' name (taboo) [4].

3.5 Solutions of the Tabu Search Algorithm

Tabu Search:

Algorithm 1. Tabu Search pseudo-code for the considered problem

```
determine start solution s
sA = s
repeat
    for i = 0 to E
        Chech the neighboring solutions
        Chech Tabu List
        if true then omit solution
        Else Add to Tabu List
        if (f(s') < f(sA)) then sA = s'
return sA
```

where:
s – current solution
S_A – best solution
f(s) – objective function
E – number of iteration

The analysed problem may be presented as a binary sequence:

$$1\ 0\ 0\ 0\ 0\ \ldots,$$

in which each bit corresponds to the order assigned to the working month. The value "1" means that the order is taken and proceeded, the value "0" means that the order is rejected. Implementation of five orders can be described as a bit sequence, e.g.:

$$1\ 0\ 0\ 0\ 0.$$

Check the neighbouring solution: In this step, the algorithm checks all of the neighbouring solutions (differing in one bit):

$$\begin{matrix} 1 & 0 & 0 & 0 & \mathbf{1} \\ 1 & 0 & 0 & \mathbf{1} & 0 \\ 1 & 0 & \mathbf{1} & 0 & 0. \\ 1 & \mathbf{1} & 0 & 0 & 0 \\ \mathbf{0} & 0 & 0 & 0 & 0 \end{matrix}$$

Check Tabu List: The algorithm checks, whether the movement between previous and actual solutions has been already made. If so, this solution is omitted and then the best solution is chosen, according to the specified criterion.

Add to Tabu List: The algorithm adds it to the Tabu List, i.e. the list of prohibited movements (TL). The algorithm remembers, which solution is the best one and repeats the steps the specified number of times.

In the case when each bit represents an order [o]- [o_1, o_2.....o_n], its parameters are:

- priority [prio]- [$prio_1$, $prio_2$, ..., $prio_n$] – depends of the contract type;
- production time [pt]- [pt_1, pt_2, ..., pt_n] – required for the order execution;
- rearm time [rt]- [rt_{12}, rt_{13}, ..., rt_{nn}] – for machine rearm during order changing;
- profit [p]- [p_1, p_2, ..., p_n] - that the company can gain thanks to its execution;

According to the assumptions, the total operational time is strictly defined. The algorithm has a limit: total time of the orders multiplied by the bit value representing the order and rearm times between accepted orders cannot exceed the maximum operational time T_{max}.

$$\mathbf{1}pt_1 + \mathbf{1}rt_{12'} + \mathbf{1}pt_2 + \mathbf{0}rt_{23'} + \mathbf{0}pt_3 + \cdots + \mathbf{1}rt_{24'} + \cdots + \mathbf{1}pt_{4'} + \cdots + \mathbf{0}pt_n \le T_{max} \quad (8)$$

In addition, in following case there are two time margins:

- The first one: 306 working hours, what means 20 working days multiplied by 2 shifts, no additional costs are include;
- The second one: 459 working hours, what means 20 working days multiplied by 3 shifts, additional costs are included.

The additional cost in working at three shifts per day is 50.000 euro to start the night shift and higher payment for workers, as was said in case study part of article. According to the adapted objective function, the criterion for choosing the best solution is profit of the orders multiplied by priority values and by values of bits representing the order data.

$$F(c) = \mathbf{0}p_1 prio_1 + \mathbf{1}p_2 prio_2 + \mathbf{1}p_3 prio_3 + \cdots + \mathbf{0}p_n prio_n \rightarrow MAX \quad (9)$$

In following case priority orders have their prio parameter set to 1000000 and non priority set to 1 that guarantee the privilege for orders that should to be considered first.

The algorithm implements also calculation of the weekend shift. If the calculated time is longer then – which means that the weekend shift works – total profit is decreased by 50 000 euro and additional employers salaries. The algorithm was set to 50 iterations.

3.6 Comparison of Results

To compare the results of algorithm methods and current schedule process, there were analysed data from 10 month of fulfilling the orders by company (Table 2).

Table 2. Amount of fulfilled orders

Month	Amount of all orders	Amount of fulfilled orders		
		Company method	Greedy algorithm	Tabu Search
1	18	12	11	12
2	23	14	14	16
3	12	12	12	12
4	17	6	7	6
5	26	14	14	14
6	33	21	21	21
7	22	22	22	22
8	26	12	14	13
9	36	29	34	31
10	33	10	10	11

The company fulfilled 152 orders in total, in the period of 10 months. With using the greedy algorithm, they would fulfil 159 orders and with Tabu Search, 158 orders. But these are only information about number of orders that could be done. Next step of

Table 3. Amount of fulfilled orders

Month	Profit per month [euro]			Amount of overtime [hours]		
	Company method	Greedy algorithm	Tabu Search	Company method	Greedy algorithm	Tabu Search
1	245037	243037	245037	19	12	19
2	209376	209376	228465	42	42	50
3	117623	117623	117623	132	132	132
4	175006	114667	175092	0	61	2
5	336470	336470	336470	0	0	0
6	384738	384738	384738	0	0	0
7	109274	109274	109274	0	0	0
8	86455	92384	94637	0	148	117
9	203867	206348	208365	0	9	0
10	256098	256098	276405	163	163	172
Total:	**2123944**	**2070015**	**2176106**	**356**	**567**	**492**

analysis was to check the profit, which is the criterion of optimization in this case. The profit and overtime hours (third shift) that need to be done in each case were shown in Table 3.

The company decided to perform least overtime hours, while greedy algorithm and Tabu Search proposed more. As mentioned above, considering matrix of rearmings that gives bigger variability of potential solution, the problem becomes very complex and notification of optimum solution using company manual method may be impossible.

4 Conclusion

In the paper scheduling of production process was considered. The aim was to verify current method and improve it by using the intelligent methods. Data about experiment results are shown in Fig. 2.

The criterion of optimization was profit earned by company. Greedy algorithm failed the experiment, giving total results worse than the method currently used in company. According to studies, the biggest profit is guaranteed by using Tabu Search as a method of production scheduling. It allows company to earn average 5.216 more per month. The biggest difference between potential profit was observed in the tenth month and it is more than 20 thousands euro. If the company would decide to use greedy algorithm method in scheduling, they can lose profit because of its' limitations. The most meaningful result is that Tabu Search method was always at least as good as other ones and frequently it gave even better profit. This case study proves the legitimacy of using intelligent method in production process scheduling, as it can give measurable results of improvement. The last, but not least is the fact, that once prepared algorithm need just adding new orders' data and can give the schedule to company, without all decision-making process, which takes a lot of time and engineers work.

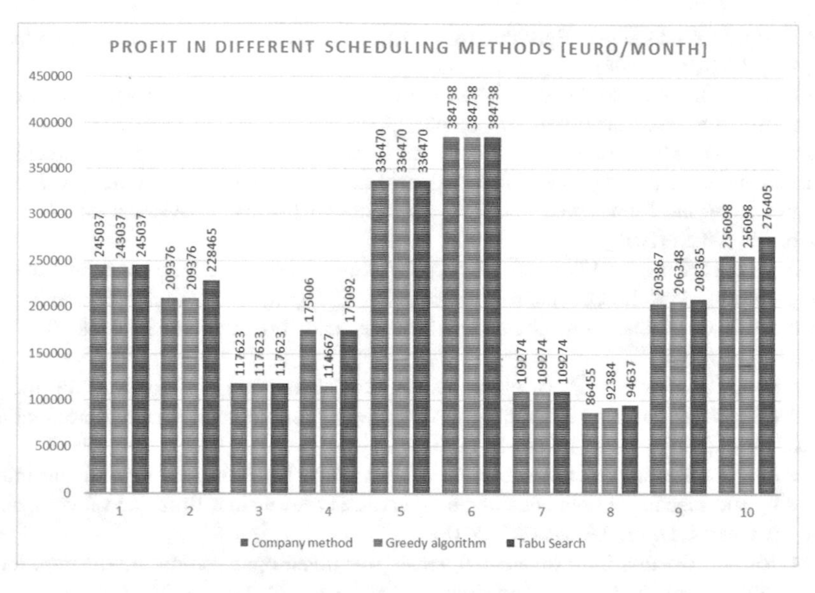

Fig. 2. Profit in different scheduling methods

References

1. Musiał, K., Kotowska, J., Górnicka, D., Burduk, A.: Tabu search and greedy algorithm adaptation to logistic task. In: Saeed, K., Homenda, W., Chaki, R. (eds.) Computer Information Systems and Industrial Management. CISIM 2017. Lecture Notes in Computer Science, vol. 10244. Springer, Cham (2017)
2. Zwolinska, B., Grzybowska, K., Kubica, L.: Shaping production change variability in relation to the utilized technology. In: DEStech Transactions on Engineering and Technology Research (ICPR 2017). Destech Publications Inc. (2017)
3. Hua, M.: A greedy algorithm for interval greedoids. Cent. Eur. J. Math. **18**(1), 260–267 (2018)
4. Yin, P.Y., Day, R.F., Wang, YC.: Neural Comput. Appl. **29**(5) (2018)
5. Kotowska, J., Markowski, M., Burduk, A.: Optimization of the supply of components for mass production with the use of the ant colony algorithm. In: Burduk, A., Mazurkiewicz, D. (eds.) Intelligent Systems in Production Engineering and Maintenance – ISPEM 2017. Advances in Intelligent Systems and Computing, vol. 637. Springer, Cham (2018)
6. Burduk, A., Musiał, K.: Genetic algorithm adoption to transport task optimization. In: Graña, M., López-Guede, J., Etxaniz, O., Herrero, Á., Quintián, H., Corchado, E. (eds.) International Joint Conference ICEUTE 2016, SOCO 2016, CISIS 2016. Advances in Intelligent Systems and Computing, vol. 527. Springer, Cham (2017)
7. DeVore, R.A., Temlyakov, V.N.: Some remarks on greedy algorithms. Adv. Comput. Math. **5**, 173–187 (1996)
8. Kahraman, C., Engin, O., Kaya, I., Öztürk, R.E.: Multiprocessor task scheduling in multistage hybrid flow-shops: a parallel greedy algorithm approach. Appl. Soft Comput. **10**, 1293–1300 (2010)
9. Cordeau, J.F., Gendreau, M., Laporte, G.: A tabu search heuristic for periodic and multi-depot vehicle routing problems. Networks **30**(2), 105–119 (1997)

10. Brandão, J.: A tabu search algorithm for the open vehicle routing problem. Eur. J. Oper. Res. **157**(3), 552–564 (2004)
11. Grabowski, J., Wodecki, M.: A very fast tabu search algorithm for the permutation flow shop problem with makespan criterion. Comput. Oper. Res. **31**(11), 1891–1909 (2004)
12. Górnicka, D., Markowski, M., Burduk, A.: Optimization of production organization in a packaging company by ant colony algorithm. In: Intelligent Systems in Production Engineering and Maintenance ISPEM. Advances in Intelligent Systems and Computing, Springer, Cham (2018)
13. Sobaszek, Ł., Gola, A., Kozłowski, E.: Application of survival function in robust scheduling of production jobs. In: Ganzha, M., Maciaszek, M., Paprzycki, M. (eds.) Proceedings of the Federated Conference on Computer Science and Information Systems (FEDCSIS), New York (2017)
14. Cordone, R., Hosteins, P., Righini, G.: A branch-and-bound algorithm for the prize-collecting single-machine scheduling problem with deadlines and total tardiness minimization. Inf. J. Comput. **30**(1), 168–180 (2018)
15. Stanzani, A., Pureza, A., Silva, B.J.V.D., Yamashita, D., Ribas, P.: Optimizing multiship routing and scheduling with constraints on inventory levels in a Brazilian oil company. Int. Trans. Oper. Res. **25**, 1163–1198 (2018)
16. Darvish, M., Coelho, L.C.: Sequential versus integrated optimization: production, location, inventory control, and distribution. Eur. J. Oper. Res. **268**(1), 203–214 (2018)

Intuitive Methods of Industrial Robot Programming in Advanced Manufacturing Systems

Kamil Krot and Vitalii Kutia(✉)

Wroclaw University of Science and Technology, Lukasiewicza Street 5, 50-371 Wroclaw, Poland
{kamil.krot,vitalii.kutia}@pwr.edu.pl

Abstract. In this article a brief review of the modern industrial robot programming methods is given. It is noted that there are a lot of research conducted to improve robot programming process, make it shorter, easier, cost-effective and user friendly. These goals can be achieved by implementing of new advanced achievements of the IT sphere into industrial robotics. Industrial robot programing by demonstration alongside with the use of virtual and augmented reality is one of the most promising technologies that can significantly reduce the integration costs and time for industrial robot integration into a production process.

Keywords: Industrial robots · Augmented reality · Human-robot interaction
Intuitive robot programming · Manufacturing systems

1 Introduction

Nowadays industrial enterprises are in great demand for flexible automated systems and innovative solutions that help to establish efficient production workflow and at the same time minimize the negative impact on workers. All this contributed to the rapidly increasing quantity of industrial robots used in different applications at production sites because of their high productivity, ability to perform production operations continuously with high repeatability and precision. A tendency to the cost decreasing of industrial robots has become an important factor in increasing their use not only in large companies but also in small and medium-sized enterprises (SMEs) [1].

Industrial robots are advanced machines and require appropriate programming and maintenance skills to achieve all the benefits from their use in manufacturing systems. Despite the continuous technical improvement and decreasing prices for industrial robots, they are still rarely used in SMEs. This is caused by their relatively high costs of their integration into production processes. In addition, small and medium-sized enterprises are oriented primarily on the production of small parties of products, so robotic cells at these enterprises require frequent reconfiguration and re-programming [2, 3].

Schraft et al. [4] noted the following requirements for successful implementation of robotic workplaces in small productions: shortening staff training time for the robot programming to one day, an ability to quickly and easily change robot's program, a significant reduction of programming time. To meet these requirements it is necessary

© Springer Nature Switzerland AG 2019
A. Burduk et al. (Eds.): ISPEM 2018, AISC 835, pp. 205–214, 2019.
https://doi.org/10.1007/978-3-319-97490-3_20

to develop and implement new practical methods of industrial robot programming that would be distinguished by convenience, flexibility and versatility. These methods should make it possible for industrial robots programming directly by production line engineers or service staff, without the involvement of additional specialists in industrial robotics.

In this article the short characteristics and perspectives of modern methods of industrial robot programming are presented.

2 Conventional Approaches to Industrial Robot Programming

Industrial robot programming includes the description of the robot movements and implementation of the user's logic algorithm.

There are a lot of industrial robot programming methods that are usually divided on on-line methods and off-line methods.

On-line Programming. On-line robot programming methods are characterized by the fact that the real robot cell is actively used during the programming process. One of the on-line techniques is programming by teaching. The basic idea of the programming by teaching method is guiding the robot through the defined trajectory to perform the necessary tasks. All movements are recorded in the memory of the robot controller, and then the robot can reproduce these movements many times. There are two variations of this method: walk-through and lead-through programming.

In walk-through programming, the human operator directly moves the manipulator manually through the required trajectory in the robot workspace. However, this method can be used only for special type robots with low inertia. In addition, walk-through programming cannot be used if the robot works in environment with harmful or dangerous conditions for humans.

In lead-trough programming, the robot is also guided in manual mode by the movement control device. The operator with the help of teach pendant or special joystick makes the necessary movements of the robots TCP. The TCP coordinates are stored in the robot controller continuously during the movement or in specified positions.. Unlike walk-trough programming, this method does not require any special design of the robot and its drives, but requires high qualifications of the operator.

Off-line Programming. Off-line programming methods do not involve the real robot in the process of the robot task defining, thus the non-productive time of the robot approaches to a minimum.

Off-line methods allow to define the robot tasks using high-level languages or special simulation software packages and express them in text or graphic form. First of all, the tool used is expected to be able to verify the correctness of the created program based on the model of the real scene and the real robot, and model its execution. Meeting this requirement will avoid the need to engage the real robot and the real scene in the programming process.

Programming languages used in off-line methods are characterized by a large number of available instructions, thanks to which they enable programming of very complex tasks. Conditional instructions allow the use of sensory data and the dependence of the task execution on the sensor data. Text or graphic form of task representation allows for easy and cheap modification of the task.

The development of off-line methods enabled by the rapid development of computer technologies that made it possible to create advanced means of robot programming (simulation software and CAD oriented environments).

The basic difficulty of off-line methods is the need to model the robot and its working environment and to describe the interrelations between all elements. It is a time-consuming process that require detailed understanding of the robot tasks. The task definition itself does not have to be trivial, and depending on the level of development of the programming environment, it may require a lot of programming skills from the operator. Therefore, off-line methods are best suited to situations where non-productive time of the robot is significantly high. Off-line programming can be used for such processes as machining, welding, water cutting, plasma, gas and laser cutting, scanning, layer gluing, 3D printing, painting, coating, etc.

Hybrid Programming. Hybrid (mixed) programming is the optimization of robot programming combining the advantages of on-line and off-line methods. The robot program usually consists of two parts: specified positions (coordinates) and program logic (task algorithm, communication, calculations). The program logic and the main part of the traffic instructions can be very effectively developed off-line using CAD data. Motion instructions requiring indication of the position within the robot workplace can be programmed on-line. In this way, the advantages of the both programming methods can be used.

In recent years, significant progress has been made in the development of collaborative robots (cobots) designed to work directly with a human in a common workspace while performing manufacturing operations. Such robots have additional security functions (integrated power sensors, current overload detection, etc.).

Models of collaborative robots appear among the market proposals of the leading manufacturers of industrial robots. In particular, models of collaborative robots are offered by such manufacturers as ABB, KUKA, FANUC, COMAU, Yaskawa/Motoman, STÄUBLI and others. Also, in the last decade, industrial robotic arms of the UR brand (UR3, UR5 and UR10) from Universal Robots have become the world's leaders in the market of collaborative robots.

Most collaborative robots can be easily programmed with a playback method without having a thorough knowledge of programming. However, it has been actively underway to develop new devices and software for the implementation of more intuitive techniques and interfaces for industrial robots programming.

3 Intuitive Programming of Industrial Robots

There are a lot of research developing new intuitive programming devices and techniques aiming to eliminate the disadvantages of traditional methods of industrial robots programming. Intuitive robot programming can be understood as an approach that enables ordinary people to program a robot with minimal special skills and training, guided by their own experience and perception.

According to [5], recent progress in intuitive programming of industrial robots includes: development of interactive teach pendants, programming by demonstration, the use of multimodal interfaces for human-robot interaction, virtual and augmented reality etc. (Fig. 1).

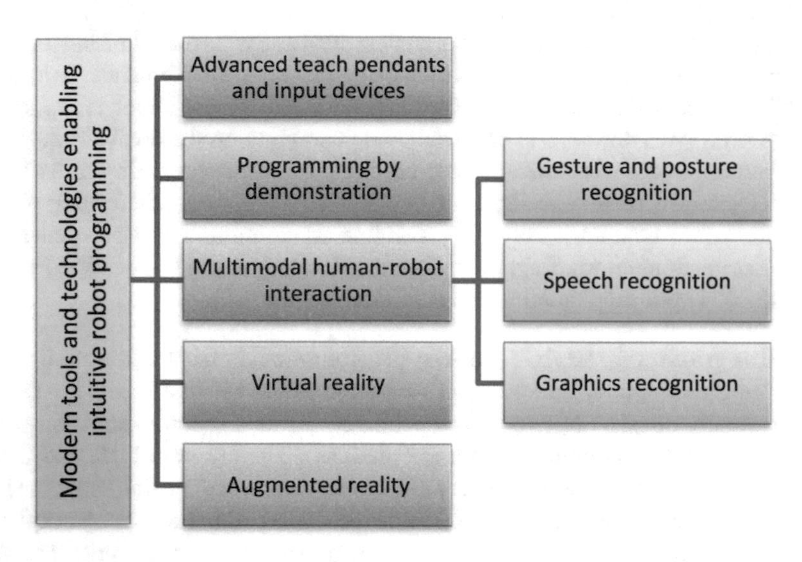

Fig. 1. Main research directions of development new technologies applied to industrial robot programming

Advanced Teach Pendants and Input Devices. A teach pendant is the main input device enabling on-line programming and configuration of an industrial robot. Most of traditional teach pendants are not intuitive and require a lot of practice to use them quickly and properly [6]. So, the industrial robot manufacturers continuously develop and introduce new design and ergonomic variations with additional user-friendly features, such as 3D joystick (ABB), 6D mouse (KUKA) or wireless teach pendant (COMAU). Bischoff et al. proposed the style guide for intuitive icon-based programming interfaces operated via touch screen and speech input [7].

The Flexible Graphical User Interface concept is presented in [8] which implements an abstract level between the technical realization of robot operations and the service oriented operation. Due to this concept the robot operator can use the teach pendant in a more intuitive way with decreasing task setup and completion time and the reducing number of interactions.

A series of studies on the use of mobile devices as a means of industrial robot programming was presented. For example, Jan et al. [9] proposed smart phone based teaching pendant with a user friendly interface giving the user possibilities of remote control, operation and monitoring of the industrial robot. Mateo et al. [10] developed an Android based application for industrial robot programming that makes it easier to program tasks for industrial robots like polishing, milling or grinding. The visual programming concept implemented in the application allows non-skilled in programming operators to create programs and monitor the tasks while it is being executed by overlapping real time information through augmented reality.

Moreover, even industrial robot manufacturers are interested to develop the user interfaces supporting robot programming based on usual mobile devices (tablets and smartphones). For example, COMAU Robotics developed the PickApp, an Android based user-friendly application to perform pick and place operations [11].

Abbas et al. [12] presented an idea of augmented reality based teaching pendant on smartphone which could help user to program industrial robot more intuitively.

Programming by Demonstration. The concept of Programming by Demonstration (PbD) consists in that the user shows the robot an example (or several examples) how to perform a specific task, unlike to the on-line robot programming. In the case of the programming with a teach pendant, the user must manually move the robot to a number of positions along the intended path and determine the speed and type of movement needed to achieve the position. The PbD aims to combine the simplicity of the teach-in with the learning and interaction with the user at all stages of the demonstration of task execution, trajectory adjustment and program playback [13].

Programming by Demonstration can be divided into: PbD with the use of a real robot (by guiding the robot through the desired trajectory) and PbD without the use of real robots (demonstration of task execution by means of special input devices) [14]. The first method has been used over the last few decades. Successful application of this method in some cases requires the using of force/tactile sensors (for example, when teaching a robot machining and grinding operations). Implementation of the second method requires the use of special systems for tracking human movements (handheld pointing devices, optical tracking systems, gesture recognition systems, etc.).

Recent research in PbD supported by the other technologies (like VR and AR) for human-robot interaction shows that this method has the most promising future. However, existing PbD concepts or other natural human input (graphic, speech or gestures) still don't have enough robustness for industrial applications [15].

Virtual Reality. Virtual Reality (VR) technologies have recently become widespread as useful industrial tools that cover the design, manufacturing and test applications [16]. The application of virtual reality opens up new opportunities for more convenient offline programming of industrial robots. Virtual reality technologies is a tool that ensures the operator's interaction with the industrial robot 3D model in a virtual working environment in a similar way to the real environment. The main advantage of using VR is to provide intuitive interfaces due to its scalable ability to simulate the entire robot work space. However, the accurate modeling of the environment in VR is quite long-term and non-trivial process, since it requires the development of special content to represent the actual working scenario [5].

Augmented Reality. Augmented Reality (AR) is a modern technology that comes from VR. AR combines real perception of the environment by the user with virtual elements created by computer software. AR differs from VR in that it does not require a model of the entire environment, but only complements the real world with additional computer-generated elements. Azuma [17] defined three main characteristics of the AR systems: combination of real and virtual elements, ability of interaction with virtual objects in real time and registration of virtual objects in real environment.

The concept of AR has been known for several decades, but only very recently it has begun to be implemented at the enterprises. This is primarily caused by the rapid development of hardware and software technologies that are needed to deploy AR throughout the entire company. Different types of devices (smartphones, tablet PCs, head-mounted displays, monitors, eyeglasses, projectors, etc.) can be used to deploy AR experience depending on the use case.

As reported in [18], AR technologies are successfully used by manufacturers of industrial products, automotive, aerospace, high-tech and software companies.

AR in robotics provides user friendly tools for human-robot interaction. There are a lot of very interesting AR-based projects in the field of industrial robotics, for example [1–3, 5, 15, 19–28] and many others. However, AR applications in robot programming are still limited to research projects and still not reached production [19].

Girbacia et al. [27] designed a methodology and a prototype system for off-line programming of an industrial robot using AR technology. The short algorithm of the methodology is presented in Fig. 2.

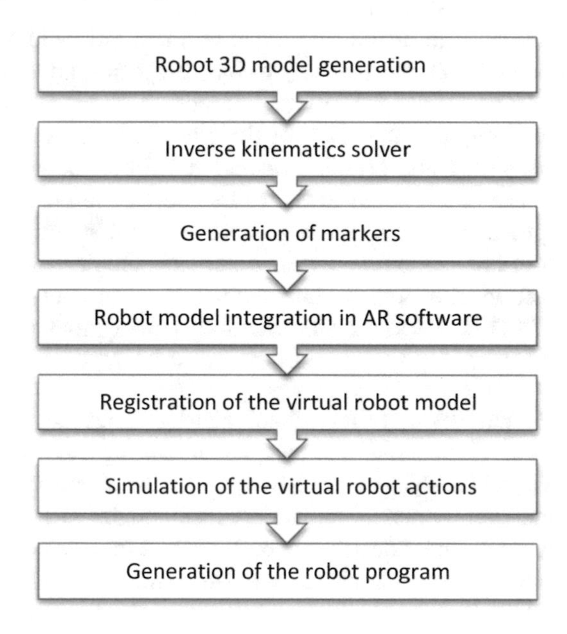

Fig. 2. Methodology for industrial robot programming using AR

Akan et al. [3] proposed the incremental multimodal language, which uses AR environment and makes it possible to manipulate, pick or place the objects in the scene. The aim of their approach is to shift the focus of industrial robot programming from coordinate based programming paradigm, to object based programming scheme.

Gaschler et al. presented an AR system for defining virtual obstacles, specifying tool positions, and robot tasks [15].

Fang et al. presented a novel Euclidean distance-based method to assist the users in the interaction with the virtual robot and the spatial entities in an AR environment. They also described the two case studies on the proposed AR-based interface for intuitive human-robot interaction in a robotic pick-and-place application and a path following application [5].

Significant contribution to the development of intuitive methods of industrial robot programming was made by Lambrecht et al. [20–22]. In [21] they presented an intuitive robot programming system that implements a multimodal approach that combines markerless gesture recognition and AR-based simulation on common mobile devices. The user is enabled to draw poses and trajectories into the workspace of the robot supported with simultaneous visual feedback in AR. The user can also easily change the robot program by gestural manipulation of poses and trajectories. Within a task-oriented implementation of the robot program a pick and place task was implemented through the PbD principle.

A variety of mentioned above projects shows that the use of AR technology in industrial robot programming allows to create a virtual working environment for real robot with the necessary elements, which is especially useful when designing or modification of robotic workcells (Fig. 3, a). AR supported robot programming is also useful when robotizing existing production sites, when a virtual robot can be augmented to the real workcell (Fig. 3, b).

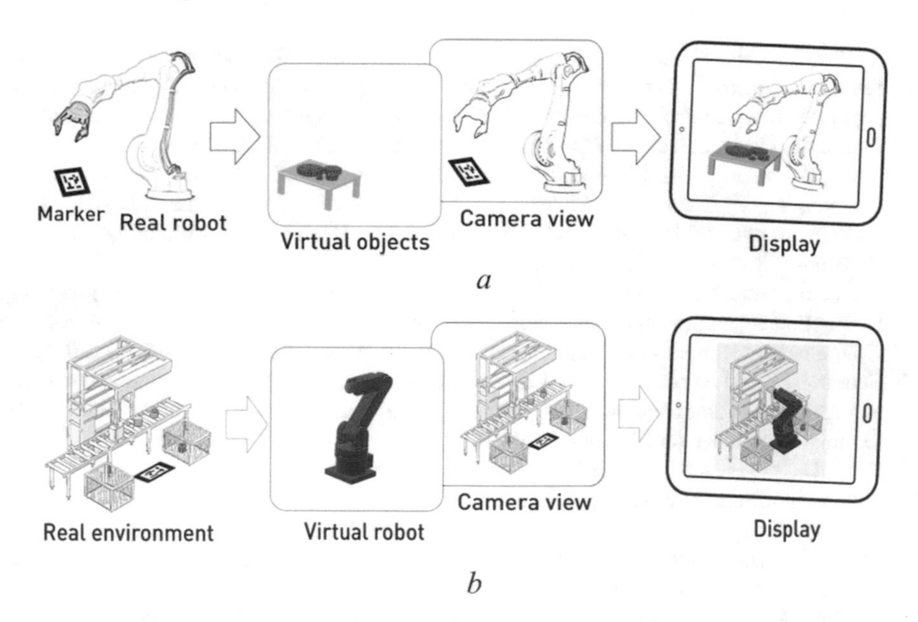

Fig. 3. General design approaches of AR application in industrial robot programming: (*a*) simulation of the virtual work environment with the real robot, (*b*) simulation of the virtual robot actions in the real working environment

The use of AR gives new opportunities to create human-robot collaborative (HRC) manufacturing where industrial robots would work in joint space with the human workers [29]. Emerging multimodal interfaces can be successfully used for human-robot interaction in such productions. AR technology is the basis in development of modern worker support systems for HRC manufacturing where the robot commands and worker instructions can be augmented for human workers [30]. This aims to make the HRC manufacturing more safe, intuitive and flexible.

4 Summary and Conclusions

In this article modern techniques of industrial robot programming are considered. There are a lot of research dedicated to the improving of industrial robot programming process, make it shorter, easier, cost-effective and user friendly. These goals can be achieved by implementing of new intuitive programming paradigms into industrial robotics. Industrial robot programing supported by VR and AR is one of the most promising techniques that will be extremely developing in the next few years. The use of these techniques will make it available to program industrial robots by persons without special programming skills in more intuitive way. This will significantly reduce the costs and time for integration of industrial robots in manufacturing enterprises.

References

1. Stadler, S., Kain, K., Giuliani, M., Mirnig, N., Stollnberger, G., Tscheligi, M.: Augmented reality for industrial robot programmers: workload analysis for task-based, augmented reality-supported robot control. In: The 25th IEEE International Symposium on Robot and Human Interactive Communication (RO-MAN), New York, pp. 179–184 (2016)
2. Akan, B., Ameri, A.E., Çürüklü, B.: Augmented reality-based industrial robot control. In: Larsson, T., Kjelldahl, L., Jää-Aro, K.-M. (eds.) Proceedings of the SIGRAD 2011. Evaluations of Graphics and Visualization – Efficiency, Usefulness, Accessibility, Usability, Stockholm, Sweden, pp. 113–114 (2011)
3. Akan, B., Ameri, A., Çürüklü, B., Asplund, L.: Intuitive industrial robot programming through incremental multimodal language and augmented reality. In: 2011 IEEE International Conference on Robotics and Automation, ICRA 2011, Shanghai, pp. 3934–3939 (2011)
4. Schraft, R.D., Meyer, C.: The need for an intuitive teaching method for small and medium enterprises. In: Joint Conference of the International Symposium on Robotics (ISR) and the German Conference on Robotics (ROBOTIK), Munich, pp. 95–105 (2006)
5. Fang, H.C., Ong, S.K., Nee, A.Y.C.: Novel AR-based interface for human-robot interaction and visualization. Adv. Manuf. **2**(4), 275–288 (2014)
6. Neto, P., Pires, J.N., Moreira, A.P.: 3D CAD-based robot programming for the SME shop-floor. In: 20th International Conference on Flexible Automation and Intelligent Manufacturing, FAIM 2010, San Francisco (2010)
7. Bischoff, R., Kazi, A., Seyfarth, M.: The MORPHA style guide for icon-based programming. In: Proceedings of the 11th IEEE Symposium on Robot and Human Interactive Communication, pp. 482–487, Berlin (2002)
8. Dániel, B., Korondi, P., Sziebig, G., Thomessen, T.: Evaluation of flexible graphical user interface for intuitive human robot interactions. Acta Polytech. Hung. **11**(1), 135–151 (2014)

9. Jan, Y., Hassan, S., Pyo, S., Yoon, J.: Smartphone based control architecture of teaching pendant for industrial manipulators. In: 2013 4th International Conference on Intelligent Systems, Modelling and Simulation (ISMS 2013), Bangkok, pp. 370–375. IEEE (2013)
10. Mateo, C., Brunete, A., Gambao, E., Hernando, M.: Hammer: An android based application for end-user industrial robot programming. In: 2014 IEEE/ASME 10th International Conference on Mechatronic and Embedded Systems and Applications (MESA), Senigallia, pp. 1–6 (2014)
11. PickApp 1.1: Intuitive interface for robot programming on your Android tablet. http://www.comau.com/EN/our-competences/robotics/software/pickapp. Accessed 20 Apr 2018
12. Abbas, S.M., Hassan, S., Yun, J.: Augmented reality based teaching pendant for industrial robot. In: 2012 12th International Conference on Control, Automation and Systems (ICCAS), Jeju Island, Korea, pp. 2210–2213 (2012)
13. Friedrich, H., Münch, S., Dillmann, R., Bocionek, S., Sassin, M.: Robot programming by demonstration (RPD): supporting the induction by human interaction. Mach. Learn. **23**(2/3), 163–189 (1996)
14. Münch, S., Kreuziger, J., Kaiser, M., Dillmann, R.: Robot programming by demonstration (RPD) - using machine learning and user interaction methods for the development of easy and comfortable robot programming systems. In: Proceedings of the 24th International Symposium on Industrial Robots (ISIR 1994), pp. 685–693 (1994)
15. Gaschler, A.K., Springer, M., Rickert, M., Knoll, A.: Intuitive robot tasks with augmented reality and virtual obstacles. In: 2014 IEEE International Conference on Robotics and Automation (ICRA). IEEE, Hong Kong (2014)
16. Aron, C., Marius, I., Cojanu, C., Mogan, G.: Programming of robots using virtual reality technologies. In: Talaba, D., Amditis, A. (eds.) Product Engineering, pp. 555–563. Springer, Dordrecht (2008)
17. Azuma, R.T.: A survey of augmented reality. Presence Teleoperators Virtual Environ. **6**(4), 355–385 (1997)
18. Campbell, M., Kelly, S., Jung, R., Lang, J.: The State of Industrial Augmented Reality 2017. PTC (2017). https://www.ptc.com/-/media/Files/PDFs/Augmented-Reality/State-of-AR-Whitepaper.pdf. Accessed 20 April 2018
19. Andersson, N., Argyrou, A., Nägele, F., Ubis, F., Campos, U.E., de Zarate, M.O., Wilterdink, R.: AR-enhanced human-robot-interaction - methodologies, algorithms, tools. Procedia CIRP (2016)
20. Lambrecht, J., Kleinsorge, M., Rosenstrauch, M., Krüger, J.: Spatial programming for industrial robots through task demonstration. Int. J. Adv. Rob. Syst. **10**(5), 254 (2013)
21. Lambrecht, J., Krüger, J.: Spatial programming for industrial robots: efficient, effective and user-optimised through natural communication and augmented reality. AMR **1018**, 39–46 (2014)
22. Lambrecht, J., Walzel, H., Krüger, J.: Robust finger gesture recognition on handheld devices for spatial programming of industrial robots. In: 2013 IEEE RO-MAN. The 22nd IEEE International Symposium on Robot and Human Interactive Communication, Gyeongju, Korea, 26–29 August 2013, pp. 99–106 (2013)
23. Fang, H.C., Ong, S.K., Nee, A.Y.C.: Interactive robot trajectory planning and simulation using augmented reality. Robot. Comput. Integr. Manuf. **28**(2), 227–237 (2012)
24. Ong, S.K., Chong, J.W.S., Nee, A.Y.C.: A novel AR-based robot programming and path planning methodology. Robot. Comput. Integr. Manuf. **26**(3), 240–249 (2010)
25. Pai, Y.S., Yap, H.J., Singh, R.: Augmented reality–based programming, planning and simulation of a robotic work cell. Proc. Inst. Mech. Eng. Part B J. Eng. Manuf. **229**(6), 1029–1045 (2014)

26. Pai, Y.S., Yap, H.J., Md Dawal, S.Z., Ramesh, S., Phoon, S.Y.: Virtual planning, control, and machining for a modular-based automated factory operation in an augmented reality environment. Sci. Rep. **6**, 27380 (2016)
27. Girbacia, F., Duguleana, M., Stavar, A.: Off-line programming of industrial robots using co-located environments. AMR **463–464**, 1654–1657 (2012)
28. Bischoff, R., Kazi, A.: Perspectives on augmented reality based human-robot interaction with industrial robots. In: Proceedings of 2004 IEEE/RSJ International Conference on Intelligent Robotics and Systems. IROS 2004, Sendai, Japan, pp. 3226–3231 (2004)
29. Fogal, D., Rauschecker, U., Lanctot, P., et al.: Factory of the future. White paper. International Electrotechnical Commission, Geneva (2015)
30. Liu, H., Wang, L. (eds.): An AR-based worker support system for human-robot collaboration. In: 27th International Conference on Flexible Automation and Intelligent Manufacturing, FAIM 2017, Modena (2017)

Manufacturing Activities Modelling for the Purpose of Machining Process Plan Generation

Jan Duda and Jacek Habel(✉)

Cracow University of Technology, Al. Jana Pawła II 37, 31-864 Cracow, Poland
{duda,habel}@mech.pk.edu.pl

Abstract. The idea of modelling the manufacturing activities is presented in the paper. The main goal was to realise the procedure of machining process plan generation using semi-generative method of CAPP system build. This method requires to use the expert system. The manufacturing knowledge is determine by machining process plan template, in the form of hierarchical networks, and production rules where the definition of manufacturing activities is mandatory. Moreover the classification and formal description of manufacturing activities is also included. This formalization let us to define manufacturing capabilities of manufacturing system resources and is also used in workpiece features identification and transformation. The new definition of machining process as an ordered set of activities is introduced. During the machining process the intermediate states of workpiece are generated. Also the manufacturing system model in the form of objects and vector of its state is given. The described activities cause the intended discrete change of manufacturing system and workpiece state.

Keywords: CAPP · Manufacturing activities · Machining process planning

1 Introduction

The new trends, like industry 4.0, force to introduce the higher level of automation in product life cycle. One of its phase is a manufacturing process preparation. Different kind of CAx systems are used. The most common are CAD/CAM systems, which, due its integration on digital mock-up level, offers wide range of features like: 3D modelling, 2D workshop documentation, CNC machine tool programming with NC program generation etc. PLM systems are the next level, where very good example is a newest version of Dassault 3D-Experience [9]. Even those kind of systems have still some weaknesses, which should be eliminated. It is a stage of manufacturing process planning, where the CAPP systems can be used.

The idea of CAPP system has a long, almost forty years, history [3]. Generally speaking, it is a computer system, where the implemented procedure is able to create and store the machining process plan (mostly those kind of system were implemented). In the eighties of last century, two general methods were developed: variant and generative [6]. Both of them have advantages and disadvantages. Variant method is easier to implement and the main advantage is a fast process planning using the process plan template (based on similarity of stored process plans in database). The biggest

© Springer Nature Switzerland AG 2019
A. Burduk et al. (Eds.): ISPEM 2018, AISC 835, pp. 215–224, 2019.
https://doi.org/10.1007/978-3-319-97490-3_21

disadvantage is no possibility to create the process plan for part which is not similar to other, already stored in database. This is possible with generative method. But here is necessary to implement difficult algorithms (e.g. clustering) and AI methods, what cause a lot of problems [8]. Moreover there is a problem with structure creation of machining process (how to divide process into operations, setups etc.) because there is no process template. Some solution could be the hybrid method, which takes the advantages from both previous ones. Those solution, called semi-generative method, is an original idea created by authors of this paper (more details can be find in [2, 5]).

Nowadays manufacturing systems are more and more automated. The general structure of it can be easily defined. Such system consists of a set of object. Some of them are crucial and fulfil main role in production process, creating value added. Those kind of resources are for example machine tools or manufacturing stands. Other, even they don't create value added, are still required, e.g. robots, warehouses. There is also a big set of exchangeable equipment like workholders or cutting tools. All of them can be described by a set of attributes, which define their state.

Manufacturing process is realised in the manufacturing system, where some events occur. Those events can be defined as a sequence of manufacturing activities, which consists the manufacturing process. The wide automation of process planning requires the exact recognition of all activities, which appear in the real manufacturing process. The differentiation of these activities and their formalized description makes it possible to build the activities database and knowledge base, which allows the correct planning of the structure of the machining process. The activities occurring in the machining process should be formalized in order to select the right elements of the process structure in the semi-generative process planning method.

2 The Idea of Process Planning with Semi-generative Method

The CAPP system is a missing component in integrated CAD/CAM systems. Those kind of system should realize the following tasks (not included in CAD/CAM):

- Raw material or semi-finished product selection,
- Creating the structure of machining process, with division on operations, setups and machining cuts,
- Machine tool and its equipment (e.g. workholders) selection,
- Cutting tool and cutting parameters selection with material considering of both workpiece and tool,

The worked-out semi-generative method combine best practices of variant and generative methods. The new machining processes can be added to the common database. Like in variant method, we can use the machining process template (*MPT*). For the process plan generation the manufacturing knowledge is used, like in generative systems. The most important here is a connection between machining process template and manufacturing knowledge, organized in the form of production rules [2].

2.1 Definition and Formal Description of Manufacturing Activities

Manufacturing process is an ordered sequence (set) of manufacturing activities which are present in manufacturing system work. Let the definition of manufacturing activity will be as follow: The manufacturing activity causes the required changes of objects' state vector in manufacturing system.

If we assume that each object in manufacturing system is described by state vector than each vector changes is caused by manufacturing activity. The most often are manufacturing activities causes the object location changes. Change of location (translation, rotation etc.) is possible thanks to free relocation and kinematic dependences which are present in kinematic structure of devices in manufacturing system (realized by power transmission and control system). Manufacturing activities are different because can realize different functions in manufacturing system.

The following manufacturing activities are distinguished (see Fig. 1):

- Transformation Activity D_{TR} of workpiece state causes the change of geometrical and technological parameters (e.g. dimensions, accuracy) or physical (e.g. hardness). This activity, through interaction of cutting tool *CT* on workpiece *WP*, creates the temporary pair *TR* and dependence *F* (cutting force). Transformation has influence only on nominal parameters *Nm* of workpiece.
- Identification Activity D_{ID} (comparison) of parameters causes receiving the real characteristic of measured object. This activity, through interaction of measuring tool *MT* on workpiece *WP*, creates the temporary pair *ID* and dependences *I*. Identification receive and store the real value of parameters *Re* of object.
- Set-up Activity D_{SU} (locating & clamping) causes arise or break locating constraint (contact of two base surfaces) and clamping constraint (apply or remove clamping force). This activity, through interaction of object 2 (e.g. robot gripper) on object 1 (e.g. workpiece), creates the temporary pair *SU* and dependences *LC*. Set-up activity has influence only on *COC* parameters (character of object contact) of pair *SU*.
- Relocating Activity D_{RC} cause the relative change of position or orientation of objects in manufacturing system. This activity, through interaction on dependences *R* (relocating) in objects chain (or pair), cause the changes in *MGP* parameters (mutual geometrical position) of given pair.

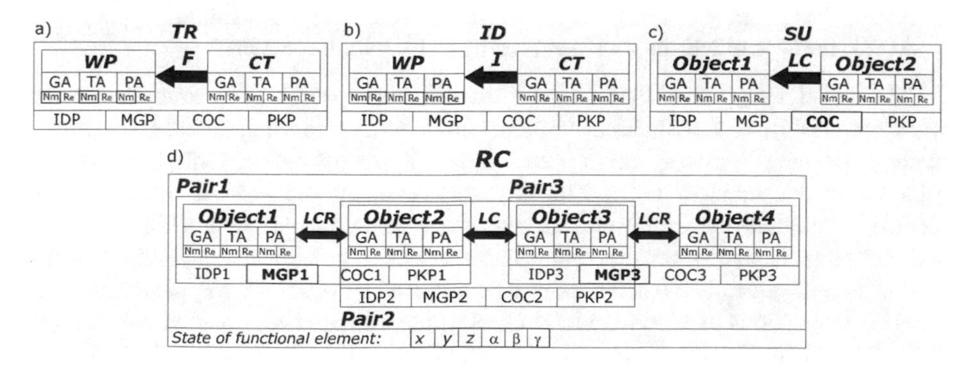

Fig. 1. Schemes of activities: (a) transformation, (b) identification, (c) set-up, (d) relocating

In conclusion, to execute the manufacturing activity, it have to cover minimum two objects and these objects have to be connected by minimum one dependence. The manufacturing activities cause the changes of manufacturing system objects characteristic (state vector).

To calculate worktime standard of operation we usually consider the main and auxiliary activities. The main activities creates value added of product (transformation D_{TR}). The auxiliary activities don't create value added but still are necessary. There are the activities associated with setup change time and auxiliary time (activities for each workpiece in batch). Using elementary activities we can define more complex activities. For example all auxiliary activities D_{AU} can be defined as a sequence of D_{SU} and D_{RC}. Here can be distinguished the following list of auxiliary activities associated with workpiece and cutting tool:

- Supply (load or unload) of workpiece or cutting tool on machine tool,
- Change of workpiece setup or change of cutting tool on machine tool,
- Change of workpiece or cutting tool position on machine tool,

Each manufacturing activity can be described by a set of parameters, which can be grouped by three blocks of information:

- The model of activity *AD* – defines general parameters which distinguish model from others. Gives general overview of activity.
- The method of realization *BD* – each model of activity can be realized in different ways. The possible ways can be defined here.
- The parametrization of realization *CD* – when we introduce a real values for a specific example we get parametrization of realization.

Let's consider for example, the activities belonging to the class of transformation of shape change during the machining. The model AD_{TR} is described by the following sets of parameters:

- *ADT1*: the machining method and its kinematics, e.g. turning, milling, drilling,
- *ADT2*: the model of workpiece feature, e.g. cylindrical surface, pocket, slot,
- *ADT3*: the model of volume transformation of manufacturing feature, e.g. with constant or variable value of machining allowance,
- *ADT4*: the accuracy of machining method, e.g. roughing or finishing,
- *ADT5*: the type of cutting tool, e.g. turning cutting tool, twist drill.

The model AD_{TR} defines the elementary machining capability. For example the same workpiece feature (e.g. cylindrical surface) can be processed using different machining methods (turning, grinding, milling) with different type of cutting tools (turning tool, grinding wheel, shoulder mill), with different accuracy and volume transformation model (for roughing machining allowance is bigger and variable, for finishing is smaller and constant). Each combination defines new model of transformation. But this information is still general. Each model can be realized in different ways, what defines possible realizations of given model. For example, different tool pathways strategies could be applied. Finally, when real parameters are added (for example: diameter and length of cylindrical surface, values of feed and cutting speed) we got parametrized realization of model, what is equivalent with single machining cut.

To define all types of manufacturing activities, with division on *AD*, *BD* and *CD*, the proper classifications are required. Those classifications, in the form of trees, which defines attributes domain, are described in [1, 2, 4].

2.2 The Machining Process Template and Manufacturing Knowledge

The machining process template is created for the part families. The family includes parts with similar machining process plan. It means that includes all possible process elements (like operations, setups, machining cuts) which can be applied to any part of this family. For a new part, during the process plan generation, only necessary process elements are selected [2].

The machining process template defines general process plan structure, but also defines the manufacturing knowledge (by the heuristic how to create the process structure). In general, the structure of manufacturing knowledge *WT* can be described by manufacturing activities models and *MSPO* structure (which defines connections between activities), what presents following equation [2]:

$$WT \rightarrow \{AD_{TR}, AD_{AU}, AD_{ID}\}, MSPO \quad (1)$$

where:

- AD_{TR}, AD_{AU}, AD_{ID} – models of manufacturing activities: transformation, auxiliary and identification,
- *MSPO* – the model of structure of machining process plan.

The *MSPO* model is represented in the form of nested networks. On the first level, the division into operations is defined. The example network presents Fig. 2.

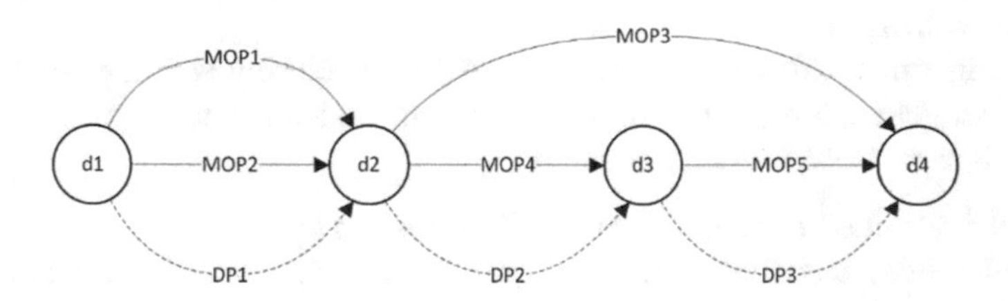

Fig. 2. The example network of *MSPO* on first level

The network consists of:

- Nodes – represents both: place of decision making and intermediate state of workpiece during the machining process,
- Directed edges – represents necessary design activities, after which the node is changed. On this level there are two types of edges:
  - *MOP* – the model of manufacturing operation which is represented by the network of second level. It defines the operation design algorithm.

– *DP* – the apparent activity, which cause only the change of node in network.

Each operation model *MOP* can be defined also by a network of lower level. The example network presents Fig. 3. This network defines the operation design algorithm.

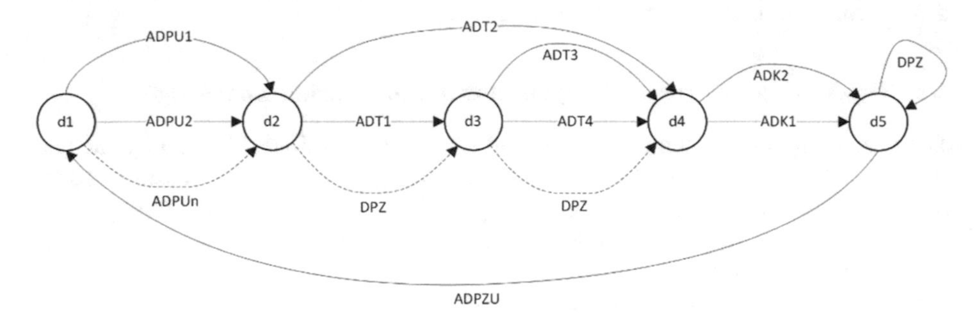

Fig. 3. The example network of *MPO* on second level

There are the same rules. Nodes represents intermediate states of workpiece. But here is more types of edges, equivalent to models of manufacturing activities:

- *ADPU* – the model setup activity. It must be first in the second level network.
- *ADT* – the model of transformation activity.
- *ADK* – the model of identification activity.
- *ADPZU* – the model of setup change. Cause the return back and starts operation design from the beginning.
- *DPZ* – the apparent activity, which cause only the change of node in network.

Using such *WT*, the generation of machining process for a new part is possible. During process planning are generated intermediate states of workpiece, structure of machining process plan and all necessary manufacturing resources (machine tools, equipment, cutting tools).

The main aim of this work is utilization of manufacturing activities for machining process plan generation (with semi-generative method for automated manufacturing systems). It is connected with necessity of:

- Formal description of automated manufacturing system,
- Definition of manufacturing activities occurred in given manufacturing system,
- Identification, based on manufacturing activities set, of required resources.

3 Formal Description of Automated Manufacturing System State

The automated manufacturing system could be presented as a set of elementary and complex objects which are organized in the form of hierarchical tree [4]. Any complex object *O* (e.g. manufacturing system, machine tool, robot) consists of objects $o_i (i = 1, 2, \ldots, n)$, which are constitute the ordered set of objects:

$$O = \{o_1, o_2, \ldots, o_n\} \tag{2}$$

Each object o_i can be described by a set of attributes A. The main three subsets of attributes can be distinguished: AG – geometric, AT – technological and AF – physical:

$$A = \{AG, AT, AF\} \tag{3}$$

The set of attributes A have to be defined separately for each kind of object o_i. Each attribute (parameter) from AG, AT and AF subsets, should be recorded twice: first as a nominal value (*zd*, required) and then as a real value (*rz*, measured). State of object o_i can be defined with the following state vector (set of attributes):

$$So_i = \left[AG_{zd}, AT_{zd}, AF_{zd}, AG_{rz}, AT_{rz}, AF_{rz}\right]^{\mathrm{T}} \tag{4}$$

There may be some links w_i between objects, which constitute the set of links W:

$$W = \{w_1, w_2, \ldots, w_n\} \tag{5}$$

Each two connected objects, at least by one link, determine the pair of objects and they remain in a relationship with each other, which can be described by state vector. Each complex object consists of pairs p_i, which are determine the object's set of pairs Po_i:

$$Po_i = \{p_1, p_2, \ldots, p_n\} \tag{6}$$

To describe the mutual state of objects in a pair, the following subsets of parameters (attributes) must be specified:

- *IDP* – identifiers of objects in pair and type of pair,
- *WPG* – describing the mutual geometrical location,
- *CZO* – describing the nature of the objects' surfaces contact,
- *PPK* – describing the parameters of the kinematic pair.

These subsets determines a set of parameters of pair state Sp_i, which can be recorded using the pair state vector:

$$Sp_i = [IDP, WPG, CZO, PPK]^{\mathrm{T}} \tag{7}$$

A complex object can be represented as a chain of object pairs. If different configurations of a complex object are possible, then the object may have several variants of this chain. Chains l_i determine the set of object's chains Lo_i:

$$Lo_i = \{l_1, l_2, \ldots, l_n\} \tag{8}$$

To describe the state of the object's chain l_i, have to be specified the position of the coordinate system of the final object of the chain (executive element) relative to the coordinate system of the initial object (base).

The state space of the manufacturing system is described as the state of all simple objects, object pairs and the chains of the pairs in it.

If we describe the manufacturing system with the help of the state space, we can define the functions defined in this space. These functions can be implemented using operators. Each of these operators causes a change in the state in the manufacturing system. Thus, manufacturing activities can be described by means of operators (operator = action, in a mathematical sense, on a given set of parameters causing a change in state).

The domain of manufacturing activity D is the state space of the manufacturing system S_{SW}. Every manufacturing activity causes its change. In general, it could be presented in the following form:

$$S_{SW}(j) + D \rightarrow S_{SW}(j+1) \tag{9}$$

As a result of manufacturing activity D, the j-th intermediate state of the manufacturing system S_{SW} passes into another intermediate state $j + 1$. The manufacturing system state is described by a state of all objects (elementary objects So_i, pairs of objects Sp_i and chains of objects Sl_i). The formal description of the activities only shows how they affect the state of the manufacturing system.

4 Process Plan Definition as a Set of Manufacturing Activities

According to the previous chapters, the new definition of manufacturing process can be presented. The manufacturing process is a intended and ordered series of discrete manufacturing activities occurred in manufacturing system with reference to workpiece. The manufacturing process is realized in manufacturing system.

During the manufacturing process, the change of workpiece state is present. There are following states of workpiece:

- Initial state of workpiece – it is a raw material,
- Final state of workpiece – it is a final product,
- Intermediate states of workpiece – represents all states during machining process, what is equivalent to *WIP* (work in process).

The manufacturing activities determines the discrete change of intermediate state of workpiece. Therefore the manufacturing process can be described by the set of manufacturing activities and hierarchical process structure *SPO*, which defines the sequence of activities occurred in the process:

$$PTO = \{D_{TR}, D_{AU}, D_{ID}\}, SPO \tag{10}$$

Where:

- $\{D_{TR}, D_{AU}, D_{ID}\}$ – the set of manufacturing activities,
- *SPO* – the structure of machining process plan.

The manufacturing process, realized in given manufacturing system, is described by the set of activities realized by given resources. As a result, the discrete change of workpiece state from raw material to finished product take place.

5 Practical Use of Manufacturing Activities Models

The formal description of manufacturing activities gives opportunity to define manufacturing capabilities of any complex object in manufacturing system. First should be defined general models *AD* and possible realizations *BD*. Using these sets, we assign to resources (e.g. machine tool) what kind of activity models *AD* can be applied and with which realization *BD*. In that way, the manufacturing capabilities are defined [1].

Second aspect of practical use is a utilization to define the manufacturing knowledge. The model of knowledge for manufacturing process planning will be created on the basis of concept of description of activities occurring in the manufacturing system. The set of activities occurring in the manufacturing system *D* can be presented as:

$$D = \{D_I, \{D_{II}, \{D_{III}, \ldots, \{D_R\}, \ldots, \}\}\} \tag{11}$$

where: $D_I, D_{II}, D_{III}, \ldots, D_R$ – subsets of activities on the I, II, …, Rth level of classification.

The subsequent levels of classification distinguish the types, groups and variations of manufacturing activities. Such classification of activities forms the basis for the definition of:

- manufacturing knowledge (used to define both machining process plan template and production rules) [2],
- procedures for the identification and transformation of workpiece manufacturing features [7],
- manufacturing capabilities of manufacturing system resources [1, 4],
- connections between manufacturing activities and resources selection, cutting parameters selection, using table knowledge with catalogue values, etc.

The assumed formal description is the base for the development of source database necessary to generate the machining process plans and covering the constituent data (models of activities *AD* and the methods of their realization *BD*) for the definition of activities occurring in the manufacturing system.

The operation of this CAPP system can be described as follows. The input is the data about the part, for which the machining process must be prepared. This description must be oriented on manufacturing features [7]. CAPP system is equipped with procedures for the recognition of manufacturing features, which are further used to determine the structure of the manufacturing process using the knowledge base. The second stage is the generation of stock and intermediate states of part, which can occur during the manufacturing. Next stage covers the selection of methods and resources for realization of activities with the use of data about the processing capabilities of manufacturing system. The detailed description of CAPP system operation is given in [2, 5]. All stages of operation of CAPP system described above use data and manufacturing knowledge related to the manufacturing activities.

6 Summary

The result of this work is a developed method of manufacturing activities definition. It can be used to:

- Manufacturing knowledge definition,
- Describe the manufacturing capabilities of any automated manufacturing system,
- Record the machining process plan, as a result of generation by semi-generative method, as a ordered set of manufacturing activities with definition of the way of realisation and parametrization.

This method makes possible the realization of machining process generation with Expert System use and possibility to:

- Automated workpiece features recognition [7],
- Manufacturing resources selection (machine tools, equipment, cutting tools) [4],
- Further analysis of machining processes to build new manufacturing knowledge.

Acknowledgements. This work is a part of project called CyberTech, titled: "The expert system of machining process planning of aircraft components". This project is realized within the "Innolot programme – innovations in aviation", financed by Polish government.

References

1. Duda, J., Habel, J., Kwatera, M., Samek, A.: Data base with open architecture for defining manufacturing system capabilities. In: Proceedings of 9th FAIM International Conference, pp. 1083–1094 (1999)
2. Duda, J., Habel, J., Pobożniak, J.: Repository of knowledge for manufacturing process planning. In: Proceedings of 15th FAIM International Conference, pp. 171–178 (2005)
3. Maraghy, E.: Evolution and future perspectives of CAPP. Ann. CIRP **42**(2), 739–751 (1993)
4. Habel, J.: Manufacturing system resources modelling for CAx systems. In: Weiss, Z. (ed.) Virtual Design and Automation, New Trends In Collaborative Product Design, pp. 391–398. Publishing House of PUT, Poznań (2005)
5. Habel, J.: The idea of integrated manufacturing process planning system. In: Proceedings of 16th FAIM International Conference, pp. 185–192 (2006)
6. Kumar, S.L.: State of the art-intense review on artificial intelligence systems application in process planning and manufacturing. Eng. Appl. Artif. Intell. **65**, 294–329 (2017)
7. Pobożniak, J.: Interacting manufacturing features in CAPP systems. In: Advances in Manufacturing, pp. 249–258 (2018)
8. Chang, P.-T., Chang, C.-H.: An integrated artificial intelligent computer-aided process planning system. Int. J. Comput. Integr. Manuf. **13**, 483–497 (2000)
9. The Dassault 3D Experience. http://www.3ds.com/products-services/3dexperience/

Model of Application of Cluster Analysis in Storage Area Designing

Peter Trebuňa, Jana Kronová(✉), and Miriam Pekarčíková

Technical University of Kosice, Kosice, Slovak Republic
{peter.trebuna, jana.kronova, miriam.pekarcikova}@tuke.sk

Abstract. The article deals with an application of cluster analysis in stock area designing. There was designed the model of application of cluster analysis in stock area designing, which was depicted on practical example and with that was verified the applicability of the designed model. Cluster analysis forms the clusters according to the similarity of finished products. According to the resulting clusters is divided the stock of finished products, for which is created 2D and 3D layout model of the stock.

Keywords: Storage area design · Cluster analysis

1 Introduction

Cluster analysis belongs to multivariate statistical methods. "Cluster analysis is a general logic process, formulated as a procedure by which groups together objects into groups based on their similarities and differences" [5, 13].

Having a data matrix *X* type *n x p*, where *n* is the number of objects and *p* number of variables (features, characteristics). Next there is a decomposition *S(k)* of set *n* objects to *k* certain groups (clusters), i.e.

$$S^{(k)} = \{C_1, C_2, C_3, \ldots C_k\}, [1, 4, 6, 7]: \tag{1}$$

$$C_i \neq \varnothing, i = 1, \ldots, k, \tag{2}$$

$$\bigcup_{i=1}^{k} C_i \text{ comprises all the space.} \tag{3}$$

If that set of objects $o = \{A_1, A_2, \ldots, A_n\}$ and any dissimilarity coefficient of objects *D*, then a cluster is called a subset of *p* sets of objects *o* to which it applies [1, 4, 6, 7]:

$$\max_{i,j} D(A_i; A_j) < \min_{k,l} D(A_k; A_l), \tag{4}$$

where $A_i, A_j, A_l \in o$ a $A_k \notin p$. This means that the maximum distance of objects belonging to the cluster must always be less than the minimum distance any object from the cluster and object outside cluster.

© Springer Nature Switzerland AG 2019
A. Burduk et al. (Eds.): ISPEM 2018, AISC 835, pp. 225–233, 2019.
https://doi.org/10.1007/978-3-319-97490-3_22

The input for the clustering of the input data matrix and output is a specific identification of clusters. The input matrix X of size $n \ x \ p$ contains the i-th row of characters x_{ij} object A_i, where $i = 1, 2, \ldots, n$ and $j = 1, 2, \ldots, p$ [9, 12].

Classification of cluster analysis methods is shown in Fig. 1.

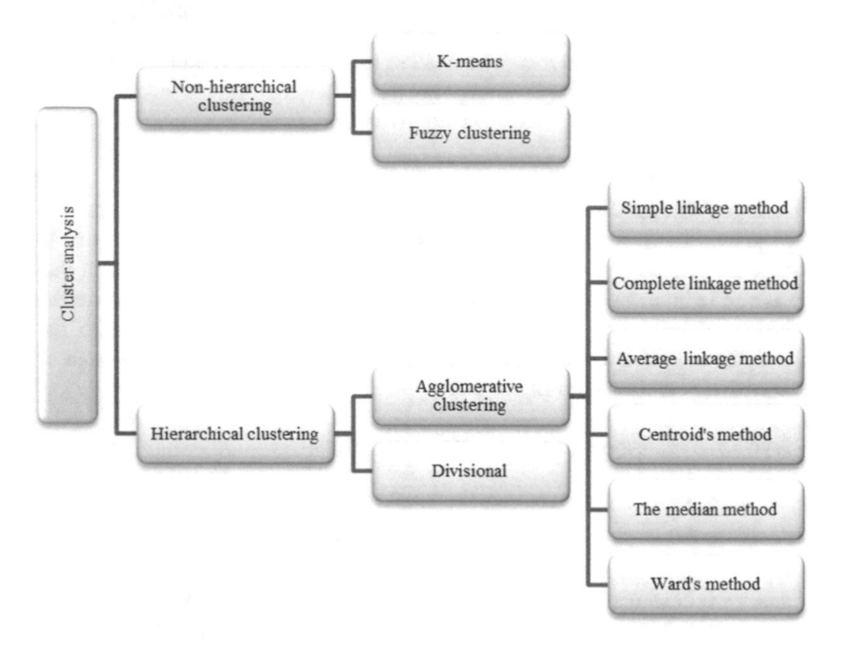

Fig. 1. Classification of cluster analysis methods.

Cluster analysis is a summary term for a group of methods that aim to either group the objects into clusters or clusters create a hierarchy of objects. Hierarchical cluster analysis methods analyzed classify objects into a hierarchical system of clusters. After this is important effective programming or utilization available software. [10] Between the hierarchical methods of cluster analysis method are simple linkage, complete linkage method, average linkage method, centroid method, median method, Ward's method. Non-hierarchical methods do not create hierarchical (tree) structure and the objects are categorized into the number of disjunctive clusters specified in advance. Between the non-hierarchical cluster analysis methods of cluster analysis method are k-means, fuzzy clustering [3].

In this paper was used hierarchical cluster analysis method - Ward's method. Ward's method is also marked as a method of minimizing the increases of errors of sum squares. It is based on optimizing the homogeneity of clusters according to certain criteria, which is minimizing the increase of errors of sum squares of deviation points from centroid. This is the reason why this method is different from previous methods of hierarchical clustering, which are based on optimization of the distance between clusters [2, 4].

The loss of information is determined at each level of clustering, which is expressed as the increase of total sum of aberrance square of each cluster point from the average *ESS* value. Then comes to an connection of clusters where there is a minimal increase in the errors of sum of squares [2, 5].

The accruement of *ESS* function is calculated according to [2, 11]:

$$\Delta ESS(A_i, A_j) = \frac{1}{2} d_{ES}(A_i, A_j), A_i, A_j \in 0, i, j = 1, 2, \ldots, n. \quad (5)$$

The article will deal with the application of methods of cluster analysis as a tool for creating groups of similar inventory items (clusters) to speed up reactions on customer requirements and efficiency to the arrangement in the warehouse. The proposed algorithm of application of cluster analysis in storage area designing will be verified on a practical example and reorganization measures of the storage space will be proposed.

2 Model of Application of Cluster Analysis in Storage Area Designing

Proceeding of cluster analysis application in in-plant production processes modeling is divided into three consequential stages:

I. stage – preparation,
II. stage – realization = cluster analysis,
III. stage – evaluation and proposition.

First stage of cluster analysis application in material and technical supply is preparatory, in which is needed to create a list of stock item, which will be a subject of analysis. In the case of item analysis of output stock are input data the information from expedition plan, which is the subject of this article, it means, the input for cluster analysis are the information from following movements of company processes. Selected data are analysed from the view of their development in tracking period and after that is created the input data matrix in the form "stock item tracking period" in table form.

Then follows second stage, cluster analysis itself applied on input data matrix "stock item tracking period". As a result of cluster analysis is creation of groups (clusters) of stock items.

Clusters of stock items in the form of matrix "stock items clusters are in the third stage analysed, whether they fit to conditions of the stock. If it is possible to regard the selection of cluster as optimal solution, then is the process of clustering ended and the matrix "stock item clusters" is possible to use as a basis for proposition of configuration reorganization of stock items. If the output matrix is not from the view of stock conditions optimal, process of clustering is repeated from the point of selection of optimal number of clusters (Fig. 2).

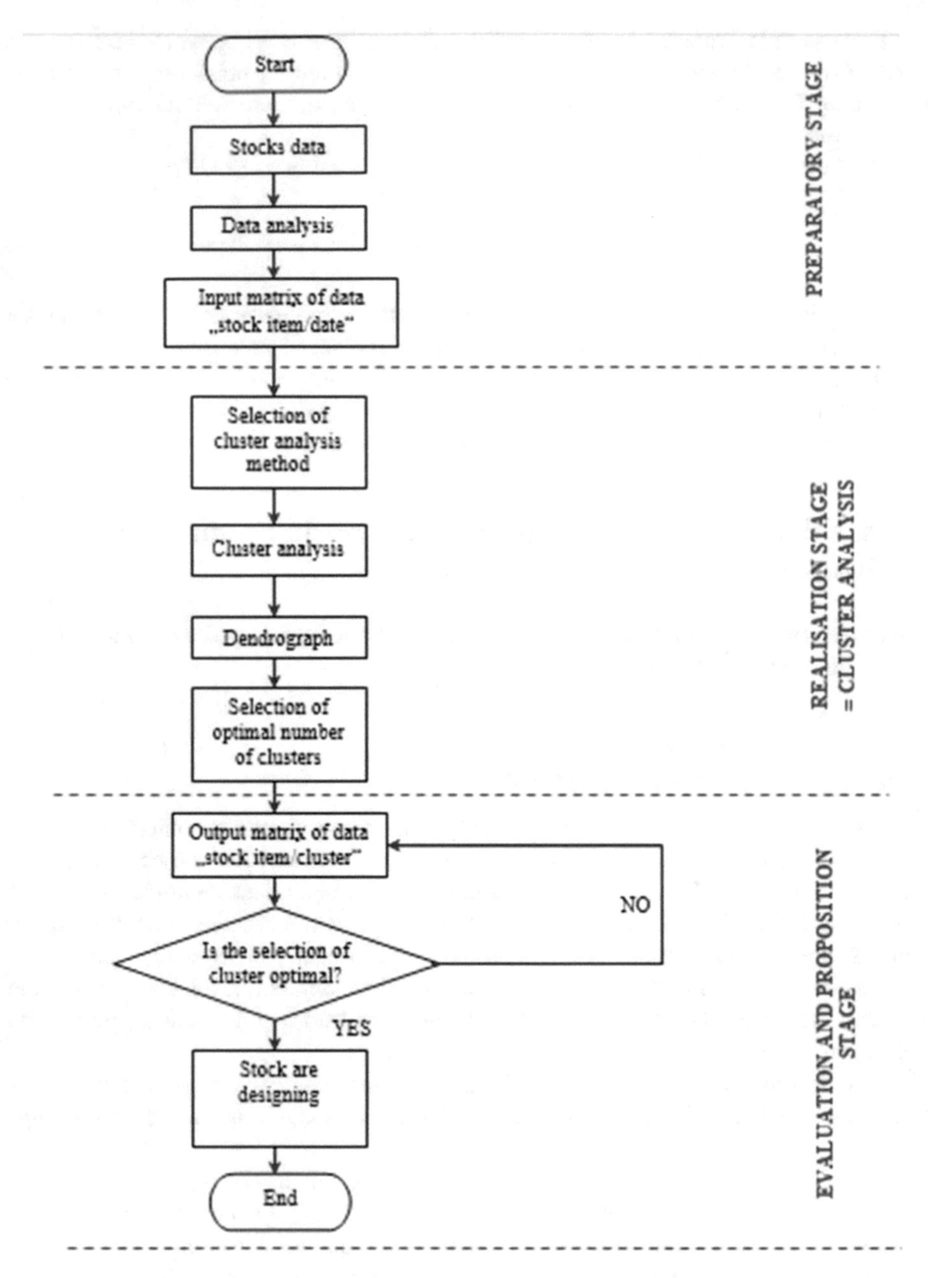

Fig. 2. Model of application of cluster analysis in storage area designing.

3 Example of the Application of Cluster Analysis in Storage Area Designing

3.1 Preparatory Stage

Proposed model of application of cluster analysis in stock area designing is in following part applied on practical example. To perform the analysis are needed of data on the expedition of finished products particular company for the year. [8] Fig. 3 shows the evolution of customer requirements resp. expedition during the year.

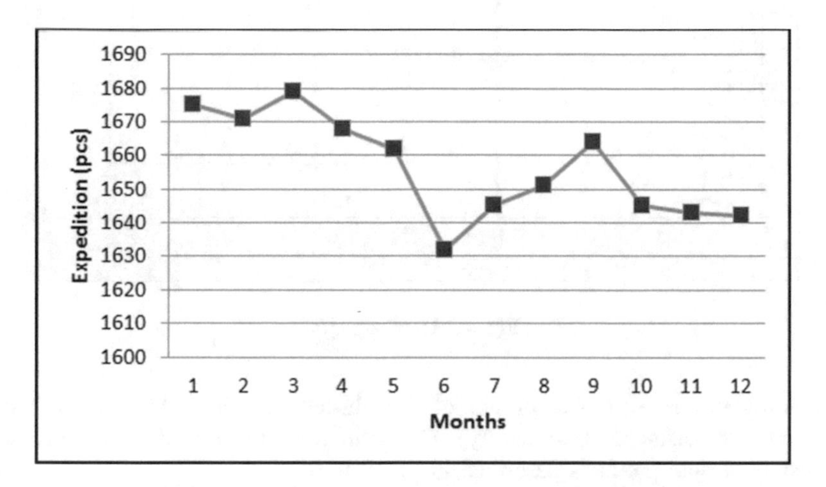

Fig. 3. Evolution of expedition during the year.

It can be seen that the evolution of customer requirements resp. expedition of finished products in the year has variable course. It can be seen that the evolution of means to customer i.e. evolution dispatch of products has variable course, the average expedition represents 1,66 pcs of products per month, the greatest demand resp. export was observed in March of that year and the smallest demand resp. export was observed in June of that year. In some months of was created a single customer's request for the selected product.

3.2 Realisation Stage – Cluster Analysis

Based on the input data on a monthly expedition of finished products for the customers in year was performed cluster analysis. In the program STATISTICA 10 was realised the cluster analysis of final products, as method of cluster analysis was used the Ward method. The result of the process clustering is a dendrograph showing the different clusters according to the distance (dissimilarity) show in Fig. 4.

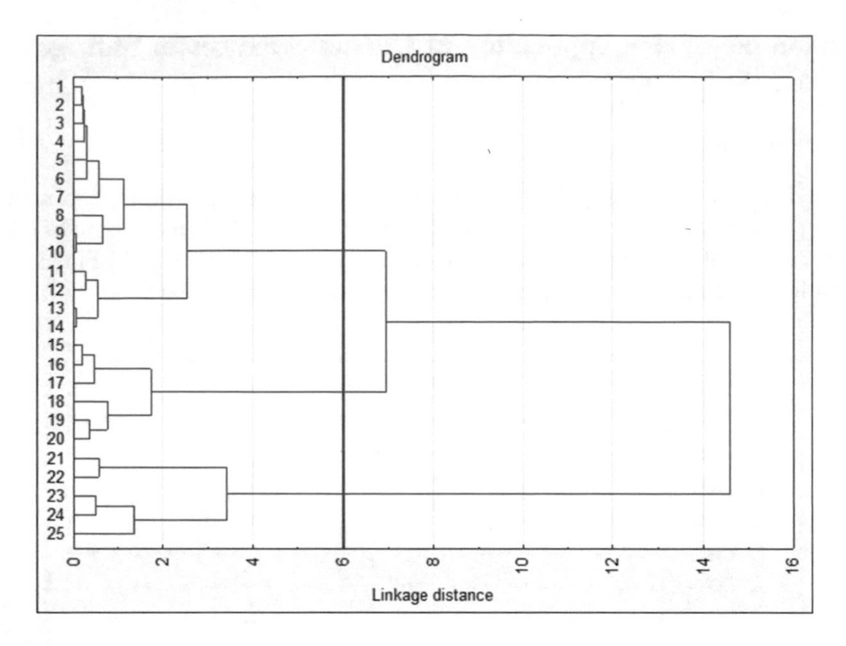

Fig. 4. Dendrograph.

Dendrograph is a graphic output of the cluster analysis, which shows clusters according to the distance (dissimilarity). The optimal clusters on judgment of the solver are clusters of final products described in the Table 1.

Table 1. Cluster of final products.

Parameters/months	1	2	3	4	5	6	7	8	9	10	11	12	Clusters
1	0	0	0	12	12	0	0	0	0	0	0	0	**1**
2	0	0	0	0	0	0	0	0	12	0	0	0	**1**
3	0	17	0	0	0	0	0	0	0	0	0	0	**1**
4	18	0	0	0	0	0	0	0	0	0	0	0	**1**
5	0	0	0	0	0	0	0	0	0	0	15	17	**1**
6	0	0	23	0	0	0	0	0	0	0	0	0	**1**
7	0	0	0	0	0	0	22	23	23	22	0	0	**1**
8	34	29	28	30	23	0	0	0	0	0	0	0	**1**
9	19	19	19	20	19	20	19	19	20	19	20	19	**1**
10	21	21	21	21	21	21	21	21	21	21	21	21	**1**
11	48	47	48	47	47	47	48	47	48	47	48	47	**1**
12	40	40	40	39	39	39	39	39	39	39	39	40	**1**
13	31	31	32	31	33	33	32	31	33	31	34	31	**1**
14	32	32	33	32	32	33	32	34	32	33	32	32	**1**

(continued)

Table 1. (*continued*)

Parameters/months	1	2	3	4	5	6	7	8	9	10	11	12	Clusters
15	77	78	77	77	78	77	77	78	77	77	78	77	**2**
16	71	72	71	72	71	72	71	72	71	72	71	72	**2**
17	63	63	61	63	63	63	62	63	63	62	63	63	**2**
18	112	112	112	112	112	112	112	112	112	112	112	112	**2**
19	98	98	98	98	98	99	98	98	99	98	98	98	**2**
20	87	87	89	87	87	90	87	87	89	87	87	87	**2**
21	138	139	139	138	139	138	139	139	138	138	139	139	**3**
22	155	155	155	156	155	155	155	156	155	155	155	155	**3**
23	229	229	229	229	229	230	229	229	229	229	229	229	**3**
24	215	215	215	215	215	215	215	215	215	215	215	215	**3**
25	187	187	189	189	189	188	187	188	188	188	187	188	**3**

It was confirmed that the use of cluster analysis in sorting of stocks of finished products is justified because the clusters are formed on the basis of similarities in our case similarity of expedition to customers, which is the main criterion for the formation of groups of products towards our customers. Products with the greatest expedition should be placed closest to the exit. This criterion will then be taken into account when designing the layout. The percentage share of clusters on total expedition is graphically shown in Fig. 5.

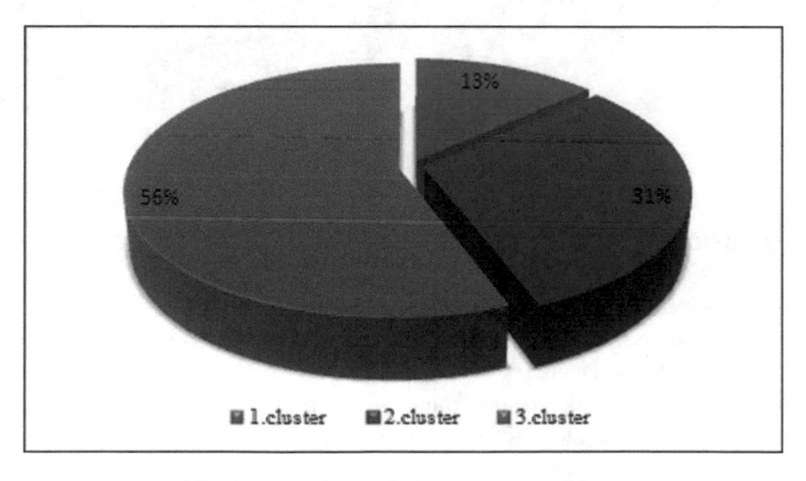

Fig. 5. The share of clusters on expedition.

3.3 Evaluation and Proposition Stage

From the graphic illustration of Fig. 3 follows that largest share of the expedition have products 3.cluster, then 2.cluster and the smallest share has 1.cluster. It is therefore appropriate stored of products 3.zhluku closer towards the exit in the warehouse of finished products.

Draft of layout of distribution warehouse is shown in Fig. 6 (2D layout – sw. MS Visio 2013), Fig. 5 (3D layout – sw. visTABLE®). In the layout is marked clusters by the percentage of share in the expedition.

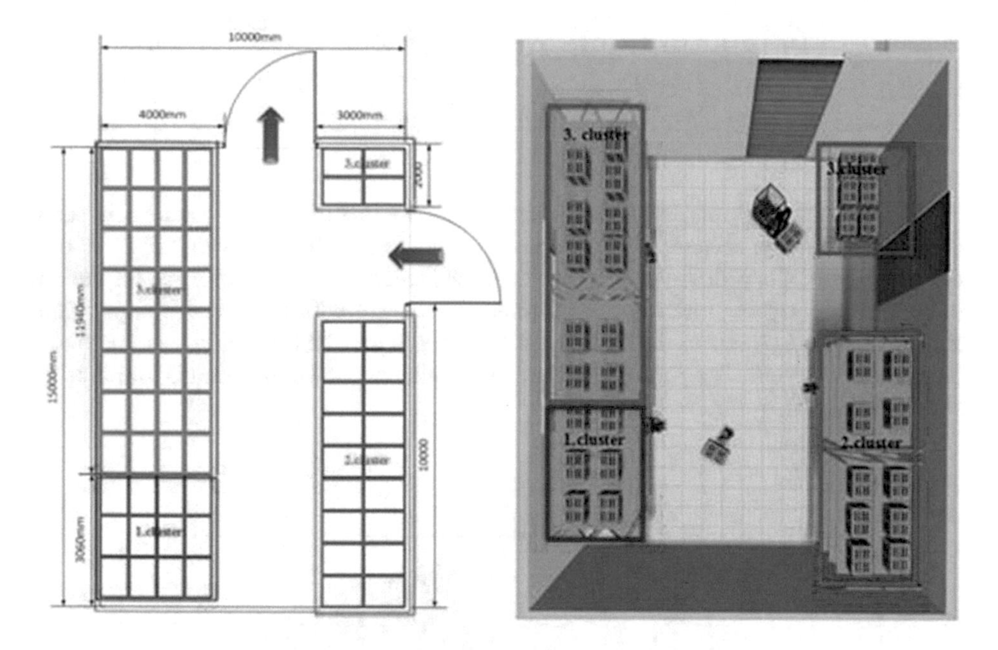

Fig. 6. Layout of storage area with clusters.

4 Conclusion

At the beginning of this article is introduced brief characteristic of cluster analysis and also the possibilities of its application. The article is concerned with an application of cluster analysis in the stock area designing. There was an algorithm designed for cluster analysis application of stock area designing, which usefulness was verified on practical example. At the end of the article is typified graphical design of stock area with distribution to clusters according to results of analysis. The main advantage of cluster analysis application in stock area designing is selection of criterion or more precisely criterions for cluster analysis realisation in dependence on conditions of concrete project.

In the paper was applied for stocks of finished products. This procedure is applicable in all types of warehouses, in the input warehouse, in warehouse of auxiliary materials and in warehouse of finished products. When creating groups (clusters) of stock items using cluster analysis are necessary information about the process, which immediately followed, i.e. for the analysis of stock items in output warehouse are input data of information on the expedition of products from previous years. Described procedure of the application of cluster analysis in sorting of stocks is a systematic and logical.

Acknowledgements. This article was created by implementation of the grant project VEGA 1/0708/16 "Development of a new research methods for simulation, assessment, evaluation and quantification of advanced methods of production". KEGA 030TUKE-4/2017 "Implementation of innovative instruments for increasing the quality of higher education in the 5.2.52 Industrial Engineering field of study".

References

1. Bacher, J., Poge, A., Wenzig, K.: Cluster Analysis - Application-Oriented Introduction to Classification Methods. Oldenbourg, Munchen (2010)
2. Everitt, B.S., Landau, S., Leese, M., Stahl, D.: Cluster Analysis. Wiley, London (2011)
3. Han, J., Kamber, M.: Data Mining – Concepts and Techniques. MK Publisher, San Francisco (2006)
4. Kaufmann, L.: Finding Groups in Data: An Introduction in Cluster Analysis. Wiley, Hoboken (2005)
5. Straka, M.: Logistics of Distribution, How Effectively to Put Product Into the Market, 1st edn. EPOS (Original in Slovak), Bratislava (2013)
6. Trebuňa, P., Halčinová, J.: Experimental modelling of the cluster analysis processes. Procedia Eng. **48**, 673–678 (2012)
7. Palumbo, M., Lauro, C.N., Greenacre, M.J.: Data Analysis and Classification. Springer, Berlin (2010)
8. Straka, M., Malindzak, D.: Distribution Logistics. TU Kosice, Kosice (2005)
9. Abonyi, J., Feil, B.: Cluster Analysis for Data Mining and System Identification. BV. Berlin, Berlin (2007)
10. Aldenderfer, M.S.: Cluster Analysis. Sage Publications, Thousand Oaks (1944). ISBN 0-8039-2376-7
11. Backhaus, K., Erichson, B., Plinke, W., Weiber, R.: Multivariate Analysemethoden. Springer, Munchen (2006)
12. Gan, G., Ma, Ch., Wu, J.: Data Clustering: Theory, Algorithms and Applications, PA, Alexandria (2007)
13. Rud, O.P.: Data Mining. Computer Press, Bratislava (2001)

Module for Prediction of Technological Operation Times in an Intelligent Job Scheduling System

Łukasz Sobaszek[1], Arkadiusz Gola[1(✉)], and Edward Kozłowski[2]

[1] Faculty of Mechanical Engineering, Institute of Technological Systems of Information, Lublin University of Technology, Lublin, Poland
{l.sobaszek, a.gola}@pollub.pl

[2] Faculty of Management, Department of Quantitative Methods in Management, Lublin University of Technology, Lublin, Poland
e.kozlovski@pollub.pl

Abstract. This paper presents a model for the prediction of technological operation times in the framework of an intelligent job scheduling system. The developed prediction module implements ARMA/ARIMA time series models. In addition, the paper introduces the mathematical prediction model and its implementation to the particular test case. The scheduling made use of dispatching rules: LPT, SPT, FCFS and EDD. The validation of the model appears to confirm the effectiveness of the proposed solution and substantiate further research works in this direction.

Keywords: Intelligent scheduling system · Real production scheduling
Operation times uncertainty

1 Introduction

The mounting pressure of the competition forces the enterprises to maximise their efforts in constant upgrading and optimising the effectiveness of conducted processes [1, 2]. Therefore, scheduling production jobs remains the crucial element in the field of production organisation and management. Sequencing jobs on technological machines affects both the completion time of production jobs, as well as the degree to which the production capabilities of a given enterprise are utilised [3, 4].

Although there is a considerable amount of academic literature in the field of scheduling production jobs, the proposed solutions are based on fixed operation times [5, 6]. In practical applications, however, the real job completion times are uncertain, being dependent on a number of various factors, such as: the training and experience of the personnel, the size of production, or even which working shift carries out production tasks (day or night) [7]. Consequently, the real processing times of particular jobs show certain discrepancy from the nominal schedule, which has a direct impact on the disparity between the target schedule and its actual execution.

The principle objective of this paper was therefore to present the assumptions of a prediction module for scheduling jobs under processing time uncertainty. The module,

© Springer Nature Switzerland AG 2019
A. Burduk et al. (Eds.): ISPEM 2018, AISC 835, pp. 234–243, 2019.
https://doi.org/10.1007/978-3-319-97490-3_23

which is to become an integral element of a larger intelligent job scheduling system, is an autoencoding system capable of unsupervised learning, equipped with tools for the prediction of production job processing times based on real historical data.

2 Variable Processing Times as the Source of Production Process Uncertainty

Although the list of potential disruptions of the production process is lengthy, this is the uncertainty regarding the processing times of technological operations that most frequently surfaces in real scheduling conditions. Uncertain processing times prevent us from defining a precise makespan. What is more, any change of operation time will have a bearing not only on the subsequent stages of the technological process, but also on other jobs. Figure 1 shows a scenario where the processing time of one of the operations of job 3 is changed (tardiness), as a result of which all other operations marked with an asterisk move, and the time of completion of all jobs is longer. In the real production system, this could lead to enforcing (often very high) contractual fines.

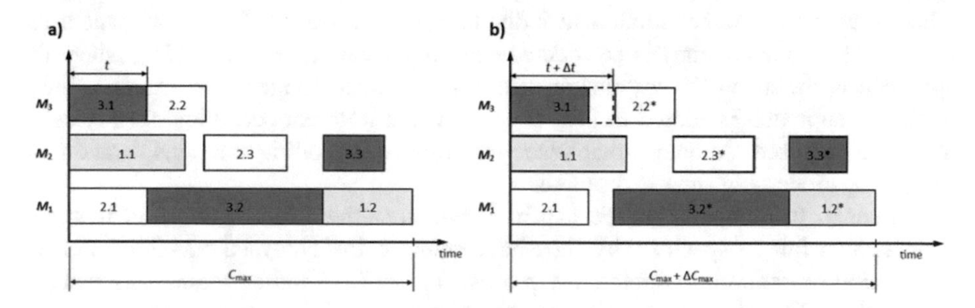

Fig. 1. Consequences of elongating technological operations processing times: (a) nominal schedule, (b) tardiness of one operation

The analysis of the impact of job tardiness is widely regarded as both an essential and a current issue. In recent years, there has been a growing body of literature presenting solutions to analysing the effect of production job processing time errors on scheduling.

3 Processing Times Uncertainty in the Literature

The analysis of the body of literature on the problem of job scheduling under technological times uncertainty indicates that there are several established methods for solving such problems, which are as follows:

- distribution function,
- fuzzy logic,
- other variable technological operation time analysis methods.

One of the major trends in analysis of processing times is the application of the distribution function. However, no specific subgroups can be classified in this field as researchers consider various typical distribution functions – from the uniform distribution to mixed distributions. It is, nevertheless, infrequent to approach job-shop problems with multiple-machines [8], which are relatively common in the industrial conditions, and the majority of analytical works focus on single-machine or two-machine problems [9, 10].

Processing time uncertainty is also analysed by means of fuzzy logic. The body of literature on fuzzy numbers in robust scheduling in job-shop systems is quite extensive. In works [11, 12], authors approach scheduling under uncertainty in a two-machine environment, executed by means of an improved version of a simple local search. The solution is tested for typical test scenarios. Another work [13] attempts to verify the efficiency of fuzzy-logic-based solutions with the bee colony approach. The fuzzy number theory finds implementation in other scheduling problems as well, however, the number of published works is significantly lower [14].

The two abovementioned methods for solving scheduling problems under processing time uncertainty in job-shop systems are currently the most popular in such applications. The literature offers other alternatives to these methods, however, the other solutions are rather limited to individual case scenarios. In several papers the processing time uncertainty is considered deterministically, such as in [15], where the approach is based on the arithmetic mean of registered longest and shortest times. A similar approach is shown in [16] or [17], where different scenarios of processing times are analysed. Another notable solution proposes limiting the considerations to shortening processing time by compression [18].

Although the majority of the papers above implement the distribution function, their authors fail to specify why they have selected the analysed functions and the distribution of random variables, *i.e.* processing times. Another problem is that the proposed solutions do not satisfactorily explore the necessity to identify the distribution and prediction of its future values.

4 Module for Prediction of Variable Processing Times

The presented literature review indicates the importance of studying the character of processing time variability to determine the individual characteristics of its distribution function. Consequently, we propose an alternative approach in the form of an intelligent job scheduling system, whose idea was introduced in the previous work [19]. Figure 2 presents the idea of job scheduling in production and the position of the scheduling module with regard to planning processes and production execution. The presented module aims to support the work of planning teams in the reactive scheduling phase of production jobs scheduling. The key objective of the module is to absorb the negative effects of uncertainty resulting from the variable production job processing times, and to obtain information on the character of these changes.

In the process of developing schedules and predicting processing times, planners employ tabular data or make use of their own experience. In practical applications, the distinction into the predictive and the reactive phase may not be clear-cut either,

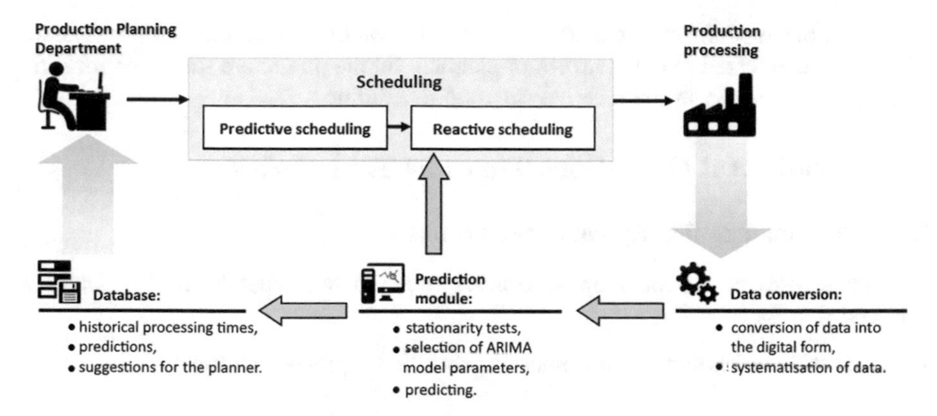

Fig. 2. The framework of intelligent job scheduling with a processing time prediction module

however, in the case of our solution, it is of considerable importance that the prediction module should be executed in two separate stages. The predictive stage enables implementing randomness in the early stages of scheduling, by means of data obtained from the execution of the predictive module, which are stored in the database. More importantly, the real-time data analysis enables employing the proposed solution in the reactive scheduling phase as well. As a result, the schedule is updated according to the predictions, which may take place during the night shift, thus producing the schedule more resembling the real conditions. Moreover, the module includes optimisation tools, which are capable of producing a schedule answering to the actual needs of an enterprise and its defined objective function.

What is the key element, however, is to obtain suitable data from the production stage. This is in this stage that the information showing the variable character of the processed jobs may be obtained. Depending on the character of production, the data may be obtained from:

- in the case of conventional, largely non-automated production: the foreman, job sheets or CCTV material,
- in the case of computer-aided and automated production: CNC machine tools, ERP software or MES systems.

In the former case, the gathered data requires conversion into the digital form, however, in modern enterprises which broadly implement IT-solutions (as per modern standards) the data may be easily managed and shared through local networks. Proper infrastructure enables gathering and storage of data from alternative sources, such as cutting edge machine tools management and control systems.

The converted data are applied in the proposed module, where used in the form of the time series are analysed by means of different tests in order to define their character regarding stationarity, and to identify the key parameters of ARMA or ARIMA models. By applying the model comparison criterion AIC, the optimal model is determined for the prediction of values of considered operations.

The entire module will represent a DSS-type system (Decision Support System), which is implemented to aid the work of planners (in the predictive stage), or constitute a scheduling solution in the reactive stage of scheduling.

5 Technological Operations Times as Time Series

5.1 Mathematical Description of the Problem

In order to present the job-shop scheduling problem, we must define the following datasets:

- set M of m technological machines (workstations) processing jobs:

$$M = \{M_1, \ldots, M_m\}, \tag{1}$$

- set J of n orders (jobs) to process:

$$J = \{J_1, \ldots, J_n\}, \tag{2}$$

Processing job J_j on machine M_i is referred to as operations, hence what needs defining is:

- matrix MO of order $m \times n$ representing the rank of jobs on particular machines:

$$MO = \begin{bmatrix} o_{11} & o_{12} & \cdots & o_{1n} \\ o_{21} & o_{22} & \cdots & o_{2n} \\ \cdots & \cdots & o_{ij} & \cdots \\ o_{m1} & o_{m2} & \cdots & o_{mn} \end{bmatrix} \tag{3}$$

where: o_{ij} – ranking of jobs i on the machine j taking the value of: $o_{ij} = 0$ – when the operation i is not processed on the machine j, $o_{ij} = \{1, \ldots, m\}$ – when the operation i is processed on the machine j; otherwise $o_{ij} = 0$),

- matrix of processing times PT containing data regarding processing times of particular technological operations:

$$PT = \begin{bmatrix} pt_{11} & pt_{12} & \cdots & pt_{1n} \\ pt_{21} & pt_{22} & \cdots & pt_{2n} \\ \cdots & \cdots & pt_{ij} & \cdots \\ pt_{m1} & pt_{m2} & \cdots & pt_{mn} \end{bmatrix}, \tag{4}$$

where: pt_{ij} – processing time of job i on the machine j, while:

$$\bigwedge_{o_{ij}=0} pt_{ij} = 0.$$

What is more, we need to define the matrix of real processing times, *RPT(n)*, which provides the source of data for the proposed intelligent job scheduling system. It can Take the following form:

$$RPT(n) = \begin{bmatrix} pt_{11} + tc_{11}(n) & pt_{21} + tc_{21}(n) & \ldots & pt_{1n} + tc_{1n}(n) \\ pt_{21} + tc_{21}(n) & pt_{22} + tc_{22}(n) & \ldots & pt_{2n} + tc_{2n}(n) \\ \ldots & \ldots & pt_{ij} + tc_{ij}(n) & \ldots \\ pt_{m1} + tc_{m1}(n) & pt_{m2} + tc_{m2}(n) & \ldots & pt_{mn} + tc_{mn}(n) \end{bmatrix}, \quad (5)$$

where: pt_{ij} – processing time of job j on the machine i, $tc_{ij}(n)$ – error of completion time of job j on the machine i predicted by the technologist.

Matrix $RPT(n)$ includes the following dependence:

$$\bigwedge_{o_{ij}=0} (pt_{ij} + tc_{ij}(n)) = 0.$$

Therefore, the selected prediction tools should determine the processing time errors $tc_{ij}(n)$, which are the difference between the operations times predicted by the technologist and the real time of completing job i on the machine j. Included in the schedule, they make it robust and enable producing a more accurate, real time of completion of all jobs.

5.2 Implementation of ARMA/ARIMA Models for Prediction Processing Times

Let the time series $\{tc_{ij}(s)\}_{1 \le s \le n}$ represents the times between operation time assumed by technologist and real operation time. For each $1 \le j \le m, 1 \le i \le n$ these series were identified by the ARIMA models. Below we present some technical aspects regarding the ARIMA models.

Let (Ω, Σ, P) be a probability space and $\{\omega_t\}_{t \in \mathrm{N}}$ be a sequence of identically independent random variables with normal distribution N(0,1).

Definition 1. The time series $\{x_t\}_{t \in N_0}$ is strongly stationary if for every $m \in N$, any $t_1 < t_2 < \ldots < t_m$ and every $\tau \in N$ the joint distributions of the probability m of random sequences of elements $x_{t_1}, x_{t_2}, \ldots, x_{t_m}$ and $x_{t_1+\tau}, x_{t_2+\tau}, \ldots, x_{t_m+\tau}$ are identical.

Definition 2. The time series $\{x_t\}_{t \in N_0}$ is weak stationary if mean and covariance do not vary with respect to time.

Any strictly stationary time series is a weak stationary. The stationary time series usually are modelled by ARMA models. The elements of $ARMA(p, q)$, $p, q \in \mathrm{N}$ model are presented by formula:

$$\begin{aligned} x_t = {} & \alpha_1 x_{t-1} + \alpha_2 x_{t-2} + \ldots + \alpha_p x_{t-p} + \omega_t - \theta_1 \omega_{t-1} - \theta_2 \omega_{t-2} \\ & - \ldots \theta_q \omega_{t-q},\ t > max(p, q) \end{aligned} \quad (6)$$

where $\alpha_1, \ldots, \alpha_p, \theta_1, \ldots, \theta_q$ are parameters.

To verify the stationary of time series we often employ the unit root tests: the augmented Dickey-Fuller test, the Kwiatkowski-Phillips-Schmidt-Shin test, Philips-Perron test, *etc.* [20]

Let the differential operator Δ be defined as $\Delta x_t = x_t - x_{t-1}$, while $\Delta^{k+1} x_t = \Delta^k x_t - \Delta^k x_{t-1}$ for $k \in N$.

Definition 3. The time series $\{x_t\}_{t \in N_0}$ is integrated of order d ($\{x_t\}_{t \in N_0} \in I(d)$) if the series $\{\Delta^k x_t\}_{t \geq k}$ for $0 \leq k < d$ is non-stationary, while the series $\{\Delta^d x_t\}_{t \geq d}$ is said to be stationary.

Definition 4. The time series $\{x_t\}_{t \in N_0}$ is $ARIMA(p, r, q)$, $p, r, q \in \mathrm{N}$ if and only if the time series $\{\Delta^d x_t\}_{t \in N_0}$ is $ARMA(p, q)$.

Of course, the stationary time series $\{x_t\}_{t \in N_0}$ can be identified by different ARMA models (regarding the different autoregressive and moving average orders). Below we propose to use the *Akaike Information Criterion* (AIC) to choose the best model.

6 Preliminary Tests

The proposed solution was verified by means of simulation performed in LiSA software (*Library of Scheduling Algorithms*). The scheduling concerned 8 production orders processed on a stock of 5 machine tools. The processing times, the job order, and the number of jobs assigned to machines were randomly generated with LiSA software, on the basis of the following assumptions:

- processing times cannot exceed one shift (maximum processing time is 7.5 h),
- machine loading is specified at 75%,
- machine routing is predetermined and not subject to change.

In order to analyse the processing times uncertainty and to verify the proposed solution it was resolved that the machines subject to processing time uncertainty constraint are M_1, M_2, M_3 and M_4. The processing times variability was randomly generated – for each operation 100 time errors were generated from the range of $tc_{ij} = \langle -10\% \cdot pt_{ij}; 25\% \cdot pt_{ij} \rangle$. Subsequently, the data was used to define the change in processing times and the predicted makespan. The predicted values were employed in updating the schedule, which was carried out with the use of methods implemented during nominal scheduling (where the processing times were derived from the technological data).

The analytical work has lead to producing the following predicted times of technological operations presented in Table 1.

The conducted verification test works were based on 4 popular dispatching rules:

- *FCFS* – First Come First Service,
- *EDD* – Earliest Due Date.
- *LPT* – Longest Processing Time,
- *SPT* – Shortest Processing Time.

These dispatching rules provided the basis for developing both schedules: the nominal and the robust one. The schedules were produced under the objective function C_{max}, (*make-span, total production time*).

Table 1. Value of C_{max} of produced schedules

Job number	Operation times [h]									
	Nominal times (technology)					Real times (prediction)				
	M_1	M_2	M_3	M_4	M_5	M_1	M_2	M_3	M_4	M_5
Job J_1	6	5	7	5	5	6.49	5.34	7.4	5.37	5
Job J_2	5	–	7.5	7.5	1	5.4	–	8.05	8.14	1
Job J_3	–	1	1	5	7.5	–	1.09	1.07	5.43	7.5
Job J_4	4	4	3	5	–	4.35	4.26	3.25	5.38	–
Job J_5	7	2	5	–	–	7.36	2.11	5.41	–	–
Job J_6	5	6	–	7.5	7.5	5.36	6.55	–	8.05	7.5
Job J_7	1	7	7.5	5	–	1.06	7.56	7.98	5.37	–
Job J_8	–	–	3	4	5	–	–	3.24	4.31	5

The optimisation of the schedules was attempted with the implementation of the Branch and Bound method, which is classified as one of the exact scheduling algorithms. This allowed us to perform the preliminary verification of the proposed module of the intelligent scheduling system. The results obtained from the tests are collated in Table 2.

Table 2. Value of C_{max} of produced schedules

Dispatching rule	Makespan – C_{max} [hours]		
	Nominal schedule (technology)	Robust schedule (prediction)	Optimal schedule (prediction)
LTP	48	51.41	42.05
SPT	46	48.55	
FCFS	43.5	45.59	
EDD	49	50.73	

The analysis of results from the preliminary tests appears to indicate that prediction of real processing times elongates the completion time of all jobs (makespan). However, it ought to be remarked that the sum of real delay of operation times amounted to 9.38 h, whereas the delay of the schedules was on average equal to 2.45 h. What may be inferred is that the schedule is to a certain extent capable of compensating for the delay, and that the prediction-based scheduling does not directly translate to a significant deterioration of schedule parameters. It seems that the implementation of suitable optimisation tools produces a robust schedule (based on predicted times) of a lower objective function value (42.05 h) than in the case of the nominal schedule, based on times obtained from the technology specifications.

The preliminary tests appear to verify the proposed solutions positively. Further investigation based on real industry data should be conducted, with regard to both predicting processing times by means of ARMA/ARIMA models, and implementing the data in producing robust schedules. In future studies, the proposed solutions should be verified in execution of the production process by means of advanced simulation software.

7 Conclusions

One of the serious challenges to face in the field of production management is the uncertainty rooted in the dynamic character of the conducted production processes. Although the literature offers a wide range of solutions for optimising planning and scheduling production, their practical applicability is significantly limited on account of several factors, *e.g.* owing to the deterministic character of the analysed data, which reduces the scope of applications of these models to barely the theoretical rather than the practical.

This paper was an attempt to present the assumptions of the processing times prediction model to become an element of a larger intelligent production job scheduling system. The work assessed the applicability of ARMA/ARIMA models for the prediction of processing time changes and their potential implementation in the robust scheduling of production jobs. The results from the tests show that by employing the stochastic character of the job processing time distribution in the scheduling process enables developing a robust schedule that would account for the processing times of individual jobs obtained from the predictions based on the real time series. The presented solution is an important milestone on the way to developing an intelligent autoencoding production scheduling system capable of unsupervised learning based on the theoretical assumptions of robust scheduling.

References

1. Relich, M., Muszyński, W.: The use of intelligent systems for planning and scheduling of product development projects. Procedia Comput. Sci. **35**, 1586–1595 (2014)
2. Gola, A.: Economic aspects of manufacturing systems design. Actual Probl. Econ. **156**(6), 205–212 (2014)
3. Zwolińska, B., Grzybowska, K., Kubica Ł: Shaping production change variability in relation to the utilized technology. In: 24th International Conference on Production Research (ICPR 2017), pp. 51–56 (2017)
4. Jasiulewicz-Kaczmarek, M., Bartkowiak, T.: Improving the performance of a filling line based on simulation. In: Materials Science and Engineering, vol. 145 (2016)
5. Sitek, P., Wikarek, J.: A hybrid programming framework for modeling and solving constraint satisfaction and optimization problems. Sci. Prog. **2016**, 13 (2016). Article ID 5102616

6. Sobaszek, Ł., Gola, A., Kozłowski, E.: Application of survival function in robust scheduling of production jobs. In: Ganzha, M., Maciaszek, M., Paprzycki, M. (eds.) Proceedings of the 2017 Federated Conference on Computer Science and Information Systems (FEDCSIS), pp. 575–578 (2017)
7. Kłosowski, G., Gola, A., Świć, A.: Application of fuzzy logic in assigning workers to production tasks. In: Advances in Intelligent Systems and Computing, vol. 474, pp. 505–513 (2016)
8. Deepak, K., Yi, M., Gang, C., Mengjie, Z.: Dynamic job shop scheduling under uncertainty using genetic programming. In: Intelligent and Evolutionary Systems, vol. 8, pp. 195–210 (2016)
9. Chung-Cheng, L., Kuo-Ching, Y., Shih-Wei, L.: Robust single machine scheduling for minimizing total flow time in the presence of uncertain processing times. Comput. Ind. Eng. **74**, 102–110 (2014)
10. Daniëls, F.M.J.: On minimizing the probabilistic makespan for the flexible job shop scheduling problem with stochastic processing times. Eindhoven University of Technology, Eindhoven (2013)
11. Gonzalez-Rodriguez, I., Puente, J., Varela, R., Vela, C.R.: A study of schedule robustness for job shop with uncertainty. In: Lecture Notes in Computer Science, vol. 5290, pp. 31–41 (2008)
12. Gonzalez-Rodriguez, I., Vela, C.R., Puente, J., Hernandez-Arauzo, A.: Improved local search for job shop scheduling with uncertain durations. In: Proceedings of the Nineteenth International Conference on Automated Planning and Scheduling, pp. 154–161 (2009)
13. Kai, Z.G., Ponnuthurai, N.S., Quan, K.P., Tay, J.C., Chin, S.C., Tian, X.C.: An improved artificial bee colony algorithm for flexible job-shop scheduling problem with fuzzy processing time. Expert Syst. Appl. **65**, 52–67 (2016)
14. Hamed, A.: Apply fuzzy learning effect with fuzzy processing times for single machine scheduling problems. J. Manuf. Syst. **42**, 244–261 (2017)
15. Al-Hinai, N., ElMekkawy, T.Y.: Solving the flexible job shop scheduling problem with uniform processing time uncertainty. Int. J. Mech. Aerosp. Ind. Mechatron. Manuf. Eng. **6** (4), 848–853 (2012)
16. Sotskov, Y.N., Sotskova, N.Y., Lai, T.-C., Werner, F.: Scheduling under Uncertainty – Theory And Algorithms, Minsk. Belorusskaya nauka (2010)
17. Shafia, M.A., Pourseyed, A.M., Jamili, A.: A new mathematical model for the job shop scheduling problem with uncertain processing times. Int. J. Ind. Eng. Comput. **2**, 295–306 (2011)
18. Karimi-Nasab, M., Seyedhoseini, S.M.: Multi-level lot sizing and job shop scheduling with compressible process times: a cutting plane approach. Eur. J. Oper. Res. **231**, 598–616 (2013)
19. Sobaszek, Ł., Gola, A., Świć, A.: Preditive scheduling as a part of intelligent job scheduling system. In: Advances in Intelligent Systems and Computing, vol. 637, pp. 358–367 (2018)
20. Kosicka, E., Kozłowski, E., Mazurkiewicz, D.: The use of stationary tests for analysis of monitored residual processes. Eksploat. i Niezawodn. – Maint. Reliab. **4**(17), 604–609 (2015)

Pricing and Ordering Decisions in a JELS-Model for Items with Imperfect Quality

Ina Bräuer and Udo Buscher(✉)

Faculty of Business and Economics, TU Dresden, Dresden, Germany
{Ina.Braeuer,Udo.Buscher}@tu-dresden.de

Abstract. This paper presents a joint economic-lot-size (JELS) model for coordinated inventory replenishment decisions between buyer and vendor. The vendor's production process is imperfect and produces a certain number of defective items with a known probability density function. The items are delivered to the buyer in equal-sized lots. The buyer faces a linear, price sensitive demand and aims to maximize her expected profit. A mathematical model is developed to determine (a) the selling price, (b) the order quantity, and (c) the number of shipments to the buyer together with this the vendor's production quantity. A numerical example for an independent as well as integrated policy is provided and its results are discussed.

Keywords: JELS · Imperfect quality · Pricing · Inventory Inspection

1 Introduction

The idea to coordinate the vendor's production policy with the buyer's procurement policy from an inventory perspective probably originates from [5]. [1] took up this idea and introduced the term JELS for the first time. Since then this approach has attracted a great deal of attention and the basic model has been developed into many different directions. For a recent comprehensive review the reader is referred to [4].

A significant extension, which is also examined in this paper, is the consideration of imperfect quality. [13] is assumed to be the first taking the effect of defective items into the classical EOQ-model into account. One important contribution goes back to [15]. They assume that lots arriving from the supplier contain defective items with a known probability density function. With the help of a reliable screening process all defective units can be detected and sorted out. [11] provide a comprehensive overview of extensions of EOQ-based models with imperfect quality items.

[7] integrates this consideration into the JELS and derives an analytic solution to minimise system-wide costs. For comparatively slight modifications to

© Springer Nature Switzerland AG 2019
A. Burduk et al. (Eds.): ISPEM 2018, AISC 835, pp. 244–253, 2019.
https://doi.org/10.1007/978-3-319-97490-3_24

this model see [6,8]. Other authors increase the number of actors and thus model more complex supply chain structures (see e.g. [10,16]).

[14] make an interesting extension of the JELS-model by incorporating a price-sensitive demand function instead of assuming constant demand. This allows the buyer to influence the demand and thus also his profit by setting the price. Recently, an additional complementary product was added to this setting (cf. [3]). In a similar approach from [20], demand depends not only on price but also on environmental performance. However, these approaches are based on completely reliable production processes.

[18] present a lot-size model with a simple lot-for-lot replenishment policy in a three-echelon setting that includes both a price-sensitive demand function and imperfect quality items. Unfortunately, this paper suffers from some shortcomings that are addressed in [2]. To the best of our knowledge the present paper is the first to integrate additional pricing decisions into a JELS-model with imperfect quality products.

The paper is organized as follows: In Sect. 2, the problem is defined and notation as well as assumptions are introduced. Section 3 describes the formulation of the model. Section 4 derives the independent policies of the vendor and the buyer as well as the integrated solution. The paper is concluded in Sect. 5 with a numerical example, some remarks and future research directions.

2 Problem Definition and Notation

Consider a two stage supply chain including a vendor and a buyer. The vendor manufactures the product in lot sizes which are a multiple n of the order quantity Q. The transportation lots are equal-sized batches and are equivalent to the order sizes. The behaviour of the inventory level over time is depicted in Fig. 1. The objective is to determine the number of shipments n, the selling price p_{b} as well as the buyer's order size Q so that the total profits of the vendor and the buyer are maximized. The notation can be found in Table 1. The following assumptions are considered to develop the model:

1. The model deals with a single vendor and a single buyer for a single product.
2. The buyer faces a linear demand $D(p_{\mathrm{b}})$ as a function of the selling price p_{b}.
3. Vendors production rate r_{pv} is finite, with $r_{\mathrm{pv}} > D(p_{\mathrm{b}})$ for every p_{b}.
4. The inventory is continuously reviewed. The buyer orders a lot of size Q when the on-hand inventory reaches the reorder point.
5. The vendor manufactures a production batch nQ at one set-up, whereas the size of shipment delivered to the buyer is Q.
6. Successive deliveries are scheduled to arrive at the buyer's stock if previous shipment has just depleted.

7. The number of perfect units is at least equal to the demand during screening time.
8. The defective units are sold as single batch at the end of the screening period.
9. Backorders and shortages are not allowed.
10. The time horizon is infinite.

Table 1. Table of symbols.

Q	Buyer's order quantity
n	Number of shipments
p_{b}	Buyer's selling price of good quality items [EUR/unit]
$\hat{p}_{\mathrm{b}}$	Buyer's selling price of defective items [EUR/unit]
p_{v}	Vendor's selling price [EUR/unit]
c_{sb}	Buyer's screening cost [EUR/unit]
c_{fb}	Buyer's fixed cost per order [EUR/order]
c_{fv}	Vendor's fixed cost per production lot (set-up costs) [EUR/production lot]
c_{tv}	Vendor's transportation cost per transportation lot [EUR/order]
h_{b}	Buyer's inventory holding cost for good quality items [EUR/unit/time unit]
$\hat{h}_{\mathrm{b}}$	Buyer's inventory holding cost for defective items [EUR/unit/time unit]
h_{v}	Vendor's inventory holding cost [EUR/unit/time unit]
r_{sb}	Buyer's screening rate [unit/time unit]
r_{pv}	Vendor's production rate [unit/time unit]
D	Demand rate [unit/time unit] with $D = D(p_{\mathrm{b}}) = a - bp_{\mathrm{b}}$
a	Potential demand
b	Price sensitivity
t_{sb}	Buyer's screening time per order [time unit]
T_{b}	Buyer's cycle length [time unit]
T_{v}	Vendor's cycle length [time unit]
λ	Percentage of defective items
$f(\lambda)$	Probability density function (pdf) of λ
α	Minimum value of a uniform pdf
β	Maximum value of a uniform pdf
M	Expected value of λ
N	Expected value of $(1 - \lambda)$
K	Expected value of $(1 - \lambda)^2$
$E[\cdot]$	Expected value operator

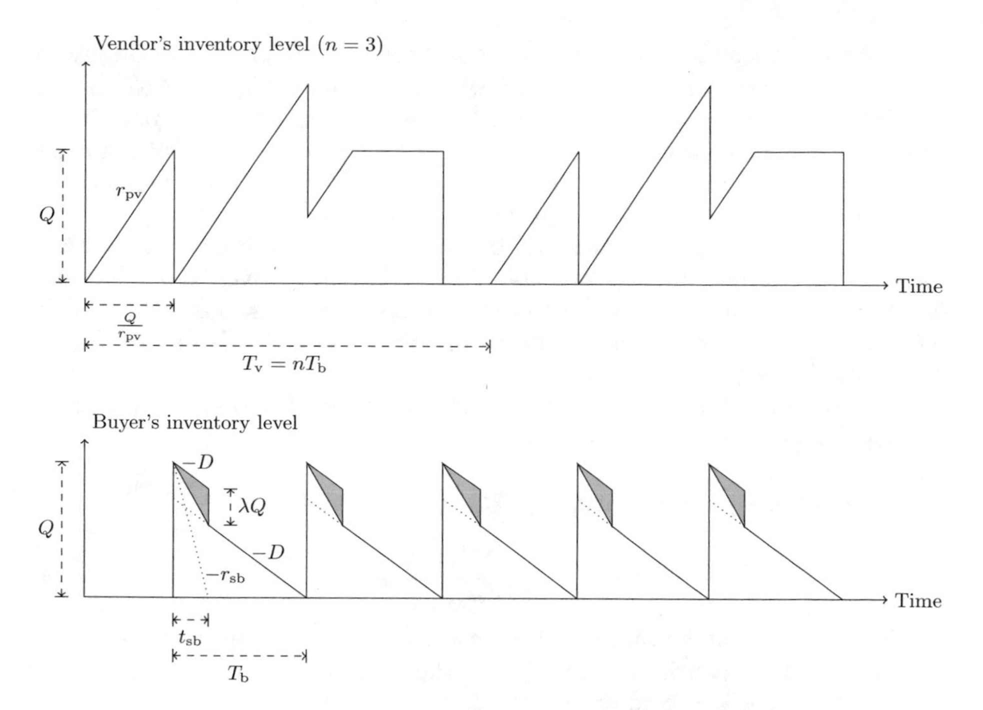

Fig. 1. Behaviour of the inventory level over time.

3 Mathematical Model Formulation

3.1 Buyer's Profit Function

First of all, the buyer has to be considered in more detail. She orders Q units from his supplier. Once the inspection process is finished λQ units are identified as defective and can be sold immediately at a reduced price $\hat{p}_b$. Buyer's total revenue per cycle consists of revenues for good and non-defective units $TR_b = p_b(1-\lambda)Q + \hat{p}_b\lambda Q$. With regard to costs, the buyer has to pay a price p_v per purchased unit and each single unit has to be screened at a cost of c_{sb}. Additionally, each order leads to ordering costs of c_{fb}. The corresponding costs per order cycle are $C_{b1} = (p_v + c_{sb})Q + c_{fb}$.

Inventory holding costs have been neglected so far. Following the average cost (AC) approach, the cost of capital (interest rate times the marginal cost for ordering one item) is added to the out-of-pocket holding cost rate. The latter comprises handling costs, which differ depending on the product's condition. Following [19] we assume that good and defective items are treated in a different way. In consequence we use different inventory holding cost parameters for good (h_b) and defective ($\hat{h}_b$) items with $h_b > \hat{h}_b$.

Inventory holding costs consist of two parts. To determine inventory holding costs for defective items (first part) we calculate the time-weighted inventory $I_{b1} = 0.5(t_{sb}\lambda Q)$ first, which is the grey area in Fig. 1. The time t_{sb} that is necessary to screen all units of one order is given by Q/r_{sb}. Substituting t_{sb} in I_{b1} and multiplying the new expression with $\hat{h}_{\mathrm{b}}$ results in $C_{\mathrm{b2}} = 0.5\lambda Q^2 \hat{h}_{\mathrm{b}}/r_{\mathrm{sb}}$.

The second part of inventory holding costs concerns the good items. After the screening process is finished, $(1-\lambda)Q$ good quality units are left and the corresponding time-weighted inventory is $(1-\lambda)QT_{\mathrm{b}}/2$. Substituting T_{b} by $(1-\lambda)Q/D(p_{\mathrm{b}})$ results in $I_{\mathrm{b2}} = 0.5(1-\lambda)^2 Q^2/D(p_{\mathrm{b}})$.

Moreover, units are assumed to be of good quality during the screening time t_{sb} as long as they are not identified as defective. Thus, the area I_{b1} also has to be considered to calculate the inventory holding costs for good items. Multiplying the summed up areas with h_{b} leads to $C_{\mathrm{b3}} = 0.5h_{\mathrm{b}}\left(\lambda Q^2/r_{\mathrm{sb}} + (1-\lambda)^2 Q^2/D(p_{\mathrm{b}})\right)$.

We receive the total profit function per cycle by subtracting the total costs C_{b1}, C_{b2} and C_{b3} from the revenue TR_{b}:

$$TP_{\mathrm{b}}(Q, p_{\mathrm{b}}) = p_{\mathrm{b}}(1-\lambda)Q + \hat{p}_{\mathrm{b}}\lambda Q - (p_{\mathrm{v}} + c_{\mathrm{sb}})Q - c_{\mathrm{fb}} - \frac{\lambda Q^2 \hat{h}_{\mathrm{b}}}{2r_{\mathrm{sb}}} - \left(\frac{\lambda Q^2}{2r_{\mathrm{sb}}} + \frac{(1-\lambda)^2 Q^2}{2D(p_{\mathrm{b}})}\right) h_{\mathrm{b}}. \tag{1}$$

Since the percentage rate of defective items λ is a random variable with a known probability density function (pdf), the total profit function is no longer deterministic. The expected total profit $E[TP_{\mathrm{b}}(Q, p_{\mathrm{b}})]$ is:

$$E[TP_{\mathrm{b}}(Q, p_{\mathrm{b}})] = p_{\mathrm{b}}(1-E[\lambda])Q + \hat{p}_{\mathrm{b}}E[\lambda]Q - (p_{\mathrm{v}} + c_{\mathrm{sb}})Q - c_{\mathrm{fb}} - \frac{E[\lambda]Q^2 \hat{h}_{\mathrm{b}}}{2r_{\mathrm{sb}}} - \left(\frac{E[\lambda]Q^2}{2r_{\mathrm{sb}}} + \frac{E[(1-\lambda)^2]Q^2}{2D(p_{\mathrm{b}})}\right) h_{\mathrm{b}}. \tag{2}$$

To get the expected total profit per time unit, we have to divide the expected total profit per cycle by the cycle length. In contrast to previous literature, [12] rightly point out that the cycle length itself is an expected value $E[T_{\mathrm{b}}] = (1-E[\lambda])Q/D(p_{\mathrm{b}})$.

The expected total profit per unit time $E[TPU_{\mathrm{b}}(Q, p_{\mathrm{b}})]$ can be determined by dividing $E[TP_{\mathrm{b}}(Q, p_{\mathrm{b}})]$ by $E[T_{\mathrm{b}}]$:

$$E[TPU_{\mathrm{b}}(Q, p_{\mathrm{b}})] = p_{\mathrm{b}}D(p_{\mathrm{b}}) + \frac{\hat{p}_{\mathrm{b}}E[\lambda]D(p_{\mathrm{b}})}{(1-E[\lambda])} - \frac{(p_{\mathrm{v}} + c_{\mathrm{sb}})D(p_{\mathrm{b}})}{(1-E[\lambda])} - \frac{c_{\mathrm{fb}}D(p_{\mathrm{b}})}{(1-E[\lambda])Q} - \frac{E[\lambda]Q(h_{\mathrm{b}} + \hat{h}_{\mathrm{b}})D(p_{\mathrm{b}})}{2r_{\mathrm{sb}}(1-E[\lambda])} - \frac{E[(1-\lambda)^2]Qh_{\mathrm{b}}}{2(1-E[\lambda])}. \tag{3}$$

To simplify notation we define $M = E[\lambda], N = 1 - E[\lambda]$, and $K = E[(1-\lambda)^2]$. If we additionally substitute the demand function $D(p_{\mathrm{b}})$ by $a - bp_{\mathrm{b}}$, then $E[TPU_{\mathrm{b}}(Q, p_{\mathrm{b}})]$ can be written as follows:

$$E[TPU_{\mathrm{b}}(Q, p_{\mathrm{b}})] = (a - bp_{\mathrm{b}}) \cdot \left[p_{\mathrm{b}} + \frac{1}{N}\left(\hat{p}_{\mathrm{b}}M - p_{\mathrm{v}} - c_{\mathrm{sb}} - \frac{c_{\mathrm{fb}}}{Q} - \frac{MQ(h_{\mathrm{b}} + \hat{h}_{\mathrm{b}})}{2r_{\mathrm{sb}}}\right)\right] - \frac{KQh_{\mathrm{b}}}{2N}. \tag{4}$$

3.2 Vendor's Cost Function

After deriving the profit function for the buyer, we now turn to the vendor. The order size Q equals the transportation lot size and represents a given parameter for the vendor. His task is merely to determine the (integer) number n of orders that are combined into one production lot nQ, which is produced with one set-up. During the vendor's cycle time $T_\mathrm{v} = n(1-\lambda)Q/D(p_\mathrm{b})$, n shipments are delivered to the buyer (see Fig. 1).

The vendor's task is to minimise his cost, which consists of inventory holding costs, set-up costs, and transportation costs. The planning situation on the supplier's side corresponds to the one already considered by [9]. However, due to defective items being sorted out by the buyer, we have to consider a slightly different cycle time for the vendor with $T_\mathrm{v} = n(1-\lambda)Q/D(p_\mathrm{b})$. This results in the following vendor's total cost function per time unit:

$$TCU_\mathrm{v}(n) = \frac{Q}{2}\left[n-1-(n-2)\frac{D(p_\mathrm{b})}{(1-\lambda)r_\mathrm{pv}}\right]h_v + \frac{c_\mathrm{fv}D(p_\mathrm{b})}{n(1-\lambda)Q} + \frac{c_\mathrm{tv}D(p_\mathrm{b})}{(1-\lambda)Q}. \quad (5)$$

Again, incorporating λ as a random variable, we consider the expected total cost per time unit:

$$E[TCU_\mathrm{v}(n)] = \frac{Q}{2}\left[n-1-(n-2)\frac{(a-bp_\mathrm{b})}{Nr_\mathrm{pv}}\right]h_v + \frac{(a-bp_\mathrm{b})}{NQ}\left[\frac{c_\mathrm{fv}}{n}+c_\mathrm{tv}\right]. \quad (6)$$

3.3 Joint Profit Function

In a joint optimisation of both supply chain members we consider the total profit function of the supply chain. For this purpose, we first establish the vendor's profit function, which results from the revenues less the costs. The vendor's expected total revenue per unit time is $E[TRU_\mathrm{v}] = (p_\mathrm{v}(a-bp_\mathrm{b}))/N$. With this, we obtain the following expected vendor's profit function per unit time:

$$\begin{aligned} E[TPU_\mathrm{v}] = {} & \frac{p_\mathrm{v}(a-bp_\mathrm{b})}{N} - \frac{Q}{2}\left[n-1-(n-2)\frac{(a-bp_\mathrm{b})}{Nr_\mathrm{pv}}\right]h_v \\ & - \frac{(a-bp_\mathrm{b})}{NQ}\left[\frac{c_\mathrm{fv}}{n}+c_\mathrm{tv}\right]. \end{aligned} \quad (7)$$

Since the vendor's revenues are equal to the buyer's purchasing costs, these two factors are shortened, so that the expected total profit of the supply chain per unit time can be determined as follows:

$$\begin{aligned} E[TPU] &= E[TPU_\mathrm{b}] + E[TPU_\mathrm{v}] \\ &= (a-bp_\mathrm{b})\cdot\left[p_\mathrm{b} + \frac{1}{N}\left(\hat{p}_\mathrm{b}M - c_\mathrm{sb} - \frac{c_\mathrm{fb}}{Q} - \frac{MQ(h_\mathrm{b}+\hat{h}_\mathrm{b})}{2r_\mathrm{sb}}\right)\right] - \frac{KQh_\mathrm{b}}{2N} \\ &\quad - \frac{Q}{2}\left[n-1-(n-2)\frac{(a-bp_\mathrm{b})}{Nr_\mathrm{pv}}\right]h_v - \frac{(a-bp_\mathrm{b})}{NQ}\left[\frac{c_\mathrm{fv}}{n}+c_\mathrm{tv}\right]. \end{aligned} \quad (8)$$

4 Model Solution

4.1 Independent Policy

In this section we look at a situation in which buyer and vendor optimise their objective functions independently. For this, we assume that the buyer initially determines both lot size Q and price p_b optimally for herself. The optimal order quantity $Q^*(p_\text{b})$ can be obtained by taking the first derivative of the buyer's profit function with respect to Q and by equating that expression to zero, which provides

$$Q^*(p_\text{b}) = \sqrt{\frac{2c_\text{fb}}{\frac{Kh_\text{b}}{a-bp_\text{b}} + \frac{M(h_\text{b}+\hat{h}_\text{b})}{r_\text{sb}}}}. \tag{9}$$

Similarly, the optimal price $p_\text{b}^*(Q)$ for a given order quantity Q can be determined by taking the first derivative of the buyer's profit function with respect to p_b and by equating this expression to zero, which leads to

$$p_\text{b}^*(Q) = \frac{a}{2b} + \frac{1}{2N}\left(p_\text{v} + c_\text{sb} - \hat{p}_\text{b}M + \frac{c_\text{fb}}{Q} + \frac{MQ(h_\text{b}+\hat{h}_\text{b})}{2r_\text{sb}}\right). \tag{10}$$

To avoid extremely complex expressions, it is advisable to iteratively determine the optimum values Q^* and p_b^* by inserting them in alternating steps.

The number of shipments n is the only decision variable for the vendor. However, it must be noted that n is an integer. Since revenue of the vendor is independent of n, it is sufficient to minimise the vendor's cost function. It is easy to show that this cost function is convex in n. Therefore, the optimal integer value n^* must meet the following condition: $E[TCU_\text{v}(n^*-1)] \geq E[TCU_\text{v}(n^*)] \leq E[TCU_\text{v}(n^*+1)]$. Applying the procedure presented in [17] results in the following expression:

$$n^*(Q, p_\text{b}) = \sqrt{\frac{2c_\text{fv}(a-bp_\text{b})}{Q^2h_\text{v}(N - \frac{a-bp_\text{b}}{r_\text{pv}})} + 0,25}\,\Big\updownarrow. \tag{11}$$

Rounding the root expression to the nearest integer value (symbolised by the arrows) provides the optimum number of transports for given values of Q and p_b.

4.2 Integrated Policy

From a formal point of view, the procedure for determining the optimum values for an integrated planning is similar to those for an independent planning, except that the common profit function is now the basis for derivation. For the buyer this results in:

$$Q^*(p_\text{b}, n) = \sqrt{\frac{2(a-bp_\text{b})(c_\text{fb} + c_\text{tv} + \frac{c_\text{fv}}{n})}{Kh_\text{b} + h_\text{v}\left[(n-1)N - (n-2)\frac{(a-bp_\text{b})}{r_\text{pv}}\right] + \frac{M(a-bp_\text{b})(h_\text{b}+\hat{h}_\text{b})}{r_\text{sb}}}}. \tag{12}$$

$$p_{\mathrm{b}}^*(Q,n) = \frac{a}{2b} + \frac{1}{2N}\left[c_{\mathrm{sb}} - \hat{p}_{\mathrm{b}}M + \frac{c_{\mathrm{fb}}}{Q} + \frac{MQ(h_{\mathrm{b}}+\hat{h}_{\mathrm{b}})}{2r_{\mathrm{sb}}} + \frac{c_{\mathrm{fv}}}{nQ} + \frac{c_{\mathrm{tv}}}{Q} - \frac{Q}{2r_{\mathrm{pv}}}h_{\mathrm{v}}(n-2)\right]. \tag{13}$$

Since the number of transports n only influences the vendor's profit function, the optimal value does not differ from the independent planning (see Eq. (11))

5 Numerical Example and Conclusion

In this section, the analytic solution procedure developed in Sect. 4 is applied to solve the following numerical example: $p_{\mathrm{v}} = 100$ EUR/unit, $\hat{p}_{\mathrm{b}} = 50$ EUR/unit, $c_{\mathrm{sb}} = 0.5$ EUR/unit, $c_{\mathrm{fb}} = 3,000$ EUR/order, $c_{\mathrm{fv}} = 10,000$ EUR/production lot, $c_{\mathrm{tv}} = 500$ EUR/order, $h_{\mathrm{b}} = 20$ EUR/unit and time unit, $\hat{h}_{\mathrm{b}} = 5$ EUR/unit and time unit, $h_{\mathrm{v}} = 10$ EUR/unit and time unit, $r_{\mathrm{sb}} = 100,200$ units/time unit, $r_{\mathrm{pv}} = 80,000$ units/time unit, $a = 55,000$, $b = 100$, $\alpha = 0$ and $\beta = 0.1$.

With the given parameters α and β of the uniform distribution we calculate $M = 0.05$, $N = 0.95$ and $K = 1/0.1\int_0^{0.1}(1-\lambda)^2 d\lambda = 0.90\overline{3}$. The obtained solutions for both policies are summarised in Table 2.

Table 2. Decision variables and profits.

	Q^*	p_{b}^*	n^*	$E[TPU_{\mathrm{v}}]$	$E[TPU_{\mathrm{b}}]$	$E[TPU]$
Independent policy	2,699.64	327.17	3	2,289,206.63	4,939,531.04	7,228,737.67
Integrated policy	3,326.56	274.93	3	—	—	7,501,453.00

Some interesting findings can be summarized as follows: (a) As expected, the system-wide profit can be increased by a cooperation along the supply chain in comparison to an individual optimisation. (b) The order quantity is always larger for joint optimisation than for individual planning and increasing defect rates always lead to larger order quantities for given values of n (see here and in the following Fig. 2). However, above a certain defect rate, a higher value of n, which is coupled with a smaller quantity Q^*, proves to be advantageous. (c) The integrated policy leads to a lower selling price than individual planning. It can be observed that the price decreases as the defect rate increases. Interestingly, individual planning shows an increase in price with higher defect rates.

This paper addresses optimal pricing and ordering decisions in a vendor-buyer supply chain with consumer demand that depends on the price. The considered new model also takes into account that the seller delivers defective products, which, however, are discovered by the buyer through a screening process. We analysed two different settings: an independent policy in which the vendor and

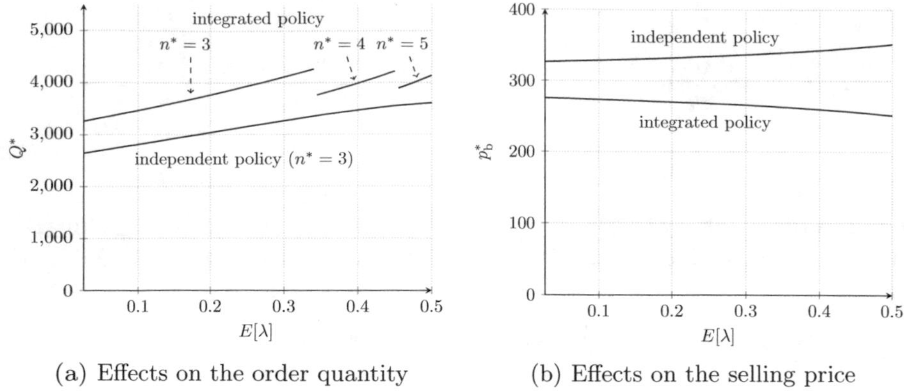

(a) Effects on the order quantity (b) Effects on the selling price

Fig. 2. Comparison of independent and integrated policy for a variation of the expected defect rate.

the buyer optimise their profit functions independently and an integrated policy, where both actors cooperate. An example finally served to illustrate the presented method.

Nevertheless, the presented model can be extended in different ways. One possible extension is to give the vendor the opportunity to increase production quality through investments and thus influence the defect rate. In addition, game-theoretical approaches could be used to adequately reflect the power structures within the supply chain.

References

1. Banerjee, A.: A joint economic-lot-size model for purchaser and vendor. Decis. Sci. **17**(3), 292–311 (1986). https://doi.org/10.1111/j.1540-5915.1986.tb00228.x
2. Bräuer, I., Buscher, U.: A note on "Pricing and ordering decisions in a supply chain with imperfect quality items and inspection under buyback of defective items". Int. J. Prod. Res. (2017). https://doi.org/10.1080/00207543.2017.1399223
3. Dehghanbaghi, N., Sajadieh, M.S.: Joint optimization of production, transportation and pricing policies of complementary products in a supply chain. Comput. Ind. Eng. **107**, 150–157 (2017). https://doi.org/10.1016/j.cie.2017.03.016
4. Glock, C.H.: The joint economic lot size problem: a review. Int. J. Prod. Econ. **135**(2), 671–686 (2012). https://doi.org/10.1016/j.ijpe.2011.10.026
5. Goyal, S.K.: An integrated inventory model for a single supplier-single customer problem. Int. J. Prod. Res. **15**(1), 107–111 (1976). https://doi.org/10.1080/00207547708943107
6. Goyal, S.K., Huang, C.K., Chen, K.C.: A simple integrated production policy of an imperfect item for vendor and buyer. Prod. Plann. Control **14**(7), 596–602 (2003). https://doi.org/10.1080/09537280310001626188
7. Huang, C.K.: An integrated vendor-buyer cooperative inventory model for items with imperfect quality. Prod. Plann. Control **13**(4), 355–361 (2002). https://doi.org/10.1080/09537280110112424

8. Huang, C.K.: An optimal policy for a single-vendor single-buyer integrated production-inventory problem with process unreliability consideration. Int. J. Prod. Econ. **91**(1), 91–98 (2004). https://doi.org/10.1016/S0925-5273(03)00220-2
9. Joglekar, P.N.: Note-comments on "A quantity discount pricing model to increase vendor profits". Manag. Sci. **34**(11), 1391–1398 (1988). https://doi.org/10.1287/mnsc.34.11.1391
10. Khan, M., Jaber, M.Y., Guiffrida, A.L.: The effect of human factors on the performance of a two level supply chain. Int. J. Prod. Econ. **50**(2), 517–533 (2012). https://doi.org/10.1080/00207543.2010.539282
11. Khan, M., Jaber, M.Y., Guiffrida, A.L., Zolfaghari, S.: A review of the extensions of a modified EOQ model for imperfect quality items. Int. J. Prod. Econ. **132**(1), 1–12 (2011). https://doi.org/10.1016/j.ijpe.2011.03.009
12. Maddah, B., Jaber, M.Y.: Economic order quantity for items with imperfect quality: revisited. Int. J. Prod. Econ. **112**(2), 808–815 (2008). https://doi.org/10.1016/j.ijpe.2007.07.003
13. Porteus, E.L.: Optimal lot sizing, process quality improvement and setup cost reduction. Oper. Res. **34**(1), 137–144 (1986). https://doi.org/10.1287/opre.34.1.137
14. Sajadieh, M.S., Jokar, M.R.A.: Optimizing shipment, ordering and pricing policies in a two-stage supply chain with price-sensitive demand. Transp. Res. Part E **45**(4), 564–571 (2009). https://doi.org/10.1016/j.tre.2008.12.002
15. Salameh, M.K., Jaber, M.Y.: Economic production quantity model for items with imperfect quality. Int. J. Prod. Econ. **64**(1–3), 59–64 (2000). https://doi.org/10.1016/S0925-5273(99)00044-4
16. Sana, S.S.: A production-inventory model of imperfect quality products in a three-layer supply chain. Decis. Support Syst. **50**(2), 539–547 (2011). https://doi.org/10.1016/j.dss.2010.11.012
17. Szendrovits, A.Z., Drezner, Z.: Optimizing multi-stage production with constant lot size and varying numbers of batches. Omega **8**(6), 623–629 (1980). https://doi.org/10.1016/0305-0483(80)90003-1
18. Taleizadeh, A.A., Noori-daryan, M., Tavakkoli-Moghaddam, R.: Pricing and ordering decisions in a supply chain with imperfect quality items and inspection under buyback of defective items. Int. J. Prod. Res. **53**(15), 4553–4582 (2015). https://doi.org/10.1080/00207543.2014.997399
19. Wahab, M.I.M., Jaber, M.Y.: Economic order quantity model for items with imperfect quality, different holding costs, and learning effects: a note. Comput. Ind. Eng. **58**(1), 186–190 (2010). https://doi.org/10.1016/j.cie.2009.07.007
20. Zanoni, S., Mazzoldi, L., Zavanella, L.E., Jaber, M.Y.: A joint economic lot size model with price and environmentally sensitive demand. Prod. Manuf. Res. **2**(1), 341–354 (2014)

Probabilistic Fuzzy Approach to Assessment of Supplier Based on Delivery Process

Katarzyna Rudnik(✉) and Ryszard Serafin

Institute of Processes and Products Innovation, Opole University of Technology, Ozimska 75, 45-370 Opole, Poland
{k.rudnik,r.serafin}@po.opole.pl

Abstract. In the article, a new tool for the assessment of suppliers' performance is proposed. This tool uses a probabilistic fuzzy approach based on the assessment of a delivery process. The approach uses a probabilistic fuzzy system of MISO type, in which the knowledge base is described as fuzzy if-then rules with the probabilities of fuzzy events in the antecedents and consequents of rules at the same time. The system identification with limiting the number of elementary rules based on a measure of the minimal support of rules is presented. Various cases of suppliers assessments in a real-life company are illustrated by the analysis of the probabilistic fuzzy knowledge base.

Keywords: Probabilistic fuzzy system · Expert system
Probability of fuzzy event · Assessment of supplier · Delivery process

1 Introduction

According to current research, approximately half of the financial resources that enterprises earn are allocated to goods and services provided by external suppliers. Managing the delivery costs and improving the efficiency of the delivery processes implemented must be based on monitoring and managing the performance of business partners. In the literature, the Supplier Performance Measurement (SPM) is a frequently described as the Supplier Management Methodology. The assessment of suppliers is a key factor in the company's success and is of great importance in the case of mass production, especially in the problems related to the need to reduce stocks. Delivery in accordance with orders is then required, and any process disruption may result in problems in the implementation of production plans. In traditional suppliers' assessment methods an evaluation procedure is carried out before the commencement of delivery, supplemented in the later period by systematic verification of delivery processes [1].

The inadequate analysis of the criteria related to the measurement of supply efficiency may destabilize the process and inflate the costs envisaged at the stage of negotiations with suppliers [2]. Studies show [3] that for around 75% of enterprises, the measurement of supplier performance is very important or even critical for their business. However, it turns out that slightly more than half of the enterprises (56%) have formally developed procedures for measuring the performance of suppliers, but in practice, only less than half of the suppliers undergo performance measurement. These

© Springer Nature Switzerland AG 2019
A. Burduk et al. (Eds.): ISPEM 2018, AISC 835, pp. 254–266, 2019.
https://doi.org/10.1007/978-3-319-97490-3_25

results clearly indicate that the measurement of supplier performance should be one of the key factors in the implementation of strategic priorities in business operations of enterprises.

Due to the above, in the article, a new tool for the assessment of suppliers' performance is proposed. This tool uses a probabilistic fuzzy approach based on the assessment of a delivery process. There are various solutions that take into account the uncertainty of information in terms of both fuzzy and probabilistic [4–8]. Mamdani probabilistic fuzzy rules are described in [9]. This idea is being developed in more articles [10–12]. Mentioned approaches use probability measure to define relative frequency of fuzzy events in conclusion. This idea is more favorable as far as knowledge base interpretation is concerned. With this goal, the paper prefers original fuzzy sets [13]. Moreover, in order to describe more broadly the uncertainty of delivery process, the IF-THEN rules contain marginal and conditional probability of fuzzy events, held in the antecedents and consequents of rules. This presentation of fuzzy rules with the probability of proper fuzzy events directly in the model was proposed by Walaszek-Babiszewska [14, 15]. The approach enables an easy analysis of cumulated knowledge and direct rule modification with the use of both experts knowledge and empirical data. In the article, various cases of suppliers assessments in a real-life company have been illustrated by the analysis of the probabilistic fuzzy knowledge base.

The paper is organized as follows. In the second Section, the related works are described. The inference system with a probabilistic fuzzy knowledge base is given. Moreover, the method of the system identification is presented. In Sect. 3, the application of the system to suppliers assessment based on delivery processes is described, the results are presented and discussed. The conclusions are reported in Sect. 4.

2 Related Works

2.1 Inference System with Probabilistic Fuzzy Knowledge Base

In an uncertain environment of incomplete probabilistic information, the state of reality can be described using a fuzzy inference system with the empirical probability distribution of fuzzy events. In that case, we use a probabilistic fuzzy model of MISO system with N inputs, based on Mamdani-Assilians type fuzzy models, in the following form [14, 15]:

$$\begin{aligned} \mathrm{R}^{(i)}&: \left[\mathrm{If}\left(X_1 \text{ is} A_1^i\right) \text{ And } \left(X_2 \text{ is} A_2^i\right) \text{ And } \ldots \text{ And } \left(X_N \text{ is} A_N^i\right)\right] w_i \text{ Then} \\ &\left(Y \text{ is} B_1^i\right) w_{1/i} \ldots \text{Also} \left(Y \text{ is} B_j^i\right) w_{j/i} \ldots \text{Also} \left(Y \text{ is} B_j^i\right) w_{J/i} \end{aligned} \quad (1)$$

where $(X_1, X_2,\ldots, X_N)$ is an input vector of linguistic random variables (criteria of the delivery assessment in the form of a description), Y is an output linguistic random variable (the linguistic assessment of the delivery for a supplier), $\left(A_1^i, A_2^i, \ldots, A_N^i\right)$ are linguistic values (descriptions) for the input variables in ith file rule ($i = 1,2,\ldots,I$) and B_j^i is a linguistic value for the output variable ($j = 1,2,\ldots,J$) in ith file rule. Weights of rules: w_i, $i = 1,2,\ldots,I$ and $w_{j/i}$, $j = 1,2,\ldots,J$; $i = 1,2,\ldots,I$ represent the marginal probability of fuzzy events in the antecedent and conditional probability of a fuzzy event in the consequent part of rules, respectively. The rule-based knowledge representation (1) and the database constitute a knowledge base. The database includes a collection of linguistic values of the system variables and fuzzy sets definitions for linguistic values. In order to define fuzzy sets, numeric variables space is divided into intervals of equal width. For each interval, the membership degree of variable values to a fuzzy set is determined.

The block of inference makes first the infer and then it aggregates according to generalized *modus ponendo ponens* with using weights of rules [14–16]. The inference method uses parametric triangular norms (e.g. a family of Yager t-norms with a shape parameter), which enable an easier optimization of inference parameters in order to improve the fit of inference results to empirical data [17]. The numerical value of the output can be determined as the centroid of output fuzzy set calculated e.g. by the Center Of Area (COA) method.

2.2 Identification of the MISO System with Probabilistic Fuzzy Knowledge Base and Inference Parametric Operators

In the construction of the knowledge base in the form (1) it is used an approach to mining of fuzzy association rules [18], which fulfill the assumption that rules support (s) is greater than a minimal support ($min\ s$):

$$\begin{aligned} \alpha \Rightarrow \beta &: \text{If}(X_1 \text{ is } A_1^i) \text{ And } (X_2 \text{ is } A_2^i) \text{ And} \ldots \text{ And } (X_N \text{ is } A_N^i) \\ &\text{Then } Y \text{ is } B_j^i\ (s{:}\ s > min\ s;\ c), \end{aligned} \tag{2}$$

Each association rule is connected with two statistic measures specifying the importance and the strength of the rule: s*upport s* – called a significance factor, is the probability of simultaneous occurrence of sets α and β in the collection of sets and co*nfidence c* called certainty or reliability is a conditional probability, $P(\beta| \alpha)$. The measure confidence is not limited. In that case, generation of rules can be understood as discovering rules of a high frequency of fuzzy events occurrence, simultaneously in the antecedents and consequents. The occurrence frequency parameter characterizes adjustment of rules. The choice of minimal support ($min\ s$) value is one of the stage in the system identification process (Fig. 1).

Some algorithms for building a knowledge base of the analyzed MISO system, have been implemented [19, 20]. The research [19, 20] indicated that the modified FP-Growth algorithm, which calculates the rule support in accordance with the value of the fuzzy set power is adequate for more than four system inputs. In this algorithm the support is computed as follows:

$$\begin{aligned} s\left(A_1^i, \ldots, A_N^i, B_j^i\right) &= P\left(A_1^i \times \ldots \times A_N^i \times B_j^i\right) \\ &= \frac{\sum_{(a_{l_1},\ldots,a_{l_N}, b_k) \in \aleph^{N+1}} \mu_{A_1^i \times \ldots \times A_N^i \times B_j^i}(a_{l_1}, \ldots, a_{l_N}, b_k)}{O} \end{aligned} \tag{3}$$

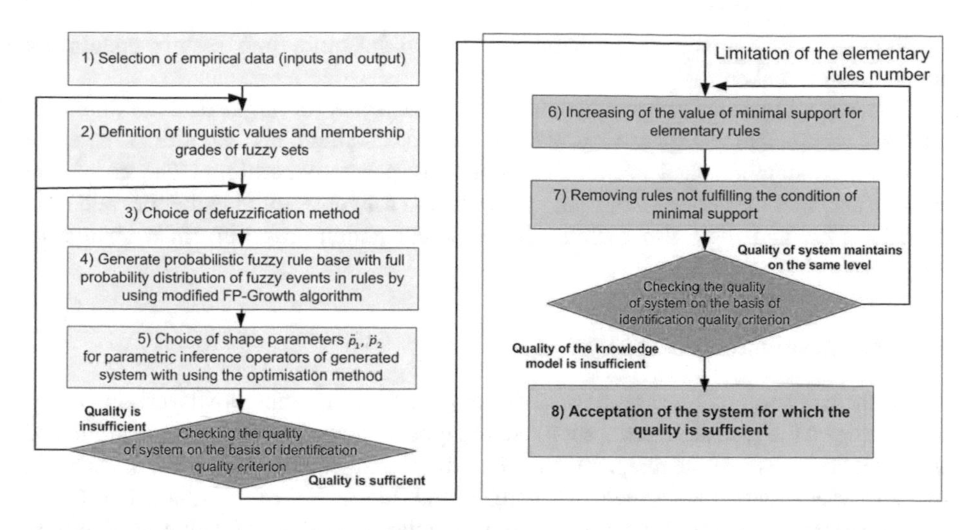

Fig. 1. Identification of system with probabilistic fuzzy knowledge base and inference parametric operators (cf. [20]).

where O is the number of nonzero degrees of membership $(a_{l_1}, \ldots, a_{l_N}, b_k) \in \aleph^{N+1}$ to fuzzy set $A_1^i \times \ldots \times A_N^i \times B_j^i$ $(\mu_{A_1^i \times \ldots \times A_N^i \times B_j^i}(a_{l_1}, \ldots, a_{l_N}, b_k))$, $a_{l_1}, \ldots, a_{l_N}, b_k$ are disjoint intervals in the space $\aleph$ of inputs and output respectively.

The whole idea of the system identification is presented in the Fig. 1. The identification consists of two main stages: choosing system parameters and limiting the number of elementary rules.

3 Assessment of the Supplier Based on Delivery Process

3.1 Problem Statement

The delivery of a material supply is the task of a logistic system operating in the enterprise and is a key element in achieving production goals. In the case of repeated deliveries over long periods of time, it is possible to assess suppliers using probabilistic measures based on data from the history of deliveries collected in the IT management system. This is a part of the probabilistic system concept for controlling supply processes, where decisions regarding cooperation with suppliers depend on the results of data analysis describing the history of a number of completed deliveries. The delivery process is burdened with uncertainty resulting from random events occurring in the supply chain. As a result, the effectiveness of deliveries decreases, assessed on the basis of inconsistencies between the expected (ordered) and realized delivery. This inconsistency may concern parameters such as assortment, quality, quantity, timeliness, and others. In this case, the uncertainty modeling can be performed with the use of the description of fuzzy variables based on fuzzy logic.

The problem in question consists in the development of a tool for the multi-criteria assessment of supplier based on deliveries process, which specifies the result of the assessment in a linguistic and probabilistic form at the same time, easy to understand for a decision-maker.

The tool will allow determining probabilistic measures for linguistic assessments of delivery parameters and a delivery. It is also the assessment of a supplier at the same time, which supports decision making regarding corrective actions. The corrective actions may include actions modifying the ordering schedule with the number and time of deliveries, and may also include more radical actions resulting in a change of suppliers.

3.2 The Assumptions of the System

There are many criteria for assessing deliveries [21]. The importance of the criteria varies depending on the type of material supplied, the type of production/services, location and financial conditions of the recipient, etc. Over the years, global trends in the assessment of deliveries also change, which changes the hierarchy of criteria importance. Continuing the research [22], the article presents the assessment of deliveries based on the following criteria and their linguistic values:

- $X1$ – [Delivery] – amount of a missing material in the delivery [0 pc.; 3 pc.] – three linguistic values: $L(X1)$ = {'Zero', 'Low', 'High'} = {'Z', 'L', 'H'},
- $X2$ – [Quality] – percentage deviation of the quality of a delivery [0%; 30%] – three linguistic values: $L(X2)$ = {'Zero', 'Low', 'High'} = {'Z', 'L', 'H'},
- $X3$ – [Time] – delivery delay expressed in days [0 days; 3 days] – three linguistic values: $L(X3)$ = {'Zero', 'Low', 'High'} = {'Z', 'L', 'H'},
- $X4$ – [Price] – deviation of the actual delivery price from the assumed one [PLN -0.04; PLN 0.3] – four linguistic values: $L(X4)$ = {'Positive', 'Zero', 'Low negative', 'High negative'} = {'P', 'Z', 'LN', 'HN'},
- $X5$ – [Payments] – deviation of the actual payment date from the assumed date [0 days; 14 days] – two linguistic values: $L(X5)$ = {'Zero', 'Negative'} = {'Z', 'N'},
- $X6$ – [Transport/packaging] – deviation of the quality of packaging and transport services [0%; 100%] – three linguistic values: $L(X6)$ = {'Zero', 'Low', 'High'} = {'Z', 'L', 'H'}.

The parameters constitute input fuzzy variables for the analyzed decision system. The output variable Y is the assessment of a delivery with five linguistic values $L(Y)$ = {'Very Low', 'Low', 'Medium', 'High', 'Very High'} = {'VL', 'L', 'M', 'H', 'VH'}. The definitions of fuzzy sets for the above mentioned linguistic values of inputs and the output have been shown in the Fig. 2.

Firstly, the system model with full marginal and conditional probability distribution of fuzzy events has been generated. The quality criterion of the system identification is the minimization of the Root Mean Square Error (RMSE) for training data set, calculated as:

$$\mathrm{RMSE} = \sqrt{\mathrm{MSE}} = \sqrt{\frac{1}{L}\sum\nolimits_{l=1}^{L}\left(\hat{y}^l - y^l\right)^2}, \tag{4}$$

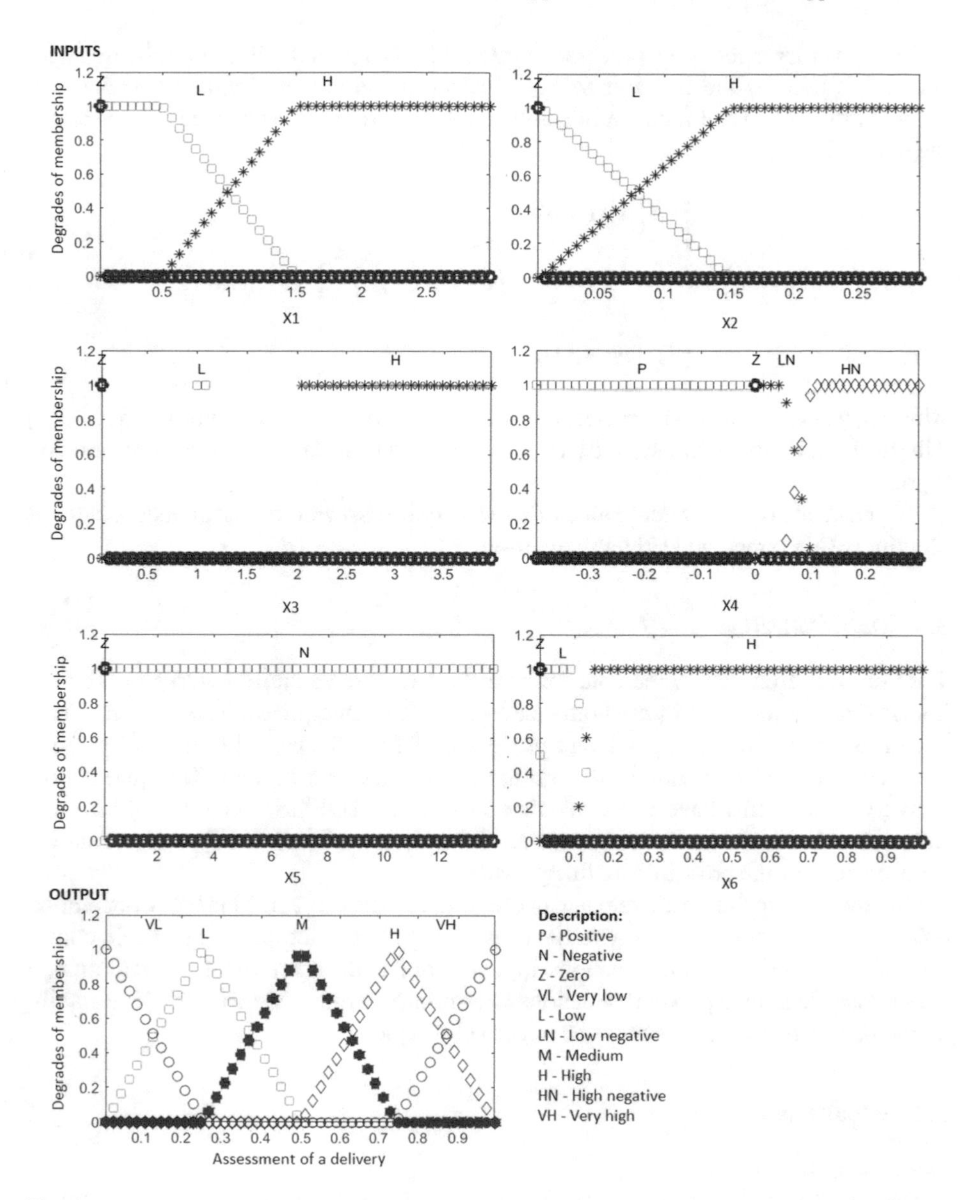

Fig. 2. Membership degrades of the values intervals of input and output variables according to the defined fuzzy sets.

where: L – the number of input-output measurements in the set of experimental data; $\hat{y}^l$ – the expected value calculated in the system based on the lst data set; y^l – the real value for the lst data set.

In the research to generate rules the modified FP-Growth algorithm [19, 20] with product t-norm has been used. The influence of fuzzy inference operators on the RMSE has been analyzed based on the optimization with Sequential Quadratic Programming method. In order to make the process of system identification flexible, a family of Yager t-norms is used as the operator to aggregation of individual simple antecedents in complex rules ($\ddot{T}_{Y1}$) and fuzzy implication ($\ddot{T}_{Y2}$). According to Yager [23], $\ddot{T}_{Y1}$ is calculated as follows:

$$\ddot{T}_{Y1}\left(\left\{\mu_{A_n^i}(x_n)\right\};\ddot{p}_1\right)=\begin{cases} T_d(\left\{\mu_{A_n^i}(x_n)\right\}) for\, \ddot{p}_1=0 \\ MAX\left\{0,1-\left(\Sigma_{n=1}^{N}\left(1-\mu_{A_n^i}(x_n)\right)^{\ddot{p}_1}\right)^{\frac{1}{\ddot{p}_1}}\right\} for\, \ddot{p}_1\in(0,\infty), \\ T_m(\left\{\mu_{A_n^i}(x_n)\right\}) for\, \ddot{p}_1=\infty \end{cases} \quad (5)$$

where $\ddot{p}_1$ marks the shape parameter, T_d – drastic t-norm, T_m – Zadeh t-norm, $\{x_1, \ldots, x_N\}$ – linguistic variables of inputs, and $\{A_1^1, \ldots, A_N^1\}$ – linguistic values for the corresponding inputs.

To calculate the numerical value of a delivery assessment, the Expanded Center of Gravity (ECOG) method [19] has been used.

3.3 Data Collection

The company from which the data for analysis have been acquired works in the food industry and is included in medium-sized production enterprises. The data that have been used for analysis of the delivery processes relate to the period from January 2014 to April 2015 and were obtained from the IT management system. The quantitative delivery assessments have been calculated using the TOPSIS method [24] with the following vector of criteria weights [0.25 0.30 0.10 0.15 0.15 0.05]. These data are training data for the probabilistic fuzzy system.

In practice, suppliers are evaluated in the company once a year. This type of operation only gives the opportunity for a verification of suppliers and repairing possible consequences of disturbances in the supply. In a system that does not operate in real time, it is not possible to take preventive actions to eliminate further disturbances. The stabilization of supply processes cannot be effectively implemented.

3.4 Results and Discussion

Case study 1

In the first case study, a probabilistic fuzzy system has been used to assess the supplier, whose deliveries deviate from the expectations of the customer. The supplier realizes deliveries once a week, supplying the raw material for production in large quantities, of a size of several tons. Delivery time is 4 days from the date of receipt of the order. Data training with 60 deliveries have been used for analysis. According to the assessment of the company itself, this supplier is characterized by an average opinion on contracts.

In the first case study, the system model with full marginal and conditional probability distribution of fuzzy events contains 104 elementary rules, which is 52 file rules. The optimization of inference parameters has proved that the proper values of shape parameters for parametric Yager t-norms are: $\ddot{p}_1 = 25000.75$ for $\ddot{T}_{Y1}$ as a representation of logical conjunction AND in antecedents of the rules and $\ddot{p}_2 = 25000.75$ for $\ddot{T}_{Y2}$ as a fuzzy inference operator. For these parameters, the RMSE of the system model with full marginal and conditional probability distribution of fuzzy events is the lowest (0.0091).

In the last stage of system identification, the number of knowledge base rules has been limited. For this purpose, the assumption of the minimal support of rules has been increased (*min s*) and the rules for which values of supports are lower than the minimal support value have been removed from the knowledge base. The result of this identification stage is presented in the Fig. 3. The Figure shows a relative plateau at the level of the RMSE for the minimal support smaller than 0.0041. After this value, the RMSE increases significantly. The number of elementary rules decreases steadily when the minimal support increases. For *min s* = [0.0038, 0.039] the adjustment error rate is the smallest (0.0088) preserving a smaller number of rules (68 elementary rules, 35 file rules).

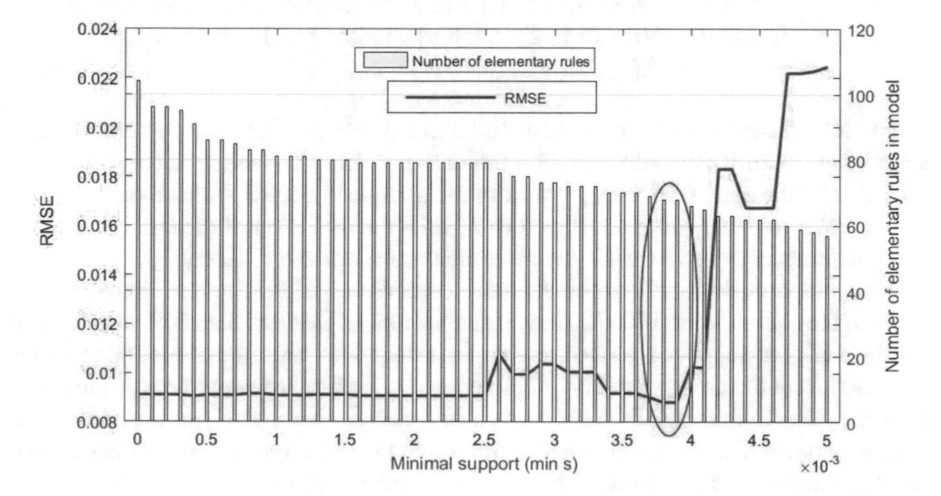

Fig. 3. The dependency of the RMSE and the number of elementary rules in knowledge base from the minimal support value (case study 1).

The example of the final model rules with the higher support is as follows:

1: IF ($X1$ IS Z) AND ($X2$ IS H) AND ($X3$ IS Z) AND ($X4$ IS LN) AND ($X5$ IS Z) AND ($X6$ IS Z) [0.1836] THEN (Y IS L) [0.5719] ALSO (Y IS VL) [0.4281]
2: IF ($X1$ IS Z) AND ($X2$ IS H) AND ($X3$ IS Z) AND ($X4$ IS Z) AND ($X5$ IS Z) AND ($X6$ IS Z) [0.0882] THEN (Y IS L) [0.5914] ALSO (Y IS VL) [0.4086]
3: IF ($X1$ IS Z) AND ($X2$ IS L) AND ($X3$ IS Z) AND ($X4$ IS LN) AND ($X5$ IS Z) AND ($X6$ IS Z) [0.0771] THEN (Y IS L) [0.6129] ALSO (Y IS VL) [0.3871]
4: IF ($X1$ IS Z) AND ($X2$ IS H) AND ($X3$ IS Z) AND ($X4$ IS LN) AND ($X5$ IS Z) AND (X6 IS H) [0.0650] THEN (Y IS L) [0.5505] ALSO (Y IS VL) [0.4495]
5: IF ($X1$ IS H) AND ($X2$ IS H) AND ($X3$ IS Z) AND ($X4$ IS Z) AND ($X5$ IS Z) AND ($X6$ IS Z) [0.0409] THEN (Y IS L) [0.5056] ALSO (Y IS VL) [0.4944]

where: $X1$ – the amount of a missing material in the delivery; $X2$ – the percentage deviation of the quality of a delivery; $X3$ – the delivery delay expressed in days; $X4$ – the deviation of the actual delivery price from the assumed one; $X5$ – the deviation of the actual payment date from the assumed date; $X6$ – the deviation of the quality of packaging and transport services; Y – the assessment of supplier's delivery.

The joint probability of the simultaneous occurrences of fuzzy events in the antecedents of the above 5 rules is 0.46. Thus, it can be noticed that less than 50% of deliveries for the analyzed supplier are rated low or even very low. This is mainly due to a high deviation from the desired quality of deliveries and a slight overstatement of the actual delivery price. In individual rules, there is also a high deviation from the desired quality of transport and packing services and shortages in the delivered material. Other criteria values in the above rules do not differ from the desired values.

In order to broaden the supplier's assessment, a larger number of rules are presented in a tabular form, and a list of the rules for the assessment criteria is presented (Table 1). These are elementary rules, sorted from the highest to the smallest probability of the simultaneous occurrence of fuzzy events in the antecedents of rules (w_i). The most negative values of the inputs and outputs of the model are marked with an intensive red color. The knowledge base clearly illustrated that the supplier's weakest point is the quality of its deliveries: 80% of deliveries were deliveries with a high deviation from the proper quality, the remaining deliveries had low deviations from the proper quality. There was no delivery of a good quality. In addition, 20% of deliveries had high deviations from the number of delivered goods, 27% of deliveries had high deviations from the proper quality of packaging and transport, and 18% of deliveries had high deviations from the delivery date. The best-assessed supplier's parameter is the deviation from the expected payment date. All rules in the entire knowledge base clearly indicate that the supplier should receive a low or very low rating.

Table 1. Knowledge base with the general rules for the assessment of deliveries and the probabilities of occurrence of the fuzzy events in the model (case study 1).

	X1	X2	X3	X4	X5	X6	Y	$w_{j/i}$	w_i
1	Z	H	Z	LN	Z	Z	VL	0,43	0,18358
2	Z	H	Z	LN	Z	Z	L	0,57	
3	Z	H	Z	Z	Z	Z	VL	0,41	0,08819
4	Z	H	Z	Z	Z	Z	L	0,59	
5	Z	L	Z	LN	Z	Z	VL	0,39	0,07708
6	Z	L	Z	LN	Z	Z	L	0,61	
7	Z	H	Z	LN	Z	H	VL	0,45	0,065
8	Z	H	Z	LN	Z	H	L	0,55	
9	H	H	Z	Z	Z	Z	VL	0,49	0,04093
10	H	H	Z	Z	Z	Z	L	0,51	
11	L	H	Z	Z	Z	Z	VL	0,47	0,0358
12	L	H	Z	Z	Z	Z	L	0,53	
13	L	H	Z	LN	Z	H	VL	0,46	0,03436
14	L	H	Z	LN	Z	H	L	0,54	
15	Z	H	H	LN	Z	Z	VL	0,42	0,03066
16	Z	H	H	LN	Z	Z	L	0,58	
17	H	H	Z	Z	Z	H	L	0,44	0,02743
18	H	H	Z	Z	Z	H	VL	0,56	
19	Z	H	H	LN	Z	H	VL	0,43	0,02576
20	Z	H	H	LN	Z	H	L	0,57	
21	H	H	Z	LN	Z	H	L	0,44	0,02509
22	H	H	Z	LN	Z	H	VL	0,56	
23	Z	H	H	Z	Z	Z	VL	0,41	0,02188
24	Z	H	H	Z	Z	Z	L	0,59	
25	H	H	Z	LN	Z	Z	VL	0,49	0,02036
26	H	H	Z	LN	Z	Z	L	0,51	
27	H	L	Z	LN	Z	Z	VL	0,46	0,01917
28	H	L	Z	LN	Z	Z	L	0,54	
29	Z	H	Z	Z	Z	H	VL	0,43	0,0167
30	Z	H	Z	Z	Z	H	L	0,57	
31	Z	H	Z	Z	N	H	VL	0,43	0,0167
32	Z	H	Z	Z	N	H	L	0,57	
33	Z	H	H	Z	Z	H	L	0,49	0,0167
34	Z	H	H	Z	Z	H	VL	0,51	
35	H	H	L	LN	Z	Z	L	0,41	0,0167
36	H	H	L	LN	Z	Z	VL	0,59	
37	H	H	H	Z	Z	Z	L	0,41	0,0167
38	H	H	H	Z	Z	Z	VL	0,59	
39	L	H	L	Z	Z	Z	L	0,49	0,0165
40	L	H	L	Z	Z	Z	VL	0,51	
41	L	L	Z	Z	Z	Z	VL	0,43	0,0147
42	L	L	Z	Z	Z	Z	L	0,57	
43	L	L	Z	LN	Z	Z	VL	0,44	0,0138
44	L	L	Z	LN	Z	Z	L	0,56	
45	Z	L	H	LN	Z	Z	VL	0,38	0,013
46	Z	L	H	LN	Z	Z	L	0,62	
47	Z	H	Z	HN	Z	Z	VL	0,41	0,0126
48	Z	H	Z	HN	Z	Z	L	0,59	
49	L	H	Z	HN	Z	H	L	0,5	0,012
50	L	H	Z	HN	Z	H	VL	0,5	
51	Z	L	Z	Z	Z	Z	VL	0,35	0,0118
52	Z	L	Z	Z	Z	Z	L	0,65	
53	Z	L	H	Z	Z	Z	VL	0,35	0,0115
54	Z	L	H	Z	Z	Z	L	0,65	
55	H	L	H	LN	Z	H	L	0,49	0,0115
56	H	L	H	LN	Z	H	VL	0,51	
57	L	L	Z	LN	Z	H	VL	0,43	0,0111
58	L	L	Z	LN	Z	H	L	0,57	
59	Z	H	H	LN	N	Z	VL	0,43	0,0108
60	Z	H	H	LN	N	Z	L	0,57	
61	L	H	Z	LN	Z	Z	VL	0,47	0,0100
62	L	H	Z	LN	Z	Z	L	0,53	
63	L	H	H	LN	Z	Z	L	0,49	0,0085
64	L	H	H	LN	Z	Z	VL	0,51	
65	H	H	H	LN	Z	Z	L	0,49	0,0082
66	H	H	H	LN	Z	Z	VL	0,51	
67	Z	L	Z	HN	Z	Z	L	1	0,0063
68	H	L	Z	Z	Z	Z	L	1	0,0049

Probability of occurrence of the fuzzy event in the model							
	X1	X2	X3	X4	X5	X6	Y
P	0,00	0,00	0,00	0,00	0,00	0,00	
Z	0,64	0,00	0,78	0,36	0,97	0,73	
N	0,00	0,00	0,00	0,00	0,03	0,00	
LN	0,00	0,00	0,00	0,61	0,00	0,00	
HN	0,00	0,00	0,00	0,03	0,00	0,00	
L	0,16	0,20	0,03	0,00	0,00	0,00	
H	0,20	0,80	0,18	0,00	0,00	0,27	
VL							0,49
L							0,51
M							0,00
H							0,00
VH							0,00

The map of the knowledge base (Table 1) allows for assessment the supplier based on the delivery process, which was in the past. The probabilistic fuzzy system also allows for the assessment of future deliveries of a given supplier based on the inference process [15, 16, 19] with the optimal parameters values ($\ddot{p}_1 = 25000.75$ for $\ddot{T}_{Y1}$, $\ddot{p}_2 = 25000.75$ for $\ddot{T}_{Y2}$). The result of defuzzification method for output fuzzy set is close to result of TOPSIS method but not the same. Figure 4 presents a comparison of results of TOPSIS

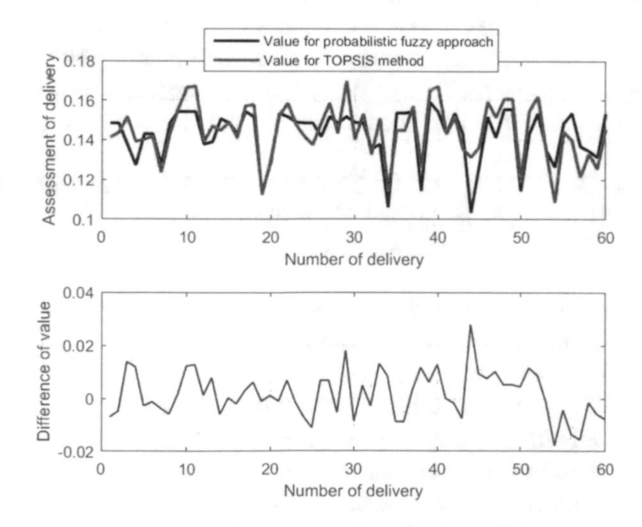

Fig. 4. Comparison of the results for TOPSIS method and the values computed on the basis of the model with a probabilistic fuzzy knowledge base (case study 1).

method with the values computed using the created knowledge model with 68 elementary rules.

Case study 2
In the second case study, a probabilistic fuzzy system has been used to assess the supplier, whose deliveries only slightly deviated from the expectations of the customer. The supplier realizes deliveries for the company every 2 weeks, supplying the raw material for production in small quantities (several dozen kilograms). Delivery time is 3 days from the order date. Data training with 34 deliveries have been used for analysis. According to the assessment of the company itself, this supplier is characterized by a good opinion on contracts.

Due to the fact that the criteria values have been not varied the knowledge base with full marginal and conditional probability distribution of fuzzy events contains only 4 elementary rules (2 file rules) in the following form:

1: IF (*X*1 IS Z) AND (*X*2 IS L) AND (*X*3 IS Z) AND (*X*4 IS Z) AND (*X*5 IS Z) AND (*X*6 IS Z) [0.6875] THEN (*Y* IS VH) [0.5918]
ALSO (*Y* IS H) [0.4082]

2: IF (*X*1 IS Z) AND (*X*2 IS H) AND (*X*3 IS Z) AND (*X*4 IS Z) AND (*X*5 IS Z) AND (*X*6 IS Z) [0.3125] THEN (*Y* IS VH) [0.5918]
ALSO (*Y* IS H) [0.4082]

where: *X*1 – the amount of a missing material in the delivery; *X*2 – the percentage deviation of the quality of a delivery; *X*3 – the delivery delay expressed in days; *X*4 – the deviation of the actual delivery price from the assumed one; *X*5 – the deviation of the actual payment date from the assumed date; *X*6 – the deviation of the quality of packaging and transport services; *Y* – the assessment of supplier's delivery.

The system identification with the optimization of inference parameters values has proved that the proper values of shape parameters for parametric Yager t-norms are: $\ddot{p}_1 = 6250.94$ for $\ddot{T}_{Y1}$ as a representation of logical conjunction AND in antecedents of the rules and $\ddot{p}_2 = 6250.94$ for $\ddot{T}_{Y2}$ as a fuzzy inference operator. Finally, after limiting the number of rules, a very simple model with one file rule has been obtained:

1: IF (*X*1 IS Z) AND (*X*2 IS L) AND (*X*3 IS Z) AND (*X*4 IS Z) AND (*X*5 IS Z) AND (*X*6 IS Z) [0.6875] THEN (*Y* IS VH) [0.5918]
ALSO (*Y* IS H) [0.4082].

The adjustment error rate (RMSE) for this system is $8.25 \cdot 10^{-4}$.

The analysis of the obtained knowledge base indicates that delivery process of the mentioned supplier is rated very high (with the probability 0.5918) and high (with the probability 0.4082). Almost all criteria are fulfilled (there are no deviations from the desired values). Only the delivery quality criterion has a low deviation (with the probability 0.6875). Therefore, it can be concluded that the future deliveries for this supplier will very well meet the requirements of the entrepreneur. However, you should take into account the risk of lower quality of deliveries.

4 Conclusions

In a modern organizational culture of an enterprise, all economic decisions should be made on the basis of data analysis. This trend is part of the proposed methodology for assessing suppliers using a probabilistic fuzzy knowledge base. The presented approach allows analyzing uncertainties of a delivery process in probabilistic and fuzzy categories. The use of fuzzy logic with the rule-based knowledge base makes it possible to express incomplete and uncertain information in a natural language, characteristic for a human way. The presented model is friendly for a human interpretation which is of great importance especially while making decisions about the selection of a supplier. Additionally, the use of probabilities of events expressed in linguistic form makes it possible to adapt the model using empirical data from the IT system of the enterprise. Using this methodology, in addition to the direct benefits of inventory management and warehouse management, increases the level of the enterprise innovation, which is now an important image element.

References

1. Skowronek, C., Sarjusz-Wolski, Z.: Logistics in the Enterprise. edn. III changed, Polskie Wydawnictwo Ekonomiczne, Warszawa (2003). (in Polish)
2. Zeydan, M., Colpan, C., Cobanoglu, C.: A new decision support system for performance measurement using combined fuzzy TOPSIS/DEA approach. Int. J. Prod. Res. **47**, 4327–4349 (2011)
3. Aberdeen Group: The Supplier Performance Measurement Benchmark Report, September 2005
4. Merigó, J.M.: Fuzzy multi-person decision making with fuzzy probabilistic aggregation operators. Int. J. Fuzzy Syst. **13**(3), 163–174 (2011)
5. Yan, L., Ma, Z.M.: A Fuzzy probabilistic relational database model and algebra. Int. J. Fuzzy Syst. **15**(1), 244–253 (2013)
6. Chen, C., Xiao, T.: Probabilistic fuzzy control of mobile robots for range sensor based reactive navigation. Intell. Control Autom. **2**, 77–85 (2011)
7. Yager, R.R., Filev, D.: Using dempster-shafer structures to provide probabilistic outputs in fuzzy systems modeling. In: Trillas, E., Bonissone, P.P., Magdalena, L., Kacprzyk, J. (eds.) Studies in Fuzziness and Soft Computing. Combining Experimentation and Theory, vol. 271, pp. 301–327. Springer-Verlag, Berlin, Heidelberg (2012)
8. Amiri, M., Ardeshir, A., Zarandi, M.H.F.: Fuzzy probabilistic expert system for occupational hazard assessment in construction. Saf. Sci. **93**, 16–28 (2017)
9. Meghdadi, A.H., Akbarzadeh-T, M.R.: Probabilistic fuzzy logic and probabilistic fuzzy systems. In: Proceedings of 10th IEEE International Conference on Fuzzy Systems, vol. 2, pp. 1127–1130. Melbourne, Australia (2001)
10. Tang, M., Chen, X., Hu, W., Yu, W.: Generation of a probabilistic fuzzy rule base by learning from examples. Inf. Sci. **217**, 21–30 (2012)
11. Almeida, R.J., Verbeek, N., Kaymak, U., Sousa, J.M.C.: Probabilistic fuzzy systems as additive fuzzy systems. In: Information Processing and Management of Uncertainty in Knowledge-Based Systems Communications in Computer and Information Science, vol. 442, pp. 567–576 (2014)

12. Sozhamadevi, N., Sathiyamoorthy, S.: A probabilistic fuzzy inference system for modeling and control of nonlinear process. Arab. J. Sci. Eng. 21 March 2015. https://doi.org/10.1007/s13369-015-1627-8
13. Zadeh, L.A.: Fuzzy sets. Inf. Control **8**, 338–353 (1965)
14. Walaszek-Babiszewska, A.: Construction of fuzzy models using probability measures of fuzzy events. In: Proceedings of the 13th IEEE International Conference on Methods and Models in Automation and Robotics, MMAR 2007, pp. 661–666. Szczecin, Poland (2007)
15. Walaszek-Babiszewska, A.: Fuzzy Modeling in Stochastic Environment; Theory, Knowledge Bases, Examples. LAP LAMBERT Academic Publishing, Saarbrűcken (2011)
16. Walaszek-Babiszewska, A., Rudnik, K.: Stochastic fuzzy knowledge-based approach to temporal data modeling. In: Pedrycz, W., Chen, S.M. (eds.) New Volume on Time Series Analysis, Modeling and Applications. A Computational Intelligence Perspective, pp. 97–118. Springer, Heidelberg (2013)
17. Rudnik, K., Pisz, I.: Probabilistic fuzzy approach to evaluation of logistics service effectiveness. Manag. Prod. Eng. Rev. **5**(4), 66–75 (2014)
18. Kuok, C.M., Fu, A.W., Wong, M.H.: Mining fuzzy association rules in databases. SIGMOD Rec. **17**(1), 41–46 (1998)
19. Rudnik, K.: Inference System with Probabilistic-fuzzy Knowledge Base: Theory, Conception and Application. Oficyna Wydawnicza Politechniki Opolskiej, Opole (2013). (in Polish)
20. Rudnik, K., Deptuła, A.M.: System with probabilistic fuzzy knowledge base and parametric inference operators in risk assessment of innovative projects. Expert Syst. Appl. **42**(1718), 6365–6379 (2015)
21. Dickson, G.W.: An analysis of vendor selection systems and decisions. J. Purch. **2**, 5–17 (1966)
22. Gierulski, W., Luściński, S., Serafin, R.: Probabilistic measures of the mass production of logistics chain. Logistics 4/2015, CD no 2, 3363–3373 (2015). (in Polish)
23. Yager, R.R.: On a general class of fuzzy connectives. Fuzzy Sets Syst. **4**, 235–242 (1980)
24. Hwang, C.L., Yoon, K.: Multiple Attribute Decision Making Methods and Applications: A State-of-the-Art Survey. Springer-Verlag, New York (1981)

Processing of Design and Technological Data Due to Requirements of Computer Aided Process Planning Systems

Kamil Krot(✉) and Jacek Czajka

Wrocław University of Science and Technology, ul. Łukasiewicza 5, 50-371 Wrocław, Poland
{kamil.krot,jacek.czajka}@pwr.edu.pl

Abstract. The paper presents an overview of the ways of data exchange between CAD and CAPP systems. The exchange was described using neutral data exchange formats and applications programming interfaces. New possibilities in the area of data exchange with the use of solutions enabling full specification of the designed part based on the 3D CAD model have been described. It was presented how such data can be exported from the CAD system to the STEP 242 file format and how to further use this data in planning production processes in the CAPP system. This study proposes the configuration of an IT environment that streamlines and organizes the work related to product development departments of manufacturing enterprises. The mechanisms and interfaces that automate the flow of data between CAD/CAPP systems have been proposed.

Keywords: Process planning · CAPP · CAD · Product data processing

1 Introduction

The traditional approach to the development of technical documentation in the area of product development assumes that process planning is executed on the basis of CAD 2D drawings. Such documents are usually prepared in the computer aided design systems - CAD. Work in these systems usually takes place sequentially; at first CAD 3D models are prepared, in the next step CAD 2D drawings are developed. CAD 2D drawings can be based on CAD 3D models for both individual parts and assemblies, which greatly speeds up the creation of documentation. In popular CAD systems, there are functions supporting the preparation of 2D drawings based on a 3D model such as: automatic projection, inserting views, simple and aligned sections, hatching, scaling, importing dimensions and annotations from a 3D model. 2D drawings are parametrically linked to the 3D model. If one of them is modified, the system also updates the related document. The drawing for the part should contain all the information and requirements necessary to produce it. Despite many functions automating the creation of 2D documents based on 3D models, it is still a time-consuming task, costly and requiring specialist knowledge in the field of mechanical design. The modern approach to design in CAD systems allows to save product and manufacturing information (PMI) directly on the 3D model [1]. In such a model, it is possible not only to insert classical dimension

© Springer Nature Switzerland AG 2019
A. Burduk et al. (Eds.): ISPEM 2018, AISC 835, pp. 267–274, 2019.
https://doi.org/10.1007/978-3-319-97490-3_26

values in 2D sketches and 3D design operations, but also to add tolerance dimensions. Similarly, on the 3D model of the part it is possible to add surface finishing and datum identifier symbols, position and shape tolerances, material definitions, weld symbols and much more about the presented part. Such solutions allow for a full description of the parts with information necessary for its production, disregarding the phase of preparing the 2D drawing [2]. Taking into account the solutions described above, the further part of the article describes possible ways of using data related to the 3D model of parts directly in Computer Aided Process Planning systems.

2 Process Planning in CAPP – Identification of a Problem

Production enterprises wanting to control and automate business processes taking place in the company implement IT systems that improve the management of various departments of the company. Usually, ERP systems (Enterprise Resources Planning) are implemented in the hope that they will provide functions supporting the work of all company departments, ensure their integration and provide access to current data at various decision-making levels. The implementation of the ERP system, however, does not exhaust the topic of computerization of the enterprise. It turns out that there are many areas without proper support. One of them is the product development area and more specifically process planning, which is based on CAD data and data collected in ERP systems. Many companies decide to implement additional software that will fill functional gaps. These deficiencies are as follows:

- lack of a common repository of technological knowledge in the form of: templates of reference processes, previously made processes for re-use, operation templates, production resources, mechanisms supporting decision making,
- lack of standards in the scope of circulation and approval of documentation, usually data are stored locally on users' computers and documents are created using popular office packages,
- it is not possible to draw sketches for operations based on design data and it is not possible of storing them in relation to the product structure,
- no mechanisms for automatically entering route data into ERP systems.

The function of the integrator of these areas can be performed by the CAPP system, which will enable the management of data related to the production technology in connection with the product structure and data from the ERP system. Additionally, an appropriately configured IT environment will provide the opportunity for concurrent work, which will result in shortening the time of preparation of the project documentation. The following approaches are used in the CAPP systems [7–9]:

– variant - based on similarity of manufactured parts and processes templates,
– generative - the system automatically generates a technological process on the basis of principles and knowledge stored in the form of rules,
– semi-generative - combining the benefits of the variant and the generative approach.

In manufacturing enterprises, there is a need to implement CAPP solutions. There is a general tendency to gather engineering knowledge and make it available in design processes. Such activities additionally bring benefits in the form of an increase in the quality of the developed processes through total or partial independence from the individual knowledge and experience of the employees of the technological planning departments.

3 Data Exchange Between CAD and CAPP Systems

During the development of CAD and CAPP systems many methods and tools have been developed for sending geometrical data, dimensions, tolerances, surface finish parameters, and position and shape tolerances [3]. Data transfer from CAD systems can be implemented in two main ways: with the use of dedicated data exchange formats and through direct communication with the CAD system in which the considered part geometry model is active. The first approach assumes the ability to read data connected with the geometry of the part from files with an ordered, standardized structure.

For this purpose, the following file formats can be used: DXF (Data eXchange Format) Parasolid, IGES (Initial Graphics Exchange Specification) or STEP (STandard for the Exchange of Product model data). The use of 3D part model data from such file formats requires a thorough knowledge of the file structure and the assumption that the files are unified. This assumption often turns out to be problematic, because apparently consistent file formats differ from each other depending on which CAD system has been generated [4–6]. Despite the many years of development, the IGES format has not been formalized in the form of a record, which hinders its analysis and processing [7]. As a result IGES development was discontinued and work was focused on the STEP format. The STEP is a continuation after the development of the IGES standard [7]. The popular STEP format is AP203, later AP214 and recently developed AP242. AP214 is considered an extension of AP203. This format was created as a result of combining two standards AP 203 and AP 214 and was approved by ISO in 2014 [1]. Regarding the development of CAPP systems, high hopes are associated with the latest version of the STEP protocol - AP242 due to the fact that in addition to the geometry, information about production and assembly are stored, such as: assembly tolerances, surface finishing and other information about the production process. The STEP 242 format is a neutral standard, so it is not related to a specific CAD system. It turns out that there is no need to generate 2D drawings, store them, analyse them or even print them. Production companies can exchange such files and complete the task of process planning more quickly.

It is a file format independent of the system so it provides flexibility in data exchange. It can successfully replace the existing approaches in the field of data exchange: neutral and native formats. Additionally, it can be used for data archiving.

Software developers of popular CAD systems (SolidWorks, Inventor, Solid Edge, Siemens NX and others), as part of development work have added functionality in the possibility of adding product and manufacturing information - PMI directly on the 3D CAD model. In addition, the export of this data is possible to the STEP 242 format. Assuming that the PMI data are complete, it is possible to skip the 2D drawings preparation.

In addition to the neutral formats in CAD systems, there are still direct formats available (Direct CAD Interfaces) that allows to load files in the form of a specific system. Data exchange takes place without conversion and without need to be saved as an additional file. These formats are unlikely to apply to data exchange with CAPP systems due to the fact that the file structure is not widely available by CAD software vendors.

Another way to access the part data is based on direct communication with the CAD system. It is possible through programming interfaces built into CAD systems. Application Programming Interface - API allow access to data by offering appropriate functions. This approach requires the use of external software tools supporting coding processes that allows to apply the methods and properties provided by the programming interface. Written computer programs enable capturing of any parameters and data of the CAD model, regardless of the method of design and parameterization of the geometry or order of design operations. The data set obtained in this way can be processed practically in any way. One of the application areas is the possibility of using data collected from the CAD system to process planning in the CAPP system. Such a method requires the collection of two data classes: geometrical data that characterizes the geometry of the designed part and dimensional and precision data, which indicate the dimensions and acceptable tolerances in which the part is to be made. The use of API programming interfaces for CAD and CAPP integration entails the need to low-level data processing. It is necessary to accurately identify the way the geometry is representing and the data structure characterizing the dimensions, tolerances, surface finishing values or shape tolerances. The next task is to link these data structures and use it to processes planning in the CAPP system. This task is quite complex and often leads to limited functionality in terms of the completeness of the description of the geometrical data needed for manufacturing. The development of the data exchange interface in the API environment is possible for one particular CAD system or even one version of it, which is a big limitation. The code can not be used in other CAD systems due to differences in the syntax of the methods of access to objects. Along with the update version of the CAD program may change the method of access to objects or attributes, which results in that the programming code will need to be updated. The advantage of this approach is that the interface refers directly to the CAD model, analyzes it and transfers data to the CAPP system. With this approach, it is easier to control design changes and make appropriate adjustments of processes in the CAPP system.

Based on the above considerations, it can be concluded that the greatest possibilities in the field of communication and data exchange between CAD systems and CAPP are offered by the STEP 242 format. It can be the basis for the operation of systems supporting the design of technological processes due to the fact that it contains all the necessary data for these activities. This format is so neutral in the event of a change in the CAD model will be required to perform a re-export data which can be considered a disadvantage of this approach.

4 Process Planning in CAPP Systems Based on PMI Annotations Saved in CAD Systems

During the development of the product, the constructor should consult his ideas with the technology department, who analyzes the technological character of the structure and proposes possible design changes. These changes may result in production costs, delivery time and the decisions on the implementation of production using the available production resources in the company or in cooperation. Without access to design data and set requirements for a product, it is difficult to make such decisions and choose the most economical variant of the manufacturing process. This allows to detect problems in the early stages of product development and avoid inserting costly changes while already at the production stage. The modern approach to design assumes that the work of product development departments does not include the stage of creating 2D drawings. This is possible due to new CAD system functions that have MBD (Model Based Definition) modules that allow to store data about the designed part directly on the 3D model and production requirements, without the need to create and store 2D technical drawings. The use of PMI annotations brings benefits in the form of reduced design costs (shorter time) and, as a result, speed up the time of product launch on the market. An annotation mechanism is used, which is defined on specific views directly on the 3D model. The use of this approach to design can completely eliminate the need to create 2D documentation. There is a significant reduction in the time of preparation of design and technological documentation due to the possibility of concurrent work on the project. Both the designer and the process planner can work on the same 3D models. The process planner can also apply his comments to the 3D model in the form of a comments. In this way, it is possible to precisely identify problem areas in the project and avoid misunderstandings. Another factor is the ability to correctly interpret 2D documentation, which can be very troublesome with complex shapes. Incorrectly interpreted design documentation may result in errors in the selection of manufacturing technology. If such documentation is transferred to production, it can bring about large losses. The benefit of using this part description is that all design and technological documentation is available in one file, which simplifies product data management in the company. CAD systems offer 3D file viewers, which have a number of useful functions to view the annotations attached to the 3D model. Parts with annotations can be saved in formats that do not require a license for the CAD program in which the model was prepared and even a specialized browser. The part with PMI annotations can be saved as a 3D PDF file and opened with the help of popular free browsers for this format. This facilitates cooperation between companies at the stage of determining design or technological requirements for products. Popular CAD systems can save geometric data in the form of the Step 242 (Managed Model Based 3D Engineering) neutral format.

The two basic methods of data exchange between CAD systems and CAPP are discussed above: neutral and direct formats and through interfaces. Interfaces work as add-inns to CAD system and are developed in internal programming languages available in the API environment.

Figure 1 presents the concept of an integrated product development environment based on interfaces enabling data exchange between CAD and CAPP systems. Two ways of communication were proposed: through a dedicated interface developed in the CAD system's API environment and using the STEP 242 neutral file format. Other neutral file formats and direct methods were discarded due to a number of limitations described above.

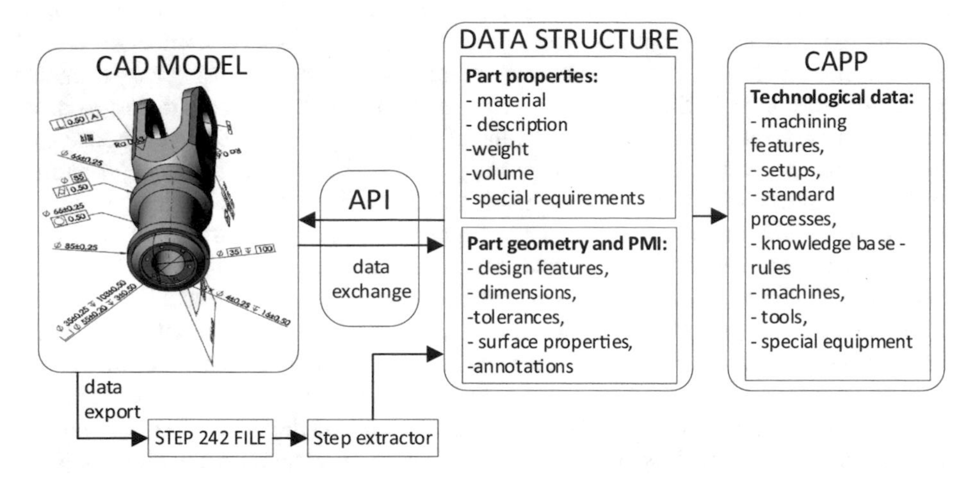

Fig. 1. Data exchange between CAD and CAPP using API and STEP AP 242.

```
...
#36=DIMENSIONAL_CHARACTERISTIC_REPRESENTATION(#44,#37);
#37=SHAPE_DIMENSION_REPRESENTATION('',(#42),#584);
#38=LENGTH_MEASURE_WITH_UNIT(LENGTH_MEASURE(0.05),#586);
#39=LENGTH_MEASURE_WITH_UNIT(LENGTH_MEASURE(-0.05),#586);
#40=TOLERANCE_VALUE(#39,#38);
#41=PLUS_MINUS_TOLERANCE(#40,#44);
#42=(
LENGTH_MEASURE_WITH_UNIT()
MEASURE_REPRESENTATION_ITEM()
MEASURE_WITH_UNIT(LENGTH_MEASURE(8.),#586)
QUALIFIED_REPRESENTATION_ITEM((#43))
REPRESENTATION_ITEM('nominal value')
);
#43=VALUE_FORMAT_TYPE_QUALIFIER('NR2 1.2');
#44=DIMENSIONAL_SIZE(#48,'diameter');
...
```

Fig. 2. Part of STEP 242 CAD model including dimensional value and acceptable dimensional deviations

In the first step, the part model is analyzed and the data is prepared in the form of a list of geometric features along with the annotations assigned to them. On the basis of such data, the CAPP system, guided by the knowledge base rules, may propose a set of data in the form of technological operations. Technological operations can be grouped in the form of a classifier for different types of geometric features. Such a system could propose several solutions (processes) and leave the final decision to the user.

As a result of exporting the part model from the CAD system to the STEP 242 format, a text file with an ordered structure is obtained. For example, for holes, both the nominal dimension and tolerance values are available in explicit form - Fig. 2

This form of data can be used to automate traditional technology design using CAPP systems. Figure 3 presents an example of PMI for counterbore holes.

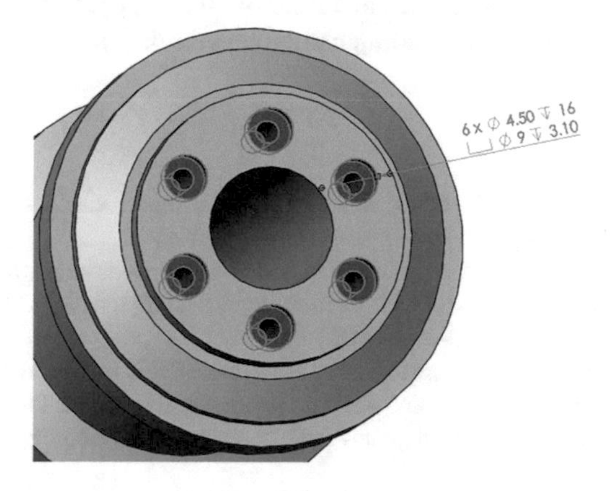

Fig. 3. An example of PMI annotation for the pattern of counterbore holes.

When adding annotations, the CAD system automatically detects geometrical features and suggests a description. In the case of the above there are 6 holes with a diameter of 4.5 mm and a depth of 16 mm. The counterbore has a diameter of 9 mm and a depth of 3.1 mm. By analyzing the model using the API interface or by interpreting the STEP 242 file format, it is possible to obtain a hole pattern with the parameters necessary to select the manufacturing technology.

5 Conclusion

Process planning is a rather complex issue. Includes selection of: material, operations, machine tool, tools, machining parameters, tooling, time and material norms, cost analysis, etc. Currently, in manufacturing companies, process planning takes place largely manually and is based on the experience of process planners. This way of working is time-consuming, requires a lot of work and it is easy to make mistakes of a different kind. Access to information necessary while creating processes is also difficult due to the lack of central knowledge bases and resources databases. The presented approach gives the opportunity to quickly create new technological documentation by associating technological features with the process models available in the CAPP system. This will increase the efficiency of work in the area of product development, giving the opportunity to re-use the data available in the system. Implementation of the CAPP system with a central database gives the opportunity to technological knowledge acquisition from process planners and writing it to knowledge base. The use of such accumulated knowledge means that the quality of process plans may be at least partly independent from the knowledge and experience of single process planners. In addition, the system gives you the opportunity to optimize production costs by analyzing various process variants.

This study proposes the configuration of an CAPP system that streamlines and organizes the work of product development departments of manufacturing enterprises. Mechanisms and interfaces have been proposed that automate the data exchange between CAD and CAPP systems. In future work, it is planned to implement the presented concepts in the form of computer system modules.

References

1. Lipman, R., Lubell, J.: Conformance checking of PMI representation in CAD model STEP data exchange files. Comput. Aided Des. **66**, 14–23 (2015)
2. Witherella, P., Herronb, J., Ametac, G.: Towards annotations and product definitions for additive manufacturing. In: 14th CIRP Conference on Computer Aided Tolerancing (CAT). Procedia CIRP, vol. 43, pp. 339–344 (2016)
3. Jeon, S.M., Lee, J.H., Hahm, G.J., Suh, H.W.: Automatic CAD model retrieval based on design documents using semantic processing and rule processing. Comput. Ind. **77**, 29–47 (2016)
4. Waiyagan, K., Bohez, E.L.J.: Intelligent feature based process planning for five-axis mill-turn parts. Comput. Ind. **60**, 296–316 (2009)
5. Rameshbabu, V., Shunmugam, M.S.: Hybrid feature recognition method for setup planning from STEP AP-203. Robot. Comput. Integr. Manuf. **25**, 393–408 (2009)
6. Marjudi, S., Mohd Amran, M.F., Abdullah, K.A.S., Widyarto, N.A., Majid, A., Sulaiman, R.: A review and comparison of IGES and STEP. In: Proceedings of World Academy of Science, Engineering and Technology, vol. 62, pp. 1013–1017 (2010). ISSN: 2070-3724
7. Schuh, G., Prote, J.-P., Luckert, M., Hünnekes, P.: Knowledge discovery approach for automated process planning. In: The 50th CIRP Conference on Manufacturing Systems, Procedia CIRP, vol. 63, pp. 539–544 (2017)
8. Kumar, S.P., Jerald, J., Kumanan, S.: Automatic feature extraction and CNC code generation in a CAPP system for micromachining. Procedia Mater. Sci. **5**, 1986–1997 (2014)
9. Chwastyk, P., Kolosowk, M.: CAD/CAPP/CAM integration system in design process of innovative products. In: Proceedings of the 23rd International DAAAM Symposium, vol. 23, no. 1 (2012). ISSN: 2304-1382

Rationalization of Decision-Making Process in Selection of Suppliers with Use of the Greedy and Tabu Search Algorithms

Joanna Kochańska(✉), Kamil Musial, and Anna Burduk

Faculty of Mechanical Engineering, Wroclaw University of Science and Technology, Wybrzeze Wyspianskiego 27, 50-370 Wroclaw, Poland
{joanna.kochanska,kamil.musial,
anna.burduk}@pwr.edu.pl

Abstract. The paper raises the issue of evaluation and selection of suppliers in a large manufacturing company. Till now, the decision-making process in the company was based on the human factor. The presented work was aimed at developing new suppliers selection system. That would make it possible to minimize costs of ordered materials. Solving the problem of optimization is offered by Tabu Search and greedy algorithms. The tuning process of algorithms and the verification of algorithm performance are reported in the paper.

Keywords: Suppliers selection · Decision-making processes
Greedy algorithm · Tabu Search · Meta-heuristics
Intelligent optimization methods of production systems

1 Introduction

Changes in demand or in manufacturing area caused by the continuous improvement of processes and products models require flexibility in cooperation with suppliers [19]. Correct management of ordering processes means striving for possibly high quality and possibly low cost. Nowadays most of decision-making processes are based only on human factor. Such an approach often results in skipping the best solution, while generating financial losses [18].

The research was carried-out on the example of a large manufacturing company. The paper examines a case where parts are delivered by different suppliers. They offer similar products and replacements with limited compatibility that differ in price. The aim of the study was to build a model that would allow obtaining the lowest cost of the ordered materials by selecting the suppliers depending on their price.

Finding the most profitable solution to this type of problem can be possible thanks to the use of artificial intelligence methods [13]. They allow finding solutions close to optimal in a relatively short time, which makes them more and more popular [11]. Application of intelligent algorithms makes it possible to reduce time, labour input and human errors in processes. They can therefore be used to support decision-making processes [4, 5, 9, 15, 16, 18].

© Springer Nature Switzerland AG 2019
A. Burduk et al. (Eds.): ISPEM 2018, AISC 835, pp. 275–284, 2019.
https://doi.org/10.1007/978-3-319-97490-3_27

2 Case Study

2.1 Decision-Making Problem

In the paper we consider decision-making problem, inspired by a real company, which is a global manufacturer and supplier of tools and components for production of electric machines. Up to now, all decision-making processes were based only on the human factor. Over time an observation was made that such an approach often generates financial losses. Then a possibility was noticed to reduce costs by an implementation of new decision-making methods.

One of the areas subjected to observation was the cooperation with the suppliers. The process being analyzed was the selection of suppliers of production support materials (materials used for machine operation, tools, consumables, etc.). They are not processed, but enable proper functioning of the production in the company.

The company cooperates with six main suppliers of production supports materials. The suppliers don't have similar product assortment to offer. Some products are the same, some the same but compatible only with another product from the specific supplier. Also suppliers offer various prices. In order to entice the company to purchase in larger quantities, the suppliers offer quantity discounts.

Planning the budget, the purchasing department evaluates the suppliers (taking into account their prices, quality and potential discounts). Based on the historical data and estimated demand, requirement for each analyzed month is calculated. Considering the above, supplier for each subsequent product is selected (Fig. 1).

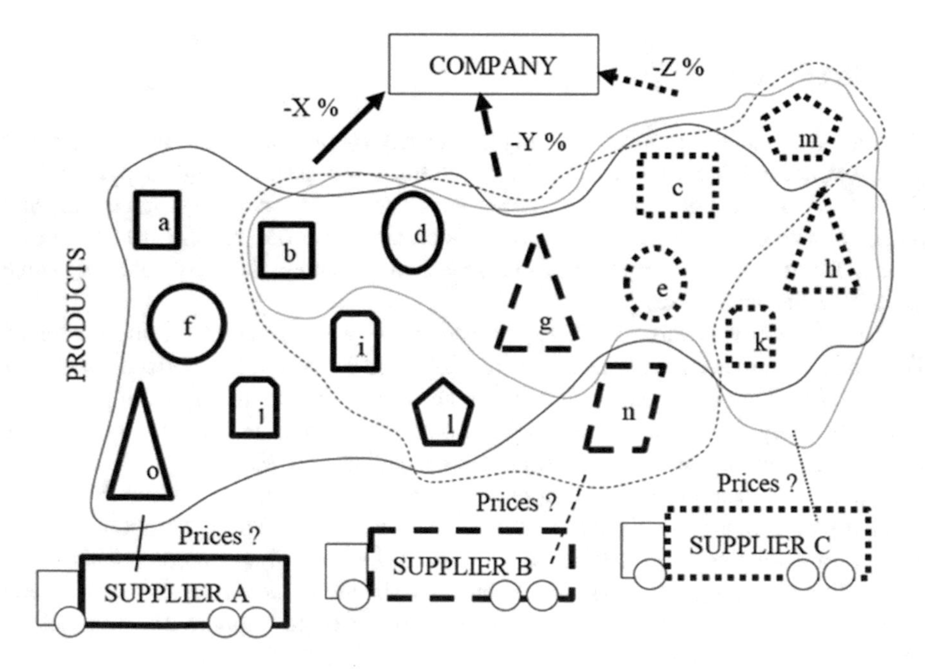

Fig. 1. Decision-making process in selection of suppliers.

The decision is made quickly but based only on the subjective opinions – there are no standardized selection methods.

From the point of view of a company oriented to profit maximisation, selection of suppliers is an important activity. In order to maintain the competitive advantage it is necessary to minimize all costs, also those of production support materials. Therefore, material resources management requires selecting the suppliers, taking into consideration both compatibility of offered products and their prices (including also possible discounts).

2.2 Basic Example

In the paper, twelve selected months were analyzed. As a basis for the research, historical data were accepted. For the needs of the research, one selected group of production support materials was analyzed. For maintenance of one machine park consisted of 54 machines 25 different products are necessary. Company cooperates with 6 suppliers (A, B, C, D, E, F) that deliver following products. Some of them are offered by all of suppliers, some by ex. 3 suppliers (A, E, F), some only by 1. Additionally, products have various versions and have to match each other. For instance, in case of order product no. 1 from supplier A, it is necessary to order also product no. 3, 17 and 22 from the same supplier because of compatibility requirement. Following situation occurs in case of ex. seals, screws, adapters ect. The products have also different prices depending on supplier. Additionally the suppliers offered various discounts with specific size of order. The requirement in twelve months (M1 – M12) has been estimated.

Considering the above, the problem of decision-making in selection of suppliers is complicated and it is impossible to find the optimal or near-optimal solution based only on human factor.

2.3 Parameters

Each product [p] of each supplier [s] was given a price [pr]. The products compatibility matrix [com] has been accepted. Moreover, each supplier offers a discount depending on the size of order [d]. Also the discount matrix has been accepted. Next, on the grounds of demand and historical data, requirement of each product for next 12 months [r] has been estimated. The purpose is selection of suppliers and products in the way guaranteeing possibly lowest cost and acceptable quality level.

Within elaboration of new decision-making method, the following parameters were accepted:

- product $[p] - [p_1, p_2, \ldots, p_n]$
- supplier $[s] - [s_1, s_2, \ldots, s_n]$
- product price from supplier $\left[pr_{p,s}\right] - \left[pr_{11}, pr_{12}, \ldots pr_{19}, pr_{21}, \ldots, pr_{n,m}\right]$
- supplier discount above specific order size $[d_s] - [d_{1,\ldots,}d_n]$
- requirement of product $[r] - [r_1, r_2, \ldots, r_n]$
- products compatibility matrix for each supplier $[com] - [com_{1,1}, com_{1,2,}\ldots, com_{n,n}]$.

3 Optimization Methods

3.1 Selected Algorithms

Many kinds of decision-making problems can be solved using meta-heuristic algorithms [4, 5, 11, 13, 16, 18]. Application of them allows finding a near-optimal solution in a reasonable time without the transformation into mathematical formulations [1, 8, 17]. For the considered problem of decision-making process in selection of suppliers it was decided to use intelligent algorithms. For the examined case, the greedy and Tabu Search algorithms were considered. Both the greedy and the Tabu Search algorithm can be used at solving NP (nondeterministic polynomial) optimization problems [3–5, 18]. Although none of them gives a guarantee of finding the optimal solution, the solutions proposed by these two algorithms are fully acceptable [2, 20].

3.2 Greedy Algorithm

The greedy algorithm solves optimization problems taking locally optimal decisions. It makes the optimal choice at each stage without assessment of its effect on further steps [14] – it seems to be the best at the analyzed moment. The search procedures used to find optimal solutions are adopted to generate alternative moves. With the use of all the locally optimal solutions, the algorithm is able to approximate a global optimal solution in a relatively short time [10]. The greedy algorithm is often used for minimization or maximization of the target value. Main advantages of this type algorithms include fast action, moderate memory requirements and easy implementation [14, 21].

Unfortunately greedy algorithm can easily miss good solution depends on data set. As was mentioned before, the greedy algorithm adapts the solution locally best. The simple example below shows the basic disadvantage of rationale of the greedy algorithm. To each order, an execution time is assigned (Table 1). The total available time of T_{max} = 100 min should be used in a possibly optimum way.

Table 1. Execution times of orders.

o - Order number	Time [min]
1	55
2	50
3	48
4	6
5	2
6	1

It can be seen that the only optimum solution is acceptance of the orders No. 2, 3 and 5, which gives exactly 100 min (the optimum solution):

$$t_2 + t_3 + t_5 = 100\,[min]. \tag{1}$$

Operation of the greedy algorithm is as follows:

1. $t_1 \leq 100$ (acceptance of the order No. 1)
2. $t_1 + t_2 > 100$ (omitting the order No. 2, going to the next one)
3. $t_1 + t_3 > 100$ (omitting the order No. 3, going to the next one)
4. $t_1 + t_4 > 100$ (omitting the order No. 4, going to the next one)
5. $t_1 + t_5 \leq 100$ (acceptance of the order No. 5)
6. $t_1 + t_5 + t_6 \leq 100$ (acceptance of the order No. 6)

The solution according to the simple greedy algorithm:

$$t_1 + t_5 + t_6 = 58\,[min]. \tag{2}$$

The above calculations show a serious disadvantage of the greedy algorithm, especially visible when times for individual orders are long in comparison to the total available time.

In investigating issue, Greedy algorithm multiply size of product order by lowest price and sort descending. then complete order following this criterion. Each next chosen product is checked by compatibility with chosen previously. If chosen product does not meet this requirement is omitted and next one is checked. If specific product has been chosen in one version is omitted if seen again in another version.

3.3 Tabu Search

The Tabu Search algorithm solves optimization problems searching the solution space created by all possible solutions by means of specific sequence of movements [6, 7, 12]. These movements are used to change the current solution to the new one. The algorithm checks the neighboring solutions trying to find a similar, but improved solution [18]. There are also taboo (forbidden) movements [7, 12]. The algorithm avoids oscillation around the locally optimal solution with the use of information stored in a Taboos List (TL) [18]. Thanks to the memory functions, the algorithm does not consider solutions, that has been already visited or has violated a rule [18].

Algorithm 1. Tabu Search pseudo-code for the considered problem

```
Algorithm Tabu Search (S0, var S, max_m, max_it)
 Set S = S0 and n_iter = 0
 Repeat
           m = 0
           best = 0
           it = it + 1
                Repeat
                        m = m+1
                        Execute Check_the_neighboring_solution (S,Sm)

                        Execute Check_Tabu_list (S,Sn)
                                if (f(Sm) > best then (best = f(Sm) and (m2= m))
                until (m = max_m)

         Execute Add_to_Tabu_list (S,Sm2)

         S= Sm2

         If best > solution then solution = best
 Until n_iter = max_it
```

The analyzed problem may be presented as a binary sequence:

1 0 0 1 1 0 1. . .,

in which each bit corresponds to the order of specific product and specific version. The value "1" means that the order is accepted, the value "0" means that the order is rejected. Implementation of five orders can be described as a bit sequence, e.g.:

1 0 0 0 1.

Check the Neighbouring Solution

The algorithm checks all of the neighbouring solutions (differing in one bit):

1 0 0 0 **0**
1 0 0 **1** 1
1 0 **1** 0 1
1 **1** 0 0 1
0 0 0 0 1.

Check Tabu List

In this step the algorithm checks, whether the transition between previous and actual solutions has been already proceed. If so, this solution is omitted and then the best solution is chosen, according to the specified criterion:

$$1\ 1\ 1\ 0\ 1.$$

Add to Tabu List

The algorithm adds it to the Tabu List, i.e. the list of prohibited movements (TL). The algorithm remembers, which solution is the best one and repeats the steps the specified number of times.

In the case when each bit represents a product in its specific version $[p] - [p_1, p_2, \ldots, p_n]$, its parameters are:

- product price $[pr] - [pr_1, pr_2, \ldots pr_n]$
- supplier discount above specific order size $[d_s] - [d_{1 \ldots} d_n]$
- requirement of product $[r] - [r_1, r_2, \ldots, r_m]$

According to the assumptions, the final price has to minimalized.

$$\mathbf{0}pr_1r_1 + \mathbf{1}pr_2r_2 + \mathbf{1}pr_3r_3 + \ldots + \mathbf{0}pr_nr_n \rightarrow \textbf{MIN} \quad (3)$$

The algorithm has a limit: in following approach it was assumed that each bit is the product in its specific version with minimal available price. In following issue the binary string consists of 42 bits wherein the solution given by Tabu Search algorithm is checked in terms of correctness following the compatibility matrixes.

$$\mathbf{0}p_1 + \mathbf{1}p_2 + \mathbf{1}p_3 + \ldots + \mathbf{0}p_n \rightarrow \textbf{ACCEPTABLE} \quad (4)$$

If solution given is acceptable – necessary product are compatible, company requirement is met – then solution is multiplied by company destination order size. Next, order sizes from specific suppliers are summed and discount is calculated. Below final objective function has been presented:

$$(\mathbf{0}pr_1r_1 + \mathbf{1}pr_2r_1)d_2 + \mathbf{1}pr_3r_2d_4 + \ldots + \mathbf{0}pr_nr_md_5 \rightarrow \textbf{MIN} \quad (5)$$

3.4 Comparison of Algorithms Results

Potential costs obtained with use of both considered algorithms for the analyzed months compared with historical data of the company are illustrated in the diagram (Fig. 2).

Information concerning the number of: products ordered, products not used (if ordered) and products necessary but not ordered is also collected (Table 2).

The results obtained by the greedy algorithm confirmed the earlier concerns (see Sect. 3.2). It is not possible to rationalize considered decision-making process with the use of the greedy algorithm because it misses good solutions in such cases. Potential

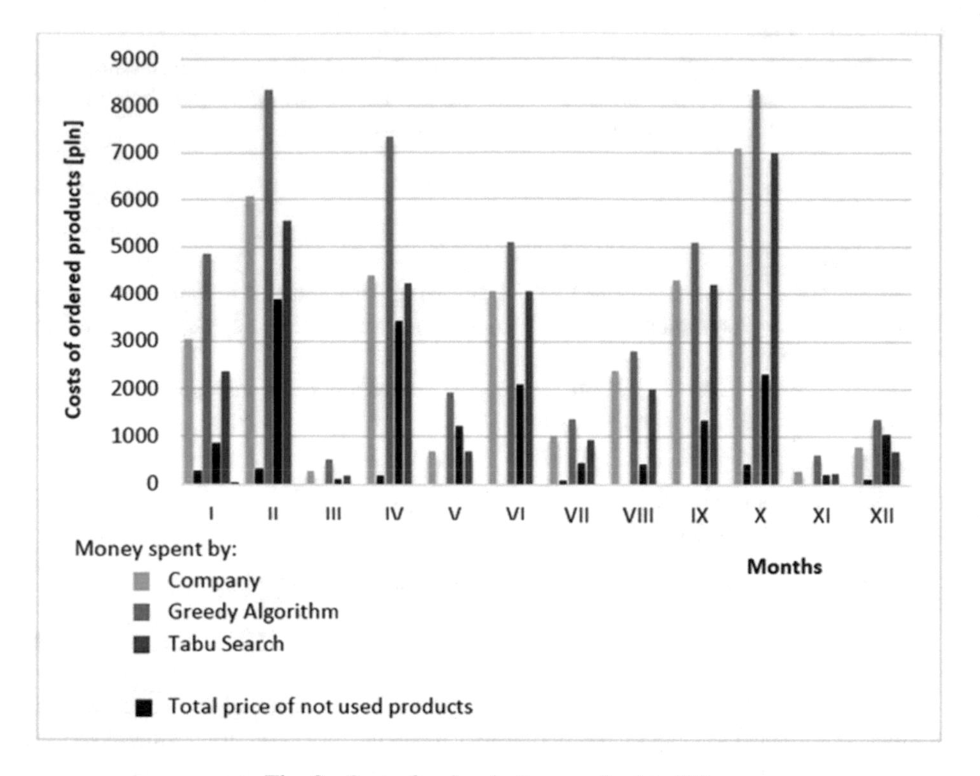

Fig. 2. Cost of orders in the months I to XII.

Table 2. Products data in the months I to XII.

	I	II	III	IV	V	VI	VII	VIII	IX	X	XI	XII	∑
Products ordered													
C	732	487	954	1062	695	844	920	316	609	498	587	449	**8153**
G	826	922	1340	1904	1345	1180	1813	427	945	524	995	277	**12498**
T	577	432	942	1018	695	844	896	298	604	458	488	372	**7624**
Products not used (if ordered)													
C	167	55	12	47	0	0	32	18	6	49	103	79	**568**
G	284	490	398	893	834	433	978	129	380	238	556	283	**5896**
T	12	0	0	3	0	0	8	0	1	9	4	2	**39**
Products necessary not ordered													
C	1	1	2	4	2	1	2	1	1	2	2	1	**20**
G	23	0	0	4	184	97	53	0	38	163	45	376	**983**
T	0	0	0	0	0	0	0	0	0	0	0	0	**0**

cost of the orders suggested by the greedy algorithm was in each case much more higher than the real one. Moreover, the algorithm suggested to order many unnecessary products, omitting the necessary ones.

Based on the data collected, the Tabu Search algorithm is able to rationalize the process. In each of the analyzed months, it suggested a solution better than the real one. The calculated costs were not much lower, however the algorithm avoided ordering unnecessary products, ensuring the order of all necessary ones at the same time.

4 Summary

In the paper the decision-making problem in selection of suppliers was considered. The presented work suggested new suppliers selection system with the use of the intelligent algorithms – Tabu Search and/or greedy algorithm. The research was carried out on an example of a real production company. Based on the results obtained by both considered algorithms, only the Tabu Search can be successfully used for solving analyzed problem. It suggested solutions that reduce the total cost of orders, guaranteeing the order of all necessary products and avoiding unnecessary ones. Based on the above, it can additionally reduce storage costs, destruction costs and also transport costs. The obtained results demonstrate high potential of the Tabu Search algorithm that can be successfully used for solving similar or even more difficult problems.

References

1. Antosz, K., Stadnicka, D.: The results of the study concerning the identification of the activities realized in the management of the technical infrastructure in large enterprises. Eksploat. Niezawodn. (Maintenance and Reliability) **16**(1), 112–119 (2014)
2. Bentley, P.J., Wakefield, J.P.: Finding acceptable solutions in the Pareto-Optimal Range using multiobjective genetic algorithms. In: Chawdhry, P.K., Roy, R., Pant, R.K. (eds.) Soft Computing in Engineering Design and Manufacturing. Springer, London (1998)
3. Bożejko, W., Uchroński, M., Wodecki, M.: Parallel tabu search algorithm with uncertain data for the flexible job shop problem. In: Rutkowski, L., Korytkowski, M., Scherer, R., Tadeusiewicz, R., Zadeh, L.A., Zurada, J.M. (eds.) ICAISC 2016. LNCS, vol. 9693, pp. 419–428. Springer, Cham (2016)
4. Burduk, A., Musiał, K.: Genetic algorithm adoption to transport task optimization. In: Graña, M., López-Guede, J., Etxaniz, O., Herrero, Á., Quintián, H., Corchado, E. (eds.) ICEUTE/SOCO/CISIS 2016. AISC, vol. 527, pp. 366–375. Springer, Cham (2017)
5. Burduk, A., Musiał, K.: Optimization of chosen transport task by using genetic algorithms. In: Saeed, K., Homenda, W. (eds.) CISIM 2016. LNCS, vol. 9842, pp. 197–205. Springer, Cham (2016)
6. Brandão, J.: A tabu search algorithm for the open vehicle routing problem. Eur. J. Oper. Res. **157**(3), 552–564 (2004)
7. Cordeau, J.F., Gendreau, M., Laporte, G.: A tabu search heuristic for periodic and multi-depot vehicle routing problems. Networks **30**(2), 105–119 (1997)
8. Dorigo, M.: Optimization, learning and natural algorithms. Ph.D. thesis. Politecnico di Milano, Italy (1992)

9. Gola, A., Kłosowski, G.: Application of fuzzy logic and genetic algorithms in automated works transport organization. In: Omatu, S., Rodríguez, S., Villarrubia, G., Faria, P., Sitek, P., Prieto, J. (eds.) DCAI 2017, AISC, vol. 620, pp. 29–36. Springer, Cham (2018)
10. Goyal, A., Lu, W., Lakshmanan, L.V.: CELF++: optimizing the greedy algorithm for influence maximization in social networks. In: Proceedings of the 20th International Conference Companion on World Wide Web, pp. 47–48 (2011)
11. Górnicka, D., Markowski, M., Burduk, A.: Optimization of production organization in a packaging company by ant colony algorithm. In: Burduk, A., Mazurkiewicz, D. (eds.) ISPEM 2017. AISC, vol. 637, pp. 336–346. Springer, Cham (2018)
12. Grabowski, J., Wodecki, M.: A very fast tabu search algorithm for the permutation flow shop problem with makespan criterion. Comput. Oper. Res. **31**(11), 1891–1909 (2004)
13. Ho, W., Xu, X., Dey, P.K.: Multi-criteria decision making approaches for supplier evaluation and selection: a literature review. Eur. J. Oper. Res. **202**(1), 16–24 (2010)
14. Kahraman, C., Engin, O., Kaya, I., Öztürk, R.E.: Multiprocessor task scheduling in multistage hybrid flow-shops: a parallel greedy algorithm approach. Appl. Soft Comput. **10**, 1293–1300 (2010)
15. Kalinowski, K., Skołud, B.: The concept of ant colony algorithm for scheduling of flexible manufacturing systems. In: Graña, M., López-Guede, J., Etxaniz, O., Herrero, Á., Quintián, H., Corchado, E. (eds.) ICEUTE/SOCO/CISIS 2016, AISC, vol. 527, pp. 408–415. Springer, Cham (2017)
16. Kotowska, J., Markowski, M., Burduk, A.: Optimization of the supply of components for mass production with the use of the ant colony algorithm. In: Burduk, A., Mazurkiewicz, D. (eds.) ISPEM 2017, AISC, vol. 637, pp. 347–357. Springer, Cham (2018)
17. Krenczyk, D., Skołud, B.: Transient states of cyclic production planning and control. Appl. Mech. Mater. **657**, 961–965 (2014)
18. Musiał, K., Kotowska, J., Górnicka, D., Burduk, A.: Tabu search and greedy algorithm adaptation to logistic task. In: Saeed, K., Homenda, W., Chaki, R. (eds.) CISIM 2017, LNCS, vol. 10244, pp. 39–49. Springer, Cham (2017)
19. Tachizawa, E.M., Thomsen, C.G.: Drivers and sources of supply flexibility: an exploratory study. Int. J. Oper. Prod. Manag. **27**(10), 1115–1136 (2007)
20. Tuncer, A., Yildirim, M.: Dynamic path planning of mobile robots with improved genetic algorithm. Comput. Electr. Eng. **38**(6), 1564–1572 (2012)
21. Zhang, Z., Schwartz, S., Wagner, L., Miller, W.: A greedy algorithm for aligning dna sequences. J. Comput. Biol. **7**, 203–214 (2000)

Risk Assessment for Potential Failures During Process Implementation Using Production Process Preparation

Robert Błocisz(✉) and Lukasz Hadas

Faculty of Engineering Management, Poznan University of Technology, 11 Strzelecka Str., 60-965 Poznan, Poland
robertblocisz@gmail.com, lukasz.hadas@put.poznan.pl

Abstract. Production preparation process is one of most important tool which is using during design or redesign of the production process. Production process can be changed by different causes. 3P (production preparation process) is universal tool which can be used in almost all cases of process modernization. 3P is connecting and engaging not only R&D department but also others company departments. The main aim of this research was to identify and assess the risk for potential failures during process implementation using Production Process Preparation (3P). The research was carried out in an enterprise from the automotive industry, which is an international supplier of components to many car brands. The prepared questionnaire contained questions (ordered in the individual categories) selected on the basis of the authors experience and consultations with the experts at least 10 years of experience. The study took into account the seniority of experts as a process engineer and experience as a manager. The results of the surveys were the basis for identifying potential failures with the highest risk in the Production Process Preparation.

Keywords: Production preparation process (3P) · Automotive industry
Risk management · Risk assessment

1 Introduction

One of the basic tasks for manufacturing engineer is production processes planning, calculation of the required performance, cycle time, the development of specifications for machines, tools and implementation related with the validation each implementation of a new product or production process launching in the production area can be treated as a new project. The manufacturing engineer who is responsible for the product then becomes the "Project Manager" for whole operation. It is he who reports status of implementation and the advancement of individual activities to superiors.

Process production preparation (3P) [1, 2] which is universal and advanced manufacturing tool is very useful during the production process planning.

3P that is Production Process Preparation, is one of tools for creating an optimal process which is based on Lean Manufacturing principles. 3P in the Lean world is perceived as one of the most powerful tool associated with the transformation of the production process. It is used only by companies which have huge experience in

© Springer Nature Switzerland AG 2019
A. Burduk et al. (Eds.): ISPEM 2018, AISC 835, pp. 285–295, 2019.
https://doi.org/10.1007/978-3-319-97490-3_28

implementing of other Lean tools. While Kaizen and other Lean tools are involved in improvement of the production process, 3P focuses on reducing waste during product and process designing [3]. In the literature, the 3P tool is described by 3 phases (1–3) [1]. However, currently in business models related to project management, there is model of 3P elaborated of 5 phases [4]:

1. Information phase
2. Creative Phase
3. Redefine Phase
4. Implementation Phase
5. Continuous Improvement Phase

These phases will be discussed in the context of practical use of them and their realization in the next chapter of this article.

2 Practical Aspects of 3P Application

During launching a new product in production area, we can often meet standard problems and frequently committed mistakes in different types of projects. We can include to them:

- Expenses for investments and costs significantly exceeding the assumed budget,
- Low level of customer service and dates of realization significantly exceeding customer requirements,
- Defective components on every stage of production process and their "re-work" at each production steps,
- Problems with fulfillment of requirements related to environmental protection, ergonomics of the operator's workplaces or health and safety.

To avoid the mentioned problems it is worth to organize a multi-day workshop based on using the 3P tool.

On workshop besides managers and manufacturing engineers from the technology department (process improvement) and R&D (production process preparation) department responsible for developing a new production process should be also invited representatives of cooperating departments such as finance and logistics. Collaboration of employees from the finance and technology department is mainly associated with the purchase of equipment and release of payments for individual suppliers. Employees from logistics cooperate with technologists in the field of materials management, material purchase, organization material shipments to suppliers for trials, acceptance of orders and organization prototypes shipments to the customer, collection and transport of equipment from abroad and break points organization.

During these short workshops the most important is to focus on design for the future production process system in which waste generated during production will be as small as is possible, while the flow of material and information will be the best. For this purpose, it is necessary to develop and present alternatives for individual steps of the production process as well as to the whole production process. To verify the correctness of individual alternatives, it is worth applying the PDCA (Plan-Do-Check-Act) principle [5], which allows determining and documenting the best solution. To guarantee the quality, traditional quality control tools can be used [6].

Basic idea of alternative solutions is to show at least two potential solutions for the same problem related to construction of product or manufacturing process. Problems in construction of the product appear mostly during validation tests on prototype parts. That is why it is so important to use simulating programs based on the finite element method (FEM) [7]. Due to application of appropriate boundary conditions, it is possible to predict the weakest element of the product structure and implement alternative solution. The same situation applies to improvement of the manufacturing process.

Tools used in 3P are universal and can be used in various situations related to production process [3]:

- Launching new product in production area, or whole range of new products related to relevant business line,
- Changes in construction of existing product,
- Increase or decrease volume of products produced in the plant,
- Introduction of new technology for production process or new production equipment,
- Transfer of production within plant or relocation of production between plants,
- Restructuring or modernization of current production process.

All cases of 3P rules application should be executed parallel with the product design process (Product Design Process - PDP). Practically every mentioned reasons of 3P usage should be agreed or at least given to customer information (product recipient). Product validation should be executed out in consultation with the client after each of mentioned above changes.

The first phase of 3P application mainly focuses on information collection about the product. Before you start designing the production process you need to know such information as [1, 3]:

- Customer specification about durability of the product, quality aspects and other requirements,
- The volume assumed by the client and potential volume increase,
- The number of variants of product/part,
- Drawings of product and components,
- Design of prototype components which will be used for building the product and design for final product prototype,
- Environmental, health and safety requirements,
- Preliminary map for flow of materials and components during the production process,
- Information about efficiency, load and condition of machines assumed in preliminary process map and other resources necessary to plan and perform the production process.

The second phase of 3P usage is called Creative Phase. During this phase, a Lean concept should be created for the entire production process. It is necessary to develop appropriate assessment criteria for the designed product. Mostly these criteria are related with reduction of costs and customer requirements mentioned in specification. Therefore, it is worth dividing the project team into small groups which will be responsible for the individual steps of the production process. Good practice is to create several alternative concepts for individual production process steps. The main goal should be to build quality into production processes and to protect against possibility of making a mistake. The best solution is to use POKA-YOKE [8]. It is important to convert the concepts of developed solution into the prototypes of final tooling as soon as possible, to check the correctness of operation. Due to this it is easy to find out how a given concept will behave after implementation into the production process and if it is worth to implemented it into the production process. In this way we are saving not only money, but also time which is a source of advantage over competition in the situation of market pressure related to the efficient implementation of new products with short life cycle. The next action which is worth to implement in this phase is to create a model of the production process. It is important to show flow of material in all process steps and also between individual workstations which are in it. Moreover, in model of production process should be marked stations which require restructuring and which are completely new. All assumed changeover should by based on SMED methodology [7]. Programs to create simulations of the production process or Excel, which show individual steps of the production process by inserted shapes can be used for this purpose [3].

3 phases 3P called Redefine Phase involves preparation of documents necessary to approve a new production process and defining and describing of the following parameters [1, 3]:

- Material flow map during realization of a new production process,
- Cycle time and tact time for individual operations,
- Process capability,
- Layout of equipment and work stations necessary for implementation of a new production process,
- Plan and schedule of implementation individual steps of the production process (reports on current implementation status should be send to all interested team members who currently can not participate in meetings),
- Organization of complementary meetings, where current plan of implementation is presented.

Phase 4 is implementation phase which involves changing concept into reality. The implementation of the plan related which launching of the production process. To do it the best is to organize a meeting which will be one or twice a week. A good solution is to organize meeting for individual task groups to not waste time other groups. On the project duration, it is worth to made table in the form of matrices where names and surnames will be placed on one X axis, and weeks on the Y axis. In this way manage tasks for individual employee and their implementation time will be easier.

The fifth phase of using 3P is Continuous Improvement Phase which goal is continuous improvement of the product or process. All works are focus on improving the process and solving production and quality problems. All action should be realized in teams. Beside person managing the producing, the team should be attended by a quality engineer, product engineer and manufacturing engineer responsible for individual production processes. In this way, not only the interpersonal relationships get strength and get improved, but also efforts are made to determine and give status quo, for process or product. It is known that errors are made, but it is important to catch them as soon as possible and not to commit more them according to the "Lessons learned" principle.

3 Identification of Critical Risks for the Basic Steps During Implementation of New Production Process

3.1 Methodology of Research

In order to explore and assess the risk of negligence or failure of using parameter defining the process in third phase and describing application of the 3P tool during launching new product in production area or modification of current production process, a survey was conducted among experts from the automotive industry. It was a group of manufacturing engineers from the R&D department and manufacturing engineering responsible for special processes. The group of 23 respondents consisted of engineers with at least several years of experience in implementing new products and launching modifications into production area. The purpose of research was to identifying significance of potential failures for parameters defined in to 3P method which can have influence for final result of project implementation.

The prepared questionnaire contained questions in the following categories: material flow map during realization of a new production process, cycle time and tact time for individual operations, process capability, layout of equipment and work stations necessary to implement a new production process, plan and schedule of implementation individual steps of the production process, organization of complementary meetings, where current plan of implementation is presented (Table 4). Questions about failures in individual categories (Table 4) were based on the author's experience and consultations with group of experts with at least 10 years of experience. The respondents evaluated according to the scale determined for occurrence (Table 1) and severity (Table 2) for prepared failures.

Table 1. Rating for the risk occurrence.

Rating for the risk occur	
Very low or equal zero risk for adjustment/process launching	**1**
Appear in about 25% o case of adjustment process/process launching	**2**
Appear in about 50% o case of adjustment process/process launching	**3**
Appear in about 75% o case of adjustment process/process launching	**4**
Appear in more than 90% of cases of adjustment process/process launching	**5**

Obtained data allowed calculating the indicator which is modeled on the RPN - Risk Priority Number which shows risks of individual mistakes. This is not exactly the RPN coefficient used in the FMEA [9] methodology because one component is

Table 2. Rating for severity of appeared failures.

Rating for severity	
No impact on the production process, no impact on process launching schedule	**1**
The process fulfils 80% of assumptions, no impact on process launching schedule	**2**
The process fulfils 60% of basic requirements/assumptions, no impact on process launching schedule	**3**
The implemented process does not meet basic assumptions it fulfils only 40% of them, acceptable overrun of the schedule in the security buffer	**4**
The process does not fulfill basic assumptions it realized only 20% of them, overrun of the process launching schedule	**5**

missing, which is detection, but for simplicity in later part of article we will use abbreviation RPN (according to FMEA methodologies). Obtained results will be used for further analysis. The maximum note of assessment which is the product of the points awarded from both of the ratings is 25. The limit value was assumed on 6 points level. The matrix below (Fig. 1) presents the product of risk of occurrence and effect for potential failures in the ranges:

- High risk for project requires immediate defining and implementation of corrective actions - red color,
- Risk appear in each project, defining corrective actions - yellow color,
- No risk for launching project - green color.

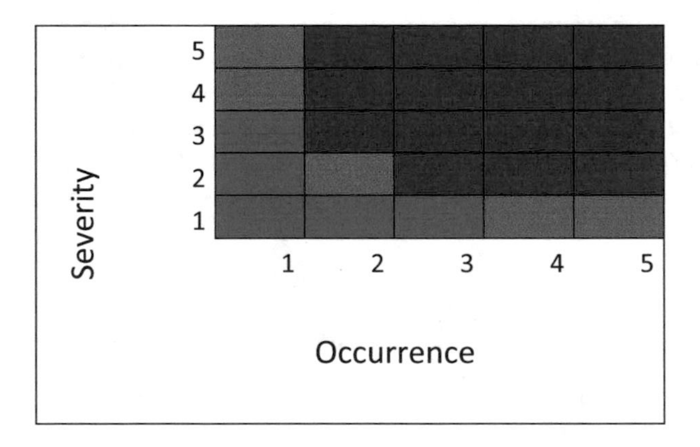

Fig. 1. Chart of colors describing RPN values.

As a limit value assume 6 point, which is the product of risk launching of the process, which may occur in 25% of cases (2 points) or 50% of cases (3 points), resulting in launching a process fulfill 80% of its basic assumptions without exceeding the assumed schedule (2 points) or a process which fulfill only 60% of the basic assumptions without exceeding assumed schedule.

Survey questionnaire had questions regarding the self-assessment of the expert's competence in tested area (Table 3), seniority and currently occupied position were added to the survey questionnaire. In that way we have gained the opportunity to analyze the distribution of experts opinion obtained depending on their seniority and position.

Table 3. Respondent self-assessment

Respondent self-assessment	
Employee without experience, no projects launched	1
Specialist with experience smaller than 2 years, many adjustment implemented in the processes	2
Specialist with experience bigger than 2 years, more than 5 products implemented into the production area	3
Specialist with experience bigger than 10 years in launching and adjustment of production processes	4
Expert, project manager	5

3.2 Research Findings

Based on results of the survey, was calculated the coefficient of significance for potential failures. Below is a chart (Fig. 2) which presents average values of SxO for failures included in the survey and a line showing the limit value which was assumed to be significant for the product of SxO.

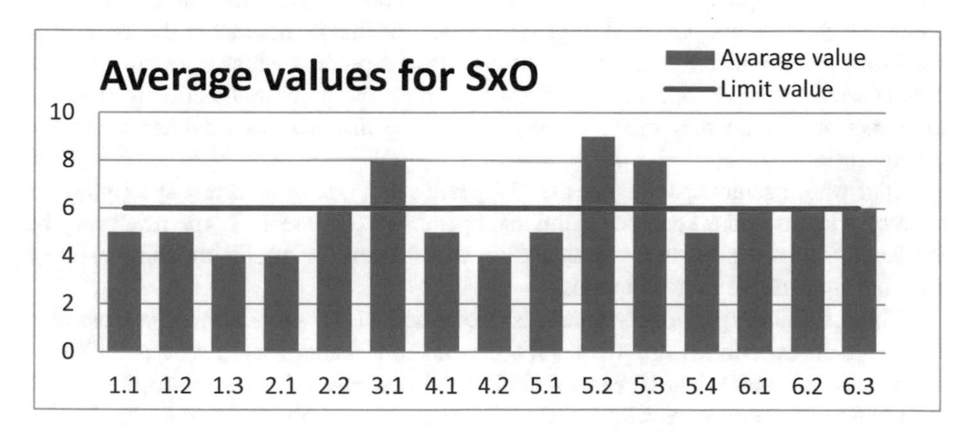

Fig. 2. The average value of the risk for potential failures (errors).

During analysis the data obtained from the survey as a potential failures burdened the highest risk (average SxO $\geq$ 6), one can designate:

- 2.2 Non-fulfillment of assumed the cycle time,
- 3.1 Non-fulfillment of process capability the Ppk or Cpk < 1.33,

- 5.2 Problems with implementation arising from lack of machines availability in the production area,
- 5.3 Overrun of assumed schedule,
- 6.1 Lack of organization meetings for status updating/no possibility to report current status,
- 6.2 No possibility to notify the problem to the supervisor during direct encounter,
- 6.3 No possibility to update current status of adjustment implementation/new process implementation.

Subsequently prepared a summary (Table 4) with percentage share of the response where a product (SxO) was greater than or equal to 6 for potential failures concerning individual parameters with division into two groups due to seniority.

In the table bolded results where 50% or more of answer exceeded 6 points in the 25 points scale. Depending on the respondents seniority, assessment of significance for potential failures was different, however, the majority of responses coincide with other and the difference falls within 5% points.

Employees with shorter seniority see a risk for successful process implementation in:

- 1.3 No identification for needless operations (storage of semi-finished products, to excessive internal transport)
- 4.1 Excessive transport between successive workstations of the process.
- 4.2 Incorrect layout of machines and devices in the production cell.
- 5.1 Incorrect established plan of production process implementation.
- 5.4 Lack of possibility for checking implemented solution.

Failures 1.3, 4.1, 4.2 and 5.4 affect the assumptions for process, not for process launching. These points mostly have influence for longer cycle time, additional operations which were not predicted in process. Effect of this is increase in the number of operators; thereby increase in costs or the release of a process which does not meet the assumed requirements. Section 5.1 affects launching the process. Incorrectly assumed implementation plan may cause delay in launching the process, establish a risk for entire project.

Staff whose seniority is higher than 10 years see a risk in incorrect specifying the necessary inputs and outputs for individual operations (Failure 1.2). The result may be the launching of a process that was badly defined from the beginning and, generally does not meet required assumptions.

Differences in expert assessments arise from the fact that employees with more than 10 years of experience are frequently persons on managerial positions (40% of employees with more than 10 years seniority) with average of self-assessment at 4.1 in a 5-point scale. Employees with less than ten years of sonority declared evaluation 2.5 and are usually manufacturing engineers responsible for the implementation of processes, being part of the launch of a new product in production area. Significant part of respondents (>80%) from both groups sees potential risk for launching the process in:

- 5.2 Problems with implementation arising from lack of machines availability in the production area.
- 5.3 Overrun of assumed schedule.

Table 4. Percentage of responses where the product SxO $\geq$ 6.

Risks for basic parameters during implementation of the production process adjustment/the new production process	All respondents	Seniority > 10 years	Seniority < 10 years
1. The map of material flow during the implementation of the new production process			
1.1 No identification for crucial operations of the production process	34,78%	40,00%	25,00%
1.2 Incorrect determination for required inputs and outputs for individual operations	47,83%	**53,33%**	37,50%
1.3 No identification for needless operations (storage of semi-finished products, to excessive internal transport)	43,48%	40,00%	**50,00%**
2. Cycle time			
2.1 Incorrect determination of cycle time and takt time	34,78%	26,67%	50,00%
2.2 Non-fulfillment of assumed the cycle time	**73,91%**	**73,33%**	**75,00%**
3. Process capability			
3.1 Non-fulfillment of process capability the Ppk or Cpk < 1.33	**60,87%**	**60,00%**	**62,50%**
4. The layout of devices and work stations in production area necessary for the implementation of the new production process			
4.1 Excessive transport between successive workstations of the process	47,83%	33,33%	**75,00%**
4.2 Incorrect layout of machines and devices in the production cell	47,83%	46,67%	**50,00%**
5. Plan and schedule for implementation of individual parts of the production process			
5.1 Incorrect established plan of implementation of the production process	47,83%	46,67%	**50,00%**
5.2 Problems with implementation arising from lack of machines availability in the production area	**82,61%**	**86,67%**	**75,00%**
5.3 Overrun of assumed schedule	**82,61%**	**80,00%**	**87,50%**
5.4 Lack of possibility for checking implemented solution	50,00%	41,67%	**62,50%**
6. Organization of supplementary meetings for the presentation of implementation plan current status			
6.1 Lack of organization meetings for status updating/no possibility to reporting about current status	**52,17%**	**53,33%**	**50,00%**
6.2 No possibility to notify the problem to the supervisor during direct encounter	**52,17%**	**53,33%**	**50,00%**
6.3 No possibility to update current status of adjustment implementation/new process implementation	**60,87%**	**66,67%**	**50,00%**

More than half of the respondents from both groups sees risk in:

- 6.1 Lack of organization meetings for status updating/no possibility to report current status,
- 6.2 No possibility to notify the problem to the supervisor during direct encounter,
- 6.3 No possibility to update current status of adjustment implementation/new process implementation.

Showed in that way how important is the plan and the schedule of implementation for individual parts of production process and the organization of complementary meetings where current status of implementation is presented.

4 Conclusion

In beginning of project connected with process launching important is to put particular attention on all above potential failures for the parameters used to define the process in the third phase of the 3P tool, treating them as critical in order to avoid possible problems for the above-mentioned failures. To do this, you can use many tools designed for project management. Moreover the proper use of the tool refers not only to the third phase but to all phases of using the 3P tool, which can give many benefits not only to the project managers but also to all persons who design and implement processes and products. Thanks to the use of 3P, the duration of the project launching can be shortened, and planed target could be achieved faster. It means that the cost of new project will be low, the product will be fabricated in the required quality and achieve this quality in a short time after the start of high-volume production. Therefore, the return on investment will start sooner and the ROI will be more beneficial for the whole project. Using 3P, you can create a strong process focused on the client, meeting his requirements and needs, not on whims. It is important to show and give the client what he needs, not what he thinks that he need. The answer to the research problem which with of the potential failures for the parameters describing the 3P method can have a significant impact on the success of the project is an important advice for project managers. Due to the high risk of fail the project, the failures identified in the study require special attention in the process of project management.

References

1. Mascitelli, R.: Lean Design Guidebook - Everything Your Product Development Team Needs to Slash Manufacturing Costs, 1st edn., pp. 189–205. Technology Perspectives, Northridge (2004)
2. Bicheno, J., Holweg, M.: The Lean Toolbox: A handbook for Lean Transformation, 5th edn., pp. 191–193. PICSIE Books, Buckingham (2016)
3. https://peterpaul.com/capabilities/process-methods/3-p-process. Accessed 21 Mar 2018
4. https://www.slideshare.net/opexcreative/3-p-production-preparation-process. Accessed 21 Mar 2018

5. Broniewska, G.: Cykl PDCA odzwierciedleniem klasycznego cyklu zorganizowanego działania (eng. The PDCA cycle reflects the classical cycle of organized activity). Probl. Jakościowe **39**(7), 36–39 (2007)
6. Wysocki, R.: Efektywne zarządzanie projektami (eng. Effective Project Management), 6th edn., p. 54. HELION, Gliwice
7. Dudek-Burlikowska, M., Szewieczek, D.: The Poka-Yoke method as an improving quality tool of operations in the process. J. Achiev. Mater. Manuf. Eng. **36**(1), 95–102 (2009)
8. Antosz, K., Kużdżał, E.: Doskonalenie procesu przezbrajania maszyn montażowych z wykorzystaniem metody SMED (eng. Improving the retooling process of assembly machines using the SMED method). Technol. Automat. Montażu **1**, 49–53 (2015)
9. Braaksma, A.J.J.: A quantitative method for failure mode and effects analysis. Int. J. Prod. Res. **50**, 6904–6917 (2012)

Strategy of Improving Skills of Innovation Managers in the Area of Advanced Manufacturing Technologies

Kamil Krot(✉), Emilia Mazgajczyk, Małgorzata Rusińska, and Anna Woźna

Wrocław University of Science and Technology, Wrocław, Poland
kamil.krot@pwr.edu.pl

Abstract. In the times of rapid market and technological changes industry is subjected for intensive development and flexibility capacity building especially in area of innovation introduction. Typically the focus is concentrated on technological and operational questions but at the bases of all there are strategic challenges that need to be answered. The key for success lies in hands of managers and CEOs, who have to cope with innovation management. This article presents new initiative of building advanced manufacturing capacities in SME of Central Europe, concentrated on managers skills. The multi-layer educational program development within the InnoPeer AVM project will be oriented on new international approach to building the potential related to AVM in local, small enterprises, as well as in leading companies.

Keywords: Innovation management · Advanced manufacturing · Industry 4.0

1 Introduction

Central Europe industry has rich achievements, however, its urgent modernisation and re-industrialisation is necessary. This is directly related to changes occurring in the economy, determined as Industry 4.0. The Industry 4.0. project, determined also as the fourth industrial revolution, is related to digitisation of companies and integration of advanced manufacturing technologies with other systems functioning in organisations. Digitisation of companies, caused by several market requirements, is related to several technological improvements shown in Fig. 1.

Companies in Central Europe, in particular small and medium-sized enterprises (SME), are obliged to carry out innovative, advanced manufacture (AVM – Advanced Value Manufacturing) in order to increase their competitiveness and to reach access to international chains of advanced manufacturing. At present, the questions related to AVM are discussed mainly as technological questions, but it is also managers for innovations and SME owners who face enormous strategic challenges. In connection with this, an urgent need has appeared to develop common basics of knowledge for the entire Central Europe with new international approach to building the potential related to AVM in local, small enterprises, as well as in leading companies.

© Springer Nature Switzerland AG 2019
A. Burduk et al. (Eds.): ISPEM 2018, AISC 835, pp. 296–305, 2019.
https://doi.org/10.1007/978-3-319-97490-3_29

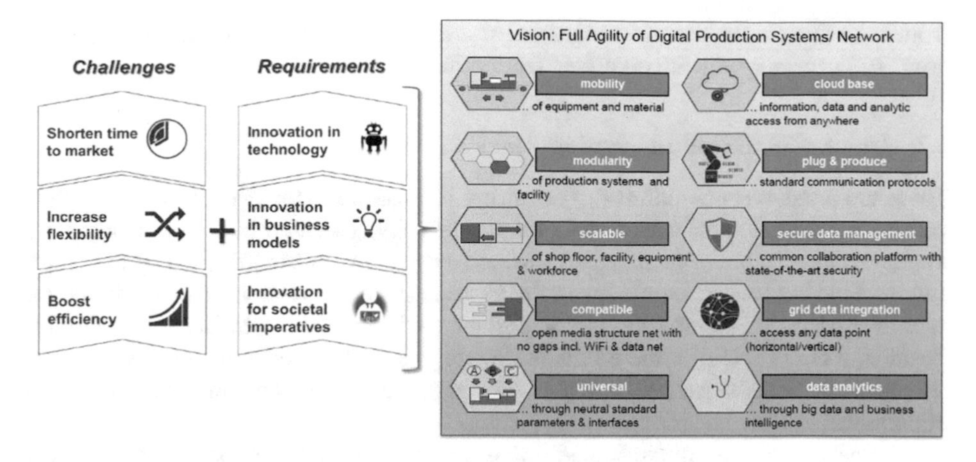

Fig. 1. Challenges, requirements and technologies related to implementation of advanced manufacturing technologies [1].

The multilayer program includes 3 scopes (AVM technologies, human resource management in relation to implementation of AVM and adaptation of a business model to AVM specificity) and will be executed with use of proved, innovative and most effective formats and training methods on two progression levels by practical training like webinars from living laboratories, practical tests in a model factory and AVM strategy camps. Moreover, the innovative compilation of “Training cases” (style of the Harvard Business School) based on real cases and challenges encountered by Central Europe companies will make it possible to improve knowledge in this scope. Participants of the entire training program shall be obliged to write an elaboration concerning their company, which shall permit their certification as the “AVM Manager with InnoPeer certificate”. Pilot trainings will involve target companies and managers for innovation from all the project partner regions.

2 Innovation Based Industry Development

2.1 Management of Innovations

Until recently, innovation was the feature distinguishing a company. At present, this is the advantage required in order that the company can survive and reach high market rank. Therefore, innovation management is the biggest challenge for SMEs [2]. If a company is managed by a person with proper qualifications, it can be very profitable for the company. A fundamental step towards development is implementation of innovations that can concern both processes, organisation and business model, as well as marketing or products [3]. At the same time, innovations are tools to build competitive advantages and image of the company.

Sources of innovations can be: performing internal research and development activity or gaining innovations from outside, but also carrying out marketing actions or stimulating creativity among the employees. In SMEs, innovations are often acquired from scientific research works carried out at universities or research institutes; "know-how" or licences can be also acquired from other companies [4].

2.2 Implementation of Innovations

There are numerous possibilities of acquiring innovations. However, success of innovations requires that changes should be properly implemented in the company. The implementation process must run concurrently on many planes [5]. First of all, small and medium businesses should break the awareness barrier related to underestimation and misunderstanding of role and importance of innovations in business activity. Besides, a big problem of the SMEs is low degree of knowledge about this problem among their personnel. This is why breaking the knowledge/ignorance barrier is so important.

The necessary and sufficient condition is convincing small and medium businesses, how much important is implementation of changes not only to products and manufacturing technologies, but also in the spheres of organisation and operation of companies, their management or even in marketing activities. They include, among others, changes of the company strategy, decision taking, controlling and organising the innovative activity [6].

2.3 Development Strategies of Companies Oriented Toward Innovations and New Technologies

The topical literature delivers numerous more or less complex models of creating and executing projects of technological innovations. These project must, on one hand, result from progress of management sciences and consider experiences of the leaders and, on the other hand, be adapted to specific conditions determined by location of the company, its size, specificity of the industry, as well as economic and staff capacities.

The following kinds of innovations can be distinguished [3]:

- product innovation – consisting in implementation of a new or significantly improved product or service,
- process innovation – consisting in implementation of a new or significantly improved method of manufacture or delivery within the process; changes within technology of equipment and/or software can be considered,
- marketing innovation – consisting in implementation of a new marketing method,
- organisation innovation – consisting in implementation of a new method of organisation in many areas, related to both principles of the company operation and organisation of workplaces.

It should be stressed here, that innovations are most often related to technology and technique of manufacture, as well as to introduction of new products. However, as was mentioned before, SMEs can mostly duplicate the solutions borrowed from big companies [7].

3 Specificity of New Technologies Within Advanced Manufacturing Technologies

Advanced manufacturing is understood as utilisation of modern technologies in order to deliver existing products and new products to the market. It is focused on improvement of the processes of design and manufacture in all the areas, together with integration of IT systems in the entire supply chain. It focuses its attention on minimisation of production time and costs by integration of the newest technological and IT solutions based on data acquisition and processing. Improvement of products, processes and company results is possible thanks to acquisition and analysis of the information coming from all stages of product life cycle, manufacturing and manufacture-related processes taking into consideration the applied processes and tools. In the market, newer and newer technologies are available, making possible integration of entire production lines and even plants into intelligent manufacturing systems with high ability to take decisions independently on the grounds of artificial intelligence modules, as well as of decisive and self-learning algorithms. These changes are called the fourth industrial revolution based on creation of cyber-physical systems and modification of manufacturing methods [8]. In the following paragraphs the most important AVM technologies are summarized.

Internet of Things - is a widely understood idea based on the concept of communication between devices (M2M – Machine to Machine). Solutions within IoT are based on the possibilities of communication, as well as data exchange, processing and acquisition by devices via computer network, with no necessity of human intervention. The network of linked intelligent machines, acting in the intended way on the basis of utilisation of sensors, advanced analyses and intelligently understood decisions, deeply changes the way how assets communicate with the company [9].

Big Data - this idea is still evolving and is still anew considered, since it remains the driving force of digital transformation, including artificial intelligence, scientific data and Internet of Things [17]. The unceasing development of technology ensures the possibility of acquiring and collecting larger and larger amounts of data, but their small part only is used in order to draw conclusions, take decisions, learn and improve [10]. The principle is the analysis and comparison of data in order to formulate new observations, work principles and relationships, but also to anticipate trends and future behaviours.

Cloud Computing - means applications and IT infrastructure delivered in form of an internet service. The cloud supplies solutions that permit technology and tools to be continuously adapted to current needs of the company. The data is stored in the virtual space, on the servers situated beyond the local network, where calculations are run [11]. Moreover, the problems related to hardware limitations, additional costs of operation and maintenance of own server rooms no longer exist [13].

Cybersecurity - developed as a result of using advanced tools for recording, collecting, transmitting, analysing and archiving industrial data. The occurrences like stealing of sensitive data, destroying the information stored in business applications or failures caused by introducing a malicious code to the corporate system cost the world-wide companies ca. USD 400 milliards per year [12]. The basic actions in data security assurance are: automatic preparation of backup copies, protection of applications, protection of websites and servers.

Additive manufacturing - technologies that generate a three-dimensional object on the grounds of its virtual 3D model by precise adding material layer by layer. Technologies are based on variable physical and chemical processes, e.g. laser sintering or melting of powder grains, gluing based on a chemical reaction of powder grains, deposition of material particles, selective hardening of photosensitive materials, laminar cutting to size and joining. An advantage of the additive technologies are big possibilities to manufacture structures with complex internal and external geometry, often impossible to be made by machining or casting [14].

Virtual and augmented reality - the possibility to represent virtually, in real time, functioning of a device or results of actions performed by an employee gives wide possibilities, accompanied by minimisation of the risk of damage or wastage of resources [15]. Use of this technology in the manufacturing industry ensures a support of engineers during design and service works, reduces costs of trainings, especially in the case of advanced, work loaded manufacturing systems that cannot be stopped any time in order to perform a training.

Simulation tools - software, based on real functioning principles in form of mathematic equations, facilitates performing operations in virtual environment imitating manufacturing processes in order to verify the assumed parameters and correctness of assumptions. The obtained results are used for optimisation of manufacturing processes, minimisation of generated costs, increasing effectiveness of processes, better distribution of resources, and also for planning new production [16].

4 Evaluation of Knowledge and Application of Solutions Within Advanced Manufacturing Methods in Lower Silesian Region

4.1 Method

So far, within the described Project, the tasks related to identification of knowledge and awareness levels concerning advanced manufacturing techniques were realised. Recognised was academic environment, small and medium-sized businesses in individual regions of Central Europe and intermediary institutions between the a.m. subjects.

Regional partition was conditioned by the project partners. In the project, eight regions were considered: Upper Austria (leader and one partner of the project), Euganean Venice – Italy (two partners), Western Hungary (two partners), District Tübingen – Germany (two partners), Lower Silesia – Poland (one partner), Emilia-Romagna – Italy (one partner) and Upper Bavaria – Germany (one partner). Evaluation of companies and academic environments was carried out in three areas:

- technologies related to advanced manufacturing,
- human resources and organisation methods,
- business model and management with business processes.

The results were used to map AVM competences of three actors based on prepared check list that was filled in by Project Partners with information gained from reports, interviews with experts, databases and available documentation analyses. Figure 2 presents comparison of AVM technologies utilisation between SMEs and Academia, with division on current situation and future needs.

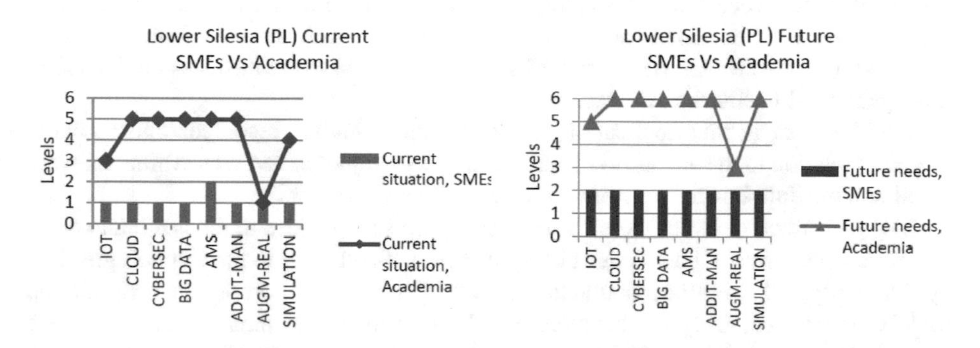

Fig. 2. AVM and future needs comparison between SMEs and Academia in Lower Silesia region

It is clearly visible that the gap between SMEs and Academia is wide, what is caused by different kind of intermediaries that are involved in supporting the development of SMEs AVM competences along the three knowledge dimensions, that strongly need support for further development.

4.2 Results

Results gained are based on data from an international survey on AVM/I4.0 activities of SMEs collected during January 2018 and February 2018. The structure of the survey consists of questions related to three AVM/I4.0 dimensions, namely technology, human resources and organization, and business model. Finally some data on firms performance were collected. We utilized a 5-point Likert-type scale ranging from 'strongly disagree' to 'strongly agree' to measure the items of all the variables. The data were collected by means of questionnaires distributed by email to SMEs, which meet the criteria described above. The respondents are CEO or employees who are knowledgeable about AVM activities. After two reminders, finally a total of 163 complete answers were obtained. Of the 163 answers used in the current analysis 28 come from Upper Austria (AU), 31 from Lower Bavaria (DE), 30 from Hungary, 33 come from Veneto (VE), 13 from Emilia Romagna (IT_EmRo) and 28 from Lower Silesia (PL).

Below, results of identification of small and medium businesses in academic environment of the Lower Silesian Region and intermediary institutions in the Region are presented.

In the academic environment, high level of utilisation of the technologies related to advanced manufacturing, like cloud computing, cybersecurity, Big Data analysis, additive technologies in manufacture were noticed. However, in relation to the Internet of Things and Augmented Reality, so good examples could not be found and the levels of knowledge and applications were evaluated low in these areas. The area of human resources and of organisation methods was evaluated fairly good. This included, among others, organisation of training courses related to advanced manufacturing techniques AVM, projects extending competences and skills of human resources, publications related to this subject matter and presence of special business units directed to human resources. Effectiveness of transferring results of development works and know-how from the academic environment to small and medium businesses was evaluated fairly low. A similar advancement level was observed in relation to the business model and to management of business processes.

Another area in that application of technologies, human resources and business model in relation to the advanced manufacturing techniques AVM was evaluated, were small and medium businesses from the Lower Silesian Region.

The performed analyses were no more so optimistic as those in the academic environment. Low levels of knowledge and application of solutions in the technologies like cybersecurity, IoT, additive manufacturing technologies, cloud computing or simulation tools were noticed. Only the technologies directly related to manufacture that were identified at medium level. This concerned the questions of robotization and automation of manufacturing processes.

The situation presents much better in the area of human resources and organisation methods in small and medium businesses. Companies oriented to innovations, employing highly skilled staff of managers, were identified. However, no significant orientation was noticed towards research and development works carried out inside the companies in specialised R+D departments.

The following chart (Fig. 3) presents on x-axis the current levels of implementation of each Industry 4.0 technology in the Lower Silesia region and, on y-axis to what extent the sampled SMEs are willing to invest on them in the next 2–3 years.

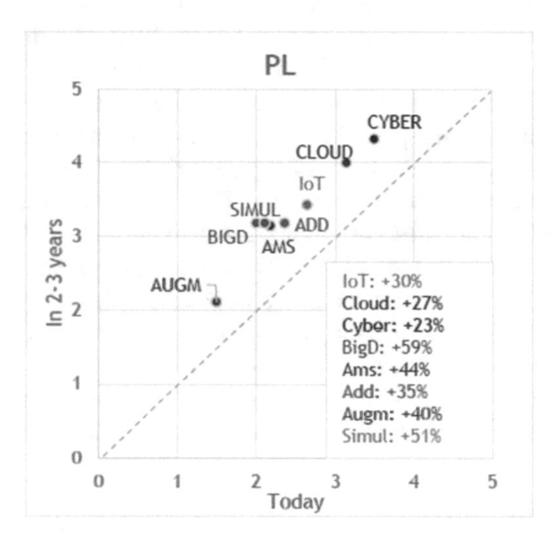

Fig. 3. Levels of implementation Industry 4.0 technologies in Lower Silesia region SMEs

The vertical distance between each points and the bisector of the first quadrant represent the relative increase of implementation of each Industry 4.0 technologies in the future, also indicated in percentage.

The subsequent actions foreseen within the InnoPeer AVM project concern polling small and medium-size companies in the region and examining the applied techniques related to advanced manufacturing. Next, it is planned to indicate a few companies that would like to share their experiences related to creation, development and implementation of innovative solutions. These cases will be converted to training materials and made accessible to other entities in other regions of Central Europe. This action is aimed at increasing skills of innovation managers in the regions involved in the project and contributing to an improvement of functioning of small and medium businesses.

4.3 Detailed Results for Lower Silesia in Reference to Whole Sample

Considering the whole sample, approximately the 70% of the SMEs has a turnover between 2 and 10 M€, while 16% has a turnover between 25 and 50 M€. They mainly produce metal products and machinery or equipment in general. The percentage of Research and Development (R&D) expenses on sales is distributed in the sample among all the three possible ranges, namely 47% of the companies spend between 0,1% and 3%, 22% of them between 3,1% and 5% and the others more than 5% on sales. The ROI trend from 2015 to 2016 is stable or increasing.

Entirely of sampled SMEs in Lower Silesia has a turnover between 2 and 10 M€, and for whole sample approximately the 70% of the SMEs has the same turnover. They are equally distributed among various industries, i.e. a third produces machinery and equipment n.e.c, some of them metal products and the 14% of the companies manufactures motor vehicles or trailers (Fig. 4).

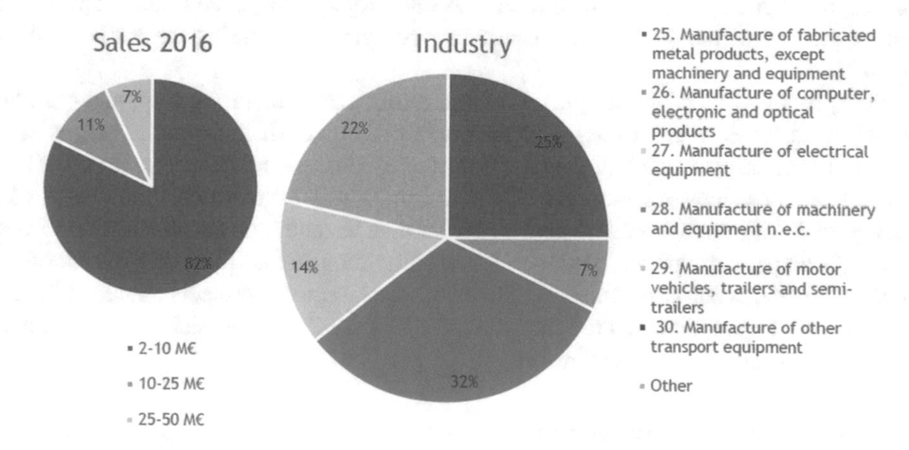

Fig. 4. Data analysis - turnover and industry sectors.

Companies with a stable ROI trend (Return On Investment) constitute half of the sample and the ROI trend of the other half is increasing (Fig. 5).

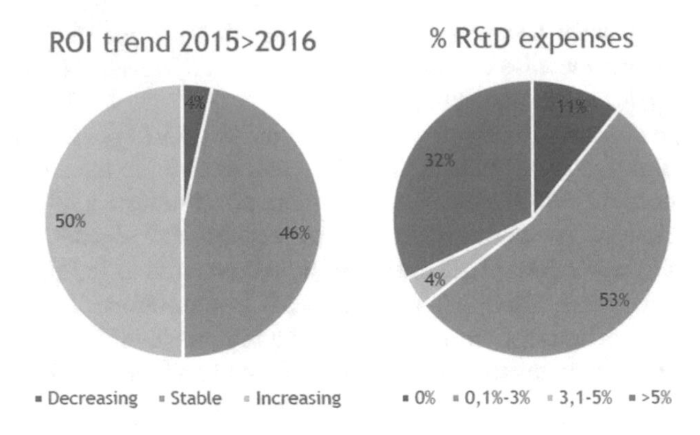

Fig. 5. Data analysis - ROI trend and R&D expenses.

The majority of firms have a percentage of R&D expenses on sales between 0,1 and 3%, but a high number of the remaining companies have this percentage higher than 5% (Fig. 5).

5 Summary and Conclusions

Conducting innovative, advanced manufacturing is a great challenge for companies in Central Europe. However, advanced value manufacturing (AVM) is not only a technological issue. Innovation managers and owners of small businesses face the challenges related to implementation of AVM. In relation to the issues of Industry 4.0, development of SMEs, as well as improvement of their machinery resources and technological possibilities, are urgently needed. The InnoPeer AVM Project is oriented to developing and testing the first comprehensive, transnational program of increasing the AVM qualifications.

The program related to many levels of managing a manufacturing company shall be prepared on three levels: basic, advanced and practical. All the training stages will include living lab webinars, practical training activities in a model factory, as well as AVM strategy camps. Participants of the trainings will be certified InnoPeer AVM managers. Pilot trainings will involve target companies and innovation managers from all the participating regions. Moreover, the training program will be available for other interested persons and companies. Within the project, regional projects related to AVM capacity will be prepared and the 'InnoPeer AVM Board' will be established, aimed at further promoting of the project results.

Acknowledgment. The presented results are part of "InnoPeer AVM" project No. CE1119. Project is supported by the Interreg CENTRAL EUROPE Programme funded under the European Regional Development Fund.

References

1. Beitinger, G.: La empresa digital. In: Basque Industry 4.0, San Sebastian (2016)
2. Havlíček, K., Thalassinos, E., Berezkinova, L.: Innovation management and controlling in SMEs. Eur. Res. Stud. **16**, 57–70 (2013)
3. Şimşit, Z., Vayvay, Ö., Öztürk, Ö.: An outline of innovation management process: building a framework for managers to implement innovation. Procedia Soc. Behav. Sci. **150**, 690–699 (2014)
4. Tont, D., Tont, M.: An overview of innovation sources in SMEs. Oradea J. Bus. Econ. **1**(1), 58–67 (2016)
5. Skibiński, A., Sipa, M.: Sources of innovation of small businesses: Polish perspective. Procedia Econ. Financ. **27**, 429–437 (2015)
6. McAdam, R., Keogh, W., Reid, R., Mitchell, N.: Implementing innovation management in manufacturing SMEs: a longitudinal study. J. Small Bus. Enterp. Dev. **14**(3), 385–403 (2007)
7. Delgado-Verde, M., Martin-de Castro, G., Navas-Lopez, J.: Organizational knowledge assets and innovation capability: evidence from Spanish manufacturing firms. J. Intellect. Cap. **12**(1), 5–19 (2010)
8. Sun, H.: Current and future patterns of using advanced manufacturing technologies. Technovation **20**(11), 631–641 (2000)
9. Zorzi, M., Gluhak, A., Lange, S., Bassi, A.: From today's Intranet of Things to a future Internet of Things: a wireless- and mobility-related view. IEEE Wirel. Commun. **17**(6), 44–51 (2010)
10. Chen, Y., Alspaugh, S., Katz, R.: Interactive analytical processing in big data systems: a cross-industry study of MapReduce workloads. Proc. VLDB Endow. **5**(12), 1802–1813 (2012)
11. Marston, S., Li, Z., Bandyopadhyay, S., Zhang, J., Ghalsasi, A.: Cloud computing — the business perspective. Decis. Support Syst. **51**(1), 176–189 (2011)
12. Urban, P., Jaczyński, R.: Zarządzanie ryzykiem. PWC, Polska (2014)
13. Aikat, J., Akella, A., Chase, J., Juels, A., Reiter, M., Ristenpart, T., Sekar, V., Swift, M.: Rethinking security in the era of cloud computing. IEEE Secur. Priv. **15**(3), 60–69 (2017)
14. Grote, K., Antonsson, E.: Springer Handbook of Mechanical Engineering. Mechanical Engineering for the Globally Working Engineer. Springer, Heidelberg (2009)
15. Berg, L., Vance, J.: Industry use of virtual reality in product design and manufacturing: a survey. Virtual Real. **21**(1), 1–17 (2017)
16. Abar, S., Theodoropoulos, G., Lemarinier, P., O'Hare, G.: Agent based modelling and simulation tools: a review of the state-of-art software. Comput. Sci. Rev. **24**, 13–33 (2017)
17. Miorandi, D., Sicari, S., De Pellegrini, F., Chlamtac, I.: Internet of Things: vision, applications and research challenges. Ad Hoc Netw. **10**(7), 1497–1516 (2012)

SYNERGY Project: Open Innovation Platform for Advanced Manufacturing in Central Europe

Maria Rosienkiewicz, Joanna Helman, Mariusz Cholewa(✉), and Mateusz Molasy

Faculty of Mechanical Engineering, Wroclaw University of Science and Technology, Wrocław, Poland
mariusz.cholewa@pwr.edu.pl

Abstract. The main aim of this paper is to present a concept of enhancement Open Innovation in Central Europe, through designing a dedicated Synergic Crowd Innovation Platform (SCIP) supporting the services and methods of cooperation. The paper presents basic terms related to Open Innovation and discusses current trends and phenomena such as crowdsourcing, crowdfunding, social product development, microworking and living labs. Their implementation to the SCIP will allow to build an effective cooperation environment, with particular emphasis on the area of Advanced Manufacturing.

Keywords: Open Innovation · Social Product Development
Crowd Innovation Platform · Crowdfunding · Crowdsourcing · Microworking

1 Introduction

According to the research performed by Chesbrough, it can be stated that the innovation process is currently facing a "paradigm shift" – the way how companies innovate new ideas and bring them to market is undergoing a fundamental change [1]. Analysis of current trends shows that increasing number of companies is implementing elements of Open Innovation (OI) into their activities [2]. Industry is gaining benefits from using i.a. crowdsourcing, crowdfunding, microworking, Living Labs and makeathons. The usage of those new phenomena, tools and methods in industrial companies is necessary if they want to keep up with worldwide competition. Business models based on Open Innovation paradigm are still under development and therefore there is a need to continue research towards building a dedicated environment, where companies could fully benefit from the new approaches. This task requires also standardization, which is crucial in order to ensure transferability among different industrial organizations and regions. In Central Europe, where advanced manufacturing is a strong branch of economy, it is especially important to support industrial companies with providing the "innovation-friendly ecosystem" equipped with Open Innovation solutions. This is a challenge particularly for scientific organizations, which should deliver state-of-the-art models and transfer technology and knowledge to industrial companies. Therefore, the main aim of this paper is to present a concept of enhancement Open Innovation in Central Europe, through designing a dedicated platform and following services.

© Springer Nature Switzerland AG 2019
A. Burduk et al. (Eds.): ISPEM 2018, AISC 835, pp. 306–315, 2019.
https://doi.org/10.1007/978-3-319-97490-3_30

In the first sections of the article the issue of open innovations has been defined and current initiatives related to this topic. Further sections describes the project as a step towards the Quadruple Helix Innovation Model.

2 Defining Open Innovation

Literature analysis shows that there are a number of definitions related to Open Innovation. The most common is the one introduced by H. Chesbrough. According to this definition, Open Innovation (OI) "is the use of purposive inflows and outflows of knowledge to accelerate internal innovation, and expand the markets for external use of innovation, respectively" [1]. When analysing Open Innovation in the context of Advanced Manufacturing it can be noticed that it will be influenced and will influence on both advanced technologies and manufacturing processes. Bearing in mind these dependences, Authors propose the framework for Open Innovation in Advanced Manufacturing illustrated in Fig. 1.

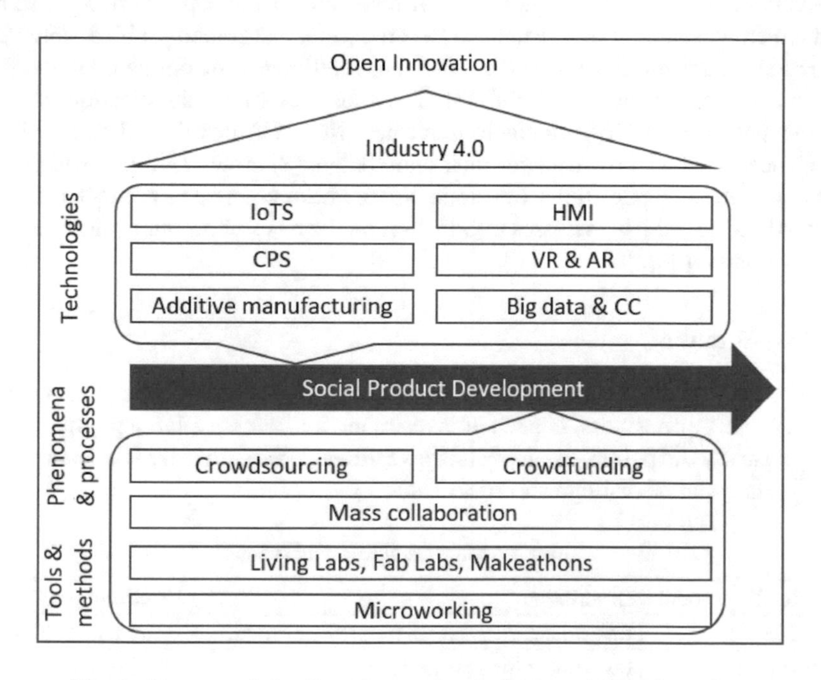

Fig. 1. Framework for Open Innovation in Advanced Manufacturing.

From a technological point of view the Open Innovation paradigm in Advanced Manufacturing cannot spread efficiently without a substantive support of ICT-based solutions which are the pillars of Industry 4.0. Internet of Things and Services (IoTS), Virtual and Augmented Reality (VR & AR), Cyber-Physical Systems (CPS), Human-Machine Interface (HMI), Big data and Cloud computing (CC), etc. enable the Open Innovation paradigm to develop and influence innovation processes among industrial entities. From a manufacturing point of view especially interesting is the impact of OI on the New Product Development Process (NPDP). Changes which can be observed in recent years indicate that traditional approach to product development in increasing number of companies is being replaced or enhanced by Social Product Development.

Together with implementation of Additive Manufacturing, these trends cause a tremendous shift in manufacturing companies and cause a need to develop new and efficient business models. As presented in Fig. 1, to efficiently build an Open Innovation ecosystem, crowdsourcing and crowdfunding should be considered and implemented.

2.1 Crowdsourcing

According to Brabham the term "crowdsourcing" was introduced by Jeff Howe in a paper "The Rise of Crowdsourcing" in 2006. "Howe illustrated the phenomenon of crowdsourcing with a number of cases. Four of these cases—Threadless.com, InnoCentive.com, Amazon's Mechanical Turk, and iStockphoto.com—have become early exemplars of the crowdsourcing model in research on the topic" [3]. According to Brabham crowdsourcing can be defined "as an online, distributed problem-solving and production model that leverages the collective intelligence of online communities to serve specific organizational goals" [3]. The most common classification of crowdsourcing, proposed by Howe, includes four categories – (1) crowd wisdom or collective intelligence, in which crowd shares their knowledge, (2) crowd creation, which can be observed when a company turns to customers to create or co-create a product or a service, (3) crowd voting, which possesses information on the basis of crowd's judgements, and (4) crowdfunding [4].

2.2 Crowdfunding

Crowdfunding, another pillar of Open Innovation, can be defined as "the process of taking a project or business, in need of investment, and asking a large group of people, which is usually the public, to supply this investment" [5]. Currently four main types of crowdfunding can be distinguished (see Table 1).

Table 1. Main crowdfunding types

Crowdfunding model	Description	Platform example
Reward-based	Investment in exchange for gifts or products	Kickstarter, Indiegogo
Equity-based	Investment for a percentage stake	Seedrs, Crowdcube
Lending-based	Peer-to-peer lending	Just Giving
Donation-based	Charitable giving	The Funding Circle

An analysis of crowdfunding phenomenon is very important from scientific point of view due to a number of reasons. First of all, in recent years, crowdfunding has influenced significantly a process of new product development (NPD). Through presenting the product idea to the public, the crowdfunding has created an opportunity for engineers and designers, to receive an immediate market feedback. It can be quickly and with little effort an investment verified if potential customers are interested in the new product. What is more, the new product is easily linked to potential investors. As it has been underlined by Forbes and Schaefer "a crowdfunding campaign can be launched with minimal cost, no proof of sales and, in three of the models, no release of equity" [5]. Thus, the crowdfunding revolutionizes traditional approach to the process of new product development.

2.3 Social Product Development

In their paper Forbes and Schaefer discuss number of definitions describing Social Product Development and select one of them as the most suitable [6]. Accordingly, the SPD can be explained as "the use of social computing technologies, tools, media, influencing the product lifecycle at any stage through the use of a defined and qualified crowd" [7]. In this paper Authors decided to follow this definition as it is appropriate in the context of Advanced Manufacturing.

2.4 Microworking

Microworking (known also as crowdworking) has been defined as the individual behaviors associated with microtask crowdsourcing work [8]. It's a new form of working beyond organizational boundaries, created mostly by social media technologies, in which engagement in work is posted by organizations or individuals on a web-based, third-party platform in exchange for monetary remuneration [9]. Workers are only hired for one particular task, even if that task takes only seconds or minutes. The idea of breaking down tasks to their lowest common denominator is nothing new itself. In fact, it is paradigmatic Taylorism from the beginning of 20th Century. The new is automatic management of workers through computer code and mostly engagement of the part-time and temporary labor. Computer code may perform a variety of supervisory tasks: assigning tasks to workers, speeding up work processes, determining the timing and length of breaks, monitoring quality, ranking employee, and more [10].

The main advantage of microworking is the fact that workers have huge flexibility to set their own working schedules. On the other hand, the huge disadvantage is the fact that workers don't receive the employee status. Moreover, there is a risk of cancellation of a task while a worker is in the midst of completion [10].

2.5 Living Labs

Direct form of Open Innovation enhancement represents Living Labs (LLs), which can be defined as "as user-centered, open innovation ecosystems based on systematic user co-creation approach, integrating research and innovation processes in real life communities and settings" [11]. "European Network of Living Labs (ENoLL) is the international federation of benchmarked Living Labs in Europe and worldwide. Founded in November 2006 under the auspices of the Finnish European Presidency, the network has grown in 'waves' up to this day" [11].

3 Open Innovation 2.0

Literature analysis shows that currently Open Innovation paradigm can be divided into two "phases" – Open Innovation 1.0 (OI 1.0) and 2.0 (OI 2.0). The criteria that distinct OI 1.0 from OI 2.0 include: integration of external knowledge in own innovation process (outside-in perspective), co-creation of knowledge (co-creation perspective) and externalization of internal knowledge (inside-out perspective). When analyzing the OI from these three criteria point of view, it can be stated that OI 1.0 should be linked to (1) licensing-in and spin-in (in terms of outside-in perspective), (2) consortia, joint-venture (e.g. R&D), industrial clusters (in terms of co-creation perspective), (3) licensing-out and spin-out (in terms of inside-out perspective), whereas OI 2.0 should be allied to (1) crowdsourcing, crowdfunding and lead user method (in terms of outside-in perspective), (2) innovating with communities like Living Labs, Fab-Labs (in terms of co-creation perspective), (3) online platforms and innovation challenges and competitions (in terms of inside-out perspective) [12]. When analyzing existing online platforms related to Open Innovation and matching them with appropriate phenomena, tools and methods crucial from Advanced Manufacturing point of view it can be noticed that a number of platforms are focused on crowdsourcing (e.g. ninesigma.com, innocentive.com, threadless.com) and crowdfunding (e.g. kickstarter.com, indiegogo.com, www.ulule.com, crowdcube.com, experiment.com), whereas less popular are platforms dedicated to Social Product Development (e.g. quirky.com), microworking (e.g. mturk.com, microtask.com, crowdsource.com) and Living Labs (enoll.org).

Each of the platform is dedicated to another goal. None of them offers full set of tools and processes to effectively support Advanced Manufacturing and especially NPDP. Therefore, within the SYNERGY project, the Authors decided to develop a platform which would fill in this gap.

4 SYNERGY Project as a Step Towards the Quadruple Helix Innovation Model

Project "SYnergic Networking for innovativeness Enhancement of central european actoRs focused on hiGh-tech industrY" (acronym SYNERGY) is financed within Interreg Central Europe programme. One of the programme's priority is Cooperating on innovation to make Central Europe more competitive with its specific objective to improve sustainable linkages among actors of the innovation systems for strengthening regional innovation capacity in Central Europe.

4.1 SYNERGY Aims and Methodology

As McAdam and Debackere underline, in recent years, the effectiveness of the Triple Helix Innovation Model - linking government, universities, and industry - has been questioned, due to the reason that "regions have failed to meet expected levels of innovation, GDP development, and employment" [13]. To improve regional innovating ecosystems, another helix has been included in the model – the 'media-based and culture-based public' and 'civil society' [14]. The Quadruple Helix Innovation Model has been established by involving societal-based innovation users.

Following the Quadruple Helix Innovation Model, the SYNERGY project partners represent not only entities based on Triple Helix Innovation Model – four higher education and research institutions (Poland, Slovenia and Germany), one SME (Austria), two business support organizations (Croatia, Italy) and government as an associated partner (Poland), but also the project aims at involving society.

To overcome common challenges including: administrative barriers for innovation, low global innovation performance and technology transfer, low New Member States participation in Research and Development, inefficient funding for local innovative initiatives, SYNERGY goal is to enhance innovativeness in EU regions through strengthening linkages and beyond border cooperation to create synergy between SMEs, industry, research, intermediaries and policy makers.

The project scope is mainly oriented on Advanced Manufacturing with a special focus on the most promising modern industrial technologies in 3 Key Project's Areas (KPAs): (1) Additive Manufacturing, (2) Micro- and nanotechnology-related processes and materials and (3) Industry 4.0.

The Fig. 2 presents an overall methodology of the project implementation. The concept of the SYNERGY project can be divided into 6 main steps illustrated below.

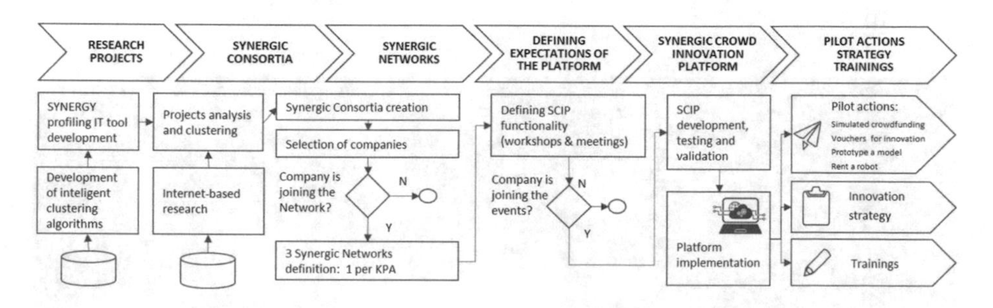

Fig. 2. Methodology of SYNERGY project implementation

The first one is an analysis of running and finalized projects in 3 Key Project's Areas. Then based on the results of the analysis, institutions that were realizing these projects, will be clustered into Synergic Consortia. The next step is to create living linkages among regional actors through innovative Synergic Networks within 3 KPAs. Members of the transnational Synergic Networks during workshops and meetings will define expectations of the new the Synergic Crowd Innovation Platform (SCIP). The SCIP will become a space for enhancement open innovation, crowdsourcing, crowdfunding and micro-working among project partners' regions. The last phase of the project will be a set of pilot actions that will test the functionality of the SCIP. Additionally, "a ready to adopt" generic regional Crowd Innovation strategy will be prepared and presented to target groups during trainings.

Before mentioned steps comes from the project structure, that consists of 4 thematic work packages: Synergic profiling (WP T1), Synergic Networking (WP T2), Synergic Crowd Innovation Platform (WP T3) and Launching Synergic Platform (WP T4).

4.2 SYNERGY Work Plan

The main aim of the WP T1 was to build and deliver a method and tool to search, profile, cluster and reach innovation-oriented organizations based on their activities and experience gained from successful project realizations in order to enhance networking, matchmaking and linking regional actors from research, industry and intermediaries operating within 3 KPAs. The main goal of the WP T2 is to enable international cooperation to create 3 KPAs innovative Synergic Networks and to define needs for successful transregional cooperation based on crowd innovation through regional and international "Simulated Sharing" networking workshops and Design Thinking idea meetings. The main output of the project that fulfil the needs of enhancement Open Innovation in Central Europe is the Synergic Crowd Innovation Platform – SCIP. The WP T3 and T4 aim to develop and test this platform.

The user's interface of the SYNERGY profiling IT tool developed within the WP T1 is presented in the Fig. 3. It is available via https://synpro.e-science.pl – at the moment, in the validation phase, with the restricted access for consortium members only.

Fig. 3. SYNERGY – SynPro IT tool

The tool is composed of "Projects", "Organizations" and "Map" modules. Presently, in the pilot phase, there are 208 projects and 112 organizations registered. The intelligent algorithms, based on graph theory, are being developed currently in order to efficiently and automatically cluster the projects and organizations according to selected criteria. The structure of the registered projects divided into 3 KPAs as well as organizations grouped by 9 countries can be found in the Fig. 4.

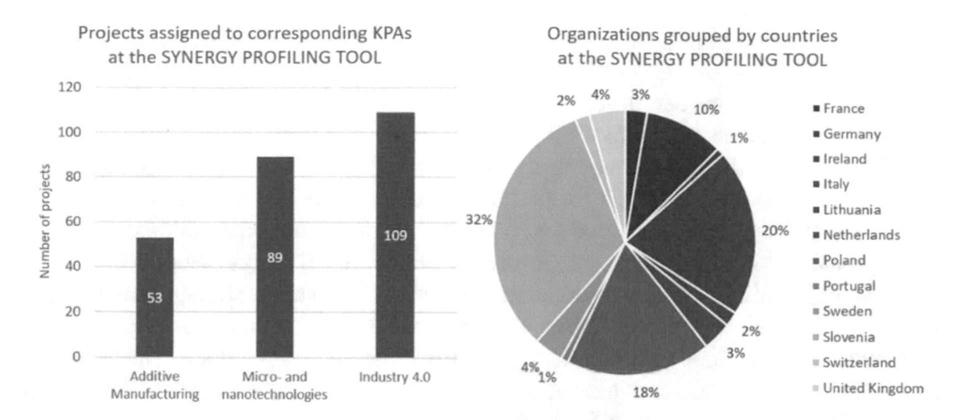

Fig. 4. Statistics regarding projects and organizations introduced to the SynPro IT tool

Wroclaw University of Science and Technology, the Project Leader, on the 24th of May 2018 organized the first regional "Simulated Sharing" networking workshop and the Design Thinking idea meeting. 17 participants from industrial companies, clusters, academia and intermediaries took part in the event. Participants were successfully discussing the barriers for efficient cooperation of science and business and ways to overcome them. Moreover, the initial and desired functionality of the Synergic Crowd Innovation Platform was defined. The project partners will organize parallel events in their regions and next, a common, international workshop is foreseen in September 2018. It will be a ground for establishment of the 3 Synergic Networks (one per KPA). The living Synergic Networks focused on 3 KPAs will enable researchers and representatives of industry to meet and share experiences in their fields to start transregional cooperation in Central Europe. Members of Synergic Networks will not only set up linkages among each other, but also will define needs for successful transregional cooperation reflecting requirements of all regional innovation actors and thus will define common functionality of the Synergic Crowd Innovation Platform. Moreover, the new services for regional actors from research, industry, intermediaries and local authorities operating within 3 KPAs will be defined and collected. It is expected that these services, implemented within the SCIP, will enable the members of the platform to exchange resources (HR, equipment, best practices), set up cooperation, build innovation friendly environment and enhance crowd innovation initiatives in Central Europe.

The basic idea of the SYNERGY project is to set up a platform ensuring crowd-funding and crowdsourcing for innovative solutions for the Central European society. As a part of functionality of the SCIP, a crowdfunding mechanism will be developed. It is planned to transfer best practices from American portal www.experiment.com. Within this approach every euro is contributed towards innovation and helps push the boundaries of knowledge. The people fund directly to the researchers and idea-givers, so there is no middleman or overhead involved (compared 50–60% when receiving a grant). Anyone can start a new project, as long as the results can be shared openly. Scientists, researchers, idea-givers will share progress, data, and results directly with their backers [15]. The SYNERGY will use the platform also to implement microworking, being an approach where community solves smaller tasks which are then reassembled into an overall result at the end. After the platform is launched a number of pilot actions is foreseen to test its functionality – e.g. "Simulated crowdfunding", vouchers for research and innovation projects, "Rent-A-Robot".

The platform will then be improved according to the results coming from the testing pilot actions. Also, on the basis for performed research on best practices and success stories related to crowd innovation and feedback coming from pilot actions, a strategy for regional public authorities, public organizations and Synergic Networks will be elaborated. Trainings will also be organized on Crowd Innovation Strategy in the project partners' regions for regional public authorities, public organizations and Synergic Networks. Finally, three pilot actions will be implemented on promoting the SCIP and its services, such as: "Crowd innovation for companies", vouchers for developed solutions of the research projects, "Design and prototype model".

Synergic Crowd Innovation Platform is the most important output of the whole project because it will be an environment where all new services enhancing crowd innovation will be available. Entities from project partners' and associated partners' networks will be invited to meetings. Once the project has ended SCIP will be transferred into a spin-off at the Wroclaw University of Science and Technology (project leader). An agreed percentage (app. up to 10%) of the budget of the crowdfunding initiatives on the platform will be spent on the SCIP sustainability and on covering current costs. It is assumed that both the project and the associated partners will be using and disseminating such a platform after the duration of the project.

5 Summary

In the paper the new phenomena, tools and methods of Open Innovation were discussed. The concept of Open Innovation enhancement in Central Europe through implementation of the SYNERGY project was presented. A role of the project in terms of Quadruple Helix Innovation Model was mentioned. The special focus was put on description of the process of the Synergic Crowd Innovation Platform development and activities supporting this process – in order to underline the new, innovative and holistic approach of creating an effective cooperation environment, with particular emphasis on the area of Advanced Manufacturing. It is assumed that the variety of services planned as components of the SCIP and active participation of its potential users in the platform's development process should result in an efficient enhancement and dissemination of Social Product Development in Central European regions. Helpful with this will be the newly-developed match-making open source online IT instrument.

An analysis of already performed activities leads to a conclusion that for the successful realization of the project, it is crucial to efficiently communicate it to the target audience. It is especially important in terms of launching the platform, which effective functioning will be possible only if an appropriate number of users is registered. Similarly, only through reaching appropriate target groups the events can be effectively organized and executed. What is more, the research performed within the paper leads to assumption that technological development causes an increased "computerisation" of jobs, which means that increasing number of jobs will become liable to digitalization. The phenomena of crowdsourcing that has recently emerged, can be defined also as a new form of organization of work. It can be observed that platforms built around it have evolved as an innovative instruments and new form of work organization.

Future research will focus not only on further development of the platform and set of complementary services, but also on development of new business models for Open Innovation in Advanced Manufacturing sector in Central Europe.

References

1. Chesbrough, H.: Open Innovation: The New Imperative for Creating and Profiting from Technology. Harvard Business Press, Brighton (2006)
2. Open innovation 2.0 yearbook 2016, European Commission, Directorate-General for Communications Networks, Content and Technology – sprawdzić poprawne cytowanie
3. Brabham, D.C.: Crowdsourcing. The MIT Press Essential Knowledge Series, pp. 18–19 (2013)
4. Sloane, P.: A Guide to Open Innovation and Crowdsourcing: Advice from Leading Experts in the Field. Kogan Page Publishers, London (2011)
5. Forbes, H., Schaefer, D.: Guidelines for successful crowdfunding. In: Complex Systems Engineering and Development Proceedings of the 27th CIRP Design. Procedia CIRP, vol. 60, pp. 398–403 (2017)
6. Forbes, H., Schaefer, D.: Social product development: the democratization of design, manufacture and innovation. In: Complex Systems Engineering and Development Proceedings of the 27th CIRP Design. Procedia CIRP, vol. 60, pp. 404–409 (2017)
7. Bertoni, M., Larsson, A., Ericson, Å., Chirumalla, K., Larsson, T., Isaksson, O., Randall, H.: The rise of social product development. Int. J. Netw. Virtual Organ. **11**, 188–207 (2012)
8. Deng, X., Galliers, R.D., Joshi, K.D.: Crowdworking - a new digital divide? Is design and research implications (2016). Research Papers 148
9. Deng, X., Joshi, K.D., Galliers, R.D.: The duality of empowerment and marginalization in microtask crowdsourcing: giving voice to the less powerful through value sensitive design. J. MIS Q. **40**(2), 279–302 (2016)
10. Cherry, M.A.: Beyond Misclassification: The Digital Transformation of Work (February 18, 2016), Legal Studies Research Paper No. 2016-2, Comparative Labor Law & Policy Journal, Forthcoming; Saint Louis (2016)
11. ENOLL. http://enoll.org/about-us/. Accessed 15 Mar 2018
12. Puechner, P.: Presentation: "NUCLEI - Transnational Brokers Coaching: Open Innovation" 28.06.2017, Munich (2017)
13. McAdam, A., Debackere, K.: Beyond 'triple helix' toward 'quadruple helix' models in regional innovation systems: implications for theory and practice. R&D Manag. **48**(1), 3–6 (2018)
14. Carayannis, E.G., Barth, T.D., Campbell, D.F.: The Quintuple Helix innovation model: global warming as a challenge and driver for innovation. J. Innov. Entrep. **1**, 1–12 (2012)
15. Experiment. https://experiment.com/. Accessed 18 Mar 2018

The Application of Augmented Reality Technology in the Production Processes

Andrzej Szajna[1], Janusz Szajna[1,2], Roman Stryjski[2], Michał Sąsiadek[2(✉)], and Waldemar Woźniak[2]

[1] DTP Ltd., ul. A. Wysockiego 4, 66-002 Zielona Góra, Poland
[2] Faculty of Mechanical Engineering, University of Zielona Góra, ul. prof. Szafrana 4, 65-516 Zielona Góra, Poland
m.sasiadek@iizp.uz.zgora.pl

Abstract. Augmented reality technology (AR), also known as mixed reality, enables the deployment of virtual computer images to the real-life images being seen. The article presents the concept of using Augmented Reality tools in industry. Examples are given which illustrate the production process support and the monitoring of production lines with the use of AR glasses; these provide the user with all the information necessary in front of his/her eyes while the control, the creation of notes and the reporting, is done using gestures and speech.

Keywords: Augmented Reality · Monitoring and inspection of production line Assembly of control cabinets · AR glasses

1 Introduction

Modern production technologies, the increasing degree of automation of production processes and the high level at which they are organised, put increasing demands on manufacturing services. The response to the growing requirements to shape reliability in production, is the emergence of new methods and techniques supporting the work of specialists from departments such as production maintenance and manufacturing [10].

One tool, in particular, which is beginning to be increasingly introduced, in order to maintain the efficiency and quality of production processes, is the use of Augmented Reality Technology (AR). This paper presents proposals for the application of this technique in the assembly of control cabinets and in the monitoring of production lines.

2 Augmented Reality Technology in the Support and Monitoring of Production Processes

Issues related to the proper functioning and effective diagnostics of production processes are an important economic factor in the modern, industrialised economy. Monitoring and management systems for technical facilities are used in small, medium and large enterprises.

© Springer Nature Switzerland AG 2019
A. Burduk et al. (Eds.): ISPEM 2018, AISC 835, pp. 316–324, 2019.
https://doi.org/10.1007/978-3-319-97490-3_31

Solutions of this type allow the life-span of individual parts to be utilised to their fullest extent by optimising their replacement dates and controlling their operating processes; this, then translates into measurable financial savings and impacts, ultimately, on the financial condition of an enterprise. Due to the significant diversity and complexity of production processes, the streamlining of monitoring processes and the supporting of operative personnel is a problem with a multi-faceted task structure. One of the factors contributing to the improvement of this situation is the collection of knowledge on the basis of which, specific operating strategies are developed. Progress in information processing also allows the development of ever more advanced IT diagnostic techniques, enabling effective registration and the analysis of signals while eliminating the human factor, thus ensuring an increase in the objectivity of the results obtained. Another important element in IT systems, is the possibility of interactive visualisation, which significantly improves the effectiveness of the activities carried out. One such technique is Augmented Reality Technology (AR), which has a significant impact on the human/machine communications process. This technology is characterised by the projection of computer-generated virtual objects onto the image of a fragment of the real environment; such a complex visualisation is presented to the user of this system. Augmented Reality has its origins in military technologies. The first work on its use was carried out in the 1960's of the 20th century, when the first HMD (*Helmet-Mounted Display*) display was made [5]; of course, at that time, it had not yet been assigned the name Augmented Reality.

The name "Augmented Reality" was used for the first time in the 1990's of the 20th century by scientists working for Boeing on the system supporting identification of the beams and individual electric wires in the aircraft they were constructing. Currently, this technology is used with increasing frequency in applications, both military and civil.

3 Structure and Operating Principles of a System Using Augmented Reality Technology

As part of the R&D work carried out at the DTP Ltd. in Zielona Góra and at the Institute of Computer Science and Production Management at the University of Zielona Góra (UZ), systems using **Augmented Reality Technology** to support technical personnel in the assembly process of complex technical equipment and in the monitoring of production lines, have been developed.

By implementing the functionality in the form of interchangeable instructions, delivered in *online mode*, it is possible for the system operator to carry out assembly or inspection work with neither training nor with the help of an instructor.

The main elements of the structure of the system developed are: an on-site server (or Cloud), a projection device, a digital camera that records an image analogous to that seen by the system operator, a computer, proprietary image processing and analysis software and, optionally, a keypad.

The projection device, depending on the requirements of the system operator, may be a traditional computer monitor, a tablet or projection glasses [8]. In addition, the structure of the hardware of the system includes a fragment of the Internet network infrastructure, which is the operator's communication medium with the instructional data.

The system works in real time. Virtual elements are generated by proprietary, computer software on the basis of prepared logic, digital instructions and real-time data from cameras and measuring devices. The image displayed by the projection device is enriched with virtual objects in the form of 3D models [9], markers and descriptions, or virtual screens (dashboards). The operator sees the exact instructions, locations and activities necessary to perform the assembly or inspection work. The positions of virtual objects are calculated relative to co-ordinate systems whose beginnings are determined by markers [10, 11] on pictograms, specially prepared for this purpose which are placed on a technical device or in a specific site near the location where the assembly takes place. In the future, markers will be replaced with advanced image recognition techniques that apply Artificial Intelligence (AI), that is, deep convolutional neural networks, which the DTP - UZ team are currently working on.

The algorithm of real image enrichment with virtual objects is as follows:

- acquisition of a technical device image with markers placed on it,
- analysis of the registered image, in order to locate the markers,
- calculation of markers' co-ordinates and their orientation towards the camera, that is, the system operator,
- identification of markers by comparing them with established patterns,
- calculation of the co-ordinates of virtual objects relative to reference systems and the orientation of individual markers,
- rendering the virtual objects and displaying them on the AR glasses, on the observed by system operator real environment.

The result of the algorithm is the seen image of the technical device enriched with virtual objects (see Fig. 1). The system operates according to client-server, IT software technology. The server's software consists of an application on which a programme for synchronising instructional data has been installed with the status of the technical device used, along with the database and set of instructions dedicated to the selected assembly or inspection processes. With regards to the client, there is a user programme operating within the presentation and communications platform, which is a dedicated application; there is a web browser in the Cloud version. Such a structure has the advantage over stationary solutions since users have greater flexibility in accessing the system and to the work assigned thereto. The system's software is also adapted to working in computing Clouds.

Fig. 1. Virtual objects, such as, arrows, markers and prompts, embedded on the actual observed scene – a control cabinet

4 Method for Supporting the Assembly Process

Supporting the assembly process with the use of the system developed, is achieved through a visual presentation to the operator: location of the device components, their correct orientation, assembly operations, descriptive prompts in textual form, measurement values, location of measurement points, 3D models of the device's components.

The methodology also assumes application of the system in such areas as:

- recognition of the technical condition of the device, that is, technical diagnostics and identification of the causes of possible failures,
- repairs (i.e. activities aimed at restoring the functional properties of the device that have been damaged, during which the regeneration or replacement of the damaged components are carried out),
- maintenance operations, that is, activities recommended by the manufacturer of the technical device in order to maintain the components of the device in such a condition as to allow trouble-free operation,
- training of technical staff, by improving their skills and by their acquisition of the qualifications necessary to undertake works related to the assembly, recognition of technical condition, repair and maintenance of the technical device.

Visualisation is carried out through a projection device that is part of the user's communication interface with the system. The choice of the type of projection device depends on the conditions in which the system is used, the complexity of the assembly operations planned and the individual preferences of the operator. If the operator is to have full freedom of movement during the work, the most preferred choice, associated with the projection device are the projection glasses, since they do not engage his/her hands or affect his/her mobility, which is beneficial in the case of devices of increased dimensions. Projection glasses are, in simplified terms, two miniaturised projectors mounted over each glass in a casing thus allowing them to be worn in front of the eyes in a manner analogous to traditional, eyeglasses or sunglasses.

The technical device is filmed using a camera that can be positioned statically in relation to the place where the assembly process is taking place or can be moved dynamically with the motion of the system operator. The camera, the position of which can change dynamically, is usually mounted on the back of the tablet or on the projection AR glasses. The user's interaction with the system takes place through simple gestures or speech. The optional keypad allows the user an alternative way to move around after each point of the programme or the instructions prepared. Individual points in the instructions can be attended to without feedback regarding the course of the assembly process or requiring return signals. The instructional data synchronisation programme, located in the server part of the system, is used to collect feedback. The instruction data synchronisation programme reads information about the status of the assembly or inspection process, based on the data provided by the software of the measuring system and inputs an appropriate entry into the database. This entry is read by an instruction programme that displays information supporting the operator, in such as, information about the correctness of the activity being performed; this allows the user to go to the next point in the instruction set. In the method developed with the application of augmented reality, the important issue is the interaction between the content of the instructions, *enriched with virtual objects*, and the user, as well as the physical technical device.

The physical and logical elements involved in the assembly, inspection and training processes influence the decisions made by the operator. The operator, based on the projection of the image of the technical device and enriched with the content of the instructions of the virtual objects marked, interprets the status of the assembly or inspection process and then decides on further action, as described in the instructions. The operator's actions, through feedback, in the course of the assembly process, are visually signalled, which, in turn, checks them in order to assess whether they have been carried out correctly; where works are not coupled with elements which generate feedback, the accuracy of the activities performed is recognised by the operator, based on a comparison of the physical condition of the device with a visual presentation, consisting of virtual objects and a textual description.

5 An Example of the Application of Augmented Reality Technology in the Assembly of Control Cabinets

In the following example, the assembly of control cabinets and monitoring of the production line is supported by the DTPoland AR Smart Wiring 4.0 system, consisting of EPLAN Smart Wiring software, that is, a control cabinet's CAD design, a wire label reader and visualisation software mounted on the Microsoft HoloLens Augmented Reality glasses (see Fig. 2).

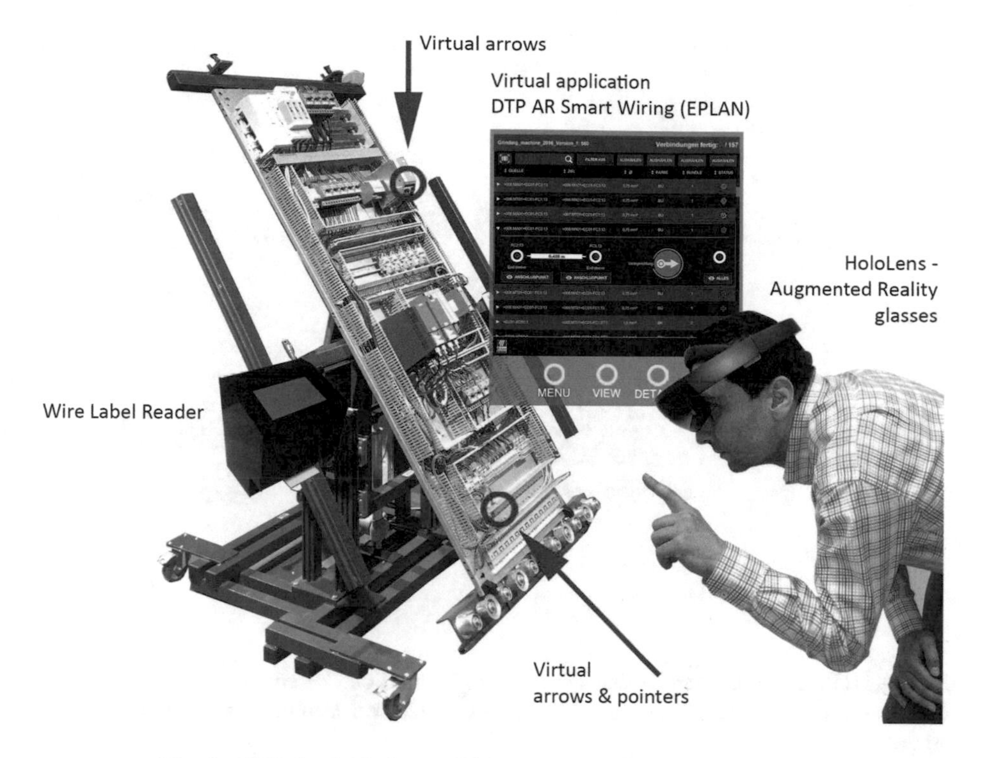

Fig. 2. DTPoland AR Smart Wiring 4.0: visualization of the system

The system is designed to support the assembly of wires between components in the control cabinet. The process is carried out as follows (see Fig. 3):

- onto the technician's work place, the following elements are delivered: the cabinet (with components mounted, without wires), set of wires, the CAD input file,
- the wires are pre-processed: cut to the appropriate length and marked with an alphanumeric label which identifies the connection between the components of the cabinet,
- the CAD file is loaded by the Smart Wiring programme,
- the technician collects one wire from the set supplied; he/she applies it to the wire reader, which identifies the wire's label and sends it to the Smart Wiring programme,
- the Smart Wiring finds the components to be connected with the particular wire in the cabinet's design and transmits their co-ordinates to the software in the Microsoft HoloLens glasses,
- the software in the glasses uses the co-ordinates provided, along with the markers in the cabinet, in order to indicate the wire's assembly points through virtual arrows, markers, virtual screens or textual prompts,
- on the virtual tablet, the technician marks with a simple gesture or voice command that mounting of this specific wire is completed. Otherwise, he/she is able to report issues and/or create a note; alternatively, the technician can use a tablet to perform these actions,
- the technician collects the next wire.

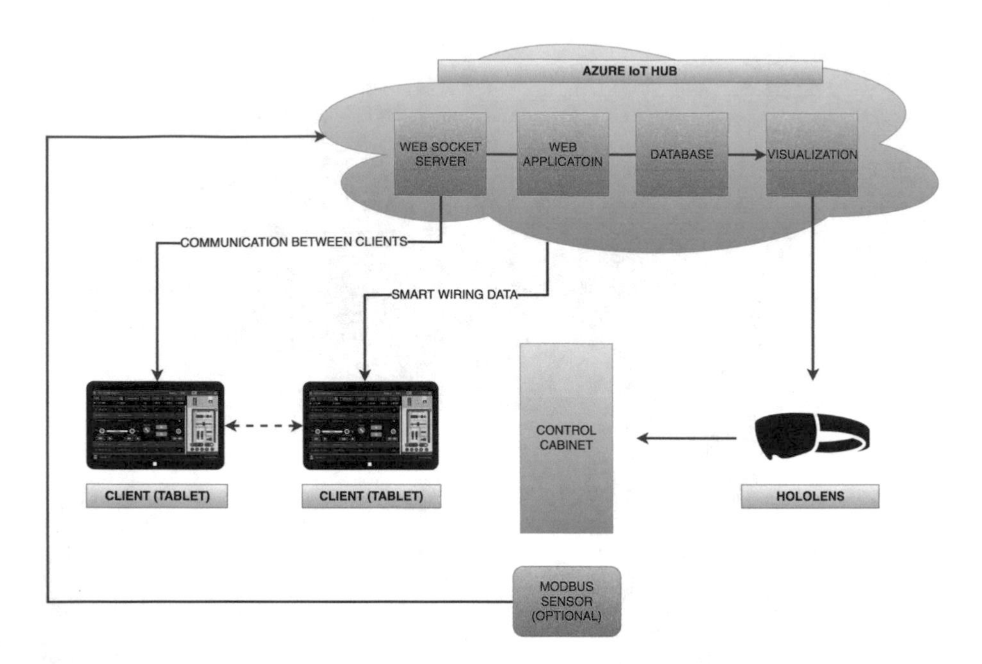

Fig. 3. DTPoland AR Smart Wiring 4.0: diagram of the Cloud based system, using tablets and Augmented Reality Technology in the Assembly of Control Cabinets and monitoring of the production line

DTPoland AR Smart Wiring 4.0 (see Figs. 2 and 3) is a prototype extension of EPLAN Smart Wiring with the AR system, including Microsoft HoloLens glasses software and the world's first, wire labelling reader which is, currently, in the course of being patented.

The wire reader is a device that consists of a micro-computer, a camera and a specially designed lighting set. The purpose of this device is to read tiny alpha-numeric identifier on the thin wire as e.g. 1,5 mm^2. Since the inscriptions on the wires are printed by a dot matrix printer and are, therefore, exposed to being deformed, blurred, shifted or damaged (see Fig. 4) none of the well-known OCR- ***Optical Character Recognition***- methods provided reliable results; this was a serious technical problem. The use of artificial intelligence and deep convolution neural networks, subjected to an advanced machine learning process, solved the problem and gave good recognition of the markings (see Fig. 5).

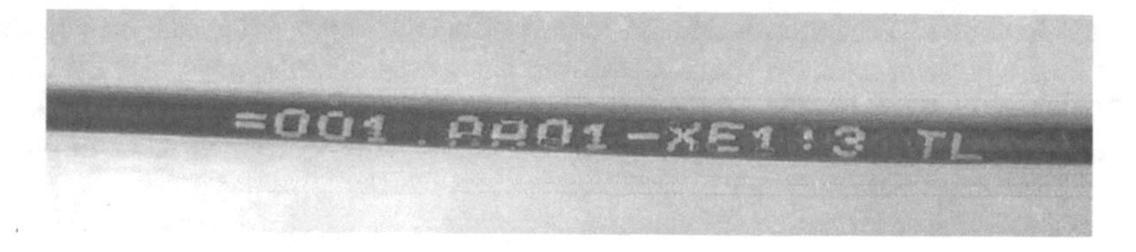

Fig. 4. Inscriptions on the wires are printed by a dot matrix printer and are, therefore, exposed to being deformed, blurred, shifted or damaged.

Fig. 5. Artificial Intelligence: correction of badly printed character

The software, created for the Microsoft HoloLens glasses, synchronises the diagram of the cabinet installed, as downloaded from the Smart Wiring program, with the image seen through the glasses by referencing it to the markers on the cabinet being assembled. Then, using co-ordinates obtained from the cabinet diagram, it marks the mounting points with the virtual objects and embeds them on the actual seen environment (see Fig. 1). The development of the software which synchronises the diagram with the image being viewed, goes towards the elimination of physical markers and the application of artificial intelligence in recognising the individual components from the cabinet diagram in the seen environment. This will allow to eliminate one production step – the need to attach the physical markers to every control cabinet.

6 Summary

The solution presented in this article, covering the methodology and the video support system of the assembly process, using the technology of augmented reality, is a significant step towards improving human/machine interaction and time needed to perform the job. Additionally, the system operator will be able to perform specific and advanced tasks without a need of training or any help of an qualified instructor. Thanks to the solution proposed, it is possible to, (a) shorten assembly times, particularly in those cases where the technical equipment is complex, (b) reduce costs related to technical staff training, (c) reduce staff by broadening the scope of work performed by a single person and (d) support training in complex assembly operations. The use of this system, developed in industrial conditions, can have a significant impact on ensuring the continuity of production by reducing the risks associated with shortages of personnel responsible for assembly, particularly in plants employing a small number of people and producing a wide range of products.

The solution developed is open, which means that the system and the method proposed can be supplemented with new elements (e.g. sensors) and instructional data, as well as the possibility of using the solution in other areas of application, such as the ongoing maintenance of technical devices and their repair.

Acknowledgments. The authors would like to thank the DTP engineering team, engaged in the project: Prof. Krzysztof Diks, Dr. Tomasz Kozlowski and also Marek Adaszyński, Piotr Charyna, Filip Chmielewski, Krzysztof Ciebiera, Joanna Cieniuch, Grzegorz Gajdzis, Anna Mroczkowska, Marcin Mucha, Sebastian Pawlak, Wojciech Regeńczuk, Eugeniusz Tswigun, Andrzej Warycha, Czesław Zubowicz, Mariusz Życiak.

References

1. Haller, M.: Emerging Technologies of Augmented Reality. Idea Group Publishing, Hershey (2006)
2. Yu, D., Sheng, J.J., Luo, S., Lai, W., Huang, Q.: A Useful Visualization Technique: A Literature Review for Augmented Reality and its Application, Limitation & Future Direction. Visual Information Communication. Springer (2010)
3. Ma, D., Gausemeier, J., Fan, X., Grafe, M.: Virtual Reality & Augmented Reality in Industry. Springer, London (2011)
4. Burdea G. C., Coiffet P.: Virtual Reality Technology. 2nd edn. Wiley-IEEE Press (2003)
5. Rash, C.E.: Helmet-mounted displays: sensation, perception, and cognition issues. U.S. Army Aeromedical Research Laboratory (2009)
6. Meni, E.: Boeing's working on augmented reality which could change space training ops. Boeing Frontiers, vol. 10 (2006)
7. Zhou, F., Duh, H.B.L., Billinghurst, M.: Trends in augmented reality tracking, interaction and display: a review of ten years of ISMAR. In: IEEE International Symposium on Mixed and Augmented Reality 15–18. Cambridge, UK (2008)
8. Cakmakci, O., Rolland, J.: Head-worn displays. J. Display Technol. A Rev. **2** (2006)
9. Dugelay, J.L., Baskurt, A., Daoudi, M.: 3D Object Processing, Compression, Indexing and Watermarking. Wiley (2008)
10. Sun, R., Sui, Y., Li, R., Shao, F.: The design of a new marker in augmented reality. IPEDR, vol. 4. IACSIT Press, Singapore (2011)
11. Uchiyama, H., Marchand, E.: Object detection and pose tracking for augmented reality: recent approaches. In: 18th Korea-Japan Joint Workshop on Frontiers of Computer Vision (FCV) (2012)
12. Mittal, S.: Powering the industry 4.0 revolution in manufacturing with Windows 10 and Microsoft Cloud/Windows Industry Lead, 24 Apr 2017

The Application of Software Tecnomatix Jack for Design the Ergonomics Solutions

Miriam Pekarčíková(✉), Peter Trebuňa, Jana Kronová, and Gabriela Ižariková

Technical University of Kosice, Kosice, Slovak Republic
{miriam.pekarcikova,peter.trebuna,jana.kronova,
gabriela.izarikova}@tuke.sk

Abstract. The article is focused on using the Tecnomatix Jack software module to solve logistics in selected company. By creating simulations of the information and gained knowledge, by selecting appropriate analyzes, Tecnomatix jack offers the possibility of organizing the workplace both ergonomically and logistically. The first part of article is oriented on theoretical knowledge of given problem. It´s focused on linking logistics and ergonomics in company and characterizes Tecnomatix Jack software. The second part deals with practical aspect of the issue and discusses the current state of workplace and proposals for improving workplace ergonomics solution for the planned new hall.

Keywords: Logistics · Ergonomics · Tecnomatix jack · Joinery

1 Introduction

Materials handling, manual work and techniques at various levels of automation and mechanization and also mental work represent the main activities of people in business logistics. They are connected with ergonomics. In such activities, accidents or diseases result from working activities. These have an impact on the efficiency of work done by people and operations in the fields of business logistics. Ergonomics complexly solves human activity and its connection to the environment and machine. It examines people's interrelationships with the working environment and technology, monitors the relationships within these subsystems and strives to achieve the maximum level of humanization and labor protection. It solves the problems associated with work tools, workplaces, workflows, creates harmonious working conditions, maximum work performance and employee satisfaction [1–4]. Nowadays, more attractive is utilization various simulation software for design of workplaces. From financial and time perspective, this way of designing is more acceptable than real state. Another advantage is the possibility of simulating the worker himself in the created work environment. Through various ergonomic analyzes, it is possible to figure out the worker's workload and then remodel the workplace. In this way, it is possible to prevent accidents and illnesses that could affect the production process [5–7]. One of the ways is optimization of workplace. Effective optimization process requires comprehensive coordination of potential possibilities. In digital factory it can be managed all the phases of this process with minimum costs and by visual demonstration and evaluation of results with the software support. Essential

© Springer Nature Switzerland AG 2019
A. Burduk et al. (Eds.): ISPEM 2018, AISC 835, pp. 325–336, 2019.
https://doi.org/10.1007/978-3-319-97490-3_32

changes can be check directly and after validation implemented into the real system. (see Fig. 1).

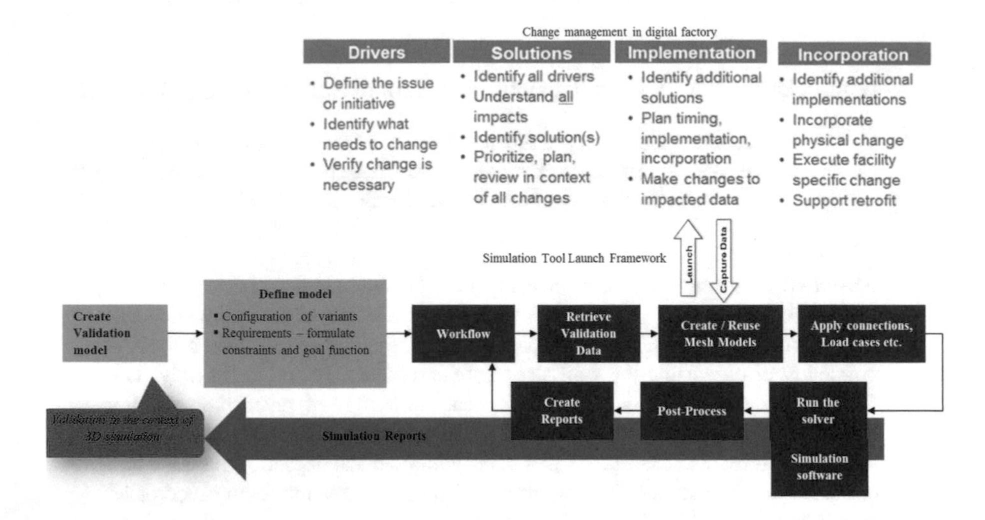

Fig. 1. Optimization process in the perception of digital factory

1.1 Importance, Aims and Benefits of Ergonomics in Business Logistics

One of the most important factors of working environment is consistency of individual activities. Therefore, the emphasis is nowadays oriented on achieving the co-operation of individual components affecting processes in organizations. Terms such as logistics and ergonomics are everyday topics of all companies. Their interconnection has a benign influence in many areas of companies. Ergonomics is aimed at increasing the efficiency of human work and has two clear basic objectives and indicators for evaluating the quality of solution: a positive effect on worker health and economic impact [8]. The main role of ergonomics is fact that Ergonomic uses a system approach for solving the problem of a man's position in the work system. It is based on the knowledge that man and environment is interconnected by material, energy and information ties. These interactions together create a new quality with specific properties and values [9].

1.2 Ergonomic Risks in the Work Process of Business Logistics

There are many factors of working environment that affects employees during the work process. Most of these factors interact each other [10]. In general, they are physical, chemical, biological, psychological and socio - economic factors. According to the effects of these factors on human health, they can be divided into: cumulative pathogenic, acute pathogenic, traumatogenic and terminal [11, 17]. Within ergonomic studies, it is necessary to identify risk factors in the working process. Risk factors can generally be divided into [12]:

- modifiable - we can influence their direct effect on health and performance of employees by preventative measures (physical, chemical, biological, psychosocial and habits).
- non-modifiable - we cannot influence their direct effect on health and performance of employee (age, gender, body type, body dimensions). An important factor that influences human ergonomics during the working process is human position. It is divided into working in a sitting position and working in a standing position.

1.3 Tecnomatix Jack Description

Tecnomatix Jack (TJ) is one of many Siemens PLM software products. It focuses on a human in terms of ergonomics. The program was created with support of NASA at the University of Pennsylvania in the 1980s [18, 19].

Table 1. The overview of analyzes used in Tx Jack software [13–16]

Method	Characteristic
Low Back Spinal Force Analysis	Analyses and evaluates the effect of the load on spine during the work performing
Static Strength Prediction	Analyses worker's workload from static forces
NIOSH Lifting Analysis	Analyses and evaluates symmetrical and asymmetrical load lifting
Metabolic Energy Expenditure	It helps to estimate the worker's metabolic energy expenditure for a particular activity in conjunction with his physical disposition
Fatigue Recovery Analysis	Calculates the amount of time needed to relax depending on the job and compares it with the required rest time for the activity
Ovako Working Posture Analysis (Owas)	It analyses and evaluates the relative discomfort of the work position with emphasis on the position of the back, arms and legs
Rapid Upper Limb Assessment	It analyses the upper limbs, assessing the working position in terms of upper limb disease
Manual Handling Limits	Analysis in connection with the evaluation and optimization of practical actions for manual handling - lifting, printing, drawing, transmission
Predetermined Time Analysis	Allows predicting time required to perform work by dividing work assignments into a set of movements using the MTM-1 method

TJ allows improving ergonomics of designed products and also improving the tasks [18, 19]. TJ gives the way to deal with problems thus simplify the visualization of solution. Program creates anthropometric and biomechanically accurate human figures that are both visual and realistic [18, 19]. In the TJ program, it is possible:

- to monitor and evaluate workers during their work,
- to focus on the optimal position at work,

- to evaluate conditions related to overloading of workers to look at positions with physiological limits resulting in hand, spine or muscle pain [13].

The TAT module evaluates risks in different intervals of load, position, and muscle utilization, frequency of movements and duration of work activity. This module contains a set of analyzes (see Table 1) which are the tools for reducing human workload in a work environment during work activities [18, 19].

2 Case Study of Implementation of Tx Jack in Selected Company

2.1 Short Characteristic of Selected Company

For analyzing the current situation of given problem, concrete company operating in the Slovak Republic was chosen. The company has been operating on the market for 10 years. The company focuses on production of furniture based on customer's requirements. It currently provides following services:

- production and installation kitchens,
- built-in cabinets also into the atypical spaces,
- high quality double beds,
- office furniture,
- custom made living room furniture,
- tables of atypical shapes,
- other services.

The company currently has only 2 employees. The area of joinery workshop is 92 m^2. In these spaces, furniture parts are also manufactured and assembled. The way in which the production workspaces are distributed is inefficient, so it is needed to make some modifications. Forasmuch as workshop is small and the joinery has more and more contracts, they have decided to build a new production hall. The total area will be 270 m^2. The owners decided for a workshop with two entrances - gates and floors that they would use for the administration. They also plan to buy new work desks or other machines to save the time in the future to produce furniture.

2.2 Creating a Simulation Model of Workplace

The first step in analyzing ergonomics of human at work is to create a simulation in TJ. It must contain objects that are actually present in work environment and characters of people in it. Each object and character in simulation is designed according to real dimensions, allowing us to accurately calculate analyzes in the TJ program [13]. The model of workplace was created in TJ. It corresponds to a real measured dimension. Also it was needed to choose a specific product, in our case it was kitchen unit. The reason was the repetitive work in the production of individual boxes from which the kitchen unit is composed.

Analysis of the Current State of Workshop
Before the analysis, a model of joinery was created in TJ software. All dimensions related to the workplace were measured in cooperation with the owner of joinery. Also, the placement of machines in the workshop was defined on the basis of the real situation in the workshop (see Fig. 2). Specific objects in 3D models with real dimensions measured in the workshop were modeled in the McNeel Rhinoceros program developed by Robert McNeel & Associates, which is used for industrial design, architecture and automotive design. Subsequently, STL objects were embedded into Solid Edge program where they were converted to .jt format. After these operations, they were imported into the TJ. After the objects were imported, they were then placed in the desired locations. After creation of working environment according to the real model of workplace, a human figure was created. The next step was to create a board from which furniture is made. It was chosen to create the own model provided by TJ. The dimensions of board (100 × 180 cm) were selected on the base of consultation with the joiners. After creating a workshop in program, the scene working environment (see Fig. 3) was saved. This action is important because simulation succeed to the creation of the working environment.

Fig. 2. Model of workplace - current state

The simulation was focused on the manipulation and work of a man with boards from which furniture is made. A whole series of commands were used for simulation. These were aimed at lifting, laying and moving the load-plate in the workplace. The individual commands were repeated, because work is performed in a standing position and worker uses work desks and machines that do not require a change of position. The simulation itself is executed on a single board and simulates the gradual movement of worker from machine to machine. In a real environment, work is carried out on several places and human moves around the workplace. For all commands, it was necessary to set the direction of simulation, whether human (Human Place) or Object Place option.

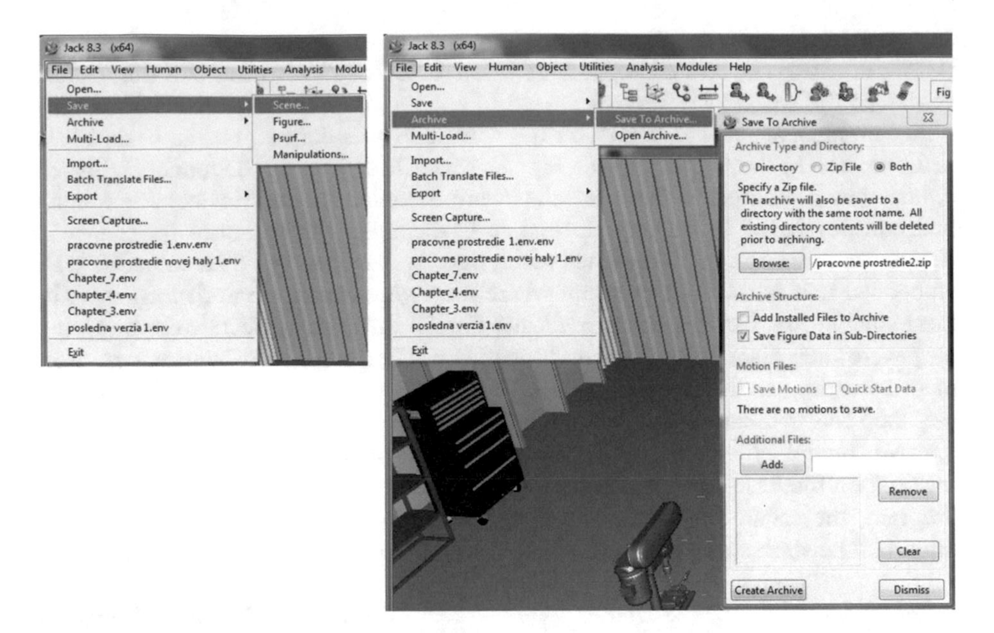

Fig. 3. Archiving of working environment

At the same time, it was necessary to adjust the position of a human at each step, because at the end of each part of the simulation the human position changed. When creating a working environment, the simulation was stored for needs of the next operation in TJ program. The next step after creating simulation was to start the particular analysis.

3 Application of Analysis RULA and LBA in Software Tx Jack

Two methods were used to analyze man in the work environment: Low Back Analysis and RULA, because these two methods examine the load on the lumbar spine and on the hands and posture of a man at work. The joiners work with different sizes and types of boards, so both parameters were averaged and used as a starting point for the analysis. Consequently, possible solutions to optimize the workplace were suggested. The simulation was focused on the movement of a person with a work load. The simulation itself was oriented on the worker's movement during the work activity with respect to the location of individual objects in the environment. At the end of the worker's simulation, a RULA analysis was performed. This analyze evaluated the load on each part of the body. The key moment for launching the analysis was lifting the board from the tables during manipulation because this activity is the most frequent. The size of these parts is different and their weight varies from 2 to 10 kg, depending on the type of material. This is determined by the customer. On the base of the analysis, this activity represents risk category no. 7, which is the highest possible degree of risk. It is necessary to take measures to reduce it (see Fig. 4).

Fig. 4. Evaluation of RULA analysis

Another selected analysis was the Low Back Analysis. This analysis was chosen to evaluate the load force on the lumbar spine in different postures. The boards from which the individual pieces of furniture are produced must be constantly lifted and manipulated. During work, the man drove the parts of cabinet an average of 160 different shapes

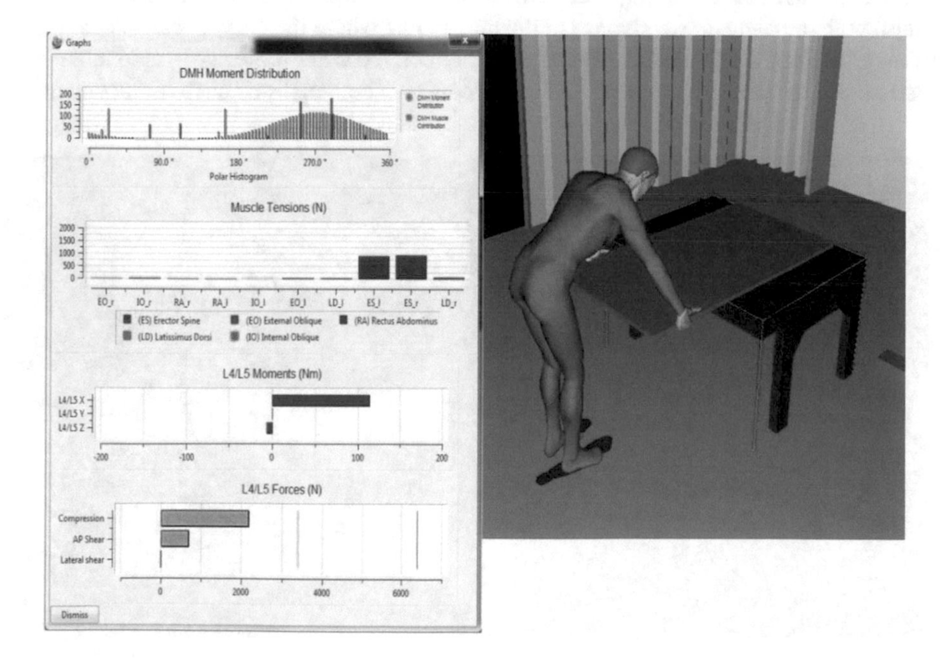

Fig. 5. Evaluation of LBA analysis

and weights. The analysis evaluated the degree of load on the lumbar spine at the lifting load as acceptable (see Fig. 5).

4 Proposal of a New Workplace Layout

4.1 Design of a New Workplace Layout

The main reason why machines, furniture and other aids were deployed in the existent workshop was an area. Despite the efforts of owners, there were still situations when the obstacle occurred at work and it was necessary to bypass it. There were also minor injuries due to lack of space and the placement of furniture. Owners of joinery have decided to invest in the construction of a new workshop, which should be three times larger than an old workshop. Based on the facts, it was necessary to make a new machine layout. Each machine has plenty of work space for work. There is the possibility of adding machines, because the hall will be more spacious than the original one. Two smaller work tables were strategically placed in the middle of the workshop. These can be used in any activity where there is a need for manual work but also for laying, depositing or transferring the material and then grabbing it from the other side of the table. The intention was to place more tables side by side and then make a transition. This will allow shortening time that was previously needed to bypass in the old workshop. A warehouse of material and finished products was solved similarly. In the original workshop, both warehouses were in one place. The material stock is located by the lumbermill. The reason is the fact that the process of manufacturing particular parts begins by cutting the material to the required sizes and shapes. In the warehouse, after a consultation with the owners, we suggested placing a table where the carvers can sketch parts of atypical shapes and then they can carve them on the work table. The same table was also placed in the warehouse of finished products. The benefit of such a placement of

Fig. 6. Model of a new hall

machines and furniture in a workshop is saving time spent on circumventing obstacles and smoother manufacturing processes. The new workplace was also modeled in TJ. The creation process was the same as in the case of current workshop. The only change that was made was the exchange of tables for new ones. The new ones will be 30 cm higher than the original tables. After creating the model of workplace, simulation of human work was carried out. Even in this case, the simulation was mainly concerned with lifting the load and moving it over the work area to the machines and the work table (see Fig. 6).

After finishing the simulation, RULA and Low Back Analysis were re-launched to evaluate the load on the upper limbs and the lumbar spine when lifting the load. This result of the RULA analysis showed that the proposed higher table and thus the grasping of the board at a different body position reduced the risk from the original 7 to 3 (see Fig. 7). Such a result is better than the original one, but is not sufficient for long-term work.In the case of new table dimensions, the results of the Low Back Analysis were better than in the first analysis (see Fig. 8). Load on the hip part of the spine has reduced due to the more upright position of the human figure when the lifting plate.

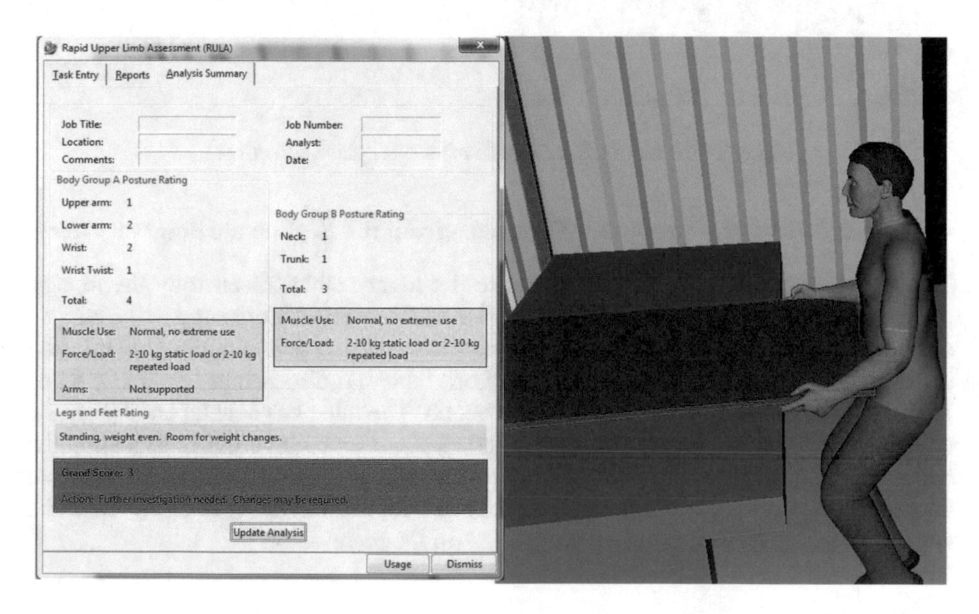

Fig. 7. Result of RULA analysis after layout of workplace

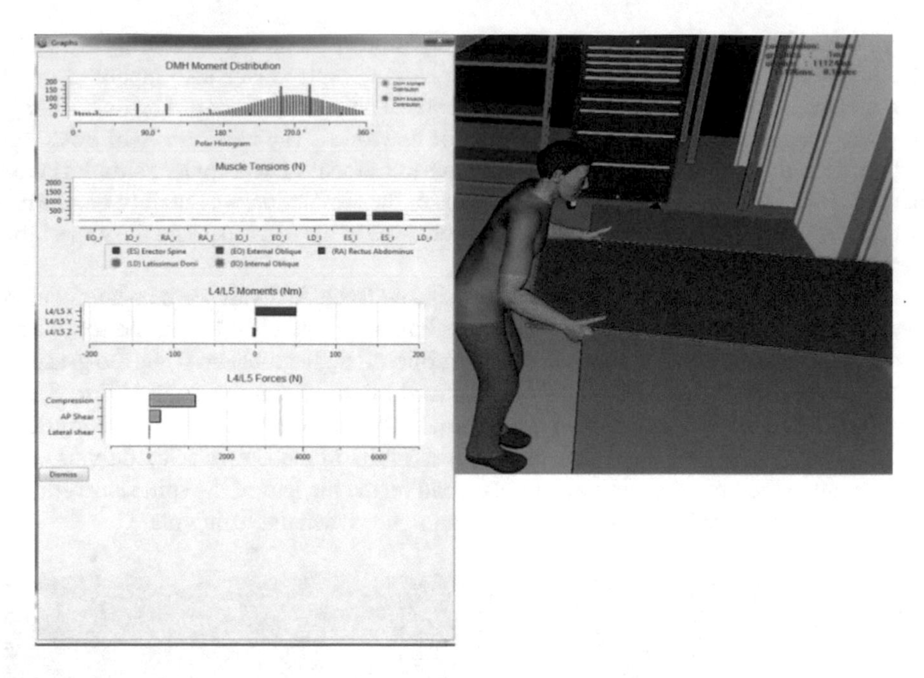

Fig. 8. Result of LBA analysis after workplace adjustment

4.2 Design of Improvement the Workplace from the Ergonomic Point of View

The RULA analysis has shown that, despite the higher tables planned by the joinery owners, the risk of Category 7 has fallen to risk of Category 3. Compared to the current situation, this is an improvement, but this risk is also not acceptable. A possible solution is the purchase of positioning tables). The work table has dimensions (HxWxD) 859 - 1,059 x 1,500 x 750 mm. The worktop is a beech. The table has a metal shelf over the entire width. Load capacity is 150 kg. The price of such a table is 350 €. Adjusting the height of the desk allows us to do different work and at the same time and also it can adapt to the human height. At the same time it provides the possibility of further material storage, respectively storage of working tools on the metal shelf.

5 Conclusion

Based on the knowledge gained from the Ergonomics study, two analyzes were selected to evaluate the load on the spine and the upper limbs of the human during the lifting, stacking and transporting of loads. The RULA analysis, which was one of two analyzes, showed a high level of risk in the work activities performed by the employee. Conversely, the second LBA analysis showed that the load on the lumbar spine and muscle strain is at acceptable levels. However, both analyzes did not evaluate the activities of a person during the whole simulation, but in part in the specific actions. It is important to note that the analyzed and evaluated operations were repeated during the whole work and the dimensions of the individual machines and furniture that were part of the examination of production process examination had the same height (deviation of approx. 3 cm). After modification of work environment, the level of worker's risk at work positions has declined, but still shows a value which is necessary to minimize. That's why one of the suggestions is to buy a positioning table. This ensures the actual height of the table for certain tasks, while the height of the table can be adjusted to the height of concrete worker. From a logistics point of view, suggestions have shown the possibility of more efficient distribution of machines in the workshop, improving material handling and time for workers to move around the workshop. At the same time, the obstacles that prevented smooth transition to individual machines were removed. From ergonomic point of view, the benefit is in reducing number of injuries caused by the layout of the tables and their dimensions and hence the sickness absence. In the case of expansion of production capacity in a given company, it is also possible to think about application 5S method in the future to eliminate potential risks arising from the working process.

Acknowledgements. This article was created by implementation of the grant project VEGA 1/0708/16 "Development of a new research methods for simulation, assessment, evaluation and quantification of advanced methods of production". KEGA 030TUKE-4/2017 "Implementation of innovative instruments for increasing the quality of higher education in the 5.2.52 Industrial Engineering field of study".

References

1. Kłos, S., Trebuna, P.: The impact of the availability of resources, the allocation of buffers and number of workers on the effectiveness of an assembly manufacturing system. Manag. Prod. Eng. Rev. **8**(3), 40–49 (2017)
2. Straka, M., Kačmáry, P., Rosová, A., Yakimovich, B., Korshunov, A.: Model of unique material flow in context with layout of manufacturing facilities. Manuf. Technol. **16**(4), 814–820 (2016)
3. Holman, D., Jirsák, P., Lenort, R., Staš, D., Wicher, P.: Understanding of the lean productivity - The source of scm productivity growth in the metallurgical industry. In: METAL 2015 - 24th International Conference on Metallurgy and Materials, Conference Proceedings, pp. 1939–1944 (2015)
4. Rosova, A., Malindzakova, M.: Material flow - starting point for recovery of inputs in the production company. Int. Multidiscip. Sci GeoConference Surv. Geol. Mining Ecol. Manag. SGEM **3**(5), 815–822 (2014)
5. Bučková, M., Krajčovič, M., Edl, M.: Computer simulation and optimization of transport distances of order picking processes. Procedia Eng. **192**, 69–74 (2017)
6. Straka, M., Fill, M.: Information system as a tool of decision support. Periodica Polytech. Transp. Eng. **45**(1), 48–52 (2017)
7. Saniuk, S., Saniuk, A.: Industry 4.0 - Technical-economic development perspective for the metallurgical production. In: METAL 2017 - 26th International Conference on Metallurgy and Materials, Conference Proceedings, pp. 2288–2292 (2017)
8. Schneider, E., Paoli, P., Brun, E.: Noise in figures, European Agency forf Safety and Healt at Work. Office for Official Publications of the European Communitie, Luxembourg (2005)

9. Cox, T., Rial-Gonzáles, E.: Working on stress, Magazine of European Agency for Safety and Health at Work, p. 28 (2002)
10. Sockoll, I., Kramer, I., Bodeker, W.: IGA –Report, Effectiveness and Economic Benefits of Workplace Health Promotion and Prevention. IGA, Germany (2009)
11. Hatiar, K., Sakal, P., Cook, T.M., Bršniak, V., Sekera, B., Výboch, J.: HCS model 3E – Microsolution of Macro – problems – Sustainable Development. In: Sustainability accounting and reporting on micro – economical and macro-economical level, International scientific conference, Brno , pp. 5–30 (2007)
12. Gilbertová, S., Matousek, O.: Ergonomie: optimalizace lidské činnosti. Ergonomics: optimization of human activity, 1st edn. Grada, Prague (2002)
13. Jack maual, Jack User Manual Version 8.0.1 (2014)
14. Stanney, K.K., Maxey J. and Salvendy, G.: Handbook of Human Factors and Ergonomics, 4th edn. pp. 637–656. Wiley, New York, (2007)
15. Trasser, H.: Principles, Methods and Examples of Ergonomics Research and Work Design. Industrial Engineering and Ergonomics, p. 363. Springer, Heidelberg (2009)
16. Walichnowski, A.: Analiza sposobów przecinania elementów składowych podłóg klejonych warstwowo. Analysis of ways to cut components of glued laminated flooring. Politechnika Gdańska, Gdańsk (2012)
17. Preventive ergonomics programs in enterprises. http://www.ibp.sk/starsie/zborník_permon_15102003/hatiar_permon.doc. Accessed 15 Feb 2018
18. Human Simulation and ergonomics. http://plm.automation.siemens.com/en_us/products/tecnomatix/manufacturing-simulation/human-ergonomics/jack.shtml#lighview-close. Accessed 15 Feb 2018
19. Jack and Process Human Simulate. https://www.plm.automation.siemens.com/en_us/products/tecnomatix/manufacturing-simulation/human-ergonomics/jack.shtml. Accessed 15 Feb 2018

The Concept of an Integrated Company Management System Combining the Results in Favour of Sustainable Development with the Company Indicator System

Piotr Cyplik[1(✉)], Michał Adamczak[1], Katarzyna Malinowska[2], and Jerzy Piontek[1]

[1] Poznan School of Logistics, Estkowskiego 6, 61-755 Poznan, Poland
{piotr.cyplik,michal.adamczak, jerzy.piontek}@wsl.com.pl

[2] Faculty of Engineering Management, Poznan University of Technology, Strzelecka 11, 60-965 Poznan, Poland
katarzyna.malinowska@doctorte.put.poznan.pl

Abstract. It is required by the development of world economies to respect social and environmental aspects. In this case, it will be a kind of development perceived by societies. The above assumption emanation is a sustainable development concept which is accompanied by forming various tools to analyse and assess the sustainability degree of particular economies. These are macro tools. Sustainable development is not formed at the entire economics level but is created by particular companies and supply chains. Therefore, there is a need for forming integrated company management systems combining the results in favour of sustainable development with the internal indicator system. In this paper, the authors used the systematic literature review and modelling method. The authors described their own integrated company management system combining the results in favour of sustainable development in social, economic, environmental and institutional political areas. The actions are to have an influence on the company results as recorded in the operational activity by the SCOR model and in the financial area by the DuPont model. The integrated company management system combining the results in favour of sustainable development with the internal indicator system is to make it possible to assess the undertaken actions in favour of sustainable development and to select value adding actions in particular company activity aspects. The companies are conscious that the actions in favour of sustainable development have an impact on their results. Main aim of the paper is to present a conception of integrated company management system joining the results of sustainable development activities with the internal key performance indicators (KPI) system.

Keywords: Sustainable development · Integrated management system
Supply chain development in green growth

© Springer Nature Switzerland AG 2019
A. Burduk et al. (Eds.): ISPEM 2018, AISC 835, pp. 337–349, 2019.
https://doi.org/10.1007/978-3-319-97490-3_33

1 Introduction

The industrialisation of world economies and the increase in consumption caused the increase in using natural resources. In order to tend to the continuous gross domestic product (GDP) increase in accordance with neoliberal economic tendencies, it was necessary to increase demand and consumption at the expense of using ever larger amounts of natural resources. The presented approach is inseparably linked with the environment devastation and, as a consequence, it leads to the life quality decrease. Is therefore the consumption increase expressed by GDP the increase an effective method of the economic development achievement [6]? One of the answers to this question is the sustainable development concept. This concept is focused on all aspects of human life and was based on three pillars: social, economic, environment. When the report entitled "Our Common Future" was published by the Brundtland Commission, the sustainable development ideas started functioning as meeting growing human needs accompanied by preserving the natural environment state at the same time [7, 11]. Due to its significance, this concept was very quickly adopted as a binding action direction of world economies. Numerous tools to assess the degree of adaptation of activities to sustainable development were also developed. Eurostat functions as a European statistical office and keeps records of the values of indicators related to sustainable development for the sake of the EU countries. Every two years Eurostat publishes a report on how the European economic, ecological and social systems are balanced. The system of indicators used by Eurostat is made of more than 100 indicators divided into 10 categories. Each theme was assigned a headline indicator. The themes with their headline indicators are presented in Table 1.

Table 1. Headline indicators in Eurostat sustainable development metrics system.

Theme	Headline indicator
Socio-economic development	Growth rate of real GDP per capita
Sustainable consumption and production	Resource productivity
Social inclusion	People at-risk-of-poverty or social exclusion
Demographic changes	Employment rate of older workers
Public health	Healthy life years and life expectancy at birth, by sex
Climate change and energy	Greenhouse gas emissions
	Share of renewable energy in gross final energy consumption
Sustainable transport	Energy consumption of transport relative to GDP
Natural resources	Common bird index
	Fish catches taken from stocks outside safe biological limits: Status of fish stocks managed by the EU in the North-East Atlantic
Global partnership	Official development assistance as share of gross national income
Good governance	No headline indicator

Source: [4]

American authors present an approach to sustainable development indicators and measures that is quite different from the European one. The U.S. Interagency Working Group on Sustainable Development Indicator (SDI Group) published a report in which one defined a set of indicators used to depict sustainable development in the United States of America. By analogy to the European system, the presented system was largely influenced by the structure of measures and indicators in the U.S. statistical area. The sustainable development measurement system as presented by the SDI Group is constructed of 40 selected indicators and measures divided into 3 areas: economic, ecological and social areas [10].

As mentioned before, not only economies of particular countries but also each company needs to face the challenge of getting involved in actions in favour of sustainable development. The real challenge for the organization is including a method of assessment a activities doing for sustainable development in the company system of business self- improvement [9]. The fact that companies are active in favour of the society or the natural environment is mostly not a result of their ecological or corporate social responsibility and their belief that it is right to be ecologically responsible. More than 6000 companies and 135 countries joined the UN Global Compact initiative by committing to reconciling their activity with 10 rules of human and employees' rights, the natural environment and corruption [3].

Decisions to take up initiatives in favour of sustainable development are made under such environment influence as social pressure and legal rules. Nevertheless, each company manager would like to treat sustainable development as a commercial activity area and expect business results from each initiative. There is a question whether projects in favour of sustainable development might generate documentable rates of return. Certainly, not each sustainable development project might undergo the ROI analysis. The lack of such possibility is implied by three reasons. Firstly, the beneficiaries of numerous sustainable development initiatives are outside the company and this makes it difficult to obtain data about the initiative effects. Secondly, most advantages of such actions are non-material and therefore, they are difficult to be measured. Thirdly, the period of the return on investment in favour of sustainable development is definitely longer [5]. It is difficult to traditionally specify the return on investment in sustainable development but this does not mean that such actions might minimise dangers, improve the company reputation and contribute to improving its business results. First of all, one should find a method to measure the effects and influence of balanced activity projects on the company results. This requires the synchronisation of the activity in the sustainable development area with the company mission and value system in order to create ideas of a "common value" that would be advantageous to the society and bring economic benefits to the company [2, 5, 7]. There is a common belief among companies that actions in favour of sustainable development are a sad duty imposed by the law in force and social expectations. This belief proves that the integration of such actions with the company strategy and objectives is not a norm. In practice, companies get involved in various actions in favour of the society and natural environment that do not fall into place. These actions are frequently conducted independently by various company employees who represent various business areas and management levels. Higher-level management staff member do often enough get involved in these actions [8]. Such projects are merely a form of execution of the duty to conduct actions in favour of the society and the natural environment. Such projects always generate costs in the company and almost never bring the company profits. The benefits to dedicated beneficiaries might also be often higher. In this article, the authors present an integrated company management concept that combines the results in favour of sustainable development with the company indicator system. Such a system would make it possible to combine actions oriented to sustainable development with the company results and thereby will enable the research on their costs and influence on the operational activity.

2 Sustainable Development Strategy in the Company

A coherent sustainable development strategy is necessary to make stakeholders and companies achieve a real value from the activity conducted in favour of the society and natural environment. An interdisciplinary sustainable development strategy should be integrated with the company mission and values and its business strategy. The whole as a coherent system gives rise to constructing balanced activity assessment indicators and measures. This whole is a starting point to build a coherent portfolio of actions in favour of the society and the environment at the same time [5, 8]. The sustainable development activities might be conducted on 3 surfaces. The first one is charity activity that is mostly a mere expenditure for the company. The second surface regards the company operational activity improvement. The actions on this surface are focused on improving the process efficiency and effectiveness. Their beneficiaries are external stakeholders and the mere company. Such actions might result in: reducing the resource use, reducing waste or the emission of harmful substances as well as increasing the company income and/or decreasing the company costs. As regards to the third surface, the projects are intended to combine the company strategy with solutions to specific social or ecological problems. Such projects result in improving the company results as a consequence of efficient social and ecological programmes. This might be exemplified by activities in favour of a local community on a new market for the company in order to recognise it and to make it easier to perform future expansion [8]. The conducted sustainable development actions need to be coherent independently of the surface they are related to. The actions might interact with each other, complete themselves or even transfer from one area to another (see Fig. 1).

In order to assess the balanced activity influence on the company results the balanced activity measures should be combined with the system of business measures and indicators. Due to the type of such actions, it is not easy to dimension them and to combine their effects with business measures and indicators. This task will be facilitated and new advantages to the company from the balanced activity will be noticed if one only focuses on such actions that support the company strategy. This chance is noticed by more and more managers which is confirmed by the survey research performed by the Harvard Business School team. The research was performed on a sample of 142 managers from various industries. The managers represented companies that used relatively advanced sustainable development practices. The majority of managers notice benefits from such actions to their company. The benefits are: company reputation improvement and increase in the staff motivation. More than 30% of the

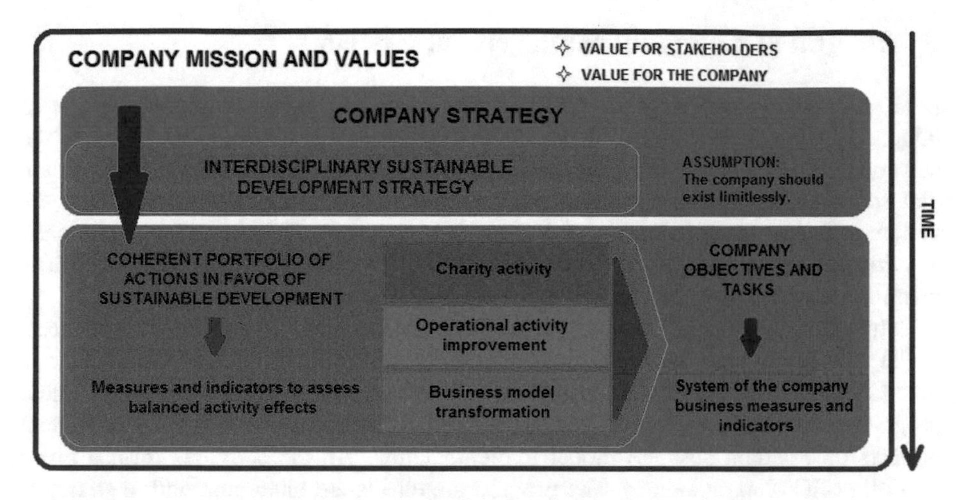

Fig. 1. Integration of the sustainable development strategy with the company business strategy.

investigated respondents notice or expect an income increase and cost reduction as a result of the CSR actions from the second or third surface. 13% of the respondents also notice an increase in income from the charity activity [8]. Another sustainable development advantage is an increase in the organisation innovation as indicated by the research conducted by Shulili Du, Xueming Luo who investigated the correlation between the activity in favour of sustainable development and the number of newly implemented products. The research included 128 companies between 2001 and 2004. The research results indicated that 30% of the companies with the highest involvement in CSR activities averagely introduced 47 new products into the market. Relatively 30% of the companies with the weakest involvement in sustainable development actions introduced almost 4 times less products into the market [3]. The benefits gained by the companies are not a result of unconsidered actions that support the society and ecology. The benefits are a result of considered long-term strategies. It should be noticed that a longer period of time is required by the return on initiatives in favour of sustainable development than standard company investments. If one assumes that a company will exist in an unlimited period of time, one might develop a sustainable development strategy that will support long-term plans and support benefit planning [2, 7]. The work for sustainable development has several times been an inspiration to improve the business activities. As far as the future prospect is concerned, one might start perceiving the society and environment support activity as chance to develop the company and to form its position on the market. While designing and implementing the sustainable development strategy, the managers should consider both the company strategic objectives and social and image benefits [2, 7]. The managers should carefully coordinate sustainable development programmes with the company strategic plans and investments. The business and sustainable development strategies should form a coherent long-term plan to create value for the company and stakeholders.

3 Integrated Company Management System

It is required by analysing the CSR activity influence on the company results to integrate the conducted action effects with the system of measures and indicators in the company. In Fig. 2 there is a concept of an integrated company management system that combines the results of activities in favour of sustainable development with the system of the company indicators. It is provided by the concept to combine 4 company activity assessment prospects: results of the activity in favour of sustainable development, brand value, financial results and operational results.

The activity in favour of sustainable development might take place in four areas: social, economic, environmental and institutional political. As part of the areas, one runs CSR programmes that are coherent with the company business strategy. These programmes might be performed as charity activities, operational activity improvement projects or a certain business model implementation. All the activities form a given portfolio of CSR programmes. The programme effects are integrated with a strategic, tactical and operational company management level. In order to combine CSR programme effects with the system of measures and indicators in the company, it is necessary to plan and dimension the effects of each CSR activity in progress and then include them in the applied business indicators as a next result influence factor. One should individually dimension CSR programmes in the case of each initiative. In this way one might specify an influence of each sustainable development activity on the company results. As the charity activity is mainly linked to expenses in favour of external stakeholders, it predominantly influences the brand value. In particular cases, such activities might result in increasing the company income. The sustainable development programmes intended to improve the operational activity should result in improving internal process and customer service reliability, reactivity and elasticity. They also reduce the use of resources and costs and thereby, they influence the increase in sales and profit. The benefits from such actions are transferred into all SCOR model attributes both in the case of a single company and the entire supply chain. The sustainable development programmes, which are intended to change or implement a new business model, have a broader view of the entire company business. Their aim at preparing a background for the company future movements and to increase their chance of success. The activity in favour of sustainable development on this surface aims at strengthening the brand, collected necessary capital and increasing sales. This activity also influences the assets turnover and total assets. Obviously, there are projects where the possibly longest period of time is required by the return on investment and the same effects are burdened by the largest uncertainty. On the other hand, these projects bring the most spectacular benefits to the company.

The brand value might be specified by comparing the expected income from the branded product with the expected income from a corresponding generic product. The probability of choosing a given brand by a customer is increased by a high degree of the brand consciousness in consumers' minds, its positive image with its correlated associations and the brand high quality. The above factors have also an influence on increasing recipients' and suppliers' loyalty. The above factors also decrease the efficiency of competitors' actions, increase the brand market share and ensure the possibility to obtain higher margin rates. Thus, the brand value influences the return on sales. All the actions, which contribute to increasing the brand value, result in increasing the company position on the market and indirectly to its financial results.

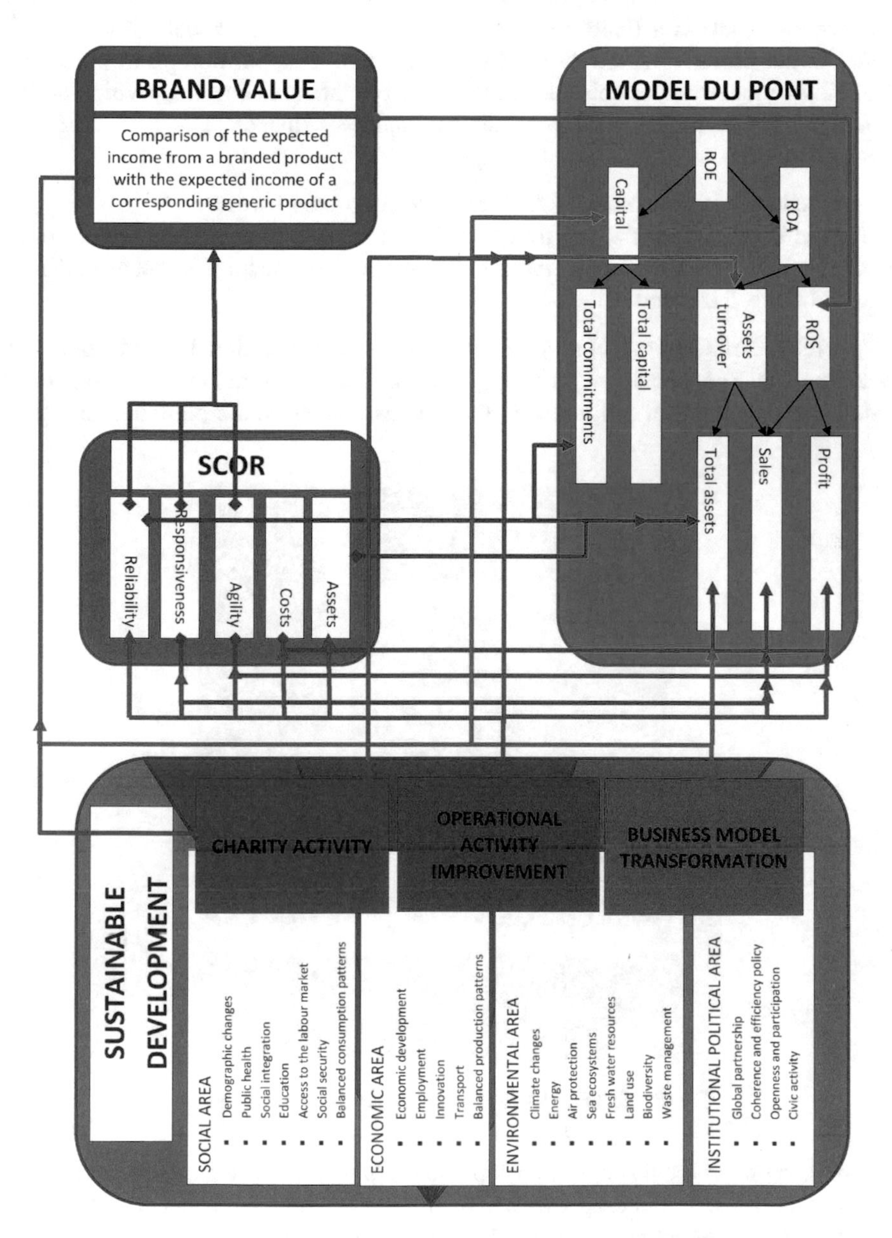

Fig. 2. The concept of an integrated business management system combining the results of activities for sustainable development with the system of enterprise indicators.

4 Operational Activity in the Integrated Management System

The activity in favour of sustainable development on the second one of the defined surfaces is based on improving the operational activity. One of the operational activity improvement tools is a Supply Chain Operations Reference Model (SCOR). SCOR model is the product of Supply Chain Council Inc. a global non-profit consortium. The SCOR model was established for the sake of standardising, improving and evaluating and comparing supply chain activities and performance [1]. The SCOR model scope embraces:

- interactions with customers – from an order to a paid invoice,
- flow of material and services from "supplier of supplier" to "customer of customer",
- market actions – their scope embraces the scope from customers' needs understand to fulfilling their requirements.

Thereby, the SCOR model scope embraces the entire operational activity undertaken by partners that cooperate within supply chains. The SCOR model might support the sustainable development fulfillment in three areas. The areas are presented in Fig. 3.

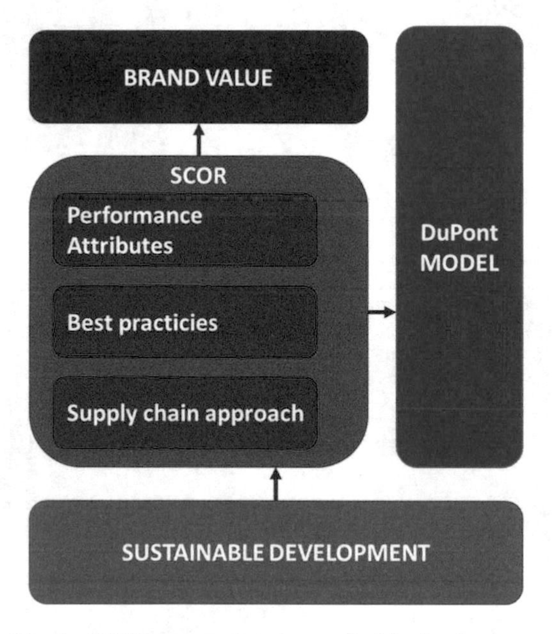

Fig. 3. SCOR approaches to sustainable development.

The first one of the distinguished approaches is performance attribute. Performance attributes on level 1 influence two out of three elements of the integrated company management system: brand value and DuPont model. Performance attribute on level 2 has an influence on sustainable development. Performance attributes on level 1 in SCOR model are divided into five dimensions. The detailed definitions of surfaces are presented in Table 2.

Table 2. Performance attribute in SCOR model.

Performance attribute	Definition
Reliability	The ability to perform tasks as expected. Reliability focuses on the predictability of the outcome of a process. Typical metrics for the reliability attribute include: On-time, the right quantity, the right quality
Responsiveness	The speed at which tasks are performed. The speed at which a supply chain provides products to the customer
	Examples include cycle-time metrics
Agility	The ability to respond to external influences, the ability to respond to marketplace changes to gain or maintain competitive advantage. SCOR Agility metrics include Flexibility and Adaptability
Costs	The cost of operating the supply chain processes. This includes labor costs, material costs, and management and transportation costs. A typical cost metric is Cost of Goods Sold
Asset Management Efficiency (Assets)	The ability to efficiently utilize assets. Asset management strategies in a supply chain include inventory reduction and in-sourcing vs. outsourcing
	Metrics include: Inventory days of supply and capacity utilization

Source: [1]

The SCOR model performance attributes are a key element of the concept of the integrated company management system that combines the activity results in favour of sustainable development with the company indicator system. The first three attributes: reliability, responsiveness and agility are an external supply chain assessment prospect. These attributes are thereby the chain assessment from the customer's view point. Therefore, one might indicate that the measures and indicators assigned to these attributes influence the brand value. In Table 3 there are level-1 metrics from the customer's view point.

The brand value is created not only by the products value but also by the logistic customer service as implied by the supply chain functioning. The brand value increases if the chain offers its customers high reliability of supplies, short order fulfillment cycle time and high agility. The above features determine the company competitive advantage and are therefore a brand value creation element. It is worth noticing that the upside adaptability and downside adaptability has an impact on: Revenue, COGS, SGA indicators from Profit & Loss and on inventory value from balance sheet. Thus, there is an indirect influence of measures and indicators from the category on the financial measures of the company and supply chain.

Costs and assets performance attributes depict the inventory (internal) prospect of the supply chain. The detailed definitions of measures and indicators at level 1 are presented in Table 4.

Performance attributes in resources perspective have an influence on the company financial results analysed by means of the DuPonta pyramid. Costs metics directly

Table 3. Level-1 metrics in customer perspective.

Performance attribute	Definition	Calculation
Reliability	Perfect Order Fulfillment	[Total Perfect Orders]/[Total Number of Orders] x 100%
Responsiveness	Order Fulfillment Cycle Time	[Sum Actual Cycle Times for All Orders Delivered]/ [Total Number of Orders Delivered] in days
Agility	Upside Flexibility	the minimum time required to achieve the unplanned sustainable increase when considering Source, Make, and Deliver components
	Upside Adaptability	the maximum sustainable percentage increase in quantity delivered that can be achieved in 30 days
	Downside Adaptability	The reduction in quantities ordered sustainable at 30 days prior to delivery with no inventory or cost penalties
	Overall Value-at-Risk	VaR = Probability of Risk Event (P) x Monetized Impact of Risk Event (I)

Source: [1]

Table 4. Level-1 metrics in resources perspective.

Performance attribute	Definition	Calculation
Costs	Total Cost to Serve	The sum of the supply chain cost to deliver products and services to customers
Assets	Cash-to-Cash Cycle Time	Cash-To-Cash Cycle Time = [Inventory Days of Supply] + [Days Sales Outstanding] – [Days Payable Outstanding] in days
	Return on Fixed Assets	Return on Supply Chain Fixed Assets = ([Supply Chain Revenue] – [Total Cost to Serve])/[Supply-Chain Fixed Assets]
	Return on Working Capital	Return on Working Capital = ([Supply Chain Revenue] – [Total Cost to Serve])/([Inventory] + [Accounts Receivable] – [Accounts Payable])

Source: [1]

influence the company profit. Assets metrics are linked with the total assets in the DuPonta model.

It is also necessary to notice that there is a relationship between the values of the measures at level 1 and the sales value. While the supply chain is being improved, it gets a competitive advantage on the market which is revealed by obtaining new customers and the increase in sales. This will be in turn reported in the DuPonta model.

The SCOR metrcis on a second level as defined in the SCOR model might support the sustainable development fulfillment by improving the operational activity of companies

Table 5. Metrics of processes on 2-level which has an influence on sustainable development.

Process on 2-level	Metrics
Make-to-Stock Make-to-Order	AM.3.22 Recyclable waste as % of total waste
Issue Material	AM.3.19 Packaging as % of total material
Produce and Test	AM.3.5% of production materials reused
	AM.3.6% of products consisting of previously used components
	AM.3.14 Hazardous materials used during production process as a % of all materials
Release Product to Deliver	RL.3.1 # of complaints regarding missing environmental documentation
Waste Disposal	RL.3.57 Waste Processing Errors
	RS.3.141 Waste accumulation time
	AM.3.15 Hazardous waste as % of total waste
	RL.3.57 Waste Processing Errors
	RS.3.141 Waste accumulation time
Package	AM.3.4% of packaging/shipping materials reused internally
Release Finished Product to Deliver	RL.3.14% of products meeting specified environmental performance requirements
	RL.3.15% of products with proper environmental labeling (if required)
Source Return Defective Product	AG.3.41 Current source return volume

Source: [1]

and all supply chains. This support mainly regards the economic area of sustainable development. Having analysed the measures and indicated them as recommended in the SCOR model, the authors included selected measures in the developed integrated company management system. The list of measures is presented in Table 5.

One selected the measures and indicators recommended to be used by companies and supply chains that supported sustainable development. This selection was based on their influence on measures as defined in the measurement systems of sustainable development at the macro level (particularly the level developed by the European Union and measured by Eurostat). As reference to this system the measures indicated in Table 4 have an influence on measures in the 'Ressource use and waste' category.

- components of domestic material consumption,
- domestic material consumption by material,
- generation of hazardous waste, by economic activity,
- emissions of hazardous substances by source sector.

The SCOR model provides its users recommended practices within each process. The SCOR model is an operational activity improvement tool. In the integrated company management system, the improvement of areas responsible for supporting

Table 6. Recommended practices.

Process on 2-level	Practices
Issue Sourced/In-Process Product	BP.171 Mixed Mode/Reverse Material Issue
Waste Disposal	BP.012 Lot Tracking
Source Return Defective Product	BP.098 Mobile Access of Information
	BP.129 Return Policy included with Shipping Document
Request Defective Product Return Authorization	BP.129 Return Policy included with Shipping Document
Return Defective Product	BP.167 Electronic Returns Tracking
Disposition MRO Product	BP.110 Product Development/Engineering/Disposition Collaboration
	BP.169 Beyond Economic Repair (BER) Management
Deliver Return MRO Product	BP.067 Returns Inventory Reduction

Source: [1]

sustainable development is the most significant. Selected recommended practices are presented in Table 6.

The implementation of the SCOR model in supply chains makes it possible to use the best practices that support the sustainable development fulfillment. The actions conducted in supply chains are characterised by a higher efficiency and might lead to a synergy effect. The actions are oriented to diminishing the business activity conduct influence on the natural environment and need to be comprehensively taken up by entire supply chains ranging from companies gaining raw materials to distributors. The efforts made by single companies are destroyed by breaking the supply chain. The coordination of prodevelopment actions between the companies is more efficient if the companies use a uniform system of defining processes, communication between the processes and finally assessing the process functioning. All these possibilities are provided by the SCOR model. Therefore, it is an absolutely necessary element of the integrated company management system that combines the results in favour of sustainable development with the system of the company indicators.

5 Summary

Contemporary customers' needs and the social consciousness of the environment protection necessity indicate new trends and challenges for business people. Nowadays, it is insufficient to provide a product to a customer in accordance with their expectations. It is also required by the social pressure and legal rules to conduct business activities with respect to social and environmental aspects. Numerous companies are not ready to face this challenge at the same time. As an attempt to fulfill the social and legal requirements, they take up numerous initiatives that are not in line with their business objectives. This is basically caused by the fact that sustainable development is not perceived as a company development chance. If one perceives the activity in favour of the society and natural environment as a chance, it will be possible to take a broader look at such activities. In this case, one might start considering to combine the sustainable development objectives with the company business objectives. In the authors' view, the integration of prosocial and proenvironment actions with the company strategy and business objectives is a necessary condition for contemporary companies to be successful. The authors proposed a concept of the integrated company management system that combines its activity results in favour of sustainable development with the company indicator system. This concept combines actions in favour of sustainable development with the company results as recorded within the operational activity by the SCOR model and in the area financed by the DuPonta model. This concept makes it possible to assess the influence of effects of the activity in favour of the society and natural environment on the company results. In order to increase the sustainable development strategy integration effects with the company business objectives, the authors encourage to expend the proposed concept to the entire supply chain. The coherent action of all the companies in the supply chain not only for business objectives but also for sustainable development will make it possible to reinforce the synergy effect and contribute to increasing profits of all stakeholders.

References

1. APICS, Supply Chain Council, Supply Chain Operations Reference Model, Revision 11 (2014)
2. Capatina, A., Micu, A., Cristache, N., Micu, A.E.: The impact of a trend pattern for sustainable marketing budgets on turnover dynamics (a case study). Contemp. Econ. **11**(3), 287–301 (2017). https://doi.org/10.5709/ce.1897-9254.243
3. Du, S., Luo, X.: Good Companies Launch More New Products. Harvard Business Review, April 2012
4. Eurostat. http://epp.eurostat.ec.europa.eu/portal/page/portal/sdi/indicators. Accessed 12 Dec 2017
5. Kuehn, K., McIntire, L.: Sustainability a CFO Can Love. Harvard Busieness Review, April 2014
6. Mousumi, R.: Managing the resources for sustainable development – where we stand and where we go from here? Adv. Manag. **3**(5), 5–7 (2010)
7. Ramusm, C.A.: Encouraging innovative environmental actions: what companies and managers must do. J. World Bus. **37**(2), 151–164 (2002). https://doi.org/10.1016/S1090-9516(02)00074-3
8. Rangan, V.K., Chase, L., Karim, S.: Czym naprawde jest CSR [original: The Truth About CSR], Harvard Busieness Review, January–February (2015)
9. Rudnicka, A.: How to manage sustainable supply chain? Issue Maturity, LogForum **12**(4), 203–211 (2016). https://doi.org/10.17270/J.LOG.2016.4.2
10. U.S. Interagency Working Group on Sustainable Development Indicators, Sustainable Development in the United States: An Experimental Set of Indicators, Washington, D.C., December 1998
11. Zaman, G., Goschin, Z.: Multidisciplinarity, interdisciplinarity and transdisciplinarity: theoretical approaches and implications for the strategy of post-crisis sustainable development. Theor. Appl. Econ. XVII **12**(553), 5–20 (2010)

The Concept of Intelligent Chlorine Dosing System in Water Supply Distribution Networks

Ryszard Wyczółkowski[1(✉)], Mariusz Piechowski[2], Violetta Gładysiak[3], and Małgorzata Jasiulewicz-Kaczmarek[4]

[1] Faculty of Organization and Management, Silesian University of Technology, Roosevelta Street 26, 41-800 Zabrze, Poland
ryszard.wyczolkowski@polsl.pl

[2] IQ-Software, Korona Street 3/31A, 60-652 Poznań, Poland
mariusz.piechowski@iq-software.pl

[3] ZW-K Śmigiel Sp. z oo., dr Skarzyńskiego Street 6A, Śmigiel, Poland
prezes@zwk-smigiel.pl

[4] Chair of Ergonomics and Quality Management, Poznan University of Technology, 11 Strzelecka Street, 60-965 Poznań, Poland
malgorzata.jasiulewicz-kaczmarek@put.poznan.pl

Abstract. The appropriate concentration of chlorine in water distribution systems is one of the most important parameters affecting the standard of drinking water supply Maintaining an even chlorine concentration in the whole water supply network is hindered by its physicochemical properties of this process. Parallel reactions related to chlorine decomposition in the water supply network (bulk reactions and reactions on the walls) cause a significant decrease in chlorine content in water with time (with growing water age) which makes it difficult to maintain optimal chlorine concentration at network points far from the source. For this, various solutions to improve the situation are implemented: for example, additional chlorination points for water, located at a certain distance from the place where the network is supplied with water. In order to introduce improved chlorination of water in its own business, the Water Supply and Sewage Plant in Śmigiel is considering introducing new chlorine dosing systems that allow automatic control of the chlorine dose over time. However, before the investment begins, its technical effectiveness is analyzed. This paper presents the first phase of simulation works in order to show whether the implementation of the new investment could be economically efficient.

Keywords: Water distribution modeling · Water quality
Genetics optimization

1 Introduction

The appropriate concentration of chlorine in water distribution systems is one of the most important parameters affecting the standard of drinking water supply [9]. The method of dosing chlorine in water supply networks should ensure a constant concentration of chlorine in water (in time and along the pipes), allowing the elimination of

© Springer Nature Switzerland AG 2019
A. Burduk et al. (Eds.): ISPEM 2018, AISC 835, pp. 350–359, 2019.
https://doi.org/10.1007/978-3-319-97490-3_34

microorganisms, pathogenic or adversely affecting the taste and quality of the water. The optimal concentration of chlorine cannot be too low, because it threatens contamination of supplied water or too high, due to unpleasant smell, worse taste [5] (maximum allowable chlorine concentration values are determined by sanitary standards).

Maintaining an even chlorine concentration in the whole water supply network is hindered by its physicochemical properties of this process. Parallel reactions related to chlorine decomposition in the water supply network (bulk reactions and reactions on the walls) cause a significant decrease in chlorine content in water with time (with growing water age) which makes it difficult to maintain optimal chlorine concentration at network points far from the source.

For this, various ways to improve the situation are used: for example, additional chlorination points for water, located at a certain distance from the place where the network is supplied with water (and chlorine). In the case of two or more chlorination points, it will be necessary to rationally choose time pattern of intensity of chlorination of water, to obtain the best level of chlorine concentration without exceeding the limit values at any point of network considered. It is noteworthy that such action requires taking into account the amount of water entering the network, which is depended on consumption of water varying over time and the topology of the network. These two parameters cause different speed of water distributed in the network and thus different age of water as well as the concentration of chlorine.

In order to introduce improved chlorination of water in its own business, the Water Supply and Sewage Plant in Śmigiel is considering introducing new chlorine dosing systems that allow automatic control of the chlorine dose over time, which will be the first step in the implementation of optimal chlorination of water. The project is part of the company's policy related to sustainable development, where the overriding goal is to improve water quality and customer satisfaction. To implement such a concept of optimal water chlorination, the following components are needed:

1. prediction module of water consumption in the network, allowing to determine the profile of water consumption in the time horizon of a few hours,
2. a module that allows to simulate the flow of water in a given network,
3. a system allowing to control the chlorine dosage to the network so as to achieve the desired time-varying intensity of chlorine in the water pumped into the network.
4. the method of optimization of controlling the dose of chlorine,

However, before the investment begins, its technical effectiveness is analyzed. In this purpose, a number of numerical experiments were planned to be carried out to show the possible improvement in the quality of chlorination of water and the economic aspect of the investment. This paper presents the first stage of these experiments.

2 The Goals and Assumptions of the Numerical Experiment

The problem of forecasting water consumption in the network has been successfully developed in many research projects, e.g. [7]. The presented results show that time series describing water consumption, especially for municipal (non-industrial) purposes, have many deterministic components, thanks to which it is possible to effectively predict the future values of the series. The occurring random components are important for predictions in short periods of time (e.g. in the minutes), which is important, for example, in network diagnostics [6, 8]. It seems that for the purpose of chlorination, a longer forecast step (hourly) is sufficient, where the impact of the random component is clearly smaller.

The problem of modeling water flows in the water supply network was also the subject of many works. Modelling always begins with the so-called mental model [4]. Currently, effective software is available to allow simulation of the water supply system. One of them is the EpaNET software [12]. Due to the availability it is widely used in research works. The advantage of the EpaNet is the ability to simulate a water flows, determine the age of water and simulate the distribution of other substances in the water supply network, for example, pollutants or additives used to disinfect the water distribution network (e.g. sodium hypochlorite) [14]. An additional value for the investigators is the ability to integrate the software with other computing platforms.

Considering the above, before making decision about implementing a new method of water decontamination in the network, the problem of possibilities for proper operation and the utility of the presented concepts could be considered. In addition, the method of controlling the concentration of added chlorine should also be considered.

To answer the first question, a numerical experiment was planned, consisting in simulation of the water disinfection process (chlorination) on the numerical model of the network (Fig. 1).

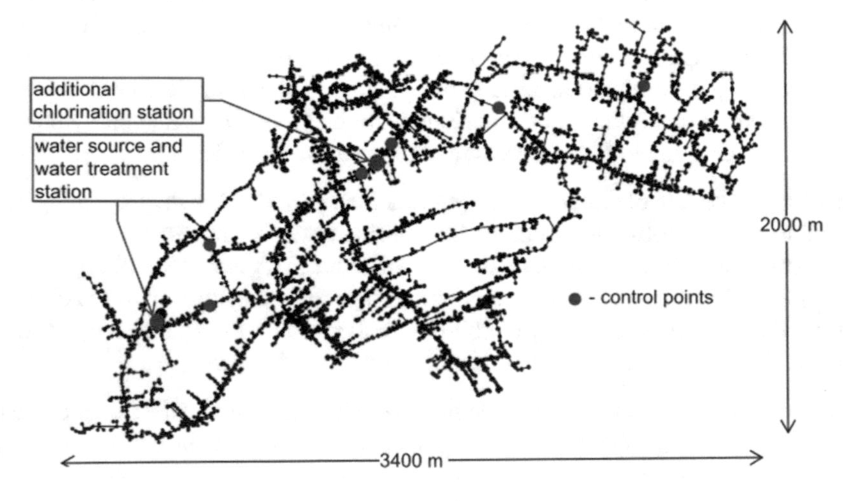

Fig. 1. Research object – a simulated water supply network

As the model of the considered water supply network was not yet available, the hypothetical network model was built in the EpaNET environment. This model was developed as a result of merging and transforming fragments of hydraulic models of other real networks, which allowed to consider the conditions of the experiment as realistic.

The figure shows the mains supply, the place of supplying chlorine, and the points in which during the experiment the current concentration of this agent was determined.

It was assumed that in the supply of the network, the pumped water will have a given concentration of chlorine, and the chlorination auxiliary point will be of the type flow paced booster and will adds a fixed concentration to that resulting from the mixing of all inflow to the node from other points.

The starting point for the considerations was the assumption that initially both chlorination points will be set to the maximum permissible dose of 0.3 mg/l. The map of obtained chlorine concentration in the network (for the selected hour of the day) is shown in Fig. 2. Obviously, with such assumptions on the chlorination process, in a certain area of the network, directly after the additional chlorination point, acceptable concentrations will be exceeded. The direct aim of the experiment was to find such chlorination time pattern for both points at which the concentration of chlorine at the control points of the network does not exceed the permissible value at any time and at the same time the concentration will be as high as possible.

Fig. 2. Map of chlorine concentration in the network - assumed parameters

The additional goal was to determine whether, given the decreasing concentration of chlorine in water over time, the benefits obtained by optimizing the chlorine process would be noticeable.

3 The Concept of the Chlorine Concentration Control System

In order to ensure the possibility of any controlled chlorine dosing, the existing water treatment station will be supplemented with two chlorine concentration analyzers and a dosing system that supplies chlorine to water (Fig. 3).

The chlorine analyzer proposed for the application provides automatic, fast and reliable measurement of free chlorine and total chlorine measurement, which is simple, fast and accurate [1]. The device is practically maintenance-free, with minimal reagent consumption and very low maintenance requirements, ensures high cost-effectiveness - in continuous operation with a measurement every 2.5 min, operator intervention is required every 30 days.

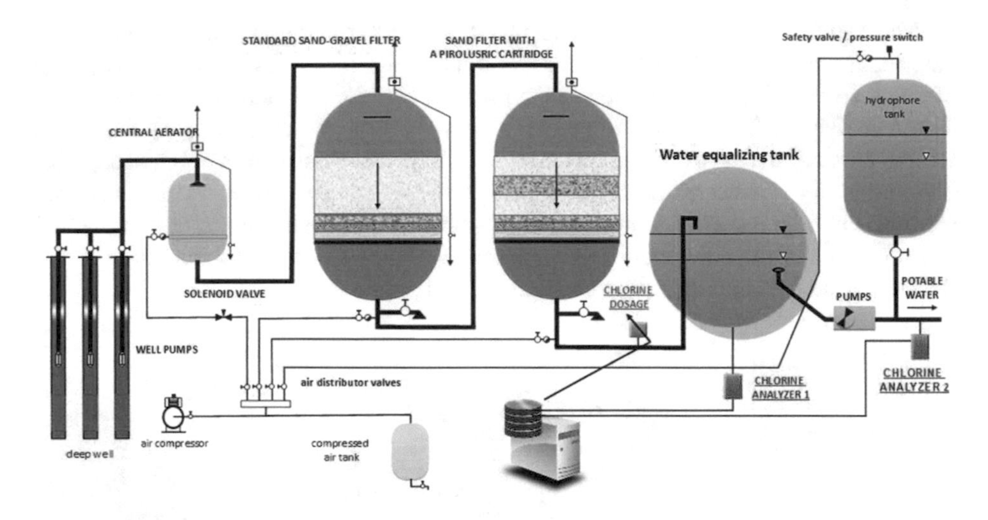

Fig. 3. Diagram of water treatment station.

Two such devices will be used in the water treatment station, enabling precise chlorine measurement. On the basis of these measurements, only the necessary amount of chlorine is dispensed for disinfection, which ensures safe quality of drinking water in the water supply network, minimizing the side effects of this disinfection method as well as the specific taste and smell.

The first device controls chlorine dosage on the clean water equalizing tank to ensure optimal chlorine dosing to the tank, where higher chlorine concentration (0.5 mg/l) is acceptable due to the longer residence time of the water in the tank. The second device provides feedback to verify the process of optimal chlorination. At the same time, the analyzer placed directly at the entrance to the water network plays a controlling role, signaling possible exceeding of the permissible concentration. The computer records and analyzes all measurement results, and on their basis and the measurement of the actual water consumption calculates the optimal size of the dose of chlorine compounds used for disinfection. Based on the actual parameters and calculated optimal pattern of chlorine concentration, the system is able to predict the necessary dose of chlorine compounds given to the equalizing reservoir and the time at which such a dose should be administered to ensure the optimal level of chlorination. At the next stage of the project, measurements are planned within the water-pipe network, on the basis of which measures will be taken to enable a more precise control of the whole system, which will further optimize the disinfection process and improve water quality.

4 Optimization Method and Obtained Results

The genetic algorithm was used for optimization - the GA library [13] from the R package [11] was used. The Epanetreader library [3] was used to connect EpaNET to the R calculation package.

The choice of the genetic algorithm as the optimization method was justified by the following considerations:

- the ease of using different forms of the objective function [10],
- possibility of multi-criteria optimization,
- simple inclusion and subsequent addition of various input parameters to the optimization process.

In the conducted simulations, it was assumed that the hourly pattern of chlorination for both sources will be optimized. Thus, the chromosome describing the subject in the population consisted of 48 (2 × 24) real numbers, being multipliers for the assumed maximum chlorination value for sources (0.3 mg/l). It has been assumed that the value of individual genes will be in the range <0.1>. It was assumed that the population will count 40 individuals, and the optimization process will be carried out so long that there will be no improvement in the obtained solution. For each simulation, the EpaNET bulk flow decay constant was set on -0.85 day^{-1} as in [2].

The value of the adaptation function for a given individual in the population was determined as $x = -\sum_{i=1:24}\sum_{j=1:n}\left|S_{ij} - 0.3\right|$, where S_{ij} - concentration of chlorine in the selected j-th checkpoint at the i-hour of the day.

To determine the objective function, for each individual (i.e. the proposed chlorination pattern), a simulation was carried out in the EpaNET program and the chlorine concentration was determined at a given hour and given points. The obtained learning curve for the genetic algorithm is shown in Fig. 4.

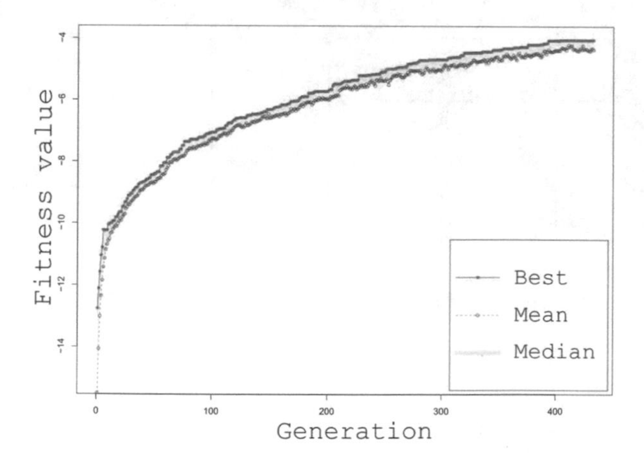

Fig. 4. Obtained learning curve for the genetic algorithm.

Exemplary chlorine concentration maps for chlorination pattern obtained in the process of optimization are shown in Figs. 5 and 6.

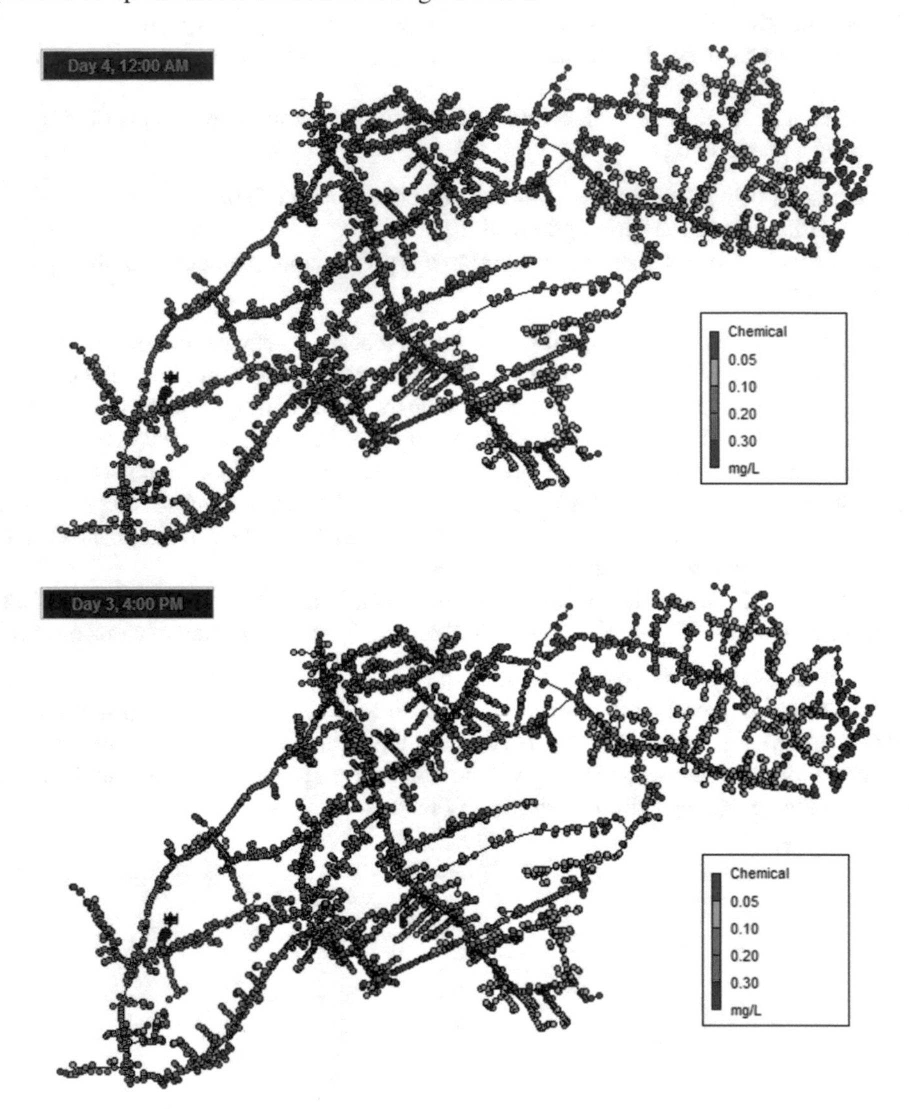

Fig. 5. Exemplary chlorine concentration maps for chlorination pattern obtained in the process of optimization

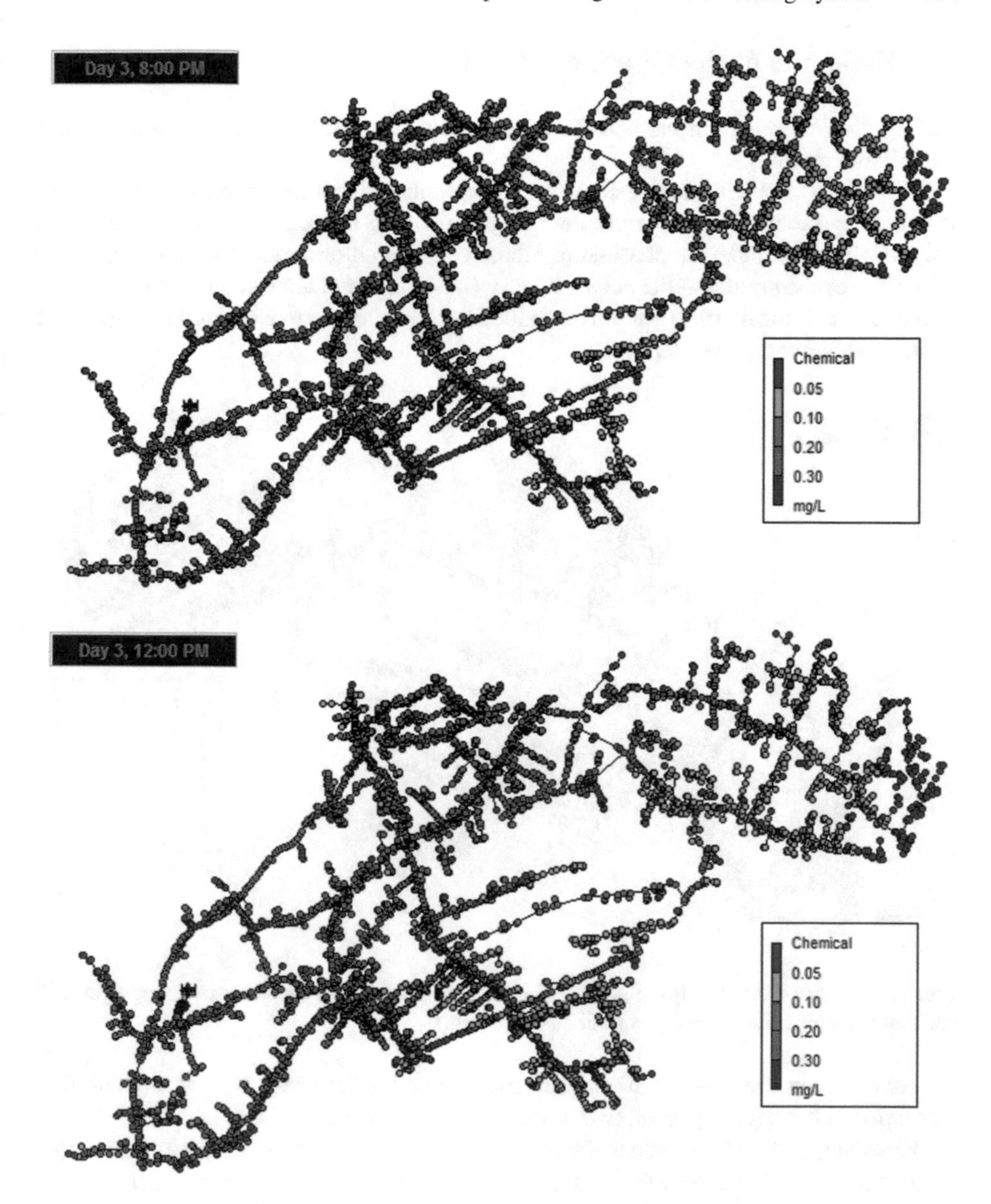

Fig. 6. Exemplary chlorine concentration maps for chlorination pattern obtained in the process of optimization

It is evident that the permissible chlorine concentration value has not been exceeded at any time. The distribution of chlorine concentration in the network does not change during the day - it is almost constant. The areas at the edge of the network, characterized by the minimum concentration of chlorine, have also decreased, but chlorine concentration changes with increasing distance from sources is still visible.

5 Discussion of the Obtained Results

The applied optimization method (GA) confirmed its usefulness. The obtained result met all assumptions.

A separate matter is whether simpler assumptions and methods would not be enough to solve the presented example problem. Figure 7 shows an example of a concentration map obtained for much simpler assumptions: constant chlorination with the maximum allowable concentration at the network entry (in source) and also a constant concentration at the additional point, but with a value chosen so as not to exceed the limit value at the next, nearest network point.

Fig. 7. Chlorine concentration maps for the selected hour of the day (constant, appropriately calculated chlorination intensity at both dosing points).

As ones can see, the results are not much worse - the areas with lower chlorine saturation at the edges of the network are little bigger only.

However, it should be noted that with two chlorination points, a perfect improvement of concentration throughout the network is not possible due to physical constraints on the chlorination process itself. It seems that the answer to the challenge posed could be the use of more chlorination points (e.g. mobile) combined with the optimization of the additional chlorine injection places. This assumption justifies the use of complex computer optimization methods, like genetics optimization.

Additional activities should also be considered, e.g. the possibility of periodically, automatically changing the network configuration (closing selected valves) so as to enforce better water circulation, especially at points far from the source.

References

1. CL17 Colorimetric Chlorine Analyzer. https://uk.hach.com/chlorine-analysers/cl17-colorimetric-chlorine-analyser/family?productCategryId=24929231577
2. Dandy, G., Blaikie, M., Commaine, C., Frankish, D., Osborne, D., Thompson, M.: Towards optimal control of Chlorine levels in water distribution systems. In: Proceedings of Australian Water Association Regional Conference, AWA South Australian Branch, Glenelg, Adelaide (2004)
3. Eck, B.J.: epanetReader: A Package for Reading EPANET Files into R. In: World Environmental and Water Resources Congress, pp. 487–496 (2016)
4. Grzybowska, K.: Application of an electronic bulletin board, as a mechanism of coordination of actions in complex systems – reference model. LogForum **11**(2), 151–158 (2015). https://doi.org/10.17270/J.LOG.2015.2.3
5. Jonkergouw, P.M.R., Khu, S.-T., Savic, D.A., Zhong, D., Hou, X.Q., Zhao, H.-B.: A variable rate coefficient chlorine decay model. Environ. Sci. Technol. **43**(2), 408–414 (2009). https://doi.org/10.1021/es8012497
6. Karwot, J., Kaźmierczak, J., Wyczółkowski, R., Paszkowski, W., Przystałka, P.: "Smart water in smart city": a case study. In: Hydrology and Water Resources, 16th International Multidisciplinary Scientific GeoConference, SGEM 2016, 28 June–6 July 2016, Albena, Bulgaria, vol. 1, Sofia: STEF92 Technology 2016, pp. 851–858 (2016). https://doi.org/10.5593/sgem2016b31
7. Kozłowski, E., Kowalska, B., Kowalski, D., Mazurkiewicz, D.: Water demand forecasting by trend and harmonic analysis. Archiv. Civ. Mech. Eng. **1**(18), 140–148 (2018). https://doi.org/10.1016/j.acme.2017.05.006
8. Moczulski, W., Wyczółkowski, R., Ciupke, K., Przystałka, P., Tomasik, P., Wachla, P.: A methodology of leakage detection and location in water distribution networks - the case study. In: Sarrate, R. (ed.) 3rd Conference on Control and Fault-Tolerant Systems, SysTol 2016, Barcelona, Spain, 7–9 September 2016, Piscataway, pp. 331–336. IEEE (2016). https://doi.org/10.1109/systol.2016.7739772
9. Musz, A., Kowalska, B., Widomski, K.M.: Some issues concerning the problems of water quality modelling in distribution systems. Ecol. Chem. Eng. **16**(S2), 175–184 (2009)
10. Oleśków-Szłapka, J., Pawłowski, G., Fertsch, M.: An optimization approach for scheduling and lot sizing problems in electromechanical industry using GA-based method. In: Burduk, A., Mazurkiewicz, D. (eds.) Intelligent Systems in Production Engineering and Maintenance – ISPEM 2017. Advances in Intelligent Systems and Computing, vol. 637. Springer, Cham (2018). https://doi.org/10.1007/978-3-319-64465-3_14
11. R Development Core Team. R: A language and environment for statistical computing. R Foundation for Statistical Computing, Vienna, Austria. http://www.R-project.org
12. Rossman, L.: EPANET2 users manual. Water Supply and Water Resources Division, National Risk Management Research Laboratory. United States Environmental Protection Agency Cincinnati USA (2000)
13. Scrucca, L.: GA: A package for genetic algorithms in R. J. Stat. Softw. **53**(4), pp. 1–37 (2013). http://www.jstatsoft.org/v53/i04/
14. Tiruneh, A.T., Fadiran, A.O., Nkambule, S.J., Zwane, L.M.: Modeling of chlorine decay rates in distribution systems based on initial chlorine, reactant concentrations and their distributions. Am. J. Sci. Technol. **3**(3), 53–62 (2016)

The Framework of IT Tool Supporting Layout Redesign in a Selected Industrial Company

Izabela Kudelska(✉), Agnieszka Stachowiak, and Marta Pawłowska

Poznan University of Technology, Strzelecka 11, 60-965 Poznan, Poland
{izabela.kudelska, agnieszka.stachowiak}@put.poznan.pl, marta.moczulska@doctorate.put.poznan.pl

Abstract. Facility layout is a problem that has been discussed in the literature for a long time, and together with growing complexity of products and advance it manufacturing process it has been gaining its importance, however in the same time becoming more and more difficult to solve. The goal of the research is to develop a tool supporting the process of designing the facility layout. In their study, the authors used the method of observation and interview. These methods allowed for a more detailed analysis of the company as well as the processes occurring in it and thoroughly analyze its needs and requirements, and model it at proper level of detail. Hence, the first step is the analysis of the literature on facility layout models. The next part presents the subject of the research. Presentation of the company is to bring the specificity of production process which is crucial for the design process and subsequent reorganization of the facility. In the last part, the authors present the concept of a computer tool that is to support the process of the facility layout. This concept is modeled using UML diagrams. The work ends with a summary and directions of further research.

Keywords: Facility layout · Shipyard organization · MAT · UML diagram

1 Introduction

Organization of a company at all its levels has a huge impact on its efficient functioning. Requires decisions concerning resources and their allocation in strategic and operational perspective. One of the examples of such decisions is the one concerning facility layout. Facility layout is determined by many factors, including product parameters, manufacturing technology, safety, ergonomics, costs, transport, etc., which makes it complex and difficult, however, in the same time important, as influencing many aspects of company's performance.

Layout development, both as a new project or redesign of the existing one, is difficult and time-consuming. During the process, the project team should thoroughly analyze all the related factors, mentioned in the paragraph above. In the case of redesigning already existing production systems, attempts to experiment can seriously disrupt the proper functioning of the production process system, which in effect will lead to disruption of the entire enterprise.

© Springer Nature Switzerland AG 2019
A. Burduk et al. (Eds.): ISPEM 2018, AISC 835, pp. 360–369, 2019.
https://doi.org/10.1007/978-3-319-97490-3_35

The general methodological approaches offer a wide range of models, methods and techniques of design: from traditional, based on graphic models, to mathematical models and methods and models based on AI implementation, among them optimization-based models and optimization methods are particularly important. To select the best approach, assessment of algorithms is required, and the criteria most often implemented include the accuracy of the solutions obtained, the speed of calculation and the technical requirements for computers. One of the basic criteria is the computational complexity of the algorithm, which is the time of operation of this algorithm expressed in the form of a function of the size of the task being solved. Most of these tasks are combinatorial problems, in which a set of permissible solutions consists of subsets of a certain n-element set or of n-element permutations [1].

Facility layout problem is a difficult task. This particularly applies to large-size problems and all discrete tasks in which there are links between objects.

The purpose of the research is to develop a tool supporting the process of facility layout and it was inspired by specific working environment – the shipyard.

2 Research Design and Methodology

The goal of the research is to develop a tool supporting the process of designing the facility layout. To achieve the goal, the authors have identified the following steps:

S1: analysis of the literature on the subject with regards to methods for facility layout design

S2: company analysis (industry, size)

S3: analysis of the production process of the company.

S4: the framework of a computer tool supporting the facility layout design.

The decomposition of the general goal into specific steps allowed structuring the research, and resulted in work planning and scheduling.

Thus, the first step is the analysis of the literature on facility layout models. The next part presents the subject of the research. Presentation of the company is to bring the specificity of production process which is crucial for the design process and subsequent reorganization of the facility. In the last part, the authors present the concept of a computer tool that is to support the process of the facility layout. This concept is modeled using UML diagrams. The work ends with a summary and directions of further research.

In their study, the authors used the method of observation and interview. These methods allowed for a more detailed analysis of the company as well as the processes occurring in it and thoroughly analyze its needs and requirements, and model it at proper level of detail.

3 Literature Review

3.1 Facility Layout Design

Facility layout is a problem that has been discussed in the literature for a long time, and together with growing complexity of products and advance it manufacturing process it has been gaining its importance, however in the same time becoming more and more difficult to solve. The literature [2] presents numerous approaches and methodologies, and their evolution but stresses that in most cases all the solutions are based on technology and material movement but in the same differ with approach, evolving with development of science as presented in the Table 1.

Table 1. Selection of layout planning/design methods

Author	The essence of layout planning
Immer [3]	Basic steps in layout planning are related to materials handling, the flow should be represented and depicted in terms of the output produced by equipment. The goal was minimizing the distance traveled between work centers, the solutions were evaluated basing on the judgment, intuition and experience of the layout analyst
Reed [4]	"Layout planning chart" is the single most important phase of systematic planning in plant layout. This chart incorporates the flow process, standard times for each operation, machine selection, manpower selection and materials-handling requirement
Nadler [5]	Ideal system approach to layout and materials handling systems in terms of a philosophy of overall development
Armour and Buffa [6]	CRAFT: a heuristic model for minimizing transportation costs when presented with flow data
Lee and Moore [7]	CORELAP: heuristic model, constructs a layout for a facility by calculating the total closeness ratio (TCR) for each department, where TCR is the sum of the numerical values assigned to the closeness relationship between departments
Tompkins and Reed [8]	COFAD: aims to minimize the extent of materials handling. This model's objective is to select a layout that resulted in a minimal-cost handling system
Seehof and Evans [9]	ADELP: automated layout design program, the same basic data input requirement and objectives as CORELAP. The basic difference between CORELAP and ALDEP was that CORELAP selected the first department to enter the layout and broke ties using the total closeness ratios, while ALDEP selected the first department and broke ties randomly. CORELAP attempts to produce the best layout whereas ALDEP produces many layouts, rating each layout but leaving decision to the facilities designer

(continued)

Table 1. (*continued*)

Author	The essence of layout planning
Edwards et al. [10]	MAT: modular allocation technique. This is a construction technique which utilizes the distances between all pairs of locations and the loads transported between all pairs of facilities per unit time to approximate the optimal allocation of facilities to locations
Apple [11]	Detailed sequence of steps to produce a plant layout, the steps are not necessarily performed in the sequence given
Scriabin and Vergin [12]	FLAC: facility layout by analysis of clusters, emulates the visual methods used by industrial engineers in solving layout problems.
Kumar et al. [13]	Augmented transition network of natural language processing as the heuristic of an expert system to determine the practical limitations of alternative layouts
Malakooti and Tsurushima [14]	An expert system and multiple-criteria decision-making
Abdou and Dutta [15]	EXSYS: an expert system approach to define appropriate layouts of machining facilities under specific combinations of manufacturing and materials handling systems. The expert system shell and relationship chart were used to construct the knowledge base
Raoot and Rakshit [16]	Fuzzy set theory to solve facility layout problems
Shih et al. [17]	AILAY: AI-based layout planning. AILAY's efficiency and effectiveness depends heavily on the intelligence of the evaluation and search process
Bozer et al. [18]	MULTIPLE: multi-floor plant layout evaluation, developed for designing multi-floor layouts
Tompkins et al. [19]	Layout as quadratic assignment problem, NP-complete
Eneyo and Pannirselvam [20]	Simulation model as a substitute for experimenting online to provide detailed insight into the evaluation of new facility designs or flow control policies
Ficko and Palcic [21]	Application of genetic algorithms for solving the layout planning problem using the modified triangle method
Kanduc, Rodic [22]	Use of simmulated annealing for facility layout problem solving
Solimanpur, Elmi [23]	A tabu search approach for cell scheduling problem with makespan criterion
Berlec, Potocnik, Govekar, Starbek [24]	A method of production layout planning based on self-organising neural network clustering

Source: based on [2]

The evolution went from simple calculation methods to application of simulation and AI in facility layout planning leaving decision-makers with numerous approaches to choose from. However, when searching for similarities, the general model for layout design could be structured as presented in the Fig. 1.

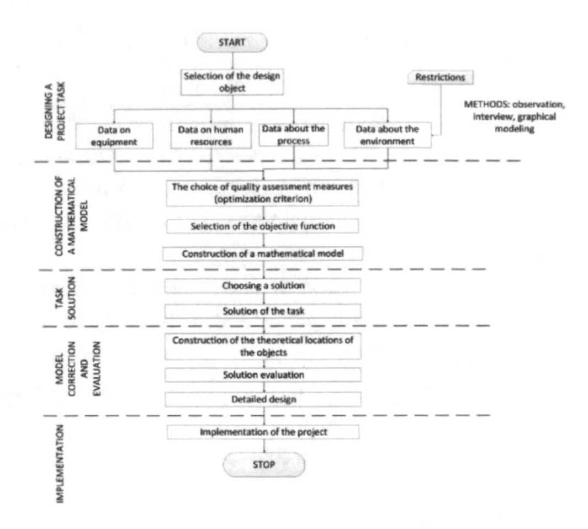

Fig. 1. Layout design/redesign general procedure [25].

The methods listed above represent selected approaches that developed over the last 70 years but the authors believe that there are problems not covered with solutions developed and there should be customized approaches tailored for specific problems, namely production processes determined by specific industry conditions and constraints.

4 The Framework of Tool Supporting Facility Layout Design

4.1 Brief Characteristic of a Company

The subject of the research is an industrial company. It produces furniture for freezing and heating equipment, as well as eaves and complete utility rooms made of stainless steel. All of these products are intended for passenger ships. For the client's needs, the company develops customized sets of furniture.

All the products must meet the USPHS (United States Public Health Service) and HHS (The United States Department of Health and Human Services) requirements. These guidelines refer to:

- the size of rooms and their equipment,
- finishing of the furniture and equipment,
- the assembly method of equipment elements,
- technical requirements of individual devices.

The company's offer is very wide and includes, among others: furniture for freezing (fridges, swimming pools, countertops), neutral furniture (tables, sinks, racks, cabinets, prams), furniture for heating (food Bain Marie, plate dispensers).

The following operations can be distinguished in the production process: cutting, bending, deburring, welding, bonding, soldering, assembly, filling, machining, grinding.

The division of labor analysis indicates a production system typical for shipbuilding industry. Elements typical for this type of system include extended production cycles, and brigade-based work system. The brigade-based work system means that the individual stages of the machining processes vary in terms of loads. The load depends on the production schedule, not planned in the long time horizon, which contributes to its variability. This leads to the situation in which employees cooperating within one technology-oriented cell are often delegated to various tasks. An example of such relocation was observed at welding and assembly stations. In addition, the system also includes product-oriented cells.

The company has, among others, the following workstations: special lasers, guillotines, presses, assembly tables, welding stations, grinding stations.

However, as a result of the reorganization in the company, a decision was made to eliminate one of the production facilities. The workstations from the facility were to be transferred to another already existing production facility.

The problems of designing the facility layout are examples of tasks with many variants of solutions. Arranging and N workstation results in N! possible solutions.

4.2 Framework of the Solution

There are no mathematical modeling methods to find an optimal or even satisfactory solution for the facility layout. There are many methods based on calculations, but they are not suitable for specific requirements of shipyard industry.

However, to illustrate the mathematically the layout modeled, the authors decided on the Modular Allocation Technique (MAT). This method is a representative of approximate, stepwise methods, with a limited number of locations.

The data for calculations are:

- N – the number of workstations, which is equal to the number of their potential locations,
- S – connections between locations resulting from transport process,
- L – connections between workstations.

The MAT is implemented to model the standard workstations allocation optimization task, which is equivalent to the zero-one quadratic programming model, as presented in the formula (1).

$$f(G) = \sum_{p=1}^{N-1} \sum_{q=p+1}^{N} S_{g(p)g(p)} * L_{pq} \quad (1)$$

The algorithm of the method is based on 2 types of data: connections between workstations and the flow of material (in the form of production batches) between them. The first type of data results from the production process technology. The data is available in the company, technology is defined and recorded. However, the second type of data needed to perform the MAT analysis could not be identified nor estimated in the company due to specific characteristics of shipyard industry.

Since it was impossible (because of lack of the data) to implement any specific calculation algorithm and the method to be applied was expected to be characterized by a short calculation time and the ability to evaluate the quality of the obtained solution, the authors have decided to implement the heuristic method.

Hence, a scheme was developed, which was a substantive basis for planning the facility layout. Its simplified algorithm is shown in Fig. 2.

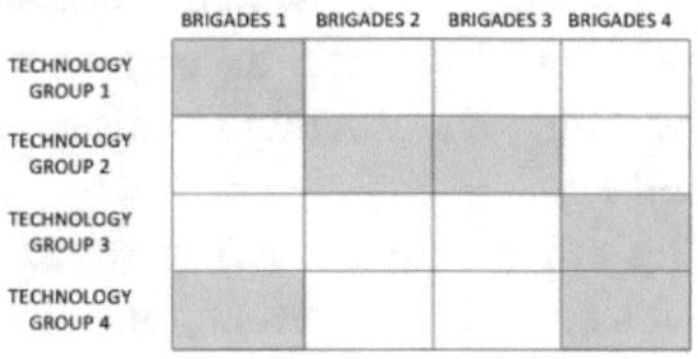

	BRIGADES 1	BRIGADES 2	BRIGADES 3	BRIGADES 4
TECHNOLOGY GROUP 1				
TECHNOLOGY GROUP 2				
TECHNOLOGY GROUP 3				
TECHNOLOGY GROUP 4				

Fig. 2. Work distribution among brigades

Figure 2 presents technological cells at vertical axis, while the production brigades are listed along the horizontal axis. The matrix designed in this way is the basis for developing the sequence of assigning brigades according to the technology of the production process. It is important from the designer's point of view to get to know the organization of production and technology thoroughly. Therefore, the IT tool should allow entering the information required and collected into the database.

The information is acquired with the interview with staff involved in production processes and observation of material flows on facility floor. Basing on data collected, conceptual diagrams were developed. The first diagram is a use case diagram. This diagram illustrates the area of application and also allows the framework design of the future facility layout to be developed (Fig. 3).

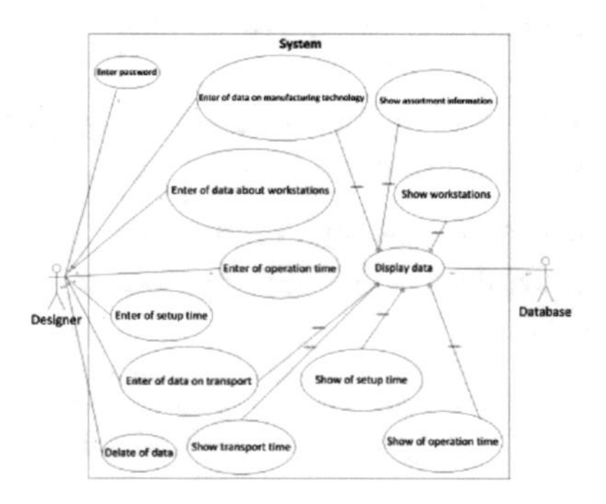

Fig. 3. Use case diagram

Users' interactions with use cases are presented in the Fig. 3. On the one hand, the user has specific roles towards the system, and on the other - use cases determine the services provided by the system to the actors (users). In the case of formulating a project task, a future user (designer) should be able to enter information about the technology (route). After searching for this data, the system should provide information related to the lead-time and workstation. In this way, the process of collecting data and analyzing them is supported.

The idea of collaboration between objects and sending messages between them is presented using the sequence diagram (Fig. 4). It encompasses actors and objects used to search for information, edit and enter it.

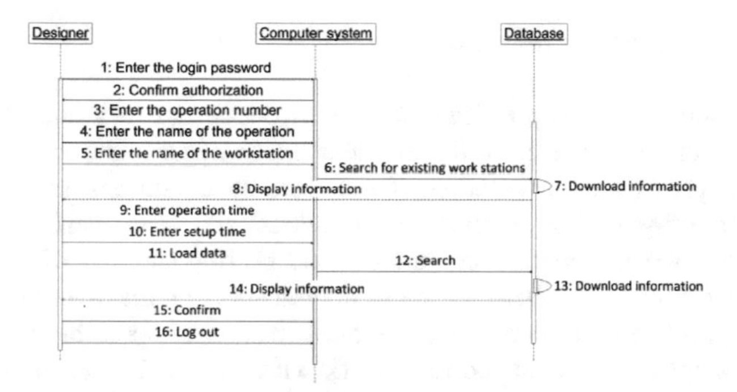

Fig. 4. Diagram of sequence

The user of the tool enters data on the route (technology) of the product. Information about performed operations is entered. The data on the machines used are important. After completing the information, the data is loaded. This stage is necessary to meet the assumed criteria according to which the optimal balancing of production capacities should take place and creation of potential possibilities to increase the workload of positions.

5 Functional Requirements of the Developed Tool

The large size of the tasks being solved and the complicated connections between the workstations being allocated results in the need to use information technologies. The purpose of developing a tool is primarily to improve the process of formulating a project task, and building a mathematical model.

The expected benefits should be the facilitating of basic information retrieval, as well as the introduction of data and proposed job positioning solutions. That is why the analysis of the requirements of future users was also an important step. Based on the interview, the following parameters were adopted:

- Editability,
- Simplicity of service,
- Cost,
- Availability.

The tool is to be easy to use so that the designer can use it without unnecessary specialized training. In addition, the process of formulating a project task and building a mathematical model (resulting from Fig. 1) should be short. The data entered should be saved in a database that would be accessed all the time and which would be constantly updated. An important feature is also the editability of the tool and editing of the data entered. Users should have access to the data themselves, including (technology) routes, and modify these data without major problems.

The user enters data on workstations and routes.

6 Conclusions and Outlook

Problems encountered in the practice of facility layout design are complex complicated. They refer not only to the size of the problem but to many additional conditions and restrictions imposed on the solution. These can be conditions resulting from the technology (multi-workshop service), construction and equipment of the building (power sources, special foundations, limited load bearing capacity of ceilings), health and safety conditions, and others. All these factors affecting the layout of workstations cause that the problem is not only difficult, but sometimes impossible to solve. In such situations it is convenient to support the work with information technology.

Implementation of the presented tool will certainly shorten the time of calculations and analyzes, which in effect will allow for a faster and more effective implementation of the solution in practice.

References

1. Santarek, K.: Podstawy metodyczne projektowania rozmieszczenia komórek produkcyjnych. Państwowe Wydawnictwo Naukowe, Warszawa (1987)
2. Gopalakrishnan, B., Turuvekere, R., Gupta, D.P.: Computer integrated facilities planning and design. Facilities **22**(7/8), 199–209 (2004)
3. Immer, J.R.: Layout Planning Techniques. McGraw-Hill, New York (1950)
4. Reed Jr., R.: Plant Layout: Factors, Principles and Techniques. Irwin, Homewood (1961)
5. Nadler, G.: What systems really are. Mod. Mater. Handl. **20**(7), 41–47 (1965)
6. Armour, G.C., Buffa, E.S.: A heuristic algorithm and simulation approach to relative location of facilities. Manag. Sci. **9**(2), 294–309 (1963)
7. Lee, R.C., Moore, J.M.: CORELAP: computerized relationship layout planning. J. Ind. Eng. **18**(3), 195–200 (1967)
8. Tompkins, J.A., Reed Jr., R.: An applied model for the facilities design problem. Int. J. Prod. Res. **14**(5), 583–595 (1976)
9. Seehof, J.M., Evans, W.O.: Automated layout design program. J. Ind. Eng. **18**(12), 690–695 (1967)
10. Edwards, H.K., Gillet, W.E., Hale, M.E.: Modular allocation technique (MAT). Manag. Sci. **17**, 161–169 (1970)
11. Apple, J.M.: Material Handling System Design. Ronald Press, New York (1972)
12. Scriabin, M., Vergin, R.C.: A cluster-analytic approach to facility layout. Manag. Sci. **31**(1), 33–49 (1985)
13. Kumar, S.R.T., Kashyap, R.L., Moodie, C.L.: Application of expert systems and pattern recognition methodology to facilities layout planning. Int. J. Prod. Res. **26**, 905–930 (1988)
14. Malakooti, B., Tsurushima, A.: An expert system using priorities for solving multiple criteria facility layout problem. Int. J. Prod. Res. **27**, 793–808 (1989)

15. Abdou, G., Dutta, S.P.: An integrated approach to facilities layout using expert system. Int. J. Prod. Res. **28**, 685–708 (1990)
16. Raoot, A.D., Rakshit, A.: A fuzzy approach to facilities layout planning. Int. J. Prod. Res. **29**, 835–857 (1991)
17. Shih, L.C., Enkawa, T., Itoh, K.: An AI-search technique-based layout planning method. Int. J. Prod. Res. **30**(12), 2839–2855 (1992)
18. Bozer, Y.A., Meller, R.D., Erlebacher, S.J.: An improvement type layout algorithm for single and multiple-floor facilities. Manag. Sci. **40**(7), 918–932 (1994)
19. Tompkins, J.A., White, J.A., Bozer, Y.A., Frazelle, E.H., Tanchoco, J.M.A., Trevino, J.: Facilities Planning. Wiley, New York (1996)
20. Eneyo, E.S., Pannirselvam, G.P.: The use of simulation in facility layout design: a practical consulting experience. In: Medeiros, D.J., Watson, E.F., Carson, J.S., Manivannan, M.S. (Eds) Proceedings of the 1998 Winter Simulation Conference, Washington DC, December 13–16, ACM Press, New York, pp. 1527–1532 (1998)
21. Ficko, M., Palcic, I.: Designing a layout using the modified triangle method, and genetic algorithms. Int. J. Simul. Model. **12**(4), 237–251 (2013)
22. Kanduc, T., Rodic, B.: Optimisation of machine layout using a force generated graph algorithm and simulated annealing. Int. J. Simul. Model. **15**(2), 275 (2016)
23. Solimanpur, M., Elmi, A.: A tabu search approach for cell scheduling problem with makespan criterion. Int. J. Prod. Econ. **141**(2), 639–645 (2013)
24. Berlec, T., Potocnik, P., Govekar, E., Starbek, M.: A method of production fine layout planning based on self-organising neural network clustering. Int. J. Prod. Res. **52**(24), 7209–7222 (2014)
25. Lis, St., Santarek, K.: Projektowanie rozmieszczenia stanowisk roboczych. Państwowe Wydawnictwo Naukowe, Warszawa (1980)

Virtual Commissioning as the Main Core of Industry 4.0 – Case Study in the Automotive Paint Shop

Jolanta Krystek, Sara Alszer(✉), and Szymon Bysko

Institute of Automatic Control, Faculty of Automatic Control, Electronics and Computer Science, Silesian University of Technology, Gliwice, Poland
{jolanta.krystek,sara.alszer,szymon.bysko}@polsl.pl

Abstract. The paper focuses on presentation of innovative paint shop concept for automotive industry, based on Industry 4.0 idea. Work begins with introduction of Virtual Engineering and Virtual Commissioning methods. Using virtual environments to test and verify new solutions allows to check them efficiency before implementing these ideas in the real plant. The main goal of the research is to improve paint shop efficiency and productivity. The simulation station, used for the purpose of research, is presented from hardware and software point of view. For the considered case study new selected heuristics, based on modification of existing ones, are proposed. The obtained results are compared and presented in the paper.

Keywords: Industry 4.0 · Automotive industry · Sequencing
Virtual Commissioning · Virtual Engineering

1 Introduction

In 2011, the German government and industrial centers initiated the Industry 4.0 revolution with the main idea of creating Smart Factory. For customers it means that in the future they will be able to order fully personalized products – this is the realization of the mass customization idea. In turn, from the producers' point of view, Smart Factory concept is the necessity to change the approach to production, e.g. centralizing all data about product and its production process only one time using modern and specialized IT tools. The main elements of the new industrial area include: cyber-physical systems, smart robots, new quality of connectivity, big data, energy efficiency and virtual industrialization, which is a subject of the paper.

The main current application areas of Smart Factory are following industries: automotive, mechanical engineering, aerospace and ship building as well as electronics and consumer goods. The circle of interests is still growing, because Smart Factory implementation results directly in economic and production indicators. Improvement and the investment in innovation solutions return after short period many additional benefits.

In the automotive industry, virtual methods are currently particularly popular. Virtual plants and products are designed to prepare physical production using simulation, verification and physical mapping. This approach is successfully applied to body shop and assembly line systems. Although this is an initial phase of building new lines or

© Springer Nature Switzerland AG 2019
A. Burduk et al. (Eds.): ISPEM 2018, AISC 835, pp. 370–379, 2019.
https://doi.org/10.1007/978-3-319-97490-3_36

modifying existing ones, because there is still a vast bridge between theoretical considerations and practical application. However, the development of Industry 4.0 forces practitioners and scientists to strengthen cooperation and to agree on contentious issues, as well as finding a common point in looking at problems occurring in the real world. Especially Virtual Commissioning is very interesting both from the academia and industry perspective. On the one hand, the innovative approach requires knowledge of process modeling principles, which are an integral part of academic thinking, and on the other hand, it is necessary to combine these theoretical methods with engineers' practice. Such approach allows bridging the qualification gap between academia and industry and providing complex analysis and solution to real problems. This is the reason, why this paper focuses on the capabilities of Virtual Commissioning and its adaptation to solve sequencing problem occurring in automotive paint shop.

In order to meet demands of modern industrial plants, the authors of the paper in cooperation with ProPoint Sp. z o.o. Sp. K. are currently conducting research on the problem mentioned above. All presented research results are based on actual data provided by one of the car factory. Due to protection of undisclosed information, the paper does not specify the location of the project and details of the proposed solution implementations.

The article is organized as follows. Section 2 introduces in details Virtual Engineering and Virtual Commissioning terms. Section 3 describes the analyzed case study and the possibilities of using virtual tools to solve the proposed problem. Section 4 contains experimental research and discussion of obtained results. The final Section concludes the paper and presents further researches.

2 Virtual Engineering and Commissioning

Developing and validating automated production systems using digital tools and methods increases quality and efficiency in ramp-up processes significantly.

Both methodologies, Virtual Engineering and Virtual Commissioning (Fig. 1), are state of the art for developing and validating production systems in automotive industry.

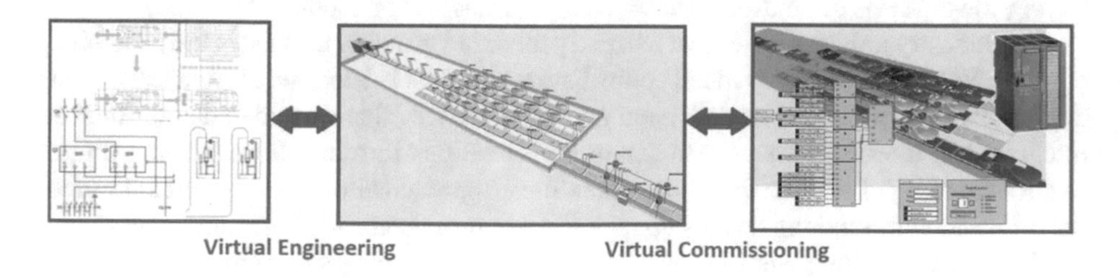

Fig. 1. Virtual Engineering and Virtual Commissioning

2.1 Virtual Engineering

The Virtual Engineering (VE) methodology consists in visualization and simulation-based validation of real system processes, cycle time and collisions for different product variants. For this purpose digital 3D design tools are used, which enable creating kinematic models of the entire manufacturing system without including its control software [1]. Today, model's quality is improved by using physics based modeling. Thanks to this, more realistic simulation results are achieved.

2.2 Virtual Commissioning

Kiefer et al. [2] defined Virtual Commissioning (VC) as a method intended for validating control engineering even before the real commissioning. According to Fratczak et al. [3] Virtual Commissioning is well established method for supporting the design and verification of the control systems based on simulation experiments. Performing the appropriate VC requires a previous modeling of this system – that model is called virtual model of the production plant [4]. The simulation is conducted in the configuration representing HiL situation (Hardware in the Loop) – a virtual plant and a real controller (usually implemented in the real PLC unit) are used.

This approach of developing control software has many benefits. Some of them are listed below [5]:

- higher quality of the control programs;
- saving of commissioning time;
- higher optimization capabilities;
- possibility for the plant owner to perform test and verification whether software fulfils his requirements and specifications;
- possibility to simulate design changes before their implementation in the real plant.

Due to many benefits of using VC, this method has become standard in the development process of automated production plants [6]. Over the last years VC has been successfully applied in the manufacturing systems, (see e.g. [7, 8]), particularly for testing lower levels of control systems, such as transport or robot cells.

The industry trend indicates that currently there is a growing interest in the possibility of using VC to test more advanced control systems. This is because, all functional tests are first carried out in VC environment and only the tested and validated solutions are integrated in the real systems. Thus, any modifications introduced on the line do not require stopping of the line for a long time. Such approach allows for significant reduction of real commissioning time by about 75% in comparison with conventional commissioning directly at the plant [9]. As a result, increasing savings of both the vendors of the control system and their industrial partners (manufactures). The risk of damaging or destroying the mechanical, electrical devices or the entire systems is also reduced. In the ideal case, the risk is completely eliminated.

These powerful virtual tools are today applied to the design and analysis of complex systems in a body shop and on an assembly line, but for a paint shop virtual methods must be improved due to complicated and difficult painting process.

3 Case Study

Based on data provided by one of the polish company, in 2010 the factory produced 533 455 vehicles. It means that daily 2320 cars leave production line – that is one car per 33 s. Achievement of such result is possible only under condition of use of effective methods of production system management. Both monitoring of process and proper organization of production play here a very important role. Considering the fact that the costs incurred for each minute of downtime can reach up to 10 000 euro, it may be asked what should be a sequence of producing cars (how to organize and plan the production) to ensure maximum throughput of the production line and offer the largest variety of cars (models and equipment). This issue, defined as the Car Sequencing Problem (CSP), has been first described by Parello et al. in 1986 [10]. Due to the growing requirements of the automotive industry, the new sequencing problem was proposed by the authors. All modifications and assumptions made for this problem have been presented in the following articles [11–13].

The currently observed tendency to transform factories into fully automated plants carries the risk of leading to so-called over-automation, which becomes the main cause of the financial problems of car companies – the costs of automation some processes often outweigh the costs of maintaining employees performing these tasks manually. Therefore, it is reasonable to simulate innovative solutions before they are actually implemented.

This section describes approaches to the proposed sequencing problem based on the use of Virtual Reality tools.

3.1 Structure of Simulation Station

The general structure of the simulation station is presented in Fig. 2. The smart buffer control system, considered as integration of advanced sequencing algorithm with buffer

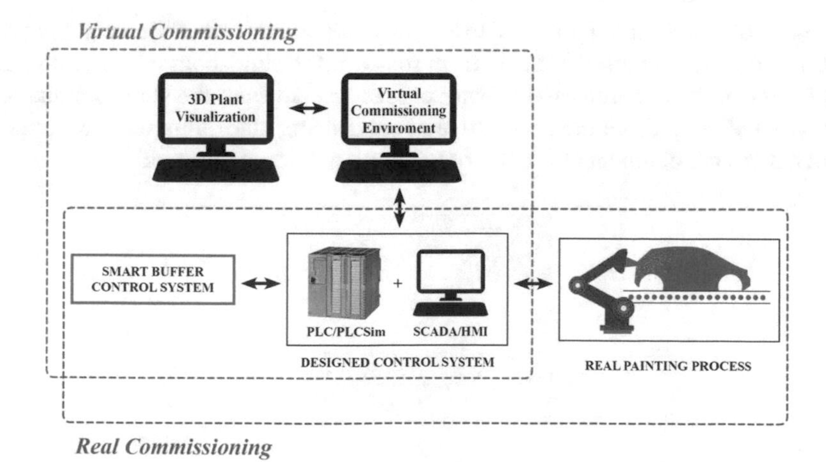

Fig. 2. Structure of the simulation station

management system, is directly connected to the designed control system which includes PLC/PLCSim and SCADA/HMI. It is connected by network to the PC where the virtual commissioning environment simulates the distributed I/O system. The 3D Plant Visualization is used in order to present the real painting process. Such approach allows transferring the software tested during the virtual commissioning into the hardware in the real plant.

3.2 Virtual Commissioning Environment

Validation of the proposed buffer control system was carried out using PLC program from a real object that had been commissioned in the past by the ProPoint company. In the conducted research, historical data were used, which consisted of sequences in which car bodies were transported between body and paint shops. The data contained information such as model and color of car bodies.

In order to develop a functional model of the buffer, WinMOD® environment was used. Based on the library of components and micro-components, created by the company as part of previous projects, a model of the transport system used in the analyzed paint shop was built. At the stage of developing the buffer model, additional models were designed: one-way skid conveyors (Fig. 3) and so called shuttle (Fig. 4).

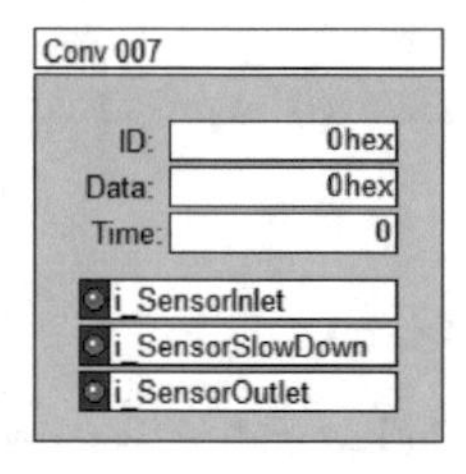

Fig. 3. A one-way skid conveyor component

Among the information provided to the conveyor component, following information input data can be distinguished: data from sensor informing about presence of skids in the area of conveyor, permission to enter or leave a skid from the conveyor and movement speed of skid. From the perspective of sequencing algorithm, the most important output data of the component is color of car body passing on the skid.

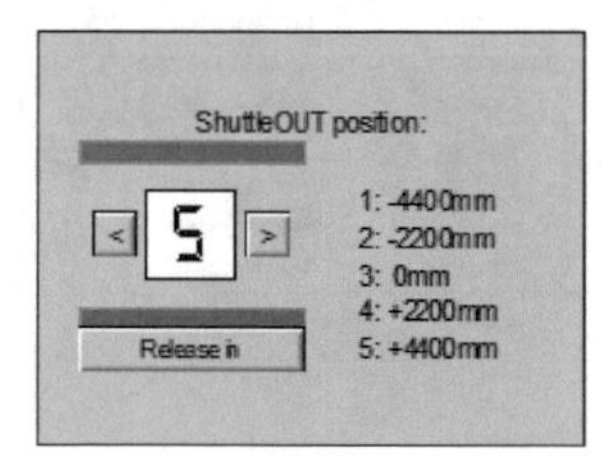

Fig. 4. Shuttle component

The second of the proposed components was designed to set the shuttle in the desired position (developed by the sequencing algorithm), so that car body could be loaded or unloaded from the appropriate position of buffer.

3.3 3D Plant Visualization

In order to facilitate analysis of transport system and buffer performance, their models were created in SIMLINE® environment with the appropriate kinematics. This allowed visualization of current state of buffer and car flow (Fig. 5).

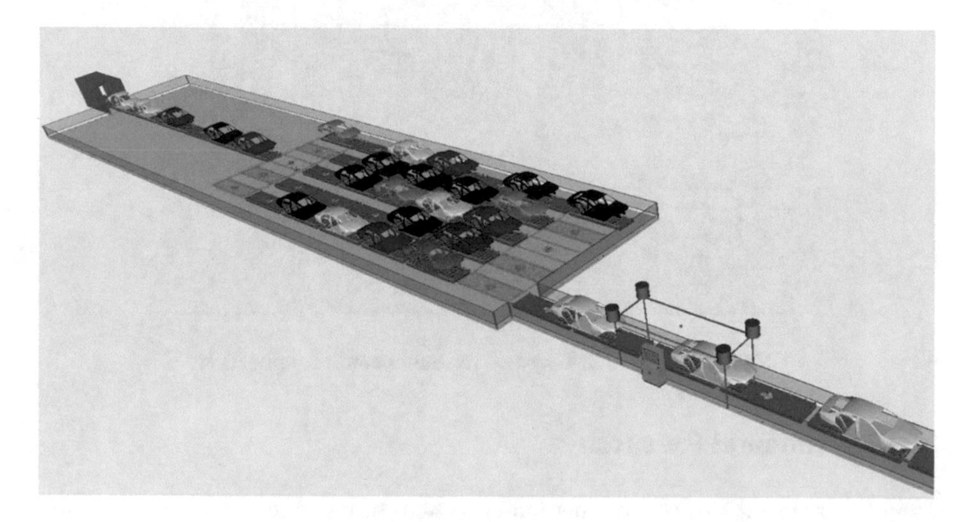

Fig. 5. Visualization of buffer and painting station

3.4 SCADA/HMI

The simple SCADA/HMI system for the buffer control application was designed in Visual Studio 2015 and consists of communication and information modules. The designed visualization window is presented in Fig. 6.

The SCADA/HMI application allows the user to choose both a loading and unloading sequencing algorithm. It is possible to activate an additional option “With Time Priorities”, which takes into account in the sequencing process time constraints resulting from production plan (Just in Time production). The system also provides basic information about the effectiveness of the selected configuration of sequencing algorithms and enables monitoring of production progress.

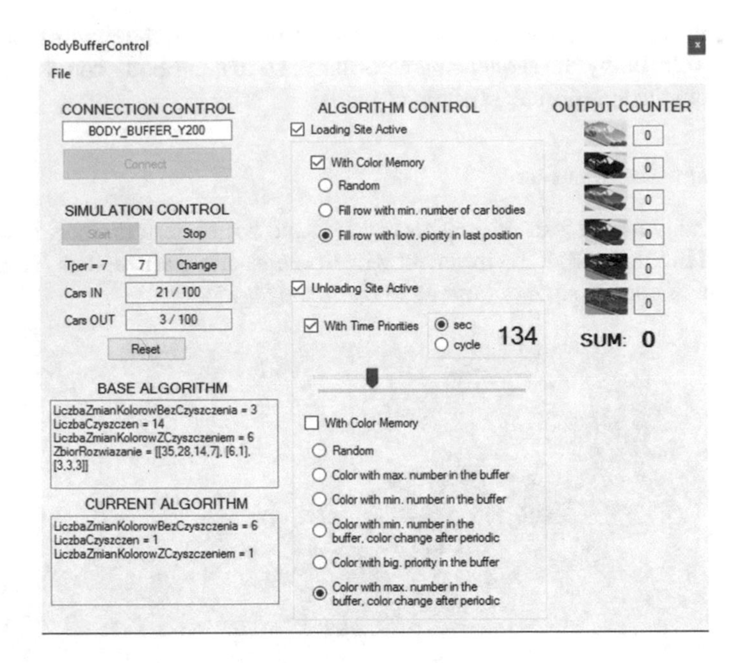

Fig. 6. SCADA/HMI system for buffer control application

4 Experimental Research

Using the designed Virtual Commissioning system series of tests were carried out in order to determine the effectiveness of the proposed sequencing algorithm understood as the number of necessary color changes. All assumptions made for the need of conducted research and selected test results are presented in this section.

4.1 Case Study Assumptions

The case study assumed consideration of buffer described in Sect. 3. The buffer consisted of 25 positions (5 × 5), intended for car body buffering, and of 2 columns used to transport the car bodies to the correct row. Following sequencing algorithms were tested:

- Conf02_03 – LP (Lowest Priority) algorithm,
- Conf12_03 – LP algorithm with Color Memory (CM) rule on buffer input,
- Conf02_13 – LP algorithm with CM rule on buffer output,
- Conf12_13 – LP algorithm with CM rule on buffer input and output.

The principle of the Lowest Priority algorithm is as follows: car located at the buffer input is directed to this row ended on the car in lowest priority color. To the buffer output is directed the car in lowest priority color. The color priority is determined based on the current buffer state. The greater the/number of a given color in the buffer, the higher the priority.

Color Memory rule ensures creating the longest possible color blocks. Thus, applying this rule at the buffer input means that the car is directed to the row ended on the car with the same color as the color of input car. In turn, unloading the buffer according to the CM rule takes place as follows: to the buffer output is directed this car, which is in the same color as the color of currently painted car.

For the purpose of the research, 5 sets of historical data were used. Each set consisted of 100 cars painted in 6 different colors. The colors distribution in each set was the same, as follows: C1: 6%, C2: 38%, C3: 29%, C4: 14%, C5: 10%, C6: 3%. In order to ensure a good quality of painting, after every 7 car bodies the necessary periodic cleaning of guns was carried out.

4.2 Experimental Results and Discussion

The results of the commissioning tests are presented in Table 1 and discussed from the perspective of practical requirements for industrial control systems.

Table 1. Experimental results – number of color changes

Data set No.	Without buffer	Manual control	Conf02_03	Conf12_03	Conf02_13	Conf12_13
Data_01	57	50	61	45	34	30
Data_02	68	64	66	59	36	33
Data_03	60	55	58	46	39	27
Data_04	61	60	63	51	31	27
Data_05	57	54	62	46	40	28

Based on the obtained results, it can be observed that the simplest algorithm (Conf02_03) is not effective enough. After algorithm modification reduction of color changes can be observed. These improvements consisted of including periodic cleaning, occurring in paint shop and ensuring correct paint quality and creating the longest possible color blocks on the buffer output. The sequencing efficiency improved almost twice for the most advanced algorithm (Conf12_13) as compared to the results obtained for manual control.

Despite efficiency improvement of car sequencing, it is also possible to achieve higher productivity of the line. Assuming that the cycle time in the paint shop is 65 s, maximum 55 body cars are painted over an hour. In the case of manual buffer control, it should be taken into account that the operator works around 75% of the full shift time – full 8 h are reduced to 6 what is caused by necessary breaks. As a result, the use of manual control allows releasing 330 car bodies from the paint shop. On the other hand, automatic mode could operate continuously all 8 h without any breaks. Thanks to that, it is possible to increase the line efficiency to 440 car bodies per shift. If the working mode is 3-shift, more about 300 body parts could be produced per day in comparison with manual buffer control.

5 Conclusions

The paper presents practical application of virtual environments tools for automotive industry. Such approaches are successfully used in body shop lines, but the possibility of using them for more advanced processes, for example process of painting, is still the subject of many studies. The main goal of the carried out research is to use above mentioned tools to implement and test novel car sequencing algorithms. Results presented in the paper provide proof that the simple heuristics are not efficient enough and there is still the need to modify them or develop complete new ones. The subject for future research is to propose more advanced algorithm and test it in the virtual environment before applying it on the real plant. Such an approach allows for significant reduction of real commissioning time and to avoid breakdowns during normal production cycle. An original and complex sequencing algorithm called Follow-up Sequencing Algorithm (FuSA), created by the authors is still developed and tested, gives promising results. However, the novel method needs to be improved in order to publish the first obtained results.

Acknowledgements. This work has been supported by Polish Ministry of Science and Higher Education under internal grants: BK-204/RAu1/2017, BKM-508/RAu1/2017 and 10/DW/201701/01 for Institute of Automatic Control, Silesian University of Technology, Gliwice, Poland.

References

1. Damrath, F., Strahilov, A., Bär, T., Vielhaber, M.: Establishing energy efficiency as criterion for virtual commissioning of automated assembly systems. In: Proceeding of Conference on Assembly Technologies and Systems, pp. 137–142 (2014). https://doi.org/10.1016/j.procir.2014.10.082
2. Kiefer, J., Olinger, L., Bergert, M.: Virtuelle Inbetriebnahme – Standardisierte Verhaltensmodellierung mechatronischer Betriebsmittel im automobilen Karosserierohbau. atp – Automatisierungstechnische Praxis **51**(7), 40–46 (2009). https://doi.org/10.17560/atp.v51i07.92
3. Fratczak, M. et al.: Virtual commissioning for the control of the continuous industrial processes – case study. In: Proceedings of the 20th International Conference on Methods and Models in Automation and Robotics, Międzyzdroje, pp. 1032–1037 (2015). https://doi.org/10.1109/mmar.2015.7284021
4. Strahilov, A., Ovtcharova, J., Bär, T.: Development of the physics-based assembly system model for the mechatronic validation of automated assembly systems. In: Proceedings of the Winter Simulation Conference, pp. 1–11, Berlin (2012). https://doi.org/10.1109/wsc.2012.6465049
5. Süß, S., Strahilov, A., Diedrich, C.: Behaviour simulation for virtual commissioning using co-simulation. In: Proceedings of the 2015 IEEE 20th Conference on Emerging Technologies and Factory Automation, Luxembourg, pp. 1–8 (2015). https://doi.org/10.1109/etfa.2015.7301427
6. Oppelt, M., Barth, M., Urbas, L.: The Role of Simulation within the Life-Cycle of a Process Plant. Technical report (2015). https://doi.org/10.13140/2.1.2620.7523

7. Hoffmann, P., Schumann, R., Maksoud, T.M.A., Premier, G.C.: Virtual commissioning of manufacturing systems a review and new approaches for simplification. In: Proceedings of the 24th European Conference on Modelling and Simulation, Kuala Lumpur, pp. 175–181 (2010). https://doi.org/10.7148/2010-0175-0181
8. Lee, C.G., Park, S.C.: Survey on the virtual commissioning of manufacturing systems. J. Comput. Des. Eng. **1**(3), 213–222 (2014). https://doi.org/10.7315/JCDE.2014.021
9. Koo, L.J., Park, C.M., Lee, C.H., Park, S.C., Wang, G.N.: Simulation framework for the verification of PLC programs in automobile industries. Int. J. Prod. Res. **49**(16), 4925–4943 (2011). https://doi.org/10.1080/00207543.2010.492404
10. Parello, B.D., Kabat, W.C.: Job-shop scheduling using automated reasoning: a case study of the car sequencing problem. J. Autom. Reason. **2**(1), 1–42 (1986)
11. Alszer, S., Krystek, J.: Contemporary aspects of car sequencing problems in a paint shop. Mechanik **7**, 527–529 (2017). https://doi.org/10.17814/mechanik.2017.7.67
12. Alszer, S., Krystek, J.: Car sequencing problem – confrontation with real automotive industry. In: CLC 2017 - Carpathian Logistics Congress - Congress Proceedings, Liptovsky Jan, pp. 240–245 (2017)
13. Alszer, S., Krystek, J.: The algorithm of buffers handling in car sequencing problem presented on an actual production line. In: 24th International Conference on Production Research (ICPR2017), Poznań, pp. 277–282 (2017)

Computer Aiding in Production Engineering

Computer Aiding Simulation of the Mechatronics Function of an Intelligent Building

Aleksander Gwiazda[1(✉)], Krzysztof Herbuś[1], Piotr Ociepka[1], and Małgorzata Sokół[2]

[1] Faculty of Mechanical Engineering, Silesian University of Technology, Konarskiego 18A str., 44-100 Gliwice, Poland
aleksander.gwiazda@polsl.pl
[2] Faculty of Architecture, Silesian University of Technology, Akademicka 7 str., 44-100 Gliwice, Poland

Abstract. The current stage of development of modern electronic and informatics solutions allow creating sophisticated solutions in different areas of human life. It is particularly related with designing complex technical means. The approach allowing creating intelligent systems is based on the Industry 4.0 concept. It enforces creating new solutions allowing integrating different cooperating components. In the paper is presented the virtual analysis of introducing concepts of mechatronics design and Industry 4.0 to creating intelligent buildings. This area could be determined as buildtronics. In this area the concept of Industry 4.0 allows developing new design methods of unconventional structural solutions of systems with characteristics of a complex character. It is realized by facilitating the integration of different subsystems of different components creating the complex technical means. However the concept of mechatronics allows creating components based on the mechatronics subsystems (mechanical, electric and control). The cooperation of these subsystems is determined as the mechatronics function of a system. Results show that the computer aided designing and modeling of sophisticated technical systems could be strongly facilitated and accelerated utilizing the virtual analysis of the mechatronic function.

Keywords: Buildtronics · Industry 4.0 · Mechatronics function

1 Introduction

1.1 Mechatronics Function

Mechatronics function is the complex function of operating of a complex technical means taking into account its mechanical, electric and control subsystems. In modern CAE environment it is possible to simulate the mechatronics function of a technical means. It reflects the complexity of a whole mechatronics system, which is a combination of its mechanical operation, electric drive functioning, computers control activity as well as communication. In modern application the communication function

© Springer Nature Switzerland AG 2019
A. Burduk et al. (Eds.): ISPEM 2018, AISC 835, pp. 383–390, 2019.
https://doi.org/10.1007/978-3-319-97490-3_37

is realized using the approach of the Internet of Things [1]. The mechatronics function could be simulated in advanced CAE environments [2]. Hence it is need to define the mechanical joints of structural components, the characteristics of drives as well as control functions to simulate this function. Mechatronics solutions allow improving power efficiency of controlled system [3]. It is gained by providing more flexible operation cycles. It leads to the higher respect for the environment. On the other hand if failures are detected the mechatronics device could take the proper action to decrease the damage rate.

The mechatronics design process let facilitating the preparation, conceptual phase of any project related to complex systems like a building. It also allows meeting specific client needs by personalizing the control functions. This is why the analysis of the mechatronics function becomes more and more important element of designing processes of complex systems.

1.2 Buildtronics Concept

The concept of a buildtronics approach is a form of transformation of mechatronics solutions to the area of intelligent buildings designing. It was assumed that an intelligent building is a combination of structural elements, drives and activators components and the control subsystem. In this way the mechatronics solution could be applied to this area. It should be stated that the drive systems and control ones relates to the intelligent component of a building. It is mainly related with such systems like: electric energy one including renewable energy sources, water system and other media, heat system, device control system as well as security system including door locks, shutters, cameras and communication devices. All these elements are linked using the Internet utilizing the concept Internet of Things. This allows facilitating the creation of intelligent connection of all various components.

Moreover it is possible to realize embedded systems among other by designing intelligent modules of such buildings linked with approach. The embedded system means that the building components and modules (windows, doors, airbricks, tiles, walls, and staircases) could be designed as mechatronic solutions including sensors, their connections and special activators. In this case it is not needed to individually design and construct all this systems one by one.

Taking into account this explanation the term buildtronics means creating intelligent and integrated solutions from the area of building constructing that include such subsystems like: structural, drive and energetic, as well as control one. The last one bases on computer and informatics solutions.

2 Virtual Analysis of an Intelligent Building

2.1 Concept of the Mechatronics Function of a Building

The analysis presented in the paper is based on the project of an estate presented in Fig. 1. The estate consists of a detached, multi-station garage and a single-family house. The gate is also a port of the system.

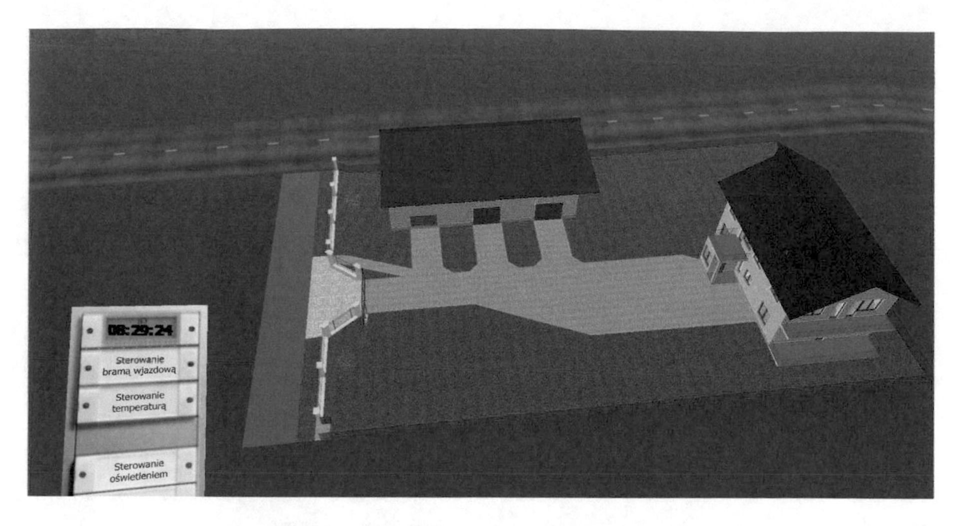

Fig. 1. Aerial view of the designed estate

The proposed system cooperates with the virtual controller system. Part of its virtual window is seen on the left side. It allows controlling all important parts of the building and all determine subsystems.

2.2 Operation of Particular Systems

The operation of all particular systems is virtually controlled like the system of the gate. The opening of the gate is controlled by a sensor mounted on a car (Fig. 2). In Fig. 3 is presented the control box of the gate. The user of the system could analyze the system of connection as well as change it.

Fig. 2. Gate operation

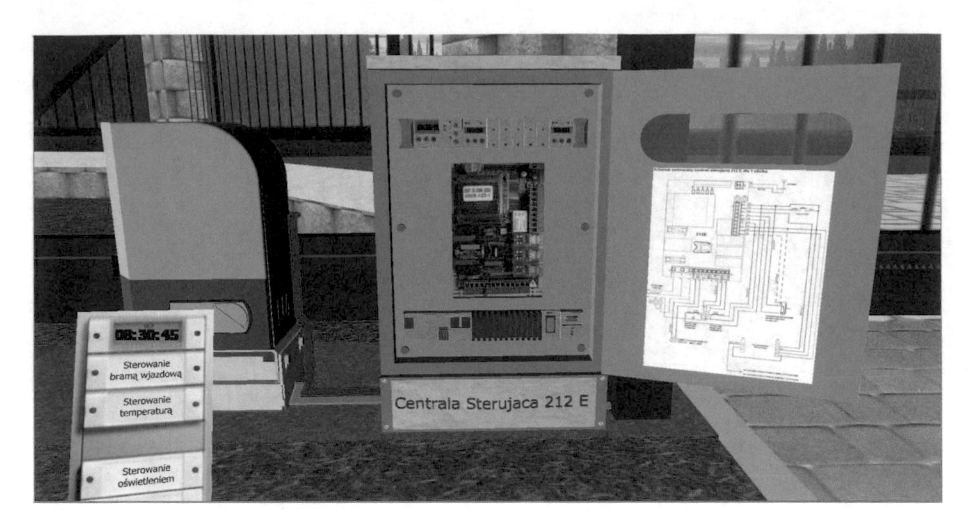

Fig. 3. Virtual gate control system

The control systems of other components of the presented intelligent house are designed in a similar manner. In a virtual controller are modeled all control units, connections and activators. In Fig. 3 are visible: control box with the scheme of connections and the activator of the horizontal (line) movement of the gate. The control box was modeled using a computer aided program for automated systems. It allows determine the control procedures realized by the given controller. The activator system was modeled in an advanced CAD program. It was realized as a 3D model that was attributed with information relating the joints of particular model components. In Fig. 4 is presented the operation of a garage door, another system of that kind.

Fig. 4. Garage doors operation

2.3 Activators Modeling

As it was stated all activators are modeled as 3D solid models including all important components related with their operation. In Fig. 5 is presented the operation of the garage doors. For better visualization the structural elements like walls have been switched of. The composition of this system consists of all elements that are important taking into account its functionality.

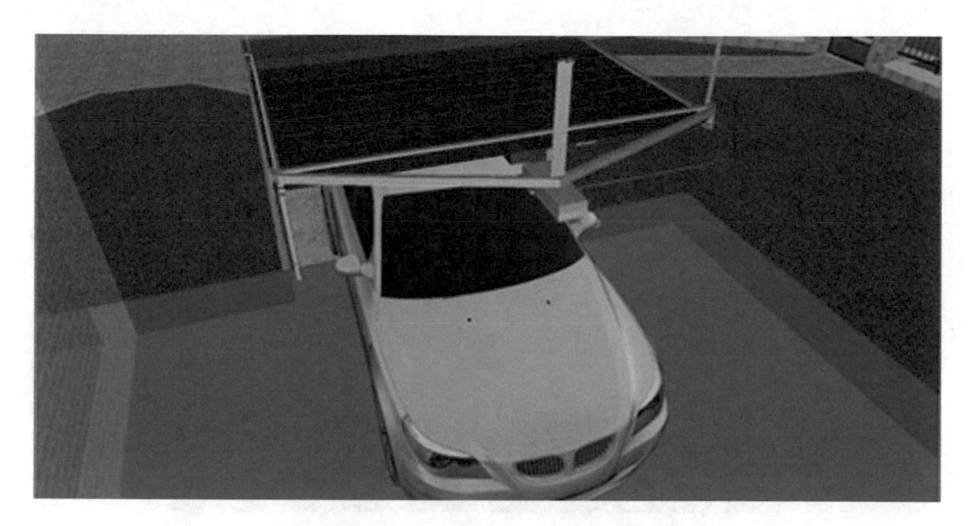

Fig. 5. Garage door activator system

The same idea is used for modeling other system. In Fig. 6 is presented the control unit of the floor heating system. Plumbing is hidden in the wall.

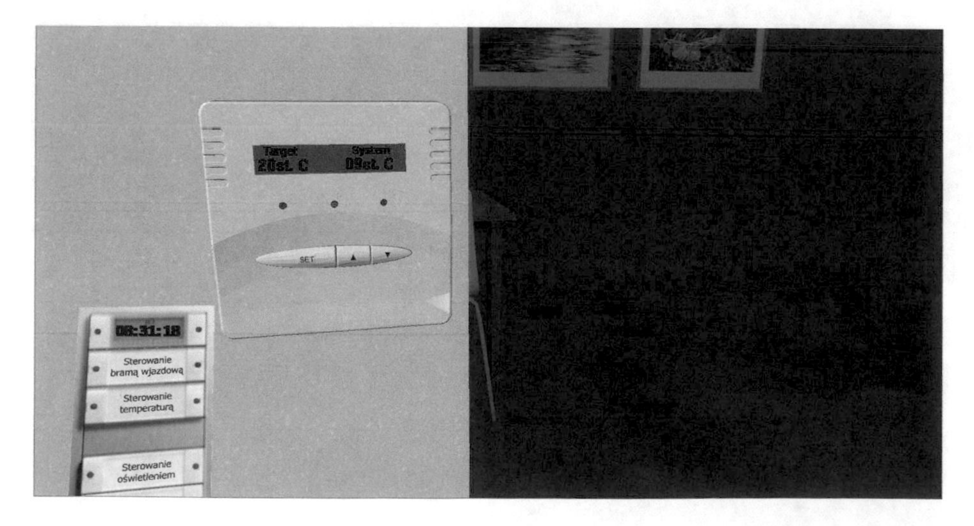

Fig. 6. Floor heating system

2.4 Central Control Module

The central control was located in a separate room and it visualizes functioning of all system of the building. In Fig. 7 is shown the operation of the security system (it is possible to look all images from the monitoring cameras). In this room is located the monitor of the central control system. So it is possible to control all operations related with all functions of this intelligent building.

Fig. 7. View of the control room

Below (Fig. 8) are presented chosen schemes of electric connections (floor heating and lighting) that are available to the user.

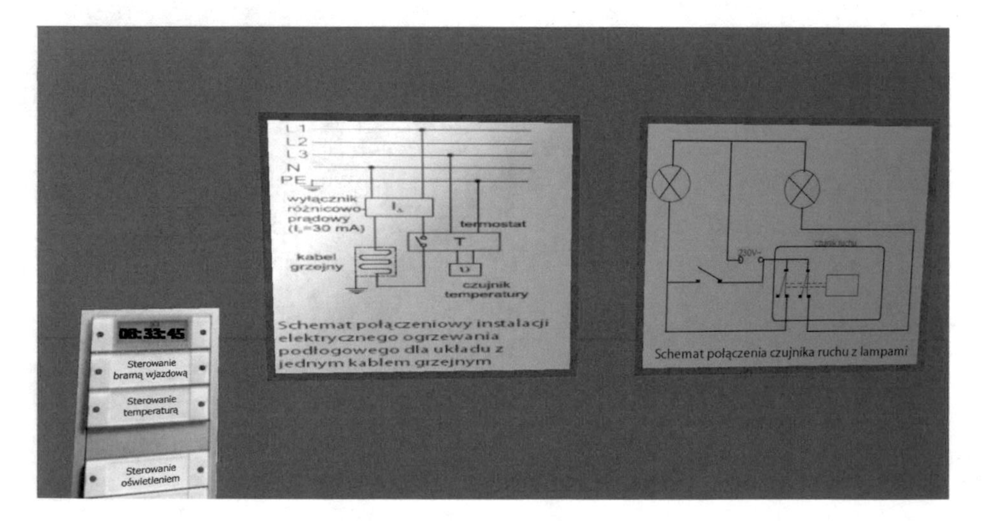

Fig. 8. Electric connections schemes

The user, at this stage, could directly from the program could change the settings of all control components as well as change the system of connections. However the change of structural components as well as the drives operation form is related with the need of changing the CAD models and their attributes.

2.5 Buildings Modeling

All virtual models of buildings were elaborated in an advanced CAD/CAE program. For the purpose of this system only the important components have been modeled. It is of course possible to utilize ready-made components from architectural libraries like windows, doors. It allows increasing the modeling phase in Fig. 9.

Fig. 9. Modeled buildings

The elaborated models must be connected with specific attributes allowing virtual controlling of all elements. The integration scheme of all virtual environments needed to obtain the reliable simulation results is presented in Fig. 10. In this scheme are also presented examples of utilized informatics platforms.

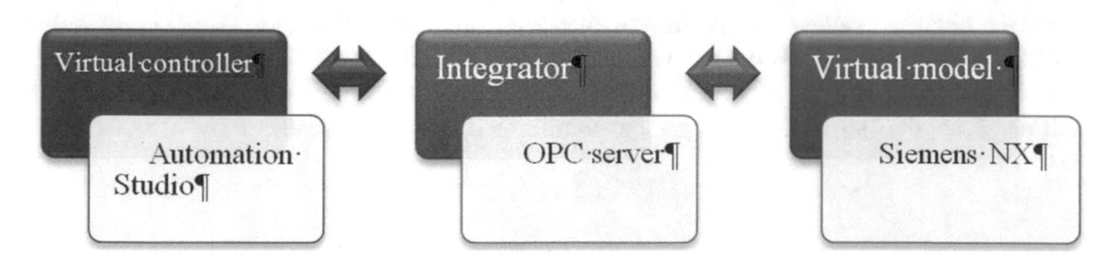

Fig. 10. Integration of the CAD model with the control system

The presented integration approach to mechatronics devices simulation has been verified in many works conducted in authors unit [4–7].

3 Conclusions

The use of mechatronic design method to describe the 3D models of the intelligent building allows preparing the simulation model for reliable analysis of their operation. The created building control panel, using the automation software, enables the verification of the building functioning in the area of its intelligent functions taking into consideration the possibility of reaching the desired sequence of actions [8, 9].

In order to speed up the work related to the preparation of the model of the building intelligent system for virtual operation, it is necessary to develop the database of intelligent building subsystems. The application of the integrator, in the form of the OPC server, enables the exchange of information between the virtual model of the building and its virtually controlled subsystems.

References

1. Kurniawan, A.: Smart Internet of Things Projects. Packt Publishing, Birmingham (2016)
2. Lückel, J.: The concept of mechatronics function modules applied to compound active suspension systems. In: Proceedings of Symposium on "Research Issues in Automotive Integrated Chassis Control Systems", International Symposium for Vehicle System Dynamics, pp. 1–6, Herbertov (1992)
3. Bolton, W.: Mechatronics: Electronic Control Systems in Mechanical and Electrical Engineering. Pearson Education, Harlow (2016)
4. Herbuś, K., Ociepka, P.: Determining of a robot workspace using the integration of a CAD system with a virtual control system. In: IOP Conference. Materials Science and Engineering, vol. 145, pp. 1–8 (2016). https://doi.org/10.1088/1757-899x/145/5/052010
5. Herbuś, K., Ociepka, P., Gwiazda, A.: Conception of the integration of the virtual robot model with the control system. Adv. Mater. Res. **1036**, 732–736 (2014). https://doi.org/10.4028/www.scientific.net/AMR.1036.732
6. Tikanmäki, A., Vallius, T., Röning, J.: Qutie - modular methods for building complex mechatronic systems. In: Mechatronics for Safety, Security and Dependability in a New Era, pp. 427–430 (2007). https://doi.org/10.1016/b978-008044963-0/50086-2
7. Sangregorio, P., Cologni, A.L., Piccinini, A., Scarpellini, A., Previdi, F.: A method for automation software design of mechatronic systems in manufacturing. IFAC-PapersOnLine **48**(3), 936–941 (2015). https://doi.org/10.1016/j.ifacol.2015.06.203
8. Gwiazda, A., Herbuś, K., Kost, G., Ociepka, P.: Motion analysis of mechatronic equipment considering the example of the Stewart platform. Solid State Phenom. **220**(221), 479–484 (2015). https://doi.org/10.4028/www.scientific.net/SSP.220-221.479
9. Bobtsov, A.A., Borgul, A.S.: Human-machine interface for mechatronic devices control. IFAC Proc. **46**(9), 614–618 (2013). https://doi.org/10.3182/20130619-3-RU-3018.00494

Practical Approach of Flexible Job Shop Scheduling Using Costs and Finishing Times of Operations

Małgorzata Olender([✉]), Krzysztof Kalinowski, and Cezary Grabowik

Faculty of Mechanical Engineering, Institute of Engineering Processes Automation and Integrated Manufacturing Systems, Silesian University of Technology, Konarskiego 18AStr., 44-100 Gliwice, Poland
malgorzata.olender@polsl.pl

Abstract. Nowadays, companies have to react dynamically to changes in the demand of a products. Therefore, the production processes should be properly planned. The problem arises when there are several options, but it is not known which version to choose. By using the KbRS tool, sample schedules for a certain production processes were shown. In addition, the program is not only focused on the achieved maximum process times, but also on the module related to the costs. Thanks to this solution, the decision-maker obtains information not only about times, but also costs. Of course, there are also specific constraints, which make a set of solutions. At this point, depending on the situation examined, which decision making people determines, which solution to choose. It can be a time-related version or a cost-related version. This is important because, manufacturers receive a signal from the market that informs about the need of a personalized products. This needs are characteristic of the so-called fourth industrial revolution, in which customers are looking for a personalized products. But also this revolution is connected with the integration: machines, people and computer systems. And next the price of personalized products are close to the price obtained in mass production. Therefore, apart from integration and security, there is need to have tools that will help planning production processes in dynamic changing needs.

Keywords: Manufacturing · Scheduling · Production planning

1 Introduction

Nowadays, customer's requirements are constantly changing. In result, companies need to be more responsive to what is happening on the market. This requires continuous changes, not only on the production line, or in value of working employees, but also in production schedules and plans. More and more personalized products force the manufacturers to react quickly and change their plans correctly. In addition to the available production resources, the production processes should be planned correctly. The problem arises, when the next personalized order arrives and the production plan must be continued. At this point, the manufacturer decides whether to resign from the new order, but then producer lose customers, or accept order for implementation and

© Springer Nature Switzerland AG 2019
A. Burduk et al. (Eds.): ISPEM 2018, AISC 835, pp. 391–400, 2019.
https://doi.org/10.1007/978-3-319-97490-3_38

try to do all production processes from the order [1–3]. Without a properly developed plan, with even the best machines, producers are not able to make task. In moment of developing a plan, decision - making person should follow the existing assumptions and constraints, which will help to make a good plan. Only, which solution is good? Which methods to follow? [4–6].

In article used the KbRS tool to schedule production. The research methodology is based on the scheduling in KbRS tool. It is a system from my Division in which can solve the problem with planning and scheduling production processes. The tests carried out and several scenarios are generated. Based on the obtained results and knowing the limitations for the existing production process/processes, solutions which can be used are showed. At the same time, these solutions differ in different versions related to the finishing process time and costs.

2 Production Planning

Production planning processes is a complicated decision - making process. The person who makes the plans has to face many problems related to the planning production and the limitations, which appear. This limitations for example are: production resources, components required for production, tools or human resources. All these elements create the whole picture of production processes. The problem is, when the schedule is planning on basis free machines and the plan should be drawn up for continue production. Depending on the problems connected with scheduling tasks, many works were created, but in general they can be divided into three categories [6–8]:

– problems related to scheduling in job shop systems,
– stochastic scheduling problems,
– issues related to the dynamism of production processes.

The development of the schedule is associated not only with the selection of machines and times, but also with deciding, whether it is possible to accept another production order with several production processes. Scheduling is important for making production orders in a timely manner, but also very important are the tools, which are helping producers to make a decisions. Currently, production is more and more often characterized by higher dynamics. This is due to the fact that customers want to have a personalized product. This causes tensions by making scheduling, because it is necessary to verify production plans of processes that are not completely repeatable. With the development of new technologies, producers need to have tools that will support decision-making people. With the increasing of personalization of the product, it is important to keep up with the customer's requirements, which change dynamically. Important point for production planning is the way of accessing production data. These data are processed in real time, which is why relevant tools that help with planning production processes and support decision - making person are so important.

In article is presented the problem related with job scheduling. Tasks are ordered on the basis of technological limitations. However, the problem concerns both on the definition of an acceptable solution and objective criteria. It usually concerns on the minimizing finishing time for all operations of the considered process but the example discussed also about cost issues of operations [6, 9–11]. To support the decision, the KbRS program was used in which schedules can be prepared. In addition, the program was developed with a module related to the costs generated depending on the chosen variant. The decision-making person got information, for example: the length of the production cycle. On this basis, this person can decide, which option is better to choose. Decision – making person can also choose production costs, which in this case are inversely proportional to times. Depending on the type of need, the information obtained supports decision making by the decision - making person at a given moment. Of course, the choice also depends on the extreme conditions, because it's define the area of the final results that can be used.

3 Example

The input data to the system is wrote in Table 1. This is a job shop system with parallel machines. Each of P1–5 processes include operations that are described by time and cost. Input data was generated randomly. Assumed: times 1–10, costs 1–10, inversely proportional to times. Each operation can be carried out on two alternative machines.

Table 1. The summary of times and costs of the operations

	t (time)					k (cost)				
P1	Op1	Op2	Op3	Op4	Op5	Op1	Op2	Op3	Op4	Op5
M1	9		7			2		4		
M2	1			2		10			9	
M3		2		9			9		2	
M4		6			6		5			5
M5			5		3			6		8
P2	Op1	Op2	Op3	Op4	Op5	Op1	Op2	Op3	Op4	Op5
M1			7	7				4	4	
M2				8					3	
M3		5			7		6			4
M4	6		4		7	5		7		4
M5	5	6				6	5			
P3	Op1	Op2	Op3	Op4	Op5	Op1	Op2	Op3	Op4	Op5
M1	5		7			6		4		
M2				8					3	
M3		1			7		10			4
M4	5	3		8		6	8		3	
M5			2		10			9		1
P4	Op1	Op2	Op3	Op4	Op5	Op1	Op2	Op3	Op4	Op5
M1	1			2		10			9	
M2	1		2	1	10	10		9	10	1

(continued)

Table 1. (*continued*)

P4	Op1	Op2	Op3	Op4	Op5	Op1	Op2	Op3	Op4	Op5
M3		9			1		2			10
M4			10					1		
M5		8					3			
P5	Op1	Op2	Op3	Op4	Op5	Op1	Op2	Op3	Op4	Op5
M1					3					8
M2		7			6		4			5
M3		3	4				8	7		
M4	5		10	2		6		1	9	
M5	1			10		10			1	

The production orders are written in Table 2.

Table 2. The list of production orders

Id of order	Z1	Z2	Z3	Z4	Z5	Z6	Z7	Z8	Z9	Z10
Process	P1	P2	P3	P4	P5	P1	P2	P3	P4	P5

On the basis of the above input data, production schedules were created. Schedules are arranged using 2 different algorithms. In the first (A1) orders are planned in full, from the first to the last operation, according to the order on the list. In the second type (A2), scheduling is done at the level of the operations, parallel for all orders. Selection of machine from alternatives is carried out according to the following:

$$\min_v (w^c u^c + w^{et} u^{et}) \tag{1}$$

where:

u^c, u^{et} is the partial score associated with the cost and ending time parameters respectively,

w^c, w^{et} is the weight of partial score associated cost and ending time parameters respectively.

While the partial scores are calculated as follow:

$$u_v^c = \frac{t_v^c}{\max_v\{t_v^c\}} \tag{2}$$

$$\mathrm{u}_\mathrm{v}^\mathrm{et} = \frac{\mathrm{t}_\mathrm{v}^\mathrm{et}}{\max_\mathrm{v}\{\mathrm{t}_\mathrm{v}^\mathrm{et}\}} \tag{3}$$

where:

$\mathrm{t}_\mathrm{v}^\mathrm{c}$, $\mathrm{t}_\mathrm{v}^\mathrm{et}$, describe cost and ending time of given operation variant v, respectively.

Parameter t_v^c, related to the cost of executing a given operation, is static. It is part of the input data and does not change during the scheduling. In contrast the t_v^{et} value is determined, when creating a schedule is at the decision-making stage associated with the selection of a given operation variant.

The experiment involved 12 simulations for different parameter weights w^c, w^{et}. In Tables 3 and 4 are presented the most important performance measures obtained in individual simulations:

- C_{max} – makespan, the completion time of a schedule,
- C_m – mean completion time in the set of orders,
- F_{sum} – total flow time of all orders,
- $F_{s\text{-}e}$ – total residence time of orders,
- CrOp – the number of critical operations,
- K_{sum} – total cost of the set of orders, determined by selected route.

Table 3. Results of scheduling using A1 algorithm

	w^{et}	w^c	C_{max}	C_m	F_{sum}	$F_{s\text{-}e}$	CrOp	K_{sum}
1	1	0	64	40,2	402	356	27	346
2	1	1	64	40,2	402	356	27	346
3	1	2	120	66,2	662	495	23	235
4	1	3	132	75	750	578	25	208
5	1	4	132	75,8	758	606	30	206
6	1	5	132	75,8	758	606	30	206
7	1	6	143	80,5	805	512	25	203
8	1	7	143	80,5	805	512	25	203
9	1	8	143	80,5	805	512	25	203
10	1	9	143	80,5	805	512	25	203
11	1	10	144	80,7	807	514	24	202
12	0	1	144	80,7	807	514	24	202

The obtained data, which are presented in Tables 3 and 4, also are illustrated in the Figs. 1 and 2.

The upward and downward tendencies of selected parameters are observed in the analyzed weight range. In addition to this interval, changing the weights does not cause any changes. In the considered case, the differences between solutions are observed in the range of weights $w^c = \{1..10\}$ with the assumed $w^{et} = 1$. The extreme values (items 1 and 12) is showed that assigning weights outside this range will not change the value of the solution parameters.

On the basis of the obtained results, a relationship can be observed. According to which as the weighting factor increases w^c in relation to weight w^{et} the time of execution of orders is extended. While the total costs of a given order are reduced. This results come from the assumed goal function (1) and the assumed assumption in the input data with the inversely proportional value of the cost to the duration of the operation. Showed measures C_{max}, Cm and F_{sum}, grow non-linear with the increase in

Table 4. Results of scheduling using A2 algorithm

	w^{et}	w^{c}	C_{max}	C_m	F_{sum}	F_{s-e}	CrOp	K_{sum}
1	1	0	55	41,6	416	396	40	331
2	1	1	55	41,6	416	396	40	331
3	1	2	81	60,3	603	566	39	243
4	1	3	89	69,8	698	653	39	211
5	1	4	92	72,4	724	679	39	209
6	1	5	97	73,2	732	687	35	204
7	1	6	108	82,6	826	743	36	202
8	1	7	108	82,6	826	743	36	202
9	1	8	108	82,6	826	743	36	202
10	1	9	108	82,6	826	743	36	202
11	1	10	108	82,6	826	743	36	202
12	0	1	108	82,6	826	743	36	202

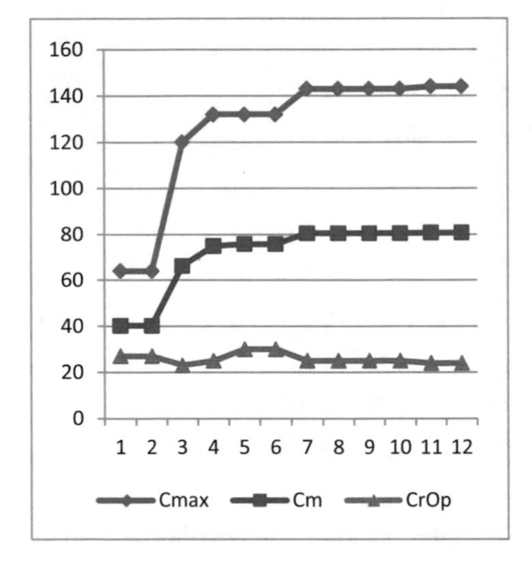

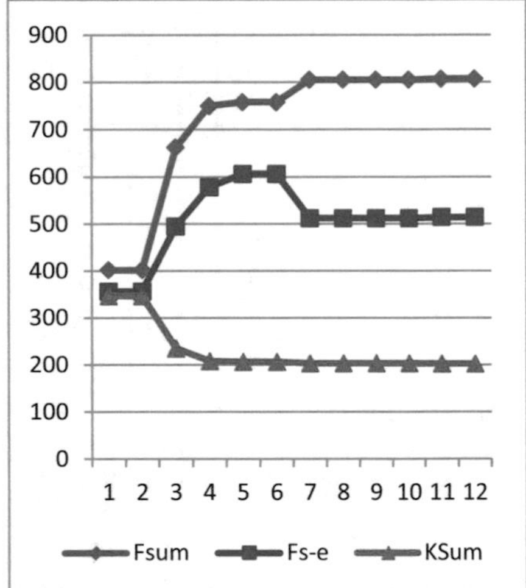

Fig. 1. Changes of performance measures in simulations using A1 algorithm

the w^c weighting parameter. Noteworthy is also the change in the values of CrOp and F_{s-e} parameters for the A1 algorithm, whose extremes (maximum values) are observed in the simulation item 5 and 6. Such behavior of parameters is strongly related to the specificity of specific input data and do not come directly from the adopted and calculated data processing.

Selected solutions subject to evaluation are showed on Figs. 3 and 8. Figures 3 and 4 are showed schedules arranged with weight coefficients with $w^{et} = 1$ and $w^c = 0$, respectively for algorithms A1 and A2. With this solution, the route selection is dominated by the time of ending the operation (the faster the better). Parameter $w^c = 0$

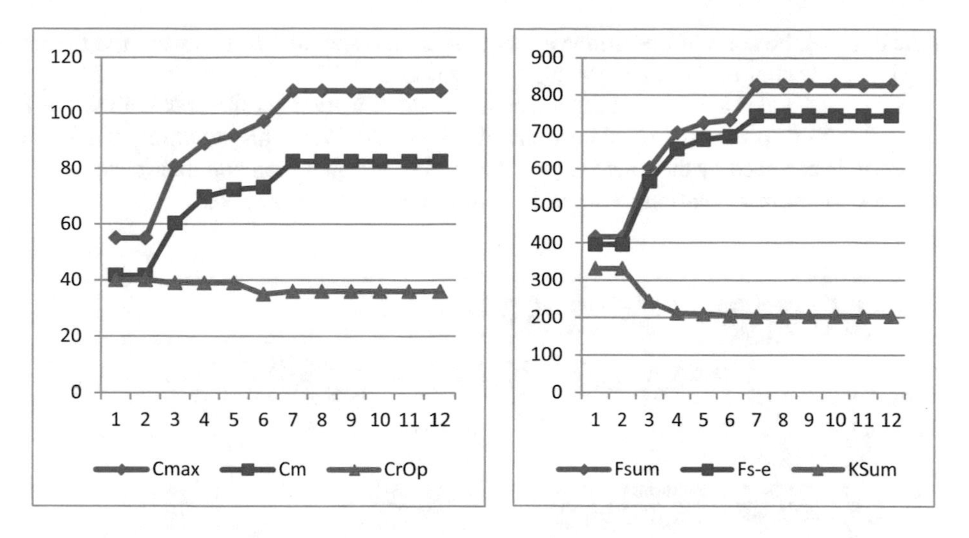

Fig. 2. Changes of performance measures in simulations using A2 algorithm

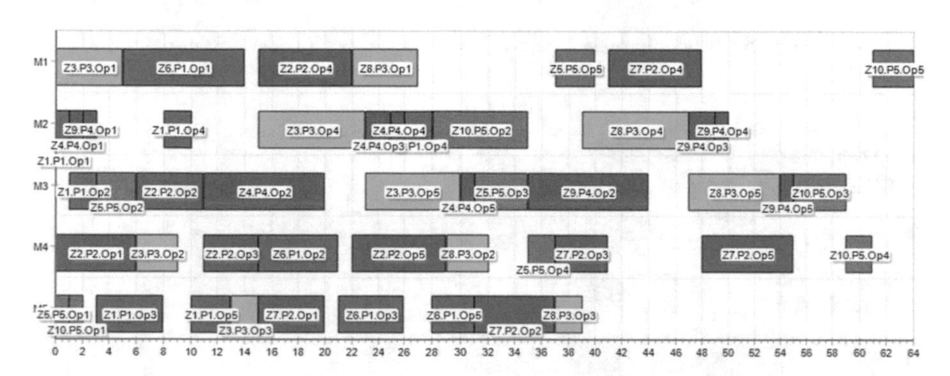

Fig. 3. Schedule by A1 with $w^{et} = 1$ and $w^{c} = 0$ (Table 3 item 1)

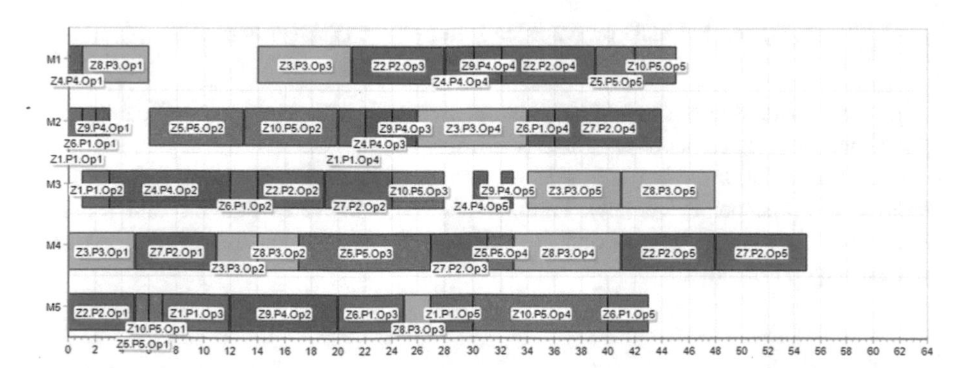

Fig. 4. Schedule by A2 $w^{et} = 1$ and $w^{c} = 0$ (Table 4 item 1)

indicates that the costs of operations are not taken into account. In this way, the fastest possible execution of the order package is obtained.

Figures 5 and 6 are showed schedules arranged at weight coefficients with $w^{et} = 1$ and $w^{c} = 5$, respectively for algorithms A1 and A2. With this solution, the route selection is balanced by the time factor of the end of the operation and their cost. In this way, a compromise solution between lead time and cost are achieved.

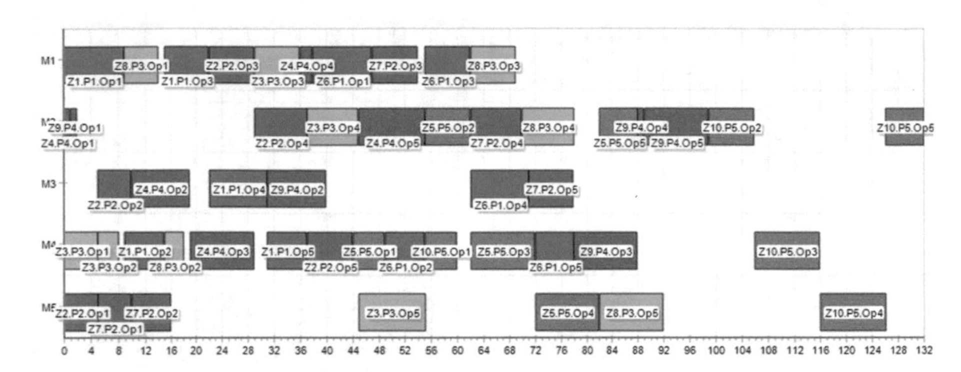

Fig. 5. Schedule by A1 with $w^{et} = 1$ and $w^{c} = 5$ (Table 3 item 6)

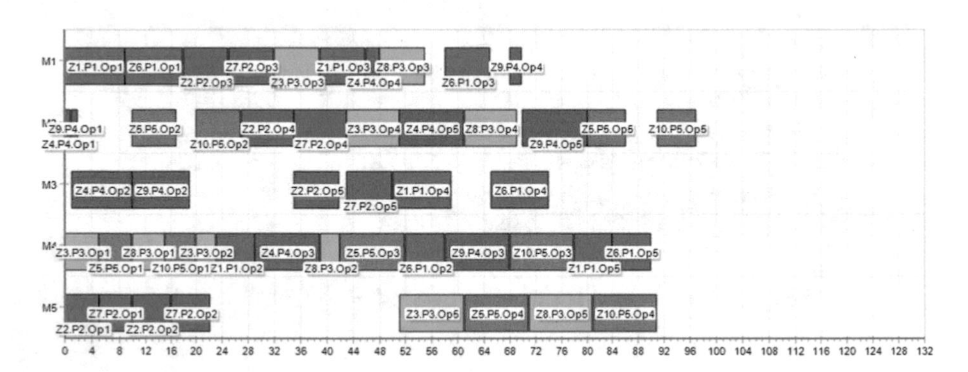

Fig. 6. Schedule by A2 $w^{et} = 1$ and $w^{c} = 5$ (Table 4 item 6)

Figures 7 and 8 are showed schedules arranged with weight coefficients with $w^{et} = 0$ and $w^{c} = 1$, respectively for algorithms A1 and A2.

With this solution, the route selection is dominated by the cost factor of the operation (the cheaper the better). Parameter $w^{et} = 0$ indicates that the end times of operations are not taken into account. In this way the fastest possible execution of the order package is obtained.

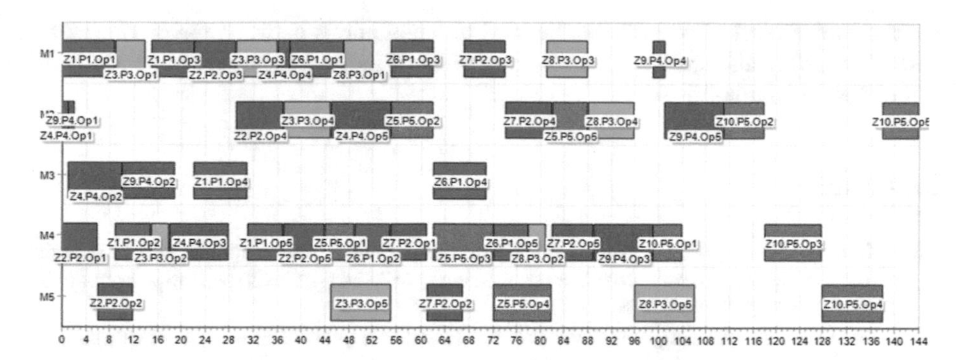

Fig. 7. Schedule by A1 with $w^{et} = 0$ and $w^{c} = 1$ (Table 3 item 12)

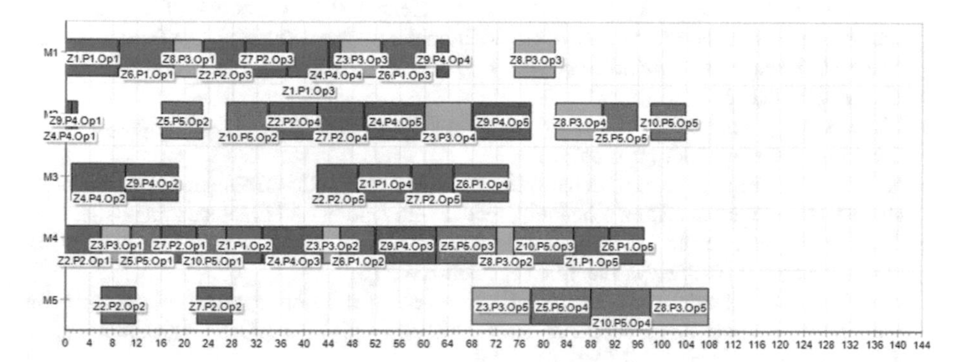

Fig. 8. Schedule by A2 $w^{et} = 0$ and $w^{c} = 1$ (Table 4 item 12)

4 Conclusion

In the paper is presented the method of scheduling in flexible job shop systems. The time and cost parameters are used to select a given variant of the implementation operation. The solutions are created with different proportions between the weighting parameters of the time operation and the cost. The adopted end of operation parameter is dynamic. It is determined at the decision-making level, when selecting a given operation variant - the selection of subsequent variants of operations depends on the decisions taken in the previous steps. For this reason, based on local optimization in individual steps, they do not guarantee globally optimal solutions.

The part of calculations presented two different algorithms, used to schedule operations - A1, according to orders, and A2, independent of orders. The results showed that algorithm A2 creates better solutions from the point of view of the C_{max}, while A1 gives the advantage at F_{sum}, $F_{s\text{-}e}$ and CrOp. The C_m and K_{sum} parameters are better only in selected simulations.

Scheduling with simultaneous consideration of time and cost parameters may be particularly important, when choosing routes depend of the urgency of the order and the acceptable level of costs. It can be used in negotiations between the producers and customer, giving an extreme. And also a number of indirect solutions from which one can choose a compromise.

References

1. Kalinowski, K., Balon, B.: Production scheduling with quantitative and qualitative selection of human resources. Adv. Intell. Syst. Comput. **747**, 245–253 (2018). https://doi.org/10.1007/978-3-319-77700-9_25
2. Kujawińska, A., Diering, M., Żywicki, K., Rogalewicz, M., Hamrol, A., Hoffmann, P., Konstańczak, M.: Methodology supporting the planning of machining allowances in the wood industry. Adv. Intell. Syst. Comput. **649**, 338–347 (2017). https://doi.org/10.1007/978-3-319-67180-2_33
3. Kalinowski, K., Krenczyk, D., Paprocka, I., Kempa, W., Grabowik, C.: Production scheduling with discrete and renewable additional resources. In: IOP Conference Series, Materials Science and Engineering, vol. 95, pp. 1–6 (2015). http://doi.org/10.1088/1757-899X/95/1/012132
4. Lisowski, B., Kozłowski, R.: Basic Issues of Production Management. Oficyna ekonomiczna, Kraków (2006). (in polish)
5. Kempa, W., Paprocka, I., Kalinowski, K., Grabowik, C.: Estimation of reliability characteristics in a production scheduling model with failures and time-changing parameters described by gamma and exponential distributions. Adv. Mater. Res. **837**, 116–121 (2014). https://doi.org/10.4028/www.scientific.net/AMR.837.116
6. Krenczyk, D., Olender, M.: Using discrete-event simulation systems as support for production planning. Appl. Mech. Mater. **809**(810), 1456–1461 (2015). https://doi.org/10.4028/www.scientific.net/AMM.809-810.1456
7. Bräsel, H., Dornheim, L., Kutz, S., Mörig, M., Rössling, I.: LiSA – A Library of Scheduling Algorithms. Otto-von-Guericke Universität Magdeburg, Magdeburg (2011)
8. Sobaszek, Ł., Gola, A., Świć, A.: Scheduling of production tasks taking into account of two-factor process uncertainty. In: Knosala, R. (ed.) Innovation in Management and Production Engineering, vol. 1, pp. 668–676. Oficyna Wydawnicza Polskiego Towarzystwa Zarządzania Produkcją, Opole (2017). (in polish)
9. Deepu, P.: Robust Schedules and Disruption Management for Job Shops. Bozeman, Montana (2008)
10. Xie, Z., Hao, S., Ye, G., Tan, G.: A new algorithm for complex product flexible scheduling with constraint between jobs. Comput. Ind. Eng. **57**(3), 766–772 (2009). https://doi.org/10.1016/j.cie.2009.02.004
11. Pinedo, M.L.: Scheduling, Theory, Algorithms and Systems. Springer, New York (2012)

The Evolution of the Robotized Workcell Using the Concept of Cobot

Wacław Banaś and Małgorzata Olender(✉)

Faculty of Mechanical Engineering, Institute of Engineering Processes Automation and Integrated Manufacturing Systems, Silesian University of Technology, Konarskiego 18AStr., 44-100 Gliwice, Poland
malgorzata.olender@polsl.pl

Abstract. In recent years, the development of the use of automation and robotics in manufacturing enterprises is even more noticeable. The possibility of robots utilization, for example in areas related to welding, painting and other, makes that producers more and more often reach for this form of solution, especially, when there is problem with employees, works are dangerous for people or when works are very monotonous. Currently robots utilization are fast, precise and work continuously. The parameters, which are obtained are more important, from the point of view of mass-customization. It means that production is personalized, but with costs that do not differ from the costs obtained in the mass production. Development of new technologies, helps not only in the planning of production processes, using databases, utilizing faster information flow, etc., but also with appropriate management and the use of the production resources on which the entire production process is based. Human works together with robots on the line. The working space should be properly managed, so that the robot could makes the specified movements, next to the working human and/or machine. At this point, integration must be done at a high level, especially that, human security is very important. Properly managed work space, programmed cooperation between human - robot, robot - machine, enable to better response to changes in product demand and quicker switch to a new product. The key for good organization of processes and integration is knowledge, how to use the potential that the company has gained.

Keywords: Production process · Safety standard · Robot integration

1 Introduction

The development of new technologies favors the introduction of changes in enterprises, not only on the level of management, but also on the level, related to the implementation of production processes. This, what used to be unattainable, now is being put into use and developed. An example, can be the area, related with the automation and robotization of a production line. In the past, individual operations were made only by employees. In the following years, part of the operations were mechanized. Then, automation began introduced in the mass production systems. It allows integrating not only robot – people cooperation. Integration is also connected also with robot - machine or machine – machine [1–3]. That integration, possible to apply, has many

© Springer Nature Switzerland AG 2019
A. Burduk et al. (Eds.): ISPEM 2018, AISC 835, pp. 401–407, 2019.
https://doi.org/10.1007/978-3-319-97490-3_39

advantages. However the used solutions, require to do many changes, in the environment of the workplace. To get the integration of workplace, firstly it should start by defining, what will be produced. Basing on production plans, schedules are developed. Then, the workplace can be prepared accordingly [4–6].

Based on the schedules one, can design a robot workcell. It is easy to simulate the program, test it and run it, when there is only one robot in the cell. In case of the example presented in Fig. 1, the problem is already more complex. Each robot movements must be coordinated with the movements of other robots, especially when robots occupy the same work space. This is very difficult to program robots on-line. In this case, many tools are used, for example, tags. However, during programming and testing, it is important to be careful, because during the collision the elements of the cell can be damaged. It should be remembered that robot movements are programmed predictable and also repetitively. The robots repeat the same movements in the same schedule, so any problems which appear should be visible at the testing stage at the first run [7, 8].

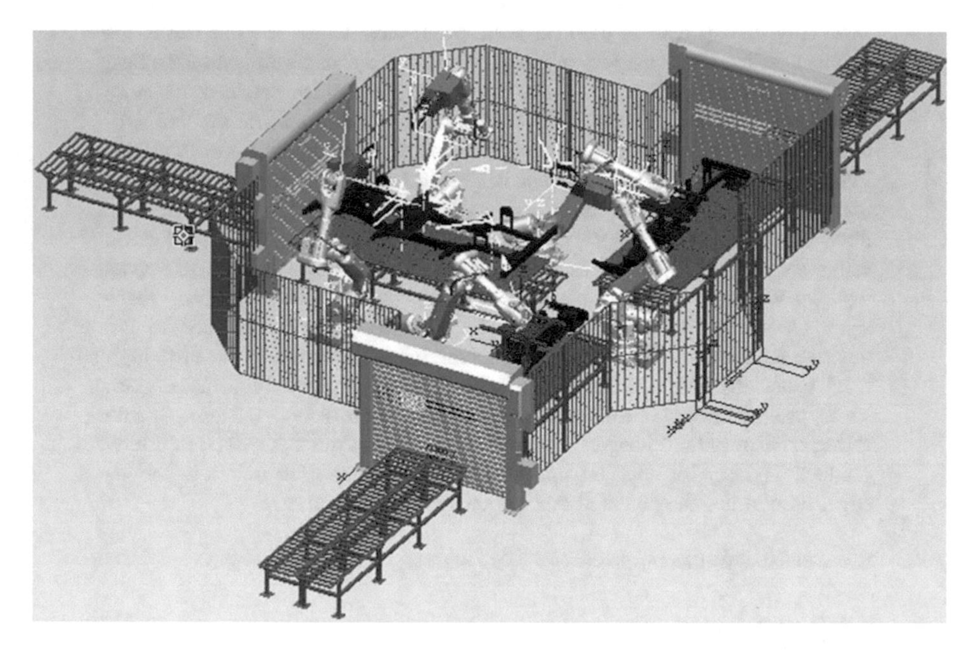

Fig. 1. Traditional robotic cell

The problem with a robotic station is associated not only with the appropriate implementation of the robot's movements, but also, among other things, with sensors and software. Because of them, it became possible to introduce people to the position in workcell without taking a lot of place for protective barriers.

In many robotized workcells, it is necessary to conduct cooperation between human and the robots. This is very dangerous in the case of traditional robots. The speed and weight of the robot can cause serious damage of the operator and sometimes even death [9, 10]. So the applied robotic cells have two separate areas (Fig. 2). One area is designated for the robot. Second area is designated for the operator. Both areas are separated by barriers. The removing of the barrier is possible only, when the operator is outside the cell. In this case robot and operator work separately and do not have contact with each other [11, 12].

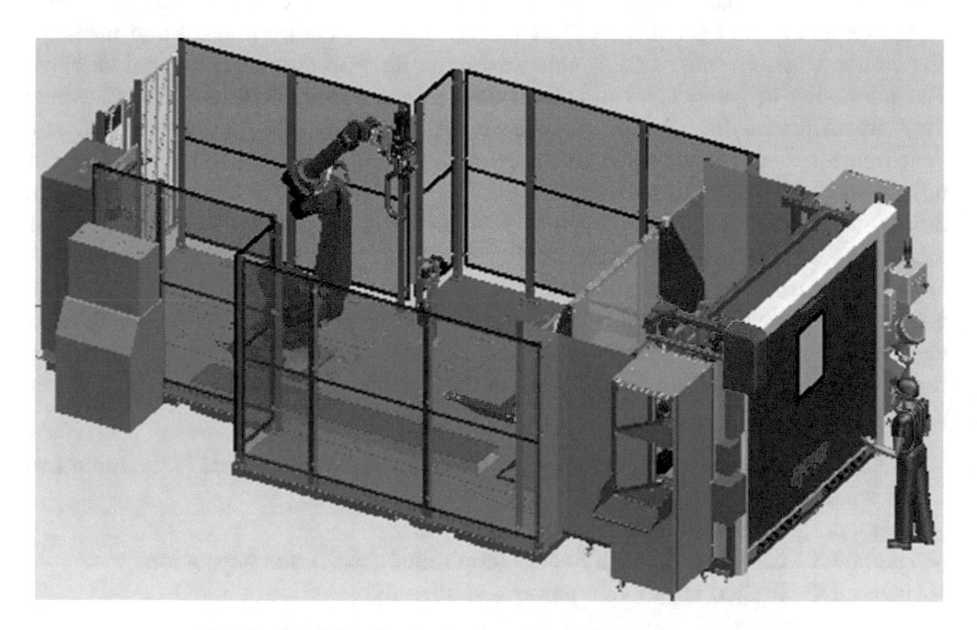

Fig. 2. Traditional robotized workcell with human support

Although that, automation and robots utilization, support production processes and ensure continuity of production (without breakdown), and constant good quality of the elements which are made. Without certainty, that people can work in robot workplace, no one would allow, to develop this area, on the production line.

This article discusses of the issue of robot and people integration on a selected production workplace. Nowadays, often in robot workplace, there are no more security barriers, because dedicated sensors and software take over this role. However, there are defined requirements which should be follow. There must be given information for example about, what maximum force the robot can work with the presence of people, when the contact is. Further, it is important to know, what kind of the tool, robot is working with, and if it is a transporting robot, what kind of the detail robot is moving?. All these aspects are an integral whole, which should be analyzed to allow such a solution on the production line, especially, when it is the human support robotized workcell.

The objective of the work is to elaborate the methods of Virtual Reality utilization to model the human – robot relation. The research methodology is based on the application of the VR.

2 Integration and Safety

Industrial robots can be used in many areas. They are used both for packaging, painting, welding and also gluing or assembly. Currently robot utilization is no longer dedicated only to mass production. They are used in the area of personalized, individual or small serial production. This allows using robots by a small and medium-sized enterprises. Because of this enterprises can become more flexible and efficient. The possibility of using robots, depends only on their parameters and a place where they are dedicated to. Due to the widespread use, it is also possible to increase development efficiency of production processes. Especially, if it based on integration of human-robot resources solutions [12, 13]. The problem appears, when people work directly with the robot. In this situation, it is necessary, to plan the movements of the robot and working area correctly. But at the same time people also have their own work area, next to the robot. Such integration is a complicated task. It is necessary to test, how far the arm will reach, which areas are dangerous for a people and in which areas the operator should moves. These issues are the basic problems that should be solved for robot-people integration. In addition, such cooperation is divided into four methods, namely [14, 15]:

- method 1 - use the function of a safe controlled stop, implemented in the robot and its controller.
- method 2 - work with manual robot guidance.
- method 3 - co-operation with limited (controlled) speed and separation.
- method 4 - limited the robot's power and strength.

The first three methods are using by classical industrial robots. Of course, the equipment of workcell robot and parameters should be selected correctly. The last method use so-called "cobots" (Fig. 3). It means that, robots support people in the implementation of production processes, but arm-in-arm, without protective barriers. Thru this solution increase the efficiency of processes and the quality of products [13, 14].

Another issue is the trend that is noticeable on the market. Cooperating robots must have appropriate parameters selected, but also should be considered [15, 16]:

- adaptation of the structure (lifting capacity, reach of the arm) robots, for various tasks and industries. Not for all branches of industry robots are a good solution, because of security reasons,
- adaptation of robots to the high requirements of narrow production sectors (food industry) or (medical and pharmaceutical industry),
- ensuring the highest levels of work safety of the robot - people,
- further development of robots (extending the use of robots and performing several tasks by robot, that were previously shared).

Another, but very important, problem related to the integration of a robot – people on a production workplace is the worker's safety, but also the work area. It is required, that the approved workplace must all legal and technical requirements and standards fulfil. It is very important to save a safety in a workcell. It is need to develop new standards and control system safety, to better cooperating in area robot – people. Like it

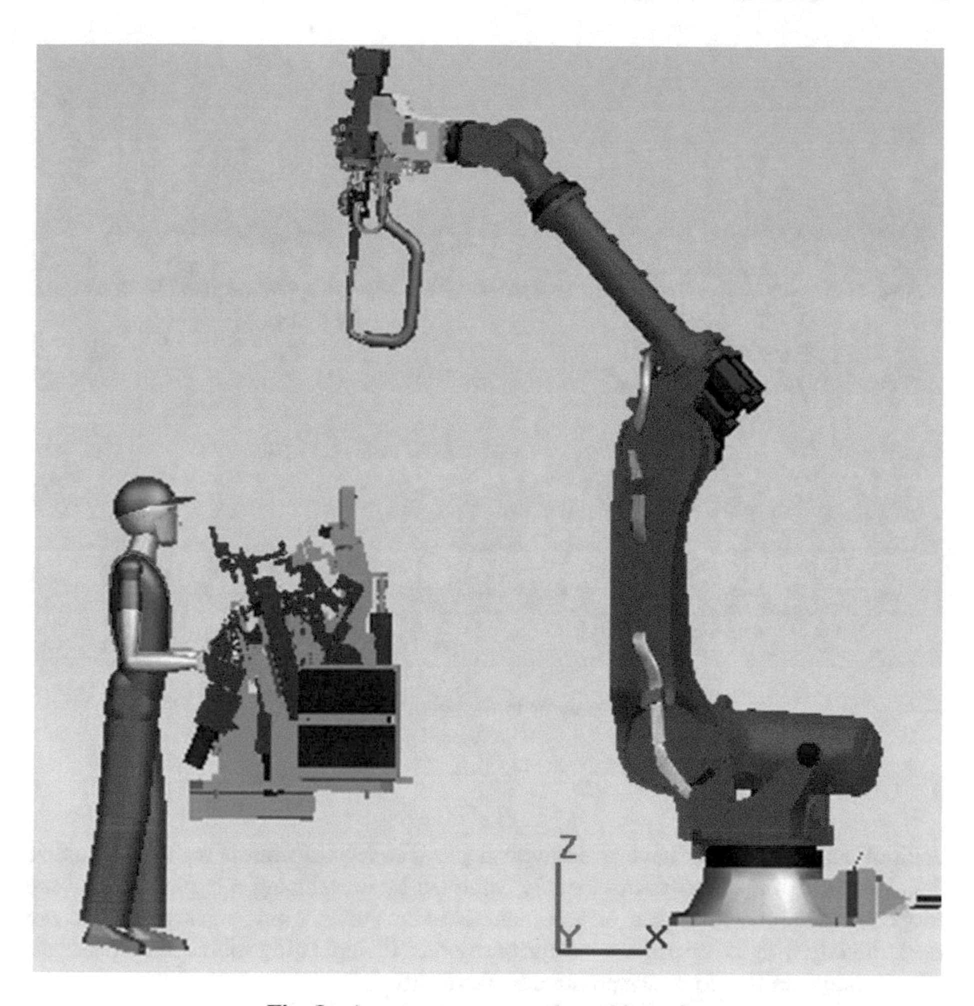

Fig. 3. An operator cooperating with a robot

is showed in the Fig. 4 this needs are very required. In this figure is showed very dangerous situation. It is possible to hit the hands of operator by the robot gripper. Hence robot must have a more complex control system.

Present robots have limited sensitivity. The robot moves on a strictly designed path, so robot does not need so much information. The cooperating robot requires more information to avoid collisions. This information can be collected for example, using a vision systems. But the use of vision systems is very restrictive. Even small changes in lighting cause errors, especially on such a large area like a robot workcell.

Fig. 4. Operator in the robot's workspace

3 Conclusion

Each of customer's want have an individual product. Manufacturers try to personalize their products. This effect is most easily achieved by introducing a human factor. The cooperating robot will appear on the manufacturers' offer. Now, most robots do not work directly with the operator for security reasons. Co-operating robots will appear for robotic purposes due to better production flexibility.

Because nowadays, time for produce new product is smaller and customers wants personalized product, it is need planning very carefully all the production operation. Due to production planning, producers must have all information about production orders. Then it should be determined connection between operations, machines and robots. Because of the area of scope robot utilization, people don't make monotonous and very dangerous tasks. Also the production is continuous, so these solutions are very interesting for production company. It is need to, develop the area connected with robotizing. Present, producers also use robots cooperated with people. But not on the level where robots directly cooperate with people, without safety barriers.

References

1. Skołud, B., Krenczyk, D., Davidrajuh, R.: Multi-assortment production flow synchronization. Multiscale modelling approach. In: MATEC Web of Conferences, vol. 112, pp. 1–6 (2017). https://doi.org/10.1051/matecconf/201711205003
2. Kalinowski, K., Grabowik, C., Paprocka, I., Kempa, W.: Interaction of the decision maker in the process of production scheduling. Adv. Mater. Res. **1036**, 830–833 (2014). https://doi.org/10.4028/www.scientific.net/AMR.1036.830
3. Krenczyk, D., Olender, M.: Production planning and control using advanced simulation systems. Int. J. Mod. Manuf. Technol. **6**(2), 38–43 (2014)
4. Bączkowicz, M., Gwiazda, A.: Optimizing parameters of a technical system using quality function deployment method. In: IOP Conference Series: Materials Science and Engineering, vol. 95, pp. 1–6 (2015). https://doi.org/10.1088/1757-899x/95/1/012119
5. Kádár, B., Pfeiffer, A., Monostori, L.: Discrete event simulation for supporting production planning and scheduling decisions in digital factories. In: 37th CIRP International Seminar on Manufacturing Systems, pp. 444–448. Production Networks, Hungary (2004)
6. Foit, K., Gwiazda, A., Banaś, W.: A multi-agent approach to the simulation of robotized manufacturing systems. In: IOP Conference Series: Materials Science and Engineering, vol. 145, pp. 739–744 (2016). https://doi.org/10.1088/1757-899x/145/5/052011
7. Krenczyk, D., Olender, M.: Using discrete-event simulation systems as support for production planning. Appl. Mech. Mater. **809**(810), 1456–1461 (2015). https://doi.org/10.4028/www.scientific.net/AMM.809-810.1456
8. Gwiazda, A., Banaś, W., Sękala, A., Foit, K., Hryniewicz, P., Kost, G.: Modular industrial robots as the tool of process automation in robotized manufacturing cells. In: IOP Conference Series, Materials Science and Engineering, vol. 95, pp. 1–6 (2015). https://doi.org/10.1088/1757-899x/95/1/012104
9. Yan, Z., Jouandeau, N., Cherif, A.: A survey and analysis of multi-robot coordination. Int. J. Adv. Rob. Syst. **10**, 1–18 (2013). https://doi.org/10.5772/57313
10. Herbuś, K., Ociepka, P., Gwiazda, A.: Conception of the integration of the virtual robot model with the control system. Adv. Mater. Res. **1036**, 732–736 (2014). https://doi.org/10.4028/www.scientific.net/AMR.1036.732
11. Sękala, A., Gwiazda, A., Kost, G., Banaś, W.: Modelling and simulation of a robotic workcell. In: IOP Conference Series, Materials Science and Engineering, vol. 227, pp. 1–7 (2017). https://doi.org/10.1088/1757-899x/227/1/012116
12. Grajo, E.S., Gunal, A., Sathyadev, D., Ulgen, O.M.: A uniform methodology for discrete-event and robotic simulation. In: Proceedings of the Deneb Users Group Meeting, pp. 17–24. Deneb Robotic, Michigan (1994)
13. Bolmsjö, G., Bennulf, M., Zhang, X.: Safety system for industrial robots to support collaboration. Adv. Intell. Syst. Comput. **490**, 253–265 (2016). https://doi.org/10.1007/978-3-319-41697-7_23
14. Roitberg, A., Somani, N., Perzylo, A., Rickert, M., Knoll, A.: Multimodal human activity recognition for industrial manufacturing processes in robotic work cells. In: Proceedings of 17th ACM International Conference on Multimodal Interaction, pp. 259–266. ACM, New York (2015). https://doi.org/10.1145/2818346.2820738
15. Głowicki, M.: Cobots - security issues in the integration of cooperating robots. Napędy i sterowanie **4**, 30–34 (2017). In Polish
16. Otrębski, T.: Robot in an integrated production environment. Complex Mach. Ind. Monit. **18**, 59–63 (2015). In Polish

Petri Nets in Modelling and Simulation of the Hierarchical Structure of Manufacturing Systems

Krzysztof Foit(✉)

Institute of Technological Processes Automation and Integrated Manufacturing Systems, Silesian University of Technology, Gliwice, Poland
krzysztof.foit@polsl.pl

Abstract. Among the methods of modeling production systems, Petri Nets occupy an important place. Due to well-formed rules, creating models is simple, though not without limitations. Some of these difficulties have been overcome by extending the existing Petri Nets formalism, but the analysis of large networks, that are a mapping of complex production systems, is still a serious difficulty. For this reason, an important issue is the possibility of scaling the network, so that there is the ability of the general look at the model and the detailed consideration of a selected fragment of the modeled process. This could reduce the amount of software and hardware resources involved in the calculations and the lower impact on the processor and memory load. However, it should be noted that it is not always necessary to accurately represent all of the component processes. On the other hand, it is important to maintain compatibility with Petri's network analysis principles in order to correctly interpret the model at different levels of detailedness. This article presents the selected problems of using the Petri Nets to modeling of production systems, taking into account comments that are available in the existing scientific sources.

Keywords: Manufacturing system · Petri Net · Modelling

1 Introduction

The modelling process, as a method of describing a selected part of a real system, consists mainly in the skillful selection and use of the method of modelling that could maintain all important information. Therefore, there is no one, universal way of modelling. Thereby, this issue comes down to deciding which information is important from the point of view of further analysis of a given phenomenon and which can be omitted without loss of accuracy. This is also related to the problem of detailedness or generality of the model.

Modeling of production processes is a complicated task. During the manufacturing process, different tasks are carried out that can be treated as discrete events, but it should be noted that each operation consists of elementary activities that are reflected in the technological documentation of the process. As in other cases, modeling of manufacturing processes will require the adoption of an appropriate scale that will provide the

© Springer Nature Switzerland AG 2019
A. Burduk et al. (Eds.): ISPEM 2018, AISC 835, pp. 408–416, 2019.
https://doi.org/10.1007/978-3-319-97490-3_40

correct level of detail, but will also reject irrelevant information. Thus, a situation arises in which particular elementary processes are nested within other processes that can be treated as a whole.

One of the commonly used methods of manufacturing systems modeling is Petri Net formalism. Petri Nets derive from graph theory and can be used to illustrate the course of a certain process in a discrete manner, i.e. mapping individual steps or stages of this process, as well as to model continuous and hybrid (discrete-continuous) processes. Initially, this formalism was referred only to the presentation of mutual connections and dependencies between the key steps of the process, but later, as a result of work carried out by various scientists, it was expanded with new possibilities. In this way the new extensions to Petri Nets have been created, like Couloured Petri Nets, high-level Petri Nets, dualistic Petri Nets, algebraic Petri Nets etc. Although there are many extensions to Petri's formalism, it is still difficult to model the multilevel hierarchical structures. There exist the methods for synthesis and detailing the network, but they still lead to the result in the form of one, big graph. Creating such a network is difficult due to the necessity of using a specific formalism, instead of intuitively connecting modules. The purpose of this article is to discuss the problem of creating and simulate of embedded network structures and propose an alternative approach to the problem. At the present stage of the discussion, the general outline of the method will be presented, without a detailed definition of formalism.

2 Petri Nets

2.1 Introduction

Petri Nets are a kind of modelling tool, formalized by Carl Adam Petri in his dissertation, submitted in 1962. The characteristic feature of the method is its dualistic representation – in the form of mathematical rules and in the form of graph.

Referring to a formal, mathematical definition, the Petri Net should be understood as a 4-tuple [1]

$$N = (P, T, F, V), \tag{1}$$

where P stands for set of places, T is set of transitions and

$$P \cap T = \emptyset, P \cup T \neq \emptyset. \tag{2}$$

Respectively F describes the flow relation of N and

$$F \subseteq (P \times T) \cup (T \times P), \tag{3}$$

while the V is set of arcs' weights

$$V{:}F \rightarrow \mathbb{N}^{+}. \tag{4}$$

According to the above definition, the Petri Net is a 2-colored, weighted, directed and finite graph [1]. The graphical representation of Petri Net consists of two main symbols: circles that represent places and bars or rectangles that represent transitions. Places and transitions are connected with directed arcs, but the ends of an arc must be connected to the place on one side and to the transition on another. The weight is often placed above the arc and has the form of a number. If the weight value is 1, there is no need to place it on the graph.

It is said that the Petri Net is marked, when it contains the tokens inside the places. In the classic net, the tokens are represented by small dots. When the transition is activated (fired) then tokens could change the place, passing through the transition in the direction indicated by the arrows placed on the arcs. The example of marked Petri Net is shown in Fig. 1.

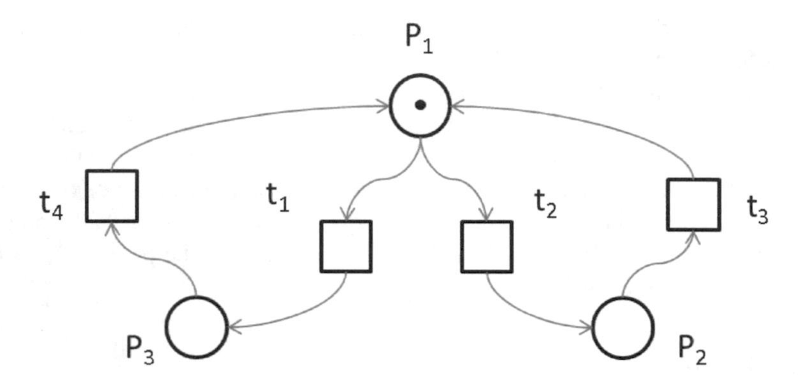

Fig. 1. The example of marked Petri Net

Analyzing the network shown in Fig. 1, two paths of token passing can be distinguished. The first one is connected with firing the transitions t_1 and t_4, while the other requires firing t_2 and t_3. As it can be seen, such approach generates a flat model, where all components are located on the same level. In the case of complicated system, the corresponding Petri Net will also be very complex and hard to analyze without the special computer software. In order to moderate this problem, several extensions of the Petri Net formalism were introduced. Some of them will be presented in the next part of the paper.

2.2 The Characteristic of the Selected Petri Net Extensions

Object-oriented Petri Nets. The concept of the Object-oriented Petri Nets is very similar to the object-oriented programming paradigm. The OOPN could be defined as the system that consists of physical objects and the relations of interconnection between them, what could be written as [2]:

$$S = (O, R), \tag{5}$$

where O is the set of physical objects in the system and R is the set of message passing relations among the physical objects. The OOPN of the physical object *i*, could be defined as [2]:

$$O_i = (SP_i, AT_i, IM_i, OM_i, F_i, C_i), \tag{6}$$

where SP_i is the finite set of places, AT_i is the finite set of activity transitions, IM_i is the finite set of input message places, OM_i is the finite set of output message places, F_i are input and output relationships between activity transitions and state/message places, C_i is a set of colour sets associated with state places, activity transitions and input/output message places [2]. The colours are used to differentiate the types of tokens that may reside is the particular state place and to subordinate the transitions response according to the associated colours of the input tokens. The example of object is shown in Fig. 2, while the example of OOPN is shown in Fig. 3.

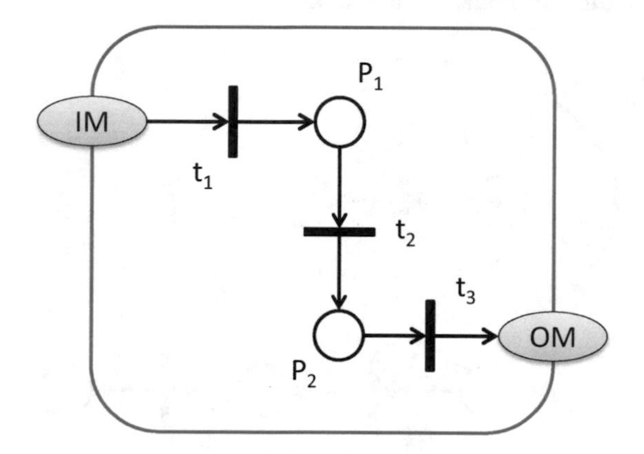

Fig. 2. The example of OOPN object

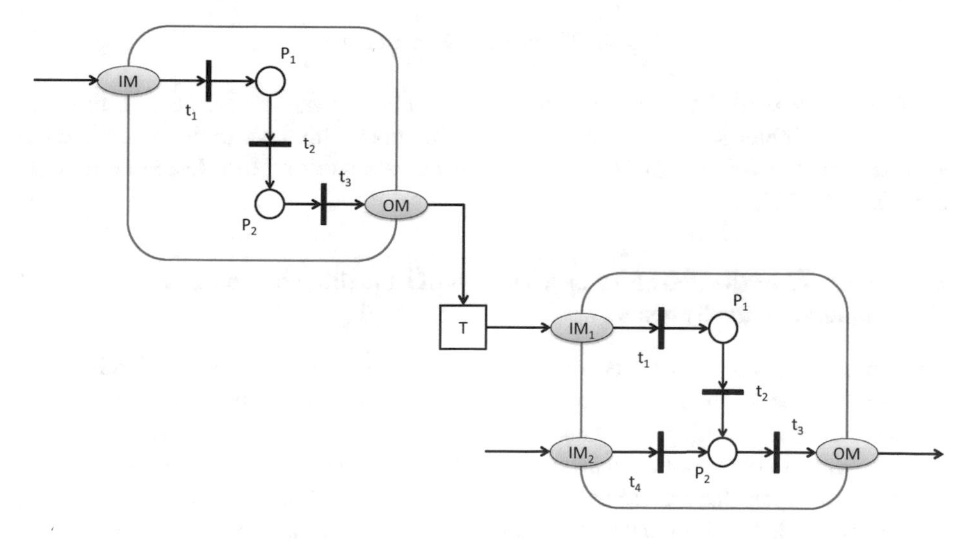

Fig. 3. The part of Object-oriented Petri Net

As it can be seen, the Object-oriented Petri Net approach can give modular models, that could be easily build, using predefined library of objects. There are some reports about using this method successfully in modelling of manufacturing systems, e.g. [2–4]. The presented approaches are slightly different, what indicates the potential of the OOPN.

Nets Within Nets Approach. The idea of "nets-within-nets" is based on the Valk approach [5, 6], who assumed that tokens could have a form of Petri Nets. This may be understood as the extension of the Object-oriented Petri Nets that could be more suitable in modelling flexible manufacturing systems and creation of agent-oriented models. The example of net within net is shown in Fig. 4.

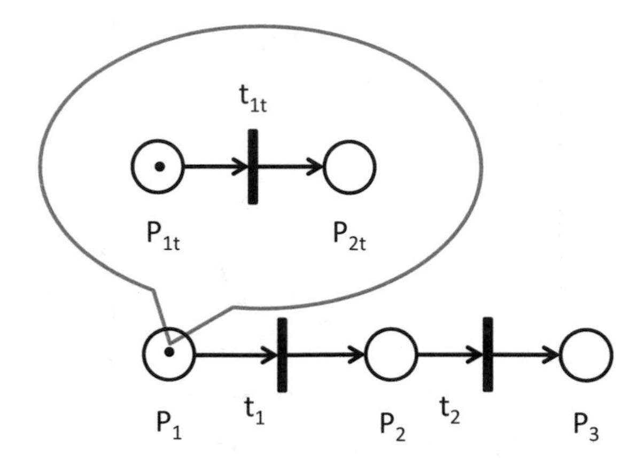

Fig. 4. The net within net example

The Valk's work has been also the starting point for the other research, like for example the Köhler's et al. studies on the use of nets within nets in the modelling of agent-oriented environments [7, 8] or the Lomazova's work on the Nested Petri Nets formalism [9–11].

2.3 Synthesis of the Petri Nets as a Method of Modelling of Complex Manufacturing Systems

The modelling of complex systems using Petri Nets approach could be long-lasting and errors-prone process. In order to avoid mistakes, such process could be divided into stages and done according to the rules of network synthesis. Despite the fact that this procedure does not allow to obtain a hierarchical model of system, the individual steps of synthesis generate the models that represents the system at various levels of detailedness.

The available literature lists the three ways of Petri Nets synthesis [12–14]:

- the bottom-up method,
- the top-down method,
- the hybrid method.

The bottom-up method consists in linking of the elementary circuits that represent the basic processes into the large network. This approach is resource-consuming, but allows the assembly of the model from blocks that can have the universal form, suitable for use in various models. In this way, it is possible to create a library of elementary networks, what greatly simplifies the modeling process and reduces the possibility of making mistakes.

The top-down method starts from the most general model of the system. In the successive steps, the model is refined by adding more and more details. The schemes for refinements involve expanding of transitions and places. The advantage of this method is the possibility to obtain a range of models that differ in the degree of detailedness.

The hybrid method consists in combination of the two methods that has been mentioned, i.e. bottom-up and top-down one. The advantage of using such approach is the possibility to alleviate the problem of complexity and find the balance between complex network and desirable qualitative properties of the obtained Petri Net.

3 The Petri Nets Formalism in Context of Computer Simulation of Manufacturing Systems

Modern manufacturing systems are characterized by high flexibility and decentralization of control systems. These properties impose high requirements on simulation tools that run on computer platforms. Referring to the Petri Nets as a tool for performing computer simulation of manufacturing systems, it is worth to say that the PN formalism, along with the extensions to the basic rules, is ready to cope with complex problems of modelling of hierarchical systems. The mentioned earlier Object-oriented Petri Nets and "Nets within Nets" approach could represent very complex systems that also may use new methodologies, like for example agent-based models. On the other hand, the offer of Petri Nets simulators on the computer software market is rather limited. Moreover the majority of programs do not support advanced features of Petri Nets, allowing only the basic analysis. In this manner is hard to use the extended formalism to represent the hierarchical structures of manufacturing systems.

3.1 The Draft of the Alternative Method of the System's Hierarchy Representation

In order to overcome the difficulties connected with the lack of the proper software, the alternative method will be presented, which does not require the use of the advanced PN formalism. Assuming that the aim is to represent and simulate the system only in visual form of Petri Net, without the need of deep mathematical analysis, the hierarchy of system could be shown in any basic software simulator that supports the Coloured Petri Nets (CPN). The method is in the early phase of development, so it is not completely formalized. It uses the tokens of different colours in order to illustrate the flow on different levels and different weight of arcs in order to ensure the proper firing of transitions. The example, shown in Fig. 5, represents the hypothetic, two-level system. The top part of the network is the most general form of the model – the highest level of generality. The bottom part of network illustrates hidden process that cannot be seen from higher level. Both parts uses different colours of the tokens – respectively black (top part) and red (bottom part). In this example, the path of the black token (P_0-P_2-P_1) corresponds with the simulation of the most general form of network. Respectively, following the black and the red token along the path P_0-P_3-P_4-P_1 illustrates the behavior of the complete network.

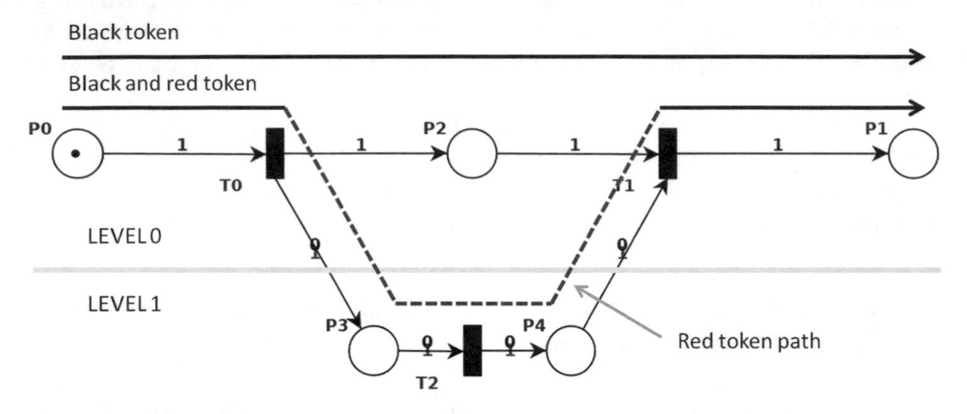

Fig. 5. The model of two-level system using CPN representation

The presented method may be useful if there is the need to observe e.g. the way of a product through the different levels of system's hierarchy. Of course the network created in the described manner will be different from the complete PN model of the system, so any mathematical analysis should be done with caution.

The presented method prospectively could be useful for representing and simulation of manufacturing systems. However, it does not solve the problems of the network complexity and demand for hardware resources, but it should be possible to implement this method on the most of existing PN simulators.

4 Conclusions

The Petri Nets formalism is the well-developed tool that could be effectively used for modelling of manufacturing systems. The existence of many PN extensions makes the analysis of complex systems easier, including the Flexible Manufacturing Systems or agent-oriented systems. The other side of using PN models for analyzing and modelling of manufacturing systems is the problem with the availability of the proper software that allow the use of extended PN formalism. Unfortunately, such kind of applications often implement only the basic PN formalism, so there is no possibility to represent the hierarchical structure. For this reason, the development of PN formalism should be accompanied by the development of appropriate software libraries in order to use them in the simulation software development process.

This paper presents the draft of the method that can partially solve this problem by using the modeler that supports the Couloured Petri Net – it is possible to use different colours of tokens to represent the token passing on different level of hierarchy, but it must be kept in mind that such solution generates the network, which is different from the complete PN model of the system. The future research will be focused on the testing of the usefulness of the presented method in modelling of manufacturing systems hierarchy as well as on the development of the mathematical formalism. The other topic for further discussion is the compatibility of this method with the PN/CPN theory.

References

1. Popova-Zeugmann, L.: Timed Petri Nets. In: Time and Petri Nets. Springer, Heidelberg (2013). https://doi.org/10.1007/978-3-642-41115-1
2. Wang, L.-C.: Object-oriented Petri Nets for modelling and analysis of automated manufacturing systems. Comput. Integr. Manuf. Syst. **9**, 111–125 (1996). https://doi.org/10.1016/0951-5240(95)00032-1
3. Wang, J., Liu, S., Liu, J., Du, Z.: An assembly process model based on object-oriented hierarchical time Petri Nets. In: AIP Conference Proceedings, p. 30023. AIP Publishing LLC (2017). https://doi.org/10.1063/1.4981588
4. Brezovan, M., Stanescu, L.: Using high-level Petri Nets with object-orientation for modeling flexible manufacturing systems. In: 2017 18th International Carpathian Control Conference (ICCC), pp. 477–482. IEEE (2017). https://doi.org/10.1109/CarpathianCC.2017.7970447
5. Valk, R.: Petri Nets as token objects: an introduction to elementary object nets. In: Proceedings of the 19th International Conference on Application and Theory of Petri Nets, pp. 1–25. Springer, Heidelberg (1998). https://doi.org/10.1007/3-540-69108-1_1
6. Valk, R.: Object Petri Nets. In: Desel, J., Reisig, W., Rozenberg, G. (eds.) Lectures on Concurrency and Petri Nets: Advances in Petri Nets, pp. 819–848. Springer, Heidelberg (2004). https://doi.org/10.1007/978-3-540-27755-2_23
7. Köhler, M., Moldt, D., Rölke, H.: Modelling mobility and mobile agents using nets within nets. In: van der Aalst, W.M.P., Best, E. (eds.) Applications and Theory of Petri Nets 2003, pp. 121–139. Springer, Heidelberg (2003). https://doi.org/10.1007/3-540-44919-1_11
8. Köhler, M., Moldt, D., Rölke, H.: Modelling the structure and behaviour of Petri Net agents. In: Colom, J.-M., Koutny, M. (eds.) Applications and Theory of Petri Nets 2001, pp. 224–241. Springer, Heidelberg (2001). https://doi.org/10.1007/3-540-45740-2_14
9. Lomazova, I.A.: Nested Petri Nets – a formalism for specification and verification of multi-agent distributed systems. Fundam. Inform. **43**, 195–214 (2000). https://doi.org/10.3233/FI-2000-43123410
10. Lomazova, I.: Nested Petri Nets: multi-level and recursive systems. Fundam. Inf. **47**, 283–293 (2001)
11. Lomazova, I.A.: Nested Petri Nets for adaptive process modeling. In: Pillars of Computer Science, pp. 460–474. Springer, Heidelberg (2008). https://doi.org/10.1007/978-3-540-78127-1_25
12. DiCesare, F., Jeng, M.D.: Synthesis for manufacturing systems integration. In: Practice of Petri Nets in Manufacturing, pp. 103–146. Springer Netherlands, Dordrecht (1993). https://doi.org/10.1007/978-94-011-6955-4_3

13. Jeng, M.D., DiCesare, F.: A review of synthesis techniques for Petri Nets with applications to automated manufacturing systems. IEEE Trans. Syst. Man. Cybern. **23**, 301–312 (1993). https://doi.org/10.1109/21.214792
14. Chrzastowski-Wachtel, P., Benatallah, B., Hamadi, R., O'Dell, M., Susanto, A.: A Top-down Petri Net-based approach for dynamic workflow modeling (2003). https://doi.org/10.1007/3-540-44895-0_23

Virtual Activating of a Robotized Production Cell with Use of the Mechatronics Concept Designer Module of the PLM Siemens NX System

Krzysztof Herbuś, Piotr Ociepka, and Aleksander Gwiazda(✉)

Faculty of Mechanical Engineering, Silesian University of Technology, Konarskiego 18A str., 44-100 Gliwice, Poland
{krzysztof.herbus,aleksander.gwiazda}@polsl.pl

Abstract. In the work the way of the virtual activating of the robotized production cell with use of the OPC server is presented. The use of CAD/CAE class systems allows simulation of the operation of an industrial robot as a part of the implementation of the modified technological process. Based on the conducted virtual tests, it is possible to verify the mechatronic function of the tested system and to predict the collision of robot elements with other elements of the system. It makes possible to eliminate collision events at the design stage of a given technological process. Robot models together with their work environment are prepared to simulate the mechatronic function of the robotized manufacturing workcell in the "Mechatronics Concept Designer" module of the PLM Siemens NX software. In the mentioned model, the "rigid body" components were created that map the geometric form of individual elements of the system, the "joint" components reflecting the nature of cooperation between the "rigid body" components, "position control" components affecting the system state and elements of the "Position sensor" type that monitors the state of the system. The OPC server is the element integrating the 3D model of the system functioning in the CAD/CAE system with the developed control application. In order to integrate the virtual model of the robotized manufacturing workcell, the objects of the "signal" type were also created, which are responsible for transmitting information about the state of the object to the control system and for reading information on the further work of the executive system.

Keywords: Virtual activating · Mechatronics Concept Designer · Integration

1 Introduction

The designed machines or technological lines could be considered as mechatronics systems, where mechanical, electronic and IT sub-systems are designed concurrently interacting with one another [1]. One of the stages of the design and construction process using the CAD/CAE systems is a virtual activating of the designed system in the context of control. In this case, the function of the mechatronic system is verified [2, 3]. The continuous development of computer-aided systems of mechatronics machines design allows testing the operation of virtual mechatronic systems using the real or virtual

© Springer Nature Switzerland AG 2019
A. Burduk et al. (Eds.): ISPEM 2018, AISC 835, pp. 417–425, 2019.
https://doi.org/10.1007/978-3-319-97490-3_41

controller. The virtual model of the mechatronic system is created in the CAD/CAE class system and the program describing the operation of the system is created in various environments used for programing the PLC controller [4–7]. In order to properly connect both environments, it is necessary to use an integrator. In the Institute of Technological Processes Automation and Integrated Manufacturing System, numerous research works are carried out related to the control of automated and robotized systems [8–12]. There are also works in the field of integration of programming environments with the use of OPC server, dynamic data exchange and ActivX technique [13–18], where the virtual model of the mechatronic system and the virtual controller are subject of integration. The paper presents an example of virtual activating of a robotized production cell in the context of its control. The OPC server is an integrating element of the virtual production cell 3D model with a virtual controller [19–24].

2 Description of the Analyzed Technological Process

The proposed method of virtual activating of the production system was presented on the example of a robotized manufacturing workcell (Fig. 1), in which the process of press welding of parts of the car's body is carried out.

Fig. 1. Robotized workcell for the press welding of car body elements

The presented production cell consists of the following subsystems:

- Input magazine (1), work Table (2).
- Ready-made products magazine (3).
- Robot for components manipulating (4 - R01).

- Robots equipped with welding guns (5 - R02, 6 - R03).
- Glue application stand (7).

During this process, four components are joined (press welded): L_PILL_OUT-external pillar plate, L_PILL_IN - inner pillar plate, L_SIDE_PRT_OUT - outer side plate, L_SIDE_PRT_IN - inner window frame plate. Figure 2 shows the positioning sequence of the elements on the work table.

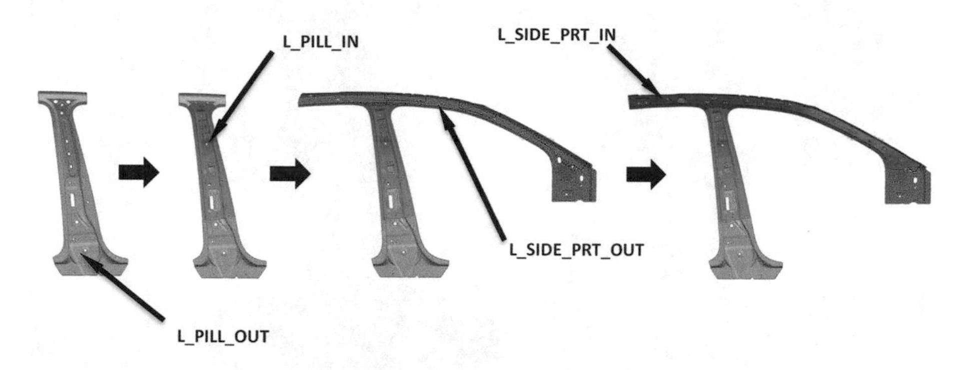

Fig. 2. Order of positioning of elements on the work table

During assembly of the elements, gluing technology was also used. Its purpose is to pre-connect the two components L_SIDE_PRT_OUT and L_PILL_OUT together just before the press welding process. This is to prevent the connected elements from moving when the clamps on the work table are not yet closed. This treatment affects the accuracy of the positioning and welding process. Figure 3 shows the place of glue application of the combined components.

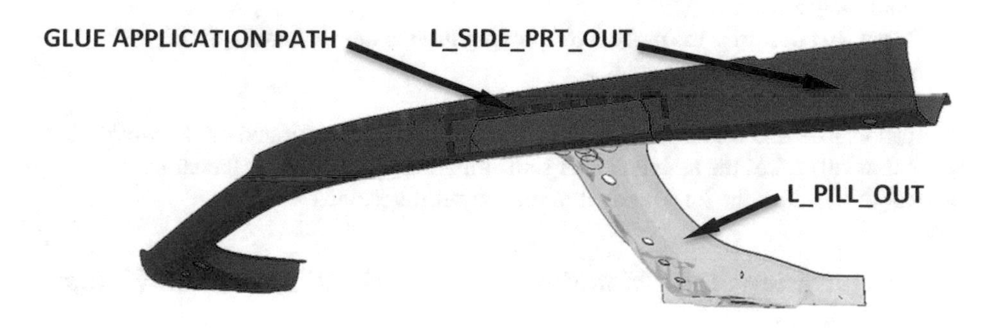

Fig. 3. Place of glue application

After proper positioning and fixing of elements on the work table, the welding process takes place. This process is carried out simultaneously by two robots. Figure 4 shows the distribution of the welding points with reference to the part of the body being assembled.

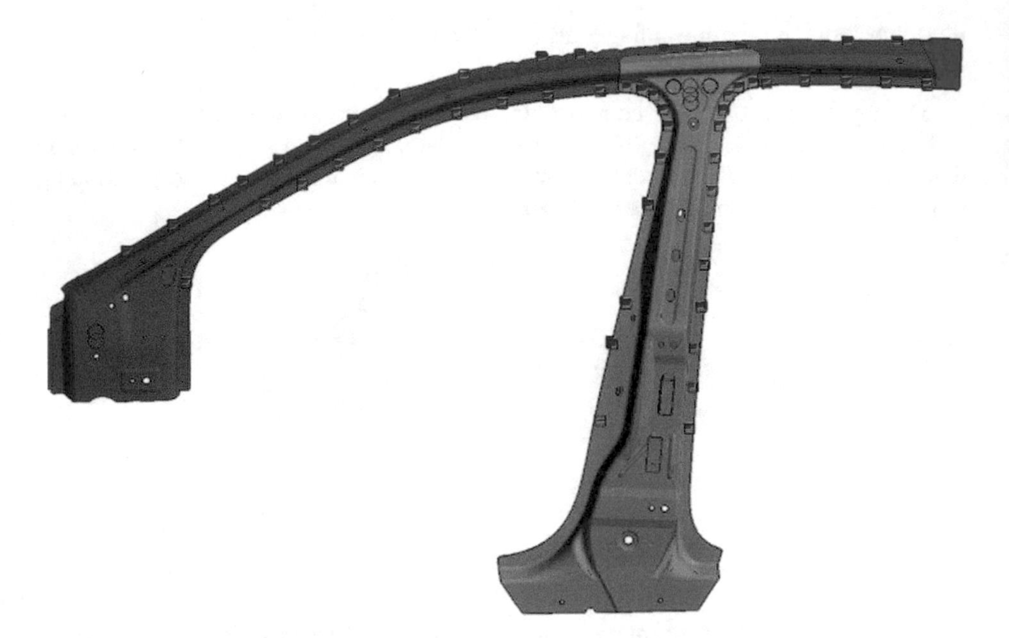

Fig. 4. Distribution of welding points on the target product

The course of the technological process in the presented production system is implemented in the following stages:

- Robot R01 takes the L_PILL_IN and L_PILL_OUT parts from the input magazine.
- Robot R01 positions the manipulated parts on the work table.
- Robot R01 gets the elements L_SIDE_PRT_IN and L_SIDE_PRT_OUT from the input magazine.
- Robot R01 performs the process of applying glue to the element L_SIDE_PRT_OUT on the external gluing machine.
- Robot R01 positions parts on the work table.
- Robots R02 and R03 simultaneously weld the details positioned on the work table.
- Robot R01 takes the assemblies of semi-finished products from the table.
- Robot R01 lays the final product to the output magazine.

3 Preparation of the Manufacturing Workcell Model for Virtual Activating

In order to virtual activate the robotized workcell, a model for integration with a virtual control system was prepared first. For this purpose, the Mechatronics Concept Designer module of the PLM Siemens NX software was used. The main executive elements of the system are drives in the form of pneumatic actuators that drive clamps located on the working table, electric motors for rotary motion controlling the operation of robot arms and motors controlling the work of welding heads. The process of preparing the

3D model of the system for integration with the virtual control system is presented on the example of the 3D model of the Kuka FORTEC KR 360 R2830 robot (Fig. 5). This process was carried out in the following steps:

- Creating the objects of the rigid body type (RBi) –which map the physical properties of elements that are involved in the simulation (mass, moments of inertia, etc.);
- Creating the objects of joint type (Ji) –which map the relationship between the objects of the rigid body type (RBi) and determine the possible type of motion of individual RBi objects during simulation;
- Creating the objects of the position control type (PCi) –which determine the values of line or angular displacements of objects of the rigid body (RBi) type in relation to objects of the joint type (Ji);
- Creating the objects of the position sensor type (PSi) – which monitor the values of linear or angular displacements of objects of the rigid body (RBi) type in relation to objects of the joint type (Ji);
- Creating the objects of the output signal type (Si_O) –which transmit information from the objects of the position sensor type (PSi) to the integrator;
- Creating the objects of the input signal type (Si_I) –which read information from the integrator and transmit it to objects of the position control type (PCi).

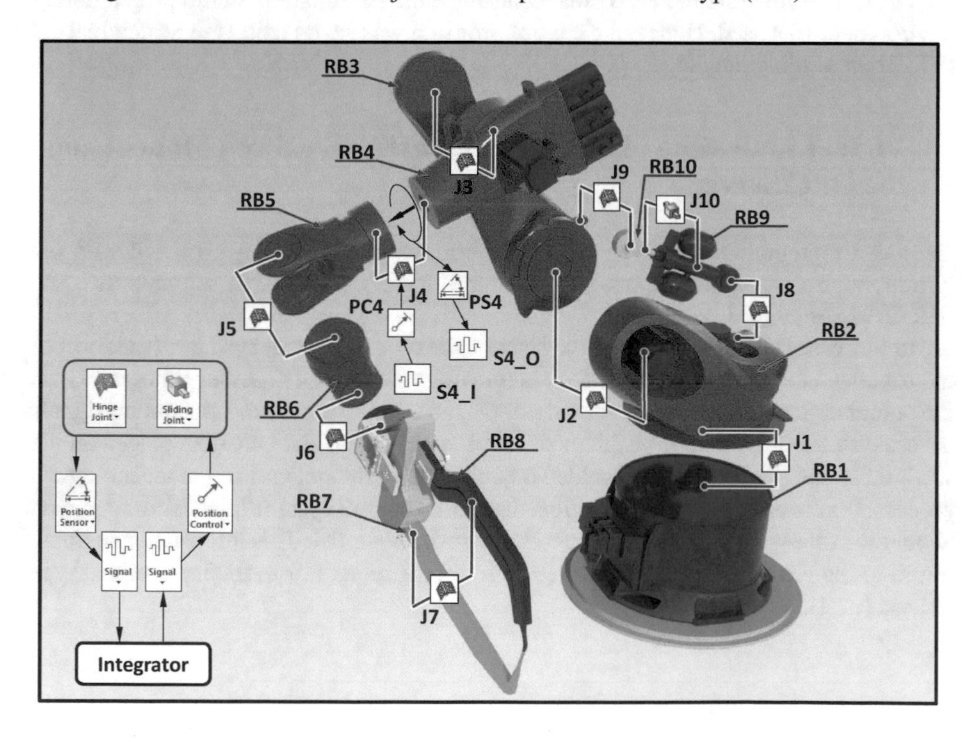

Fig. 5. Model of the robot prepared for the integration with the virtual controller

The next stage of creation the model prepared to motion simulate was associated with the imposition of constraints on the individual axes of the robot manipulator.

Imposed constrains allow faithfully reflecting the ranges of motion performed by the robot manipulator in its internal layout (Table 1).

Table 1. Range of motion of the robot in relation to individual axes

Axis number	Movement range (°)
J1	–185 to +185
J2	–130 to +20
J3	–100 to +144
J4	–350 to +350
J5	–120 to +120
J6	–350 to +350

The trajectories of robot movements as part of the technological process were modeled in the form of 3D curves in the PLM Siemens NX system. On the basis of the conducted simulation based on imposed constraints, ranges of movements and forms of assumed trajectories taking into account process points, a table of values was created in relation to individual objects of the PCi type. Then these values were implemented in the virtual control system. As a virtual controller, the Automation Studio programming environment was used. Hence, the control program was saved using the Structure Text (ST) programming language.

4 Virtual Activating of the Manufacturing Workcell Model Using the OPC Server

In order to integrate a virtual model of a robotized manufacturing workcell with the virtual control system, an integrator in the form of an OPC server was used (BR.OPC.Server_3.0_V1.14.19).

In this case, information about the state of the object (control system in the form of the robotized manufacturing workcell) is transferred to the OPC server (integrator) via the output objects of the signal type (Si_O) (Fig. 6). At the same time, the output signals describe states of particular positions of robot arms, clamp and actuators placed on the work table and the presence of welded objects at particular stages of the technological process. On the basis of this information, the virtual control system "makes" the decision about the process in relation to the robotized workcell model. This information is transferred to the virtual 3D model of the system via the integrator to the input signal type objects (Si_I).

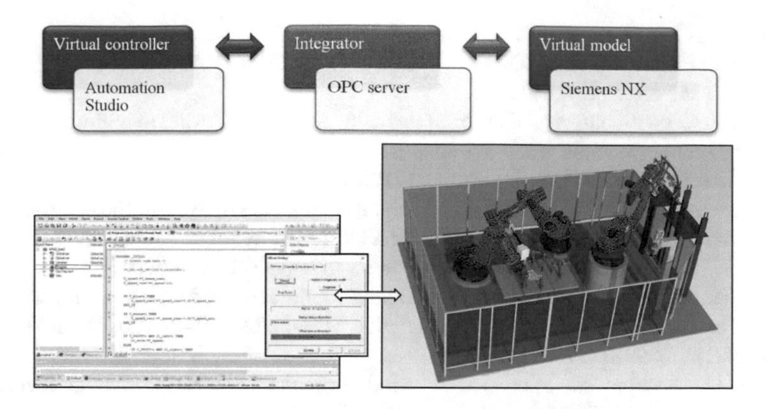

Fig. 6. Integration of the virtual model of the robotized production cell with the virtual controller

5 Conclusions

The use of the Mechatronics Concept Designer module of the PLM Siemens NX system to describe the 3D model of the robotized manufacturing workcell enables preparing the model for its virtual activating.

The application of the integrator, in the form of the OPC server, and signal objects (Si_O and Si_I) enables the exchange of signals, describing the state of the control object (PSi) and settings of input values (PCi), between the 3D virtual model of the robotized production cell and the virtual control system.

The described way of the integration of the virtual model of a robotized production cell using MCD module of PLM Siemens NX system with a virtual control system enables on minimize the risk of incorrect working of the real system.

In order to improve the process of robot control within the process under consideration, it is advisable to create a virtual control panel of an industrial robot that will enable verification of a specific robot position with a welding head or gripper.

References

1. Gawrysiak, M.: Mechatronics and mechatronics design. Wydawnictwo Politechniki Białostockiej, Białystok (1997). (in Polish)
2. Lückel, J.: The concept of mechatronics function modules applied to compound active suspension systems. In: Proceedings of Symposium "Research Issues in Automotive Integrated Chassis Control Systems", International Symposium for Vehicle System Dynamics, Herbertov, pp. 1–6 (1992)
3. Bolton, W.: Mechatronics: Electronic Control Systems in Mechanical and Electrical Engineering. Pearson Education, Harlow (2016)
4. Banaś, W., Ćwikła, G., Foit, K., Gwiazda, A., Monica, Z., Sękala, A.: Experimental determination of dynamic parameters of an industrial robot. IOP Conf. Ser. Mater. Sci. Eng. **227**, 1–8 (2017). https://doi.org/10.1088/1757-899X/227/1/012012

5. Banaś, W., Ćwikła, G., Foit, K., Gwiazda, A., Monica, Z., Sękala, A.: Modelling of industrial robot in LabView Robotics. IOP Conf. Ser. Mater. Sci. Eng. **227**, 1–8 (2017). https://doi.org/10.1088/1757-899X/227/1/012011
6. Hryniewicz, P., Banaś, W., Foit, K., Gwiazda, A., Sękala, A.: Modelling cooperation of industrial robots as multi-agent systems. IOP Conf. Ser. Mater. Sci. Eng. **227**, 1–8 (2017). https://doi.org/10.1088/1757-899X/227/1/012061
7. Banaś, W., Gwiazda, A., Monica, Z., Sękala, A., Foit, K.: Analysis of the position of robotic cell components and its impact on energy consumption by robot. IOP Conf. Ser. Mater. Sci. Eng. **145**, 1–8 (2016). https://doi.org/10.1088/1757-899X/145/5/052017
8. Gołda, G., Kampa, A.: Manipulation and handling processes off-line programming with use of K-Roset. IOP Conf. Ser. Mater. Sci. Eng. **227**, 1–8 (2017). https://doi.org/10.1088/1757-899X/227/1/012050
9. Gołda, G., Kampa, A., Paprocka, I.: Modeling and simulation of manufacturing line improvement. Int. J. Comput. Eng. Res. **6**(10), 26–31 (2016)
10. Hetmańczyk, M., Michalski, P.: The aid of a mistake proofing with the use of mechatronic systems according to the Poka-Yoke methodology. Adv. Mater. Res. **837**, 399–404 (2014). https://doi.org/10.4028/www.scientific.net/AMR.837.399
11. Hetmańczyk, M., Michalski, P.: The qualitative assessment of pneumatic actuators operation in terms of vibration criteria. IOP Conf. Ser. Mater. Sci. Eng. **95**, 1–8 (2015). https://doi.org/10.1088/1757-899X/95/1/012056
12. Michalski, P., Hetmańczyk, M.: Implementation of the safety components base on industrial networks. IOP Conf. Ser. Mater. Sci. Eng. **95**, 1–8 (2015). https://doi.org/10.1088/1757-899X/95/1/012085
13. Herbuś, K., Kost, G., Reclik, D., Świder, J.: Integration of a virtual 3D model of a robot manipulator with its tangible model (phantom). Adv. Mater. Res. **837**, 582–587 (2014). https://doi.org/10.4028/www.scientific.net/AMR.837.582
14. Herbuś, K., Ociepka, P., Gwiazda, A.: Conception of the integration of the virtual robot model with the control system. Adv. Mater. Res. **1036**, 732–736 (2014). https://doi.org/10.4028/www.scientific.net/AMR.1036.732
15. Banaś, W., Herbuś, K., Kost, G., Nierychlok, A., Ociepka, P., Reclik, D.: Simulation of the Stewart platform carried out using the Siemens NX and NI LabVIEW programs. Adv. Mater. Res. **837**, 537–542 (2014). https://doi.org/10.4028/www.scientific.net/AMR.837.537
16. Herbuś, K., Ociepka, P.: Analysis of the Hexapod work space using integration of a CAD/CAE system and the LabVIEW software. IOP Conf. Ser. Mater. Sci. Eng. **95**, 1–8 (2015). https://doi.org/10.1088/1757-899X/95/1/012096
17. Herbuś, K., Ociepka, P.: Integration of the virtual 3D model of a control system with the virtual controller. IOP Conf. Ser. Mater. Sci. Eng. **95**, 1–8 (2015). https://doi.org/10.1088/1757-899X/95/1/012084
18. Herbuś, K., Ociepka, P.: Mapping of the characteristics of a drive functioning in the system of CAD class using the integration of a virtual controller with a virtual model of a drive. Appl. Mech. Mater. **809/810**, 1249–1254 (2015). https://doi.org/10.4028/www.scientific.net/AMM.809-810.1249
19. Herbuś, K., Ociepka, P.: Determining of a robot workspace using the integration of a CAD system with a virtual control system. IOP Conf. Ser. Mater. Sci. Eng. **145**, 1–8 (2016). https://doi.org/10.1088/1757-899X/145/5/052010
20. Herbuś, K., Ociepka, P.: Integration of the virtual model of a Stewart platform with the avatar of a vehicle in a virtual reality. IOP Conf. Ser. Mater. Sci. Eng. **145**, 1–8 (2016). https://doi.org/10.1088/1757-899X/145/4/042018

21. Herbuś, K., Ociepka, P.: Verification of operation of the actuator control system using the integration the B&R Automation Studio software with a virtual model of the actuator system. IOP Conf. Ser. Mater. Sci. Eng. **227**, 1–8 (2017). https://doi.org/10.1088/1757-899X/227/1/012056
22. Herbuś, K., Ociepka, P.: Designing of a technological line in the context of controlling with the use of integration of the virtual controller with the mechatronics concept designer module of the PLM Siemens NX software. IOP Conf. Ser. Mater. Sci. Eng. **227**, 1–8 (2017). https://doi.org/10.1088/1757-899X/227/1/012057
23. Herbuś, K., Ociepka, P., Gwiazda, A.: Application of functional features to the description of technical means conception. Adv. Mater. Res. **1036**, 1001–1004 (2014). https://doi.org/10.4028/www.scientific.net/AMR.1036.1001
24. Ociepka, P., Herbuś, K.: Application of CBR method for adding the process of cutting tools and parameters selection. IOP Conf. Ser. Mater. Sci. Eng. **145**, 1–8 (2016). https://doi.org/10.1088/1757-899X/95/1/012100

Modelling and Simulation of Production Processes

A Declarative Modelling Framework for Routing of Multiple UAVs in a System with Mobile Battery Swapping Stations

Grzegorz Bocewicz[1(✉)], Peter Nielsen[2], Zbigniew Banaszak[1], and Amila Thibbotuwawa[2]

[1] Faculty of Electronics and Computer Science, Koszalin University of Technology, Koszalin, Poland
bocewicz@ie.tu.koszalin.pl
[2] Department of Materials and Production, Aalborg University, Aalborg, Denmark
{peter,amila}@mp.aau.dk

Abstract. A flow production system with concurrently executed supply chains providing material handling/transportation services to a given set of workstations is considered. The workstations have to be serviced within preset time windows and can be shared by different supply chains. The transportation and material handling operations supporting the flow of products between the workstations are carried out by a fleet of Unmanned Aerial Vehicles (UAVs). The batteries on-board the UAVs are replaced at mobile battery swapping stations (MBSs). The focus of this study is a cyclic steady-state flow of products and transportation means, i.e. a state in whose cycle workstations are serviced periodically, within preset time windows, by the same transportation means travelling the same transportation routes. Under this assumption, UAV batteries are swapped at the some locations of battery replacement depots at moments which are multiples of the cycle under consideration. Similar assumptions are made for the fleet of MBSs. To find a solution to the above problem of routing UAV and MBS fleets, one needs to determine the routes travelled by the UAVs servicing the workstations and the routes travelled by the MBSs servicing the battery swapping points, such that the total length of these routes is minimized.

Keywords: Unmanned aerial vehicles · Delivery routing problem
Declarative modelling · Mobile battery swapping station

1 Introduction

Unmanned aerial vehicles (UAVs) have the potential to significantly reduce the cost and time required to deliver commodities. In that context, they are well-suited for delivery services provided in supply chain networks, which can be seen as combinations of nodes with a given capability and capacity, connected by lanes to help commodities move between senders and receivers [1, 3–6, 8, 9]. Delivering with UAVs can be faster than delivering with traditional delivery vehicles, as UAVs are not limited by established infrastructure such as roads, and generally face less complex obstacle

© Springer Nature Switzerland AG 2019
A. Burduk et al. (Eds.): ISPEM 2018, AISC 835, pp. 429–441, 2019.
https://doi.org/10.1007/978-3-319-97490-3_42

avoidance scenarios [10, 13]. UAVs are more and more frequently considered for use in the movement of materials and products between and within departments and even between workstations [11].

One of the most important factors limiting the use of UAVs in this type of applications is the weight and limited capacity of their batteries. To eliminate the limitation imposed by the finite energy source on board each UAV, a network of shared refueling stations distributed across the field is frequently proposed [7]. In order to accomplish a mission, the UAVs can refuel, i.e. charge or swap onboard batteries, at any station and return to service. To reduce battery charging downtime, novel solutions have been proposed which enable “hot” battery swap, during which the vehicle remains powered on [9].

Solutions analogous to the approach in which a battery swapping service for electric vehicles is provided by mobile battery swapping vans [12] can be used for UAV applications. More specifically, in the case of cyclically repeated UAV missions, the operations performed along the routes, in addition to loading/unloading of goods, will also include battery replacement. When the cyclically repeated routes of UAVs and the battery discharge rate are known, it is possible to find the location of battery replacement depots and the moments at which the batteries should be replaced. The resultant network of battery swapping points, resembling a public transport network (e.g. a network of bus stops), can be operated by a fleet of mobile battery swapping stations (MBSs), which arrive at given swapping points at times specified in a preset timetable.

In this context, the routing problem for a fleet of MBS considered in this work boils down to finding a solution that will minimize the number of vehicles in the fleet and the total length of the routes they travel. In this form, our present investigation is an extension of our earlier studies [1, 2, 11, 13] in which we explored the problem of deployment (distribution) of battery swapping depots as part of a split delivery vehicle routing problem with time windows, which took into account the constraints imposed by the cyclic nature of the production flow system in which the fleet was used. The focus of the investigations reported in our previous papers were solutions which minimized the number of swapping depots used within a given production takt time.

The NP-hard character of UAV routing and scheduling problems justifies the use of a declarative modeling framework and implementation of the problems in the Oz Mozart constraint programming platform.

The remainder of this paper is organized as follows: Sect. 2 provides an example illustrating the problem of routing mobile battery swapping stations. Section 3 presents a declarative model of the problem of routing UAV and MBS fleets. The problem consists in determining cyclically travelled routes for the UAVs servicing suppliers and recipients and routes cyclically travelled by MBSs servicing battery swapping points. Section 4 shows how the problem can be solved in the Oz Mozart environment. Section 5 provides the key conclusions and indicates the main directions of future research.

2 Routing of Mobile Battery Swapping Stations

2.1 An Illustrative Example

In a flow production system consisting of four workstations, whose structure is shown in Fig. 1, two different products are being manufactured at the same time. The technological route of product J_1, marked in red, runs through workstations R_1, R_2, and R_4, with respective technological operation times of 20 u.t. (units of time) for $O_{1,1}$, 20 u.t for $O_{1,2}$, and 30 u.t. for $O_{1,3}$. In turn, the technological route of product J_2, marked in blue, runs through workstations R_1 and R_3, whose respective technological operation times are 15 u.t. for $O_{2,1}$ and 30 u.t. for $O_{2,2}$.

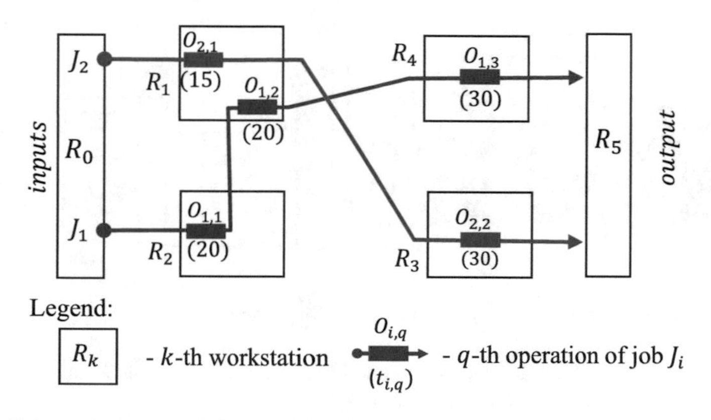

Fig. 1. Schematic layout of the considered multi-item batch flow production workshop

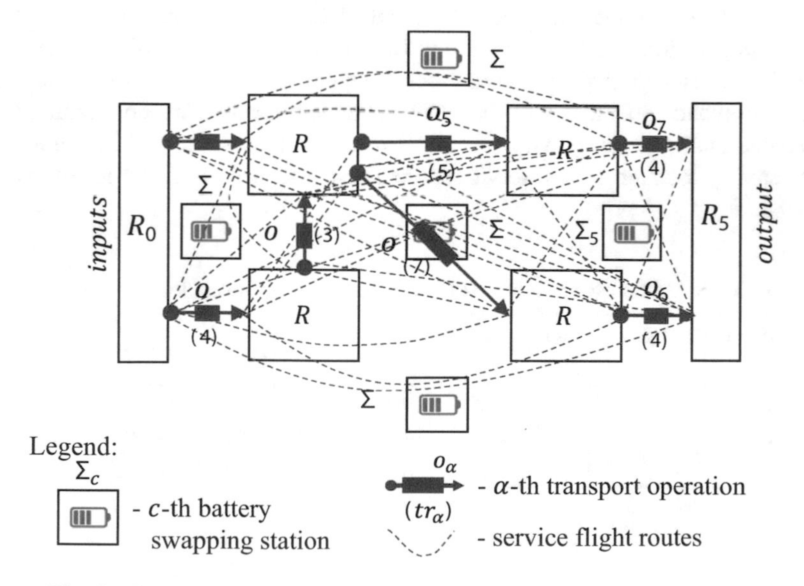

Fig. 2. Structure of possible transportation and service routes in the workshop

Let us assume that transport operations $o_1, \ldots, o_7$ are carried out by a fleet of three UAVs U_1, U_2, and U_3 (Fig. 2). Examples of trajectories of the cyclic flights of UAVs U_1 and U_2/U_3 (U_3 moves along the same trajectory as U_2) are shown in Fig. 3b. An example of the course of the trajectories for the UAV fleet is given in Fig. 3a. It is assumed that the time of each operation o_α and the times of service flights (between successive operations o_k and o_l along the trajectory of a cyclically repeated mission) are given, and are the same as in Fig. 2. This solution, in the case of a fleet consisting of three UAVs, results in a production takt time of $TP = 40$ u.t. Production takt time TP is understood here as the time that elapses between two successive items of a given product coming out from the production process.

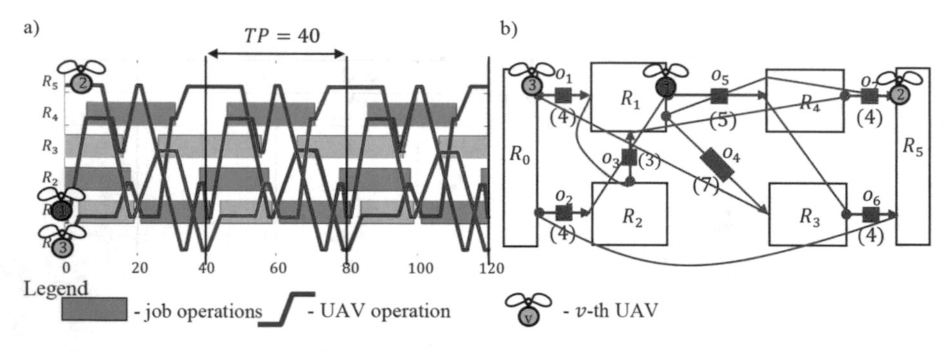

Fig. 3. Routes traversed by UAVs (a), Gantt chart of production flow taking into account material handling and service flight operations (b)

Given the takt time value determined above, the goal now is to find modifications of the previously determined trajectories, i.e. new trajectories that will incorporate battery swapping stations, such that travelling from station to station will not extend the takt time. It is assumed that the battery replacement times are the same for each swapping station and equal to 3 u.t. The newly obtained solution, which maintains the same production takt time is shown in Fig. 4. This solution uses three of the admissible battery swapping locations. The service timetables assigned to these locations are shown in Table 1.

Table 1. Schedule for servicing APP in selected battery swapping locations

Production cycle	Battery swapping stations		
	Σ_2	Σ_4	Σ_5
1	[10,13] (APP_3)	[9,12] (APP_2)	[9,12] (APP_1)
2	[50,63] (APP_2)	[49,52] (APP_3)	[49,52] (APP_1)
3	[90,103] (APP_3)	[89,92] (APP_2)	[89,92] (APP_1)

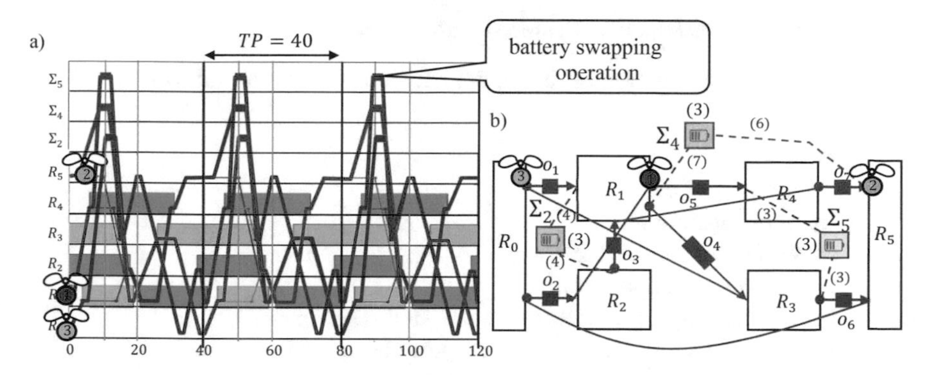

Fig. 4. Gantt chart of production flow incorporating material handling and service flight operations as well as battery swapping operations (a); cyclic trajectories travelled by UAVs and the positioning of battery swapping points (b)

2.2 Problem Statement

The problem considered assumes that there is a given set of workstations allocated at different points in space, to be served periodically (within a specified time window) by several capacitated UAVs charged from a set of spatially distributed battery swapping points serviced by a fleet of MBSs. The goal is to minimize both the number of UAVs needed and the total distance travelled by them, such that each workstation is visited with the right delivery at the right moment of time, without violating the battery capacity and battery swapping constraints. A secondary objective is to ensure, at the same time, that each battery swapping point is visited by the right UAVs at the right moments of time. Such a complex UAV and MBS routing and scheduling problem is known to be NP-hard, which, in the case of large-scale systems, such as those encountered in practice, limits the set of feasible solving methods to heuristic approaches.

3 Declarative Modelling

The declarative modeling approach is known for its ability to cope with the limited flexibility of the imperative approach. Consequently, in contrast to imperative languages, declarative languages do not specify the procedure *a priori*. Instead of determining how a searching procedure has to work exactly, only its essential characteristics are described. Following this direction, let us focus on step-by-step additive refinement of a declarative model standing behind a constraint satisfaction problem aimed at routing a multi-UAV system with Mobile Battery Swapping Stations

3.1 Constraints Satisfaction Problem Refinement

The mathematical formulation of the model considered uses the following:

Symbols:

R_k: resource k;
$O_{i,q}$: operation q of J_i;
o_α: transport operation α;
$U\tau_\mu$: movement of μ–th MBS;
Σ_c: c-th Battery swapping station;
pb_v: battery capacity of transport means v (v-th UAV);
b_α: index of transport operation which precedes o_α;
f_α: index of transport operation which follows o_α.
j_i: job i;
U_v: transport means v (v-th UAV);
$K_{\mu,\varepsilon}$: corridor linking resources R_μ and R_ε;

Sets and Sequences:

R: the set of resources R_k (workstations);
J: the set of jobs J_i, (production processes);
O_i: sequence of operations for J_i: $O_i = \left(O_{i,1}, \ldots, O_{i,q}, \ldots, O_{i,lm_i}\right)$;
p_i: route of J_i, sequence of resources on which operations $O_{i,q}$ are executed: $p_i = \left(p_{i,1}, \ldots, p_{i,q}, \ldots, p_{i,lm_i}\right)$, $p_{i,q} \in R$;
Σ: the set of battery swapping points Σ_c;
Q_k: the set of operations executed on R_k;
$\mathcal{O}$: the set of transport operations o_α;
S_k: the set of transport operations started from R_k, $S_k \subseteq \mathcal{O}$;
E_k: the set of transport operations ending on R_k, $E_k \subseteq \mathcal{O}$;
U: the set of transport means U_v (transport processes);
$U\tau$: the set of mobile battery swapping stations MBS $U\tau_\mu$
B: sequence of predecessor indices of transport operations, $B = (b_1, \ldots, b_\alpha, \ldots, b_\omega)$, $b_\alpha \in \{0, \ldots \omega\}$;
F: sequence of successor indices of transport operations, $F = (f_1, \ldots, f_\alpha, \ldots, f_\omega)$, $f_\alpha \in \{1, \ldots \omega\}$.
PB : a sequence of values characterizing the battery capacity of UAVs: $PB = (pb_1, \ldots, pb_v, \ldots, pb_l)$, pb_v – battery capacity of v-th UAV

Parameters:

m: number of resources;
n: number of jobs;
e: number of battery swapping stations,
z: number of MBSs,
lm_i: number of operations of J_i;
$t_{i,q}$: operation time of $O_{i,q}$;

$d_{a,\beta}$: travel time between resource at which operation o_α ends and resource at which operation o_β begins;
$dm_{\gamma,\delta}$: travel time between battery swapping points Σ_γ and Σ_δ,
$db_{a,c,\beta}$: flight time from workstation (resource) at which operation o_α ends to battery swapping point Σ_c to resource at which operation o_β begins;
zp : battery consumption (per unit of time) by an UAV waiting on resource (hovering);
zs: battery consumption (per unit of time) during a service flight;
zl: battery consumption (per unit of time) during a transport operation;
tw: battery replacement time (the same for all UAVs)
TP^*: maximum value of production takt time TP.
l: number of transport means;
ω: number of transport operations;
tr_α: operation time of o_α,

Variables:
TP: production takt time;
$x_{i,q}$: start time of operation $O_{i,q}$,
$y_{i,q}$: end time of operation $O_{i,q}$;
xt_α: start time of operation o_α;
yt_α: end time of operation o_α;
xs_α : the moment the resource occupied by an UAV is released after completion of operation o_α;
$x\tau_{\mu,c}$: start time of battery swapping operation at point Σ_c for the μ-th MBS,
$y\tau_{\mu,c}$: end time of battery swapping operation at point Σ_c for the μ-th MBS,
$m\tau_{\mu,\gamma,\delta}$: variable determining the route of the μ-th MBS, if $m\tau_{\mu,\gamma,\delta} = 1$, then the$\mu$-th MBS moves from station Σ_γ to Σ_δ;
lb_α: battery charge level of an UAV after the completion of operation o_α;
$\mathcal{O}_\mathcal{B}$: a subset of transport operations followed by battery replacement: $\mathcal{O}_\mathcal{B} \subseteq \mathcal{O}$;
$\Sigma_\mathcal{B}$: subset of battery swapping points ($\Sigma_\mathcal{B} \subseteq \Sigma$); each operation in set $\mathcal{O}_\mathcal{B}$ is assigned a single battery swapping station;
b_α: index of the transport operation preceding operation o_α (operations o_{b_α} and o_α are executed by the same UAV); $b_\alpha = 0$ means that o_α is the first operation of the system cycle;
f_α: index of the transport operation following o_α (operations o_α and $o_{f\alpha}$ are executed by the same UAV).

Constraints:

I. For (production processes):

$$y_{i,q} = x_{i,q} + t_{i,q}, \quad q = 1 \ldots lm_i, \forall J_i \in J, \tag{1}$$

$$y_{i,q} \leq x_{i,q+1} \quad , q = 1 \ldots (lm_i - 1), \ \forall J_i \in J, \tag{2}$$

$$y_{i,q} \leq x_{i,q} + TP \text{ , } q = 1 \ldots lm_i \text{ , } \forall J_i \in J, \tag{3}$$

$$\left(y_{i,a} \leq x_{j,b}\right) \vee \left(y_{j,b} \leq x_{i,a}\right), \text{ when } O_{i,a}, O_{j,b} \in Q_k, \forall R_k \in R, \tag{4}$$

$$TP \leq TP^*. \tag{5}$$

II. For UAVs battery swapping operations:

$$lb_\alpha = pb_v, \forall o_\alpha \in \mathcal{O}_\mathcal{B}, \text{ when } U_v \text{ executes operation } o_\alpha \tag{6}$$

$$d'_{\alpha,\beta} = d_{\alpha,\beta}, \forall o_\alpha \notin \mathcal{O}_\mathcal{B}, \tag{7}$$

$$d'_{\alpha,\beta} = db_{\alpha,c,\beta} + tw, \forall o_\alpha \in \mathcal{O}_\mathcal{B}, \Sigma_c = \varphi(o_\alpha) \tag{8}$$

$$lb_\alpha = lb_\beta - zp\left(xs_\beta - yt_\beta\right) - zs\left(xt_\alpha - xs_\beta\right) - zl\left(yt_\alpha - xt_\alpha\right), \tag{9}$$

$$\forall o_\alpha \notin \mathcal{O}_\mathcal{B}, f_\beta = \alpha,$$

$$lb_\alpha > 0, \forall o_\alpha \in \mathcal{O}, \tag{10}$$

III. For UAVs transport process operations:

$$yt_\alpha = xt_\alpha + tr_\alpha, \quad \alpha = 1, 2, \ldots, \omega, \tag{11}$$

$$b_\alpha = 0, \quad \forall \alpha \in BS, BS \subseteq BI = \{1, 2, \ldots, \omega\}, |BS| = l \tag{12}$$

$$b_\alpha \neq b_\beta \quad \forall \alpha, \beta \in BI \backslash BS, \ \alpha \neq \beta, \tag{13}$$

$$f_\alpha \neq f_\beta \quad \forall \alpha, \beta \in BI, \ \alpha \neq \beta, \tag{14}$$

$$(b_\alpha = \beta) \Rightarrow \left(f_\beta = \alpha\right), \forall b_\alpha \neq 0, \tag{15}$$

$$\left[(b_\alpha = \beta) \wedge \left(b_\beta \neq 0\right)\right] \Rightarrow \left(yt_\beta + d_{\beta,\alpha} \leq xt_\alpha\right), \alpha, \beta = 1, 2, \ldots, \omega, \tag{16}$$

$$\left[(f_\alpha = \beta) \wedge \left(b_\beta = 0\right)\right] \Rightarrow \left(yt_\alpha + d_{\alpha,\beta} \leq xt_\beta + TP\right), \alpha, \beta = 1, 2, \ldots, \omega, \tag{17}$$

$$xs_\alpha \geq yt_\alpha, \quad \alpha = 1, 2, \ldots, \omega, \tag{18}$$

$$\left[(f_\alpha = \beta) \wedge \left(b_\beta \neq 0\right)\right] \Rightarrow \left(xs_\alpha = xt_\beta - d_{\alpha,\beta}\right), \quad \alpha, \beta = 1, 2, \ldots, \omega, \tag{19}$$

$$\left[(f_\alpha = \beta) \wedge \left(b_\beta = 0\right)\right] \Rightarrow \left(xs_\alpha = xt_\beta - d_{\alpha,\beta} + TP\right), \quad \alpha, \beta = 1, 2, \ldots, \omega, \tag{20}$$

$$\left[\left(xs_\alpha < yt_\beta\right) \wedge \left(xs_\beta - TP < yt_\alpha\right)\right] \vee \left[\left(xs_\beta < yt_\alpha\right) \wedge \left(xs_\alpha - TP < yt_\beta\right)\right], \tag{21}$$

$$\forall o_\alpha, o_\beta \in S_k, \ k = 1, \ldots, m,$$

$$\left[(xs_\alpha < yt_\beta) \wedge (xs_\beta - TP < yt_\alpha)\right] \vee \left[(xs_\beta < yt_\alpha) \wedge (xs_\alpha - TP < yt_\beta)\right], \tag{22}$$

$$\forall o_\alpha, o_\beta \in E_k,\ k = 1, \ldots, m,$$

$$\left[(xs_\alpha < xt_\beta) \wedge (xt_\beta - TP < yt_\alpha)\right] \vee \left[(xt_\beta < yt_\alpha) \wedge (xs_\alpha - TP < yt_\beta)\right], \tag{23}$$

$$\forall o_\alpha \in E_k,\ \forall o_\beta \in S_k,\ k = 1, \ldots, m.$$

IV. For transport and production processes (linking UAVs with jobs)

$$x_{i,q} = yt_\alpha + c \times TP,\ c \in \mathbb{N},\ \forall o_\alpha \in E_k,\ \forall O_{i,q} \in Q_k,\ k = 1, \ldots, m, \tag{24}$$

$$y_{i,q} = xt_\alpha + c \times TP,\ c \in \mathbb{N},\ \forall o_\alpha \in S_k,\ \forall O_{i,q} \in Q_k,\ k = 1, \ldots, m. \tag{25}$$

V. Operations executed by MBSs:

$$\sum\nolimits_{i=1}^{e} m\tau_{\mu,\gamma,i} \leq 1,\ \forall U\tau_\mu \in U\tau,\ \gamma = 1 \ldots e, \tag{26}$$

$$\sum\nolimits_{i=1}^{e} m\tau_{\mu,i,\delta} \leq 1,\ \forall U\tau_\mu \in U\tau,\ \delta = 1 \ldots e, \tag{27}$$

$$(m\tau_{\mu,\gamma,\delta} = 1) \Rightarrow \left(\sum\nolimits_{i=1}^{e} m\tau_{\mu,\delta,i} = 1\right),\ \forall U\tau_\mu \in U\tau \tag{28}$$

$$(m\tau_{\mu,\gamma,\delta} = 1) \Rightarrow (x\tau_{\mu,\delta} \geq y\tau_{\mu,\delta} + dm_{\gamma,\delta}),\ \forall U\tau_\mu \in U\tau, \tag{29}$$

$$(m\tau_{\mu,\gamma,\delta} = 1) \Rightarrow (x\tau_{\mu,\delta} \leq xt'_\alpha) \wedge (y\tau_{\mu,\delta} \geq yt'_\alpha),\ \forall U\tau_\mu \in U\tau, \tag{30}$$

where: xt'_α, yt'_α times at which UAVs arrive at and depart from battery swapping station Σ_δ(determined by set of operations $\mathcal{O}_\mathcal{B}$).

For a given fleet of UAVs (set U), the following question is considered: *Does there exist a set of routes (represented by sequences B, F) and a set of battery swapping operations (sets $\mathcal{O}_\mathcal{B}$, $\Sigma_\mathcal{B}$) as well as deployment of corresponding battery swapping devices that will guarantee the existence of a production schedule ($x_{i,q}$, xt_α) satisfying the constraints imposed by the given takt time ($TP \leq TP^*$)?*

The above problem can be seen as the following Constraint Satisfaction Problem:

$$CS = (\mathcal{V}, \mathcal{D}, \mathcal{C}) \tag{31}$$

where: $\mathcal{V} = \{B, F, \mathcal{O}_\mathcal{B}, X, XT, LB\}$ is a set of decision variables, where $X = \{x_{i,q} | i = 1 \ldots n, q = 1 \ldots lm_i\}$, $XT = \{xt_\alpha | \alpha = 1, 2, \ldots, \omega\}$, and $LB = \{lb_\alpha | \alpha = 1, 2, \ldots, \omega\}$; $\mathcal{D}$ is a discrete finite set of domains of variables $\mathcal{V}$; $\mathcal{C}$ is a set of constraints describing the following relations: the execution order of job operations (1)–(3) and UAV operations (24), (25); and mutual exclusion of job operations (4) and UAV operations performed on shared resources (21)–(23). These constraints ensure cyclic routes (13)–(15) and nonempty UAV batteries (6)–(10), and determine the execution order of transport operations (11), (16)–(20) and production takt time requests (5).

To solve the problem formulated as CS (31), one must determine such values (determined by $\mathcal{D}$) of decision variables B, F (UAV routes), $\mathcal{O}_{\mathcal{B}}$ (operations and corresponding battery swapping stations $\Sigma_{\mathcal{B}}$), X, XT (production schedules and transport operation schedules), and LB (UAV battery charge levels), for which all the constraints $\mathcal{C}$ (including the mutual exclusion constraint, the cyclic operation execution constraint, etc.) will be satisfied.

Knowing the number and deployment of battery swapping stations ($\Sigma_{\mathcal{B}}$) and the service schedules assigned to them, one can consider another constraint satisfaction problem, which amounts to searching for an MBS fleet that allows servicing selected battery swapping stations, scattered over a certain (geographical) area, in preset time windows. To solve this problem, one must find an answer to the following question:

Do there exist routes (represented by $m\tau_{\mu,\gamma,\delta}$ and $x\tau_{\mu,\delta}$) of the given MBS fleet (set $U\tau$) which allow UAV batteries to be replaced at the given points $\Sigma_{\mathcal{B}}$, in time windows determined by a preset production schedule ($x_{i,q}$, xt_{α})?

The fact that the routing problem for an MBS fleet is a decision-making problem is captured in a natural way by its formulation as a constraint satisfaction problem:

$$CS' = (\mathcal{V}', \mathcal{D}', \mathcal{C}') \tag{32}$$

where:

$\mathcal{V} = \{M\tau, X\tau\}$ – a set of decision variables characterizing sections of routes along which MBS can move and a schedule for battery replacement operations at stations $\Sigma_{\mathcal{B}}$, where: $M\tau = \{m\tau_{\mu,\gamma,\delta} | \mu = 1\ldots z; \gamma, \delta = 1\ldots e\}$, $X\tau = \{x\tau_{\mu,\delta} | \mu = 1\ldots z; \delta = 1\ldots e\}$,

$\mathcal{D}'$ – a finite set of domains of variables $\mathcal{V}'$,

$\mathcal{C}'$ – a set of constraints (26)–(30) that exclude MBS collisions, as well as ensuring that the right UAVs are serviced at the right times (in accordance with the production schedule $x_{i,q}$, xt_{α} defined in CS (31))

Analogously to (31), to solve problem CS' (32), one needs to find such values (determined by domains $\mathcal{D}$) of decision variables $M\tau, X\tau$ (MBS routes and battery swapping operation schedules), for which all constraints given in set $\mathcal{C}'$ (battery replacement in time windows resulting from schedules $x_{i,q}$, xt_{α}, etc.) will be satisfied.

4 Computational Results

For the battery swapping points deployed as in Fig. 4 and the UAV service schedules assigned to these points as given in Table 1, the goal is to find an answer to the following question: what number of MBSs travelling along what routes are sufficient to service the battery swapping points $\Sigma_{\mathcal{B}} = \{\Sigma_2, \Sigma_4, \Sigma_5\}$. Assuming that

- the time of battery replacement at each station for each UAV is the same and is equal to tw = 3 u.t.,
- the time of a service flight during which a battery is swapped is 4 u.t. longer ($db_{a,c,\beta} = d_{a,\beta} + 4$) than a flight during which no such replacement takes place, and that

- battery capacity is 200, and the battery consumptions are equal to: $zl = 2$ (during a transport operation), $zs = 1$ (during a service flight), $zp = 3$ (during a hovering),

two alternative solutions can be obtained, as shown in Figs. 5 and 6. These figures show solutions to problem CS' (32) for the system from Figs. 1 and 2 and the production schedule given in Fig. 4. (Oz Mozart system, Intel Core i5-3470 3.2 GHz, 8 GB RAM, calculation time, 2 s).

In the first case, corresponding to a situation in which the available fleet consists of three MBSs, $U\tau = (U\tau_1, U\tau_2, U\tau_3)$, the solution features three routes, shown in Fig. 5b, and results in the battery swapping scenario illustrated by the Gantt diagram in Fig. 5a. Batteries are replaced at stations $\Sigma_{\mathcal{B}} = \{\Sigma_2, \Sigma_4, \Sigma_5\}$ in the time windows specified in Table 1. Each UAV has a battery exchanged once per system operation

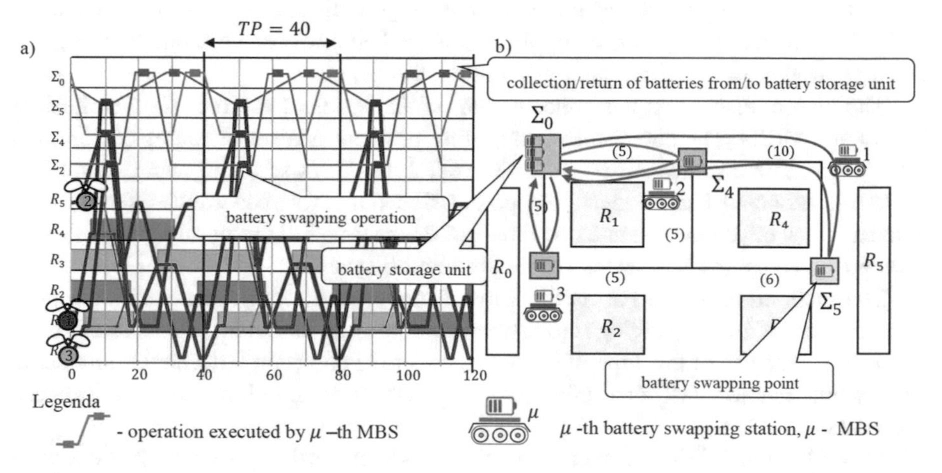

Fig. 5. MBS movement trajectories (a), Gantt chart representing battery swapping service operations (b)

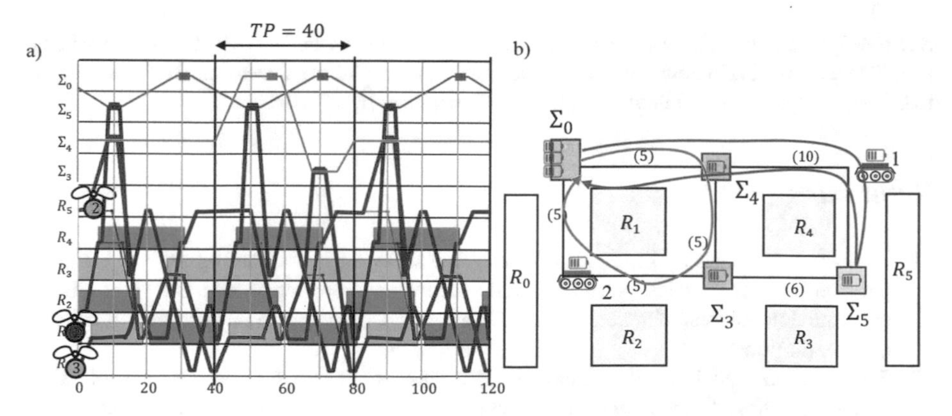

Fig. 6. Gantt chart of service activities at battery swapping stations (a), trajectories of routes traversed by MBSs in a fleet consisting of two stations (b)

cycle (once every 40 u.t.). It is easy to notice that the presented solution ensures that production can be executed in accordance with the preset schedule (production takt time: $TP = 40$ u.j.c., see Fig. 4); however, it requires the use of three mobile battery swapping stations.

Other situations are also possible in the system under consideration. An alternative solution uses a fleet consisting of only two MBSs: $U\tau = (U\tau_1, U\tau_2)$, see Fig. 6.

5 Conclusions

The possibility of flexibly arranging collision-free flight/transport routes for UAVs means that drones can find new applications in indoor settings. The flexibility of this type of solutions is particularly manifest in situations requiring frequent changes in production, in particular, changed in steady state, periodic service to and from delivery and collection depots (e.g. when supply chains are being restructured due to changes in the organization of activities such as sorting, storage, etc.).

The present discussion has concentrated on the issues of routing of given fleets of UAVs and MBSs. The solution to this routing problem is a set of routes of minimum total length. These routes are cyclically travelled by UAVs servicing workstations and by MBSs servicing battery swapping points. Schedules for these routes should guarantee the flow of production at a takt time that deviates as little as possible from the takt time governed by the bottleneck of the production system.

The multi-criterial and NP-hard nature of this problem narrows the set of approaches that can be used in solving it to heuristic methods, in particular, the class of greedy algorithms. The use of this type of strategies under the considered declarative modeling framework enables fast determination of satisfactory solutions to practical-scale problems.

In our future studies, we plan to consider cases of routing problems that allow for failure of UAVs and/or MBSs. In other words, we want to investigate the issues of restructuring UAV flight trajectories and MBS routes, i.e. to extend our approach to the problems of robust routing and scheduling.

Acknowledgements. The work was carried out as part of the POIR.01.01.01-00-0485/17 project, "Development of a new type of logistic trolley and methods of collision-free and deadlock-free implementation of intralogistics processes", financed by NCBiR.

References

1. Bocewicz, G., Nielsen, P., Banaszak, Z., Thibbotuwawa, A.: Routing and scheduling of unmanned aerial vehicles subject to cyclic production flow constraints. In: Proceedings of the 15th International Conference on Distributed Computing and Artificial Intelligence (2018). (in print)
2. Bocewicz, G., Nielsen, P., Banaszak, Z., Thibbotuwawa, A.: Deployment of battery swapping stations for unmanned aerial vehicles subject to cyclic production flow constraints. In: Proceedings of the 24th International Conference on Information and Software Technologies (2018). (in print)

3. Gorecki, T., Piet-Lahanier, H., Marzat, J., Balesdent, M.: Cooperative guidance of UAVs for area exploration with final target allocation. IFAC Proc. Vol. **46**(19), 260–265 (2013)
4. Guettier, C., Lucas, F.: A constraint-based approach for planning unmanned aerial vehicle activities. Constraint Satisf. Plan. Sched. **31**(5), 486–497 (2017)
5. Guerriero, F., Surace, R., Loscri, V., Natalizio, E.: A multi-objective approach for unmanned aerial vehicle routing problem with soft time windows constraints. Appl. Math. Model. **38** (3), 839–852 (2014)
6. Ho, H.M., Ouaknine, J.: The cyclic-routing UAV problem is PSPACE-complete. In: Pitts, A. (ed.) Foundations of Software Science and Computation Structures, FoSSaCS 2015. Lecture Notes in Computer Science, vol. 9034, pp. 328–342 (2015)
7. Suzuki, K.A.O., Kemper Filho, P., Morrison, J.R.: Automatic battery replacement system for UAVs: analysis and design. J. Intell. Robot. Syst. **65**, 563–586 (2012). https://doi.org/10.1007/s10846-011-9616-y
8. Manyam, S.G., Rasmussen, S., Casbeer, D.W., Kalyanam, K., Manickam, S.: Multi-UAV routing for persistent intelligence surveillance and reconnaissance missions In: International Conference on Unmanned Aircraft Systems (ICUAS), USA, pp. 573–580 (2017)
9. Michini, B., Toksoz, T., Redding, J., Michini, M., How, J., Vavrina, M., Vian, J.: Automated battery swap and recharge to enable persistent UAV missions. In: Infotech@Aerospace 2011. American Institute of Aeronautics and Astronautics (2011)
10. Myers, D., Batta, R., Karwan, M.: A real-time network approach for including obstacles and flight dynamics in UAV route planning. J. Def. Model. Simul. Appl. Methodol. Technol. **13** (3), 291–306 (2016)
11. Park, Y., Khosiawan, Y., Moon, I., Janardhanan, M.N., Nielsen, I.: Scheduling system for multiple unmanned aerial vehicles in indoor environments using the CSP approach. In: Czarnowski, I., Caballero, A., Howlett, R., Jain, L. (eds.) Intelligent Decision Technologies 2016. Smart Innovation, Systems and Technologies, vol. 56. Springer (2016)
12. Shao, S., Guo, S., Qiu, X.: A mobile battery swapping service for electric vehicles based on a battery swapping van. Energies **10**(1667), 1–21 (2017)
13. Thibbotuwawa, A., Nielsen, P.: Unmanned Aerial Vehicle Routing Problems: A literature review (in print)

Competence-Based Workforce Allocation for Manual Assembly Lines

Bartlomiej Malachowski(✉) and Przemyslaw Korytkowski

Faculty of Computer Science, West Pomeranian University of Technology in Szczecin, ul. Zolnierska 49, 71-113 Szczecin, Poland
{bmalachowski,pkorytkowski}@zut.edu.pl

Abstract. The paper presents an approach to multi-skilled workforce allocation based on predicted performance. The performance is estimated using Hierarchical Competence Model, which takes into account the required skills at every working station of a u-shaped manual assembly line and possessed skill of operators. The task is to find an allocation that maximises total skill improvement while guarantees required production capacity.

Keywords: Competence · Multi-skill · Learning curve · Manual assembly line Workforce allocation

1 Introduction

Manual assembly lines are organized as flow-shop production where technological operations are grouped into working station handled by human operators. Usually these kinds of assembly lines are working according to so called takt time or cycle time, the time between the exits of two consecutive units from the line. Manual assembly line can have linear or u-shaped layout. A popular special type of manual assembly lines are u-shaped lines often used in just-in-time and lean manufacturing. In the u-shaped line one operator supervises both the entrance and the exit. This kind of organization allows seamless control of production rates, maintaining a constant level of work-in-progress and a rapid response to external and internal disturbances in the materials flow. Miltenburg [10] reviewed the advantages and different variants of u-shaped line organization. Common reasons for establishing u-lines were to reduce WIP inventory, setup time, throughput time, material handling, and to improve quality.

Usually workforce allocation on manual assembly lines is performed in order to minimize makespan, lower labour costs or ensure required production quality level. In this paper we propose a different approach, we are analysing a workforce allocation that maximizes total skill improvement while guarantees predefined production capacity.

Workforce allocation problem received considerable interest in the literature. Van Den Bergh et al. [13] provides literature review. They distinguish hard and soft skill requirements constraint. In the case of a soft constraint, people with other skills could take over when there is a lack of employees with the right skill, what is not possible when hard skill constraint is in place. The same group published skill-oriented workforce

© Springer Nature Switzerland AG 2019
A. Burduk et al. (Eds.): ISPEM 2018, AISC 835, pp. 442–451, 2019.
https://doi.org/10.1007/978-3-319-97490-3_43

planning literature review [2] where skills is defined as the ability of a worker to perform certain tasks well. According to proposed there classification in this paper we investigate a case where skills reflects degree of individual technical capability, skills affects speed of work. We take into account only learning phenomena as all operators are allocated to working posts and we assume that work interruptions on certain working posts are not enough long to forgetting effect appears.

Martignago et al. [9] investigated workforce allocation on a manual assembly line taking into account requirement for single-skilled and multi-skilled operators. Developed linear programming model can be used to determine skills learning sequence. Grosse and Glock [5] analyses a slightly different system, manual order picking taking into account operators learning rates. They came to conclusions that learning effect has a significant impact on order picking efficiency and it's worth to assign operators with lowest learning rate to fast moving zones where they could gain faster experience. Pinker and Shumsky [12] analysed efficiency of cross-trained workers in servicing industry together with staffing configuration sensitivity analysis. As a result they propose to hire mixture of flexible (multi-skill) and specialised (single-skill) workers. The propositions of flexible and specialised operators depend on their learning rates, complexity and the scale of the system. Gong et al. [4] proposed a training policy for a cross-trained workforce to improve the performance of u-shaped lines. In that work competences are used as a notion of the required skills, and operator experience is not taken into account. Manavizadeh et al. [8] considered a line-balancing problem and a labor assignment policy with permanent and temporary workers. Categorical skills are used, that is, operators either possess a skill or not. The dynamics of a system, which included a model of fatigue engagement and its correlation with a resting schema, was investigated in Finggiero et al. In Firat et al. [3], skill level is interpreted as hierarchical and is measured in an ordinal scale. The skill levels are input parameters that have constant values.

The novelty of this present work is the analysis of total skill improvements of operator performance. Operator performance results from the accumulated experience in executing technological operations, which have related required skills. Our solution allows faster accumulation of competences of production line operators and the minimization of time required to achieve the required level of competence throughout the entire line cast.

2 Competence Modeling

In this paper we apply a Hierarchical Competence Model (HCM) developed by Malachowski and Korytkowski [7] and Korytkowski [6]. The HCM assumes that skills have a hierarchical structure and could be represented by a simple weighted acyclic digraph $G(S, R, \Gamma)$, see Fig. 2. The nodes of this graph are skills $= \{s_i\}$ with assigned experience level $U_k = \{u_{i,k} \in [0, 1]\}$. Leaves nodes are called elementary skills (have no prerequisite skills) and other nodes are called compound skills (partly relies on other independently defined skills). Arcs of the graph are relations between skills $R = \{r_{i,j}\}$, where $r_{i,j} = (s_i, s_j)$ and weights are skill relation force $\Gamma = \{\gamma_{i,j} \in [0, 1]\}$.

The degree of excellence in the use of skill s_i by worker p_k is called competence strength $\alpha_{i,k} \in [0, 1]$. For a compound skill, $\alpha_{i,k}$ applies only to the new set of skills, excluding those skills inherited from the component skills. The value of competence strength is calculated using the learning curve approach (Anzanello and Fogliatto [1]):

$$\alpha_{i,k} = f_i(u_{i,k}, B_i, d_k). \tag{1}$$

This means that the competence strength depends on learning curve shape $f_i()$ with parameters B_i, personal learning rate d_k and the number of executed task repetitions $u_{i,k}$. A technological operation duration time by operator p_k is denoted as $\tau_{n,k}$

$$\tau_{n,k} = \tau_{n,min} + \left(\tau_{n,max} - \tau_{n,min}\right) \cdot \left(1 - \beta_{i,k}\right), \tag{2}$$

where $\tau_{n,max}, \tau_{n,min}$ are respectively minimal and maximal duration of a technological operation at working station, and $\beta_{i,k}$ is the overall competence performance. To determine $\beta_{i,k}$ we have to find out which competence is assigned to the working station. While the performance of an elementary skill s_i is just equal to $\alpha_{i,k} \cdot \gamma_{i,i}$ in the case of a compound skill s_i it is:

$$\beta_{i,k} = \alpha_{i,k} \cdot \gamma_{i,i} + \sum\nolimits_{c_j \in C_i} \beta_{j,k} \cdot \gamma_{j,i}. \tag{3}$$

Substituting $\alpha_{i,k}$ we get:

$$\beta_{i,k} = f_i\left(u_{i,k}, B_i, d_k\right) \cdot \gamma_{i,i} + \sum\nolimits_{c_j \in C_i} \beta_{j,k} \cdot \gamma_{j,i}. \tag{4}$$

Here competence performance is a synonym for work performance. If the competence performance has a maximum value then the performance of the worker carrying out the task requiring that competence is the most efficient, i.e. it takes the minimum time. That model is valid when the structure of the skills does not change over time and skills are independent.

The HCM enables describing the current competences of a worker and the required competences at a working post. A match of possessed and required competences creates a situation where a worker is capable of working at a certain working post. Moreover HCM allows calculating the performance of a worker performing certain technological operation taking into account his/her prior experience in executing the required competences at that working post.

3 Workforce Allocation Algorithm

Having information about required skills at each working station together with information about possessed skills by all operators we can present a workforce allocation algorithm that maximises total skill improvement defined as a sum of all operators competence performance β. The algorithm is presented below in a form of a pseudo code.

```
1:shift_duration = 240
2:for i=0 to number_of_stations{
3: for j=0 to number_of_operators{
4: k = find_top_skill_index(operator[j])
5: current_beta = beta(k,j)
6: n=0
7: while n * takt <= shift_duration{
8:  n++
9:  foreach s in reqired_skills(i)
10:  u[s,j] += 1
11:  }
12: }
13: projected_beta = beta(k,j)
14: delta[i,j]=projected_beta-curent_beta
15: }
16:}
17:for k=0 to number_of_stations{
18: for i=0 to number_of_stations{
19:  max_delta[i] = max(row(i, delta))
20: }
21: v = max(max_delta)
22: m = index_of_max(max_delta)
23: n = index_of_value(v, row(m,delta))
24: assignment[m]=n
25: for z=0 to number_of_stations{
26:  delta[z,n] = 0
27:  }
28: for z=0 to number_of_operators
29:  delta[m,z] = 0
30: }
31:}
```

In lines 1 to 16 a matrix delta of dimensions $i \times j$ is calculated. The delta matrix contains information about all possible skill improvements, delta[z, n] holds value for the case when operator z would be allocated to working station n. Then in lines 17 to 30 using greedy approach one by one operator is assigned to a working post. The local optimisation uses max operator for skill improvements. When the firs operator is assigned then the procedure is repeated (lines 18–30) without taking into account the already assigned working stations (lines 25–29).

4 Case Study – Manual Assembly Line

To demonstrate the proposed above workforce allocation algorithm we used the classic Jackson u-shaped line balancing problem (Miltenburg and Winjngaard [11]) with $N = 11$, $C = 10$, i.e. a line consists of 11 technological operations and takt time equal to 10

(Fig. 1). In the optimal u-line configuration, 11 operations are assigned to five stations with one crossover station (with the first and the last operation). In the original problem processing times are constant, but in our case the actual processing times will depend on operator competences. Table 1 gives the assumed minimal and maximal duration of technological operations (t_{min} and t_{max}). The actual durations of technological operations are calculated taking into account the competences (i.e. experience) of an operator using the HCM described above.

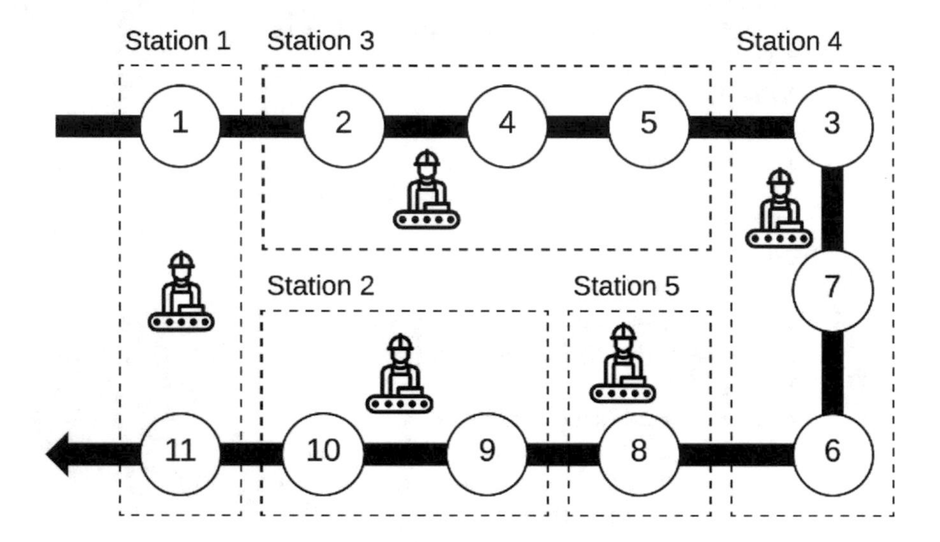

Fig. 1. Manual assembly line

Table 1. Technological operations

Technological operation	t_{min}	t_{max}	Required skill
1	6	12	s_3
2	2	4	s_4
3	5	10	s_3
4	7	14	s_5
5	1	2	s_3
6	2	4	s_4
7	3	6	s_2
8	6	12	s_1
9	5	10	s_3
10	5	10	s_3
11	4	8	s_5

Technological operations performed on the production line require five different skills from the operators. Three skills are elementary and two are compound. They are all interrelated and form a hierarchical model, depicted in Fig. 2. Table 1 shows the required skills in each operation of the production process. The relational structure of these competences shows that the elementary competences can be used alone in some other cases, while the higher-level compound skill cannot be used without its base component skills.

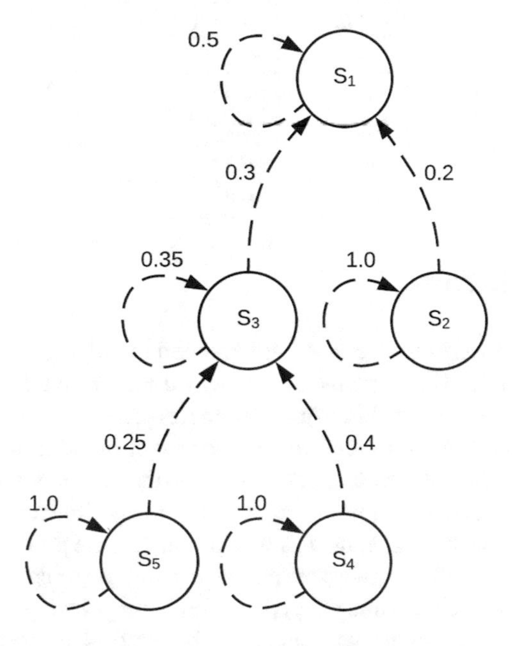

Fig. 2. Skills tree

Table 2 shows the learning curve functions used to model the skills with shape parameters of these functions (*B, Q* – growth rates, *M* – time of maximum growth, *C* – time to produce the first unit, *m* – fraction of work executed by machines). The exact meaning of shape parameters and their influence on the shapes of various learning curve functions can be found in Anzanello and Fogliatto [1].

Table 2. Skills parameters

Skills	Learning curve function	Parameters
S_1	Generalized Logistic Function	$B = 0.5, M = 100, Q = 1$
S_2	Log-linear	$C = 1$
S_3	Generalized Logistic Function	$B = 0.4, M = 35, Q = 1$
S_4	DeJong	$C = 1$, m = 0.3
S_5	Log-linear	$C = 1$

Operators working in the example production line have their own performance characteristics. Even if two or more workers have the same skills, it is natural that their learning curves can be slightly different. This is reflected in individual learning rates assigned to workers, which influence the shapes of learning curve functions. The values of these rates are presented in Table 3.

Table 3. Learning rates

Worker	Learning rate
w_1	−0.3
w_2	−0.35
w_3	−0.25
w_4	−0.3
w_5	−0.4

5 Simulation Results

The algorithm described in Sect. 3 was implemented in order to perform verification of the proposed approach. The implementation was done in Java language with several additional libraries, like Jung (http://jung.sourceforge.net/) for graph representations and DESMO-J (http://desmoj.sourceforge.net) for event-based simulation. Event-driven simulation approach was chosen in order to develop the model of the example u-shaped production line (see Sect. 4) end perform simulation experiments.

Conducted experiment was set using assumptions (ex. the production line layout and its configuration) and input data (workers characteristics, learning curve functions shape parameters, skills and their relations etc.) provided in the previous section of the article. The duration of the experiment was set to 32 h. Every 4 h the staff allocation was supposed to be reconfigured. Subsequent allocations, done after each four hours long iteration were found using the criterion of maximization of total competence gain. Table 4 presents example overall competence gains for every worker on every working stations computed in a single iteration of the algorithm. The values, that gives the highest sum of gains (and thus an optimal allocation) are emphasized in bold.

Table 4. Example projected competence performance gains computed in a single iteration of the algorithm

	Station 1	Station 2	Station 3	Station 4	Station 5
w_1	0.06551	0.11048	0.06788	**0.173178**	0.170808
w_2	0.019106	**0.082167**	0.02276	0.036316	0.032662
w_3	0.01835	0.074728	**0.022009**	0.114618	0.11096
w_4	0.012557	0.071509	0.014928	0.120226	**0.117855**
w_5	**0.013982**	0.077748	0.016159	0.029655	0.027478

During simulated time of 32 h allocations were changed 8 times. Consecutive allocations are presented in Table 5.

Table 5. Allocations in first 8 iterations of the algorithm

Iteration	1	2	3	4	5	6	7	8
Station 1	w_3	w_5	w_1	w_5	w_5	w_2	w_5	w_2
Station 2	w_1	w_2	w_5	w_1	w_3	w_4	w_4	w_1
Station 3	w_4	w_3	w_2	w_2	w_4	w_3	w_1	w_3
Station 4	w_5	w_1	w_3	w_3	w_2	w_5	w_3	w_5
Station 5	w_2	w_4	w_4	w_4	w_1	w_1	w_2	w_4

Allocation decisions are made basing on the overall competence gain. In the example every worker have the same set of skills. These skills strengthen over time according to their shape functions and personal learning rates of workers. Additionally, allocations in subsequent shifts also affect skills strengths, because on some working stations some skills can be not used at all, while on others can be used very intensively. Figure 3 presents strengths of skills of the worker w4 plotted over entire 32 h of simulation time.

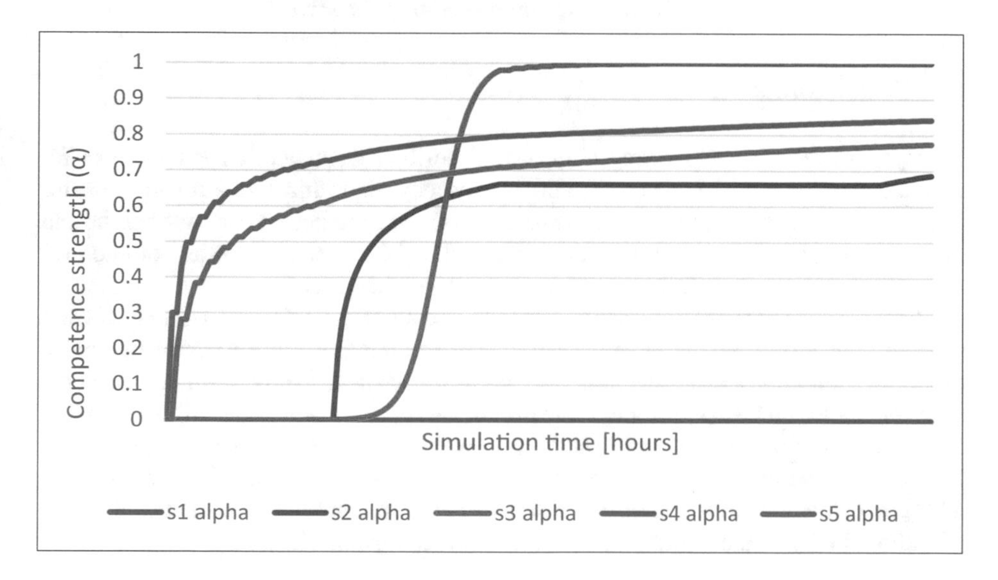

Fig. 3. Competence *strengths of skills of operator* w_4

Figure 4 presents the overall performance of the worker w4. The overall performance is computed for the top skill of the skill tree (Fig. 2), because it can be interpreted as the measure of competence required on the production line. The shape of this plot is influenced by shapes of individual skills strengths (see Eq. 3), as well as by compositions of required skills in allocated stations. Rapid changes in shape of the plot in Fig. 3 clearly indicate moments, when allocation was changed.

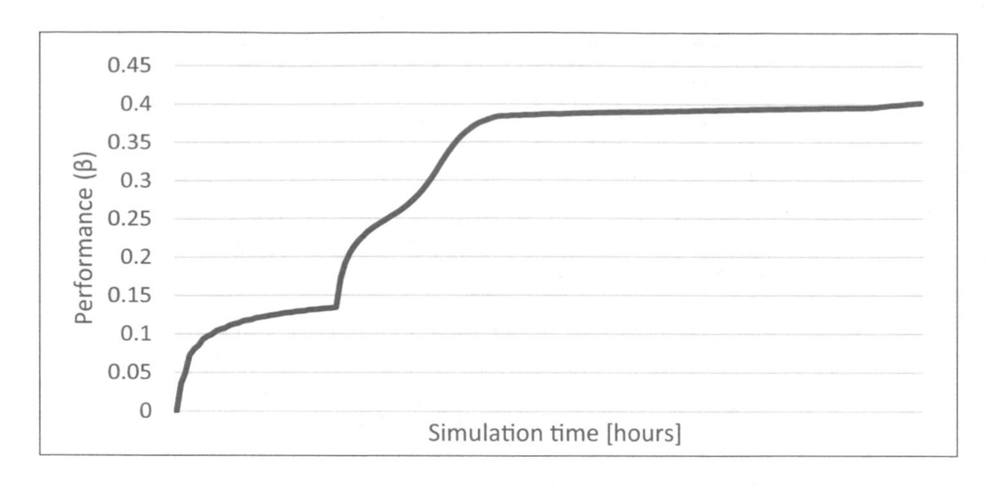

Fig. 4. Overall *performance of operator* w_4

6 Conclusions

This paper presents a workforce allocation algorithm at manual assembly lines for maximisation of total competence gain. The greedy algorithm looks for an operator allocation that enables her/him to maximise overall competence performance at the end of assignment time. Moreover thanks to Hierarchical Competence Model common for several working posts skills reinforcements are taken into account.

The proposed approach enables to take into account human factor in manufacturing planning process where the main objective is not to minimise makespan or maximise production output in a short planning horizon like it happens in environments managed in accordance with lean philosophy. The proposed algorithm will allow all e the production line operators to achieve the maximum level of competence, thanks to which the time of implementation of the employee at the new workstation will be shortened. In addition, an even distribution of competences between employees facilitates planning of staffing in the medium term and in reacting to emergency situations.

References

1. Anzanello, M.J., Fogliatto, F.S.: Learning curve models and applications: literature review and research directions. Int. J. Ind. Ergon. **41**(5), 573–583 (2011)
2. De Bruecker, P., van den Bergh, J., Beliën, J., Demeulemeester, E.: Workforce planning incorporating skills: state of the art. Eur. J. Oper. Res. **243**, 1–16 (2015). https://doi.org/10.1016/j.ejor.2014.10.038
3. Firat, M., Hurkens, C.A.J., Laugier, A.: Stable multi-skill workforce assignments. Ann. Oper. Res. **213**, 95–114 (2014)
4. Gong, J., Wang, L., Zhang, S.: A new workforce cross-training policy for a U-shaped assembly line. In: Zhu, M. (ed.) Communications in Computer and Information Science, CCIS, vol. 235, pp. 529–536. Springer, Heidelberg (2011)

5. Grosse, E.H., Glock, C.H.: The effect of worker learning on manual order picking processes. Int. J. Prod. Econ. **170**, 882–890 (2015). https://doi.org/10.1016/j.ijpe.2014.12.018
6. Korytkowski, P.: Competences-based performance model of multi-skilled workers with learning and forgetting. In: Expert Systems with Applications, vol. 77, pp. 226–235 (2017). https://doi.org/10.1016/j.eswa.2017.02.004. ISSN 0957-4174
7. Małachowski, B., Korytkowski, P.: Competence-based performance model of multi-skilled workers. Comput. Ind. Eng. **91**, 165–177 (2016). https://doi.org/10.1016/j.cie.2015.11.018
8. Manavizadeh, N., Hosseini, N.S., Rabbani, M., Jolai, F.: A Simulated Annealing algorithm for a mixed model assembly U-line balancing type-I problem considering human efficiency and Just-In-Time approach. Comput. Ind. Eng. **64**(2), 669–685 (2013)
9. Martignago, M., Battaïa, O., Battini, D.: Workforce management in manual assembly lines of large products: a case study. IFAC-PapersOnLine **50**(1), 6906–6911 (2017). https://doi.org/10.1016/j.ifacol.2017.08.1215
10. Miltenburg, J.: U-shaped production lines: a review of theory and practice. Int. J. Prod. Econ. **70**, 201–214 (2001)
11. Miltenburg, G., Winjngaard, J.: The U-line balancing problem. Manag. Sci. **40**(10), 1378–1388 (1994)
12. Pinker, E.J., Shumsky, R.A.: The efficiency-quality trade-off of cross-trained workers. Manuf. Serv. Oper. Manag. **2**(1), 32–48 (2000). https://doi.org/10.1287/msom.2.1.32.23268
13. Van Den Bergh, J., Beliën, J., De Bruecker, P., Demeulemeester, E., De Boeck, L.: Personnel scheduling: a literature review. Eur. J. Oper. Res. **226**(3), 367–385 (2013). https://doi.org/10.1016/j.ejor.2012.11.029

Development of a Simulation Platform for Robots with Serial and Parallel Kinematic Structure

Vladimír Bulej[1(✉)], Juraj Uríček[1], Ján Stanček[1], Dariusz Więcek[2], and Ivan Kuric[1]

[1] University of Žilina, Univerzitná 1, 010 26 Žilina, Slovakia
{vladimir.bulej,juraj.uricek,jan.stancek, ivan.kuric}@fstroj.uniza.sk

[2] University of Bielsko-Biala, Willowa 2, 43-309 Bielsko-Biala, Poland
wiecekd@ath.bielsko.pl

Abstract. The main aim of this article is the description of specialized software RoboSim for off-line programming and simulation of robotic systems developed at the University of Zilina during the last decade. It contains basic information about the development process of this universal software for simulation of automated workplaces equipped with one or more robots with up to 6 degrees of freedom (DOFs). The latest version can be used as a universal platform for simulation of mechanisms with serial, parallel as well as hybrid kinematic structure. This flexibility can be considered as the main advantage of the RoboSim software. Its specific feature is that there are two different methods used for calculation of inverse kinematics: heuristic and vector method as well. The versatility, openness and access to the source code create the predisposition for its deployment under laboratory conditions for controlling robotic devices developed in authors' workplace within the last fifteen years.

Keywords: Industrial robot simulation · Serial robots · Parallel robots

1 Introduction

Nowadays the complexity of products and, this way, also the complexity of production lines is very high and does increase year by year. Due to the increasing demands of the modern economy, also the development of new products must reach new levels concerning the complexity and the implemented intelligence, while saving resources and reducing the time needed for design and production [1]. These requirements put a high pressure on engineers. They can find strong support in special simulation software tools. There is a wide range of simulation software available on the market, but not all of it is suitable for each task. Simple programs, for example, are able to simulate electromechanical control of motors. High sophisticated simulation software can simulate the entire complex mechatronic multibody system, as, for instance, industrial robots, CNC machine tools, the whole production lines or groups of devices. With computer graphics and virtual reality, 3D simulation technology has been widely used not only in the design process [2]. New versions of simulation software platforms offer even more features that make simulation easier and also bring it very close to real life. Most

© Springer Nature Switzerland AG 2019
A. Burduk et al. (Eds.): ISPEM 2018, AISC 835, pp. 452–461, 2019.
https://doi.org/10.1007/978-3-319-97490-3_44

simulation tools are compatible with programming languages like C/C++, Perl, Python, Java, LabVIEW, URBI or MATLAB, however, they offer broadly varied feature sets depending on their purpose or focus areas [3, 4].

Industrial robots have several capabilities that allow them to perform a wide range of tasks, including spot and arc welding, picking up components, machine tending, spraying, assembling, carrying of different technological effectors and general manipulation with parts. As European economies start to recover from the recession, manufacturing companies are seeking ways to ramp up production in a flexible way [5]. In the field of robotics, several trends are visible both in practice and academia, for example the so-called collaborative robotics (cooperation between a robot and a human operator), mobile robotics, cooperation of multi-arm systems, as well as application of unconventional kinematic structures and simulation of robot's behaviour by suitable simulation software tools.

Simulation programs for robotized work-cells can be divided into two main categories: programs developed by industrial robots manufacturers (ABB, Kuka, Fanuc, etc.), and universal (but less capable) systems developed by software producers [6]. These software tools are used for visualization, testing, debugging and repairing programs for a single robot, robotized workcell including peripheral devices (e.g. belt conveyors, rotating heads, magazines, etc.), as well as the whole production lines composed of several work-cells equipped with several collaborating robots. The debugging option is therefore very advantageous in the pre-production stage, because the standing time of robots or a complete line can be reduced to a minimum. These programs also find their application in production planning and training operators.

After the virtual modeling, the system must be implemented in real world and tested in real life conditions in order to validate simulation results. The testing procedure implies validating different features, like control algorithms, functionality of sensor systems, failure response, etc. [6, 7]. This step can be very challenging and time consuming without using adequate tools. So, even here the positive role of simulation software seems obvious. Creating a complete virtual model of a robot or system by simulating components and control programs can significantly impact general efficiency of the whole project [1, 8]. Simulation can bring benefits, like, for example, reducing robot production costs, a possibility to simulate various alternatives without involving physical costs, or to diagnosing control program functionality, etc. [9]. The main drawbacks include the time consuming preparation phase (creating a simulation model), and the fact, that a robot can encounter many more scenarios in the real world than during simulation.

The main aim of this article is to develop the *RoboSim* software and other specialized software tools for off-line programming and simulation of robots with serial and parallel kinematic structure, or complete robotized production systems during the last fifteen years at the authors' workplace. The main motivation of the team oriented on robotics and automated production and assembly systems at the University of Žilina (*UNIZA*) was the idea to follow the main trends in robotics and, at the same time, the need to have our own software platforms for programming and simulation of the commercial as well as developed robots and mechanisms.

2 Development of Simulation Software for Robots with Serial Kinematic Structure

As already mentioned, the development of simulation software tools for industrial robots at the *UNIZA* began 15 years ago and has continued until these days. At the beginning, this was a solution to the issue of visualization and debugging of control programs for robots manufactured in the former Czechoslovakia – robots called *APR 20, APR 2.5*, as well as the robot prototype *SLR 1500*. The first task was pure visualization of robot motion within the workspace, which was later extended to simulation of various robot functions, such as linear and circular interpolation, manual control, object manipulation simulation, and collision detection. The simulation programs included the *RAMAS* controllers' editor [10, 11]. Later, also other simulation software tools for both serial and parallel robots were applied.

2.1 SLR1500 Simulation Program - Version 2004

It was the result of a diploma thesis from the year 2004 at the authors' workplace [10]. It was primarily used to simulate the movement of the *SLR 1500* robot prototype using new inverse kinematic algorithms, but it was not limited only to this one. It allows for simulation on other kinds of robot with serial kinematic structure and with five degrees of freedom as well. The program was designed as an open modular system, with the option of modifying, creating new and removing unnecessary modules (see Fig. 1).

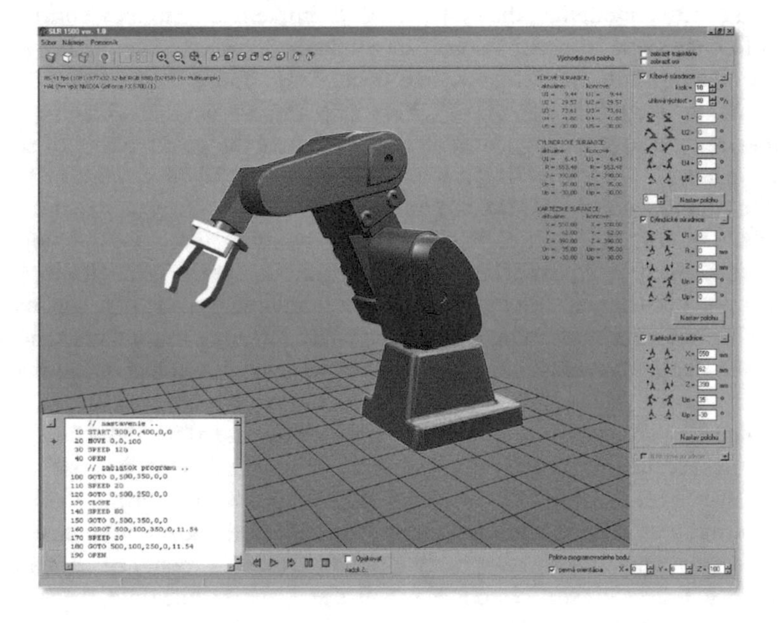

Fig. 1. The first software for simulation of industrial robots within 3D space designed at *UNIZA* in year 2004 called *"SLR 1500 ver. 1.0"* [10]

The program includes the following six basic modules:

- the core of the program and the *DirectX 8.1*,
- auxiliary functions module,
- user interface module,
- kinematics module including inverse kinematics for 5 DOFs,
- programming module.

It was planned to extend these basic modules and add a collision control module, a kinematics module for multiple degrees of freedom, a trajectory interpolation module, and a programming module of the higher programming language with a possibility of programming in all supported coordinate systems. This was done in later newly developed simulation software mentioned in the next paragraph.

2.2 *RoboSim* Simulation Program - Version 2015

In year 2008 the first version of our new 3D robot simulation software started to be developed. The main goal was to develop software suitable for simulation and programming of a single robot cell with multiple robots, manipulators, and other parts and equipment of an automated production system. For this software was set the following requirements:

- simultaneous simulation of multiple robots,
- implementation of universal algorithms of direct and inverse kinematics,
- a possibility to create and simulate control program for experimental robots developed on the author's workplace,
- the integration of conventional as well as unconventional kinematic structures (serial, parallel, hybrid),
- motion trajectory programming, linear and circular interpolation in all axes, motion control using commands read from CL data,
- detecting real-time collisions and limit states,
- a high degree of versatility, openness and modularity of the system.

Developing the Concept and Algorithms. It was necessary to focus on various areas, such as kinematics of spatial mechanisms, theoretical description of manipulated objects and their mathematical formulation, programming of automated and robotic systems, computer graphics and programming in *OpenGL* environment, as well as development and implementation of control algorithms. The algorithms used for development had to meet the requirements of speed, accuracy, versatility, ease of implementation and the possibility of parallel computing.

One of the crucial features of the whole development process was searching for a universal, direct, and inverse kinematics algorithm for serial and parallel mechanical structures, and the real-time collision control algorithm.

Forward and Inverse Kinematics. A specific feature of the *RoboSim* software is that it uses two different methods for calculating inverse kinematics: the heuristic method, and the vector method as well. The vector method is applicable to only one type of kinematic structure (in our case just the serial robot with six degrees of freedom and rotation constraints), but it is very fast and accurate (see Fig. 2). On the other hand, the heuristic method allows us to find a solution for more degrees of freedom than the number of known parameters. This method is especially useful when the robot moved fast from point to point, when there is no need of a very high orientation precision (just high position precision).

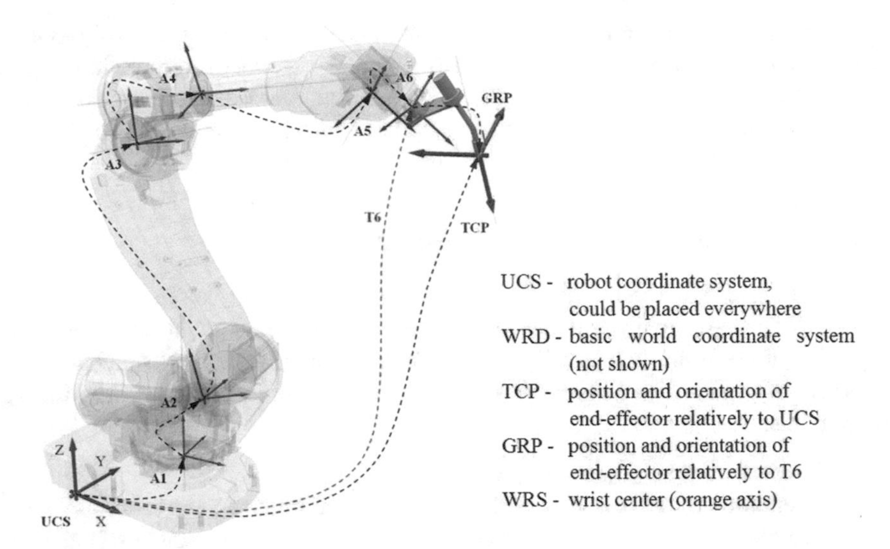

Fig. 2. Example of kinematic structure of the robot with serial kinematic structure and the definition of the partial coordinate systems related to each structural element.

The versatility, openness and access to the source code created the predisposition for its deployment under laboratory conditions for controlling the robotic devices developed in the authors' workplace within the last few years. Some of them are based on parallel or hybrid kinematic structure, and the simulation software can be used for creating and testing control programs for these machines as well. The designed simulation software is based on modular principle, which makes it possible to modify the software according to our application (the modules can be changed).

To control the robot and to achieve the most accurate trajectory, and to prevent any vibrations caused by inconstant (discontinuous) control, the cubic approximation function, not a weighting rule, is used in the program. However, the output of these inverse kinematics calculations is not displayed. For simulation and its graphical display in the program window, calculations with precision of order less than can be obtained by calculating angles from micro-streams and spreading to other axes by the weight rule would be sufficient. Exact results of inverse kinematics can only be obtained by exporting calculations to txt or csv format, or by communicating between the integrated network module and the robot control program.

The Main Screen. During the project solution, a simulation program was developed to validate the proposed control and simulation algorithms to create a universal platform for future robot simulations at authors' workplace. The user environment for the current version of the 0.9.2 program (see Fig. 3) consists of several panels, controls, and windows displaying the 3D model of the production system. Some of the panels can be moved or hidden. Camera windows, co-ordinate display, and angular rate and acceleration display can be hidden, modified, or fully disabled. The ability to customize the user environment increases the ease of use.

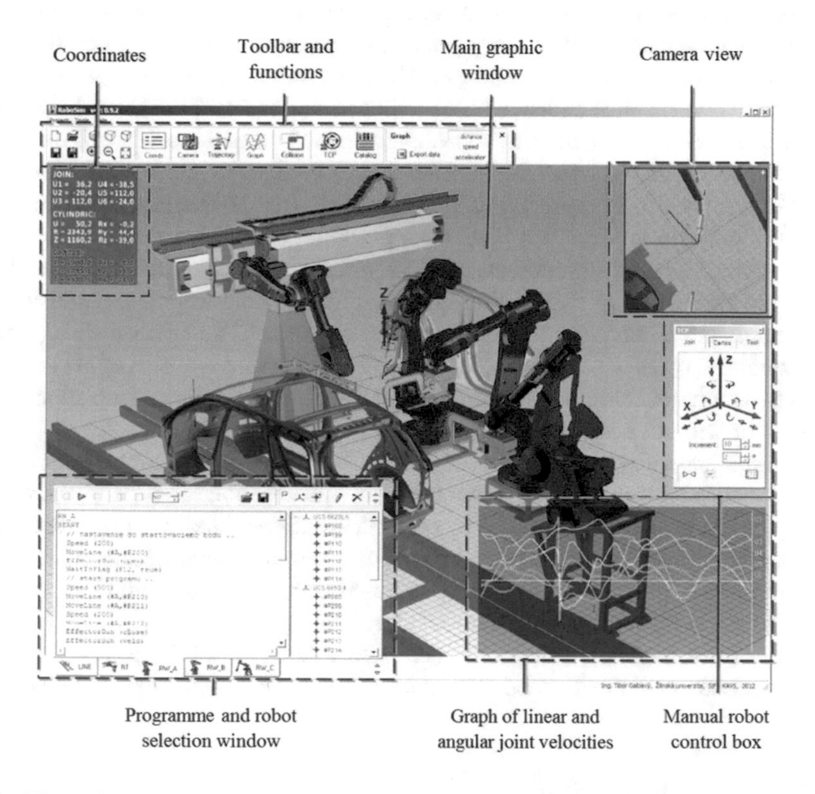

Fig. 3. The main screen of simulation software *RoboSim* developed at authors' workplace [10].

The simulation software was designed in the programming language *Delphi* with graphic library *OpenGL*. A screen of the latest version from year 2015, where cooperation of several robots is simulated, is shown in Fig. 3. You need a standard mouse to control the interface and to enter data and write a keyboard. The functions *Pan*, *Rotate* for camera control are controlled by mouse movement and by pressing the right or middle mouse button. *Zoom* is controlled by a mouse wheel or toolbar icons. The left mouse button is for *Select*. Press CTRL + move the mouse to control the TCP translation of the current robot in the XY plane, or at the tool coordinates in the UV plane and move the mouse wheel to move in the Z or W-axis. Press SHIFT + mouse to move the TCP rotation motion of the current robot around the axis perpendicular to the screen. This way of controlling robot movements is very intuitive, fast and accurate.

Programming Motion Trajectory. A simple programming language will be developed for programmable robot control, allowing simultaneous control of multiple robots. The compiler will support work with variables, subroutines, procedures, cycles, and conditional program branching. The compiler will support reading CL data from an external file and converting it to motion commands.

Collision Detection, Prediction and Avoiding. Under the term "collision" in robotics we generally understand undesirable contact between two bodies, in which at least one body encounters non-zero kinetic energy at the collision moment. The system for detection and protection against collision ("anti-collision system") means any software, hardware, or technical solution seeking to avoid collisions on real machines [12]. The off-line procedure simulates the motion of robot handling devices, so we apply specialized simulation software in this case. It is important to use a model that represents the manipulator itself and the environment in which a potential collision occurs. During the run of simulation, the control system must identify collisions which may happen. On-line systems are active during real motion, so they are also able to prevent collisions that were unpredictable in programming (an active approach). The best way is to combine several different types of sensors, as, for example, tactile sensors, ultrasound or infrared sensors and cameras. On-line system does not need an environment model and it can work with or without a model.

The *RoboSim* software works in the off-line mode, so it is a good idea to use the graphical method for suitable object description. We describe robot parts, the manipulated objects, as well as any obstacles by geometric entities. The next step is to calculate their mutual distance or detect the formation of their common intersection. If the distance between the units reaches zero value, or penetration occurs, the control system stops robot's motion action in that direction and alerts the operator.

Since the test method "General Surface - General Surface" for collision control is the only accurate one, this method is used for the *RoboSim* software (see Fig. 4). Each

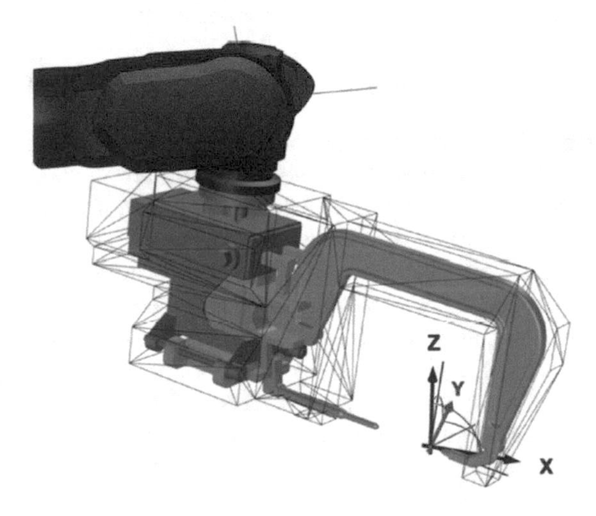

Fig. 4. An sample object description in collision detection algorithm in the *RoboSim* software

general surface or object is described as a net of polygons (triangles), whose vertices are defined by points – nodes. Finding a collision between the objects is possible only at the level of their polygons. The algorithm is checking if there occurs any intersection between the triangles of both tested objects, or not. Because of that the task of finding a collision between surfaces and objects formed by several thousand triangles is very complicated. Therefore, several various optimization and approximation methods are available.

Practical Verification. A link between the simulation program and the control must also constitute an integral part of the software. The connection is done in two ways - by exporting program files that are translated into the robot's control language, or by directing orders via a TCP-IP-based network. After the development process, functionality of the whole simulation system and its subsystems was verified on a virtual cell with multiple welding robots. Program capability was also tested on another type of kinematic structures – it was used to create, simulate and debug control programs for the prototypes of the Hexapod mechanism, developed at the authors' workplace.

3 Adjustment and Testing of Simulation Software for Mechanisms with Parallel Kinematic Structure

Another main field of study in the area of robotics at the authors' workplace is the topic of mechanisms with parallel or hybrid kinematic structure. These mechanisms are characterized, above all, by higher stiffness and higher dynamic properties (thanks to the reduced moving mass) [11]. General parallel kinematic structure (PKS) is a mechanism with closed kinematic chain, which consists of the base, platform and at least two reciprocally independent leading legs. Parallel links (struts, legs) are joined between the base and the movable platform. Motion of the platform is controlled by a change of linear or angular parameters of legs acting in a parallel system.

a) b)

Fig. 5. A prototype of CNC machine tool with PKS for 5D milling operations called *Hexapod ZU*: (a) overall view, (b) mechanism itself - both designed at *UNIZA*

The authors decided to develop, and subsequently build several prototypes of machine tool and robots with parallel and hybrid kinematic structure to verify and compare their capability and technical properties with mechanisms based on conventional, serial kinematic structure. Some of them are designed with parameters allowing for its real deployment in practice. During the period of between 2009 and 2013, some construction concepts of PKS and different kind of simulation software were designed for these types of mechanism.

After the first functional tests of the latest version of the *RoboSim* software we decided to adjust it and try to apply it for mechanisms with such kinematic structure – concretely on the designed machine tool prototype based on the Hexapod mechanism (Fig. 5).

Since the program was not primarily designed to simulate parallel kinematic structures, it was necessary to add a module for new kinematic structure, derive equations and algorithms for inverse kinematics, and customize program control for this type of robotic mechanisms as well. Interpolation and control algorithms functioned correctly. After changing the simulated program into a Hexapod simulator, it was possible to monitor identical operation on a real machine and in the *RoboSim* simulation software (see Fig. 6). The accuracy of the machine real trajectory against the simulated one still remains to be verified. The functionality and versatility of the *RoboSim* simulation program have been verified and further development will continue.

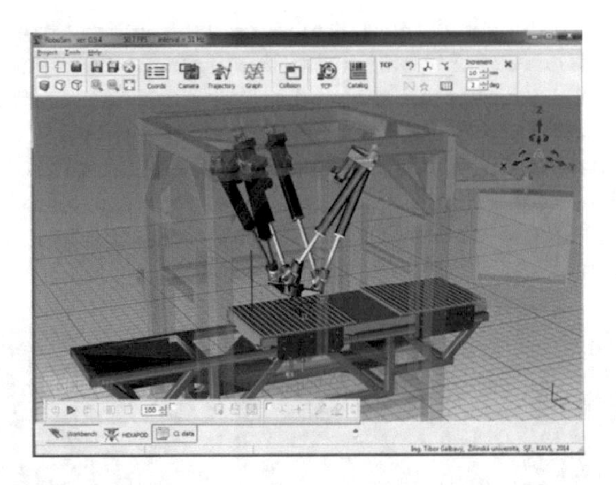

Fig. 6. Testing of the adjusted version of the *RoboSim* software on Hexapod machine tool prototype – designed at UNIZA [10]

4 Conclusions

The main aim of this article is to describe of the original *RoboSim* software developed at the authors' workplace, which enables for simulating automated manufacturing systems equipped with one or more robots with up to 6 DOFs each. Several particular development steps are described in more detail, as, for example, the development of the first simulation software *SLR 1500*, forward and inverse kinematics, description of the software, collision detection and practical verification. At the end, capability of the designed software was tested also by adjusting it to mechanisms with unconventional kinematic structure. The modified version of the software was tested on a real machine tool prototype based on the Hexapod mechanism. We can assume that the developed software can be applied mainly for off-line simulation of robots and mechanisms with serial, as well as parallel kinematic structure. There are still some features which should be improved in the near future.

Acknowledgement. The authors disclosed receipt of the financial support for the publication of this article: This work is partly supported by the project VEGA 1/0504/17 - Research and development of methods for multi-criteria diagnosis of CNC machine tools' accuracy.

References

1. Bulej, V., Stoianovici, G.-V., Poppeová, V.: Material flow improvement automated assembly lines using Lean logistics. In: Annals of DAAAM for 2011 & Proceedings of the 22nd International DAAAM Symposium, pp. 253–254. DAAAM International, Vienna (2011)
2. Pivarčiová, E., Božek, P., et al.: Analysis of control and correction options of mobile robot trajectory by an inertial navigation system. Int. J. Adv. Rob. Syst. **15**(1), 1–10 (2018). https://doi.org/10.1177/1729881418755165
3. Kumičáková, D., et al.: Utilisation of kinect sensors for the design of a human-robot collaborative workcell. Adv. Sci. Technol. **11**(4), 270–278 (2017)
4. Dodok, T., Čuboňová, N., et al.: Utilization of strategies to generate and optimize machining sequences in CAD/CAM. Procedia Eng. **192**, 113–118 (2017)
5. Korzekwa, J., Tenne, R., et al.: Two-step method for preparation of Al_2O_3/IF-WS_2 nanoparticles composite coating. Phys. Status Solidi (A) Appl. Mater. Sci. **210**(11), 2292–2297 (2013). https://doi.org/10.1002/pssa.201329320
6. Kuric, I., et al.: Development of simulation software for mobile robot path planning within multilayer map system based on metric and topological maps. Int. J. Adv. Rob. Syst. **14**(6), 1–14 (2017). https://doi.org/10.1177/1729881417743029
7. Korzekwa, J., et al.: The influence of sample preparation on SEM measurements of anodic oxide layers. Pract. Metall. **53**(1), 36–49 (2016). https://doi.org/10.3139/147.110367
8. Košinár, M., Kuric, I.: Monitoring possibilites of CNC machine tools accuracy. In: Proceedings of 1st International Conference on Quality and Innovation in Engineering and Management (QIEM), pp. 115–118 (2011)
9. Mičieta, B., Edl, M., et al.: Delegate MASs for coordination and control of one-directional AGV systems: a proof-of-concept. Int. J. Adv. Manufact. Technol. **94**(1–4), 415–431 (2018). https://doi.org/10.1007/s00170-017-0915-8
10. Galbavy, T.: Simulation of robotized workcells. EDIS-University of Žilina, Žilna, Dissertation thessis (2014)
11. Uríček, J., et al.: The calculation of inverse kinematics for 6DOF serial robot. Commun. Sci. Lett. Univ. Žilina **16**(3A), 154–160 (2014)
12. Gmiterko, A., Kelemen, M., Virgala, I., et al.: Modeling of a snake-like robot rectilinear motion and requirements for its actuators. In: 15th International Conference on Intelligent Engineering Systems, INES 2011, pp. 91–94. IEEE, Poprad (2011). https://doi.org/10.1109/INES.2011.5954726

Interactive Layout in the Redesign of Intralogistics Systems

Paweł Pawlewski(✉)

Poznan University of Technology, ul. Strzelecka 11, 60-965 Poznań, Poland
pawel.pawlewski@put.poznan.pl

Abstract. This paper describes an interactive layout as the tool for redesign of intralogistics systems. The need of interactive layout was presented in context of Industry 4.0, Layout Planning and simulation. Material handling system, material flow and distance travel of material improve productivity and operational efficiency of production systems. The paper focuses on the approach based on milkrun systems. The main foundations of interactive layout such as workstation as basic object, addressing system, using templates, core 3D simulation are described. Proposed solution was implemented in FlexSim Discrete Events System environment.

Keywords: Intralogistics · Discrete event simulation · Layout Planning

1 Introduction

Nowadays there are some serious dynamically changes in the market. It is revealed in factories by changes in produced products (continuous modernization, face lifting) and by introduction of new products. This in turn causes the changes in production processes. Production cycles and product life spans are shortens. The factories react to it by adjusting to these requirements. The challenge is reorganizing production, changing layout and matching intralogistics systems. Material handling system, material flow and distance travel of material improve productivity and operational efficiency of production systems.

The paper focuses on the concept of interactive layout as the tool for redesign of intralogistics systems – especially in assembly.

The objective of this paper is:

- to term the needs of more smart tool for design and redesign intralogistics,
- to define requirements of interactive layout as the tool,
- to present the concept of interactive layout

The main contribution of the paper is definition of the main foundations of interactive layout:

- workstation (production line or cell),
- working with workstation using mouse – it is the feature of interactivity,
- two levels of working with mouse - locally and globally
- addressing system of workstations and components which defines topography of the factory.

© Springer Nature Switzerland AG 2019
A. Burduk et al. (Eds.): ISPEM 2018, AISC 835, pp. 462–473, 2019.
https://doi.org/10.1007/978-3-319-97490-3_45

The main highlight of the paper is description of interactive layout implementation in FlexSim Discrete Events System environment which allows the practical use of this tool.

The paper is organized as follows. Section 2 defines intralogistics systems in general and discusses related literature. Section 3 describes the motivation and defines the problem. Section 4 presents the solution – the concept of interactive layout. Section 5 contains the description of implementation and Sect. 6 conclude and discuss future research directions.

2 Intralogistics Design and Redesign

The paper focuses on design and redesign of intralogistics system so firstly is necessary to define the term "intralogistics". This term is defined by The Intralogistics Forum of the Verband Deutscher Maschinen- und Anlagenbau (VDMA) (Hompel and Heidenblut 2008) as: "The organization, control, execution and optimization of in-plant material and information flows, and of goods transhipment in industry, distribution and public sector facilities.". Can find other definitions. As the best (from our point of view) definition we recognized the definition by Wynright (wynright.com 2018): "every dimension of logistics within the four walls related to implementing, managing, monitoring and optimizing materials handling and information flows." We focus on factories where assembly is the main activity. We analyze the material flow from supermarket (as entry - where there are already prepared containers with parts) by production, assembling lines and cells till the buffer with containers of finished products, which is located in front of the finished goods warehouse. The warehouse flows were not analyzed and modeled before the super-market and after the container buffer with ready products.

Intralogistics forms together with Product, Process and Layout one consistent system where each element depend on another (Fig. 1). Layout as the central point in this system plays a special role. Layout it is a floor plan of a plant that arranges equipment according to its functions. It is an integration of the physical arrangement of departments, workstations, machines, equipment, materials, common areas etc, within an existing or proposed industry.

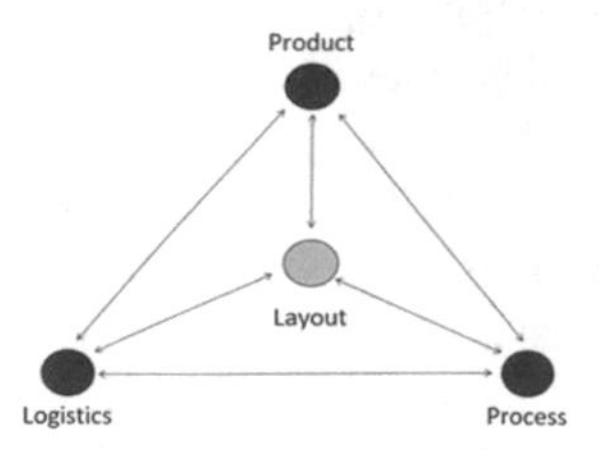

Fig. 1. Interaction between layout, product, process and logistics – based on (Muther and Hales 2015).

To design or redesign layout following goals are to achieve:

1. minimal material handling costs,
2. minimal investments,
3. minimal throughput time,
4. flexibility,
5. efficient use of space.

The same goals are set for intralogistics. Most layouts are designed properly for the initial conditions of the business, although as long as the company grows and has adapted to internal and external changes, a re-layout is necessary. Symptoms that allow us to detect the need for a relayout:

1. Congestion and bad utilization of space.
2. Excessive stock in process at the facility.
3. Long distances in the work flow process.
4. Simultaneous bottle necks and workstations with idle time.
5. Qualified workers carrying out too many simple operations.
6. Labor anxiety and discomfort.
7. Accidents at the facility.
8. Difficulty in controlling operations and personnel.

There are two main general approaches (methods):

1. The right equipment at the right place to permit effective processing,
2. Short distances and short times.

One of the first approaches to solve this problem is the Systematic Layout Planning. The Systematic Layout Planning (SLP) Design Process was developed by Richard Muther in 1961 (Muther 1961). SLP is a tool used to arrange a workplace in a plant by locating areas with high frequency and logical relationships close to each other – see Fig. 2. SLP is commonly used as the foundation of all manufacturing process design methods. This multi-step process repeatedly demonstrates its applicability to a wide number of process design projects, whether they are the design of a fully automated production facility, or the design of a flexible manufacturing job shop facility. This approach is applied to process design projects in newly constructed facilities, or process redesign efforts in well-established facilities, and can be applied in the same way to minor as to major process design efforts.

In the present, there are several methods for plant layout design (Naik and Kallurkar 2016) such as systematic layout planning (SLP) (Zhu and Wang 2009), algorithms (Deb 2005), and simulation (Jaturachat et al. 2007) can apply to design plant. In (Parveen and Ravi 2013) can find the review of metaheuristic approaches: Particle Swarm Optimization (PSO), Genetic Algorithm (GA) and Tabu Search (TS) - used to optimize the multi-objective layout problem. A number of study have attempted to explore the influence of facility layout design on operational efficiency in some manufacturing firms (Anucha et al. 2011), (Pinto and Shayan 2007), (Tao et al. 2012), (Vaidya 2013), (Yifei 2012).

In context of Industry 4.0 we have new challenges for intralogistics. The key industry 4.0 principles are (Schwab 2017):

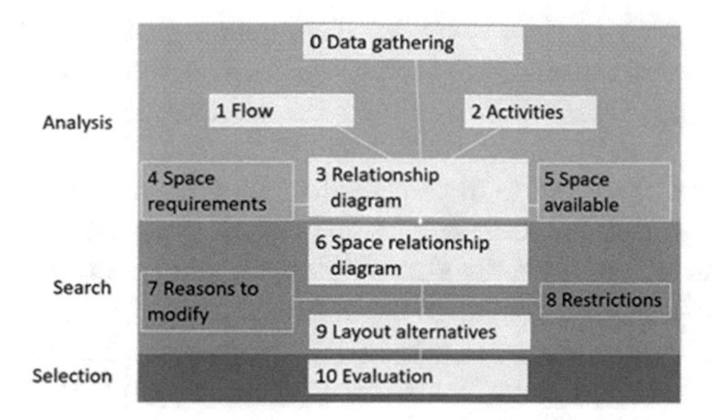

Fig. 2. Systematic layout planning – based on (Muther and Hales 2015).

1. the factory becomes digital and flexible, which means continuous and immediate communication between various workstations and tools, integrated into production lines and supply chains;
2. the use of simulation tools and data processing to collect and analyze data from assembly lines that are used for modeling and testing; this is a great value for employees who want to better understand industrial conditions and processes;
3. factories become economical in using energy and resources through the use of communication networks to exchange information in a continuous and immediate way to coordinate needs and availability.

We want to focus on two aspects: digital and simulation.

Digital means that we use digital technology for modeling, communications and to operate the manufacturing process. This arrangement of technology allows managers to configure, model, simulate, assess and evaluate items, procedures and system before the factory is constructed. The digital factory gives answers for configuration, design, screen and control of a production system (Canetta 2011).

Simulation on the one side is modeling, i.e. mapping of the real system, understanding of system behavior, virtual (and visual) assessment of possible consequences of actions, and on the other side, is - experimenting and testing ideas and alternatives before making decisions on actions and resource involvement (Beaverstock 2017). Simulation is a collection of methods and techniques to which we include discrete simulation, continuous simulations (including systems dynamics), Monte Carlo method (including static simulations in a spreadsheet), managerial games, qualitative simulation, agent simulation and others.

3 Problem Definition and Motivation

The research works were implemented in car industry enterprise in Poland. The process is performed on several welding stations, where welding operations are carried out successively and then the assembly and packaging operation of the product is carried out. It is a pipeline production. The goal of the research is to organize and maintain a management system for parts, components, subassemblies and other materials that are delivered to the enterprise from suppliers. The challenge is to find the answer to the question of how to design (redesign) such a logistic system which, slimmed down with as much waste as possible, will ensure the most effective flow of materials inside the factory.

The researchers attempted to model the flow of parts that flow in containers. The containers “flow” from the supermarket to the fields of workstations, where the welding operation was carried out, then the assembly (in containers or on the logistic trolley) “floated” to the next workstation, up to the buffer in front of the finished product warehouse. The assembly operations were modeled taking into account the operation time (described with the appropriate statistical distribution), disruptions (failures) and planned breaks. The focus was on the flow.

Factory where researches were done work in dynamic market it means that the layout “lives”, changes when some production lines are closed or new lines are introduced. Then managers, production engineers, planners, logisticians, lean specialists work on changes in layout and intralogistics.

According to the Digital Factory (Worn et al. 2000) concept some of the typical simulation application can define:

1. Production control and simulation of production flow,
2. Line balancing of assembly processes,
3. Simulation of material handling systems,
4. Industrial robotics work cells,
5. Ergonomics evaluation through simulation of hunab resources,
6. Design, validation and optimization of Digital Factory layout.

Some authors and methodologies (Beaverstock et al. 2017), (Centobelli 2016) suggest that depending on the particular goal of the simulation different tools and levels of detail are required. From the other side we meet requirements from factories to propose one consistent methodology which covers all or many from listed topics.

As mentioned in Sect. 2 – intralogistics concerns “every dimension of logistics within the four walls related to implementing, managing, monitoring and optimizing materials handling and information flows”. So we can discuss about intralogistics on some levels of details. Figure 3 shows levels of intralogistics in assembly factory.

Level 0 is formed by resources from simulation program. Modeling and analysis are performed from down to top – starting level is level 1, next level 2 and ending on level 3. Table 1 summarizes which data are necessary as input for every level and which results are achieved.

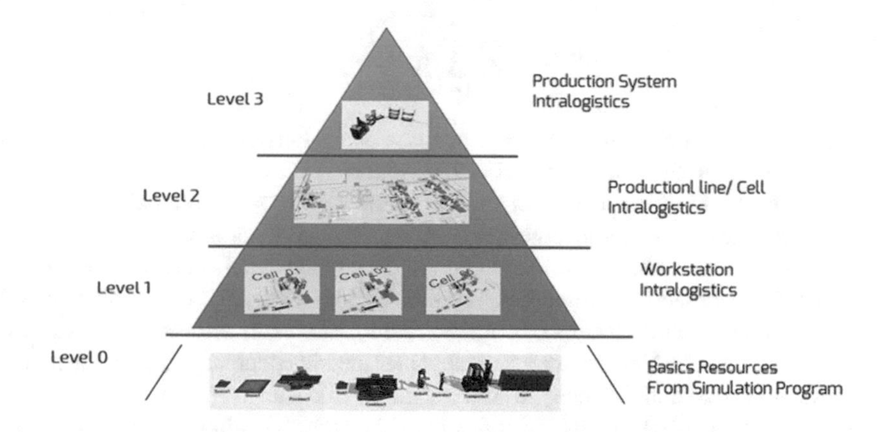

Fig. 3. Levels of intralogistics in assembly factory – own study

Table 1. Three levels of analysis – inputs and results – own study

Level	Input	Results
1	Layout with first positions of parts, BOM, operations of operator, welding time, tact time, loading/unloading time, operator speed, acceleration, deceleration, welder structure	Basic statistics, visualization in 3D, possibilities for optimization of operator's movements, good positions of totes with parts
2	High quality cells, cells connection – logic, production program	Balanced line, visualization in 3D
3	High quality line, production program, warehouse position, intralogistics concept, KPI's, objective function, decision variables	Optimized intralogistics, selection of tuggers and trolleys, milk-runs schedules, visualization in 3D

One of the challenge to solve is layouts definition and working with layout in situation where market is dynamic. Layout forms the background for intralogistics so changes in layout influence in intralogistics.

4 The Concept of Interactive Layout

As mentioned earlier (see Fig. 1), layout plays the crucial role in production and assembly system. We can find many papers about layout's problems – many of them are theoretical – the study examines how material handling system, material flow and distance travel of material improve productivity and operational efficiency. The main objectives is the minimization of the total cost of distance travelled between facilities. Stated mathematically, we have

$$\text{Min}\, C = \sum_{i=1}^{m} \sum_{j=1}^{n} f_{ij}\ d_{ij} \qquad (1)$$

Where f_{ij} is the number of flows/loads or movements between facilities I and j, d_{ij} is the distance between facilities I and j (i = 1, 2, n), (j = 1, 2, m).

In many cases this approach is not enough in the factory's operational life.

One of modern solutions to organize intralogistics systems is using the approach based on milkrun systems. (Bozer and Ciemnoczolowski 2013) described Milk-Run systems as "route-based, cyclic material handling systems that are used widely to enable frequent and consistent deliveries of containerized parts on an as-needed basis from a central storage area (the 'supermarket') to multiple line-side deposit points on the factory floor". (Meyer 2015) defined the goal of designing Milk-Run system as "the definition of cost efficient, regularly recurring transport schedules for suppliers with a regular demand". The goal of the implementation of Milk-Run system is to achieve the number of operational benefits. (Brar and Saini 2011) and many different authors in their scientific considerations, defined these benefits. The most commonly mentioned is the reduction in transportation costs. The cost of a Milk-Run system usually depends primarily on the layout of transport routes, because it affects the total distance travelled by logistics trains and the duration of deliveries.

(Harris and Harris 2004) defined 5 major steps for the transition from the traditional material flow to the lean material flow using the Milk-Run systems:

1. Develop a plan for every part (PFEP).
2. Build the purchased parts market.
3. Design delivery routes.
4. Implement pull signals.
5. Continuously improve the system.

Simulation technique is also recommended in the facility planning analysis. Simulation is an appropriate tool to help the designer to define the storage spaces of assembly system in this stochastic situation (Nica 2008). Simulation tools that commonly used in facility planning are Arena, Quest, ProModel, Witness, FlexSim. Limitation of many software is they only provide the two dimensional visualization (2D), which is not easy to visualise, understand and evaluate or 2D/3D where three dimensional visualisation is available as postprocessor – only for visualization, not for direct work with 3D objects. Only FlexSim is the real core 3D simulation tool. In situation when the location of containers with parts is crucial (because of dimension Z – see Fig. 4) core 3D simulation program is the best choice. Then we can manipulate the position of containers using mouse or by table with 3D coordinates. This software also allows designer to build virtual reality (VR) environment and can have feel of the actual setting of the factory.

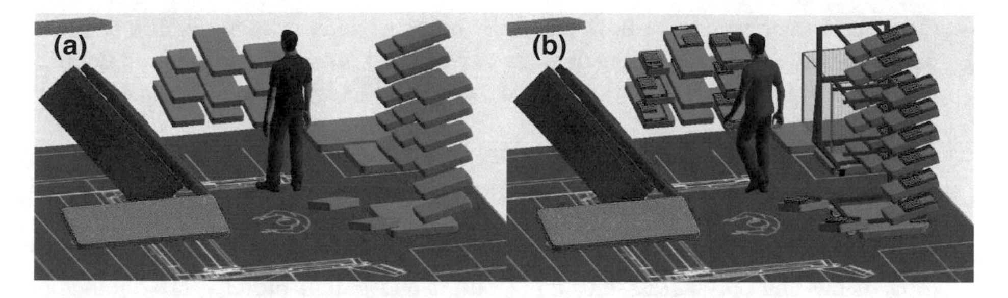

Fig. 4. Locations of containers in 3D core simulation program FlexSim – own study

Other factor is the level of objects in simulation tool – its offer the level of operator, machine, buffer, conveyor – if we want to model, design the plant we have to work with too many objects. It is difficult to work with these tools according to proposed model with three levels of intralogistics.

Engineers in factories work with layout using AutoCad to create two dimensional floor plan. If they want to change position of machine or assembling line, they redraw the drawing. As result they have many drawings).

Our concept of interactive layout can define as the list of requirements:

- Workstation as the basic object - A workstation is treated as a basic object but it can include containers, locations, worktables, assembling stations, disassembling stations, operators (as an operator we understand a human, robot, manipulator, agv), welding stations, logistics trains.
- Working with workstation using mouse it means the possibility to move whole workstation to requested position in layout (in plant) without changing the relations inside workstation.
- Working with mouse on two levels – inside workstation (locally) and outside (globally) – all positions are saved in tables accessible by user with possibility to import from or export to Excel.
- Request to address every object (workstation) and components of object by name – requirements regarding the rules for naming objects.

5 Implementation of Interactive Layout

Interactive layout was implemented in FlexSim Discrete Event System environment (see Sect. 4 and Fig. 4). We defined the workstation as basic object and we addresses the components of the workstations by name. Figure 6 shows the workstation basic object as template. We use this template to generate designed workstations automatically by indicating components and providing their quantity. Workstation is composed of:

- Plane which forms the floor of workstation – the name of this plane is the name of workstation – this name must be unique. (this plane can contain the layout of workstation in bitmap format).

- Operators – as operator we understand the objects which have possibility to move parts: human, robot (with many degrees of freedom), manipulator (without degrees of freedom), forklift and agv (automated guided vehicles). Operators work periodically – in repetitive cycles.
- Universal operators – as universal operator we understand operators which have the same behaviors, and they can form teams and the leader of team can assign the activities to these operators. These operators may be able to move: human, forklift, agv.
- Assembling station – where assembling can be performed – this station has included (not visible and not addressable) input buffer and output buffer – the number of assembled parts is set.
- Disassembling Station – where disassembling can be performed - this station has included (not visible and not addressable) input buffer and output buffer – the number of disassembled parts is set.
- Work Table – the table where some works can be performed by worker.
- Welding station – the place where welding operationsa are performed using special automate welding machine.
- Local Points which define the point on the floor – access point for Assembling station – for example.
- Conveyor which enables to move parts,
- Logistics trains formed and drived by Operator – human.
- Storage area - the place where the container is placed, in which elements and units are transported (Fig. 5).

The Fig. 6 shows the working of operator with storage area and points connected to this object.

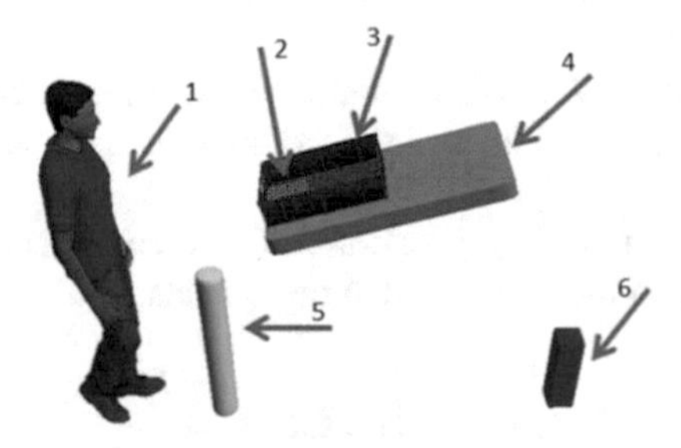

Fig. 5. Basic objects form the work environment of an operator (agent) – own study.

The legend for this Figure is as follow:

- 1 - Operator (human) – employee, he performs the tasks in a working cell,
- 2 - Parts – components needful to produce subassemblies and ready products,
- 3 - Container – object designed for storage of parts. It can have a specified type, dimensions, color, etc.,
- 4 - Storage Location – objects dedicated to storage the containers with components or products,

- 5 - Node – a virtual object, which defines the point in the model space. Each container/table/trailer has one node assigned, which informs the operator where he should go to find a specific object (e.g. welding table). It is important to include the ergonomics in a simulation model. Usually the model contractor sets manually the locations of nodes,
- 6 - InNode – a node dedicated to intralogistics workers.

Figure 6 shows the workstations with all mentioned above components. All components are named (Table 2) and its coordinate are stored in tables.

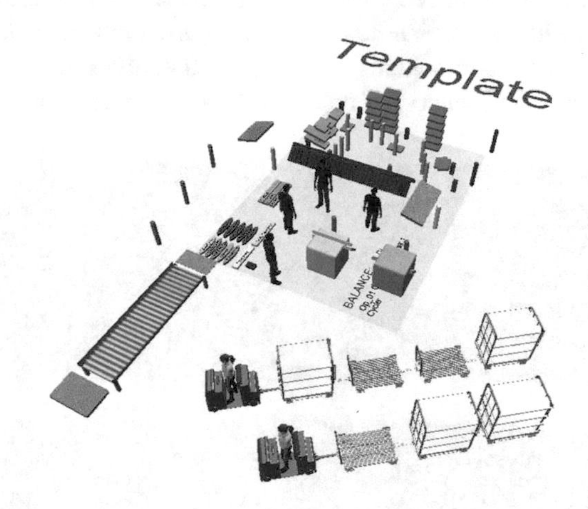

Fig. 6. Basic objects form the work environment of an operator (agent) – own study.

Table 2. List of components and its names used for interactive layout

Component name	Mark
Plane (workstation)	The name is optional
Operator	Op_xx
Universal operator	Ou_xx
Assembling station	Ass_xx
Disassembling station	Dis_xx
Work table	WT_xx
Welding station	Weld_xx
Local point	G_xx
Conveyor	Cv_xx
Logistics train	Tugger_xx
Storage area	P_xx
Worker access point to storage area	N_xx
Logistics access point to storage area	I_xx
Global – MilkRun stop	MRStop_xx
Global – global point	GG_xx

As mentioned earlier the interactive layout was implemented in FlexSim. Modeler can work with all objects in two modes:

- Table mode – where all coordinates for objects are set from table – in this mode object remembers its position in 3D and if it will be moved by mouse it will return to saved position
- Mouse mode – where modeler can move the object by mouse and this new position will be saved in table.

The modeler can switch between these modes – it form very useful and comfortable conditions for work with model. Whole workstation can be moved on layout maintaining the local properties of the components – see Fig. 7. It enables to locate workstation on plant layout and check the scenario of material flow in new situation.

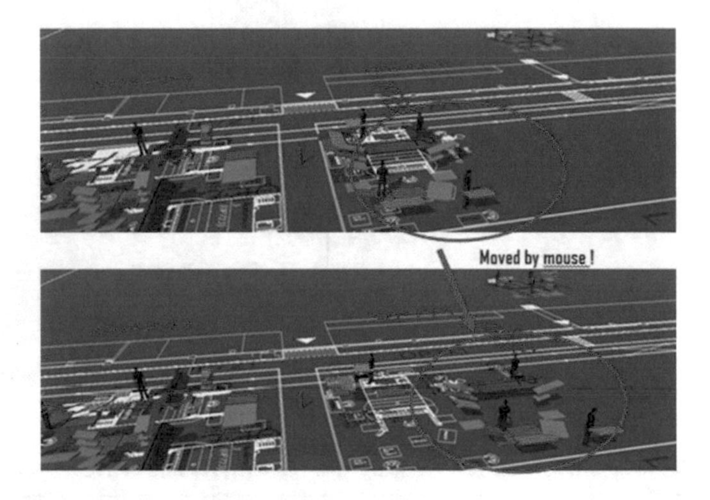

Fig. 7. Screens with visualization of moving workstation by mouse – own study.

6 Conclusion and Futher Research

This paper presents an approach for redesign of intralogistics systems using interactive layout. The initial implementation of the approach is in FlexSim simulation software as tool called LogABS. First implementations of this approach we are performing in automotive plants (two cases) and furniture factory (1 case). The reactions of industry representatives have been promising. Now we are focusing on developing the methods of collision-free and deadlock-free implementation of intralogistics processes in developed tool. The goal is to prepare a tool which can help engineers to quickly and optimally react to demands from the market.

Acknowledgements. The work was carried out as part of the POIR.01.01.01-00-0485/17 project, "Development of a new type of logistic trolley and methods of collision-free and deadlock-free implementation of intralogistics processes", financed by NCBiR.

References

Hompel, M., Heidenblut, V.: Taschen-lexikon Logistik, p. 132. Springer (2008)
http://www.wynright.com/intralogistics/. Accessed 10 April 2018
Muther, R., Hales, L.: Systematic Layout Planning, 4th edn. Management & Industrial Research Publications, Kansas City (2015)
Muther, R.: Systematic Layout Planning. Mass, Boston (1961)
Naik, S.B., Kallurkar, S.: A literature review on efficient plant layout design. Int. J. Ind. Eng. Res. Dev. (IJIERD) **7**(2), 43–50 (2016)
Zhu, Y., Wang, F.: Study on the general plane of log yards based on systematic layout planning. IEEE Comput. Soc. **4**, 92–95 (2009)
Deb, S.K.: Computerized plant layout design using hydrid methodology under manufacturing environment. IE(I) Journal-PR **85**, 46–51 (2005)
Jaturachat, P., Charoenchai, N., Leksakul, K.: Plant layout analysis and design for multi-products line production. In: IE-Network Conference, pp. 844–849 (2007)
Parveen, S., Ravi, P.S.: A review of metal-heuristic approaches to solve facility layout problem. Int. J. Emerg. Res. Manag. Technol. **2**(10), 29–33 (2013)
Anucha, W., Kajondecha, P., Duangpitakwong, P., Wiyaratn, P.W.: Analysis plant layout design for effective production. In: International Multiconference of Engineers and Computer Scientists (IMECS 2011), Hong Kong, vol. II, (2011)
Pinto, W.L., Shayan, E.: Layout design of furniture production line using formal methods. J. Ind. Syst. Eng. **1**, 81–96 (2007)
Tao, J., Wang, P., Oioao, H., Tang, Z: Facility layout based on intelligent optimization approaches. In: IEEE 5th International Conference on Advanced Computational Intelligence, China, pp. 502–508 (2012)
Vaidya, R.: Plant layout design: a review survey. Int. J. Bus. Manag. Issue **2**(1), 1–9 (2013)
Yifei, Z.: Facility Layout Design with Random Demand and Capacitated Machines. Lancaster University Management, Lancaster (2012)
Schwab, K.: The Fourth Industrial Revolution. Crown Business, New York City (2017)
Canetta, L., Redaelli, C., Flores, M.: Digital Factory for Human-oriented Production Systems: The Integration of International Research Projects. Springer (2011)
Beaverstock, M., Greenwood, A., Nordgren, W.: Applied Simulation. Modeling and Analysis using Flexsim. Flexsim Software Products, Inc., Canyon Park Technology Center, Orem (2017)
Worn, H., Frey, D., Keitel, J.: Digital factory – planning and running enterprises of the future. In: 26th Annual Conference of the IEEE Industrial Electronics Society, vol. 2, pp. 1286–1291 (2000)
Centobelli, P., Cerchione, R., Murino, T.: Layout and material flow optimization in digital factory (PDF). Int. J. Simul. Model. **15**, 223–235 (2016). https://doi.org/10.2507/ijsimm15(2)3.327
Bozer, Y.A., Ciemnoczolowski, D.D.: Performance evaluation of small-batch container delivery systems used in lean manufacturing–Part 1: system stability and distribution of container starts. Int. J. Prod. Res. **51**(2), 555–567 (2013)
Meyer, A.: Milk Run Design – Definitions, Concepts and Solution Approaches. Karlsruher Institut für Technologie, KIT Scientific Publishing (2015)
Brar, G.S., Saini, G.: Milk run logistics: literature review and directions. In: Proceedings of the World Congress on Engineering (WCE), vol. 1 (2011)
Harris, Ch., Harris, R.: Steps to Implementing a Lean Material Handling System (2004). http://www.lean.org
Nica, M., Ganea, L.M., Donka, G.: Simulation of queues in manufacturing systems. An. Oradea Univ., Fascicle Manag. Technol. Eng. **VII**, 56–61 (2008)

Technical Diagnostics at the Department of Automation and Production Systems

Ivan Kuric[1], Miroslav Císar[1(✉)], Vladimír Tlach[1], Ivan Zajačko[1], Tomáš Gál[1], and Dorota Więcek[2]

[1] University of Zilina, Univerzitna 8215/1, 01026 Zilina, Slovakia
{ivan.kuric,miroslav.cisar,vladimir.tlach,ivan.zajacko, tomas.gal}@fstroj.uniza.sk

[2] University of Bielsko-Biala, Willowa 2, 43-309 Bielsko-Biala, Poland
dwiecek@ath.bielsko.pl

Abstract. The article contains a summary of recent development in field of technical diagnostics at the Department of Automation and Production Systems, Faculty of Mechanical Engineering, University of Zilina. It covers diagnostics and monitoring of CNC machine tools, industrial robots, and production lines. Each part contains a description of basic approaches, methods, measurement tools and their implementation. In case of machine tool and industrial robot diagnostics, it is mainly laser interferometry and double Ballbar method. It also describes usage of the internet of things and machine learning as a tools to implement multiparametric diagnostics and monitoring on production lines.

Keywords: Technical diagnostics · CNC machine tool · Industrial robot Production line · Machine learning

1 Introduction

Every technical object, including basic tools of automation, has its own lifespan, in which it is capable of proper functionality, which can be extended to a certain extent by proper maintenance. The maintenance should be focused not only on repairs of devices, as it still usually is, but it also should prevent occurrence of unexpected downtimes and resulting financial losses. Such maintenance requires certain amount of information about machinery, production processes, mechanisms of faults, and impact on the environment. Therefore, correct maintenance requires implementing of technical diagnostics and condition monitoring at some extent. It should be clear that, in order to improve competitiveness of the business, it is desirable to increase machinery reliability to reduce its necessary downtime, and to predict development of its future condition. These reasons, together with the necessity to save resources, should be the main reasons for implementation of predictive maintenance and technical diagnostics in industry [1].

The importance of development in the field of technical diagnostics is undeniable, therefore it is one of the most researched topics at the Department of Automation and Production Systems, Faculty of Mechanical Engineering, University of Zilina. As the name suggests, the main interest of our department is industrial automation and its

© Springer Nature Switzerland AG 2019
A. Burduk et al. (Eds.): ISPEM 2018, AISC 835, pp. 474–484, 2019.
https://doi.org/10.1007/978-3-319-97490-3_46

practical implementation in industry. The development of technical diagnostics in our department can be divided into three main groups: machine tool diagnostics and monitoring, diagnostics of industrial robots, and diagnostics of production lines.

2 Diagnostics of Machine Tools

CNC machine tools are complex mechatronic devices which are often mistakenly considered to be absolutely accurate. In reality it always inherits various errors and inaccuracies that affect the quality of the produced parts. Therefore, machine tool precision is considered to be one of the most important diagnostic parameters, by which it is possible to evaluate its overall technical condition.

Machine tool precision is defined as the ability to produce parts of the required shape and dimensions keeping the required tolerances, and to achieve the desired surface roughness. Geometric accuracy of a machine tool is determined by actual shape and position of its individual parts, joints, and their mutual movements. The desired geometric accuracy of a machine tool can be achieved by respecting the required accuracy of production of individual machine tool parts, nodes, and its assembly. Furthermore, most of CNC machine tools are capable of compensating geometric errors resulting from inaccuracies and wear of a construction part [2].

However, geometric accuracy of a machine tool is not sufficient to ensure precise production, as it depends on the accuracy of the path shape between a work-piece and a tool, which can be affected by various additional factors. Machine tool condition on its built in state, but also on its usage and maintenance. For further improvement of machine tool reliability and safety throughout all lifetime, early detection, prediction and location of inaccuracies, faults, and errors are of high interest. Not only critical situations and collisions, but also normal everyday use causes wear that leads to decreasing accuracy of positioning, and furthermore to production of scrap [3].

Development in this area in our department is focused mainly on improving the existing procedures of machine tool precision measurement in terms of time efficiency and implementing long-term machine tools monitoring.

2.1 Measurement of Machine Tools with a Laser Interferometer

In the field of implementing laser interferometry, much effort was devoted to design devices that help to train routines necessary to effectively perform measurement in any given condition or situation. Such devices replace machine tools during training and simulate various faults and positioning errors to provide real life experience with setting up measurement and aligning the laser, which require the most time and skill of all measurement process stages [3].

Such devices are necessary, as standard training CNC machine tools commonly used in educational process in our laboratories lack the ability to set compensation tables, and thus it is impossible to present the effect of compensation to machine tool performance in laboratory conditions.

2.2 Measurement of Machine Tools with Ballbar

Ballbar type devices are probably one of the most effective tools in terms of predictive maintenance. Normally, it is implemented according to standard ISO 230-4 that describes measurement, environmental conditions and evaluation of measurement. The Ballbar system, in our case Renishaw Ballbar QC20-W, is essentially a highly accurate linear displacement sensor which measures deviations of distance between two balls that fit into magnetic bowls connected to the moving parts of the examined device. The measurement with this device is based on measurement of radius deviations during circular movement [4, 5].

Most of research at our department regarding measurements of CNC machine tool performance with Ballbar was focused on software development. For example, the B5R2SIG software can be used to export data from measurement to be analyzed in external software, and then furthermore to compare data measured by Ballbar device with profile measured on machined parts by a roundness measuring device, such as Talyrond 73. Our research confirmed that the most of the machine tool errors, measurable by Ballbar, can be at least partially diagnosed by measuring of circularity of profile machined in a controlled environment.

2.3 Machine Tool Monitoring

The laboratory of CNC programing at our department is, similarly to other educational institutions, equipped with a training machine tool from the Emco Concept series. These machine tools are characterized by a control system which is realized as a computer software with interface and behavior simulating various commercial control systems.

The fact that the control system is running on a PC as standard software offers an opportunity to access its data in various ways. Based on this, we designed software (MT Monitor) capable of reading variables that stores information from control system, such as tool position, technological parameters, or filenames of NC programs.

Such data can be filtered and analyzed in various software. For example, the analysis of workspace usage can be very valuable. Such analysis can prevent excessive usage of specific subspaces of the machine tool workspace, and thus local wear resulting to maintenance cycle shortening.

3 Diagnostics of Industrial Robots

The main focus in this field at our department is to evaluate, analyze, and improve the existing methods, or to develop new methods if it is necessary. Most of our experiments are performed on Fanuc robots, as the department is Fanuc authorized system integrator. However, the results are still applicable for any industrial robot of common construction.

Selecting of an industrial robot for individual application requires considering some basic criteria, such as payload, workspace shape and size, number of degrees of freedom, and also other characteristics. The analysis of the performance criteria offers a detailed view of the industrial robot properties that affect its performance in a given application. The importance of individual performance criteria, together with the recommended test conditions, measurement and evaluation methods are defined in international standard ISO 9283. This standard divides the performance criteria of industrial robots into four groups:

- positioning performance,
- path performance,
- minimum positioning time,
- static compliance.

The International Organization for Standardization published technical report ISO/TR 13309 to provide an overview of metrological methods and measurement methods applicable for measuring performance criteria of industrial robots. There are also a lot of various scientific publications dedicated to such topics. The most frequently measured criteria are single directional precision and repeatability. In most cases, these measurements are carried out using indicators or a laser interferometer. Other approaches implement camera systems or devices designed primarily to evaluate CNC machine tool condition. Most of the mentioned methods are highly time and skill demanding, while in other cases the biggest problem is price and availability of measuring devices [6].

Laser tracker is considered to be one of the most common measurement devices for evaluation of industrial robots performance used by most of robot manufacturers. The main benefit of implementing laser interferometers is simplicity of measurement and ability to meet almost all conditions required by ISO 9283. On the other hand, problems appear with robots with repeatability in hundredths of millimeters [6, 7].

Implementing indicators for measuring performance criteria is a simple and relatively cheap solution. Most publications dedicated to one-directional precision and accuracy of industrial robots list indicators as the most commonly used measuring device. However, without other tools for designation absolute position of robot end point, it is possible to evaluate only relative repeatability of an industrial robot.

In case of CNC machine tools, besides laser interferometer, ballbar type devices are commonly used. The Ballbar type devices are great tools for quick and easy evaluation of Cartesian kinematics, but their usefulness for alternative kinematics is quite limited. The main limitation is the lack of methodologies and compatible software. Therefore, our research is mainly focused on implementing these two measuring devices and developing appropriate methods and software [8, 9].

3.1 Industrial Robot Measurements with a Laser Interferometer

Experimental verification was performed on industrial robot Fanuc LR mate 200iC, available in the Laboratory of robotization of production processes at our department, with the laser interferometer Renishaw XL-80. This laser system allows for multiple measurement, mainly linear, angular, and straightness measurement. In case of measurement repeatability, linear measurement was used. The accuracy of linear measurement is ± 0.5 μm, and resolution is 1 nm. This type of measurement is based on relative movement of the reflector and the interferometer, which is usually static, along the laser beam emitted by the laser measuring unit XL-80. The principle of measurement limits its application to movement along the line parallel to the laser beam [6, 10].

We designed a measuring procedure based on ISO 9283 which is based on an imaginary cube (Fig. 1) that has to be placed in workspace of the examined robot. Such a cube should be as big as possible and should be placed in the most used part of robot workspace. In our case the cube edge is 470 mm long and the edges are parallel to the axes of the world coordinate system of robot (WCS). As the measurement is possible only along the line, three individual measurements are necessary, each parallel with a different axis (X, Y, Z). All measurements have a common point (P0) located in the corner of the cube. This arrangement allows for one-directional, and bi-directional measurement method.

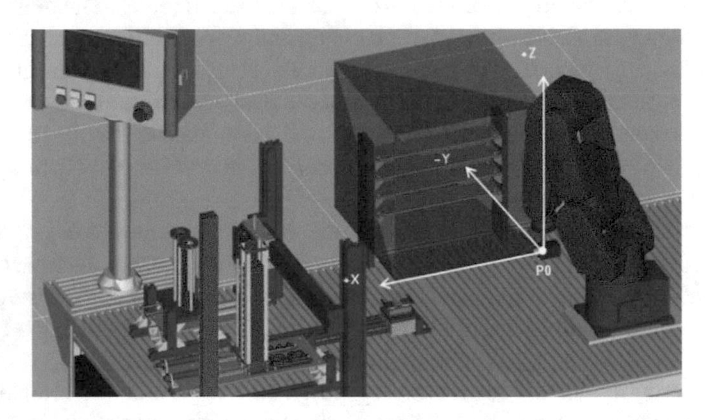

Fig. 1. A model of Fanuc LR mate 200iC workplace with position and orientation of an imaginary cube and measured axes

Measurements of industrial robots with a laser interferometer are time and skill demanding, and in comparison with measurements on CNC machines bring fewer benefits, as it is not possible to directly use a compensation table.

3.2 Industrial Robot Measurements with a Ballbar

Current trends of increasing production flexibility and shortening the time of creating a solution and its implementation to production creates more and more pressure to deploy industrial robots in applications that used to be the domain of other devices, such as machine tools or single purpose devices. Such applications require high accuracy and precision and, most importantly, performance stability over time. The most common applications include highly accurate assembly, measurement, testing, and machining. During the lifecycle, a robot's performance is changing as a result of various factors, the most important of which can be divided into the following five groups:

- environmental factors (temperature, humidity),
- parametric factors – result of production and assembly errors, impact of dynamic parameters, such as friction, hysteresis and loosen fits,

- factors related to measurement (resolution and nonlinearities of sensors),
- factors related to calculations (e.g. rounding),
- factors related to application (e.g. installation errors).

Regular diagnostic measurement, necessary to check performance criteria meeting requirements of the mentioned applications, exert pressure to develop measurement methods which would take as little time as possible. Such measurements should be performable in real industrial conditions in which the robots are deployed [5, 6].

Regular data collection is a necessary condition to implement progressive maintenance methods that allow to predict and prevent robot or production faults due to decreased performance criteria. In case of CNC machine tools, a commonly used device is Ballbar, which allows to check quickly its performance and also to evaluate errors and their sources. Results of such measurements provide a strong information base for targeting and planning maintenance [4, 5].

Measurement with Ballbar type device is not part of ISO 9283 standard, nor the technical report ISO/TR 13309. During measurement, it is not possible to fulfill measurement conditions required by the mentioned standards. The circular path on the machine tool is done as the synchronous relative motion of two perpendicular axes. Contrary, the simple circular path on industrial robot with serial kinematics requires movements in multiple joints. Therefore, measurement and result analysis also requires special approach. On the machine tool, the software for Renishaw Ballbar is capable of identifying 21 different errors describing its condition. Still, it is possible to use this device for simple check of robot condition, and with regular measurement to monitor its changes. Some errors identified by this software, such as vibrations or reversal spikes, are related to robot kinematics as well but most of them do not [6].

The basic output of this measurement is polar plot of deviations of the measured circle from the programmed ideal one. Analysis of such a plot allows to identify errors e.g. perpendicularity of both axes that manifests as an extension along 45° or 135° (Fig. 2). In case of the industrial robot, it is most probably caused by incorrect conversion of Cartesian coordinates into the angular coordinates of robot joints. The next typical error is occurrence of vibrations (Fig. 2b).

To fully understand the origin and causes of errors, it is necessary to perform tests on multiple places in the workspace, or even to use different measuring devices and methods and their combinations. It also requires detailed information about robot kinematics, exact position and orientation of tested circular path relative to the robot, and exact influence of individual drives in overall movement. Such information can be obtained by using the suitable CAE system to plan and prepare experimental verification. In our case, the Creo Parametric 2.0 was used to create a positional analysis, as shown on a model of our robot and its workplace (Fig. 3). The positional analysis can help us to plan locations and orientations of measurement in order to get different ranges of individual joint coordinates for each position, and thus to identify possible source of error in that particular joint.

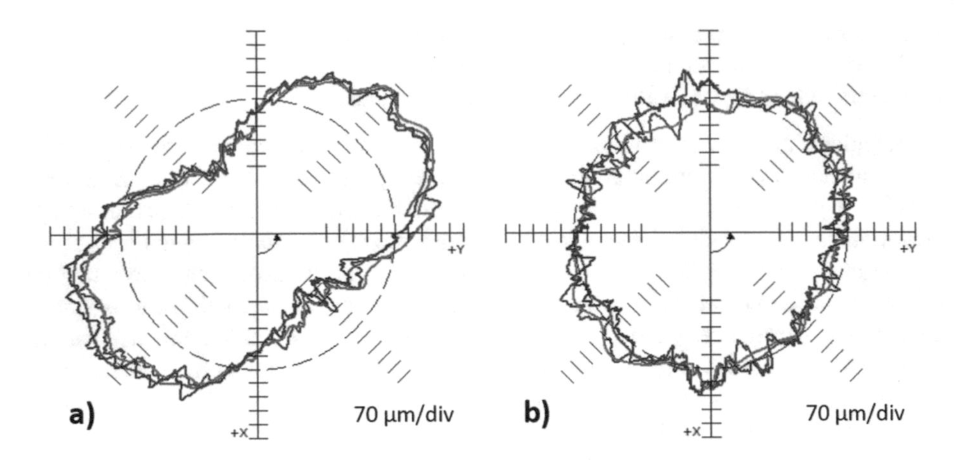

Fig. 2. An example of polar plot from measurement of industrial robot by Renishaw Ballbar QC20-W showing error in perpendicularity (a) and vibrations (b)

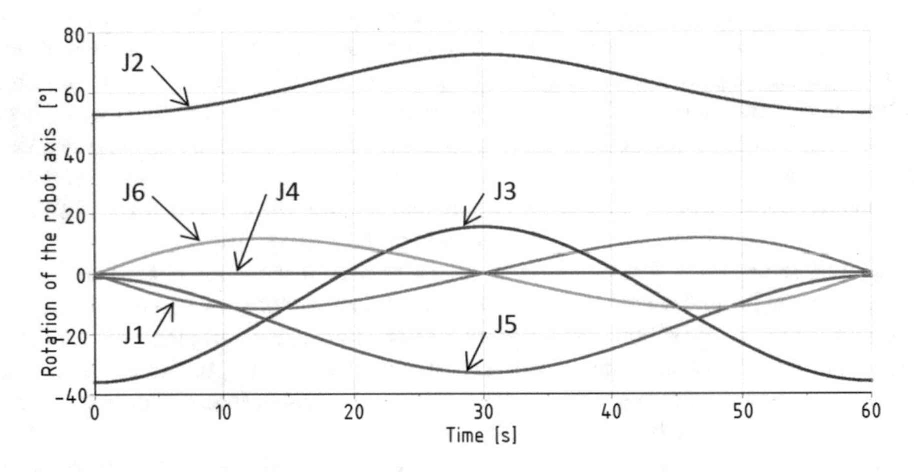

Fig. 3. Positional analysis of joints of an industrial robot during circular motion

4 Diagnostics of Production Lines

There are a lot of various elements used in general industrial automation. One of the most common types of production facilities that implements automation in engineering industry are production lines. As everything in industry, even production lines should be seen not only as tools that create products, but also as products themselves. Currently, our department cooperates with numerous companies dealing with design and manufacturing of production lines. The purpose of this cooperation is to implement modern methods of diagnostics, such as remote diagnostics and condition monitoring into newly developed production lines or to add them to the existing ones during their regular maintenance. There are a lot of various methods that can be implemented in order to increase the amount of information describing the current and future conditions of a production line and its individual parts. It is necessary to understand that a production line itself is not always simply a sum of its individual parts, and thus it requires more sophisticated and complex or even holistic approach [11].

The implementation of diagnostics into complex systems, such as production lines usually requires the following steps:

- analysis of individual production processes,
- division of the examined device into the key nodes,
- selecting appropriate diagnostic methods (physical principles, sensors, and their location),
- selecting suitable method for data acquisition (from sensors and from control system of production line),
- test run,
- analysis of data acquired in test run,
- verification or optimization of the original solution,
- deploying the final solution.

The development process usually requires several iterations and fine tuning to achieve appropriate solution that meets customer expectations. Modern technologies allow shortening of development cycles by implementing ready-to-use solutions and tools designed for rapid development [11].

One of the latest trends is to decentralize data collection by implementing a so-called intelligent industrial sensor that is meant to be connected to the Internet of Things (IoT). IoT can be used in various ways. Currently we are using Kepware KEPServerEX as a data concentrator, and PTC ThingWorx as platform to collect, process, and visualize data [8].

The collected data originates not only from sensors, that were added as diagnostic tools, but also from control system itself. The control system of any production line usually uses huge amount of variables. Therefore, it is necessary to carefully select the variables which would be reasonable to collect important data, in order to reduce traffic necessary to transmit diagnostic data. The selection and filtration can be done in one iteration of the whole optimization process. The platform used for data collection allows us to export data and process them in any other external software. We can use it to create and enhance algorithms of data processing and analysis as well. This is described in the following chapter.

4.1 Machine Learning in Technical Diagnostics

Contemporary methods of technical diagnostics commonly use state-of-the-art devices and technologies in order to ensure quality, reliability and value of data acquired by diagnostic procedures. In some cases, information of the same or even higher value can be gained by using multiple diagnostic methods, where synergic effect can take place. Simultaneous usage of multiple diagnostic method is known as multiparametric diagnostics. However, implementation of multiparametric diagnostics requires perfect knowledge of relations between the values of diagnostic parameters and condition of the examined subject. Identification of such relations is usually a complex task that requires processing of significant amount of data. Therefore, machine learning seems like an ideal tool for such kind of tasks.

It is beneficial to use machine learning and deep neural networks` algorithms in the area of multiparametric diagnostics and predictive maintenance when analyzing high volumes of data is required. It is important mainly when the dependence between measured data points and their effect on the quality of the manufacturing process is complex and non-obvious.

A typically exercised workflow when using machine learning and deep neural networks methods is the following:

- filtering and synchronizing the data,
- using simple machine learning algorithms to preliminary data analysis,
- using deep neural networks to develop a model that detects changes in condition and arising machine failures,
- implementing the model on PCs.

Filtering the data is necessary step, as the data collected from the manufacturing process can exhibit errors due environmental factors, such as interference. The most common type of error that can be easily filtered out is such a measurement error where the value is out of the physically possible range. Another frequently occurring error is missing data when, measurement was not performed or the data was not collected for any reason. If only small group of values in short time span is missing or if they are obviously incorrect, then their value can be approximated from the neighboring values. If there are more missing records, then all the values in that time span should be dropped. Then data are synchronized and resampled according to use the same time span and sampling period. Failure causes can be encoded numerically and added to the data [12, 13].

The data is then analyzed by using common regression and clustering machine learning tools to find nonobvious relationships between measurements and failure states. This step does not detect all causes of failure, but it helps to better recognize deeper relations between the measured data, collected technological parameters, and condition of the examined device acknowledged by conventional approaches. Regression and clustering results can be directly used to develop software modules for failure prediction. Sometimes it is beneficial to use some of the results of regression and clustering as additional input for deep neural networks [13, 14].

For most tasks, when it is necessary to analyze significant amount of data collected over a long time period, stamp and classical methods cannot predict all failures, but we can successfully use recurrent neural networks. The recurrent neural networks have loops that pass information from one step of the network to another and back. These loops act as memory that can store internal state and allows them to process sequential data.

The most universal recurrent neural networks that we use in monitoring, diagnostics, and predictive maintenance is the Long Short-Term Memory (LSTM) network. LSTM is a neural network capable of learning long term dependencies between input data. We use the Keras library in conjunction with TensorFlow. This approach allows us to easily test several neural network topologies and fast iteration times [13].

After the model is successfully developed and tested, it can be exported and used directly in the manufacturing process as an essential part of a control system. However, most models are too complicated and computing power demanding to run on PLCs which control the manufacturing line. In such cases, it is necessary to use either conventional computers, or dedicated industrial PCs built into the production line to collect and process data from PLCs. Running the predictive maintenance model on conventional computers has the benefit of easier monitoring and ability to update the model. This allows us to use simultaneously different models and compare their performance.

Acknowledgement. This work was supported by the Slovak Research and Development Agency under the contract No. APVV-16-0283.

References

1. Kolny, D., Więcek, D., Ziobro, P., Krajčovič, M.: Application of a computer tool monitoring system in CNC machining centres. Appl. Comput. Sci. **13**(4), 7–19 (2017)
2. Castroa, H.F.F., Burdekinb, M.: Calibration system based on a laser interferometer for kinematic accuracy assessment on machine tools. Int. J. Mach. Tools Manuf. **46**, 89–97 (2006)
3. Kosinar, M., Kuric, I.: Monitoring possibilites of CNC machine tools accuracy. In: Proceedings of 1st International Conference on Quality and Innovation in Engineering and Management (QIEM), 17–19 March 2011 (2011)
4. Majda, P.: The influence of geometric errors compensation of a CNC machine tool on the accuracy of movement with circular interpolation. Adv. Manuf. Sci. Technol. **36**(2), 59–67 (2012)
5. Kuric, I., Košinár, M., Císar, M.: Measurement and analysis of CNC machine tool accuracy in different location on work table. Proc. Manuf. Syst. **7**(4), 259–264 (2012)
6. Tlach, V., Císar, M., Kuric, I., Zajačko, I.: Determination of the industrial robot positioning performance. In: MATEC Web of Conferences, vol. 137. EDP Sciences (2017)
7. Shirinzadeh, B., Teoh, P.L.: Laser interferometry-based guidance methodology for high precision positioning of mechanisms and robots. Robot. Comput.-Integr. Manuf. **26**, 74–82 (2010)
8. Stanček, J., Bulej, V.: Design of driving system for scissor lifting mechanism. Acad. J. Manuf. Eng. **13**(4), 38–43 (2015)
9. Nubiola, A., Slamani, M., Bonev, I.A.: A new method for measuring a large set of poses with a single telescoping ballbar. Precis. Eng. **37**(2), 451–460 (2013)
10. Kuric, I., Bulej, V., Sága, M., Pokorný, P.: Development of simulation software for mobile robot path planning within multilayer map system based on metric and topological maps. Int. J. Adv. Robot. Syst. **14**(6) (2017)
11. Mičieta, B., et al.: Delegate MASs for coordination and control of one-directional AGV systems: a proof-of-concept. Int. J. Adv. Manuf. Technol. **94**(1–4), 415–431 (2018)
12. Hochreiter, S., Schmidhuber, J.: Long short-term memory. Neural Comput. **9**(8), 1735–1780 (1997)

13. Aydin, O., Guldamlasioglu, S.: Using LSTM networks to predict engine condition on large scale data processing framework. In: 4th International Conference on Electrical and Electronic Engineering (ICEEE), pp. 281–285, Ankara (2017)
14. Jozefowicz, R., Zaremba, W., Sutskever, I.: An empirical exploration of recurrent network architectures. In: Bach, F., Blei, D. (eds.) Proceedings of the 32nd International Conference on Machine Learning – (ICML 2015), vol. 37, pp. 2342–2350 (2015). JMLR.org

Use of Dynamic Simulation in Warehouse Designing

Monika Bučková[1(✉)], Martin Krajčovič[1], and Dariusz Plinta[2]

[1] University of Zilina, Univerzitná 1, 010 26 Zilina, Slovakia
monika.buckova@fstroj.uniza.sk
[2] University of Bielsko-Biała, Willowa 2, 43-309 Bielsko-Biała, Poland

Abstract. This article describes the topic of warehouse designing with using computer simulation tools. In the introduction of article is described digital factory itself and future of digitalization. The core of article includes description of methodology of warehouse designing with using tools of the digital factory. Any disturbance effects, non-delivery of goods, problems with logistics, problems with suppliers, etc., disturb the activity of a warehouse system and as a result creates situation which are hardly estimable occur. In third chapter is on a specific example described how to use computer simulation to optimize warehouse management. The main task of this example was to find out when such storing system starts failing. For this example, were designed experiments by which is it possible to find solution of problems emerging during picking of goods. The conclusion points out benefits of this methodology for practice. This way of data processing is not only safe, but it is also a good presentation tool which can be used anywhere.

Keywords: Warehouse designing · Logistics · Simulation

1 Introduction

Digital factory and its tools could be used where there is a need to increase production, effectiveness of stock management, improve order planning or decrease costs for storing. The digital factory concept involves integration of various methods, tools and information. It is necessary to work with updated information when proposing improvements. Depending on problem the company wants to deal with, it is important to recognize the information it gets from production or stock processes and customer information obtained with the aim to find out how to use purchased products. Moreover, all information obtained must be sorted in an appropriate manner, stored and process in a way companies can familiarize themselves with possibilities of further improvement as much as possible. This information can be then changed to new ideas and proposals and digital factory workers can implement this information into a real warehouse operation accordingly. Influence and impact of their proposals can be tested for a warehouse operation even before they implement them into a real operation. Every change is made in a virtual factory. Therefore, we can decide what alternative is suitable for us before we implement this change in the real system [1]. Such possibility brings the following advantages: cost reduction, better proposal processing, time saving

© Springer Nature Switzerland AG 2019
A. Burduk et al. (Eds.): ISPEM 2018, AISC 835, pp. 485–498, 2019.
https://doi.org/10.1007/978-3-319-97490-3_47

due to necessary proposal adjustments, etc. The digital future will combine the products with the software solution, which is already in progress [9].

2 Digital Factory and Warehousing

Company managers would like to use digital factory tools to support their decisions. For example, software for static warehouse display or computer simulation in other words. Computer simulation is able to remove deficiencies and analytical method incompleteness even though it is more preparation-consuming for input data, tracked time setting, conditions and rules or restrictions which must be considered when creating a model [6]. However, these difficult steps contribute to warehouse modernization and data collection that is continuously increasing. Nowadays, there is no software solution enabling both static and dynamic warehouse displaying and that would cooperate as a whole. That is why it is important to describe individual steps from a problem definition, through data collection and analysis, draft warehouse proposal, its improvement via dynamic verification up to final implementation when projecting warehouses with the use of digital factory tools.

Figure 1 displays gradually described steps of classical warehouse designing where software solutions are not used yet. These steps create a conceptual warehouse proposal which must be then further developed.

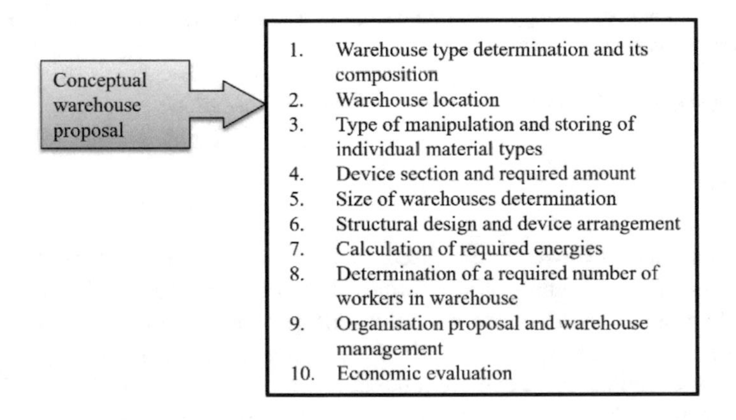

Fig. 1. Steps of conceptual proposal of warehouse

Figure 2 describes methodology of warehouse designing in a digital factory environment, extending the previous methodology by use of software tools to create complex solutions. This provided methodology does not only improve the quality of projecting but creates the opportunity to improve disposition, planning and further 3D simulation model creation directly by company workers.

In the following parts of the article steps of detailed warehouse design with the support of digital factory will be described, including dynamic verification of the proposed solution and implementation into operation.

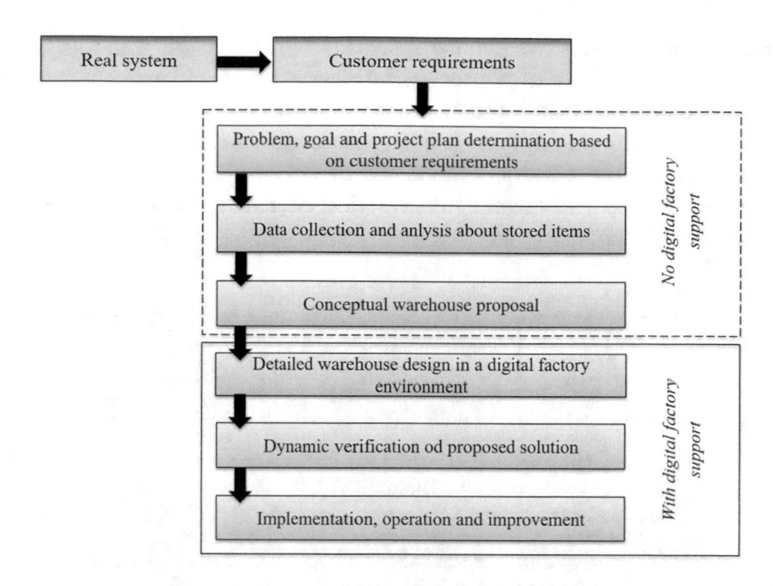

Fig. 2. Steps of methodology of warehouse designing

This methodology is suitable for warehouse, which is yet to be built. If you need to modify warehouse adapted to the methodology, it is necessary to obtain the relevant data from database systems (for example warehouse management system or another software such as SAP, ORACLE, IBM and so on), sensors or measurements.

Example of data:

- **Product Data** - parameters of product, dimensions of product, properties of product, consumption of product, packaging regulations, turnover of items, product weight,
- **Process data collection** - workflows, time calculations, snapshots, maintenance information, staff numbers,
- **Collection of source data** - warehouse data, data about operation of the warehouse and its objects, investment information, energy use, data about storage facilities, data about transport and handling equipment.

2.1 Detailed Design in a Digital Factory Environment

Any disturbance effects, non-delivery of goods, problems with logistics, problems with suppliers, etc., disturb the activity of a warehouse system and as a result situation which are hardly estimable occur. In addition to this, external influences may add to this usually in the form of deviation in quantity, material quality, power failure, effect of weather, etc., only statistical features are written about these. In other words, it is not possible to predict when, where and how this system disturbance happens. Therefore, it means these influences might cause the following:

- Selected events influencing storage processes are difficult to predict and impossible to express them stochastically.

- Relation investigation in warehouse system is time-consuming and imprecise.
- Experiment possibilities with process influences of stock management are time consuming and difficult to estimate what influences to follow and when.
- Warehouse system description and its elements could be done by relatively extensive and mathematically complicated models.
- If the level of system behaviour investigation decreases, there might be distortion and inaccurate information, etc.

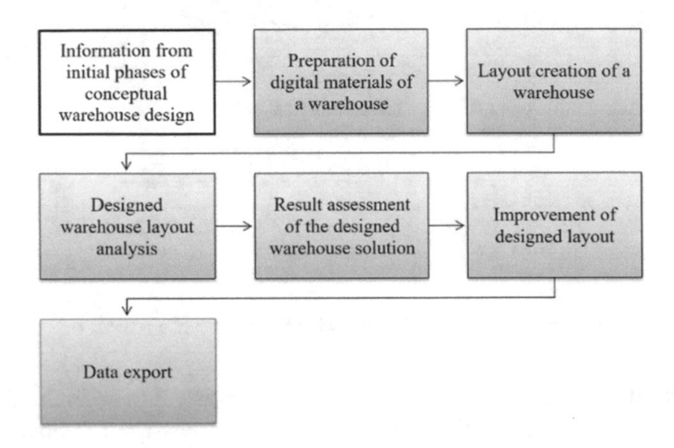

Fig. 3. Detailed proposal of warehouse designing with the use of software solution

By transferring real or planned warehouse system and its elements into virtual environment, a project architect might create a better image about the influence on the system, unexpected changes and their possible impact on it [3]. Simultaneously, it enables the project architect to verify a large number of problem solving variations which will result in a good quality of final warehouse design in early stages of the big project. In order to create detailed warehouse design in the digital factory environment, it is necessary to gradually fulfil the steps displayed in Fig. 3.

Individual steps of this proposal could be described as follows:

- **Preparation of digital materials -** for 3D model creation and selected active or passive warehouse elements, it is possible to use CAD software (CAD model adaptation with the use of software tools for detailed modelling, rendering and 3D model animation), reverse engineering, parametric 3D modelling, explicit 3D modelling or gradual intelligent 3D model development [2]. One of the active warehouse elements is for example the manipulation equipment and one of the passive warehouse elements is for example the storage equipment.
- **Creation of warehouse layout solution -** Objects from software, library of objects are inserted in the layout, which are then arranged in detail in a way areas and storage equipment would be used most effectively. These must also be operated appropriately with regard to designed logistic network. Afterwards, material flows

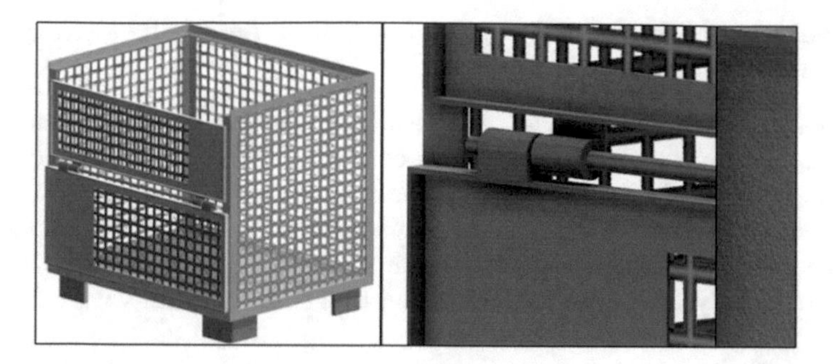

Fig. 4. Illustration of a texturing 3D model of Gitterbox by using Autodesk Inventor 2017 software.

could be designed, transport networks could be amended and also remaining data in the end (for example, product, process and cost information, etc.)

- **Warehouse layout analysis** - Designed layout solution could be analysed and evaluated thanks to software solutions, as they contain a few basic tools which in text or graphical form enable assessment of the designed solution suitability. For example, they enable diagram creation of transport road utilization, analysis of storage area structure, use of areas, etc.
- **Result assessment of a proposed solution** - software solutions offer a number of reports, apart from tools for the designed layout analysis. These reports represent an additional source for the analysis of material flows in a warehouse. Creation and browsing of software reports for static displaying requires interconnection with database systems, such as MS Access.
- **Designed layout improvement** - Optimization software tools could be used to improve layout arrangement variation. Also, it is possible to use special software modules to find a solution for a complex optimization arrangement problem. For example, module for a genetic algorithm creation.
- **Data export** - Data export is used for the following dynamic verification phase of a designed solution. 2 basic data groups could be exported. First is graphical data representing 3D warehouse model and the second is numerical data describing products, processes, links between each other and sources used in a warehouse.

Static proposal created by using software for production and logistic system projecting, so called software without using the possibility of dynamic simulation use are not sufficient to evaluate stochastic influences of external factors or experimenting [5]. It is not possible to deduce dynamic behaviour of the designed system from its results. Mathematical and analytical problem solutions are relatively difficult for evaluation via these types of software. Therefore, it is necessary to use another tool, computer simulation enabling for example interconnection of information and technical model aspects.

2.2 Dynamic Verification of Proposed Solution in the Digital Factory Concept

When creating a simulation project, it is advantageous when previous steps of detailed projecting in a company have already been completed. When 2D/3D object models or layouts in software for static warehouse design or modelling and visualization are created, the quality of model, system definition and naming depend on a model creator, his or her experience, knowledge and obtained materials. Figure 5 shows described methodology of dynamic warehouse proposal in a simulation software environment.

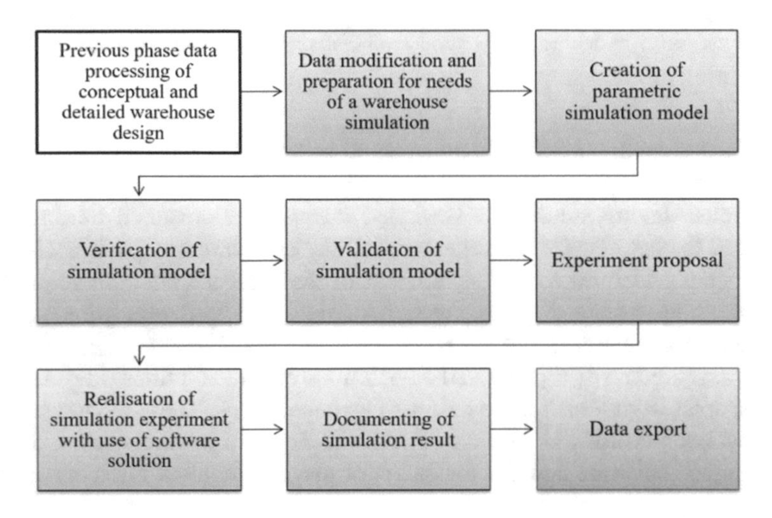

Fig. 5. Dynamic proposal of a warehouse with use of software for computer simulation

The basic entry for stated phase implementation is data exported from static warehouse design. Methodology shown in Fig. 4 consists of the following steps:

- **Data collection and amendment for a warehouse simulation -** It is necessary to provide detailed information about warehouse and manipulation equipment, operators or entities in order to prepare a warehouse simulation. Parameters used for a warehouse simulation might have a character of random quantities.
- **Parametric simulation model creation -** Simulation software offer various object parameter changes via the help of integrated tools for parameterization, thanks to which it is possible to for example change data from a clear dialogue box or insert data into a table containing object names and their parameters.
- **Simulation model verification -** Verification represents the verification of a fact whether the pc model is in accordance with the original conceptual model or directly with project goals.
- **Simulation model validation -** Validation verification of a simulation model is carried out within or after a simulation runs, implemented before simulation experiments. If a simulation model is made based on an existing system it is possible to compare real data with model outputs.

- **Experiment proposal -** For experiment implementation in the field of warehouse simulation it is necessary to know the goal of an experiment and detailed information about system in the warehouse. For example, amount of goods which needs to be stored, storage period, way and speed of preparing goods, etc. Based on these factors, the way of experimenting will be chosen.
- **Experiment implementation by using a software solution -** Thanks to software tools it is possible to use output, during and after the necessary simulation runs in a form of extensive graphs, videos or extensive report which could be open in any web browser.
- **Simulation runs result analysis and evaluation -** Statistical methods and outputs represented by tables or graphs are used to analyse data. In order to analyse data obtained from experiments, it is possible to use methods such as analysis of output diameter, analysis of the main effect of individual factors, factor effect, effect estimate (to calculate factor effect sign effect - Yates' algorithm could be used, etc.), test of effect significance, hypothesis determination and finally setting test strength.
- **The most suitable variation selection -** based on result assessment of simulation experiments, project team chooses a variation which fulfils set project goal criteria the best. If none of the generated variations provide a suitable result, it is necessary to start repeating the process from the beginning, go through simulation project goals and adjust a simulation model.
- **Simulation result documenting -** During the entire period of simulation project solving, it is necessary to process documentation which will contain information about a model structure, model development, experiment result so if necessary, a project architect or a model creator can go back to it [4].

  Simulation project report may be completed with a shortened summary of results for a company management, reports from workshop, summary of a system bottleneck and additional recommendations for their removal based on simulation experiment results.
- **Data export -** We export data from software for a warehouse dynamic design in order to store, sort it and eventually use it again. Web documents, table documents and graphical data can be exported. This data is used in planning, gradual enlargement of model parameterization and further model adjustments so it can be used in the warehouse operation process to the fullest.

Implementation of the selected practice solution variation follows up to results of a dynamic simulation. It is important to deal with simulation software implementation into the warehouse operation process simultaneously with warehouse technical solution.

2.3 Software Solution Implementation into a Warehouse Operation

Deployment of digital company tools and mainly dynamic simulation into a warehouse operation enables operation and optimization of logistic processes in warehouse management.

In individual methodology steps could been seen that for simulation model implementation into operation it is necessary to know the concept of information flows in a warehouse, having selected or established warehouse information system. Way of data import into a simulation model and a possibility to use automatic import then depends on this concept. In automatic report information does not need to be inserted in software manually, information can be loaded automatically in input tables from an external database. However, it is important to provide stability of data placement in external database, also as data uniformity in transfer between individual system components to provide correctness and speed of automatic information update. Procedure methodology of simulation software implementation into a warehouse operation is described in Fig. 6.

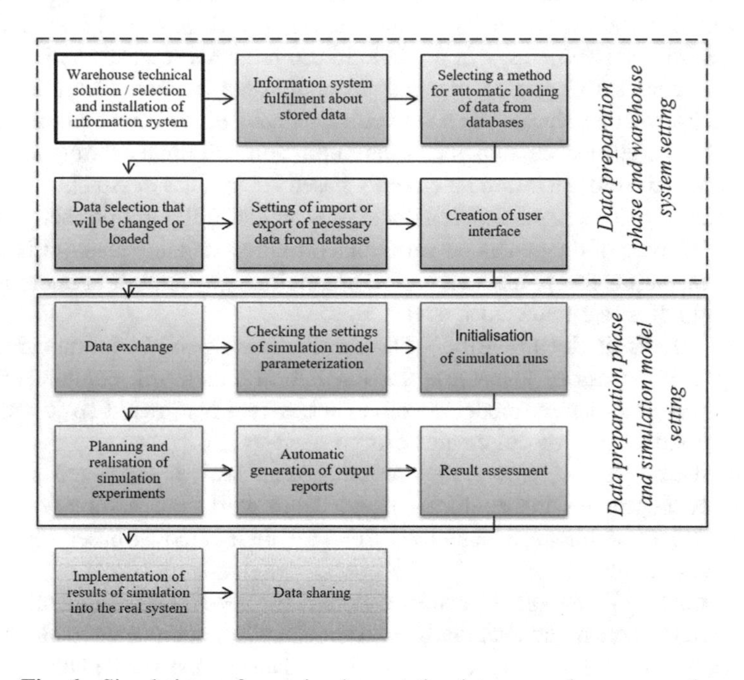

Fig. 6. Simulation software implementation into a warehouse operation

Procedure methodology of simulation software implementation into a warehouse operation is described in Fig. 7.

Data collection for needs of simulation model optimization may run for example by using information system (for example SAP, ORACLE, IBM), sensors, physical collection of data. The larger amount of data will be collected, the more need to know solution of clearing and missing data modification. Furthermore, different data, which could be gained from real-time assessment of machine's performance requires an integration of many different technologies including sensory devices, reasoning agents, wireless communication, virtual integration and interface platforms [7].

Current technologies enable interconnection of software applications with sensors, mobile devices, cameras placed on the equipment, thanks to which it is possible to gain huge amount of data. Therefore, a customer must define how he or she wants to collect data and sort it. Factories need to have an integrated telecommunication network and information technology through which they can transmit the necessary information and dates within the organization too [8].

Before the use of simulation software in operation, it is important to know how to use it. Simulation model is only a supportive instrument when deciding about next steps which could be related to logistic improvement or planning needs. Simulation model interconnection with existing warehouse systems serves as a supportive tool for managing and process optimisation in a warehouse.

All data obtained from simulation software could be used and sharing in a link to virtual reality, internet of things, expert systems, link to extended reality, decision support systems and so on.

2.4 Proposed Methodology Conclusion

Current companies face all kinds of changes in a way of storing, managing stock management, supply chain functioning due to the need of storing more and more different products and accessories. Storing premises are extending and modernisation requires high-speed mobile communications. On the other hand, customer individualism who want to use a product sale via internet with the fastest delivery possible influences a company. Such disposal solution could be simulated by for example, movement of customers or simulation of the entire storing system. Example of such simulation storing system model could be seen in next chapter.

3 Implementation Example of Dynamic Simulation Methodology When Deciding About Storing Process and Picking

For a warehouse simulation model display possibilities and 3D modelling I have prepared a simple warehouse example with semi-automated zone picking. The main concept of semi-automated zone picking is picking of small products, such as small parts, cosmetics or medication. Such storing system is suitable for distribution warehouses where medium to fast-moving goods are stored. In this particular system, consideration is being given to solution where picking is delivered directly to a person from a warehouse via conveyors where it is delivered through manipulation technique. The area where goods are received is divided into several zones with assigned operators. Products are oriented towards exactly chosen zone where there are stored in for example in stands or racks.

Example of a created warehouse model is shown in Fig. 7, it was created in Tecnomatix Plant Simulation software, version 14, in the Department of Industrial Engineering. Figure 7 also shows a 3D model of this system. The general principle is that goods collection finishes in one zone and the operator directs goods to the next zone through a conveyor.

The number of picking and empty palettes could be tracked via counter which could be placed next to objects we would like to follow in 3D. In this example, there is a set movement of 4 types of products which are mixed and placed in a palette of

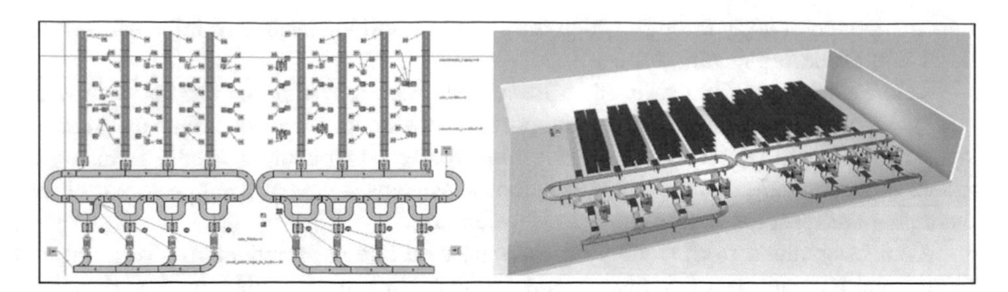

Fig. 7. Example of simulation model of zone picking

10 pieces. The products are named as Product A, Product B, Product C and Product D, and they are placed on palettes which are thanks to the manipulation technique delivered to a picking point. Procedural time is set according to a triangle division. Time of palette picking depends on the internal variable or in other words, the number of palettes per hour which are sent from one storing location.

The main task of this example was to find out when such storing system starts failing. Therefore, I have used a software tool Tecnomatix Plant Simulation, version 14 - Experiment Manager, which is suitable for implementation of such simulation studies. It studies different variations of input values and creates statistical results based on input parameter changes. Moreover, in the given example I gradually observed the way of material unloading and the speed of material picking. It was also important to observe workers' workload. Also, it would be possible to track for example use of trucks.

For each resource, there were 50 experiments carried out, with gradually increasing number of palettes per hour, coming from one storing location. In these experiments, the number of palettes and products coming from sorting workplaces was tracked and then it was workers' workload and what activity keeps them occupied at work. By monitoring of workers' activities, it is possible to improve working conditions and not only the picking system itself.

3.1 Experiment Results

In such simulation studies, it is possible to examine designed system parameters which represent a set of input values influencing resulting simulation parameters. Within experiments of the Experiment Manager, there are several simulation cycles carried out with unchanged input values and it generates output values based on given parameters. During experiments, the Experiment Manager changed given values randomly. The storing system itself behaves stochastically; the model contains information in a form of various probability distributions, which might result in unpredictable results. Therefore, the Experiment Manager outputs in the form of web document, tables or graphs are advantageous [10].

In the first column of Table 1, it is possible to see the number of experiments, in the second column - the number of pallets per hour per 1 regal, in the third column - number of all products, in the fourth column - number of products A, in the fifth column -number of products B, in the sixth column - number of products C and in the

Table 1. Example of result experimenting proportion

	Pocet paliet za hodinu na 1 regal	Pocet vyrobkov ALL	Pocet vyrobkov A	Pocet vyrobkov B	Pocet vyrobkov C
Exp 01	1	1520	630	470	115
Exp 02	2	3001	1283	904	200
Exp 03	3	4430	1922	1340	277
Exp 04	4	5880	2560	1774	357
Exp 05	5	7266	3175	2191	435
Exp 06	6	8703	3818	2624	515
Exp 07	7	10133	4464	3048	590
Exp 08	8	11588	5110	3485	671
Exp 09	9	13009	5748	3898	751
Exp 10	10	14411	6370	4323	826
Exp 11	11	15908	7031	4776	911
Exp 12	12	17333	7668	5202	989
Exp 13	13	18774	8307	5640	1066
Exp 14	14	20189	8935	6058	1148
Exp 15	15	20378	8921	6123	1159
Exp 16	16	20411	8764	6195	1158
Exp 17	17	20591	8792	6277	1159
Exp 18	18	20062	8386	6116	1139
Exp 19	19	19758	8049	6083	1108
Exp 20	20	19049	7561	6042	968

last column - number of products D. In addition, another advantage is the possibility of generated table copying into software for creating tables, such as MS EXCEL, where they can be edited even more and even evaluated. Table 1 shows experiment results, where number of palettes and products which have gone through this system are described. Based on Experiment Manager results, it can be pointed out what experiment combination would be the most suitable for improving stock management. Drawn graphs supplemented by red lines could be seen in the following figures. In the Fig. 8 on the x-axis, it is possible to see the sequence of experiments and on the y-axis is the number of all products. In the Fig. 8 on the x-axis, it is possible to see the sequence of experiments and on the y-axis is the number of all pallets.

Based on the Fig. 8(a), it is clear to see the system will reach it peak (with given setting) during the 17th experiment when 20 591 small products picked in boxes pass through it. Then, the graph of picked products shows the curve gradually increases and decreases up to the 28th experiment when the number of products starts slowly decreasing under the limit of 18 000 small products.

Next Fig. 8(b) shows graphical display of a curve of palette course in a warehouse. The system will reach its peak in the 17th experiment when more than 2056 palettes have passed through it. From the 27th experiment, where the limit of picked palettes

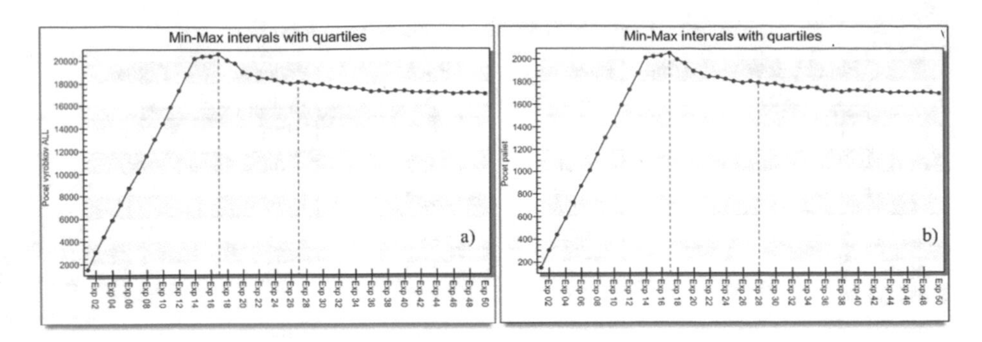

Fig. 8. Graphical evaluation of all experiments for all products (a) and all pallets (b)

dropped below 1800 pieces which reveals a gradual system congestion by palettes. When improving the system, it would be advantageous to reduce the number of palettes or set exact number which should pass the system.

Next, there is a possibility of material flow change when planning further, in a way, mutual influence would be smaller and system could handle a larger number of palettes.

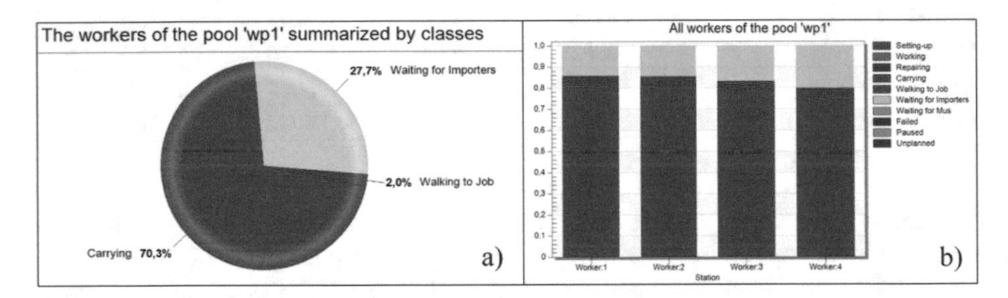

Fig. 9. On the example with name (a) could be seen a complete pie chart of workers' work and on the graph with name (b) a column graph of 4 workers' work in one storing zone

Figure 9 points out the use of workers and example of activities they did during goods picking. Figure 9(a) reveals that on average workers spent 70,3% of their time carrying and processing products, then waiting for a manipulation vehicle and other activities after. Figure 9(b) shows a column graph which was created during a simulation run. Thanks to this graph we can tell when the activity of workers increases or decreases. Workers' activity always selectively increases and during system congestion gradually decreases. Intensity increase of arriving palettes will not cause any larger system permeability and percentage ratio of waiting for palettes is caused by palette blocking on a conveyor belt.

3.2 Experiment Result Assessment

It can be seen based on this example that workload for workers in such system is still very high. The important question for system permeability is palette blocking on a conveyor belt. Therefore, conversions from detailed warehouse designing are important as these helps decide whether it is suitable to add more workers for removing empty palettes or extend positions for waiting palettes in individual unloading spots. Calculations and proposals could be tested on a model example of warehouse system even before a system failure, congestion with products, overload of people or other problem stopping a continuous warehouse operation [2].

Interesting experiment would be a human replacement by a robotic device. Thanks to Plant Simulation software, version 14, it is possible to create an interactive 3D warehouse model, optimise picking system and gain very clear results which we can easily store in web documents. This way of data processing is not only safe, but it is also a good presentation tool which can be used anywhere.

4 Conclusion

Our proposal of methodology of designing of warehouses with use of digital factory tools, described in the article, brings a lot of advantages, for example:

- Required data processing and its structure for warehouse designing in all phases of complex methodology (conceptual proposal, detailed designing, dynamic verification with use of pc simulation).
- Results from detailed warehouse designing with use of digital factory technologies (static part of proposal) can be further used in stock management optimisation.
- Dynamic verification of detailed warehouse design in the digital factory concept brings the verification possibility of proposals and midterm to long-term planning possibility of warehouse needs.
- Possibility to use pc simulation even in operation warehouse management.
- Total acceleration of a warehouse designing process.
- Clarification of a storing proposal and operational capacities with use of computer simulation.
- Visualisation improvement of a designed warehouse solution, etc.

In conclusion, PhD Thesis with name "Proposal of a Warehouse Designing Methodology with use of Digital factory tools", author: M. Bučková, 2017, was submitted in the Department of Industrial Engineering. The solution which uses Digital factory tools in warehouse designing was created in Autodesk and Tecnomatix FactoryCAD/Factory FLOW, Tecnomatix Plant, version 12.1 software. Moreover, based on research in this field and thesis results and also further research carried out at the Department of Industrial Engineering at the University of Zilina, it can be concluded that static warehouse design in connection with its dynamic verification done via pc simulation enables for example, reduction of warehouse devices in a warehouse, improves manipulation with items and warehouse logistics, improves utilisation of manipulative units, operational devices, warehouse workers, etc. For example, the introduction of automated logistics brings out significant time savings when transferring material, guaranteed delivery of material to the production line at the right time and quantity without potential human errors [8].

Acknowledgements. This work was supported by the Slovak Research and Development Agency under the contract No. APVV-16-0488.

References

1. Dilský, S.: Simulation using digital factory software tool - plant simulation. In: AIE - Advanced Industrial Engineering: Monograph, Bielsko-Biała, pp. 67–80 (2013). ISBN 978-83-927531-6-2
2. Dulina, Ľ.: Augmented reality using in modern ergonomics. In: Advanced Industrial Engineering: New Approaches in Production Management: Monograph, pp. 165–179. Wydawnictwo Fundacji Centrum Nowych Technologii, Bielsko-Biała (2015). ISBN 978-83-927531-7-9, [1,35 AH]
3. Furmann, R.: 3D laser scanning - support the implementation the digital factory. In: Digital factory management methods and techniques in engineering production, vol. V, pp. 25–28. Wydawnictwo Akademii techniczno-humanistycznej, Bielsko-Biała (2011). ISBN 978-83-62292-57-8
4. Furmann, R., Furmannová, B., Więcek, D.: Interactive design of reconfigurable logistics systems. Procedia Eng. **192**, 207–212 (2017). https://doi.org/10.1016/j.proeng.2017.06.036. ISSN: 1877-7058
5. Gregor, M., Haluška, M., Fusko, M., Grznár P.: Model of intelligent maintenance systems. In: 26th DAAAM International Symposium on Intelligent Manufacturing and Automation, DAAAM 2015, pp. 1097–1101 (2015). ISBN 978-3-902734-07-5
6. Gregor, M., Herčko, J., Grznár, P.: The factory of the future production system research. In: Proceedings of the 21st International Conference on Automation and Computing, ICAC 2015, Glasgow, UK, pp. 101–105, 11–12 September 2015. ISBN 978-0-9926801-0-7
7. Mičieta, B., Biňasová, V., Haluška, M.: The approaches of advanced industrial engineering in next generation manufacturing systems. Commun. Sci. Lett. Univ. Žilina **16**(3A), 101–105 (2014). ISSN 1335-4205
8. Mičieta, B., Herčko, J., Botka, M., Zrnić, N.: Concept of intelligent logistic for automotive industry. J. Appl. Eng. Sci. **14**, 233–238 (2016). https://doi.org/10.5937/jaes14-10907
9. Rakyta, M., Fusko, M., Herčko, J., Závodská, Ľ., Zrnić, N.: Proactive approach to smart maintenance and logistics as a auxiliary and service processes in a company. J. Appl. Eng. Sci. **14**(4), 433–442 (2016). ISSN 1451-4117
10. http://www.cardsplmsolutions.nl/en/plm-software/tecnomatix/plant-simulation-warehousing-logistics-7. Accessed 23 Feb 2017

Using Modern Ergonomics Tools to Measure Changes in the Levels of Stress Placed on the Psychophysiological Functions of a Human During Load Manipulations

Luboslav Dulina[1], Miroslava Kramarova[1], Ivana Cechova[1(✉)], and Dorota Wiecek[2]

[1] University of Zilina, Univerzitna 1, 010 26 Zilina, Slovakia
{luboslav.dulina,miroslava.kramarova, ivana.cechova}@fstroj.uniza.sk
[2] University of Bielsko-Biala, Willowa 2, 43-309 Bielsko-Biala, Poland
dwiecek@ath.bielsko.pl

Abstract. The article provides information on the CAPTIV wireless sensor system and its use in measuring the response of the human nervous system to stress caused by manual manipulation of loads of varying weight. This tool of modern ergonomics serves to gather data for subsequent analysis of stress put on psychophysiological functions of a person. In the article, you can find information on measurement results from individual sensors and the interdependence between the level of stress put on a person and the amount of the organism's response to said stress. Used in the tests were the T-Sens GSR sensors for measuring the galvanic skin conductivity and a T-Sens HRM sensors for measuring heart rate. The data obtained from the sensors were then transformed, compared and evaluated in order to find dependence between the subjective reaction to the stress of the person monitored and the objective levels of stress induced.

Keywords: CAPTIV sensor system · Heart rate · Organism activation Load manipulation

1 Measuring Psychophysiological Functions Using CAPTIV

Most of our physical and psychological processes cannot be directly perceived nor consciously influenced; these include: heart functions, blood pressure, reacting to stress, temperature, etc. [1]. The core of the CAPTIV sensor system, specifically the HR sensor, lies in measuring the differences (increases and decreases) of heart rate. The heart rate is regulated by the aforementioned autonomic nervous system. The Sympathetic part of the nervous system increases heart rate (when walking or running, when in stress or during any activity with increased physical stress and others). The parasympathetic part decreases heart rate (when the body is at rest and relaxed). When these two segments of the nervous system are in balance, they both harmonically influence the heart functions thus preserving variability of heart rate. This shows the organism is ready to take on any life events [2].

© Springer Nature Switzerland AG 2019
A. Burduk et al. (Eds.): ISPEM 2018, AISC 835, pp. 499–508, 2019.
https://doi.org/10.1007/978-3-319-97490-3_48

1.1 CAPTIV Sensor System

The captive sensor system which we use is a complex, fast and simple solution for gathering information on the physiological and vital functions of a human in his working process. It is a modern technology, where the information gathered are related to visual monitoring with wireless data transfer [3, 4]. Using hardware, high number of sensors and software, which outputs its results in visual form. The measurement is done directly during work. Real data is therefore used, instead of data from model laboratory settings. This promotes this sensor system to a tool for modern ergonomics.

The most important physiological characteristics measurable by the CAPTIV system are:

- electrical skin conductivity,
- skin temperature,
- heart rate,
- heart rate variability (HRV),
- breathing (frequency, amplitude, type of breathing),
- muscle tension.

1.2 Material Manipulation as the Most Hazardous Logistics Activity

The majority of our research up to now has been focused on ergonomics as it relates to production and installation work. However, manual manipulation is just as present in other areas of industrial activities. For instance, is forms a key part of a logistics worker. Furthermore, manual material manipulation adds no value to the products while creating the highest percentage of employee injuries [5]. Logistics is one of the main segments in every company. Logistics include various areas: information flow, transportation, storage, material manipulation and safety. For this reason, the focused the use of CAPTIV on the stresses related to a work in this particular area.

1.3 T-Sens HRM Sensor for Measuring Heart Rate Variability

Measuring the heart rate variability and researching the load on the human cardiovascular apparatus caused by activities during work processes. It's used to measure the amount of physical stress during work, measuring performance, analyzing resistance, etc. [6]. The unit of heart speed is heart beats per minute (BPM), which are the basic indicator of physical effort or psychological stress. As the consumption of oxygen by the body rises, so does the heart's activity. A healthy heart is characterized by natural heart rate variability – a key factor for a healthy heart. The rhythm of a heart changes also with changes in emotional state. The sensor is placed directly on the skin of a person into his metasternum [7].

1.4 Slovak Laws as They Relate to Heart Rate

As part of the law system in Slovakia and other EU member states, there are regulations and decrees that define the limits of physical load (energy output, heart rate and many others). In terms of comparing results output by the HR measuring sensor, it is necessary to look at the values of heart rate as well as the heart rate variability. The aforementioned limits on heart rate are same for men and women, but categorized into 5 groups by age in the Ministry of Health Decree On Details of Protection against Physical, Psychological and Sensory Overload at Work (see Table 1). The defined limiting values are related to activities performed by large muscle groups and cannot exceed the value of 150 beats per minute, not even in the short-term. In all cases of our measurements, the load manipulation included the manual manipulation of a load and activation of large muscle groups. Another important information is that heart rate is dependent on many external and internal influences and the overall health of a person. It is therefore necessary to monitor the highest acceptable value of heart rate increase over the baseline values. The maximum acceptable HR increase values are also defined in the aforementioned decree [8].

Our measurements were performed for a test case of monitoring heart rate changes dependent on the weight of the load being manipulated and on working positions. The results of these measurements can be compared to the limits of heart rate (Table 1) as defined by Slovak laws. The measured persons were women in the 30–39 age group. Therefore, the relevant data for us will be the heart rate limit values given in Table 1, row 2, marked by green colour.

Table 1. The criteria for evaluating heart rate changes during an activity involving large muscle groups (as defined by Slovak laws)

Age group	Values of heart rate change per minute			
	Absolute values		Heart rate increase over the baseline value	
	A Average values	B Limit values	C Average values	D Limit values
18-29	108	117	30	33
30-39	106	115	29	32
40-49	101	110	26	28
50-59	97	105	23	25
60-65	93	100	20	22

Explanatory notes for Table 1:

A – the value defined for assessing the results of measuring a group of people, if a custom baseline value of heart rate is not measured or applicable.
B – the value that can be bearable for the measured person in the long-term unless the C value is being exceeded, i.e. if the maximum increase over baseline HR is not exceeded.
C – the highest acceptable increase of heart rate over the baseline value that is bearable long-term in healthy individuals.
D – the highest acceptable value of heart rate increase over the baseline value, that is not to be exceed.

1.5 T-Sens GSR Sensor for Measuring Galvanic Skin Conductivity

The GSR sensor determines the organism activation ratio in relation to a stimulus based on secreting micro-particles of sweat from a human body. The stimulus may be psychological or emotional stress as well as a sudden deep breath or a surprising event.

The GSR sensor applies very small, safe and unrecognizable electrical voltage, and thus electric current, to the skin. Changes in this small current can be used to measure the activity of sweat glands of the skin beyond the limits of self-awareness. Activity of sweat glands are set by the autonomic nervous system. Sweat glands in the skin are directed by sympathetic nervous system only, which makes them a very good indicator of inner tension and stress.

The visual output of the GSR sensor reading doesn't give us information about positive or negative activation ration, only about its intensity. It is used to perform analyses to determine the degree of stress in a person as well as biofeedback, reactivity to stimuli and others. The unit of Galvanic skin conductivity is 1 μS (micro Siemens). Monitoring the organism activation ration has a great deal of influence in the context of effectivity and productivity. Due to an increase in work activities where emphasis is placed on mental performance, it is necessary to deal with this factor which significantly influences working performance and psychological comfort.

1.6 Slovak Laws as They Relate to GSR

There is no decree or regulation in Slovakia that would set limiting values for the maximum organism activation ratio during an entire work shift.

The GSR measuring sensor does not provide information on positive or negative level or organism activation, only on the intensity of this activation. However, when talking about stress on a person caused by the influence of emotions, it is not critical to know whether these emotions are positive or negative. The amount of stress on a person due to the organism activation is the same with both negative and positive emotions with the same intensity.

2 Influence of Increased Physical Stress on a Person's Health

Several studies have shown that insufficient activity of the parasympathetic branch of the autonomic nervous system - which regulates relaxation and regeneration activities of the body – is a significant risk factor for cardio-vascular disease, diabetes, obesity and other lifestyle diseases. Long-term stress with continuous sympathetic activity is also related to a more frequent occurrence of indigestion, immune or reproductive disorders, pains in the musculoskeletal system as well as psychological problems, mainly attention disorders and uneven sleep patterns, depression and burnout syndrome [9].

2.1 The Psychophysiological Functions Monitored During Manual Material Manipulation

5 sensors were placed on the measured volunteer during the CAPTIV sensor system monitoring as part of measuring the amount of stress on a person during manual manipulation of heavy loads. The first sensor was to measure heart rate. The second was from the T-Sens Respiration sense. This serves to analyze the rhythm and amplitude of breathing, i.e. the ratio of deformation of chest and stomach of the measured person. The measured parameter depends on the placement of the sensor. Other sensors were T-Sens Temperature to measure the skin temperature (which can be used to measure the temperature of the surrounding environment as well), T-Sens GSR sensor to measure the galvanic skin conductivity and T-Sens sEMG. The sEMG sensor provides information on local muscle stress. Its output is the determination of % Fmax exerted by a person during a given activity.

This article compares stresses on a human body during manual manipulation of loads of various weight in physiological and non-physiological positions. Descriptions of the researched outputs will follow; these can be obtained from the heart rate sensors or rather the HRV sensors and the GSR sensor. The GSR sensor gives up, as is stated later, information on the organism activation ratio based on a given stimulus.

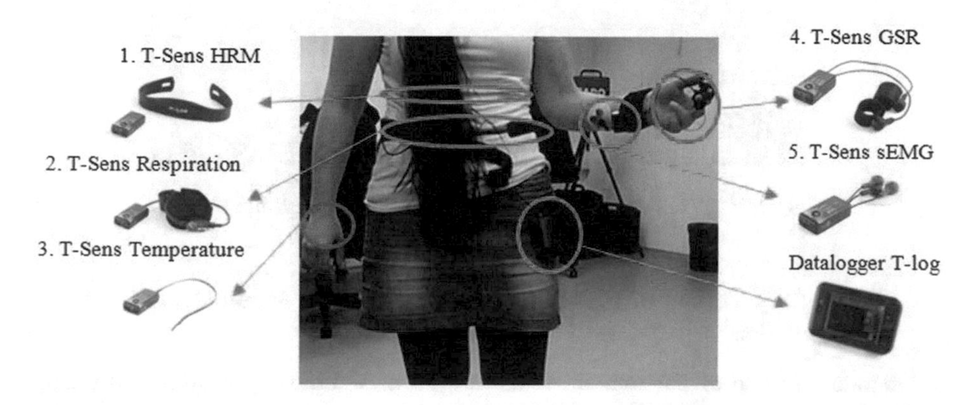

Fig. 1. Placement of sensors during the monitoring of psycho-physiological functions

2.2 The Process of Measuring Psychophysiological Functions

The CAPTIV system is a modular solution to record the activity of a worker and synchronize the sensors with video footage using T-log. Video may be captured by an analog or digital camera (and synchronization can occur with 4 videos at once). The data gathered from the sensors placed on the body of the worker (Fig. 1) serve as input for software analysis (Fig. 2).

The data gathering focused on the amount of stress on the psychophysiological functions of people was performed using manual load manipulation. The loads were 3 kg, 5 kg and 10 kg and the manual manipulation with them was done in both physiological and non-physiological positions. The weight for the loads was selected as stated because the literature and publications are not clear and consistent in terms of the weights of the loads; in some cases, the authors mention a load of 3 kg, in others they start at 5 kg. The Slovak laws allow women of the selected age group to handle loads of up to 10 kg in physiological positions. For this reason, these three weights were selected to compare the amount of stress placed on the cardiovascular system of a person, i.e. the change in their BPM.

Load manipulation was performed in a standing position and the load was only handled with both hands at the same time.

The software gives us visual evaluation of the data gathered from individual sensors and their outputs are also synchronized, which means it is easy to see changes in the monitored characteristics in relation to the visual outputs of other sensors (Fig. 2). This software then outputs a report which gives us information on the indicators being monitored throughout the selected work activity.

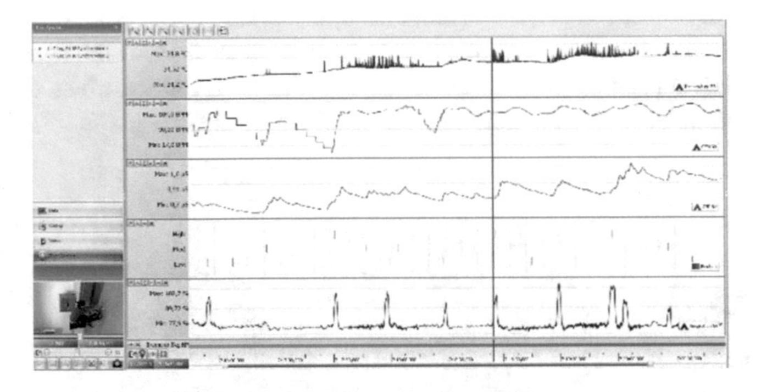

Fig. 2. CAPTIV software - video is synchronized with the sensors

2.3 Laboratory Testing of Heart Rate Changes During Manual Load Manipulation

In laboratory testing of the amount of stress on a person during manual load manipulation, the premise was that the BPM changes with the level of physical or psychological effort during a given activity.

During testing, the person tested manipulated loads of 3 kg, 5 kg and 10 kg in physiological work positions as well as non-physiological work positions. BPM was measured before the manipulation – this measurement was used as a baseline – and then BPM was also measured during the manual manipulation using the CAPTIV system.

In Figs. 3, 4, 5, 6 and 7, the changes in BPM can be seen; these occur during manual load manipulation. The baseline BPM value is in green. The limits for heart rate increase over the baseline, as defined by Slovak laws are displayed in yellow and red color. The yellow line marks the highest acceptable value of heart rate increase over the

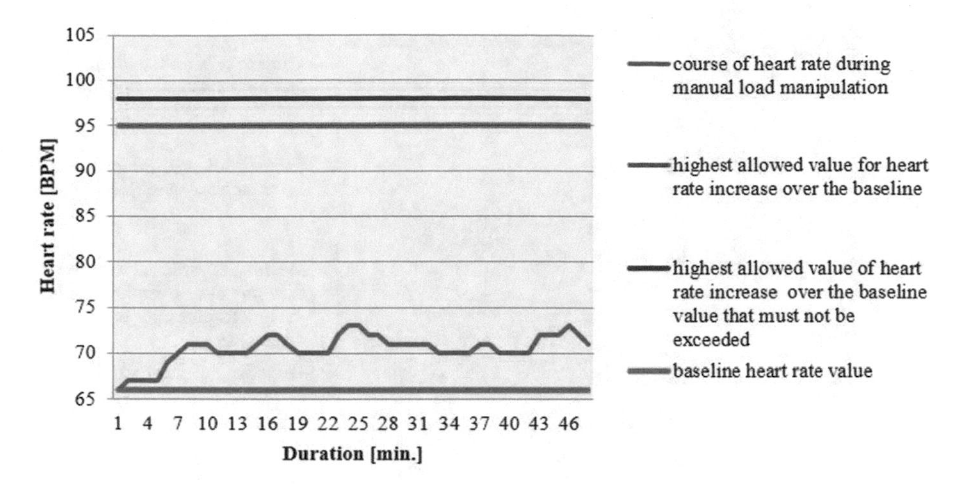

Fig. 3. BPM during manual manipulation with a load of 3 kg in physiological positions

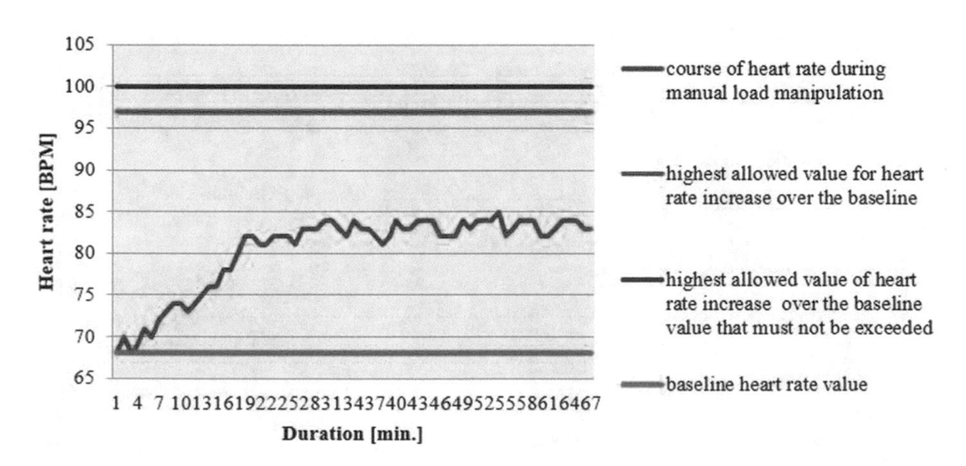

Fig. 4. BPM during manual manipulation with a load of 3 kg in non-physiological positions

baseline that is allowed long-term in healthy individuals. The red line marks the heart rate value that must never be exceeded.

Figures 4 and 5 show that - in our testing - the amount of stress placed on a human heart is influenced not only by the weight of the load, but also by the position in which the load is being manipulated. This can be concluded from the fact that the same increase in BPM was measured for a 3 kg load in non-physiological working position and 5 kg load in a physiological working position.

Looking at Fig. 7 and Table 2 shows that during manual manipulation with a 10-kg load in non-physiological positions, the heart rate almost achieved the highest

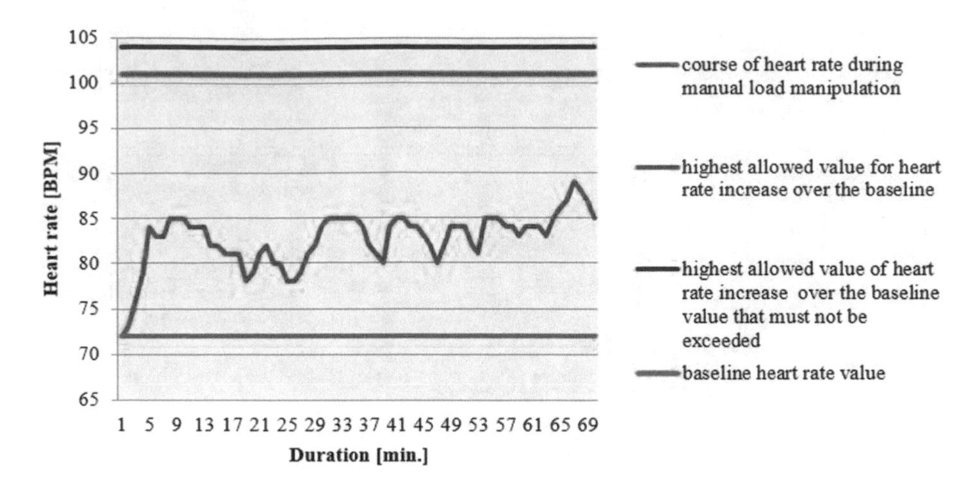

Fig. 5. BPM during manual manipulation with a load of 5 kg in physiological positions

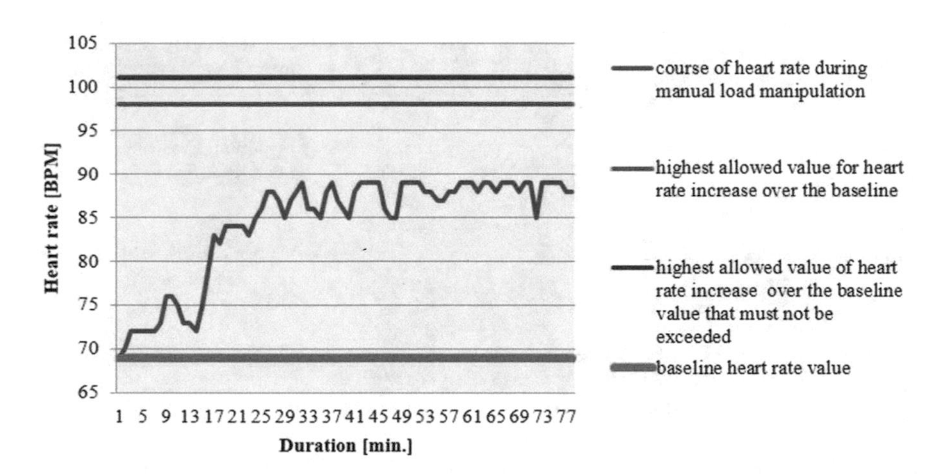

Fig. 6. BPM during manual manipulation with a load of 5 kg in non-physiological positions

allowed value for heart rate increase. This means that if several factors influencing the increase of stress on a human heart occur at once, it is probable that long-term manipulation with a 10-kg load in non-physiological positions more than twice in 1 min will create a health hazard, i.e. the maximum allowed value will be reached.

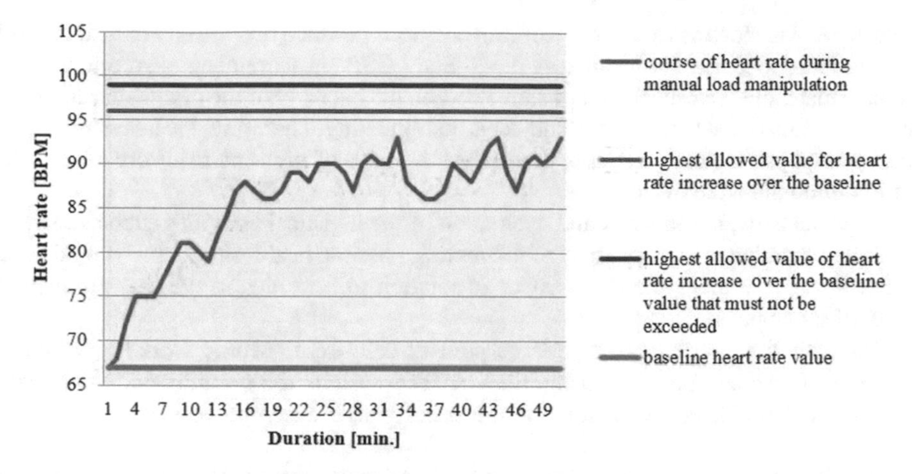

Fig. 7. BPM during manual manipulation with a load of 10 kg in non-physiological positions

The changes in BPM with varying degrees of stress placed on a person can be seen in Table 2.

Table 2. Comparison of significant BPM values at varying degree of stress

	Load weight/ type of working position				
	3kg/PP	3kg/NP	5kg/PP	5kg/NP	10kg/NP
Baseline HR	66 BPM	68 BPM	72 BPM	69 BPM	67 BPM
Maximum HR achieved	73 BPM	85 BPM	89 BPM	89 BPM	93 BPM
BPM increase over the baseline	7 BPM	17 BPM	17 BPM	20 BPM	26 BPM

3 Conclusion

Using the sensor system, it is possible to measure the effect of load and stress on a person in their working process and subsequently evaluate, analyze and eliminate causes that brought them about. Through the CAPTIV system, we are able to gain an objective view of a subjective response of the nervous system of a person doing a given activity.

In our case, the subjective reaction of a person to manual manipulation with loads of varying weight was transformed into an objective state of stress. This objective state of stress was evaluated using the comparison of BPM readings measured by the T-Sens HRM sensor with the legal regulations of the Slovak Republic and the number of BPM between the individual manipulations. The results show that the stress on human heart, and the cardiovascular system as a whole, is influenced during manual load manipulation by weight of the load as well as the working positions in which this activity of manual load manipulation occurs. The results of the T-Sens GSR measurements also allow us to deduce that there is a causal relationship between the intensity of the organism activation and the weight of the load which a subject is manipulating. Therefore, the heavier the load is in non-physiological positions, the higher the ratio of high-intensity organism activation throughout the event.

In conclusion, the analysis and evaluation of the logistical activities such as manual load manipulation is a critical part of preventing possible health risks which could arise either during the manipulation itself or in relation to activities in the non-production parts of a company.

By gathering and analyzing a large pool of data from specific workspaces (combination of load and position) using the CAPTIV system, the groundwork can be laid for more detailed legal regulations on the stress placed on employees.

Acknowledgement. This paper is supported by the following project: University Science Park of the University of Zilina – II. phase (ITMS: 313011D013) supported by the Operational Programme Research and Innovation funded by the European Regional Development Fund and by VEGA project [1/0936/16] entitled Using tools of digital factory for development of ergonomic prevention programs.

References

1. Vignais, N., Bernard, F., Touvenot, G., Sagot, J.C.: Physical risk factors identification based on body sensor network combined to videotaping. Appl. Ergon. **65**, 410–417 (2017). https://doi.org/10.1016/j.apergo.2017.05.003
2. Balog, A.: Psychoterapia, Biofeedback (Inst. 2016-08-10). http://www.mindandbody.sk/vas-doktor/psychoterapia-biofeedback-attila-balog
3. Kuric, I., Bulej, V., Saga, M., Pokorný, P.: Development of simulation software for mobile robot path planning within multilayer map system based on metric and topological maps. Int. J. Adv. Robot. Syst. **14**(6) (2017). ISSN 1729-8814
4. Krajcovic, M., Plinta, D.: Comprehensive approach to the inventory control system improvement. Manag. Prod. Eng. Rev. **3**(3), 34–44 (2012). ISSN 2082-1344
5. Krkoska, L., Gregor, M., Matuszek, J.: Simulation of human effect to the Adaptive Logistics System used in public facilities. Procedia Eng. **192**, 492–497 (2017). https://doi.org/10.1016/j.proeng.2017.06.085
6. Uusitalo, A., Mets, T., Martinmaki, K., Mauno, S., Kinnunen, U., Rusko, H.: Heart rate variability related to effort at work. Appl. Ergon. **42**, 830–838 (2011). https://doi.org/10.1016/j.apergo.2011.01.005
7. Visnovcova, Z., Tonhajzerova, I.: Biomedicinsky princip a vyuzitie elektrodermalnej odpovede v klinickej praxi. Cogn. Remediat. J. (2013)
8. Gasova, M.: Modern methods in ergonomics research and european legislation. In Advanced industrial engineering – new approaches in production management, Monograph (Dariusz Plinta), Publisher: Publ. House of Center for New Technologies Foundation, Bielsko Biala, pp. 147–164 (2015). ISNB 978-83-927531-7-9
9. Cook, D.J., Song, W.: Ambient intelligence and wearable computing: Sensors on the body, in the home, and beyond. J. Ambient Intell. Smart Environ. **1**(2), 83–86 (2009)

Yamazumi Analysis in Milk-Run Intralogistics Systems Using Simulation Tools

Kamila Kluska(✉)

Faculty of Engineering Management, Chair of Production Engineering and Logistics, Poznan University of Technology, Strzelecka 11, 60-965 Poznan, Poland
kamila.kluska@put.poznan.pl

Abstract. Simulation is a successful tool for designing new transport systems and improvements. The milk-run is popular material-handling system in manufacturing companies. Designing, modeling and analyzing of milk-run system is often a big challenge for the designer due to its complexity and dynamics. The paper focuses on Yamazumi charts as a tool for analysis and work balancing of logistics train operator. The mechanism for automatic generation of Yamazumi charts is built with use simulation technology. In article authors describes important step of methodology of designing the Milk-Run systems and initial version of tool for verification and improvement of such systems.

Keywords: Modeling · Intralogistics · Lean · Material-handling system Logistics train

1 Introduction

Nowadays the Lean Manufacturing system is implemented in production companies due to the wide range of applications and impressive potential for cost reduction, increasing flexibility, implementing clear rules of employee work management and organization of workplaces, and thus increasing competitive advantage of company on the market.

Lean management in the production company includes both improvement activities within production processes, and activities related to the design and management of the internal transport system. Production processes based on the Kanban system, requires timely, precisely matched deliveries, which do not generate large inventory of stocks and wastage. This means frequent deliveries in small batches, delivered to working cells at exact time set.

The organization of the internal transport system based on milk-run concept, allows to meet these requirements and achieve the many benefits described in the further part of the publication. Milk-run system can be implemented in internal logistics system as main concept or as a part of an enterprise's internal logistics system, as well as in external logistics system [12].

Many studies focuses on describing and comparing the milk-run system to other internal transport systems. Also, the subject of such studies is the optimization or improvement of the Milk-Run transport system in enterprises. Literature describes tools for simulation modeling of transport system, planning deliveries, designation the best

© Springer Nature Switzerland AG 2019
A. Burduk et al. (Eds.): ISPEM 2018, AISC 835, pp. 509–519, 2019.
https://doi.org/10.1007/978-3-319-97490-3_49

routes for the logistics trains, designation and allocation of transport orders for trains, analysis of various scenarios for modifying a number of variables such as number of trains, speed, loading and unloading time.

However, the results of the simulation and evaluation of the variants, under certain assumptions, are usually based on one parameter, e.g. the total distance traveled by the logistics trains. Due to the complexity of milk-run system and multitude of requirements, transport system should be evaluated based on many parameters and various analyzes. The publication proposes a milk-run system analysis using simulation modeling and Yamazumi load balancing graphs. Mentioned tools enables to include the dynamics and complexity of the system, as well as analyze the work of the system not only in terms of timely deliveries and many typical parameters, but also in terms of the distribution of various types of activities carried out by production employees and the logistics train operator.

The goal of the article is to propose the developed tools that enables preparation of Yamazumi analysis in simulation model.

The article is divided into five sections. The second section contains literature review on milk-run advantages, milk-run system optimization projects with use simulation and lean tools. The developed mechanism for Yamazumi charts generation is presented in section three. The fourth section describes the possibilities of interpreting such charts and the possibilities of introducing improvements in the internal transport system. Section five focuses presents the conclusions.

2 Optimization of Milk-Run System and Yamazumi Analysis in Literature

Milk Run is internal material-handling system, which goal is increasing frequency and simultaneous reduction of delivery costs. The mode of transport is logistics train served by operator. The logistic train consist of one tugger and specified number of trailers. The operator drives logistics train along designated route, in order to perform regular, multiple, indirect deliveries in small lot sizes, in short intervals. The milk-run operator gets containers with materials from the supermarket, delivers containers to points of use, collects ready-made products, and then returns goods to the warehouse. Logistic trains also support the transport of empty containers in the production area.

Material-handling is critical issue in production companies. The implementation of the milk-run system is becoming more and more popular, due to possibility of achieve a number of operational benefits, including:

- better efficiency of the material handling system [16],
- more efficient satisfaction of transport demand, elimination of delays and increasing the timeliness of deliveries [16],
- minimization of the number of empty runs [12],
- synchronization of means of transport, which leads to increased reliability [12],
- lower transportation cost and shorter shipment distances, as a result of reduction of total distance traveled by means of transport and less energy consumption [3],
- consolidated distribution and collection of full and empty containers [3],

- improvement of throughput of the production system,
- lower level of WIP (Work-In-Progress) materials [12],
- process transparency and a high level of reliability [10],
- better use of working time of intralogistics employees [16],
- improvement of load factors [6],
- prevention of overproduction,
- obtaining the additional space as a result of reduction total area of transport routes [6], and therefore
- reduction of traffic and lower collision rate.

Meyer [10] defined the goal of implementation of Milk Run system as "the definition of cost-efficient, regularly recurring transport schedules for suppliers with a regular demand". Bowersox et al. [2] defined Milk Run as an important element of integrated lean logistics strategy.

However, the design of an internal transport system based on logistic trains and scheduling of deliveries is much more complex, than in transport systems based on direct deliveries. This complexity results from the need to implement decision-making processes related to the choice of materials supplied in different types of containers, their quantity, frequency of deliveries, train routes and their sequences related to the delivery and collection of goods [13]. In addition, specific deliveries must be assigned to individual trains with an appropriate number of specific trailer types and a proper tugger. As in many cases, the success of milk-run system implementation lies in details and in good planning.

In production processes is many variations. It causes that the number of deliveries served by logistics trains varies per specified periods of time. To enable efficient and stable coverage of production needs, milk-run system must be well-designed, consciously managed, and adjusted to serve the peaks in the production system demand.

A commonly used tool for designing and verification of projects is simulation. Korytkowski and Karkoszka [7] described optimization of milk-run internal logistics system in assembly plant using discrete-event simulation model.

Hao and Shen [4] used hybrid simulation approach to build the software, dedicated for simulation modeling of assembly lines and material handling milk-run system. The goal of simulation experiments is improvement of internal transport system performance.

Next simulation tool is created by Kitamura and Okamoto [9]. It is used in order to find automatically the optimal routes for logistics trains in defined map.

Hosseini et al. [5] developed the model to designate an optimal solution for consolidation network, which uses cross-docks, milk-runs and direct shipment. The goal of this study is minimization of the total shipping cost.

Bae et al. [1] tried to find the best combination of input parameters in simulation model of milk-run system. Paper describes three experiments, conducted in order to analyze the results of implementation of such system in automotive company.

Kilic et al. [8] described the models dedicated for minimization of total distance traveled by logistics trains and reduction of internal transport vehicles number.

Miwa et al. [11] developed simulation model of the JIT manufacturing system, that integrates suppliers, Mizusumashi and dual-card Kanban systems in order to designate the minimal number of containers, and therefore prevent part stock-out events.

Lean methods are widely used in simulation studies. Tokola et al. [15] reviewed and categorized 26 papers that study lean methods with the use of simulation tools. Paper describes the occurrences of lean methods in literature. The most popular are: value stream mapping (VSM), setup time reduction, pull system, total productive maintenance (TPM), continuous improvement (Kaizen), Kanban, cell layouts and layout change, U-shaped lines, bottleneck control, process automation, dedicated material handler (Spiderman), visual management, line balancing, one-piece flow, standardized work charts; 5S, pokayoke, quality-at-the-source, reduction of WIP, reduced changeover time, single-minute exchange of die (SMED).

The tool used by many of these methods is Yamazumi board. It helps to quickly analyze and rebalance processes, in case of changing the tact time. It allows for a visual, fast indication of underutilized and overloaded operations.

Yamazumi means to stack up. It is load balancing graph, where usually operations are classified by VA (value-added), NVA (non-value added), or waste. It is a visual tool used within lean manufacturing to show the various work elements of process, which allows for comparison to the required customer tact time and output time. It supports the design of workstations, redesign of processes and continuous improvement.

3 Yamazumi Analysis in Simulation

The Yamazumi board is a stacked bar chart. It allows to visualize the amount of work, balance of workloads between the operators at workstation or production line in relation to the cycle time, define variations in the process, reduce unsaturation, determine added value and not added value activities.

Balancing is aimed at uniform distribution of the workload between resources while avoiding exceeding certain standards, reduction of time of no added value activities, reduction of the resources number, improving the efficiency, or decrease the cycle time of production line.

The described tool is part of the methodology for designing intralogistics systems based on the milk-run concept. The methodology involves the use of various tools, such as CalculatorForMR built using MS Excel and a simulation model built using FlexSim simulation software.

CalculatorForMR is a tool used to organize data on the material needs of workstations and to transform it into specific orders, appearing cyclically with a specific frequency, taking into account specific quantities of materials, as well as the necessary number of resources needed to fulfill such orders, e.g. logistic trailers and tuggers. The orders also includes the bus stops for train positioning, all locations for containers, the supermarket and the target warehouse for final goods and empty containers. An important basis for data is PFEP table (Plan For Every Part).

The simulation model concerns the production system and the internal milk-run transport system. It takes into account all workstations, their work logic, work resources, locations for containers with materials and empty containers, working tables and machines, relations between workstations, flow of goods, and all dependencies necessary for proper modeling of the production processes. After building the simulation model of the production system, the intralogistics system is designed based on data prepared in CalculatorForMR and facility layout. Such a system in the simulation model consists of a parking area for trailers and tuggers, a supermarket, a warehouse, milk-run stops, and a network of transport routes.

All elements of the system in the methodology are controlled by operators, robots, manipulators, etc. which periodically perform lists of tasks. These lists consist of commands and specific differencing parameters. Commands allows for fast modeling work activities, decision-making and control processes. The list of commands for the train operator is generated automatically by the program based on the data entered by the user into designated tables. The visualization of work of the logistic train operator in the described methodology is presented at Fig. 1.

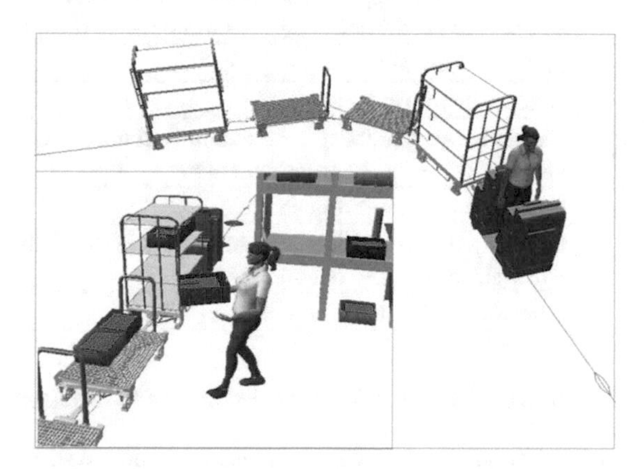

Fig. 1. Milk-run system in simulation model. Source: own study.

After simulation modeling of the internal transport system, it is necessary to verify it with the use of specific analytical tools. In the described methodology, such a tool is Yamazumi graph generator, which enables to analyze the distribution of various types of activities in the operator's designated working time (usually with reference to the tact time), and analyze of the total distance traveled, total working time and the weight of carried loads.

Preparation Yamazumi charts with use described method consist of three basic steps:

- Analysis of activities performed by logistics train operator
- Definition of Yamazumi classes and its assignment to operator activities
- Simulation experiment and analysis of charts in simulation model.

3.1 Analysis of Activities Performed by Logistics Train Operator

In case of Yamazumi analysis for the work of the logistics train, it is necessary to focus on the train operator. The preparation of the Yamazumi chart for the milk-run system in the simulation model requires analysis of the activities performed by each logistic train operator and preparation of their list. In the described tool, the operator's activity means reading and executing commands with respect to their names and parameters concerning time or quantity. The commands used to program the train operator work are presented and described in Table 1.

Table 1. Commands for logistics train operator. Source: Pawlewski [14]

Command	Description
Travel	Operator is moving to a specific location in model space
GrabTugger	Taking a logistic tugger by an operator- an operation dedicated only for intralogistics operators
ReleaseTugger	Leaving a logistic tugger by an operator in order to carrying out transport operations - an operation dedicated only for intralogistics operators
DriveTugger	Move a logistic tugger by an operator in order to drive the logistics train-an operation dedicated only for intralogistics operators
GrabTote	Picking up the tote
ReleaseTote	Unload the tote to location
WaitForEmptyTote	Checking by operator if the one empty container is in the shelf – if not operator will wait for empty container
GrabEmptyTote	Load an empty tote from shelf
CheckTrolleyWait	Checking if the one blocked container is in the shelf – if not operator will wait for blocked container
ReadyForTask	End work and wait for new tasks

3.2 Definition of Yamazumi Classes and Its Assignment to Operator Activities

The next step is to determine the Yamazumi classes, which will be assigned to the activities carried out by the train operator and displayed on the chart. It enables the program to generate personalized analyzes.

The user can define any number of classes. For the presented example four Yamazumi classes have been designated:

- 0 – idleness due to lack of work (waste)
- 1 – VA the value added activity (in the case of transport processes carried out by the train operator, the VA activities not exist)
- 2 – NVA (non-value added) an activity that does not add value. It must be performed and can't be reduced
- 3 – NVAA (non-value added attackable) an activity that does not add value, but it is possible to reduce to a minimum, or eliminate it.

While defining the Yamazumi classes for activities performed by the logistics train operator, it is necessary to take into account that the transport processes are not intended for adding the value to the finished product. Their execution enables timely delivery of materials to locations generating demand, and thus execution of value added activities on workstations. Therefore, for the preparation of proper analyzes, it is necessary to prepare many individual, useful classes of activities.

After definition of the Yamazumi classes, user can assign these classes to the commands performed by the operator of the logistic train. The commands are used for automatic programming of the train operator's work. Commands are saved in the command lists, executed cyclically during simulation experiments.

The numerical values of the Yamazumi classes are entered by the user into the table in the simulation program. The table in simulation program is presented at Fig. 2.

	Value
Travel	3
GrabTugger	2
ReleaseTugger	2
DriveTugger	3
GrabTote	2
ReleaseTote	2
WaitForEmptyTote	3
GrabEmptyTote	2
CheckTrolleyWait	3
ReadyForTask	0

Fig. 2. Yamazumi classes in simulation model. Source: own study.

3.3 Simulation Experiment and Analysis of Charts in Simulation Model

After implementation of the input data and simulation experiment, the program will automatically generate graphs. The program generates two groups of charts for logistic train operators.

The first group is Yamazumi charts, where the time unit is one second. The charts are presented at Fig. 3. The first graph shows the total share of activities of specific classes in the available working time of each operator, assigned to the group of operators named LT_01. In this case, only activities of type 2 (non-value added, without possibility of reduction) and 3 (non-value added activity that can be reduced to a minimum or eliminated) are taken into account.

The next three charts shows the distribution of specific classes of activities performed by the operator during the simulation experiment. Charts were generated separately for every operator in the LT_01 group.

The second group of graphs presents the workload of employees, resulting from carrying the elements with given weight (e.g. containers with parts, empty containers), total distance traveled as a result of train and operator movement, and total working time. The graphs presenting the workload of each employee belonging to the logistics team named LT_01, are presented at Fig. 4.

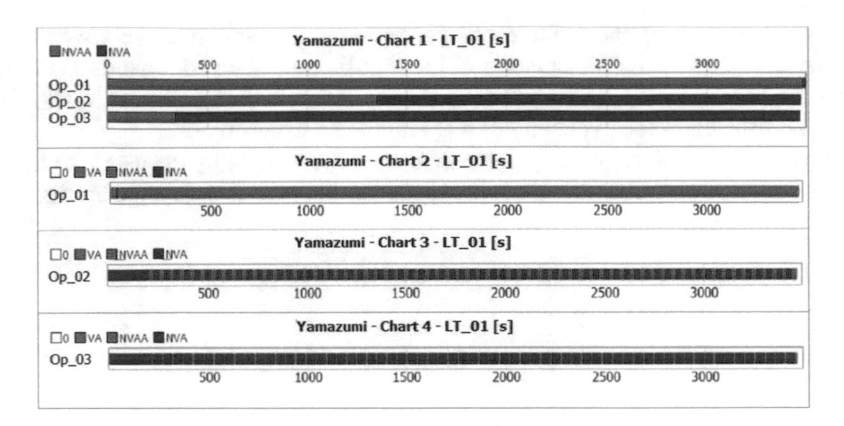

Fig. 3. Yamazumi charts. Source: own study with use simulation model

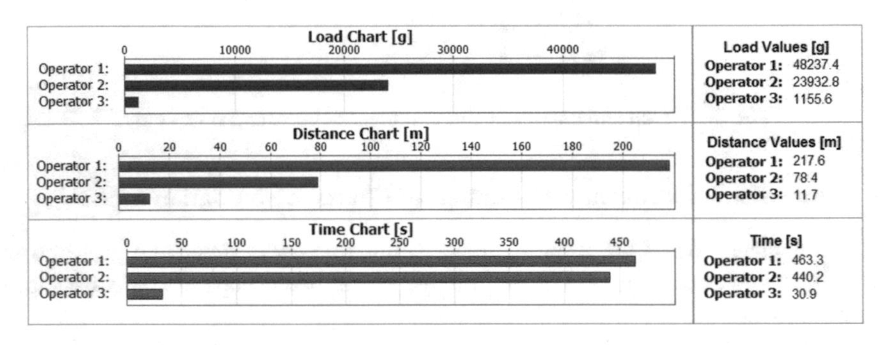

Fig. 4. The workload of logistics train operators on charts. Source: own study with use simulation model

The chart which presents the total weight carried by the operator during the experiment is generated on the basis of PFEP (Plan For Every Part) implemented in the simulation model as a knowledge base for modeling the production system. PFEP takes into account, inter alia the weight of each element and container. When operator transfers an empty container or a container filled with elements, the program can calculate the weight of the cargo carried during the transport operations.

The graph showing the total distance traveled by the train operator (distance unit is meter) and the graph showing the total working time (time unit in second), are generated during the work of operator. The data for charts are collected as a result of execution of subsequent commands form operators commands lists.

Analysis of the workload of train operators is extremely important for balancing workloads and improving their work organization.

4 Simulation Experiments and Analysis

The charts generated by the tool are the basis for analyzing and designing system improvements. Designed improvements can be introduced into the simulation model, in order to carry out simulation experiments and determine the results of introduced changes. Simulation models offer the possibility of multiple experiments, without risk the consequences in real system. Thus, it is possible to quickly find effective solutions and determine the cost, time and reasonableness of their implementation in the real system.

The result of the charts' interpretation can inform, that a specific type of activity for a given operator should be reduced. An important postulate may also be the reduction of the number of involved resources (e.g. operators) or the balancing of their workload. The next step is the definition of possible actions that may cause the above-mentioned reduction and obtain additional available operator's working time, reduction of various types of waste, or reduction of its work load. In described example, such actions are expressed with the following commands: Travel, DriveTugger, WaitForEmptyTote and CheckTrolleyWait.

In order to reduce Travel activities, user can:

- search the better arrangement of milk-run stops in relation to workstations on the production space layout,
- analyze the possibilities of positioning methods for the logistics train relative to the milk-run stop (the train can be allocated in one location, or be moved closer to the position/storage place at single workstation).

However, it should be taken into account that multiple movements of the train in relation to the workstation and bus-stop, the operator carrying loads for a shorter time, but the service time for the logistics train is extended. The impact of the train positioning method and its impact on the processes at workstation, safety of employees and total distance traveled by the train operator (which must efficiently provide containers for the working cells and pick up empty containers), is also a complex issue. Charts generated by the described tool give a great opportunity to analyze this phenomenon and determine the best way of positioning the train relative production units for a given system.

In order to reduce the duration of the DriveTugger activity, it is necessary to minimize the total distance traveled by the train (problem of route optimization) and to consider changing the way of the train positioning.

The reduction of WaitForEmptyTote and CheckTrolleyWait activities duration, requires improvement of the logistics train schedule and better work organization, and may require the improvement of the organization of work of operators at working cells.

A very important issue is the control of the total weight of components carried by the train operator. This is a frequently overlooked problem, however, due to the ergonomics of work and various regulations for this issue, special attention should be paid to it.

5 Conclusions

Yamazumi analysis is a powerful source of information about the process, the work of operators and the possibility of design and implement various improvements in order to achieve many benefits.

These charts are needed both for the analysis of the work distribution of operators working at single production cell, and for logistic train operators. The problem of designing and improving milk-run systems is more and more often raised due to its complexity and high usefulness for manufacturing companies, especially assembly lines.

Yamazumi analyzes allow for the analysis of the train operator's work, taking into account the contribution of non-value added activities that user can try to reduce, or not. It gives the opportunity to design and implement many improvements, aimed at reducing the burden on operators, reducing wastage and better distribution of work between employees.

The mechanism that generates Yamazumi charts has a very large potential for development. It can be adapted to display the described statistics on charts in another form, and to generate new statistics. Due to the possibility of assigning Yamazumi classes to particular types of activities, the use of the mechanism is fast, simple and user-friendly. This means quick, automatic generation of various charts, which are the basis for important, valuable conclusions.

Acknowledgments. The work was carried out as part of the 503228/11/140/DSMK/4160 project.

References

1. Bae, K.G., Evans, L.A., Summers, A.: Lean design and analysis of a milk-run delivery system: case study. In: Roeder, T.M.K., Frazier, P.I., Szechtman, R., Zhou, E., Huschka, T., Chick, S.E. (eds.) Proceedings of the 2016 Winter Simulation Conference (2016)
2. Bowersox, D.J., Copper, M.B., Closs, D.J.: Supply Chain Logistics Management. McGraw-Hill Publishers, New York (2002)
3. Brar, G.S., Saini, G.: Milk run logistics: literature review and directions. In: Proceedings of the World Congress on Engineering, WCE, vol. I, pp. 797–801 (2011)
4. Hao, Q., Shen, W.: Implementing a hybrid simulation model for a Kanban based material handling system. Robot Comput. Integer Manuf. **24**, 635–646 (2008)
5. Hosseini, S.D., Shirazi, M.A., Karimi, B.: Cross-docking and milk run logistics in a consolidation network: a hybrid of harmony search and simulated annealing approach. J. Manuf. Syst. **33**(4), 567–577 (2014)
6. Karagul, H., Albayrakoglu, M.M.: Selecting a third-party logistics provider for an automotive company: an analytic hierarchy process model (2007)
7. Korytkowski, P., Karkoszka, R.: Simulation-based efficiency analysis of an in-plant milk-run operator under disturbances. Int. J. Adv. Manuf. Technol. **82**(5–8), 827–837 (2016)
8. Kilic, H.S., Durmusoglu, M.B., Baskak, M.: Classification and modeling for in-plant milk-run distribution systems. J. Ind. Eng. Manag. **5**(2), 382–405 (2012)

9. Kitamura, T., Okamoto, K.: Automated route planning for milk-run transport logistics using model checking. In: Proceeding of 2012 Third International Conference on Networking and Computing, pp. 240–246 (2012)
10. Meyer, A.: Milk Run Design – Definitions, Concepts and Solution Approaches. Karlsruher Institut für Technologie KIT Scientific Publishing (2015)
11. Miwa, K., Nomura, J., Takakuwa, S.: Module-based modeling and analysis of just-in-time production adopting dual-card Kanban system and Mizusumashi worker. In: Chan, W.K.V., D'Ambrogio, A., Zacharewicz, G., Mustafee, N., Wainer, G., Page, E. (eds.) Proceedings of the 2017 Winter Simulation Conference (2017)
12. Nemoto, T., Hayashi, K., Hashimoto, M.: Milk-run logistics by Japanese automobile manufacturers in Thailand. Procedia Soc. Behav. Sci. **2**(3), 5980–5989 (2010)
13. Lashine, S.H., Fattouh, M., Issa, A.: Location/allocation and routing decisions in supply chain network design. J. Model. Manag. **1**(2), 173–183 (2006)
14. Pawlewski, P.: Script language to describe agent's behaviors, highlights of practical applications of complex multi-agent systems. In: Bajo, J., et al. (eds.) International Workshops of PAAMS 2018. Springer, Cham (2018)
15. Tokola, H., Niemi, E., Vaisto, V.: Lean manufacturing methods in simulation literature: review and association analysis. In: Yilmaz, L., Chan, W.K.V., Moon, I., Roeder, T.M.K., Macal, C., Rossetti, M.D. (eds.) Proceedings of the 2015 Winter Simulation (2015)
16. Zhenlai, YE., Yang J.: Development and application of milk-run distribution systems in the express industry based on saving algorithm. Math. Probl. Eng. 1–6 (2014). http://dx.doi.org/10.1155/2014/536459

Product Design and Product Manufacturing in Industry 4.0

Automation and Digitization of the Material Selection Process for Ecodesign

Izabela Rojek[1(✉)], Ewa Dostatni[2], and Adam Hamrol[2]

[1] Institute of Mechanics and Applied Computer Science, Kazimierz Wielki University, Bydgoszcz, Poland
izarojek@ukw.edu.pl

[2] Department of Management and Production Engineering, Poznan University of Technology, Poznań, Poland
{ewa.dostatni,adam.hamrol}@put.poznan.pl

Abstract. The research presented in this article focuses on the concept of Industry 4.0 developed in Poland and across the world. This concept applies, among others, to the broadly understood automation of processing, digitization and exchange of large volumes of data in production processes. It covers not only the manufacturing process, but the entire life cycle of the product. It applies to those areas of organization functioning, which are supported by intelligent systems facilitating decision making and automation that improves work efficiency. The method presented in the article automates the selection process of materials and connections at the product design stage, taking into account the recyclability aspects. It is an extension of earlier research by the authors in this area. The developed method has been extended with further parameters describing the properties of materials that are necessary when selecting appropriate construction materials at recycling-oriented product design stage. The method was developed with the use of the decision tree induction method for selecting environmentally friendly materials and connections allowing for high level of product recyclability. The method is a practical solution supporting ecodesign (at the stage of both construction and technology design). Thanks to the developed method, the material and material selection data are digitized and stored in the expert system.

Keywords: Material · Selection · Ecodesign · Expert system
Decision tree induction method

1 Introduction

Product design involves actions and events that occur between the moment when a problem appears, and the time when documentation is drawn, describing problem resolution which meets the functional, economic, and other pre-defined requirements. It is a complex process, very important in the life cycle of any product. Decisions made in the design stage affect the manufacturing costs and determine actions that will have to be made in the final phase of the product's life cycle, i.e. after it is withdrawn from service. The sooner the environmental impacts are identified and included in the life cycle of a product, the better the results of such actions. Such approach to design is

© Springer Nature Switzerland AG 2019
A. Burduk et al. (Eds.): ISPEM 2018, AISC 835, pp. 523–532, 2019.
https://doi.org/10.1007/978-3-319-97490-3_50

called ecodesign. It consists in the identification of environmental issues concerning the product and taking them into account at the early stage of product development. Ecodesign is also referred to as design for environment or sustainable product design [10, 11, 17, 18].

Artificial intelligence is used in various areas of computer aided production processes to acquire knowledge from experienced employees and to automate its use in a computer system [7, 9, 21–23, 26, 27]. These methods include mainly machine learning methods, such as neural networks or the decision tree induction method.

The first stage of the authors' research, which was described in their publications [23], presented an expert system for the selection of materials and connections at the design stage of the product. It supported product designers and process engineers in selecting materials and connections for the designed product to ensure its highest possible recyclability. It enabled the selection of materials compatible in terms of recycling and connections most suitable for quick and simple disassembly. The proposed method offered two ways of selecting materials. In the first one, the user defines the main material (from which one element of the product is made) and the additional material (from which another product element is made and connected to the first element), and the system provides information to what degree the two are compatible. In the second method, the user selects the material from which an element of the product is made and specifies the requested degree of compatibility of another material needed in the product. The system suggests the material which should be used to enable recycling of the product in the final stage of the product lifecycle. At this stage the authors used the matrices of compatible materials. After selecting the materials, it was necessary to choose the most appropriate material connections. For compatibility reasons, the materials may be joined by separable or inseparable connections. In the case of good compatibility, both separable and inseparable connections may be used. However, if material compatibility is low or if materials are incompatible, only separable connections should be used.

The next stage of research, reported in this paper, focused on increasing the number of parameters describing the properties of materials chosen using the first selection method. The method consists in providing the main material component, the properties described in the further part of the article and additional material. As a result, the modified expert system provides information to what degree the materials are compatible. The second method of material selection, where the user inputs the main material and the requested degree of compatibility, and the expert system suggests the appropriate additional material, has not been modified. Also, the method of selecting material connections is the same as in the previous version of the expert system.

The research presented in this article is consistent with the concept of Industry 4.0, developed in Poland and across the world. The concept puts significant emphasis on the automation of processing and exchange of large volumes of data [15]. The term Industry 4.0 means the unification of the real world of manufacturing machines with the virtual world of Internet and information technology. People, machines and IT systems automatically exchange information during the production process. Industry 4.0 deals mainly with the production process performed by manufacturing companies. However, in the literature one can find an extension of this concept, which concerns not only the production process, but the entire life cycle of the product and the associated value added chain, including pre-production, production and post-production phases. Pre- and post-production areas, which create the greatest added value, are especially important. Therefore, there was a need to develop and use new technologies, techniques and methodologies, which poses numerous challenges for engineers [12, 16]. One of such challenges is the development of artificial intelligence-supported systems for data analysis and processing in order to choose the best solutions. The tool presented in the article automates the process of material and connection selection at the product design stage, taking into account recycling. Also, certain properties of the product which are important in the final stages of its life cycle are taken into account already at product design stage.

The author's expert system for material selection in ecodesign has been tested on real data from the company.

2 Literature Review

Regulations implemented in the European Union require that designers and engineers take appropriate actions related to ecodesign [2, 8, 14, 20]. However, insufficient knowledge of ecodesign often limits the creative approach to such engineering solutions [19]. Therefore, designers need tools to aid them in such work [3, 4]. Baunman et al. identify over 150 different aid tools used in the development of eco-friendly products. The first group comprises tools which include information about toxicity and compatibility of materials, the sources of origin of raw materials, etc. Another group are IT tools which support product evaluation in its entire life cycle; they present a holistic approach to ecodesign [1, 5, 13, 19]. The tools supporting ecodesign include also solutions dedicated e.g. specifically to product designers, based on the automation of the design process taking into account environmental concerns. However, the large number and complexity of some of the solutions discourage potential users. The tools that work best are simple, easy to use and dedicated to specific users and purposes [6].

3 Methods

Artificial intelligence, in particular machine learning, is an important element of industry 4.0. It allows for the automatic processing of data and information into the expert knowledge contained in computer systems. A computer system with expert knowledge can replace a human expert in certain activities. Following Industry 4.0, the authors chose one of the machine learning methods, i.e. decision tree induction method, to acquire knowledge and experience in selecting materials in ecodesign.

In induction methods, the available data are analyzed to generate a hypothesis describing the existing dependencies. In practice, we are dealing with data organized as a set of records described by a fixed set of attributes. In such case, inductive inference consists in generating a hypothesis which describes the relations between attributes. The decision tree induction method was chosen due to such advantages as speed of classification, intelligibility, "mature methodology", and numerous practical implementations. Additionally, this method makes it possible to process data as symbols, and not only numbers. The decision tree induction method allows for an approximation of classification functions with discrete input values referring to certain notions – decision classes. Nowadays decision trees are the basic method of inductive machine learning [24]. This is due to their high efficiency, the possibility of simple software implementation, and intuitive evidence for a man. The method of knowledge acquisition is based on the analysis of already completed and tested selections, but each example must be described by a set of attributes, and each attribute may assume different values. Based on the training file with examples a decision tree is built, in which this method generalizes examples from the training file to the decision rules placed in the expert system. The method of induction of decision trees is used to acquire knowledge from examples in the DeTreex program.

An expert system is a computer program that performs complex tasks with high intellectual requirements and does it as well as a human being who is an expert in the field [25]. The most important features of the expert system include:

- gathering the fullest and most competent knowledge in a given field and the possibility of its updating in line with scientific and technical progress,
- skilful imitation of the man-expert reasoning, which he uses when solving problems of the same type,
- prompting system decisions using the inference method,
- explaining the course of "reasoning" that led to the results obtained,
- justification of the solutions received.

The basic components of the expert system are: knowledge base, inference mechanism, explanation module and user interface.

Most of the expert systems implemented so far use rules to represent knowledge. The universality of this type of knowledge representation results from many advantages of this formalism. The main advantages of the rules are their simplicity and generality. Simplicity makes them understandable even for users who are not specialists in the field of expert systems. Generality means that the field of the problem is not a limitation in the use of this method of knowledge representation. There are many examples of practical applications in such different fields as medicine, technology or economics.

Rules may represent knowledge in an explicit way, especially in comparison with traditional algorithm-based techniques. They are particularly useful for representing knowledge of a heuristic nature (depending on the expert's experience and intuition), used to simulate the decision-making processes. Experts from various fields generally have no problem with expressing a part of their knowledge with the help of this formalism, which facilitates the cooperation of knowledge engineers and experts. The rules are generally legible for end users of the application, which is very important for the success of the practical applications of expert systems. At the same time, this feature facilitates the introduction of explanatory mechanisms that are of great practical importance in some areas of application. The rules facilitate the incremental development of the knowledge base and its improvement as further experience is gained and verified in practice. However, it should be remembered that not all knowledge can be expressed by rules, it can also take the form of facts.

4 Author's Expert System for Material Selection in Ecodesign Using Decision Tree Induction Method

4.1 Data Preparation

The stages of designing an expert system supporting the selection of materials have been developed as follows:

- analysis of input data for the selection of materials (based on the analysis of material properties).
- development of a file with examples of material selections, which will then be used to build a decision tree,
- generation of the model in the form of a decision tree,
- testing the tree model on a test file,
- generating decision rules based on a decision tree,
- development of an interface supporting the selection of materials.

The training and testing files were developed based on the analysis of material properties, which include (Fig. 1): material name (e.g. S235), density, expressed in grams per cubic centimeter (e.g. 7.88), tensile strength, expressed in megapascals (e.g. 35.5), yield point elongation Re [%], expressed as a percentage (e.g. 5.5), processing temperature, expressed in degrees Celsius (e.g. 20.8), dielectric constant (e.g. 2.0), dielectric strength, expressed in kilowatts per millimeter (e.g. 22.0), Young's modulus (E), expressed in gigapascals (e.g. 4.61), water absorption, expressed as a percentage (e.g. 22.55), environmental impact (e.g. true) and the cost of recycling, expressed in PLN per kilogram (e.g. 4.25); positive value means profit from the sale of material, while the negative value is the cost of disposal.

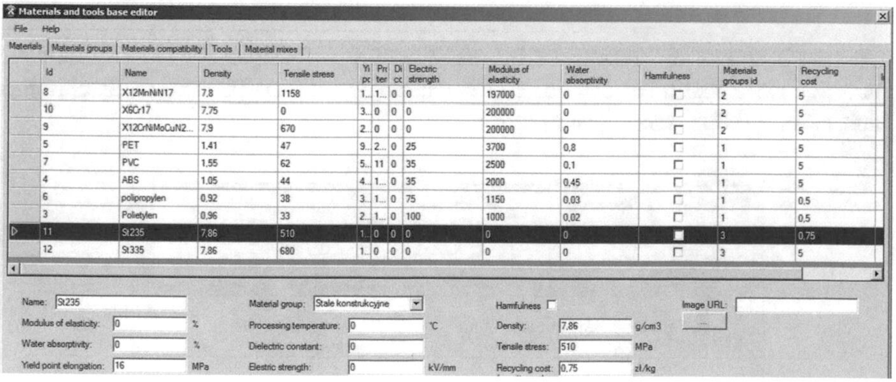

Materials and tools base editor

File Help

Materials | Materials groups | Materials compatibility | Tools | Material mixes

Id	Name	Density	Tensile stress	Yi pc	Pr ter	Di cc	Electric strength	Modulus of elasticity	Water absorptivity	Harmfulness	Materials groups id	Recycling cost
8	X12MnNiN17	7,8	1158	1...	1...	0	0	197000	0		2	5
10	X6Cr17	7,75	0	3...	0	0	0	200000	0		2	5
9	X12CrNiMoCuN2...	7,9	670	2...	0	0	0	200000	0		2	5
5	PET	1,41	47	9...	2...	0	25	3700	0,8		1	5
7	PVC	1,55	62	5...	11	0	35	2500	0,1		1	5
4	ABS	1,05	44	4...	1...	0	35	2000	0,45		1	5
6	polipropylen	0,92	38	3...	1...	0	75	1150	0,03		1	0,5
3	Polietylen	0,96	33	2...	1...	0	100	1000	0,02		1	0,5
11	St235	7,86	510	1...	0	0	0	0	0		3	0,75
12	St335	7,86	680	1...	0	0	0	0	0		3	5

Name: St235 | Material group: Stale konstrukcyjne | Harmfulness | Image URL:

Modulus of elasticity: 0 % | Processing temperature: 0 °C | Density: 7,86 g/cm3

Water absorptivity: 0 % | Dielectric constant: 0 | Tensile stress: 510 MPa

Yield point elongation: 16 MPa | Elestric strength: 0 kV/mm | Recycling cost: 0,75 zl/kg

Fig. 1. Material data, excerpt from the database [10]

Next, examples of material selections based on literature and real data were prepared, which were then used in the DeTreex program to build a decision tree. The file with examples contained 195 examples. In contrast, the test file that served to check the quality of the model in the form of a decision tree contained 20 examples. The input parameters for the development of the tree are the properties of materials, taking into account their environmental impact. The decision-making class is the compatibility of materials.

4.2 A Decision Tree for Material Selection

The file with examples of material selection was divided into a training file, based on which the tree was generated, and a test file, which was used to assess the classification accuracy of the generated tree.

The developed decision tree was built on the basis of a set of training examples and then used to create decision rules. A fragment of a generated decision tree is shown in Fig. 2. All input attributes from the training file were used to generate the tree. The additional material attribute became the root of the tree. Other material properties were individual tree nodes.

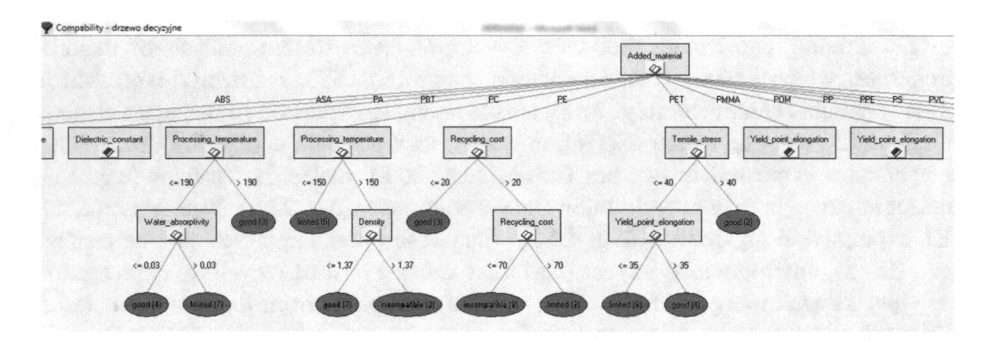

Fig. 2. A fragment of the decision tree for the selection of materials

Figure 3 shows the results of the classification of test examples (classification matrix and classification error). For 20 test examples, all have been correctly classified, which gives a 0.00% classification error.

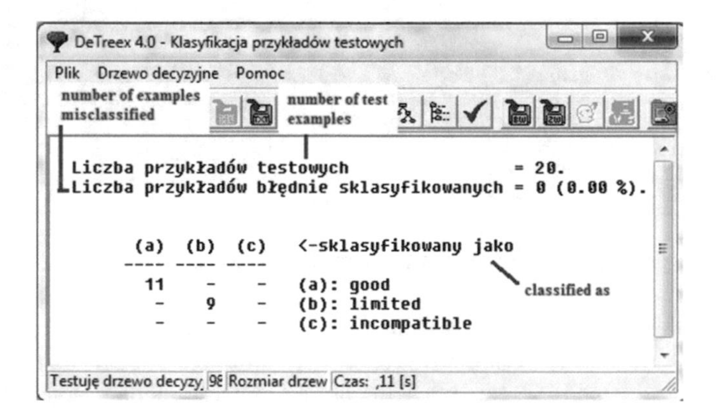

Fig. 3. Testing the correctness of the model's activity - classification of test examples

Next, decision rules were created and placed in the expert system. Examples of rules:

```
0018 : compability = "limited" if        // (5.0)
         added_material = "SAN",
         processing_temperature <= 150;

0019 : compability = "good" if           // (4.0)
         added_material = "PBT",
         processing_temperature <= 190,
         water_absorptivity <= 0.03;

0020 : compability = "good" if           // (4.0)
         added_material = "PET",
         tensile_stress <= 40,
         yield_point_elongation > 35;
```

DeTreex is a tool for acquiring knowledge using the method of induction of decision trees. Using the Detreex program, a knowledge base was created in the author's expert system.

4.3 Operation of Expert System for Material Selection

The expert system can support the product designer and/or process engineer in the selection of materials, taking into account the environmentally-friendly properties of products.

In order to support product designers and process engineers, we developed an expert system which allows for:

(a) the selection of materials according to their compatibility,
(b) the selection of additional components with regard to compatibility,
(c) choose the most appropriate material connections.

The article shows changes in the expert system that occurred in relation to previous research. The changes concern (a) selection of materials with a view to their compatibility, and they consist in the introduction of additional properties of materials, based on which selection of compatible materials is made.

When choosing the materials according to their compatibility, process engineer enters to the expert system the symbol of the main material, the properties of the main material, and the symbol of the additional material. The system will then advise the compatibility between the materials (e.g. good, limited or incompatible).

Figure 4 shows sample screens of the expert system. Figure 4a shows the expert system start window, Fig. 4b - selection of main material ASA, Fig. 4c - determination of properties of main material ASA, Fig. 4d - selection of ABS material to be combined with ASA material, Fig. 4e - the answer of the system, i.e., information on the degree of compatibility (good).

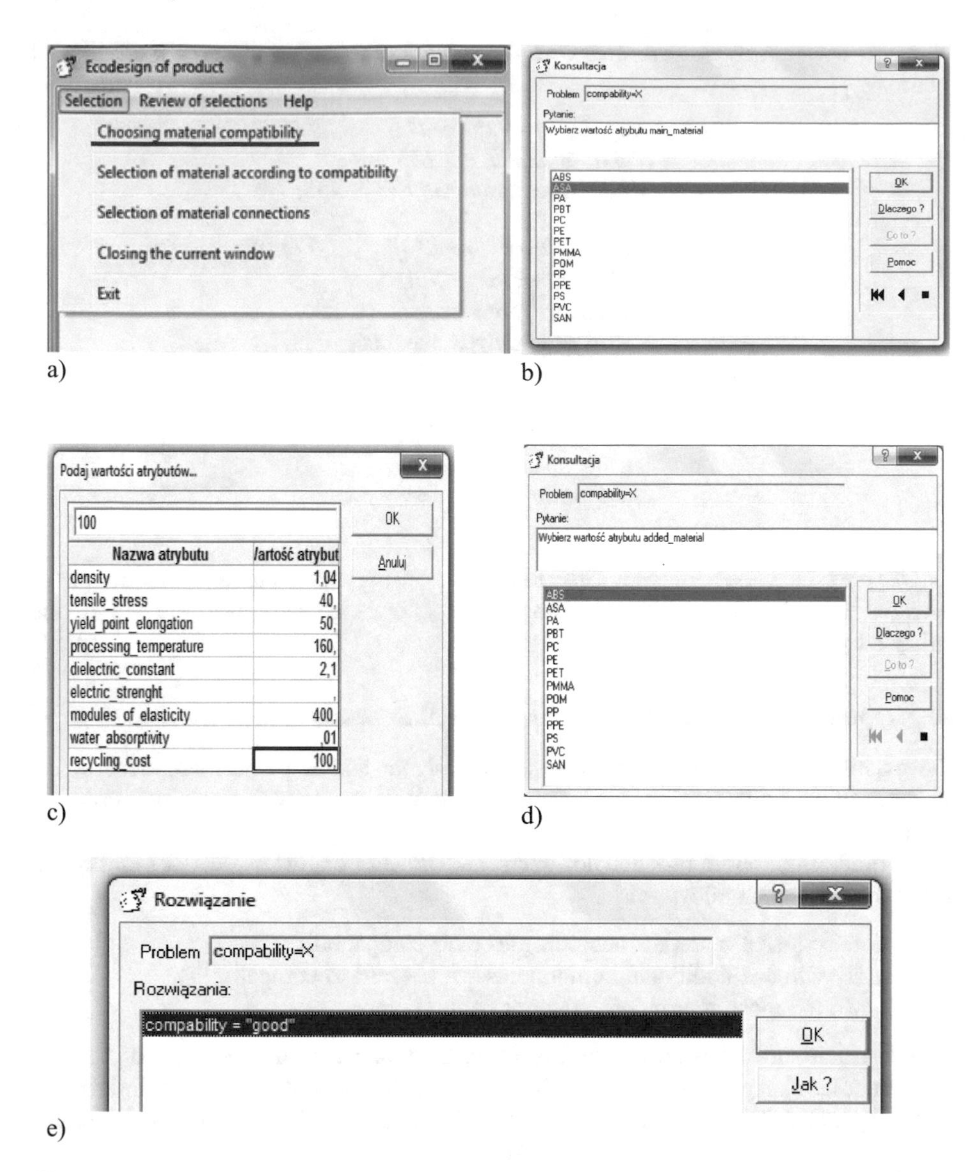

Fig. 4. Sample screens of the material selection expert system indicating the compatibility of these materials

5 Summary

In the case of sustainable product design, it is appropriate to use the decision tree induction as the classification method due to the large number of input data represented in a symbolic form. Decision trees have very good classification properties. Rules generated based on decision trees are more concise and the time needed for drawing conclusions is significantly reduced. Decision rules were then introduced to the expert system. Research proved the usefulness of classification trees and their high efficiency in supporting the material selection for ecodesign.

Classification trees, as very good data exploration algorithms, offered great opportunities to use the data contained in the databases. They made it possible to create automatic rules based on examples, which a very good solution for discovering knowledge resulting from the experience of process engineers.

The material selection expert system is an innovative solution in line with Industry 4.0.

Acknowledgements. The presented results derive from a scientific statutory research conducted by Chair of Management and Production Engineering, Faculty of Mechanical Engineering and Management, Poznan University of Technology, Poland (no. 02/23/DSPB/7716) and Institute of Mechanics and Applied Information Science, Faculty of Mathematics, Physics and Technical Sciences, Kazimierz Wielki University, Poland, supported by the Polish Ministry of Science and Higher Education from the financial means in 2018.

References

1. ATROiD. http://root.ew.eea.europa.eu/sd-online/tools/atroid-30-assessment-tool-recycling-oriented. Accessed May 2013
2. Azevedo, S.G., Carvalho, H., Machado, V.C.: The influence of green practices on supply chain performance: a case study approach. Transp. Res. Part E **47**(6), 850–871 (2011)
3. Baunman, H., Boons, F., Bragd, A.: Bragd mapping the green product development field: engineering, policy and business perspectives. J. Clean. Prod. **10**(5), 409–425 (2002)
4. Birch, A., Hon, K.K.B., Shor, T.: Structure and output mechanisms in Design for Environment (DfE) tools. J. Cleaner Prod. **35**, 50–58 (2012)
5. Boustead model. http://www.boustead-consulting.co.uk/products.html. Accessed May 2014
6. Bovea, M.D., Pérez-Belis, V.: A taxonomy of ecodesign tools for integrating environmental requirements into the product design process. Int. J. Clean. Prod. **20**, 61–70 (2012)
7. Burduk, A.: Artificial neural networks as tools for controlling production systems and ensuring their stability. Lecture Notes in Computer Science, vol. 8104, pp. 487–498. Springer, Heidelberg (2013)
8. Buyukozkan, G., Cifci, G.: A novel hybrid MCDM approach based on fuzzy DEMATEL, fuzzy ANP and fuzzy TOPSIS to evaluate green suppliers. Expert Syst. Appl. **39**, 3000–3011 (2012)
9. Czapczuk, A., Dawidowicz, J., Piekarski, J.: Artificial intelligence methods in the design and operation of water supply systems. Rocz. Ochr. Sr. **17**, 1527–1544 (2015)
10. Dostatni, E., Diakun, J., Grajewski, D., Wichniarek, R., Karwasz, A.: Multi-agent system to support decision-making process in ecodesign. In: Proceedings of 10th International Conference "Soft Computing Models in Industrial and Environmental Applications", SOCO, pp. 463–474 (2015). https://doi.org/10.1007/978-3-319-19719-7_40
11. Dostatni, E., Diakun, J., Grajewski, D., Wichniarek, R., Karwasz, A.: Functionality assessment of ecodesign support system. Manag. Prod. Eng. Rev. **6**(1), 10–15 (2015)
12. Engineers of Industry 4.0 (Not) ready for change? [In Polish: Inżynierowie Przemysłu 4.0 (Nie)gotowi do zmian?], Astor Whitepaper (2017). https://www.astor.com.pl/images/Industry_4-0_Przemysl_4-0/ASTOR_Inzynierowie_4.0_whitepaper.pdf
13. Gabi. http://www.gabi-software.com. Accessed Jan 2013

14. Govindan, K., Khodaverdi, R., Vafadarnikjoo, A.: Intuitionistic fuzzy based DEMATEL method for developing green practices and performances in a green supply chain. Expert Syst. Appl. **42**(20), 7207–7220 (2015)
15. Hermann, M., Pentek, T., Otto, B.: Design Principles for Industrie 4.0 Scenarios. http://www.snom.mb.tu-dortmund.de/cms/de/forschung/Arbeitsberichte/Design-Principles-for-Industrie-4_0-Scenarios.pdf. Accessed Jan 2016
16. Industry 4.0. The revolution is here. What do you know about it? [In Polish: Przemysł 4.0 Rewolucja już tu jest. Co o niej wiesz?], Astor Whitepaper (2016). https://www.astor.com.pl//images/Industry_4-0_Przemysl_4-0/ASTOR_przemysl4_whitepaper.pdf
17. ISO/TR 14062, Environmental management – integrating environmental aspects into product design and development (2001)
18. Jasiulewicz-Kaczmarek, M.: The role of ergonomics in implementation of the social aspect of sustainability, illustrated with the example of maintenance. In: Proceedings of 9th International Symposium "Occupational Safety and Hygiene", SHO, Guimaraes, Portugale, pp. 250–251 (2013)
19. Jung Yang, Ch., Lewis Chen, J.: Forecasting the design of eco-products by integrating TRIZ evolution patterns with CBR and Simple LCA methods. Expert Syst. Appl. **39**(3), 2884–2892 (2012)
20. Lin, R.J.: Using fuzzy DEMATEL to evaluate the green supply chain management practices. J. Clean. Prod. **40**, 32–39 (2013)
21. Rojek, I.: Neural networks as prediction models for water intake in water supply system. Lecture Notes in Artificial Intelligence, vol. 5097, pp. 1109–1119. Springer, Heidelberg (2008)
22. Rojek, I.: Technological process planning by the use of neural networks. Artif. Intell. Eng. Des. Anal. Manuf. **31**(1), 1–15 (2017)
23. Rojek, I., Dostatni, E., Hamrol, A.: Ecodesign of technological processes with the use of decision trees method. In: Pérez García, H., et al. (eds.) Advances in Intelligent Systems and Computing, vol. 649, pp. 318–327. Springer, Cham (2018)
24. Rokach, L., Maimon, O.: Data Mining with Decision Trees: Theory and Applications. World Scientific Pub. Co. Inc., Singapore (2008). ISBN 978-9812771711
25. Russell, S.J., Norvig, P.: Artificial Intelligence: a Modern Approach. Prentice Hall, Englewood Cliffs (2009)
26. Sapietova, A., Dekys, V., Sapieta M., Pechac, P.: Application of computational and design approaches to improve carrier stability. In: Proceedings of 6th Conference "Modelling of Mechanical and Mechatronic Systems", Modelling of Mechanical and Mechatronic Systems. Procedia Engineering, vol. 96, pp. 410–418 (2014)
27. Sapietova, A., Petrivic, M.: Analysis of the dynamical effects on housing of the axial piston hydromotor. Novel Trends Prod. Dev. Syst. Appl. Mech. Mater. **474**, 357–362 (2014)

Automation of the Ecodesign Process for Industry 4.0

Ewa Dostatni, Jacek Diakun(✉), Damian Grajewski, Radosław Wichniarek, and Anna Karwasz

Faculty of Mechanical Engineering and Management, Poznań University of Technology, Poznań, Poland
{ewa.dostatni,jacek.diakun,damian.grajewski, radoslaw.wichniarek,anna.karwasz}@put.poznan.pl

Abstract. One of the key aspects of Industry 4.0 is high automation of various processes conducted through product lifecycle. In the article the authors consider the issue of final stages of product lifecycle and the automation of product assessment from ecological point-of-view in the context of Industry 4.0. The automation of product assessment is conducted by set of measures, calculated from special type of product model called Recycling Model of Product, that is also showed in the paper. The Recycling Model of Product and the assessment are conducted using dedicated module, operating in CAD 3D system environment. The assessment of example product is presented in the paper as well.

Keywords: Industry 4.0 · Automation of product assessment · Ecodesign Agent technology

1 Introduction

Despite the growing environmental awareness in the society, numerous surveys [1, 2] show that restrictive formal requirements are one of the most important reasons for introducing environmental considerations in enterprises. Manufacturers still see costs, efficiency and customer satisfaction as the most important factors in developing innovative products. On the other hand, end customers are most interested in the price and quality of the product.

Research within the project "Sustainable production patterns (SPP) in enterprises – a proposal of systemic solutions supporting the implementation of SPP in SMEs" [3] showed that the increase in the turnover of environmental services and goods is affected by the tightening of legal provisions, not by the increase of public awareness.

In many countries of the European Union it is required to search for solutions reducing the negative environmental impact of products already at product design stage. Firstly, it involves appropriate production technologies, which minimize the amount of waste. Secondly, it requires the use of less hazardous materials. Finally, technologies of waste recovery and treatment must be implemented. Pro-environmental activities in a company should be analyzed and improved throughout the product development process, because only then can it bring the best results [4, 5]. It is estimated that about

© Springer Nature Switzerland AG 2019
A. Burduk et al. (Eds.): ISPEM 2018, AISC 835, pp. 533–542, 2019.
https://doi.org/10.1007/978-3-319-97490-3_51

80% of a product's sustainability performance is defined during the early stages of its development [6]. Including environmental issues in product design consists in the identification of such issues and taking them into account at the early stage of product development. It is referred to as ecodesign, design for the environment or sustainable design [7–9]. To be precise, ecodesign includes e.g. the selection of materials aimed at reducing the environmental impact of the product [10]. It usually refers to various aspects that can have a big impact on the production environment. These aspects include the choice of materials (non-toxic substances, recyclable materials), the selection of production processes (in terms of waste and emissions), as well as the determination of energy demand of products in the use phase, as well as its decommissioning (i.e. repairs and recycling) [11, 12].

At a time when industrial evolution has reached Industry 4.0, aspects of ecodesign should also be included in this concept. The article describes the assumptions of Industry 4.0 including the broadly understood ecodesign. The authors present concepts of a solution that can be applied at the design stage to automate and digitize works related to the environmental impact assessment of the designed products. The method proposed in the article will also enable the automation of the recycling stage (disassembly, selection of materials, etc.) in the last phase of the product life cycle.

2 Industry 4.0 Concept and Ecodesign

The Industry 4.0 concept is related to the beginning of the so-called the fourth industrial revolution. Industry 4.0 means the unification of the real world of manufacturing machines with the virtual world of Internet and information technology. People, machines and IT systems have access to tools that allow automatic exchange of data and information. Integration and exchange of data applies to the entire enterprise as well as its surroundings. It applies to the entire logistics chain, i.e. customer, manufacturer and supplier. The Industry 4.0 concept can also be extended to include the client's "client" (if the manufacturing company's customer is an enterprise producing components for another end product manufacturer) and/or supplier's "supplier" (if the products provided by the supplier consist of product components made by other manufacturers). Thanks to the Industry 4.0 concept the production in the entire logistics chain can be fully automated, digitized and computerized. The necessary data and information will be available anywhere, at any time, and in the required scope. The availability of data and information in the company will enable quick adaptation to customer requirements, flexible adaptation of production to market needs, and consequently, it will offer a competitive advantage. The implementation of the concept Industry 4.0 is to accelerate the transformation of enterprises into "smart factories" in which networks based on information and communication technologies connect machines, processes, systems, products, customers and suppliers [13, 14]. The main pillars of the concept are the Internet of Things, cyber-physical systems, network infrastructure, self-learning processes and dispersed artificial intelligence that will enable network connections of dispersed enterprises and contribute to the development of intelligent production systems [15]. Thanks to the networking and digitalization of data related to production processes, the time of adapting the machine to new working conditions will be shortened, which will enable shortening of innovative production cycles [16]. Currently, however, there are still no uniform and clearly defined definitions describing what exactly the term Industry 4.0 means [17–20]. Published research has shown that in 2014 term Industry 4.0 was unknown to 64% of the surveyed companies. At the same time, many companies, especially industrial giants, have already invested in technologies related to Industry 4.0 [21]. With the popularization of the Industry 4.0 concept, e.g. through the development of national Platforms of Industry 4.0 in various countries (including Poland and Germany) or organization of conferences for enterprises, one could say that the concept is recognized by a much larger percentage of enterprises. The following elements are defined in the literature as priorities for Industry 4.0 [14, 22–25]:

- development of new business models and network ecosystems using artificial intelligence; integration of customer and manufacturer within individualized production, more efficient communication between suppliers, manufacturers and customers,
- intensive data exchange between entities cooperating during the product life cycle,
- development of intelligent and uniquely identifiable products,
- development of intelligent software for total design and immediate on-line response to any production problems,
- development of intelligent production equipment,
- development of new business models,
- providing a new teaching method based on augmented reality,
- ensuring a better balance between work and private life,
- ensuring consistency of intelligent technological solutions with care for the protection of the natural environment,
- efficient use of resources and energy saving.

Summing up, it can be said that the new concept of industry applies to those areas of organization functioning, which are supported by intelligent systems facilitating decision making and automation that improves work efficiency [14, 24, 26].

There are no direct references to environment protection in the premises of Industry 4.0. However, it was noticed that environment protection and the effective use of resources will remain an important element. The studies on Industry 4.0 do not directly address the area related to, for example, recycling of products or ecodesign. In the works of Kokot and Kolenda [27, 28], it was noticed that in accordance with the idea of sustainable development during the development of Industry 4.0, aspects related to intelligent environment, intelligent water management, intelligent energy and intelligent health and life will be discussed in the near future. It can be noted that in this case the concept of Industry 4.0 is treated in a broader context and is not limited only to manufacturing enterprises.

Industry 4.0 affects the entire product life cycle and related value added chain, i.e. pre-production, production and post-production phase. When implemented in companies, the concept of Industry 4.0 will ensure data exchange within value chains, and manufacturing tools will be able to (in the majority of cases) modify their functioning to adjust to new tasks, without any direct participation of man [29]. The data collected and shared throughout the entire life cycle of the product should also include information that concern ecodesign and can be used to design better, more environmentally friendly products. Hence, it will be necessary to develop tools that will be able to automatically use such type of data and share it. The tools will have to be developed based on new information technologies and be part of the concept of the Internet of Things. This article describes an example of a tool that meets the above requirements, and the collected data, information and knowledge can be made available and serve to improve the environmental aspects of the designed products.

3 Research Methodology

As part of the work, a method of product recycling assessment at design stage was developed. The method includes the so-called Recycling Product Model (RmW) and agent system. The agent system enables the assessment of recycling parameters during product design without user intervention. The data necessary to carry out the assessment are obtained from the RmW implemented in the CAD system. The developed method automates the analysis of the designed product regarding recycling parameters. When designing a product using a 3D CAD system, the structure of the designed product is assessed on an ongoing basis in terms of meeting the ecodesign requirements.

For the purpose of comprehensively capturing the product design features which affect its environmental properties, a type of product model has been developed, called the Recycling Model of Product (RmW). The model includes the following elements: CAD 3D geometric model, extended product structure, extended material attributes, disassembly attributes (data on the disassembly process) and product categorization.

The product structure in RmW is based on the typical structure, created by the designer in the CAD 3D system, and extended by additional data defined as below.

Considering the strong impact of joints between product components on product recyclability and the fact standard 3D product models lack an explicit representation of joints as a design feature, the recycling-oriented product model has been extended by detailed data on joints.

Product components are divided in RmW into three groups as follows:

- the first group includes primarily standardized parts, whose primary purpose is to connect parts and assemblies in the product. The group comprises screws, bolts, nuts, rivets, etc.
- the second group includes those elements of the product, whose function is different than connecting parts,
- the third group includes elements that combine the function of the connecting element with another function, not related to connecting (e.g. threaded parts of bodies).

The division above is functional, as it describes the functions of all components in the model, looking at their role in joints. Due to this feature of the discussed solution, such a model can be described as functional, focused on the structure of joints in the product (in short: a functional model of joints).

Representation of joints should be made by way of grouping selected elements of the model, where each group reflects a specific joint and additional attributes unambiguously describe (categorize) the joint. For this reason, the adopted method of defining joints – as logically related groups of elements – may be described (due to the analogy with geometric constraints) as functional constraints oriented to joint structure in the product (in short: functional joint constraints).

The inclusion of the recycling-related issues into the tools that support the designer's work requires as well that the list of standard material attributes be expanded. This list has been expanded to include:

- harmfulness of material (the presence of hazardous materials precludes a positive recycling-oriented product assessment),
- material cost – the amount which can be obtained for the utilization of a unit of a given material (positive value) or which has to be paid for such an operation (negative value).
- graphic designation of the material according to the standards and other legal acts on the requirements for materials used in the product.

In the solution described here it was assumed that in the product recycling process disassembly will be non-destructive. Thus, the basic (elementary) disassembly operation will be to dismantle the connections in the product. Therefore, on a given level of product structure the components are considered disassembled when all connections have been dismantled.

The data describing the disassembly include: disconnection of a given joint during disassembly, tools used during the disconnection operation and time of disconnecting the joint.

Disconnection of a particular joint is a logical attribute (yes/no), representing the assumed (planned) way of dealing with this connection in the disassembly process (yes - the connection will be dismantled, no - the connection will not be dismantled). If the connection is disconnected, such operation is determined by the toolbox used for this purpose and the estimated time of such operation. The inclusion of this last attribute makes it possible to estimate the total time of disassembly of the entire product.

For some products (such as household appliances), the current regulations categorize particular products within types, specifying minimum requirements concerning the recycling rate. Assigning the product to a given group is the final element of the RmW, and it allows for a direct comparison of the product assessment values resulting from the legal regulations with the current values at a given stage of design or analysis. This in turn allows us to decide whether the product meets the legal criteria for the given category, and thus determines the possibility of placing the product on the market.

Agent technology was used to develop a tool supporting ecodesign. Agent technology is based on the concept of an autonomous agent that has the following characteristics [30, 31]: the ability to observe, autonomy, mobility, the ability to communicate, intelligence.

4 Automation of the Analysis of Recycling Properties of the Product

In order to ensure the possibility of using a system supporting the recycling-oriented product assessment, in a distributed design environment a multi-agent MAS system (for simplicity called agent systems) was developed, consisting of cooperating agents to solve the problem. The main task of the multi-agent system is to conduct a recycling-oriented assessment of a product designed in a CAD 3D environment, based on data from RmW, recycling product model [32, 33]. The system makes it possible to control and monitor recycling parameters of products on an ongoing basis in subsequent versions of the design, and suggest potential remodeling which may improve the parameters. It supports designers working in a distributed environment. Each designer who participates in the work on the project may work in a different node of the distributed network. He may use an IT tool (the agent system), which analyzes his work and based on the observations it conducts a recycling-oriented assessment of the product. Moreover, the system offers suggestions of construction modifications to improve the recyclability of the product.

The tasks of the system are conducted by independent software agents, and each of them is in charge of a separate phase of the environmental product assessment. The agents coordinate their work using a common, central message server and knowledge base. The agent system supporting green, recycling-oriented design features the following functions:

- analysis of product structure, including the logical correctness of the defined joints between elements – the design is the process in which data are supplemented in stages, in certain situations; when the stored data do not allow for a clear description of the product, further analysis could yield meaningless results; the aim of this function is to detect these moments and block further analysis,
- detection of changes made to the product design by the user – product design analysis is started when modifications are detected; the user will have access to current information
- calculating statistics and recyclability indicators of the entire product and its individual elements
- detecting changes in the recycling rate and recyclability indicator – thanks to the feature the user can see which changes in the product structure caused the increase, reduction or exceeding of acceptable levels and rate of recycling
- detecting and identifying elements, inseparable and incompatible elements
- detecting and identifying elements that have the greatest negative impact on the recycling rate
- creating a numerical summary of the used materials and types of joints
- detecting the use of hazardous materials and warning the user
- suggestions and tips on changing materials or joints to improve recyclability parameters of the designed product
- creating reports.

5 Example of Method Application

An exemplary product assessment is shown on the example of an electric kettle. The assessment was made using a CAD 3D model of the product (Fig. 1), on the basis of which its recycling model was created (RmW). The recycling model served as the base for an analysis of the impact of selected constructional features of the product on its recycling properties. As a result, the authors automatically obtained the measures of the product's recycling-oriented product assessment (Table 1). The results of the analysis may provide the grounds for final decisions regarding the constructional solutions applied in the product.

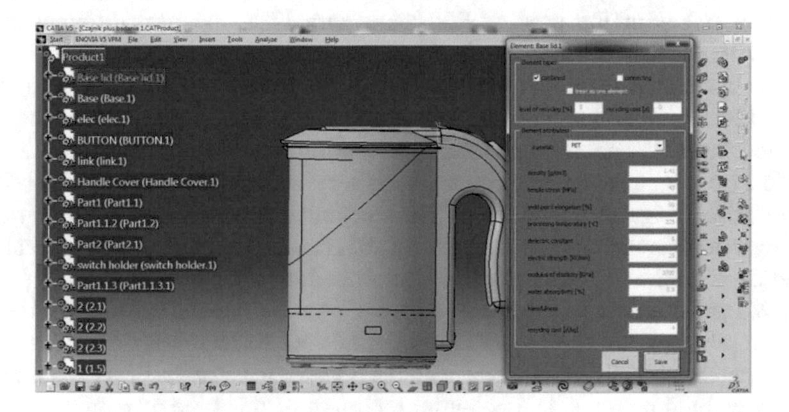

Fig. 1. The model of the kettle subjected to the recycling-oriented assessment

Table 1. Selected construction variants of the product and the values of their recycling-oriented assessment

	CWR	WPR	WRM	WRP	TD [s]	LM	LP	K_{Dem}	K_{MD}	K_{RW}	WPR%
Variant 1	21	15	3	3	38	4	3	0.05	0.09	0.04	41
Variant 2	7	1	3	3	40	4	3	0.06	1.17	1.11	100

Table 1 presents the following values of the recycling assessment: CWR – total recycling ratio; WPR - recycling rate; WRM - material diversity; WRP - joint diversity; TD - time of disassembly; LM - the number of different materials; LP - the number of different joint; KDem - disassembly costs; KMD cost of good materials (suitable for recycling and reuse); KRW - product recycling cost; WPR% - percentage of recycling rate. A detailed description of the recycling parameters can be found in [34]. The values of recycling parameters were calculated by the agent system without user intervention. In the current form of the system, the user chooses which product variant is the best in terms of recycling. In the next stage of system development, the selection of the best solution could be performed by the system. The use of the agent system for the recycling assessment of the product enables the further sharing of data with other systems or placing them in the calculation cloud. This is very important from the point of view of Industry 4.0.

6 Final Remarks

In the life cycle of a product, significant amounts of data of diverse nature are used and created. The data include product requirements, standards and legal acts related to the product, product concept sketches, 2D and 3D geometric models, technological data, operating manuals, etc. The data are integrated and managed based on PLM (Product Lifecycle Management) solutions. Due to the frequent practice of the implementing subsequent stages of the product life cycle with IT tools from different software suppliers, the standard of recording data resulting from activities related to a given stage of the life cycle is of particular importance [35]. It should be noted that computer aid systems differ at individual stages of the product life cycle. The last stage of the product life cycle is relatively poorly supported by IT solutions, in terms of both tools and types of models (in comparison, for example, with very well-supported activities related to the geometric design of the product). The Recycling Product Model (RmW) and the recycling modeling module is a proposal for standardization in this area and enables the integration of environmental aspects into product design practice.

RmW can also be applied at the stage of product decommissioning. Using the model, we can analyze the product in terms of various methods of its disassembly, determining the most beneficial one from the point of view of the profit obtained from the disassembled product. The potential recipient of this type of model is therefore also the entity performing the disassembly of the product (recycler). This is a situation in which data (as RmW) generated at an earlier stage of the product life cycle (design) are used during the decommissioning of the product, i.e. at a much later stage in its life cycle. This aspect (data exchange between stages of the product life cycle) is part of the Industry 4.0 concept, i.e. the exchange of data between entities cooperating during product development.

References

1. Dekoninck, E.A., Domingo, L., O'Hare, J.A., Pigosso, D.C.A., Reyes, T., Troussier, N.: Defining the challenges for ecodesign implementation in companies: development and consolidation of a framework. J. Clean. Prod. **135**, 410–425 (2016)
2. Kara, S., Ibbotson, S., Kayis, B.: Sustainable product development in practice: an international survey. J. Manuf. Technol. Manag. **25**(6), 848–872 (2014)
3. Anuszewska, I.: Sustainable production patterns (SPP) in enterprises – a proposal of systemic solutions supporting the implementation of SPP in SMEs, a report on the qualitative part of the study (in Polish: Wzorce zrównoważonej produkcji (WZP) w działalności przedsiębiorstw – propozycja rozwiązań systemowych wspierających wdrażanie WZP w MSP, raport z części jakościowej badania). EFS POKL (2011)
4. Rodrigues, V.P., Pigosso, D.C.A., McAloone, T.C.: Measuring the implementation of ecodesign management practices: a review and consolidation of process-oriented performance indicators. J. Clean. Prod. **156**, 293–309 (2017)

5. Rojek, I., Dostatni, E., Hamrol, A.: Ecodesign of technological processes with the use of decision trees method. In: Pérez García, H., et al. (eds.) Advances in Intelligent Systems and Computing, vol. 649, pp. 318–327. Springer, Cham (2018)
6. McAloone, T., Bey, N.: Environmental Improvement through Product Development. A guide. Danish Environmental Protection Agency, Copenhagen (2009)
7. Dostatni, E., Diakun, J., Grajewski, D., Wichniarek, R., Karwasz, A.: Multi-agent system to support decision-making process in design for recycling. Soft. Comput. **20**, 4347–4361 (2016)
8. Pigosso, D.C.A., McAloone, T.C., Rozenfeld, H.: Characterization of the state-of-the-art and identification of main trends for ecodesign tools and methods: classifying three decades of research and implementation. J. Indian Inst. Sci. **94**(4), 405–427 (2015)
9. Pigosso, D.C.A., Rozenfeld, H., McAloone, T.C.: Ecodesign maturity model: a management framework to support ecodesign implementation into manufacturing companies. J. Clean. Prod. **59**, 160–173 (2013)
10. Bovea, M.D., Gallardo, A.: The influence of impact assessment methods on materials selection for eco-design. Mater. Des. **27**, 209–215 (2006)
11. Köhler, A.R.: Challenges for eco-design of emerging technologies: the case of electronic textiles. Mater. Des. **51**, 51–60 (2013)
12. Kiurski, J.S., Maric, B.B., Oros, I.B., Keci, V.S.: The ecodesign practice in Serbian printing industry. J. Clean. Prod. **149**, 1200–1209 (2017)
13. Stadnicka, D., Zielecki, W., Sęp, J.: Industry 4.0 – assessment of implementation possibilities on the example of a selected enterprise, (In Polish: Koncepcja Przemysł 4.0 - ocena możliwości wdrożenia na przykładzie wybranego przedsiębiorstwa). In: Knosala, R. (ed.) Innovations in Management and Production Engineering (in Polish: Innowacje w zarządzaniu i inżynierii produkcji), vol. 1, pp. 472–483. Oficyna Wydawnicza Polskiego Towarzystwa Zarządzania Produkcją, Opole (2017)
14. Bembenek, B.: Industry clusters 4.0 within sustainable knowledge-based economy (Klastry Przemysłu 4.0 w zrównoważonej gospodarce opartej na wiedzy). Research Papers of Wrocław University of Economics, vol. 491, pp. 31–44 (2017)
15. Pejs, K., Patalas-Maliszewska, J.: Manufacturing enterprise module in 4.0 formula (in Polish: Model przedsiębiorstwa produkcyjnego w formule 4.0). In: Patalas-Maliszewska, J., Sasiadek, M., Jakubowski, J. (eds.) Production Engineering. Process Quality and Effectiveness (in Polish: Inżynieria produkcji. Jakość i efektywność procesów), vol. 11, pp. 53–64. Instytut Informatyki i Zarządzania Produkcją, Zielona Góra (2016)
16. Pluciński, M., Mularczyk, K.: Industry 4.0 in Polish Manufacturing Companies - Chances and Risks (in Polish: Przemysł 4.0 w polskich przedsiębiorstwach produkcyjnych – szanse i zagrożenia). In: Mazurek-Kucharska, B., Dębski, M. (eds.) Management of a small and medium-sized enterprise in Poland. Innovative strategies, tools and implementations (in Polish: Zarządzanie małym i średnim przedsiębiorstwem w Polsce), Przedsiębiorczość i Zarządzanie, vol. 17, pp. 311–314. SAN, Łódź-Warsaw (2016)
17. Heng, S.: Industrie 4.0 – Upgrade des Industriestandorts Deutschland steht bevor. In: Slomka, L. (ed.) Aktuelle Themen, pp. 2–16. Deutsche Bank Research, Frankfurt am Main (2014)
18. Industrie 4.0 – Volkswirtschaftliches Potenzial für Deutschland. Studie, BITKOM, Berlin (2014)
19. Industrie 4.0. Chancen und Herausforderungen der vierten industriellen Revolution. Studie, PWC. https://www.strategyand.pwc.com/media/file/Industrie-4-0.pdf. Accessed 10 Apr 2018

20. Wolter, M.I., Mönnig, A., Hummel, M., Schneemann, Ch., Weber, E., Zika, G., Helmrich, R., Maier, T., Neuber-Pohl, C.: Industrie 4.0 und die Folgen für Arbeitsmarkt und Wirtschaft. IAB Forschungsbericht, Aktuelle Ergebnisse aus der Projektarbeit des Instituts für Arbeitsmarkt- und Berufsforschung. http://doku.iab.de/forschungsbericht/2015/fb0815.pdf. Accessed 10 Apr 2018
21. Deutschland droht die Zukunft zu verschlafen. http://www.welt.de/135151615. Accessed 10 Apr 2018
22. Kiraga, K.: Industry 4.0 - the fourth industrial revolution according to Festo (in Polish: Przemysł 4.0 – czwarta rewolucja przemysłowa według Festo). Autobusy, vol. 12, pp. 1603–1605 (2016)
23. Roblek, V., Mesko, M., Krapez, A.: A complex view of industry 4.0. Sage Open **6**(2), 1–11 (2016)
24. Kuczmaszewski, J.: HR training for "Industry 4" (in Polish: Kształcenie kadr dla „Przemysł 4"). In: Knosala, R. (ed.) Conference materials Innowacje w Zarządzaniu i Inżynierii Produkcji, Zakopane 2018, pp. 149–154. Oficyna Wydawnicza Polskiego Towarzystwa Zarządzania Produkcją (2018)
25. Zawadzki, P., Żywicki, K.: Smart product design and production control for effective mass customization in the industry 4.0 concept. Manag. Prod. Eng. Rev. **7**(3), 105–112 (2016)
26. Rojek, I.: Technological process planning by the use of neural networks. Artif. Intell. Eng. Des. Anal. Manuf. **31**(1), 1–15 (2017)
27. Kokot, W., Kolenda, P.: What is the Internet of Things (in Polish: Czym jest Internet Rzeczy). In: Internet Rzeczy w Polsce, IAB Polska. https://iab.org.pl/wp-content/uploads/2015/09/Raport-Internet-Rzeczy-w-Polsce.pdf. Accessed 10 Apr 2018
28. Ziernicka-Wojtaszek, A.: Industrial revolution 4.0 and ecology and meteorology - inspirations and perspectives. Technical review. (in Polish: Rewolucja przemysłowa 4.0 a ekologia i meteorologia - inspiracje i perspektywy. Przegląd Techniczny). Gazeta inżynierska, vol. 25, pp. 14–15 (2017)
29. Piątek, Z.: What is Industry 4.0? - part 1 (in Polish: Czym jest Przemysł 4.0? – część 1). http://przemysl-40.pl/index.php/2017/03/22/czym-jest-przemysl-4-0/. Accessed 10 Apr 2018
30. Brenner, W., Zarnekow, R., Wittig, H.: Intelligente Softwareagenten Grundlagen und Anwendungen. Springer, Heidelberg (1998)
31. Wooldridge, M.J.: An Introduction to Multiagent System. Wiley, Hoboken (2002)
32. Dostatni, E., Diakun, J., Grajewski, D., Wichniarek, R., Karwasz, A.: Functionality assessment of ecodesign support system. Manag. Prod. Eng. Rev. **6**(1), 10–15 (2015)
33. Dostatni, E., Diakun, J., Grajewski, D., Wichniarek, R., Karwasz, A.: Multi-agent system to support decision-making process in ecodesign. In: Herrero, Á., Sedano, J., Baruque, B., Quintián, H., Corchado, E. (eds.) 10th International Conference on Soft Computing Models in Industrial and Environmental Applications. Advances in Intelligent Systems and Computing, vol. 368, pp. 463–474. Springer, Cham (2015)
34. Dostatni, E., Diakun, J., Wichniarek, R., Karwasz, A., Grajewski, D.: Product variants recycling cost estimation with the use of multi-agent support system. In: Hamrol, A., Ciszak, O., Legutko, S., Jurczyk, M. (eds.) Advances in Manufacturing. Lecture Notes in Mechanical Engineering, vol. 48, pp. 311–320. Springer, Cham (2018)
35. Rachuri, S., Subrahmanian, E., Bouras, A., Fenves, S.J., Foufou, S., Sriram, R.D.: Information sharing and exchange in the context of product lifecycle management: role of standards. Comput. Aided Des. **40**(7), 789–800 (2008)

Efficiency of Automatic Design in the Production Preparation Process for an Intelligent Factory

Przemysław Zawadzki(✉), Krzysztof Żywicki, Damian Grajewski, and Filip Górski

Chair of Management and Production Engineering,
Poznan University of Technology, Piotrowo 3 Street, 60-965 Poznan, Poland
{Przemyslaw.Zawadzki,Krzysztof.Zywicki,
Damian.Grajewski,Filip.Gorski}@put.poznan.pl

Abstract. This article presents the research on the efficiency of performance of automatic design of variant products system, including the process of product variant configuration, preparation of the 3D model in CAD system and the preparation of the assembly instructions for the manufacturing positions. The variants of the described product take the form of various combinations of individual assembly parts combined together and constitute an example of customized products, and the process of their manufacture (assembly) is the subject of the research carried out in the SmartFactory Laboratory at the Poznan University of Technology.

Keywords: Design automation · Smart factory · Industry 4.0

1 Introduction

An intelligent factory is the idea of a production system characterized by a high level of process automation and production flexibility, understood as the ability to react quickly to individual customer needs [1–3]. Intelligent factories are somehow the result of implementing various assumptions and solutions known as Industry 4.0 [4]. Therefore, it can be assumed that a smart factory will be a way to implement the assumptions of mass customisation, which is defined as production strategy designed to adapt flexibly to individual customer needs at a mass scale [5]. In the future intelligent factories will be autonomous units capable of planning, organizing and even controlling highly flexible production [6, 7]. However, despite the fact that the idea of Industry 4.0 will be crucial in implementing the assumptions of the mass customization strategy, efficient organization of production processes may prove insufficient in the effective implementation of the idea of mass customization [8, 9]. The basis for the operation of a smart factory of the future should therefore also be the efficient design of products that meet the individual requirements of customers [10, 11].

The development of tools available in CAx systems in recent years and their integration with PLM solutions has resulted in more and more manufacturing companies willing to implement solutions that accelerate and integrate the implementation

© Springer Nature Switzerland AG 2019
A. Burduk et al. (Eds.): ISPEM 2018, AISC 835, pp. 543–552, 2019.
https://doi.org/10.1007/978-3-319-97490-3_52

of complex engineering tasks in the design process [12, 13]. A special example may be the combination of the parametric CAD modelling technique with the knowledge-based design (KBE) approach, which thanks to the development of the so-called generative CAD models allows the implementation of partial or even complete automation of the design process [10]. Solutions built in this way may include a visual configuration of the structural and functional features of the product [14], as well as the preparation of technical and technological documentation [15].

In KBE systems, the recognized knowledge and gathered experience from experts of what, how and when something should be done, is processed by the computer system, allowing it to be easily used in new projects [16]. The formal description of the rules used by the constructors affects the standardization of the design process and reduces its time [17, 18]. The process of building KBE solutions is usually a task developed for the specific case [19]. Although some methodological standards are known (e.g. MOKA, KADM, MDAVP), as the literature analysis shows, they are rarely used in practice [20–22]. The scope of configuration of the designed product, i.e. what features and to what extent they may be subject to changes, depend on the expectations of the recipient, but nevertheless it is estimated that approx. 80% of the design time concerns routine tasks [19]. Their acceleration can therefore significantly affect the optimization of the entire life cycle of the product and savings.

2 SmartFactory Laboratory at PUT

The production system of the SmartFactory laboratory at Poznan University of Technology (PUT) was built to represent real production (assembly) processes, aimed at realization of customized orders, according to the Industry 4.0 concept.

The production system in the laboratory consists of the following elements (see Fig. 1):

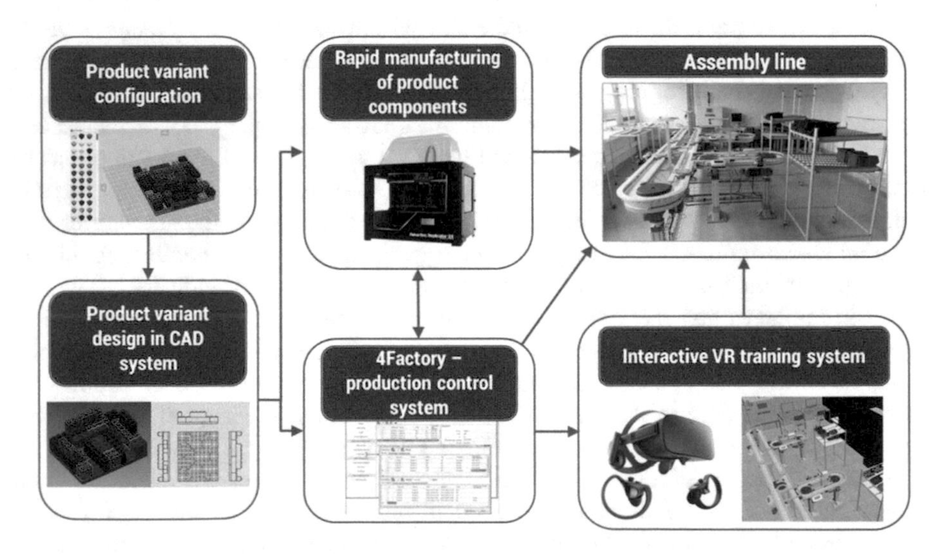

Fig. 1. Structure of the SmartFactory system

- application for product configuration,
- CAD system for product documentation,
- RM section for parts and tools manufacturing,
- 4Factory program for production control,
- assembly automatic line,
- VR training system.

The produced blocks are characterized by different colours, sizes and configurations of parts (Lego blocks) (Fig. 2).

Fig. 2. Example of product variant

Product design is carried out using an application that is an independent, commercial program that allows for the visual selection of components of the variant being built and the determination of their relative position. It is easy and intuitive to use, which is why this solution can be run even by inexperienced persons, which in turn allows the implementation of mass customization assumptions regarding customer engagement in the process of developing a new product. Unfortunately, it is not integrated with the CAD program in any way, therefore in the next step the CAD model of the variant is built from scratch and on its basis the technical documentation of the product variant is prepared. The main element is the automatic assembly line. It consists of four transporting loops, by which there are production stands. Transport of products is realized by small pallets, which can be directed to any production stand. Their automated identification in the system is realized with use of the RFID technology. The whole production system is controlled by an author-made computer program, named 4Factory.

To maintain proper efficiency of this production process, it is necessary to properly prepare operators, by frequent trainings. Performing them in real conditions is often related to stopping the production, which in turn may cause delays in realizing orders of the clients. This means that training in real conditions is economically unjustified. Using VR systems for training purposes makes it possible to study reactions and capability of realization of tasks by the operators without pausing the production, thus being significantly cheaper and enabling more flexible training scenarios, at cost of lower realism and lack of physical interaction with the production line. It was decided to build a VR training system for the SmartFactory line, in order to improve effectiveness of work of operators. The next chapters present the process of building this solution

A production process realized in the SmartFactory laboratory consists in assembly of parts at subsequent production stands. The parts, subassemblies and ready products are transported through an automated line, using small pallets. Production stands are equipped with flow racks, that allow storing containers with parts and subassemblies for assembly of customized products. The racks are additionally equipped with heads reading RFIDs, allowing their identification.

Number of assembly activities, as well as activities related to taking parts from the material racks is obviously related to the scope described in the manufacturing instruction (see Fig. 3).

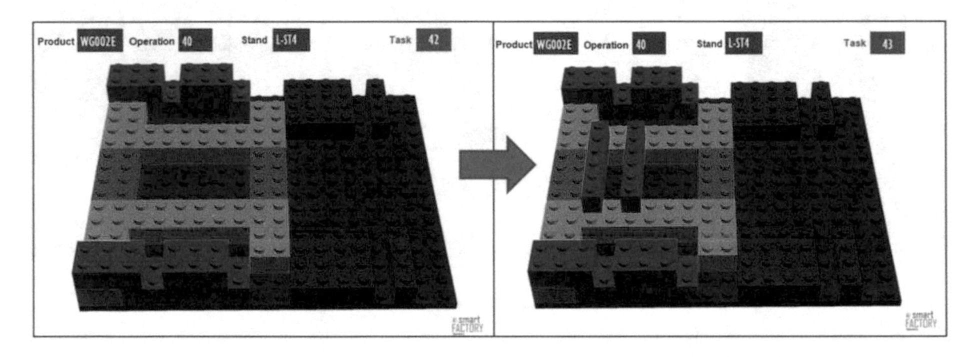

Fig. 3. Example of manufacturing (assembly) instruction

3 Aim of the Work

The aim of the research was to build and validate a solution that allows the automation of design process of product variants for the needs of the Smart Factory Laboratory of the Poznan University of Technology. The following assumptions of the new solution were adopted:

- it should support (facilitate) the configuration process of the product variant and ensure data exchange with the CAD system,
- design of the product variant is prepared in the CAD program - Autodesk Inventor, in an automatic way,
- results of the solution operation should facilitate the preparation of production documentation (assembly in that case).

The developed solution was subject to validation by carrying out the design process (configuration) for 3 selected product variants. The research involved 5 constructors with experience in using the Autodesk Inventor program and 3 people without such experience. The duration of the design process and the ease of use of the configuration process have been checked.

4 Automated Design System for SmartFactory Laboratory

4.1 Description of the CAD Solution

It turned out that the simplest solution, ensuring the integration of configuration tasks and automation of the design process was to prepare the interface for configuring the product variant directly in the CAD environment. The interface was supposed to be so simple to use that the choice of the components of the product being built and the determination of their relative position did not require the user to know the CAD program. To ensure this, it was decided to use the special features of the Autodesk Inventor iLogic package. And, to accelerate and enable the generation of 3D models of individual parts, a special generative CAD model was prepared.

4.2 Automatic Design System Preparation

For the construction of model parts, used in the product variants, one generative CAD model was developed, which in addition to the parameters describing the shape (length, width, height and colour) has been enriched with a record of the relevant relationships between these parameters. This way, the generative CAD model allowed for quick preparation and saving of the part model with any combination of features (see Fig. 4), even one that could be considered as a non-standard combination in the entire future production cycle (a part never used before in the production process).

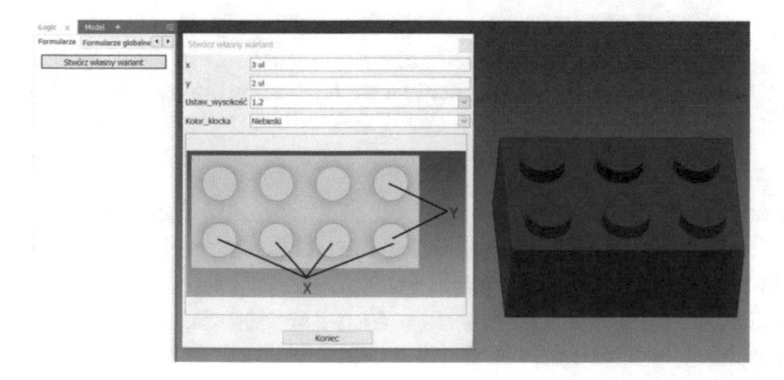

Fig. 4. Form with parameters of the block (left) and generative CAD model of the block (right)

Next, using the tool named iPart, additional parameters of the part were defined to ensure the ability to assign (and later identify) the level at which the block should be located (in the context of the whole product). This is related to instruction to be prepared later and order of the assembly sequence. By using the iPart feature, it was also possible to create a library of used parts dedicated to the design process, so that the future user could more quickly use those parts that are considered standard. Next, using the iMate feature, the ways of joining the blocks to each other were determined based on assigning geometrical bonds: the indication of the contact plane of the combined blocks and two axes determining their relative position in space. After completing the model building, 2D documentation is generated automatically.

4.3 Integration of Configuration and Design Process

Maintenance of the developed solution for the configuration and design of product variants takes place in the area of the construction of the assembly model in Autodesk Inventor, using the interface of the prepared library (see Fig. 5). The user has the freedom in the order of defining the components of the product variant, but usually the work starts with placing the base model and configuring its parameters. Then, the process of adding and placing the block begins (first on the base, then on any block), by indicating the contact plane and two axes fixing the position of the block being added. The user has the option of adding the so-called standard block (sorted by the colour and shape) or a "new" brick (not used previously in a system), defining its features using a special form.

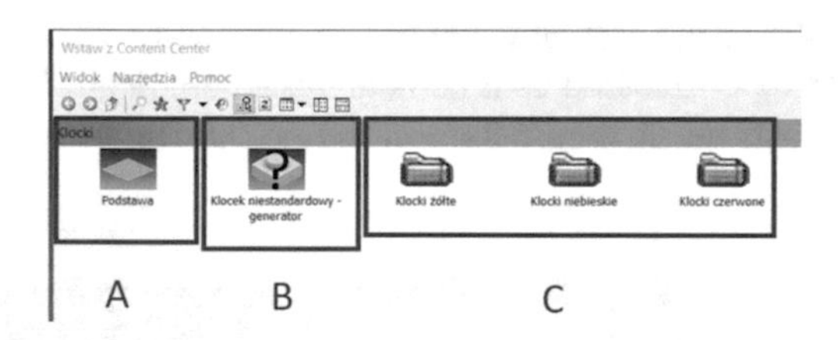

Fig. 5. Main user interface window - Content Center (Inventor library), A - basis block, B - user block generator, C - standard blocks

The advantage of the developed solution is that during the process of configuring the product variant and adding more blocks, the 3D CAD model of the product variant is built right away. After the construction of the model has been finished, the relevant 2D documentation is also ready (see Fig. 6). A special function was also prepared, allowing for hiding selected layers of the product variant, in order to make it easier to prepare assembly instructions.

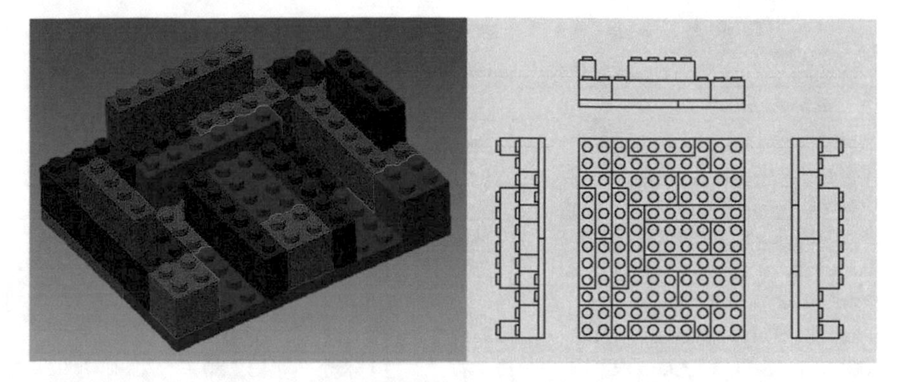

Fig. 6. Example of 3D CAD model of the product variant (left) and 2D documentation (right)

5 Efficiency Evaluation of Automatic Design System

In accordance with the assumptions, the research on the effectiveness of the developed solution was carried out with the participation of two testing groups: the first being the constructors, who have experience in using the Autodesk Inventor program, the other being people who do not have such skills. The research was to show whether the developed automation system reduces the time of the design process and whether the configuration process is user friendly. For this reason, a user manual for the configuration interface was prepared, which the testers were familiarized with.

Three variants of the product were selected for carrying out the system test (Fig. 7) In the case of tests with the participation of the first group, the time needed to develop the 3D model in the standard way was measured first, i.e. by means of standard modeling in Autodesk Inventor. Then, each tester repeated the design process of selected product variants, but using an automatic solution. The results of the implementation times of the design process of selected product variants are presented in Table 1.

Fig. 7. Three variants of the product

Table 1. Comparison of design method for first test group [min]

	Designer 1	Designer 2	Designer 3	Designer 4	Designer 5
Product variant 1					
Standard design process	12:19	14:28	13:55	11:52	13:34
Automatic design system	06:30	08:03	07:42	07:13	06:55
Product variant 2					
Standard design process	21:02	23:10	24:44	20:14	23:04
Automatic design system	11:52	13:02	15:17	12:22	13:35
Product variant 3					
Standard design process	25:02	27:11	26:12	23:16	24:35
Automatic design system	13:12	15:27	13:56	13:30	14:27

In the case of tests for the other group, the time of implementation of the product variant configuration was also measured using the developed automation system, and each of the tested persons made a subjective assessment of ease of use of the developed solution (in the scope of interface support) using 1–5 scale, where 1 means difficult in use, and 5 means easy. The results of the tests for the second group are shown in Table 2.

Table 2. Results of second test group

	User 1	User 2	User 3
Product variant 1			
Automatic design system [min]	10:55	11:53	10:26
Product variant 2			
Automatic design system [min]	13:28	14:23	13:48
Product variant 3			
Automatic design system [min]	14:25	15:55	14:12
User friendly (1–5 scale)	4	3	4

As a result of the conducted research, it has been shown that the time of the product design process with the use of an automatic system has been reduced by 43% compared to the traditional process. The conducted research also indicated further activities related to the development of the presented IT solution.

6 Conclusions

The idea of a smart factory enables the implementation of individual customer requirements while rationally using production resources. Planning the execution of orders must be characterized by close integration with regards to exchange of design and technological requirements. Only the close integration of these areas guarantees synchronization of production flow and thus can contribute to the timely delivery of customer orders.

The presented IT system for configuring products taking into account individual customer requirements is a part of the assumptions for the functioning of the production system in accordance with the Industry 4.0 concept. It has been prepared for the needs of the operation of the science-teaching laboratory SmartFactory, which aims to develop solutions for intelligent control of the production system. The system allows for the automation of activities related to the design of products which translates into reduced time of preparation and start of production. The use of a non-standard component generator enables the design of an unlimited number of product variants, and the user-friendly interface allows using it by people without much experience.

Acknowledgements. The presented results are derived from a scientific statutory research conducted by Chair of Management and Production Engineering, Faculty of Mechanical Engineering and Management, Poznan University of Technology, Poland, Poland, supported by the Polish Ministry of Science and Higher Education from the financial means in 2018 (02/23/DSPB/7674).

References

1. Brettel, M., Friederichsen, N., Keller, M., Rosenberg, M.: How virtualization, decentralization and network building change the manufacturing landscape: an Industry 4.0 perspective. Int. J. Mech. Ind. Sci. Eng. **8**(1), 37–44 (2014)
2. Shrouf, F., Ordieres, J., Miragliotta, G.: Smart factories in Industry 4.0: a review of the concept and of energy management approached in production based on the Internet of Things paradigm. In: IEEE International Conference on Industrial Engineering and Engineering Management (IEEM), pp. 697–701. IEEE (2014)
3. Kemény, Z., Beregi, R.J., Erdős, G., Nacsa, J.: The MTA SZTAKI smart factory: platform for research and project-oriented skill development in higher education. Procedia CIRP **54**, 53–58 (2016)
4. Gorecky, D., Schmitt, M., Loskyll, M., Zühlke, D.: Human-machine-interaction in the Industry 4.0 era. In: 12th IEEE International Conference on Industrial Informatics (INDIN), pp. 289–294. IEEE (2014)
5. Zawadzki, P., Żywicki, K.: Smart product design and production control for effective mass customization in the Industry 4.0 concept. Manag. Prod. Eng. Rev. **7**(3), 105–112 (2016)
6. Lee, J., Bagheri, B., Kao, H.A.: A cyber-physical systems architecture for Industry 4.0-based manufacturing systems. Manuf. Lett. **3**, 18–23 (2015)
7. Hu, F.: Cyber-Physical Systems: Integrated Computing and Engineering Design. CRC Press, Boca Raton (2013)
8. Żywicki, K., Zawadzki, P.: Fulfilling individual requirements of customers in smart factory model. In: Hamrol, A., Ciszak, O., Legutko, S., Jurczyk, M. (eds.) Advances in Manufacturing. Lecture Notes in Mechanical Engineering, pp. 185–194. Springer, Cham (2018)
9. Trojanowska, J., Żywicki, K., Pająk, E.: Influence of selected methods of production flow control on environment. In: Golinska, P., Fertsch, M., MarxGomez, J. (eds.) Information Technologies in Environmental Engineering, pp. 695–705. Springer, Heidelberg (2011). https://doi.org/10.1007/978-3-64219536-5_54

10. Górski, F., Zawadzki, P., Hamrol, A.: Knowledge based engineering as a condition of effective mass production of configurable products by design automation. J. Mach. Eng. **16** (4), 5–30 (2016)
11. Żywicki, K., Zawadzki, P., Hamrol, A.: Preparation and production control in smart factory model. In: Rocha, Á., Correia, A., Adeli, H., Reis, L., Costanzo, S. (eds.) Recent Advances in Information Systems and Technologies. WorldCIST 2017. Advances in Intelligent Systems and Computing, vol. 571, pp. 519–527. Springer, Cham (2017)
12. Dostatni, E., Grajewski, D., Diakun, J., Wichniarek, R., Buń, P., Górski, F., Karwasz, A.: Improving the skills and knowledge of future designers in the field of ecodesign using virtual reality technologies. In: International Conference Virtual and Augmented Reality in Education, Procedia Computer Science, vol. 75, pp. 348–358 (2015). ISSN 1877-0509
13. Dostatni, E., Diakun, J., Grajewski, D., Wichniarek, R., Karwasz, A.: Functionality assessment of ecodesign support system. Manag. Prod. Eng. Rev. **6**(1), 10–15 (2015)
14. Górski, F., Buń, P., Wichanirek, R., Zawadzki, P., Hamrol, A.: Immersive city bus configuration system for marketing and sales education. Procedia Comput. Sci. **75**, 137–146 (2015)
15. Górski, F., Hamrol, A., Kowalski, M., Paszkiewicz, R., Zawadzki, P.: An automatic system for 3D models and technology process design. Trans. FAMENA **35**(2), 69–78 (2011)
16. Reddy, E.J., Sridhar, C.N.V., Rangadu, V.P.: Knowledge based engineering: notion, approaches and future trends. Apm. J. Intell. Syst. **5**(1), 1–17 (2015)
17. Elgh, F., Cederfeldt, M.: Documentation and management of product knowledge in a system for automated variant design: a case study. In: New World Situation: New Directions in Concurrent Engineering, pp. 237–245. Springer, London (2010)
18. Sandberg, M., Gerth, R., Viklund, E.: A design automation development process for building and bridge design. In: Proceedings of the 33rd CIB W78 Conference 2016 (2016)
19. Verhagen, W.J.C., Bermell-Garcia, P., Van Dijk, R.E.C., Curran, R.: A critical review of knowledge-based engineering: an identification of research challenges. Adv. Eng. Inform. **26** (1), 5–15 (2012)
20. Skarka, W.: Application of MOKA methodology in generative model creation using CATIA. Eng. Appl. Artif. Intell. **20**(5), 677–690 (2007)
21. Stokes, M.: Managing Engineering Knowledge; MOKA: Methodology for Knowledge Based Engineering Applications. Professional Engineering Publishing, London (2001)
22. Zawadzki, P.: Methodology of KBE system development for automated design of multivariant products. In: Hamrol, A., Ciszak, O., Legutko, S., Jurczyk, M. (eds.) Advances in Manufacturing. Lecture Notes in Mechanical Engineering, pp. 239–248. Springer, Cham (2018)

Low-Cost 3D Printing in Innovative VR Training and Prototyping Solutions

Paweł Buń(✉), Filip Górski, Radosław Wichniarek, Wiesław Kuczko, and Magdalena Żukowska

Chair of Management and Production Engineering,
Faculty of Mechanical Engineering and Management,
Poznan University of Technology, 60-965 Poznan, Poland
pawel.bun@put.poznan.pl

Abstract. The paper presents possibility of using a low-cost Fused Deposition Modelling process, realized by MakerBot Replicator 2X machine, in comparison with a professional one – realized by Dimension BST 1200 machine – to build tooling for an innovative training simulation, in accordance with Industry 4.0 guidelines. A physical object was manufactured and then tracked to manipulate in a virtual environment – it was a manual head used during the procedure. The objects were manufactured on two different machines out of ABS material, 3D scanned for accuracy testing and finally possibilities of their use in a VR system were evaluated.

Keywords: 3D printing · Virtual Reality · Virtual training

1 Introduction

Work of a modern engineer requires lots of decision-making processes (Dostatni et al. 2015). One of modern, innovative technologies aiding these processes is the Virtual Reality (VR). It uses digitally built worlds to create a sense of presence at a user and help realize certain tasks and make decisions faster and more effectively. As the technology develops and hardware prices drop, VR becomes more and more widely applied in entertainment, professional training (Grajewski et al. 2015) (especially regarding medicine (Hamrol et al. 2013)) and in some industry branches (Wu et al. 2013), among other things.

Both historically and in the present day, VR is successfully used in engineering – mostly for extended, CAD-based virtual design, as well as industrial training. VR, as well as Augmented Reality (AR), plays an important role in Factories of the Future, within the Industry 4.0 concept (Zawadzki et al. 2016; Żywicki et al. 2018). It should be remembered that the concept of Industry 4.0 has a place for a human operator. The concept of the so called Operator 4.0, introduced in (Romero et al. 2016), assumes supporting physical and mental work performed by the operator with various techniques - Virtual Reality among others. Within the concept of Operator 4.0, several functions can be distinguished, among which the augmented operator and virtual operator are most important for the considerations in this paper. The operator in a

© Springer Nature Switzerland AG 2019
A. Burduk et al. (Eds.): ISPEM 2018, AISC 835, pp. 553–562, 2019.
https://doi.org/10.1007/978-3-319-97490-3_53

company realizing the Industry 4.0 concept utilizes VR and AR for his training, that is why it is important to study these two techniques in this context (Żywicki et al. 2018).

Early VR-focused studies (Slater et al. 1995) prove that going into artificially prepared reality allows subconscious obtaining of competences in many fields. The state of being inside VR is known as the immersion. This phenomenon is used, among other things, for curing phobias, such as arachnophobia (Parsons et al. 2008), as well as in training. To obtain immersion, it is important to engage as many senses of a user as possible in interaction with a virtual environment. It is possible to achieve in many ways, among other things by representation of movement of real objects, manipulated by a user, in an application. The more real, physical object is similar to its virtual representation (both in terms of shape and movement accuracy), the easier is to achieve full immersion state, which translates into increase of training effectiveness (Bowman et al. 2007). The physical objects which can be tracked for manipulation in virtual environments are often manufactured using 3D printing techniques (Górski et al. 2013). As in recent years many low-cost 3D printers were introduced to the market (usually working in Fused Deposition Modeling – FDM – technology), it has become more and more possible to produce additional objects tracked inside a virtual simulation due to greatly reduced costs. The paper addresses the problem of effective manufacturing of objects aiding VR simulations for training and prototyping using low-cost 3D printers to manufacture interactable objects out of thermoplastics.

2 Materials and Methods

2.1 Case and Problem Definition – VR Application for Training and Prototyping

A training procedure selected for tests of the presented approach was a medical procedure - ultrasound examination. It was selected for the initial studies on manual operation in VR, as the procedure requires a human operator to perform precise movements and be well-trained in order to work effective – just like in case of Operator 4.0. The aim was to make the examination simulation realistic and inexpensive at the same time. Therefore, it was decided to utilize a hybrid prototyping approach (Górski et al. 2013), expanding a virtual prototype with a physical device – a representation of manual head for the ultrasound examination (in real life operated by a doctor who is performing the procedure), along with a physical phantom of a patient (in form of a mannequin). The device was manufactured using two different 3D printers, a low-cost 3D printer – MakerBot Replicator 2X, and a professional 3D printer – Dimension BST 1200. The aim of the studies presented in this paper was to compare both processes in terms of accuracy of obtained products and evaluate empirically if the device manufactured using a low-cost device can be successfully used in professional training and prototyping of manual tools and procedures using Virtual Reality.

2.2 Manipulation Idea – A Tracking System

A tracking system is a device which allows real-time measurement of position and/or orientation of a given object. Usually tracked objects are special markers. There are certain devices, which allow tracking objects of any shape, by placing patterns of markers on them. This concept is utilized in the described ultrasound examination procedure, where a special object – a head for manipulation – is covered in markers recognized by a tracking system. The system used in the presented studies is PST-55 – it can track objects distant between 40 cm and several meters from the device, by means of infrared light detection. The device sends a wave of IR light, which is reflected back by the retroactive markers. If their pattern is recognized as forming a previously recorded object (shape), its position is calculated by analyzing image from two cameras built inside the device. It is important for the device to see at least 4 markers simultaneously, so the more markers on a surface of a given object, the better (Buń et al. 2015). Figure 1 presents a general concept of markers and their recognition by the software.

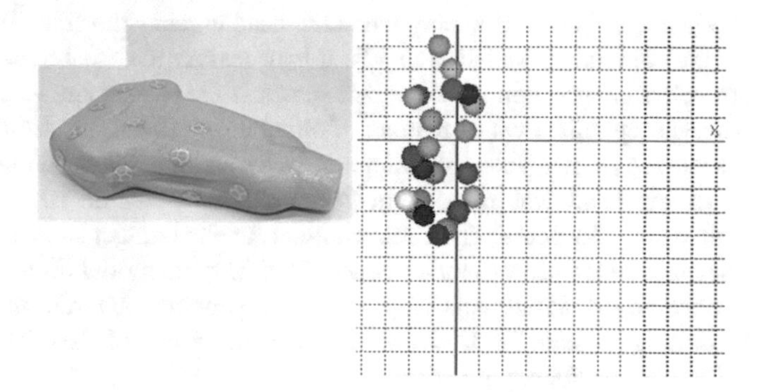

Fig. 1. Markers of the PST-55 system and their recognition in the system software (Buń et al. 2015)

Initial shape of the ultrasound examination device (see Fig. 1) was not properly recognized in the preliminary studies by the authors. That is why it was decided to change its geometry and expand it with additional elements for larger area to put markers into.

2.3 Interactable Object Preparation

The studied objects were models of heads used for ultrasound examination of human abdominal cavity. These models were prepared on the basis of commercial ultrasound examination systems. To make it more realistic, the physical objects were 3D scanned using the ATOS I optical scanner by the GOM company, with measurement field of 125 × 125 mm. Three-dimensional scanning ensures rapid time of measurement and its result is a point cloud (Gessner 2012). The initial head was re-created from the 3D scan and the modified models were prepared directly in the CATIA v5 CAD system, on the basis of 3D scans. The two models (Fig. 2) are based on real ultrasound devices, applied to examine different internal body areas (see Table 1 for details).

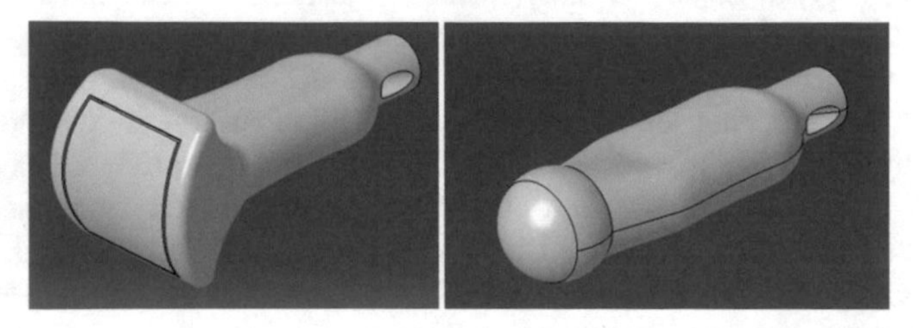

Fig. 2. Models of ultrasound examination heads with visible exit of the cable channel

To use the re-created head in a VR simulation, a contact sensor was built in the head tip and the models were modified to allow its assembly. The contact sensor usually is non-present in real ultrasound examination head. In the real examination, the image of the patient's body internals is visible only when the head touches the skin. To obtain the same effect for the computer simulation in VR, it is necessary to detect a moment when user touches the physical phantom. The contact sensor is the easiest and least expensive method to ensure acceptable level of realism of simulation. The signal from the sensor can be easily accessed in available software, e.g. by emulating a mouse click.

Main part of the head was divided in 2 parts, which can be assembled using specially designed snap fasteners. The cable channel for the contact sensor was shaped in a way to not cause discomfort to the operator. Dividing the model directly along the channel axis allowed to avoid necessity of using support material in the Fused Deposition Modelling process. Figure 2 presents location of exit of the cable channel in both shapes of heads for the examination.

2.4 Manufacturing of Ultrasound Heads

In the previous work by the authors, modified head models manufactured by 3D Printing process were used. The 3D Printing was realized using ZPrinter 310 machine (3D Systems company), out of a powder based on gypsum, joined by a binder based on methyl alcohol (old head visible in Fig. 1).

The new models of examination heads were manufactured using the Fused Deposition Modelling technology. The FDM technology consists in linear deposition of plasticized thermoplastic material, extruded by a nozzle of a small diameter, by a special head. The extrusion head can move in two axes (XY) and the table (on which the model is made) can move in the vertical (Z) axis. After each layer is made, the table goes down in the Z axis leaving a space equal to desired layer thickness between itself and a nozzle. To ensure support of geometry of manufactured object, if subsequent layer contour is significantly going beyond the previous layer, it is necessary to build special support structures. They are also made of thermoplastic material, but with different mechanical properties, which makes it possible to mechanically or chemically separate them from the actual object after the layer deposition is finished (Chua 2010).

Table 1. Comparison of manufacturing parameters for heads produced using BST-1200 and MakerBot Replicator 2X machines

Machine/parameter	BST 1200	Makerbot Replicator 2X
Parameters common for both heads		
Infill	Sparse (linear)	Sparse (hexagonal)
Layer thickness	0,254 mm	0,25 mm
Model material	Dimension P400 black (ABS)	Makerbot True Black ABS
Support material	Dimension P400-RP (ABS)	Makerbot Dissolvable Filament (HIPS)
Temp.	Model & support: 300 °C	
Chamber: 74 °C	Model: 230 °C	
Support: 250 °C		
Table: 110 °C		
Device preparation time	1 min (cleaning and assembly of a tray)	7 min (table cleaning, kapton tape deposition, covering tape with ABS dissolved in acetone for better sticking)
Wide head GE RSP6-16-RS 3D/4D Linear Probe		
Applications: Small Parts, Vascular, Pediatrics, Ortho		
Production time	6 h 20 min	5 h
Model material	75,6 g	70 g (purge walls used)
Support material	27,8 g	24 g (purge walls used)
Support removal time	12 min	2 min
Approx. cost	700 PLN	250 PLN
Spherical head GE RNA5-9		
Applications: Neonatal, Pediatrics		
Production time	3 h 48 min	3 h 21 min
Model material	47,46 g	47,4 g (purge walls used)
Support material	19,74 g	16 g (purge walls used)
Support removal time	7 min	1 min
Approx. cost	450 PLN	175 PLN

The head models were both manufactured using the Dimension BST 1200 – professional machine (worth approx. $35 000) – and the MakerBot Replicator 2X – a low-cost machine (worth approx. $2 500). Manufacturing parameters are presented in Table 1.

After manufacturing the heads, it was necessary to remove supports (Fig. 3). Relatively large difference between heads manufactured using two different machines is a result of applying two different materials. The HIPS (polystyrene) material used in the MakerBot Replicator 2X machine is removed in a much easier manner than the ABS material used in the Dimension BST 1200 machine. Moreover, it can be removed

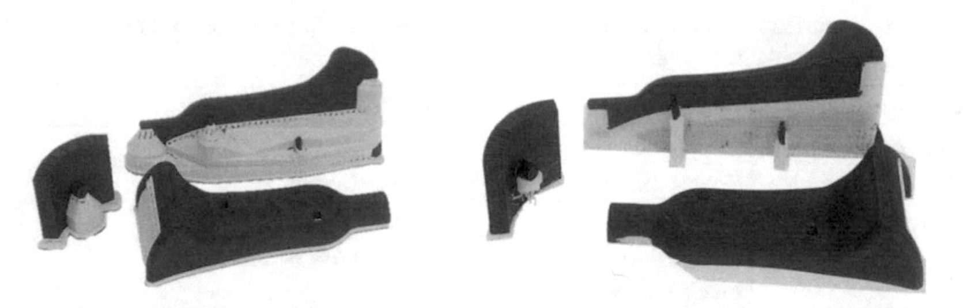

Fig. 3. Heads with support, left – BST 1200, right – MakerBot Replicator 2X

chemically (using citric acid) if the support is located in places difficult to reach with manual tools – it was not the case here, all the supports were removed mechanically.

To ensure proper fitting of parts of the head, proper assembly clearances were assumed while designing the parts (presented in Fig. 4). Size of clearance was assumed 0,4 mm total (0,2 mm per one side).

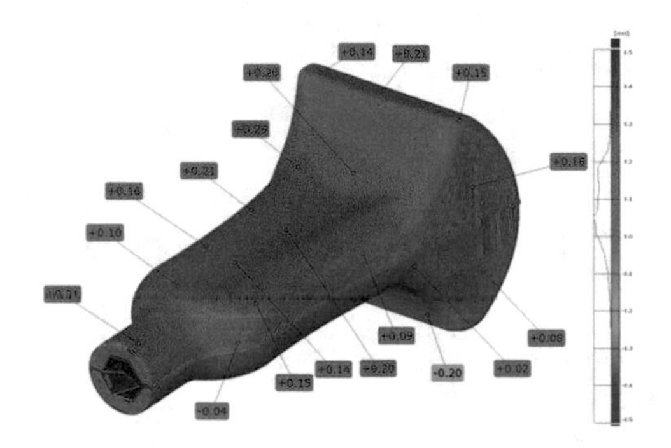

Fig. 4. Colorful deviation map for the wide head made using the Dimension BST 1200 machine, scale between −0,5 and +0,5 mm

Elements manufactured using the BST 1200 professional machine were possible to assembly directly after support removal. After assembly, there was a minimal clearance between the parts – assumed clearance size could be smaller, to ensure better snap fitting.

In case of the Replicator 2X machine, it was not possible to assembly both halves of the main head frame – there was no clearance. A layer of ABS material had to be removed on both sides of the snap fasteners, to enable joining the parts together (0,5 mm total). Removal of the material did not cause obtaining clearance – the joining was tight, although the two halves were not matched precisely, with a visible displacement.

3 Results and Discussion

The visual evaluation was performed first. In case of objects manufactured using the BST 1200 machine, the threads of material are visibly more parallel to each other, with no "waving" effect on side walls. The only visible defect is a "sew" perpendicular to layer division plane, caused by breaking material during layer contour deposition – it always occurs at the same point on the layer contour. The objects manufactured using the MakerBot Replicator 2X machine do not have the "sew", as it is possible to eliminate it by software means (starting and finishing the layer contour in each consecutive layer is shifted by a small distance). Still, visual quality of deposed threads in the low-cost machine was visibly worse than in the professional machine. Layers in contact with the support material were slightly displaced.

After initial visual evaluation, the parts were assembled together, to be used in the VR simulation. As the assembly process had different course for objects from each machine, as mentioned in the chapter 2.4, it was decided to investigate the obtained accuracy by 3D scanning. The manufactured wide heads were scanned using the GOM Atos Compact Scan 5 M optical 3D scanner, using a measurement field of 150 × 110 × 110 mm. Then, data analysis (matching scan with the nominal solid CAD model) was performed and colorful deviation maps were prepared. Selected maps are presented in Figs. 4 and 5.

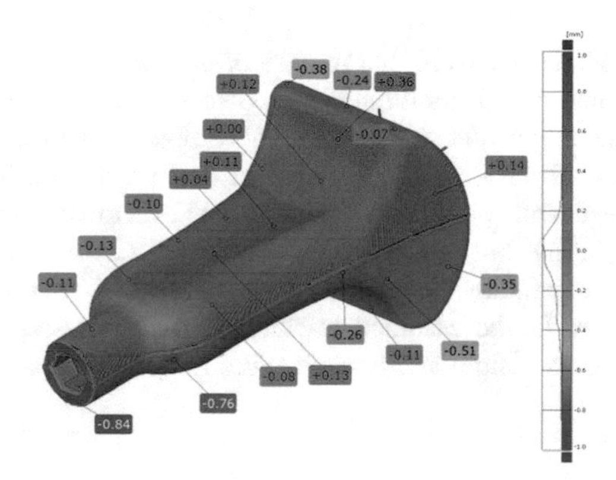

Fig. 5. Colorful deviation map for the wide head made using the MakerBot Replicator 2X machine, scale between −1,0 and +1,0 mm

Accuracy, approximated by an average deviation of the "best fit" method of matching between the CAD model and the scans, was 0,098 mm for the wide head manufactured by the BST 1200 machine and 0,147 mm for the Replicator 2X machine. For the BST 1200 machine, 100% of measured points were in the ± 0,5 mm tolerance field and 99,4% in ± 0,38 mm field. For the Replicator 2X machine, the percentage for the same tolerance fields were 95,3% and 89,7%, respectively.

It can be therefore assumed, that accuracy of the low-cost FDM process is far worse than the professional one, despite roughly the same parameters of manufacturing (leaving aside construction differences between two machines). The parts manufactured by the Replicator 2X machine require higher assembly clearances or manual processing after manufacturing to assure the proper assembly.

After the accuracy study, the heads were covered with markers of the PST-55 tracking system. Their arrangements were then introduced to the tracker's memory. Markers on the devices and their visibility in the tracker software are presented in Fig. 6.

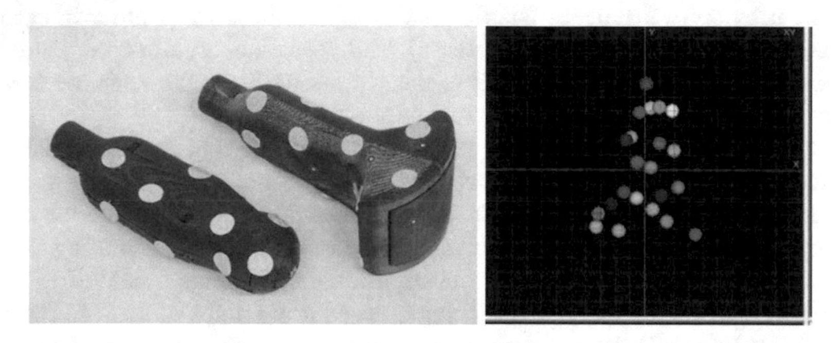

Fig. 6. Markers on 3D printed heads (left) and recognition in the tracker software for the wide head (right)

The practical tests have proven that in case of both ultrasound examination heads, method of gripping has significant influence on detection of marker arrangement in the tracker software. The authors decided to solve this problem by adding extra geometrical elements on top of each head (Fig. 7), with additional markers, impossible to cover while gripping a head with one hand. The additional elements are universal and are assembled by snap fitting, but their addition requires re-calibration of tracked marker arrangement. After expanding the heads with additional markers, recognition of devices by the tracker in the assumed tracking space increased significantly, to an acceptable level (objects visible and tracked more than 95% of manipulation time).

Fig. 7. Heads with additional 3D printed geometry, allowing better recognition by the PST-55 tracker

During the tests, the following observations were made regarding influence of manufacturing quality on use of objects in virtual simulation:

(1) Worse surface quality in objects from the Replicator 2X machine had negative influence on placement of markers of a tracking system. They are placed more easily and recognized more effectively when the surface is planar. No problems were observed for heads manufactured by the professional machine.
(2) Ergonomic quality of heads manufactured by the low-cost machine was slightly lower due to displacement of head parts and worse surface quality, but it was still on an acceptable level.
(3) The above mentioned flaws can be reduced if proper post-processing is applied, e.g. chemical and mechanical surface polishing.
(4) Despite the flaws and visible defects, it is still possible to effectively use the objects manufactured using a low-cost 3D printed in a training simulation.

4 Summary

The performed processes and tests proved, that it is possible to effectively use low-cost 3D printers in professional VR training and prototyping simulations. There is a number of problems with the low-cost process itself – the Replicator 2X machine must be supervised constantly as the nozzle is prone to clogging and breakage of material thread. Both situations require immediate action from the machine operator. The Replicator 2X (similarly to other low-cost FDM devices) also requires longer time of machine preparation, as it is not equipped in replaceable trays. Obtained quality and accuracy are also lower. Despite the defects, it is fully possible to obtain a working manipulator of dedicated, individualized shape for a simulation in Virtual Reality. The cost is significantly lower, therefore it is a recommended approach, although it requires higher qualifications of the machine operator.

It is noteworthy, that VR simulations can be successfully expanded with 3D printed accessories to further increase realism of simulations, which translates into better educational results. Both VR and 3D printing technologies have become more accessible in recent years due to emerging of low-cost processes, so the hybrid approach (joining virtual and physical objects) will become more and more widespread in authors' opinion.

Future work will consist in completing the VR simulation and conducting more tests with a group of students.

Acknowledgements. Presented studies realized in scope of a subject no. 02/23/DSMK/7647 were financed from science subvention by the Polish Ministry of Science and Higher Education.

References

Bowman, D.A., McMahan, R.P.: Virtual reality: how much immersion is enough? Computer **40**(7), 36–43 (2007)
Buń, P., Górski, F., Wichniarek, R., Kuczko, W., Zawadzki, P.: Immersive educational simulation of medical ultrasound examination. Procedia Comput. Sci. **75**, 186–194 (2015)
Chua, C.K., Leong, K.F., Lim, C.S.: Rapid Prototyping: Principles and Applications. World Scientific Publishing Co. Pte. Ltd., Singapore (2010)
Dostatni, E., Diakun, J., Grajewski, D., Wichniarek, R., Karwasz, A.: Multi-agent system to support decision-making process in ecodesign. In: Proceedings of 10th International Conference Soft Computing Models in Industrial and Environmental Applications, SOCO, pp. 463–474 (2015)
Gessner, A., Staniek, R.: Evaluation of accuracy and reproducibility of the optical measuring system in cast machine tool body assessment. Adv. Manuf. Sci. Technol. **36**(1), 65–72 (2012)
Górski, F., Hamrol, A., Grajewski, D., Zawadzki, P.: Integration of virtual reality and additive manufacturing technologies – hybrid approach to product development. Mechanik **86**(3), 173–176 (2013)
Grajewski, D., Górski, F., Zawadzki, P., Hamrol, A.: Immersive and haptic educational simulations of assembly workplace conditions. Procedia Comput. Sci. **75**, 359–368 (2015)
Hamrol, A., Górski, F., Grajewski, D., Zawadzki, P.: Virtual 3D Atlas of a human body - development of an educational medical software application. Procedia Comput. Sci. **25**, 302–314 (2013)
Parsons, T.D., Rizzo, A.A.: Affective outcomes of virtual reality exposure therapy for anxiety and specific phobias: a meta-analysis. J. Behav. Ther. Exp. Psychiatry **39**(3), 250–261 (2008)
Romero, D., Stahre, J., Wuest, T., Noran, O., Bernus, P., Fast-Berglund, A., Gorecky, D.: Towards an operator 4.0 typology: a human-centric perspective on the fourth industrial revolution technologies. In: Proceedings of International Conference on Computers and Industrial Engineering (CIE46), Tianjin, China, pp. 1–11 (2016)
Slater, M., Usoh, M., Steed, A.: Taking steps: the influence of a walking technique on presence in virtual reality. ACM Trans. Comput. Hum. Interact. (TOCHI) **2**(3), 201–219 (1995)
Wu, Y., Zhang, Y., Shen, J., Peng, T.: The virtual reality applied in construction machinery industry. In: Shumaker, R. (ed.) Virtual, Augmented and Mixed Reality, Systems and Applications, VAMR 2013. LNCS, vol 8022, pp. 340–349 Springer, Heidelberg (2013)
Zawadzki, P., Żywicki, K.: Smart product design and production control for effective mass customization in the industry 4.0 concept. Manag. Prod. Eng. Rev. **7**(3), 105–112 (2016)
Żywicki, K., Zawadzki, P., Górski, F.: Virtual reality production training system in the scope of intelligent factory. In: Burduk, A., Mazurkiewicz, D. (eds.) Intelligent Systems in Production Engineering and Maintenance – ISPEM 2017. ISPEM 2017. AISC, vol. 637, pp. 450–458. Springer, Cham (2018)

Safety Improvement of Industrial Drives Manual Control by Application of Haptic Joystick

Paweł Bachman[1(✉)] and Andrzej Milecki[2]

[1] University of Zielona Góra, Zielona Góra, Poland
p.bachman@iibnp.uz.zgora.pl
[2] Poznan University of Technology, Poznań, Poland
andrzej.milecki@put.poznan.pl

Abstract. This paper presents the research results related to the use of the haptic joystick to control the electrohydraulic and electric linear drives. The main aim of this study is to examine the extent to which the use of haptic interface with force feedback and the corresponding control algorithms can contribute to improving the quality and safety of manual control of these drives. In the research several control algorithms have been proposed and validated. The investigations were performed in simulation and on a test station, using a PC with the input/output card and Simulink software. At first, the simulation model of the hydraulic and DC drive was built and connected with real haptic joystick, which enabled the control algorithm development. Then the same joystick was used in experimental investigations. The obtained results showed that the haptic interface can greatly improve the safety of the control of hydraulic and electric drives.

Keywords: Haptic · Force feedback · Control · Hydraulic drive
DC drive

1 Introduction

Currently in many industrial companies, in mines or on construction sites, most devices are controlled manually by the users. The examples of such devices are: working machines like lifts, cranes, excavators [2, 9, 10] robots and vehicles or only single drives. In such cases usually different joysticks are used, which serve to control the output force, velocity or position of the linear or rotary actuator. Such type of control is also applied in teleoperation, in which the user controls mainly robots or vehicles, located at a considerable distance from the operator [8]. The joysticks with teleoperation are also used wherever there is a safety danger, in cases like working with explosive or dangerous materials, working under water (in high pressure conditions) or working in space vacuum. Joysticks are also used in the control of surgical robots [3]. Usually, such joysticks generate electrical signal which is proportional to the current position of its arm. The human operator observes the movement and position of the controlled device and changes the position of the joystick lever, controlling the drive movement in this way. Usually a mechanical spring is used in the joystick, which in

© Springer Nature Switzerland AG 2019
A. Burduk et al. (Eds.): ISPEM 2018, AISC 835, pp. 563–573, 2019.
https://doi.org/10.1007/978-3-319-97490-3_54

case of abandoning the lever by the operator, brings the joystick arm to a zero position. In this position the joystick output electric signal is zero, which causes the drive to stop the generation of the drive force and generally means stopping. In such control situations, the disadvantage is the lack of information about the values of forces or moments generated by the drive. To solve this, the drive current position or force is measured and fed back from the controlled object to the joystick, in which active or semi-active elements (e.g. small drives or brakes) are used. Such joysticks are robotic like devices, which enable the human operator to feel the force generated by controlled device and its environment [4, 6]. The difference between classical control and haptic control is visualized in Fig. 1.

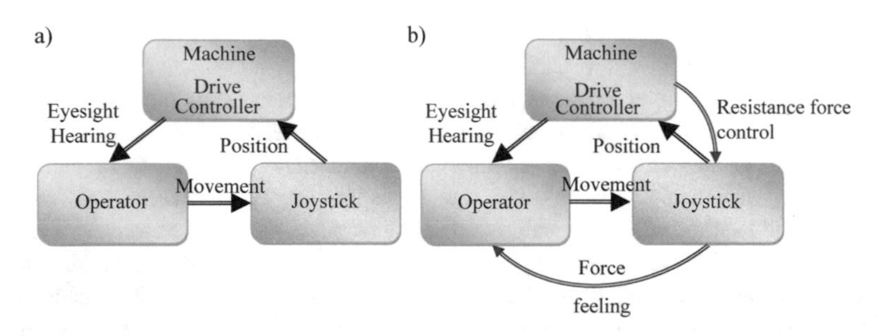

Fig. 1. Comparison of a traditional manual control interface (a) and haptic (b)

Haptic devices are used to directly control of devices and machines, where the precise feeling of touch is required. Haptic joysticks enable person-machine communication with the use of simultaneous information exchange between an operator and a machine. Thanks to the force feedback, the operator is supposed to be able to control the drive more accurately and more safely. The force feedback is very important when the observation of the controlled drive (machine) is difficult i.e. its quality is poor. In the literature about haptic control, two basic control methods are presented i.e.: force-command and motion-command [1, 5, 11], also called impedance and admittance control. Impedance control is when the operator's joystick motion is measured and taken as an input signal, but the generated force is fed back to the joystick. The drive produces a force thus makes a movement of the actuator. If the actuator acts on external object the force increases and the user can feel it in a joystick. The main objective of this strategy is to control the drive when it is in contact with the load. In admittance control the force exerted by the user on a joystick arm is measured and used as drive input signal, while drive position is fed back to the operator. In practise, impedance control is more often used for haptic interfaces than admittance control, because force measurement is easier and cheaper than position measurement.

Nowadays there are several different designs of joysticks and tangible devices already available on the market [12–15]. Another, sometimes very specific solutions are built as prototypes and described in the literature [11]. In research presented in this paper both, the designed and built by authors haptic joystick and a joystick used in computer games are applied to control of hydraulic and electric DC drives.

2 Design of Joysticks with Force Generation

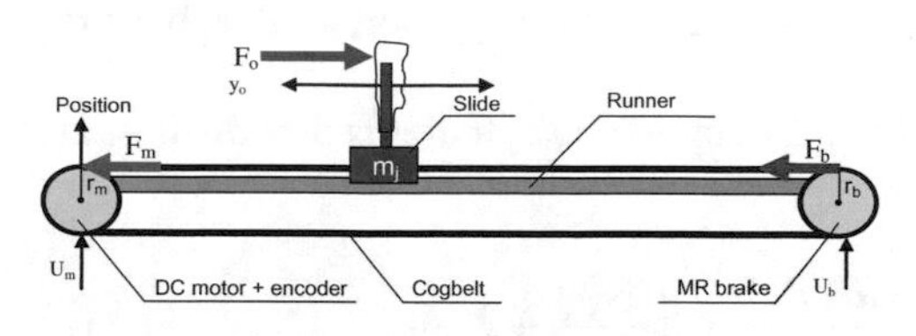

Fig. 2. Structure of linear joystick

In the research, two types of joysticks: linear and rotary have been built and applied to drives' control. The first of them is used in impedance control of an electrohydraulic servo drive and the second one in admittance control of a small electric drive. In order to generate the feedback signal from electrohydraulic drive in the joystick, (i.e. a resistance force) a small DC motor and magneto-rheological (MR) fluid brake are used. In Fig. 2 the structure of a linear joystick is shown.

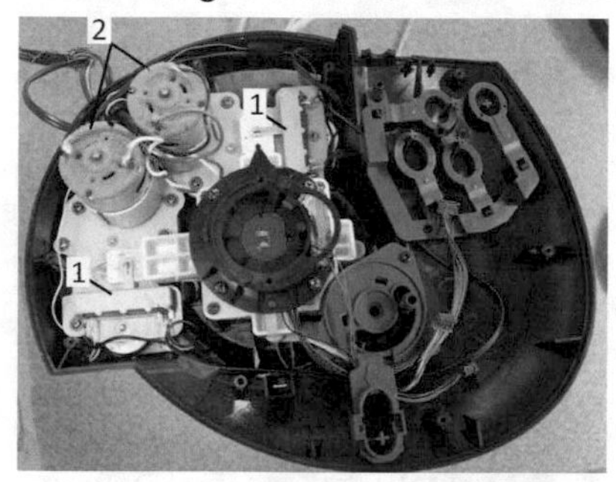

Fig. 3. Interior view of the joystick

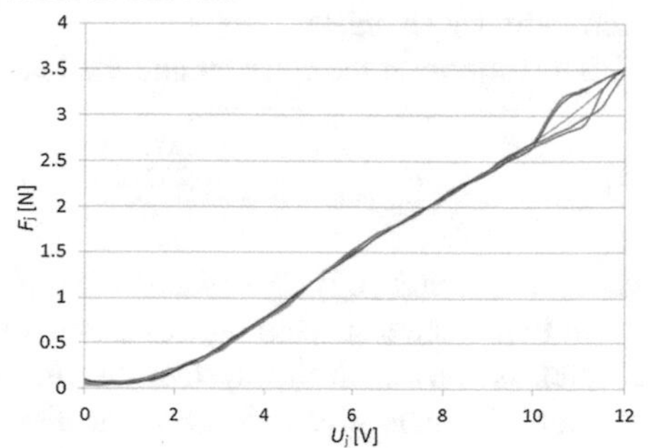

Fig. 4. Joystick's resistance force F_j in function of DC motor supply voltage U_j

In the investigations also the modified 2-axis joystick Microsoft Sidewinder Force Feedback Pro is used. In this joystick the electronics is removed and replaced by two potentiometers (1) as shown in Fig. 3, which generate output signal in a range ±10 V. The dead zone in a neutral position is in a range ±0.2 V. Two small DC electric motors are used in a joystick with a task to bring the arm to a neutral position. The force generated by the DC motor applied in a joystick is not linear, which is visible in the measured characteristic shown in Fig. 4.

3 Modelling and Simulation of the Haptic Control Systems

Typical electrohydraulic servo drive consists of proportional valve and hydraulic cylinder. The electromechanical transducer used in proportional valve is described by following second order transfer function [15]

$$G(s) = \frac{x(s)}{U(s)} = \frac{k \cdot \omega_z^2}{s^2 + 2\zeta \cdot \omega_z s + \omega_z^2} = \frac{200}{s^2 + 230\,s + 10000} \qquad (1)$$

where: x – spool displacement [m], U – coil input voltage [V], $\zeta_z = 1.15$ – damping coefficient, $\omega_z = 100$ Hz – natural frequency, $k_z = 0.02$ [m/V] – gain coefficient.

These parameters are calculated basing on data given in proportional valve producers catalogues.

The hydraulic part i.e. hydraulic amplifier and hydraulic cylinder is described in the following equation [7]:

$$K_{Qp}x(t) - K_l\Delta p(t) = \frac{V}{2E_0}\frac{d\Delta p(t)}{dt} + A\frac{dy(t)}{dt} + K_v\Delta p(t) \tag{2}$$

where: K_{Qp} – valve flow gain [m^2/s], K_l – valve flow-pressure coefficient $\left[\mathrm{m}^5/\mathrm{N}\cdot\mathrm{s}\right]$, Δp – pressure difference in a cylinder, E_0 – fluid bulk modulus [MPa], A – cylinder piston cross-sectional area, K_v – leakage coefficient, V – average contained volume of each cylinder chamber.

The motion of the cylinder and the load can be described with a following equation

$$m\frac{d^2y(t)}{dt} + D\frac{dy(t)}{dt} = A\Delta p(t) \tag{3}$$

where: m – mass [kg], D – viscous damping coefficient (friction) $[\mathrm{N}\cdot\mathrm{m}\cdot\mathrm{s}]$.

In Fig. 5 the test stand block scheme of electrohydraulic drive controlled by haptic joystick is shown. In this system the impedance control is applied. The joystick arm position signal is given to the controller, which sends the assumed force signal to electrohydraulic drive. The measured output force is fed back to the joystick. Operator assesses the drive actual position and compares it with assumed one. In this way, the system has two feedback loops: inner – force one and outer – position one. The model of the system made in Simulink is shown in Fig. 6. The parameters in this model are calculated basing on data: cylinder piston area A = 0.01 m^2, flow coefficient 1.0 m^2/s, dynamic friction coefficient 40 000 Ns/m, movable mass 500 kg and oil bulk modulus 10^9 Pa.

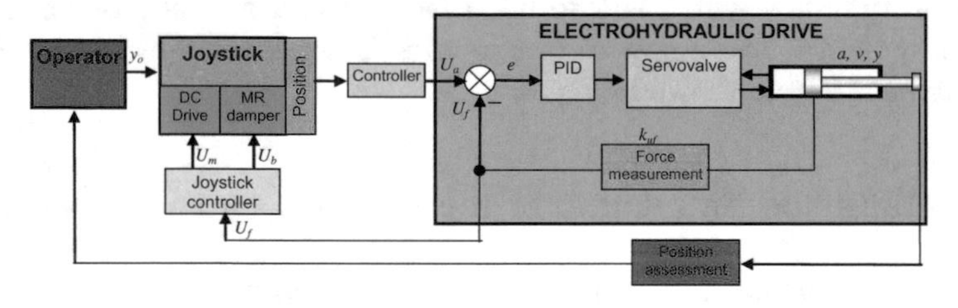

Fig. 5. Block diagram of electrohydraulic drive control by haptic joystick

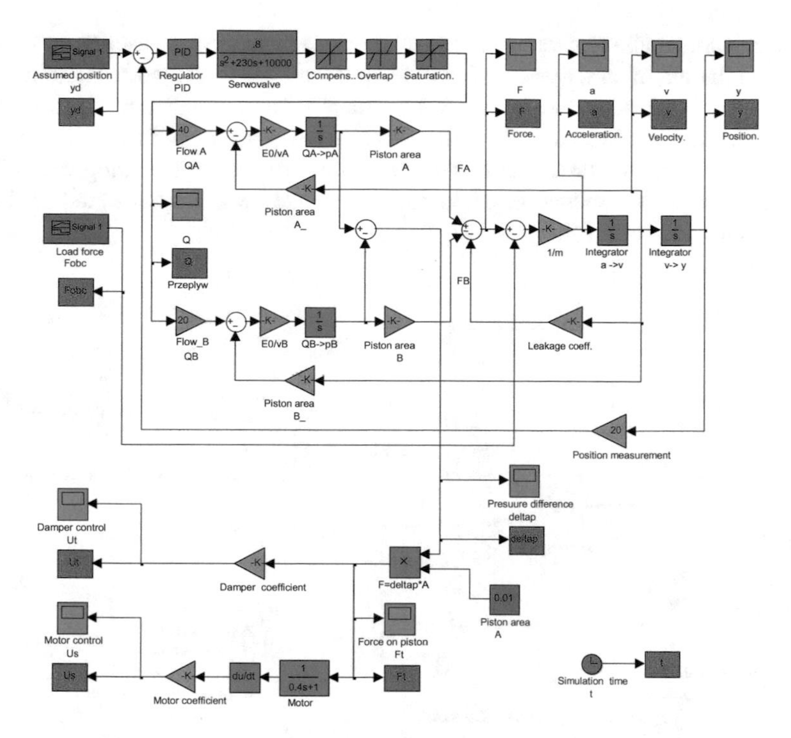

Fig. 6. Model of electrohydraulic drive with haptic joystick

The scheme of DC drive is shown in Fig. 7. It is described by following equations:

$$\frac{di(t)}{dt} = -\frac{R}{L}i(t) - \frac{K_b}{L}\omega(t) + \frac{1}{L}U_{app}(t) \quad (4)$$

$$\frac{d\omega(t)}{dt} = -\frac{K_f}{J}\omega(t) + \frac{K_m}{J}i(t) \quad (5)$$

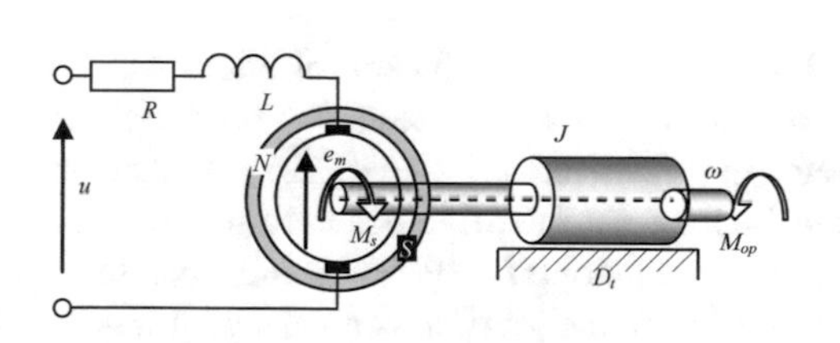

Fig. 7. Simple model of DC motor

where: $R = 3.2$ [Ω], $L = 0.233$ [H] – resistance and inductance of the motor coil, i – motor current [A], ω – angular rate of the load [rad/s], U_{app} – input voltage [V], $J = 0.1\text{kg}\cdot\text{m}^2$ – reduced inertia module, $K_b = 0.045\text{V}\cdot\text{s/rad}$ – the emf constant (depends on certain physical properties of the motor), $K_f = 0.2$ Nm $\cdot$ s – linear approximation for viscous friction coefficient, $K_m = 0.14$ Nm/A – the motor torque constant (is related to physical properties of the motor, such as magnetic field strength, the number of turns of wire etc.).

The mentioned above parameters are taken from manual or measured. In the model also a simulation of emergence of the load (force of 10 N) corresponding to the collision with the obstacle have also been simulated. The model enabled the simulation of user behaviors during control of the drive using joystick without and with force feedback. The operator task was to withdraw the drive after detecting the collision, which may cause the reduction of loses. The model was implemented in Matlab-Simulink software (Fig. 8).

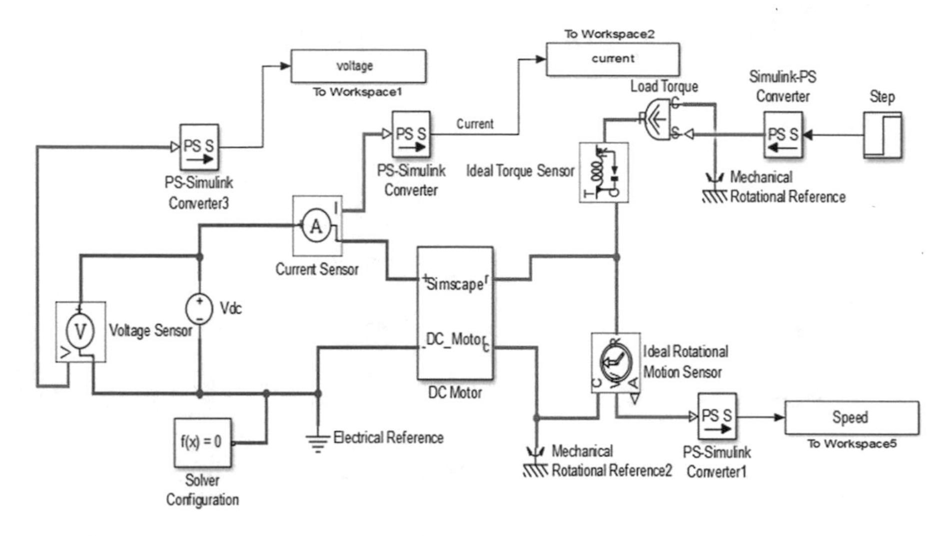

Fig. 8. The model of DC drive controlled by joystick made in Matlab-Simulink

The simulation results are shown in Figs. 9 and 10. In all control attempts, the output position y_d, force F_d and joystick voltage U_{j_out} were recorded. Three characteristic points were marked on the waveforms registered on signal: A – impact on obstacle, B – stopping of the drive and C – the moment the drive began to move away from the obstacle. The time elapsed from impact to stopping (B − A) and operator's reaction time (C − B) has been determined. In Fig. 10 the joystick control signal U_{FF} is included. The investigations have shown that thanks to the use of the force feedback the reaction time is reduced from 1.5 s to 0.7 s. To the algorithm, a modification was introduced causing the joystick to automatically "move back" after encountering an obstacle, which significantly improved the quality of control.

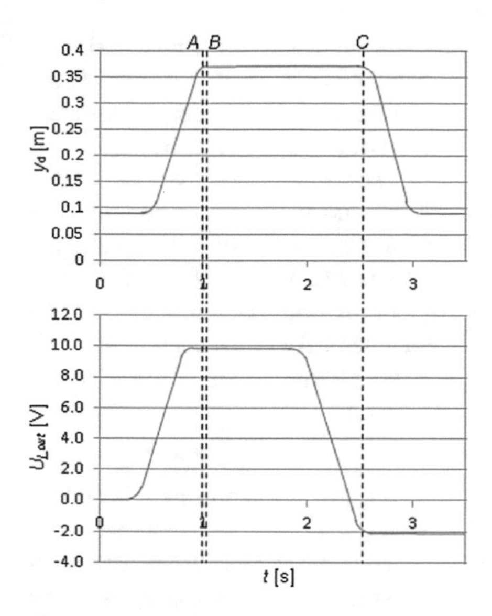

Fig. 9. Simulation results without force feedback

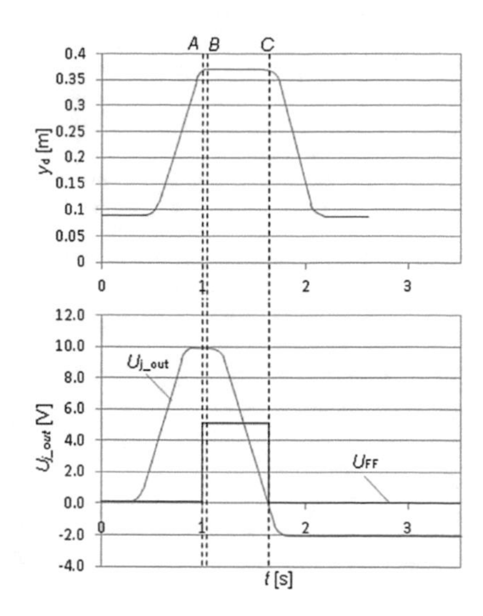

Fig. 10. Simulation results with force feedback

4 Experimental Tests

The view of the test stand for investigations of electrohydraulic servo drive is shown in Fig. 11. The investigations were focused on operator's reaction time when he couldn't see the cylinder piston rood accurately while the controlled electrohydraulic drive impacts the obstacle. In Fig. 12 following recorded signals are presented: α_j – joystick angle position, y_c – piston position, F_c – cylinder force, F_j – joystick force. The vertical dashed line described as "x" shows the point in which the drive meets the obstacle. In Fig. 12a signals recorded when the MR brake was off, are presented. In this case operator spotted the obstacle with a delay of 1 s. (vertical dashed line "y"). If the MR brake was activated, the operator could feel the cylinder movement opposite force on the joystick almost immediately. This gave him the information that the drive meets obstacle which caused the operator to stop moving the joystick and as a result the cylinder piston was stopped.

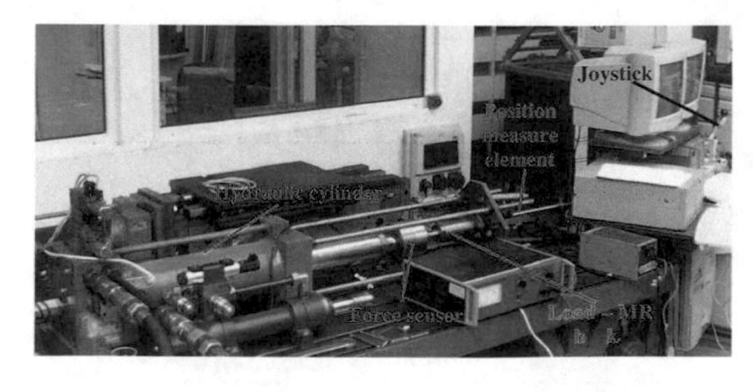

Fig. 11. Investigations stand

In Fig. 13 the test stand used for investigations of control of DC drive by haptic joystick is shown. It consists of a joystick (1), linear DC drive (2), the supply and control electronics (5). The drive and a joystick are connected to the PC (4) with an installed input/output card type RT-DAC PCI. On the movable trolley a force sensor

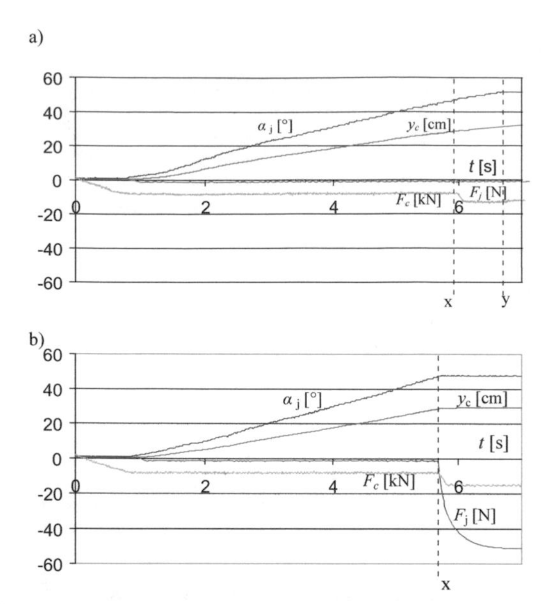

Fig. 12. Signals recorded during obstacle recognition when the MR brake was switched: (a) off, (b) on

Fig. 13. Test stand for haptic control of DC drive

(3) is installed. The whole system is controlled by PC (4) on which the Matlab-Simulink software is installed with control scheme.

The connection scheme of the test stand is shown in Fig. 14. The measured and control signals are transferred to and from the computer by input/output card and by adequate electronic amplifiers. Following signal are used in the system:

- U_{jX}, U_{jY} – joystick DC motors supply voltage in axis X and Y,
- U_{j_outX}, U_{j_outY} – output voltages from potentiometers measuring the joystick arms angle positions,
- F_d – measured force (converted to voltage) with which a drive acts on the obstacle,
- y_d – measured by potentiometer, current position of the drive output element (troley) (converted to voltage),
- U_d – DC main motor supply voltage.

The control algorithm is implemented in Matlab-Simulink software.

In experimental tests, similar investigations were carried out as during simulation. The recorded

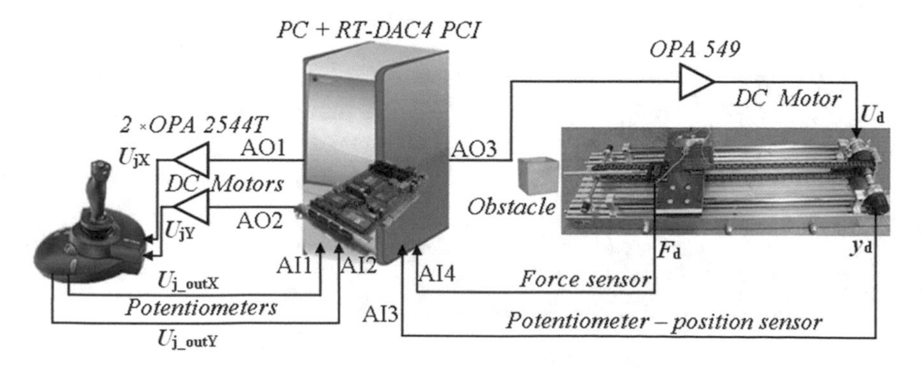

Fig. 14. Connection diagram of the test system

Table 1. Results of operator's response time measurements

		A[s]	B[s]	C[s]	B − A[s]	C − B[s]
Simulation	No force feedback	1.00	1.05	2.5	0.05	1.45
	With force feedback	1.00	1.05	1.60	0.05	0.55
Experimental	No force feedback	1.16	1.46	2.15	0.30	0.69
	With force feedback	1.00	1.24	1.53	0.24	0.29

curves are presented in Figs. 15 and 16. The algorithm automatically moved back the joystick arm after contact with the obstacle. In Table 1 the measured times showing the operator's reaction during control of the drive without and with force feedback are listed. Their comparison clearly shows that the times to reach the positions A, B and C are similar in all cases. The same concerns the times B − A, which elapsed from drive contact with the obstacle to drive stop (B − A). The most important thing is to compare operator response times to contact with an obstacle and operator's reaction time (C − B). These times with force feedback are clearly shorter (0.55 s and 0.29 s) than times when feedback was disconnected (1.45 s and 0.69 s).

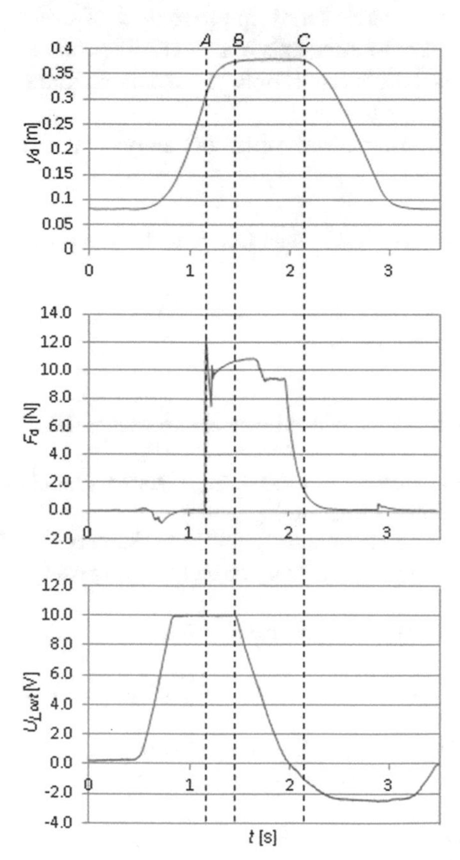

Fig. 15. Experimental test results without force feedback

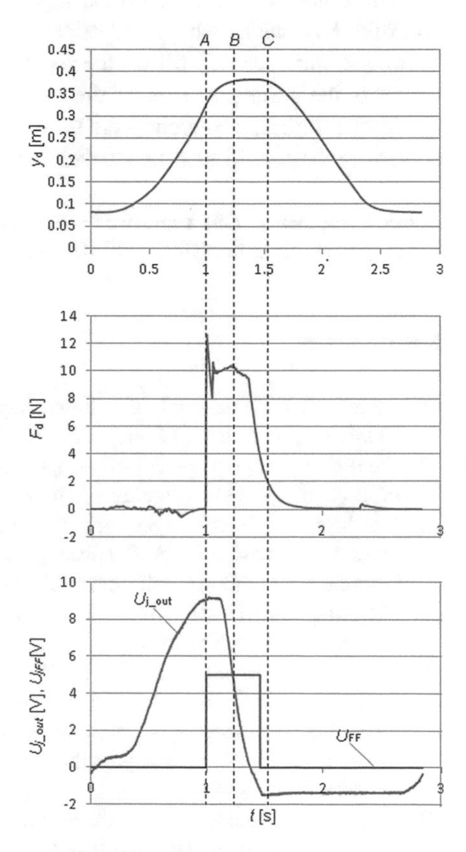

Fig. 16. Experimental test results with force feedback

The main parameter which determines the degree of the safety of the system is the operator's reaction time on the collision of the controlled drive with an obstacle.

5 Conclusion

Dozens of series of the same measurements were made, giving a similar result. In this paper only one example is included, showing the practical results. The simulation tests confirmed the possibility of designing such a control algorithm, which enables to reduce the operator's reaction time to the obstacle.

The paper proposed the application of one-axis joystick with DC motor and with MR brake which enabled force feedback to communicate and interact with a electro-hydraulic and electric servo drive. The joystick was connected to the PC, which controlled the drive and the joystick. By the usage of force feedback in a joystick, the human operator's reaction times have been shortened. Thanks to this, the operator-drive interaction was more realistic and faster, which may improve the safety of industrial machines manual control.

The obtained in the described in this paper research safety improvement of drive controlled by one axis haptic joystick, can be extent to three axis of devices like robots. However, the user will fill on the joystick, thus on his hand, only the resultant force, which is the geometric sum of the three constituent forces.

In the future, researches on how behaviour of various operators impact at the control process, will also be carried out.

Acknowledgment. This paper was supported by the Polish National Centre of Research and Development, grant no. 02/22/DSPB/1434.

References

1. Adams, R.J., Hannaford, B.: Stable haptic interaction with virtual environments. IEEE Trans. Robot. Autom. **15**, 465–474 (1999)
2. Frankel, J.G.: Development of a Haptic Backhoe Testbed. MS thesis, The Georgia Institute of Technology, G.W. Woodruff School of Mechanical Engineering (2004)
3. Goethals, P.: Tactile Feedback for Robot Assisted Minimally Invasive Surgery: an Overview. Division PMA Department of Mechanical Engineering Katholieke Universiteit Leuven, scientyfic research raport (2008)
4. Grunwald, M.: Human Haptic Perception – Basics and Applications. Birkhäuser-Verlag, Boston, Basel, Berlin (2008)
5. Harward, V.: Haptic synthesis. In: Proceedings of 8th International IFAC Symposium on Robot Control, SYROCO (2006)
6. Kern, T.A.: Engineering - Haptic devices. Springer, Berlin, Heidelberg (2009)
7. Milecki, A.: Linear electrohydraulic servo drives. Modeling and control. Poznan University of Technology (2003)
8. Mora, A., Barrientos, A.: An experimental study about the effect of interactions among functional factors in performance of telemanipulation systems. Control Eng. Pract. **15**, 29–41 (2007)

9. Oh, K.W., Kim, D., Hong, D., Park, J.-H., Hong, S.: Design of a haptic device for excavator equipped with crusher, In: 25th International Symposium on Automation and Robotics in Construction, ISARC 2008. Vilnius Gediminas Technical University Publishing House "Technika", pp. 202–208 (2008)
10. van der Zee, L.F.: Design of a haptic controller for excavators. Thesis MScEng, Electrical and Electronic Engineering, University of Stellenbosch (2009)
11. Zhuang, Y., Canny, J.: Haptic interaction with global deformations. In: Proceedings of IEEE Robotics and Automation Conference, vol. 3. IEEE (2000)
12. http://www.forcedimension.com
13. http://www.immersion.com
14. http://www.quanser.com
15. http://www.sensable.com

Scan$^{\pi}$ - Integration and Adaptation of Scanning and Rapid Prototyping Device Prepared for Industry 4.0

Paweł Krowicki(✉), Grzegorz Iskierka, Bartosz Poskart, Maciej Habiniak, Tomasz Będza, and Bogdan Dybała

Faculty of Mechanical Engineering, Wrocław University of Science and Technology, Łukasiewicza 5, 50-371 Wrocław, Poland
pawel.krowicki@pwr.edu.pl

Abstract. The article presents the design and implementation of a device that replicates 3D objects and works within the Internet of Things (IoT). The first part presents the design and construction of the replication machine. In the following, integral components are presented. Broadly discussed FDM printer adaptation and modernisation. Afterwards, a concept and realisation of 3D scanner is presented. Image processing and analysis algorithms were developed and implemented for Scan$^{\pi}$ device. After careful preparation of subdevices, such as a 3D scanner and an FDM printer, integration with the Internet of Things was carried out. Presented solution is compared with professional system delivered by Phoenix Contact. Finally, the device tests are carried out. The last chapter describes the advantages and disadvantages of the newly developed device.

Keywords: Industry 4.0 · IoT · Reverse engineering · Image processing Rapid prototyping · REST web service

1 Introduction

Nowadays, the technologies of reverse engineering and rapid prototyping are very popular and commonly used in both commercial and noncommercial applications. Mostly they operate separately, complementing each other. Focusing on the reverse engineering it is easy to show diversity of a number of applications. Use of the technology in a lot of areas was indicated by Vinesh Raja and Kiran Fernandes [1]:

- production (e.g. quality control),
- medicine (e.g. surgical planning or customized implant development),
- products customization (e.g. footwear),
- criminalistics,
- aviation,
- archeology.

Those technologies would not exist without devices which accuracy depends on the applied scanning technique [2]. Simple division of scanners indicates two types - contact and contactless. The first group consists of all kinds of pointing tools such as measuring arms or coordinate measuring machines (CMM) [3]. The operation of these devices

© Springer Nature Switzerland AG 2019
A. Burduk et al. (Eds.): ISPEM 2018, AISC 835, pp. 574–586, 2019.
https://doi.org/10.1007/978-3-319-97490-3_55

based on the data which come from sensors (e.g. encoders). Estimation of the pointing tool's position works on the principle of computation of the inverse kinematics where the data from sensors is obligatory. The second group of 3D scanners defines contactless systems. Those reverse engineering devices mostly operate on the principle of reflection of emitted medium such as light or ultrasound. In this method the shape of light pattern projected onto the object is analyzed. Another possibility gives measuring the time of flight from transmitter to the object (light reflection) and back to the receiver. In this case short light impulses are generated. These two scanning methods are based on triangulation [4] and the time-of-flight measurement [5]. Professional and amateur solutions are easily available on the market. The result of the scanning are the digital models of scanned objects.

The technology that often uses the results of reverse engineering is rapid prototyping. Digitalized objects, thanks to the mentioned technology, could be duplicated. Numerous possibilities of rapid prototyping are already in use. The best known are fused deposition modeling (FDM), stereo lithography (STL), and selective laser sintering (SLS). Application of those methods can be observed in medicine (e.g. manufacture of implants), aviation, or automotive industry. FDM devices became very popular, cheap and easy to use, therefore they can be found in many home applications.

As mentioned before, reverse engineering and rapid prototyping complement each other but mostly function separately. To enable the cooperation of two independent devices, it is necessary to use a master control device. Currently, there are few devices on the market that can scan and print in 3D. First of them is Zeus from AIO Robotics [6] which provides functionality of scanning, printing, copying and faxing. The important thing is that scanning and printing cannot take place at the same time. Scanned objects are stored in the cloud. Introduced functionality in Zeus responds to the Internet of Things integration [7]. The next all in one appliance is da Vinci 1.0 Pro 3D from XYZprinting [8]. This device does not work with IoT and, just like in the previous example, the scanning and printing processes cannot take place at the same time.

The integration of 3D printers and scanners into the Internet of Things (IoT) or Industrial Internet of Things (IIoT) extends the ability to remotely control the processes, store data in the cloud, perform calculations in the cloud (e.g. slicing), exchange models between devices or allow 3D faxing.

The article presents a comprehensive approach to the development and implementation of a scanning and printing device operating in IoT ready for Industry 4.0. The second section presents the assumptions and design of the device together with its detailed solutions. FDM printer modification and adaptation to IoT is presented in the next chapter. Then the scanner's design and software development are described. After that, IoT integrating algorithm and its implementation is explained in detail. Finally, the test results and conclusions are presented.

2 Device Development

The purpose of this work was to develop a device that would be able to scan and 3D print at the same time. Additionally, information on the current status of the device should be available online. Status information is important for introduction of an interactive tutorial and remote control. The device was designed to implement an FDM printer - TEVO TARANTULA and a specially developed 3D scanner based on triangulation. A Raspberry Pi minicomputer was used to control all peripherals. In the first stage, devices and sensors were selected and wiring diagrams were developed. Then the housing was designed. In the next steps a scanner was developed and tested.

2.1 Simplified Electronics Schematic

As already mentioned, the heart of the device is Raspberry Pi which is connected to an extension board through a specially created PCB (Printed Circuit Board). A power supply ATX 550 W was selected to power all of the subdevices. Work on the device has been divided into three parts of integration: printer, scanner and IoT.

The first part needed integration of a TEVO TARANTULA printer controller with Raspberry Pi. The FDM device was behind the door which is controlled by a limit switch. Scanner integration required device design and connections of components such as Logitech C510 camera, two linear lasers, stepper motor (400 steps per revolution and torque 0.48 Nm) connected to the rotary table, an E18-D80NK presence detector and a limit switch mounted in the door of the scanning chamber. The project incorporates an intelligent LED lighting, operated using a touch screen. The LED lighting of the device is controlled by a relay module. The project uses 4 USB ports available in Raspberry Pi for communication with the camera, the printer's controller, the touch screen and external devices (USB HUB). The built-in Wi-Fi module was used to communicate with the developed service (REST). The simplified diagram below shows the connections between the components of the device. Red lines symbolize 12 V DC (direct current), yellow 5 V, and green 3.3 V. Signal connections were marked with black lines (Fig. 1).

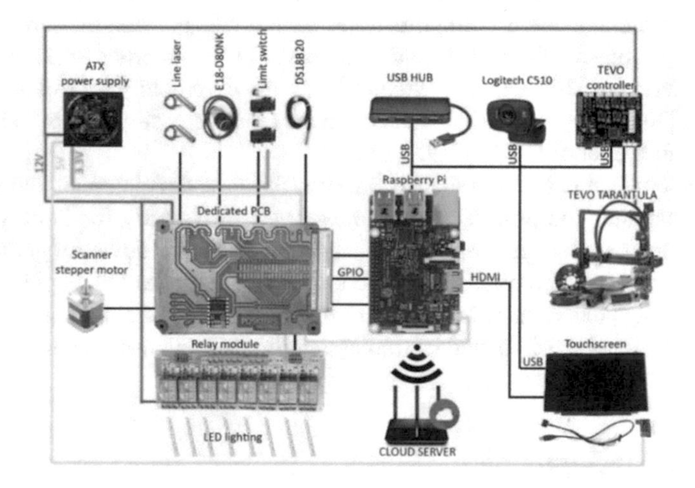

Fig. 1. Simplified schematic of connections between components of the device.

2.2 Scan$^{\pi}$ Device Design

The aforementioned heart of the device was the basis for awarding the device name Scan$^{\pi}$ which was associated with scanning and Raspberry Pi. Appropriately selected devices, sensors and actuators made it possible to design the device's construction. The final appearance of the internal and external structure looked as shown in Figs. 2 and 3.

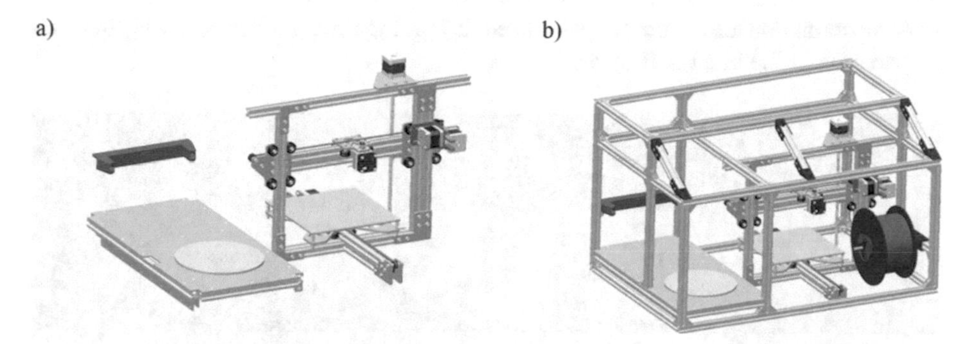

Fig. 2. The construction of the scanner and 3D printer: (a) before integration, (b) integrated.

Fig. 3. Visualizations of the final version of the housing (external case).

2.3 FDM Printer Integration

The applied TEVO TARANTULA 3D printer is equipped with a heated working platform with the most typical size of 200 × 200 mm. The maximum temperature of the heated bed (120 °C) and nozzle (260 °C) allows the use of materials such as: ABS, PLA, PETG, NYLON, WOOD. The main advantage of the open design TARANTULA is the possibility of parts modernization. Potential weak points were found during usage. The modernization concerned the frequent need to level the heated bed resulting from the deformation of the bracket (Fig. 4).

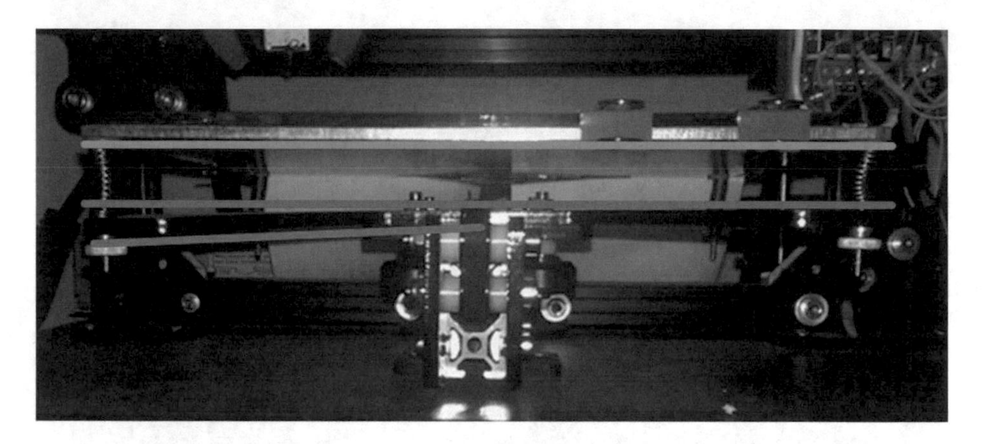

Fig. 4. The essence of the problem of the frequent need to calibrate a heated bed.

The green line indicates the level of the heated bed. The red line (parallel to green) indicates the level on which the table support should be. The purple line indicates the actual deformation of the bracket. Under the influence of the load, resulting from the pressure of the nozzle on the heating bed, the weight of the heating platform, the glass sheet and the print itself, the bracket deforms. The bracket was made of plastic (probably ABS), therefore the temperature generated by the heating platform (up to 120 °C) could have influenced easier plastic deformation of the bracket. Inaccurately leveled heated bed may result in the lack of adhesion of the first layer of the printout with the work platform, the detachment of the model from the table during printing, clogging of the nozzle or errors in the printout. Errors could also occur when layer shift is caused, skipping layers due to printer error, when the platform is not preheated [9]. Appropriate calculations were done and modifications were implemented.

A modified heated bed support made of 3 mm thick aluminum PA6 was designed and made. The results of the FEM (Finite Element Method) analysis are presented (see Fig. 5). As a result, the maximum deflection was reduced from 1.78 mm to 0.145 mm (a 92% decrease) with a weight increase of 25 g (18%). The modernization has eliminated the problem of needing frequent calibration of the table. Wider recognition of the problem, models, analyses, calculations and detailed modernization made with the design thinking method is in the master's thesis [10].

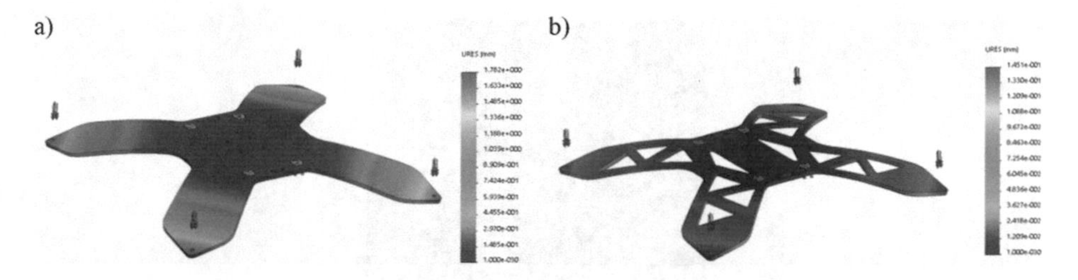

Fig. 5. FEM analysis of the: (a) original printer table bracket, (b) modernized printer table bracket.

2.4 3D Scanner Design and Implementation

The idea for the scanner was to use a linear laser, a camera and a rotary table. The measurement based on taking a photo of the illuminated object, and then performing the rotation by a given angle. Performing measurements for full rotation enables digitization of the scanned object. Computation of registered data was performed with the triangulation algorithms based on set distances between the laser and the camera (LC), and the camera and the object (OC), presented on the figure below (see Fig. 6).

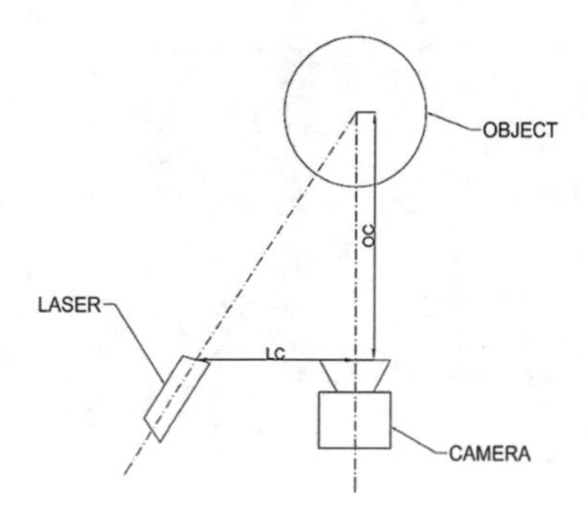

Fig. 6. Conception schematic of the developed laser scanner

Additionally, due to the fact that the device was designed to be entirely independent of external computers, a built-in Raspberry Pi manages all of the processes and computations, including execution of implemented 3D scanning algorithms. The technique required one or more line lasers for pattern projection onto the surface of the scanned object.

For image acquisition, an external web camera was used. The images are recorded in a 1280 × 720 resolution. To allow rotation of an object, a rotary turntable has been designed with the use of Igus Robolink D standard robot joint to allow scanning of heavier objects and more precise angle of rotation control. The final look of Scan^{II}'s scanning chamber is presented on Fig. 7.

Fig. 7. Scan^{II}'s scanning chamber.

Scanning software has been developed in Python, with the use of OpenCV library for image acquisition and processing, as well as Numpy library for further calculations. To fully automate the scanning process, image processing, profile's rotation transformation and writing an STL file algorithms have been devised and will be further explained. The core scanning algorithm is presented in Fig. 8, where green boxes show the aforementioned algorithms.

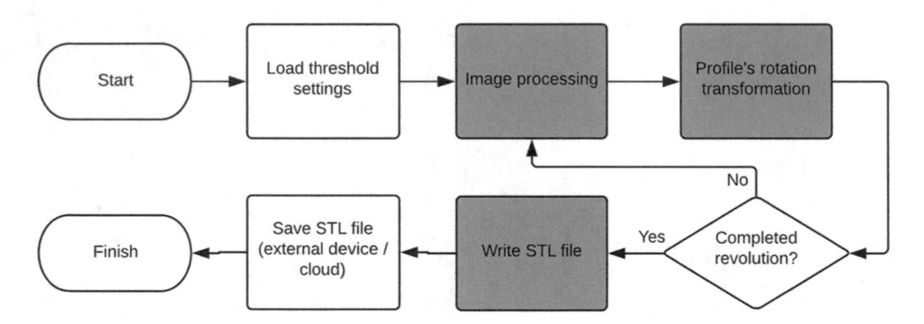

Fig. 8. Scanning algorithm.

As presented on the diagram (see Fig. 8), the scanning process starts by loading threshold settings from a file, previously set with a separately developed piece of software, which allows for correct thresholding, and in return, correct recognition of the projected pattern. Afterwards, the software proceeds into the main scanning loop consisting of consecutive image processing operations as well as rotation transformation of gathered data, in order to create a spatial point cloud correlating with the scanned objects orientation. Upon completion of full revolution of the scanned object, software uses the calculated coordinates of gathered points to create a meshed 3D model, later saved in an ASCII STL file format.

Image Processing. The algorithm (see Fig. 9) begins with image acquisition, registered with the built-in camera, which is then converted to HSV colour model, allowing for easier thresholding of the colour of the used line laser. The process of colour thresholding begins with applying the previously loaded threshold settings to an HSV mask and applying the mask to the captured image.

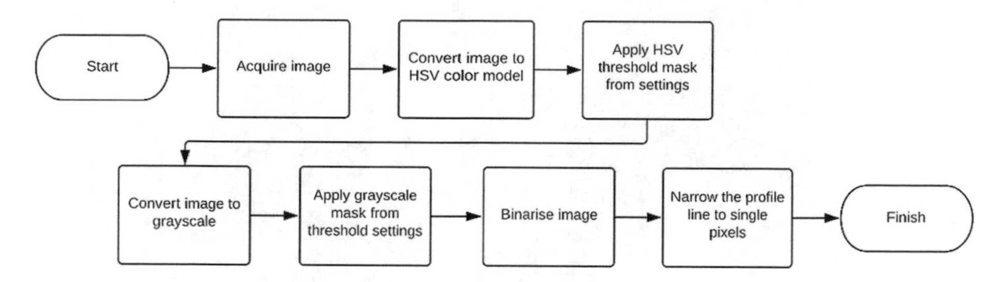

Fig. 9. Image processing algorithm.

To further improve thresholding and reduce background noise, caused by reflections or other light sources, the image is converted to grayscale, and a grayscale mask is being applied accordingly. The final image is then binarised and the received projection line is narrowed by leaving only the middle pixel in each row of pixels of the image. This way, the product of the algorithm is a set of points representing the distorted line laser's projection.

Profile's Rotation Transformation. In order to convert the flat image coordinates provided by the previous algorithm, a triangulation calculation is carried out for each of the points registered in each image, following the schematic roughly presented in Fig. 10, as a result receiving a set of spatial Cartesian coordinates.

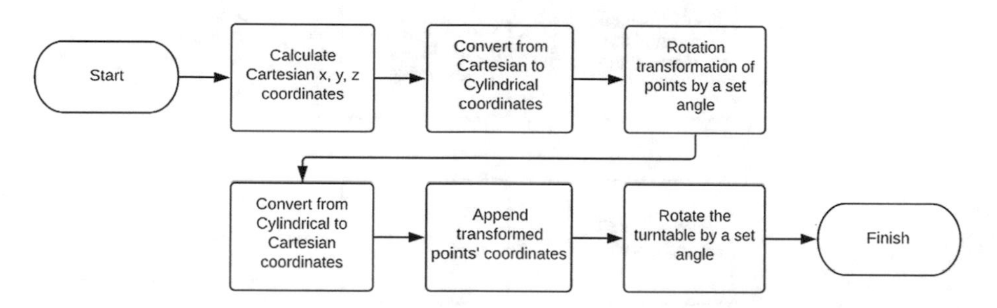

Fig. 10. Profile's rotation transformation algorithm.

In order to carry out the rotation transformation more efficiently, the points' coordinates are converted from Cartesian (Fig. 11a) to Cylindrical (Fig. 11b) coordinate system. This allows rotation by a set angle simply by changing the φ coordinate.

Upon the completion of rotation calculations, the coordinate system is changed back to Cartesian, as is required for further operations, and the points' coordinates are appended to a point cloud list. The turntable is then being rotated by a set angle to prepare the object for the following scan.

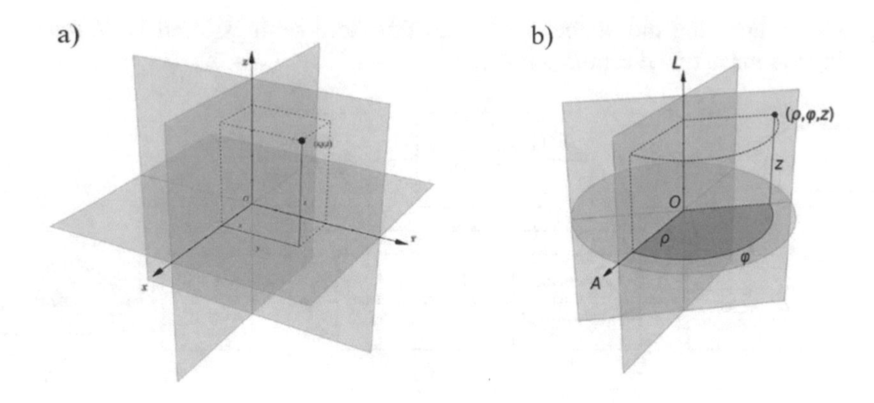

Fig. 11. Three dimensional coordinate systems: Cartesian (a), Cylindrical (b).

Writing an STL File. After the entire scan of an object is completed, a point cloud needs to be meshed in order to create outer walls of a 3D model. The model is saved in an ASCII STL format, which makes the algorithm (see Fig. 12) easier to develop.

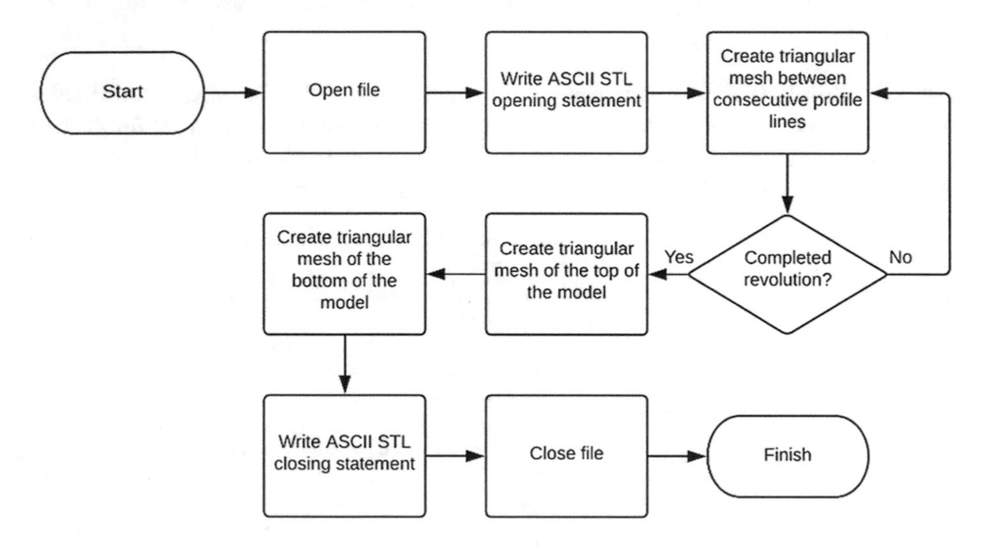

Fig. 12. Writing an STL file algorithm.

Upon opening of a file, it is initiated with a beginning line, which allows for file format recognition and provides an object's optional name. The meshing process is conducted by writing consecutive triangles in a format presented below (see Fig. 13).

A single triangle in an ASCII STL format consists of three points in Cartesian coordinate system, presented on Fig. 13 as vertices. In addition, the format requires to provide a normal to the surface created by said three points to establish the surface's orientation.

The developed algorithm creates a triangular mesh with points of two nearby lines of scanned data, until it finishes building the sides of the model, at which point it closes off the top and bottom part of the model. The file then ends with an ending statement and is closed.

```
solid name
  facet normal n_i n_j n_k
    outer loop
      vertex v1_x v1_y v1_z
      vertex v2_x v2_y v2_z
      vertex v3_x v3_y v3_z
    endloop
  endfacet              } +
endsolid name
```

Fig. 13. Sample ASCII STL format.

Spatial Scanner Implementation – Capabilities and Limitations.
Current iteration of the developed scanner is best suited for objects of simple geometry, mostly because of the limitations of the scanning method used and the developed algorithms.

Since the method used is structured light-based, it is important to notice that best results can be achieved for white or bright-coloured objects with less-reflective surfaces. Worst results were observed for highly reflective, black, or dark-coloured objects.

Considering the meshing method used in the scanning software, it can be concluded that under current implementation it is impossible to produce through holes. Additionally, due to camera's position and orientation, it is impossible to register any concave structures on the top or bottom part of scanned objects.

3 Internet of Things Integration

As of recent years, Internet of Things has become very prominent in the industry, with countless new devices being connected to the Internet each year. With internet connectivity come additional functionalities and features aiding the user in various ways. Since Scan$^{\Pi}$ is equipped with a Raspberry Pi micro-computer, which has a built-in Wi-Fi module, a global server has been set up to allow additional features such as:

- 3D models cloud database,
- Augmented reality user tutorial,
- Remote device status information.

The set-up server and provided features will be explained in following paragraphs.

3.1 REST Server

Simple Object Access Protocol (SOAP) became very popular and is strictly connected with IoT. SOAP is a kind of web service which includes REST (Representational State Transfer) architecture. To allow quick connection and good responsiveness, a REST Server has been set up, which allows for communication through standardised HTTP request protocols.

For server setup, a PythonAnywhere service was used. It allows running a Python script on a global server, which in this case allows the use of REST technology through Flask API. Flask is an API (Application Programming Interface) which allows the server to handle standard HTTP requests, of which most common ones are: GET, POST, PUT, DELETE. The requests allow the server to send, receive, change, or delete data. On the basis of these features a cloud computing server was developed.

3.2 3D Models Storage

The PythonAnywhere account includes additional disk space on the server (500 MB for free user accounts), which for this application serves as a storage space for uploaded 3D models. The files are simply contained on the server's hard drive, and can be easily accessed with previously programmed URL (Uniform Resource Locator).

The URL can be accessed from multiple sources through GET requests. This means that the file can be downloaded programmatically, directly through any web browser or other specialised software (e.g. Postman). This allows for easy access to scanned models, but also raises safety or privacy concerns, which should be further addressed, but are not for the purpose of this demonstration.

3.3 3D Models and Status Information Database

To allow the user access to information about currently stored 3D models and machine status, separate URLs have been programmed, which return all the information in a JSON (JavaScript Object Notation) format, examples of which are shown in Fig. 14.

a)


```
{
  "status": {
    "printer": "printing",
    "printer_door": "closed",
    "scanner": "scanning",
    "scanner_door": "closed",
    "scanner_object": "no",
    "total_size": "35 MB"
  }
}
```


b)


```
{
  "1": {
    "date": "06-03-2018  17:31",
    "gcode": "no",
    "image": "yes",
    "link_PNG": "http://gviazdeathka.pythonanywhere.com/file/NESSIE.png",
    "link_STL": "http://gviazdeathka.pythonanywhere.com/file/NESSIE.stl",
    "name": "NESSIE.stl"
  },
```


Fig. 14. JSON files returned for: machine status (a), stored models (b).

Machine status information is mostly used for the implemented AR (Augmented Reality) user tutorial, using feedback from the on-board sensors to walk the new user through following steps of the scanning or printing processes. The models list presented above organises the files for the machine to interpret, should the user choose to print models directly from the provided cloud storage. Presented approach is strictly connected with the Internet of Things. In order to extend the device functionality into the IIoT, some modifications should be introduced. For this purpose, the use of a PLC controller should be considered. A PROFINET controller type AXC CLOUD PRO could serve as an example. Presented solution could easily replace Raspberry Pi in the discussed project. The functionality and modularity of the PLC is a big advantage. However, one of the biggest disadvantages is the need to buy credits. For economic reasons, this solution was abandoned. Please note that the presented solution (with Raspberry Pi) should not be considered as a solution in the Industrial Internet of Things.

4 Measurements and Tests

First scanning tests were performed on an object of complicated shape (see Fig. 15). The developed scanner was compared with the Sense 3D V2 scanner from 3D SYSTEMS. Results of the measurement are presented below.

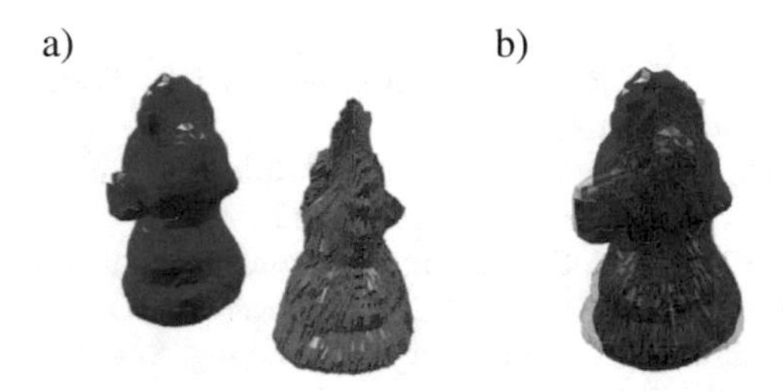

Fig. 15. Comparison between models: side by side (Sense V2 – left, developed scanner – right) (a), overlapped (b).

The difference between the models is presented in the table below (Table 1).

Table 1. Comparison between Sense V2 and the developed scanner.

	Sense V2	Developed scanner
Vertices count	1142	8946
Faces count	2354	18324
Model volume [cm^3]	100	50

Another of the tests was to check the connection speed with the server. Internet connection speed has been validated via spedtest.net and was downloading up to 190 Mb/s, sending up to 19 Mb/s and 17 ms server response time. For these conditions, the response time of the REST server per query was measured. 20 measurements were carried out from which the mean time of response was calculated. Measured mean time for the Wi-Fi connection is equal to 311 ms and 305 ms for LAN.

5 Conclusion

The aim of this work was to develop a replicating device. All scheduled tasks have been successfully completed. The preliminary device tests were carried out positively. The improved design of the printer table bracket has increased the reliability of the printer's operation. The built-in scanner algorithms should be revised and further tested. The best results of scanning can be achieved for white or bright-coloured objects with less-reflective surfaces. Integration of the Scan$^{\pi}$ device into the IoT was done properly. The use of the REST server in PythonAnyware service is free but with some restrictions. One of the considered IoT solution is applying a PLC controller which could collaborate with industrial networks. Implementation of a Phonix Contact PLC which meets requirements is associated with additional fixed costs. The developed device can be used by people without much knowledge about scanning or 3D printing. A sample Scan$^{\pi}$ application could be provided for companies where it is necessary to quickly repair damaged elements.

References

1. Raja, V., Fernandes, K.T.: Reverse Engineering. 1st edn. Springer Series in Advanced Manufacturing, London (2008)
2. Barbero, B.R., Ureta, E.S.: Comparative study of different digitization techniques and their accuracy. Comput.-Aided Des. **43**, 188–206 (2011)
3. Sadaoui, S.E., Mehdi-Souzani, C., Lartigue, C.: Combining a touch probe and a laser sensor for 3D part inspection on CMM. In: Teti, R., D'Addona, D.: 11th CIRP Conference on Intelligent Computation in Manufacturing Engineering, vol. 67, pp. 398–403. Elsevier, Naples (2018)
4. Isa, M., Lazoglu, I.: Design and analysis of a 3D laser scanner. Measurement **111**, 122–133 (2017)
5. Vázquez-Arellano, M., Reiser, D., Paraforos, D., Garrido-Izard, M., Burce, M., Griepentrog, H.: 3-D reconstruction of maize plants using a time-of-flight camera. Comput. Electron. Agric. **145**, 235–247 (2018)
6. Zeus documentation. https://www.robotshop.com/en/zeus-all-in-one-3d-printer.html. Accessed 30 Apr 2018
7. Atzori, L., Irea, A., Morabito, G.: Understanding the Internet of Things: definition, potentials, and societal role of a fast evolving paradigm. Ad Hoc Netw. **56**, 122–140 (2017)
8. Debapriyo, P., Puneet, K.A., Gourab, G.M., Debamalya, B.: A comparative analysis of different hybrid MCDM techniques considering a case of selection of 3D printers. Manag. Sci. Lett. **5**, 695–708 (2015)
9. Song, R., Telenko, C.: Material and energy loss due to human and machine error incommercial FDM printers. J. Clean. Prod. **148**, 895–904 (2017)
10. Poskart, B.: Project and construction of 3D copy device based on "Internet of Things". Master thesis 2018, Wrocław University of Science and Technology

The Application of a Vision System to Detect Trajectory Points for Soldering Robot Programming

Andrzej Milecki and Piotr Owczarek(✉)

Institute of Mechanical Technology, Poznan University of Technology, ul. Piotrowo 3, 60-965 Poznań, Poland
{andrzej.milecki,piotr.owczarek}@put.poznan.pl

Abstract. This article focuses on the application of a vision system for a point to point programing of a robotized soldering station. The station is used in the last stage of production for the soldering of elements, applying Through-Hole Technology (THT). In the soldering station a SCARA-type robot is used. The main aim of the usage of the vision system is to extract pads from the image of a printed circuit board (PCB), and to obtain the coordinates of central point of every pad. These coordinates are used in the soldering program, prepared for a robot. In the paper the PCB pictures processing algorithm is described. The results of practical investigations on pads coordinates recognition are presented and finally their application in the robot program is shown.

Keywords: Vision system · PCB · THT soldering · Robot · Image processing

1 Introduction

Progress in robotics plays a crucial role in the development of Industry 4.0, which is going to link the existing factory with virtual reality. Today, the strongest growth in the robotics industry is visible in Asia, and specifically China has the largest number of industrial robots in operation. It is predicted that in all developed countries the number of robots will increase significantly in the near future. The International Federation of Robotics estimated that the worldwide stock of operational industrial robots will increase from about 1.8 million at the end of 2016 to 3.0 million units by 2020; this means an average annual growth rate of about 14%. Electronics assembly industry uses more and more robots as well. This concerns both Surface Mounted Technology (SMT), in which dedicated robots are used, and Through-Hole Technology (THT), in which typical robots are applied.

The current requirements for soldering robots are to be simultaneously easier to use and to program. Therefore the producers of robotized soldering stations try to facilitate the programming with direct transformation of data stored in Gerber-type files into a robot program. Unfortunately, sometimes the manufactured PCBs differ from the theoretical model defined in Gerber file. The most common error is produced during drilling of holes and cutting shapes of PCB. Such pads position error could reach up to ± 0.2 mm. If PCB with such errors is soldered it may occur that the final soldering is

© Springer Nature Switzerland AG 2019
A. Burduk et al. (Eds.): ISPEM 2018, AISC 835, pp. 587–596, 2019.
https://doi.org/10.1007/978-3-319-97490-3_56

inappropriate. Therefore the auto-generated soldering program must be further adapted to the real PCB. Sometimes the producer, to prevent his product from being copied, purposely doesn't make Gerber files available. In other cases, the PCB could be ordered in a different company and ad checking them is necessary before soldering. In all these cases the application of visual system comes in handy. The vision system may be used to determine the position of PCB in the working station with the recognition of so-called "fiducial points", as well as to detect all pads with holes.

In the last ten years vision systems became a fast-growing technology which is getting more and more accurate and fast. As a result it is used in everyday life, for example in mobile phones' cameras with a face recognition system, or in smart TVs, in which gesture control is possible, or in industries, in which quality control [1], products classification [2], items sorting in cooperation with robots [3, 4] or monitoring systems [5] is executed using such systems. Every application of a visual system requires the implementation of special recognition algorithms. These algorithms can be divided into features detection [6, 7] and marker detection, which can be passive [8], active [9, 10] and coded [11]. One of the most commonly used markers has a circle-like shape [12], which is believed to be an optimal configuration, according to several publications, for example [13].

This paper presents the application of a vision system to recognize the pads position on the PCB and the coordinates of their central. The pads have circular-like shape and therefore the already known method for their recognition may be applied. The vision system is used for programing the trajectory of soldering machine with SCARA robot developed by Renex company in cooperation with Poznan University of Technology. The main aim of the vision system is to extract central point's coordinates of all pads in the image of PCB, and to send these coordinates to the soldering robot. The soldering station described in this paper is used only for soldering of THT elements. This operation is usually made in the last stage of electronic boards production.

2 PCB Image Processing Algorithm

In this chapter we will describe steps to extract central points of pads in a PCB through a visual image processing. In the first step, a smoothing filtering is applied. In most cases the Gauss blur filter (called also Gaussian smoothing) is used to reduce in the image noise and a non-necessary small details. A two-dimension representation of Gauss filter is defined by a following equation [14, 15]:

$$G(x,y) = \frac{1}{2\pi\sigma^2} e^{-\frac{x^2+y^2}{2\sigma^2}} \tag{1}$$

where: x – is the distance from the origin point (pixel) in the horizontal axis to ±kernel size, y – is the distance from the origin in the vertical axis ±kernel size, σ – σ is the standard deviation of the Gaussian distribution.

This equation produces a surface where contours are circles with a Gaussian distribution from the center point. Every pixel's new value depends of its neighborhood and the original pixel's value is bigger than neighboring pixels as their distance to the original pixel increases. The maximum distance is a relative kernel size, which value is in a range from 0.0 to 3.0. Gaussian filter is used in order to enhance image smoothness at different scales. Image for processing can be made with a digital scanner or can be taken with a visual camera. Practical investigations made by the authors have proven that the images taken with a scanner are better to recognize soldering pads on PCB than images taken directly from camera, mainly because they have the same intensity of light in the whole picture area. Moreover, scanner images do not have lens distortions, which are typical for camera images. The images were made using digital scanner HP Deskjet Ink Advant k209a-z with a resolution of 600 DPI, which proved to be sufficient for proper pads recognition. Smaller DPI gave lower quality of result. Bigger resolution required much more CPU processing time and as a result a delay occurred, which was unacceptable to the user. All image processing were made using an Adaptive Vision Studio software environment ver. 4.3.

The results of the use of a Gaussian filter are shown in Fig. 1. In the first picture the relative kernel is 1.0, but in next pictures it is 3.0. The standard deviation σ is 3.0, 9.0 and 18.0. The obtained effect is called a visual blur. Further research has shown that the acceptable results of first step of image processing are those, presented in Fig. 1c and d. This is mainly because blur is only a little visible and simultaneously the image is readable. Greater blurring causes smudging of the edges and inability to set holes dimensions. In the research described in this paper following parameters are used: kernel = 3 and σ = 3.

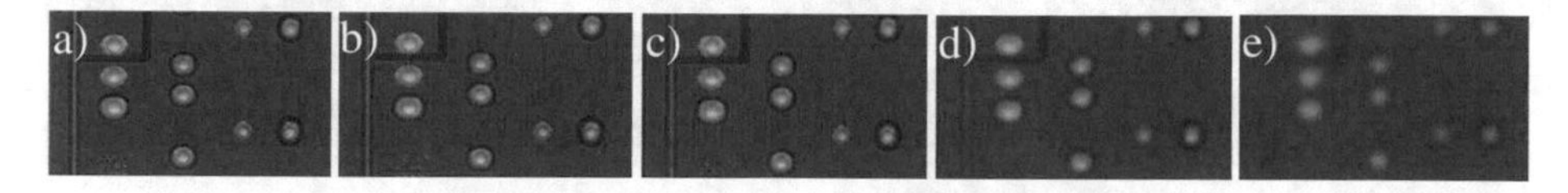

Fig. 1. Smoothing results: (a) original image, (b) Gauss kernel = 1.0, σ = 3.0, (c) Gauss kernel = 3.0, σ = 3.0, (d) Gauss kernel = 3.0, σ = 9.0, (e) Gauss kernel = 3.0, σ = 18.0

In the next step of the image processing the threshold filter is used to find pixel group, which are pads with holes. This filter operates with a different types of color space of the image e.g. intensity, RGB, HSV, HSI, HSL. In this work only HSV color space is used for filtering of the image. There are three parameters of the color space i.e.: "H" (0…360) means "hue", "S" (0…255) means saturation, and "V" (0…255), which is "the brightness" of the light. Every pixel in threshold filtration is compared with HSV parameters using min and max values for each of them, and converted either to white or to black. As a result a "black-and-white" image is obtained, which is also called "binary image".

In Fig. 2a an example PCB is presented, in which the holes diameters are equal to 0.9 mm. In Fig. 2d another PCB fragment is shown, in which the holes diameters are 1.5 mm. In this case a background color is white, which is very similar to the pads color. In the Fig. 2b the images after Gaussian and threshold filter application are shown. The investigations have proven, that good results can be achieved, when the values of filter min and max parameters are: for H – 0 and 149, for S – 0 and 79, and for V – 25 and 255. However, inside the hole pads (white circles) some dark points (pixels) can be noticed, which are in fact “noises” (Fig. 2b). This is most probably because of light reflexes. These figures result in an incorrect calculation of pads-center points. Moreover, the pads inner border curve is jagged. The parameters used for Gauss filter were: kernel = 3.0 and $\sigma = 3.0$. In order to improve the quality of the images the CloseRegion filter [16] should be applied, after the use of Gauss filter and threshold filter.

In the Fig. 2e the images after only threshold filter application is shown to visualize the impact of noise in image without Gaussian filter. In Fig. 2c and f, the output image is shown, which is obtained after application of CloseRegion filter, which performs a morphological closing on a region using selected predefined kernel. Parameters of this filter are a region defined by its width, height (x, y) and a run-length which is a radius equal to 4. The use of this filter resulted in filling of all closed areas with white color, giving a fully satisfactory outcome. However, although the pads shapes have been improved (Fig. 2f), i.e. are without noise pixels, the line on the PCB left side is more visible, which is an undesirable result. Therefore a Gauss filter was added before threshold and Close Region filter which is shown in Fig. 2g.

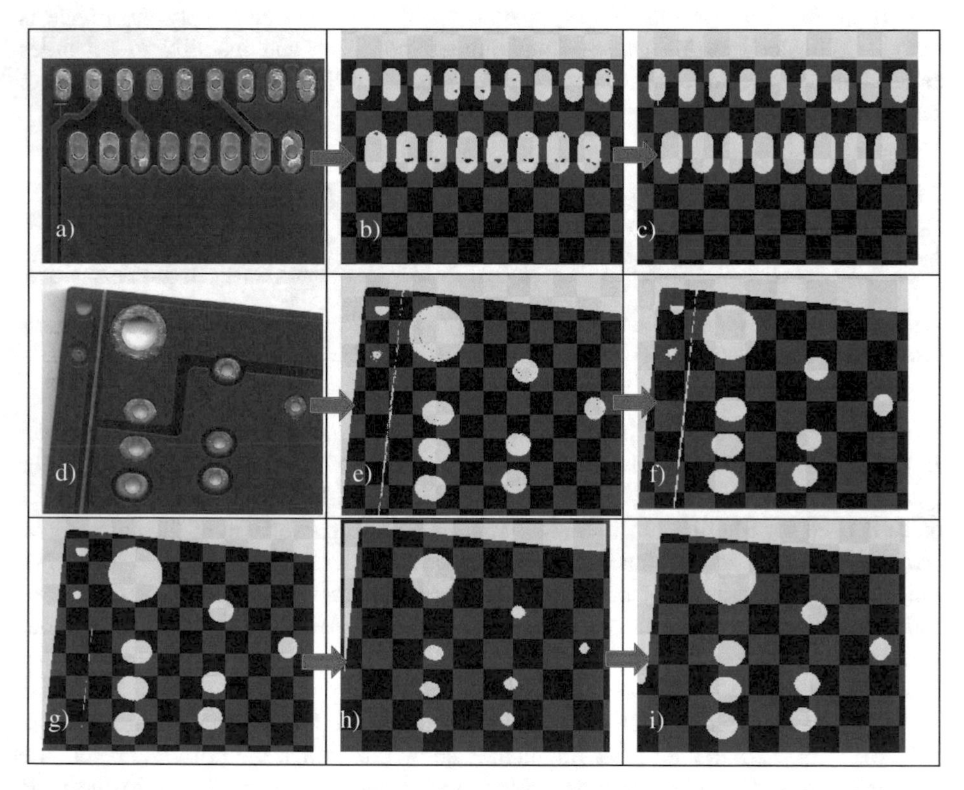

Fig. 2. PCB images obtained during processing: (a) PCB with a holes diameters 0.9 mm; (d) PCB with a holes diameters 1.5 mm on the white background; (b), (e) after application of smoothing and of the threshold filtration – binary image; (c), (f) after CloseRegion filter usage; (g) after application of Gauss filter; (h) after Erode filter application; (i) after Dilate filter – output image

In the visual system analysis of the PCB, also the filter called Erode is commonly used to delete small groups of pixels or thin line near path from the image. Erode filter is used also to remove sharpness of edges around the pads. The parameter in this filter can be chosen from: kernel-like box, ellipse or cross and aligned to axis *x* and *y*. In this investigation the type of kernel is chosen to be "box" and its radius is equal to 3.0. The result of Erode filter application is shown in Fig. 2h. As you can see, the unnecessary item (line of the left of PCB) was removed, but all pads are now smaller. This happens because the filter removes the outer layers of pixels. In order to restore the correct size of pads, a Dilate filter is used, which enlarges the areas of parts of image which were decreased in the previous step. In Fig. 2i a result of using a Dilate filter is shown. This filter gradually enlarges the boundaries of regions of foreground pixels, (white pixels creating circles). As a result, the areas of foreground circles were increased in size. Dilate filter was used with following parameters: ellipse kernel and a radius = 3.0. This is the last image filtering process used to prepare input image for computing central points of the pads.

In order to calculate pads' areas and estimate of pads' central points, every pad region in the image is split into a blob using "SplitRegion into blobs" function. This function segments a region into individual objects when the objects do not touch each other. Blobs means a group of pixels, which are stored in computer memory, like object of data. Every blob can be processed independently. In a next step, the pads features extraction is made. When pads are properly selected as white regions i.e. blobs, their areas can be computed using following equation:

$$S = \sum_{i=1}^{n} \sum_{j=1}^{m} p(i,j) \tag{2}$$

where: p – value of the pixel in position (i, j) on an image, $n \times m$ –size of region, $p(i, j) = 1$ – when the object is i.e. white, $p(i, j) = 0$ – when the object is, i.e. black.

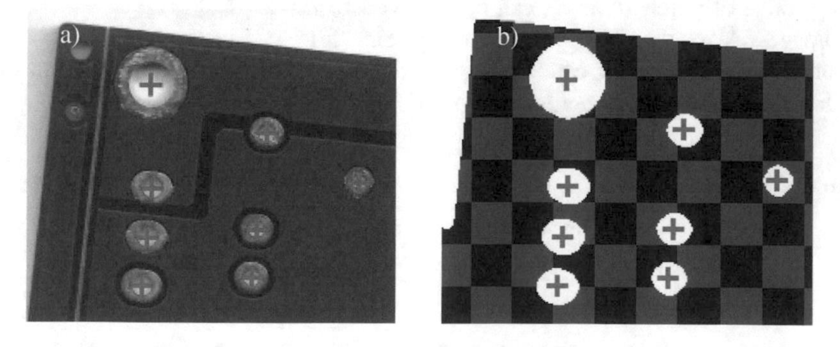

Fig. 3. Center of the mass calculation: (a) raw image result, (b) processed output image

In the next step of the algorithm, the center of the mass of every white object (blob) is calculated. This is made using the two following equations:

$$\tilde{x} = \frac{\sum_{i=1}^{n}\sum_{j=1}^{m} k}{S}, \tilde{y} = \frac{\sum_{i=1}^{n}\sum_{j=1}^{m} k}{S} \tag{3}$$

where: k – is equal to 1 when pixel is white, and 0 in other case.

The calculated coordinates x and y are used to draw the blue crosses in the middle of every pad. The results of application of all filters described above is shown in Fig. 3a and b. The blue crosses appoint a central position of all pads.

The conducted research allowed to form a complete algorithm (Fig. 4) of image processing to extract central points of pads in a PCB. First stage named "Load image" reads data from a computer. Second block is a "Gauss smooth" filter, which is used to reduce local min and max values of pixels and also to smoothen edges in the image. Next filter called "Threshold image" is used to cut off unnecessary pixels from the image and to produce a binary image. "Close region" filter removes "dark holes" from group of pixels by filling them fully with white color. Further on "Erode filter" makes every group smaller, and as a result small noise regions are removed from image, but edges of group are smoother. "Dilate filter" restores the size of the white group of pixels in the image. Next step of the algorithm is to transform these single groups into an array. This is done with the "Split region into blobs" filter. Blobs are a separate objects on image which store information about their size and position in image. This function can also filter blobs by size. This is very useful, because using this block, a small groups of pixels (noises), can be removed from the image. Also "too-big" groups of pixels, like mounting holes for screws, can be removed, thanks to setting maximum sizes of blobs. The last filter computes a center position of pads with sub pixels resolution.

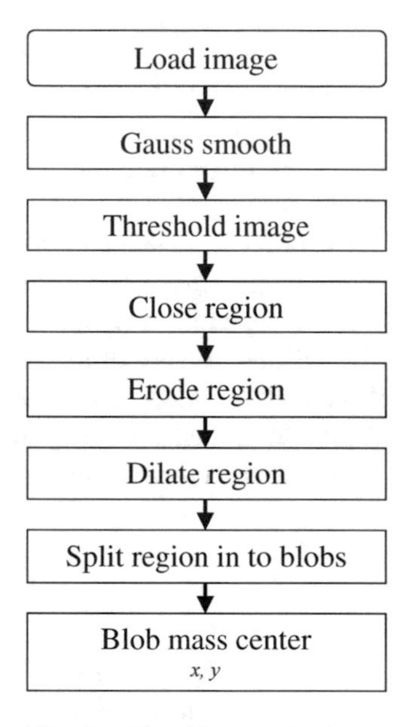

Fig. 4. Algorithm of image processing

Table 1. Comparison of usage CPU for filter image processing with size 3101 × 2785 px

	Loading	Gauss	Threshold	Close	Erode	Dilate	Center
Time [ms]	134,6	9,87	29,8	4,5	3,1	7,2	1,5

To compare the usage of processor, the execution time of all filters was measured and shown in Table 1. It is visible, that the "Loading" task is the longest one. Second longest is the "Threshold filter".

3 Results of Processing of Different PCB Sample Images

To improve the correctness of the above-described algorithm in detecting in PCB central points of holes in pads, several tests have been performed. At first, the investigations were focused on pads recognition on the bottom side of the PCB, on which all electronic elements are located. On this side of almost every PCB there is only a green solder mask without any description. The results of holes in pads recognition on two different PCB bottom sides are shown in Fig. 5.

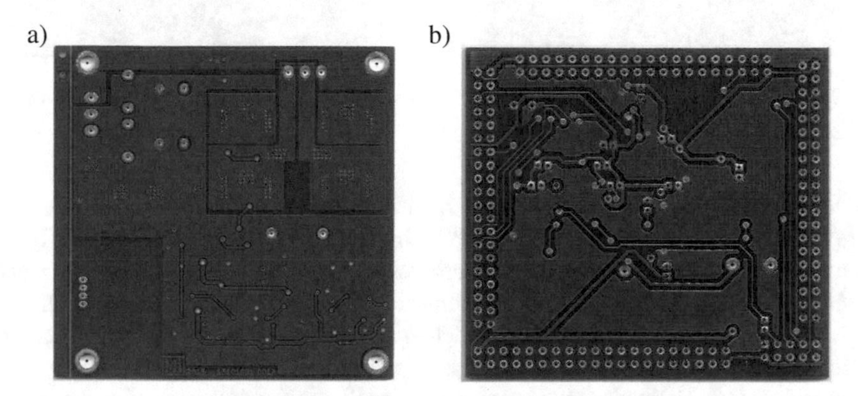

Fig. 5. Detection results of pads on bottom side of PCB: (a) pads diameter 3,0 mm, (b) pads diameter 1.5 mm

The visible red points (small circles) denote, that the holes in pads are recognized. All holes are marked properly, which means that the algorithm worked correctly and every point was detected. Another examples are shown in Fig. 6, in which two same PCB boards with white color text description are shown. Because the text is written in white color, it is very similar to pads color, and as a result in Fig. 6a, some elements of text are incorrectly recognized by the algorithm as pads. In order to avoid that type of error, a pink color of the background is used. The result of algorithm processing is shown in Fig. 6b. In this case, the software recognized properly only pads with the holes inside (red points) while the white text description is omitted.

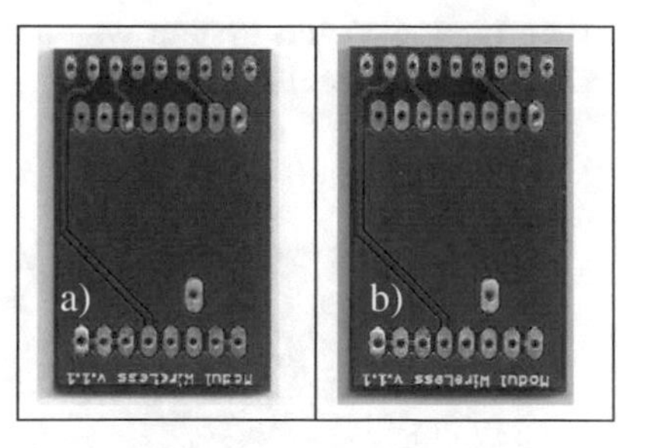

Fig. 6. Recognition of pads on small PCB with a white description on the white and pink background

In the next step, the PCB with a blue solder mask (Fig. 7a) is used to test the correctness of the proposed algorithm. The text descriptions on the PCBs have caused a big number of errors, therefore the color of the background was changed to pink. The detection results are shown in Fig. 7b. It is visible that all necessary pads are detected

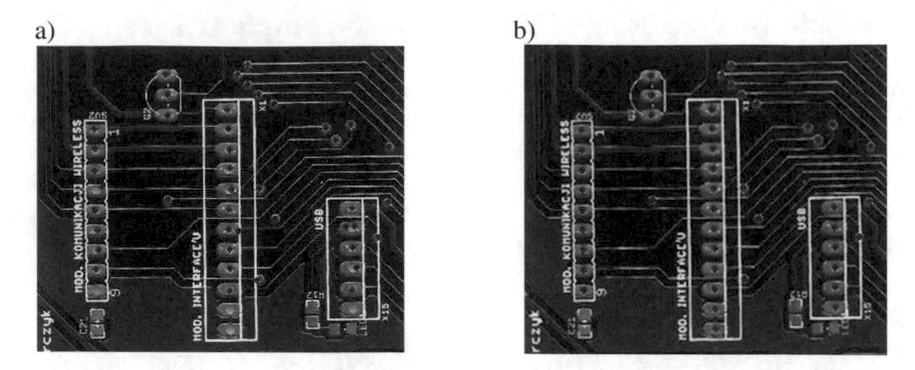

Fig. 7. Holes recognition on PCB with blue solder mask and white description on pink background: (a) white background, (b) pink background

correctly. In the next step of described here software, the coordinates of every hole were calculated and written in a file. A PCB image taken for tests is shown in Fig. 8a. All recognized points are shown as blue crosses in Fig. 8b. The coordinates from vision system given in pixels are converted to dimensions in mm. There is the possibility in a software to set on an additional filter, which enable skipping of holes whose diameter is bigger than assumed value, thus omitting a mounting holes in corners.

In order to check the accuracy of obtained results, the coordinates of every hole in a pad from Gerber type files are taken and compared with coordinates received from the vision system. Small parts of PCB were zoomed in Fig. 9 to show a difference between positions from Gerber file and positions generated from vision system; the position error in x axis is about 0.03 mm. The comparison of all obtained results showed that the maximum error of position in both axis was equal to 0.12 mm compared with references from Gerber file, but the average error of position was equal to 0.06 mm. The proposed vision system was used to extract the real coordinates of the pads in PCB. These coordinates were send to the soldering robot. So, by this way the

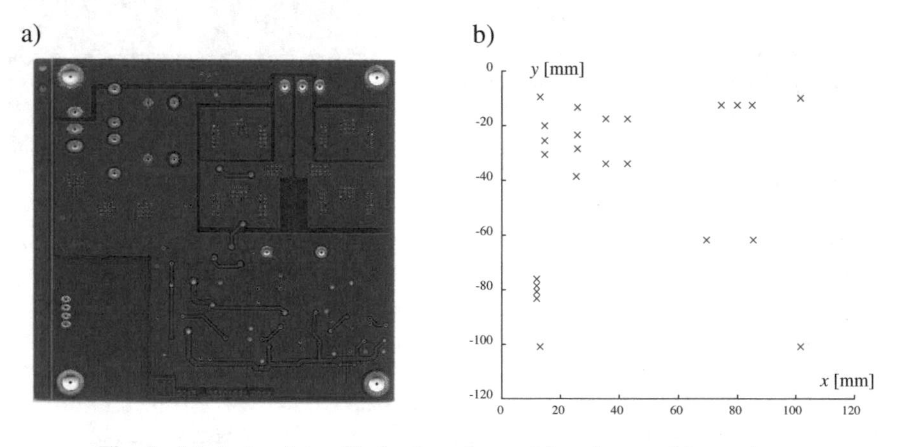

Fig. 8. Marked points of holes in pads on: (a) an image, (b) *x-y* plate

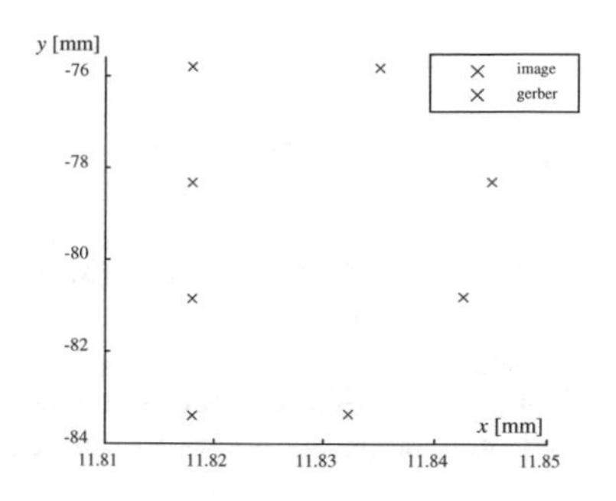

Fig. 9. Comparison of points from Gerber file with points from image processing

Fig. 10. Test stand with soldering robot

differences between pads coordinates given in the Gerber file and pads coordinates in a real PCB are not important for the soldering process. No external tool was used to check positions of all points. The user may add such parameters like: position in z axis, rotation of soldering head, position for start, soldering time etc. and send to the SCARA robot (Fig. 10) controller using serial interface with TCP/IP protocol.

4 Conclusion

In the paper the use of a visual system with the algorithm for soldering pads recognition is described. The research has shown that the recognition system worked in line with expectations. The application of the presented algorithm for pads detection can reduce the time required to prepare the control program for a soldering robot. The important advantage of the proposed software is the possibility to work in the off-line mode, without the use of the soldering station. After preparing data points, user can connect to the internet and chose station by TCP/IP protocol. New soldering program can be downloaded to the station according to industry 4.0 principle. The best results of pads recognition were achieved when the PCB images were taken from a 2D scanner. The use of industrial camera caused problems with lighting and with a perpendicular orientation of the camera above a center point of the PCB. White lights, which are commonly used in rooms, can produce shades and different intensity of light, therefore the use of camera caused errors in pads recognition. The algorithm worked very well with scanned image when background color was set properly; the best results were obtained when the background color was pink. The problem with proper recognition occurred, when there was a text description on a PCB, in a similar color to the color of pads (for example white). To avoid detection errors in such cases, a pink color of background was used. After that recognition system has worked as expected. The average error of detection point compared with Gerber file is 0.06 mm.

Acknowledgment. This paper was supported by the Polish National Centre of Research and Development, grant no. POIR.01.01.01-00-0014/15.

References

1. Nandi, C.S., Tudu, B., Koley, C.: An automated machine vision based system for fruit sorting and grading. In: 2012 Sixth International Conference on Sensing Technology (ICST), pp. 195–200 (2012)
2. Akbar, H., Prabuwono, A.S.: The design and development of automated visual inspection system for press part sorting. In: International Conference on Computer Science and Information 194 Technology, ICCSIT 08, pp. 683–686 (2008)
3. Bodhale, D., Afzulpurkar, N., Thanh, N.T.: Path planning for a mobile robot in a dynamic environment. In: IEEE International Conference on Robotics and Biomimetics (ROBIO), Bangkok, pp. 2115–2120 (2009)
4. Hong, S.M., Jang, W.S., Son, J.K., Kim, K.S.: Evaluation of two robot vision control algorithms developed based on N-R and EKF methods for the rigid-body placement. In: International Conference on Advanced Intelligent Mechatronics, Wollongong, Australia, vol. 204, pp. 938–943, 9–12 July 2013
5. Govardhan, P., Pati, U.C.: NIR image based pedestrian detection in night vision with cascade classification and validation. In: International Conference on Advanced Communication Control and Computing Technologies (ICACCCT), pp. 1435–1438 (2014)
6. Schomerus, V., Rosebrock, D., Wahl, F.M.: Camera-based lane border detection in arbitrarily structured environments. In: 2014 IEEE Intelligent Vehicles Symposium Proceedings, vol. 207, pp. 56–63 (2014)
7. Chugo, D., Hirose, K., Nakashima, K., Yokota, S., Kobayashi, H., Hashimoto, H.: Camerabased navigation for service robots using pictographs on the crossing point. In: IECON 8th Annual Conference on IEEE Industrial Electronics Society, pp. 154–4159 (2012)
8. Sampe, I.E., Amar Vijai, N., Tati Latifah, R.M., Apriantono, T.: A study on the effects of lightning and marker color variation to marker detection and tracking accuracy in gait analysis system. In: International Conference on Instrumentation, Communications, Information Technology, and Biomedical Engineering (ICICI-BME), pp. 1–5 (2009)
9. Kim, D., Choi, J., Park, M.: Detection of multi-active markers and pose for formation control. In: 2010 International Conference on Control Automation and Systems (ICCAS), pp. 943–946 (2010)
10. Fukuzawa, M., Hama, H., Nakamori, N., Yamada, M.: High-speed distance measurement between moving vehicles with NIR-LED markers. In: 11th International Conference on Computer and Information Technology, ICCIT 2008, pp. 516–520 (2008)
11. Gherghina, A., Olteanu, A., Tapus, N.: A marker-based augmented reality system for mobile devices. In: 2013 11th Roedunet International Conference (RoEduNet), pp. 1–6 (2013)
12. PONTOS—Dynamic 3D Analysis. http://www.gom.com/metrology-systems/system-overview/pontos.html. Accessed 15 May 2018
13. Mochizuki, Y., Imiya, A., Torii, A.: Circle-marker detection method for omnidirectional images and its application to robot positioning. In: IEEE 11th International Conference on Computer Vision, ICCV 2007, pp. 1–8 (2007)
14. Khler, J., Pagani, A., Stricker, D.: Detection and Identification Techniques for Markers Used in Computer Vision, in Modeling and Engineering. Bodega Bay, CA, USA, o.A (2011)
15. Haddad, R.A., Akansu, A.N.: A class of fast gaussian binomial filters for speech and image processing. IEEE Trans. Acoust. Speech Signal Process. **39**, 723–727 (1991)
16. Nixon, M.S., Aguado, A.S.: Feature Extraction and Image Processing, p. 88. Academic Press (2008)

The Idea of "Industry 4.0" in Car Production Factories

Jarosław Kurosz[1,2] and Andrzej Milecki[1,2(✉)]

[1] Volkswagen Poznań, Poznań, Poland
jaroslaw.kurosz@vw-poznan.pl
[2] Poznan University of Technology, Poznań, Poland
andrzej.milecki@put.poznan.pl

Abstract. This article is covering the topic of Industry 4.0 which is a name for the new trend of production automation and data exchange in manufacturing systems. The history of industrial revolutions is shortly sketched in the beginning. Then the term "Industry 4.0" and its area of influence is presented. The different solutions used in Industry 4.0 factory are described focusing on the methods important for car production; these methods are then shortly explained and illustrated. The idea of smart factory for a car producing sector is presented. Finally the most important disadvantage of Industry 4.0 is discussed; which is the wide open access to Internet requirement that may be dangerous for the company.

Keywords: Industry 4.0 · Car production · Production control
Manufacturing · Internet of Things (IoT)

1 Introduction

The first industrial revolution began with the development of mechanics which enabled constructing a steam engine in 1800s [1], that was used to generate high mechanical power. The availability of such power has allowed the mechanization of many industrial processes in which the manual work (mostly in textile industry) was replaced with machines. As a result the production efficiency and also quality of life were significantly improved. The electrification was a main driver of the change called "the second industrial revolution", which enabled industrialization and mass production. The introduction of microelectronics into everyday life and implementation of automation has caused the third industrial revolution. It is characterized by the digitalization, which allows for flexible production, where different, customized products are manufactured on the production lines controlled by programmable digital controllers like PLCs.

The last achievements of science in the areas like computerization, modern control, modelling, artificial intelligence and Internet of Things (IoT), that have begun their way into almost every household in high developed countries, induces the industry to formulate an answer to the question whether and how wider than before the application of these developments is possible. Nowadays, the most popular answer is the term

© Springer Nature Switzerland AG 2019
A. Burduk et al. (Eds.): ISPEM 2018, AISC 835, pp. 597–607, 2019.
https://doi.org/10.1007/978-3-319-97490-3_57

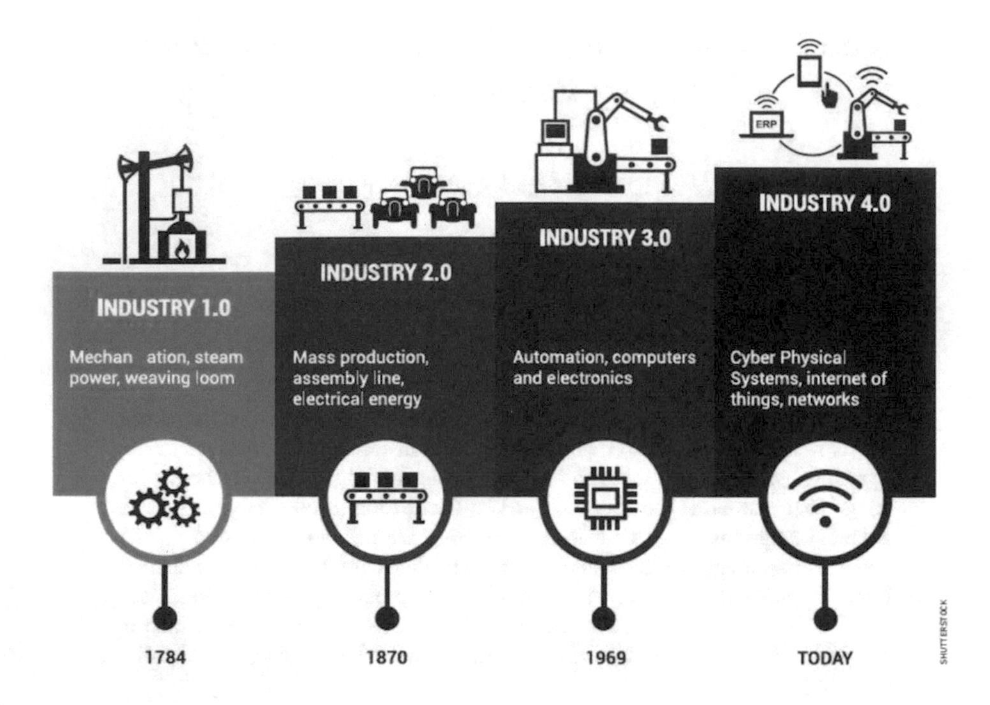

Fig. 1. History of industrial revolutions (Industry infographic [2])

"Industry 4.0", which is believed to be the foundation for the next, fourth industrial revolution [2]. The history of industrial revolutions is illustrated in Fig. 1.

2 The Main Features of "Industry 4.0" Concept

The term "Industry 4.0" has originated from a German government project in the field of development strategies [2]. The initiative "Industrie 4.0" (original German term) was revived and published during Hannover Fair [3]. One year later, the group working on Industry 4.0 presented a set of its implementation recommendations to the German federal government. In 2013 at the Hannover Fair, the final report of the Industry 4.0 was presented [4]. Although "Industry 4.0" is now an important concept for industries and universities, a generally accepted definition of this term still does not exist. For example, in [5] the Authors have defined it as an autonomous, knowledge and sensor based, self-regulating production system. In a study [6] Industry 4.0 is defined as profound transformation of business models by enabling the fusion of virtual and real words, and the application of digitization, automatization, and robotics in manufacturing.

The beginning of the fourth industrial revolution is mainly based on Information and Communications Technologies (ICT) development [7], but in the field of the technology the base of this revolution are: smart automation of cyber-physical systems, decentralized and intelligent control, advanced connectivity implemented by Internet of Things (IoT), clouds and big data processing. For industrial production systems, the implementation of new technologies results in changes of classical automation systems to self-organizing cyber physical production systems that allow flexible adjustments in product features and in production quantities. The main goal of Industry 4.0 is to fully utilize the potential of those technologies to create a so-called "smart factory".

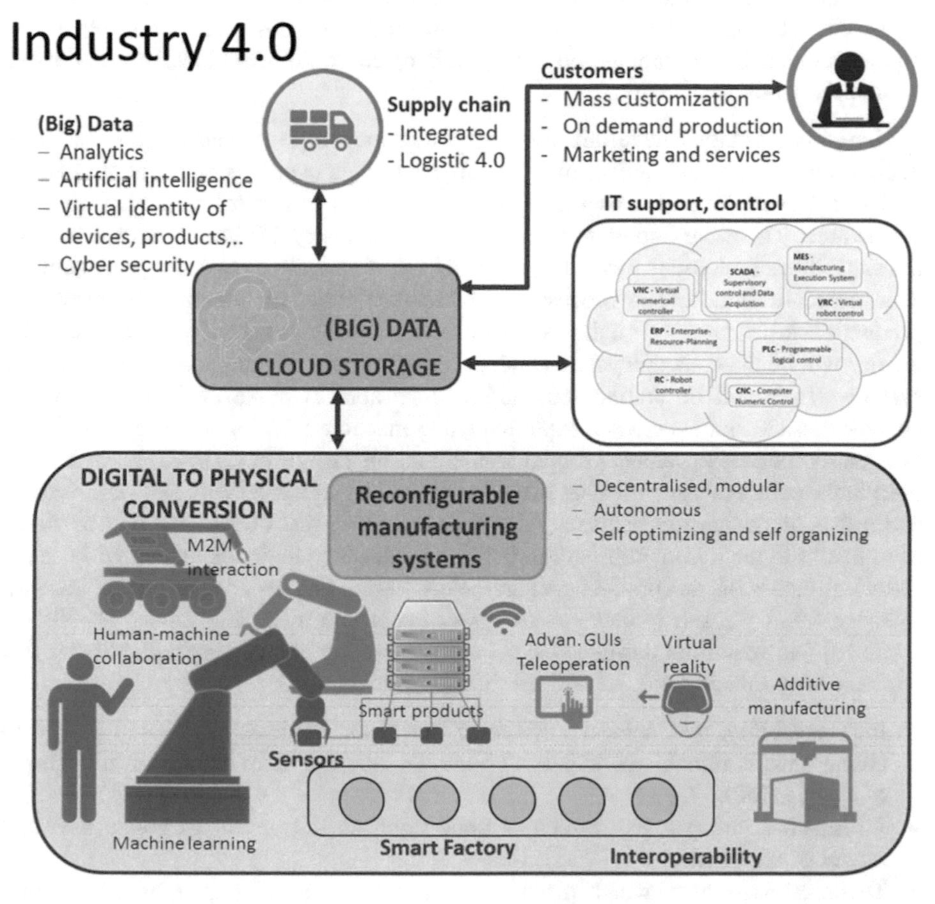

Fig. 2. The illustration of different solutions of Industry 4.0 factory [7]

Figure 2 presents the Industry 4.0 solutions in a factory producing smart products. Its core element is a production line with machine tools, robots and additive manufacturing machines. In this factory the reconfigurable manufacturing systems are implemented, with flexible production system controlled by programmable devices, machines, robots, transport system like Autonomous Guided Vehicles (AGV), etc. All that allows for manufacturing of different types of products with variable equipment. The most advanced reconfigurable manufacturing systems are able to adapt their equipment (hardware and software) to a wide spectrum of products and thus to various market requirements concerning types and quantity of the products. Thanks to these features and technological possibilities, such a factory can be named as "smart" or even "intelligent". According to Industry 4.0 requirements, the following functionalities and methods are used in the modern factory:

- Communication between machines, devices and human operators with the use of Virtual Reality, Augmented Reality and direct Human-Machine collaboration,
- Machine-To-Machine (M2M) communication and interaction,
- Machine learning application in robots and production devices,
- Decentralized, autonomous and self-optimizing control of machines, robots, devices and the whole production.

Large corporations have already been using thousands of sensors on their production lines that would deliver many gigabytes of data every second. The problem is how to retrieve information useful for the improvement of control methods from such big datasets. Therefore important innovation in the Industry 4.0 factory is expansion of the capabilities of computer systems, to be able to effectively explore large amounts of data stored in clouds. The processing of big data may, for example, improve the production in the area of quality and efficiency. Production devices in Industry 4.0 factory will be the Cyber-Physical Devices, i.e. physical elements integrated with ICT systems. They will be autonomous and will be able to make decisions basing on analytics results and experience exploited using machine learning algorithms and real-time data capture. Nowadays programmable controllers like PLC, PAC, IC or NC are used in the control systems, which in the future will be equipped with self-organization and self-optimization mechanisms. Also the production line downtimes will be shortened thanks to the introduction of Predictive Maintenance methods, which will use the historical data with the production parameters. According to certain predictions [8], Industry 4.0 may result in decrease of costs of production and logistic by 10–30%.

In [9] four important design principles for supporting companies in identifying and implementing Industry 4.0 are defined. These principles are:

- Interoperability, which means ability of machines, devices, robots and people to connect and communicate with each other via the Internet of Things or the Internet of People (IoP).
- Information transparency, which is understood as the ability to create a virtual model of the plant.
- Technical assistance, which means the ability to support humans by aggregating and visualizing information about production for solving urgent problems.
- Decentralized decisions, which means the ability of cyber physical systems to make decisions on their own and to perform their tasks as autonomously as possible.

3 Industry 4.0 in the Car Production Sector

The car production sector is considered to be highly capital- and labor intensive. The big auto manufacturers are called Original Equipment Manufacturers (OEMs). They produce some of their own parts, like auto car bodies, but they can't produce every part and component that is installed into a new vehicle.

According to the Industry 4.0 principle, production devices can connect to each other and to human through digital interfaces, providing real-time data from many sensors and actuators. A human can receive data about production at any time [10–14]. Now most automotive facilities haven't yet reached that perfect connection in which machines and humans work together seamlessly. Implementation of Industry 4.0 will enable better communication between the factory and their products, i.e. with manufactured and sold cars.

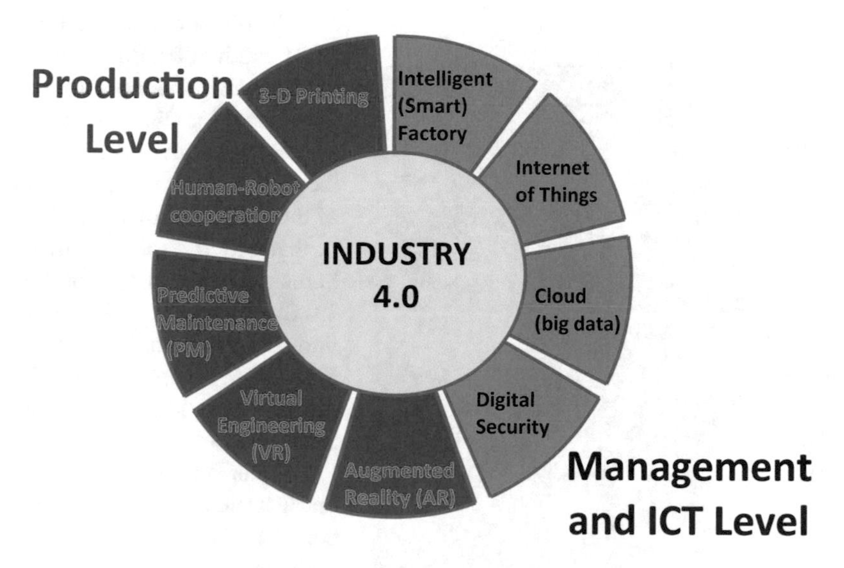

Fig. 3. Industry 4.0 principles - the areas of interest for car producing factories

Fig. 4. 3-D printing examples

According to analysis made at VW Poznan, the implementation of Industry 4.0 methods in the auto production facility should be initially focused on two levels (Fig. 3):

- Production Level
- Management and ICT Level.

In the Production Level the following implementation projects (areas) are already started: 3D printing, Human-Robot direct cooperation and envisaged: Predictive Maintenance, Virtual Engineering, Augmented Reality. In the Management and ICT related Level the following implementation of projects (areas) of Industry 4.0 seems to be most important: storage big data in private cloud and retrieval in real time useful information. In this are crucial is the data security is extremely important. The long term future is the creation of so called “Intelligent Factory”.

In the last few years the possibilities of using 3D printing devices in production of specific car components has increased significantly. These devices are now capable of printing hard, rigid, durable and precise elements that can be used in cars without any concerns. For example VW Poznań factory uses the 3D printing technology for manufacturing of prototype elements, instruments for production and spare parts (see Fig. 4).

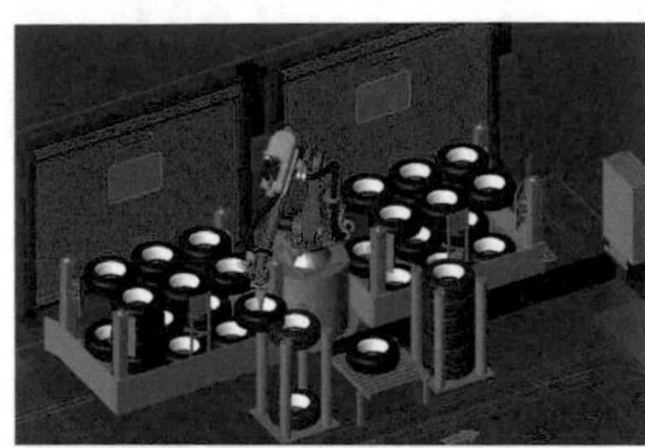

Fig. 5. Robot – human cooperation

The direct cooperation between robot and human, during which the robot assists the operator is very important for the car producing sector, as shown in the examples in the Fig. 5. This is also a vital part of the definition of Industry 4.0, as in this way the robot complements human capabilities. In the Industry 4.0 facility there will be no division into automatic and manual workplaces. People and robots will work together seamlessly - without protective fences.

Thanks to the algorithms based on historical data and thanks to constant monitoring of technical parameters the Predictive Maintenance (PdM) of production devices is now possible. Using huge historical data (Big Data) and advanced analytics, the prevention of failures is possible, which allow longer devices availability. Thanks to implementation of PdM the breakdown times and thus production costs can be significantly reduced. In the book [15], the following benefits to PdM are linked

- Maintenance costs - down by 50%,
- Unexpected failures - reduced by 55%,
- Repair and overhaul time - down by 60%,
- Spare parts inventory - reduced by 30%,
- 30% increase in machinery mean time between failures (MTBF).

These numbers seem too high, but PdM can really bring significant impact and, for the typical manufacturing plant, a 10% reduction in maintenance costs is realistic. In car factory, every 10 min. breakdown means not producing 3-5 cars. In an advanced PdM system the repair procedures can be carried out according to:

1. Error detection (automatically or by the user)
2. Automatic diagnosis – error identification
3. Repair procedure
4. Tests and reporting to the supervisory system.

Such support delivered by PdM will surly reduce significantly the repair time.

Next Industry 4.0 principle is the Virtual Engineering (VE). The 3D visualization of production line enables verification of designs and conducting a thorough analysis of possible solutions (Fig. 6). It leads to the reduction of financial risks and streamlining the decision-making and purchasing processes. In the car design, the use of VE leads to improvement of development process. It also allows faster preparation of the production of new models.

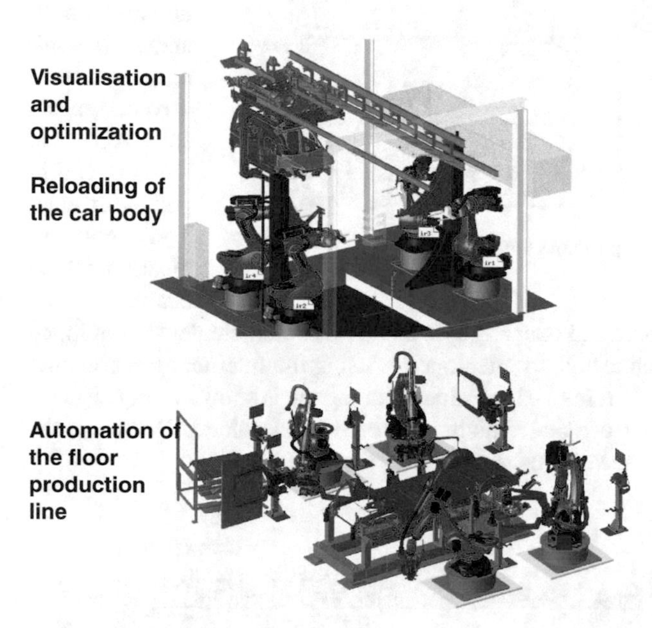

Fig. 6. Virtual Engineering examples

The amount of electronic information measured on the production lines is staggering. Storing all those digital data in one typical data centre can be too expensive. That's why cloud storage has become of interest (Fig. 7), due to lower maintenance costs and shifting Capex (business expense incurred to create future benefit) to Opex (expenditures required for the day-to-day functioning of the business). The digital data is stored in logical pools, on multiple servers, which crate cloud of data. The cloud physical environment is owned and managed by a hosting company, which is responsible for keeping the data available and accessible in a "as a service" mode.

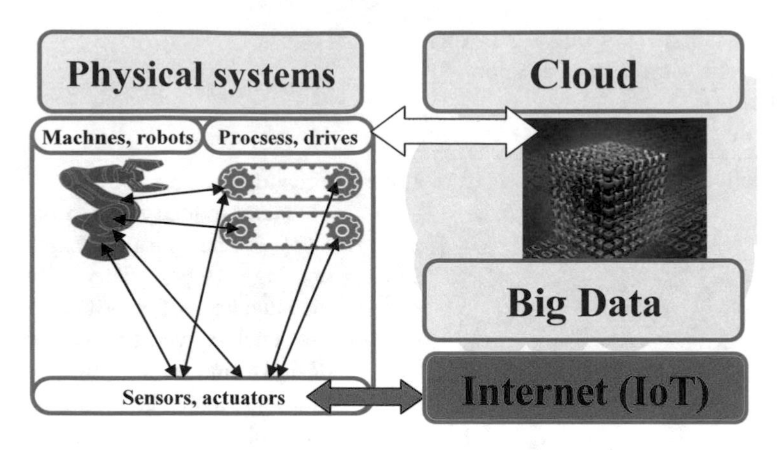

Fig. 7. IoT and cloud interaction

The Internet of Things (IoT) is the network of physical devices like production appliances (machines, robots, drives, sensors etc.), AGV and humans equipped with electronic communication, which enables all these subcomponents to connect and exchange data (Fig. 8). Each element is equipped with computing system, which is able to inter-operate using the Internet infrastructure. A typical example of such system for a car producing company is shown in Fig. 9. This system produces a big amount of data, which are stored in the cloud. However, the problem is how to assure the security of data and ICT system.

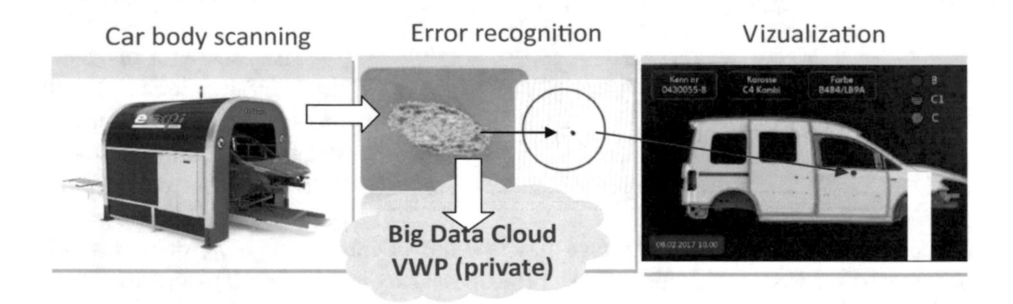

Fig. 8. Big Data generation by car body scanning for error recognition in visualization

The real problem is how to utilize the huge amount of measured and stored data sets at terabyte or even petabyte scale in order to make them usable in the production process. To this end only a new data mining technologies can be useful and offline batch data processing is typically used. However the toughest task is to a do real-time analysis on a complete big data set, which means that terabytes (or even more) of data must be processed within seconds. This is only possible when data is processed in parallel by several processors. The utilization of gathered data for generation of feedback information for the production control is illustrated in Fig. 9.

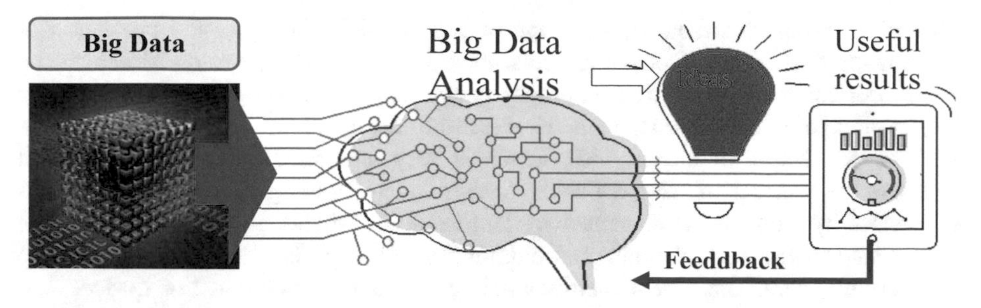

Fig. 9. Big Data analysis for delivering of usable feedback for the production control

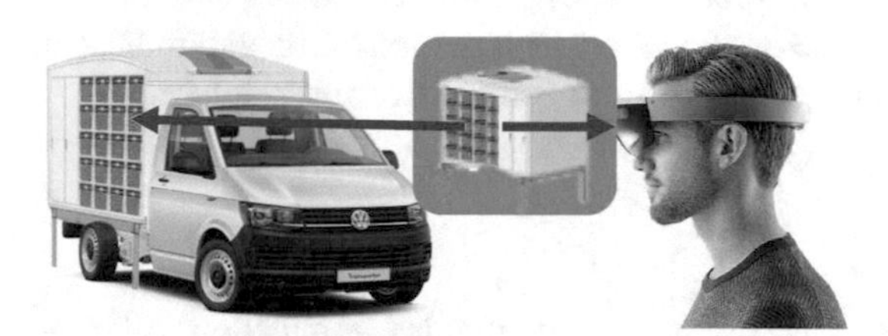

Fig. 10. AR in design

The important element of Industry 4.0 is the common use of Augmented Reality (AR) solution in chosen areas of design and production. Nowadays in VW Poznan it is used in the design of cargo space of such cars like Caddy, T5 and Crafter, The virtual picture of designed cargo space is augmented on a real car. The designer uses special glasses to observe how the result would look like (Fig. 10). Another application of AR technology is its use for training or in assembly stations. In such case the additional pictures (drawings) or instructions will be displayed and combined with the real elements.

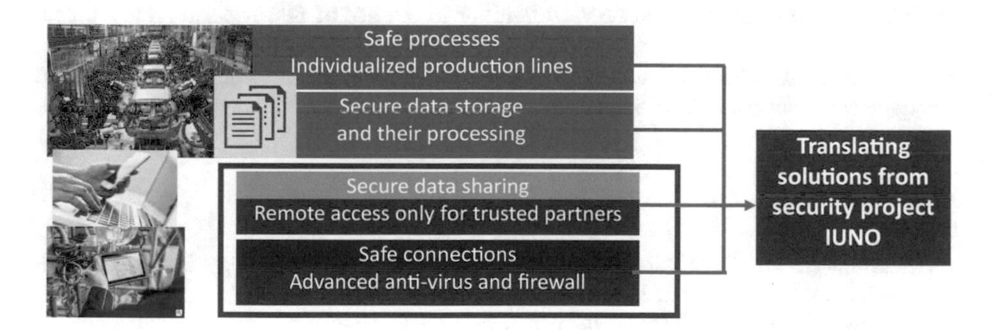

Fig. 11. Security in Industry 4.0 [16]

The concept of Industry 4.0 assumes wide access and use of Internet and that the factory network would be open for communication with the whole word. However, this would enable every internet users, in or outside the company to communicate with the facilities in the company. This exposes the plant network to the risk of data theft and to the possibility of being attacked by hackers infecting the plant automation systems.

This is really an important disadvantage of Internet 4.0. Therefore the protection of information during their processing and storage is absolutely necessary. Moreover the access control methods as well as proper networks segregation should be used. The important solutions in security area are presented in Fig. 11.

The last feature of Industry 4.0, which is very important for the car production sector is the concept of "Smart Factory", also called "Intelligent Factory", which is supervised by integrated management system for all installations. The Smart Factory is a next step in the way of traditional automation into a fully connected and flexible production system. This type of factory will integrate data from production devices, and human assets with operational data from management systems to introduce optimal control of manufacturing processes. Such control system will react (i.e. predict and adjust) in real time, adopting production process, taking into account its current state and other conditions, leading to increase in efficiency, cost minimization and finally, in better positioning of the car producing factory in the competitive marketplace.

4 Conclusion

The described above Industry 4.0 concept will surly require applications of new advanced technologies and methods in manufacturing. The integration of production equipment with human operators and with digital control and management systems will force to use faster communication networks and more secure systems. The smart factory will be based on self-adoption and devices which will utilize the control and reasoning methods based on Big Data and artificial intelligence. Implementation of smart factory in practice requires new competences of human personnel involved in the production and control of the devices and of the whole factory. In addition to excellent knowledge of computer systems, mechanical engineering, automation, the worker of the Industry 4.0 factory should be very familiar to the use of Internet and its resources.

Acknowledgment. This paper was supported by the Polish National Centre of Research and Development, grant no. 02/22/DSPB/1434.

References

1. Tunzelmann, N.: Historical coevolution of governance and technology in the industrial revolutions. Struct. Change Econ. Dyn. **14**(4), 365–384 (2003)
2. http://wbj.pl/industry-4-0-are-we-there-yet/. Accessed 07 Jun 2018
3. https://www.bmbf.de/de/zukunftsprojekt-industrie-4-0-848.html. Accessed 07 Jun 2018
4. https://www.dfki.de/web/presse/pressehighlights/industrie-4-0-mit-dem-internet-der-dinge-auf-dem-weg-zur-4-industriellen-revolution/view. Accessed 07 Jun 2018
5. Kagermann, H., Anderl, R., Gausemeier, J., Schuh, G., Wahlster, W. (eds.): Industrie 4.0 in a Global Context. acatech STUDY (2016)
6. Lasi, H., Fettke, P., Feld, T., Hoffmann, M.: Industry 4.0. Bus. Inf. Syst. Eng. **6**(4), 239–242 (2014)
7. Götz, M., Jankowska, B.: Clusters and Industry 4.0 – do they fit together? J. Eur. Plan. Stud. **25**(9), 1633–1653 (2017)

8. Rojko, A.: Special focus paper - Industry 4.0 concept: background and overview. iJIM, **11**(5), 77–90 (2017)
9. Bauernhansl, T., Krüger, J., Reinhart, G., Schuh, G.: WGP-Standpunkt Industrie 4.0. Wissenschaftliche Gesellschaft für Produktionstechnik WGP e.v. (2016)
10. http://www.huawei.com/en/about-huawei/publications/winwin-magazine/29/accelerating-success-in-the-4th-industrial-revolution. Accessed 07 Jun 2018
11. Bures, T., Weyns, D., Klein, M., Haber R.E.: 1st International workshop on software engineering for smart cyber-physical systems. In: Proceedings of International Conference on Software Engineering, pp. 1009–1010 (2015)
12. Lin, S.-W., Miller, B., Durand, J., Bleakley, G., Chigani, A., Martin, R., Murphy, B., Crawford, M.: The Industrial Internet of Things. vol. G1, Industrial Internet Consortium (2017)
13. https://www.plattform-i40.de/I40/Redaktion/DE/Downloads/Publikation/rami40-eine-einfuehrung.pdf?__blob=publicationFile&v=10. Accessed 07 Jun 2018
14. IMPULS Foundation of the German Engineering Federation (VDMA). https://www.industrie40-readiness.de/?lang=en. Industrie 4.0 readiness check tool for companies. Accessed 07 Jun 2018
15. Mobley, K., (eds.): Plant Engineer's Handbook. BH (2001)
16. https://iuno-projekt.de/. Accessed 07 Jun 2018

Mining 4.0 and Intelligent Mining Transportation

Process-Oriented Approach for Analysis of Sensor Data from Longwall Monitoring System

Edyta Brzychczy(✉) and Agnieszka Trzcionkowska

AGH University of Science and Technology, Kraków, Poland
{brzych3,toga}@agh.edu.pl

Abstract. In the paper we address possibility of industrial process analysis based on sensor data with the use of process-oriented analytic techniques. We propose in this area usage of process mining techniques. Important issue in the context of usefulness of industrial sensor data gathered in the monitoring systems for process analysis is proper level of abstraction of an event log.

We present our approach requiring creation of high-level event logs based on low-level events from longwall monitoring system in order to model and analyse the mining process in an underground mine. We use combination of unsupervised data mining techniques as well as domain knowledge to discover stages in an example process and create an event logs for further process analysis.

Keywords: Sensor data · Process mining · Event logs · Longwall face
Process-oriented analysis

1 Introduction

Just like the classical branches of manufacturing industry, mining also tries to create new solutions that improve the efficiency of processes within the Industry 4.0 concept. New mining technologies, new machines and devices, digitization, virtual and augmented reality are becoming a fact [1]. Wide automation initiatives and development in mines can be seen all over the world in various mining branches, both in underground as well as in opencast mines [13, 14, 16, 27].

Main goal of automating the longwall mining process was deemed to be the principle means to drive significant improvements in efficiency and safety [14]. At the same time, automation enabled access to very detailed data characterizing the operation of machines and devices (stored in monitoring systems), that allows also a closer look at the ongoing processes in the mines. Vast amount of data is generated that should be used and handled more efficiently in modern mining operations [7]. In this aspect, advanced data analytics and usage of hidden knowledge from data to support process analysis and management should be applied [3].

In analysis of monitoring data, especially for reliability and predictive maintenance purposes, various wide known data mining techniques could be used: association rules induction [12], classification techniques (support vector machines [26], decision trees [6], artificial immune systems [4]), regression [10], time series analysis [19], clustering methods [15] and many others [9]. They are rather data-oriented techniques, that do not

© Springer Nature Switzerland AG 2019
A. Burduk et al. (Eds.): ISPEM 2018, AISC 835, pp. 611–621, 2019.
https://doi.org/10.1007/978-3-319-97490-3_58

handle well with characteristic aspects of processes (e.g. process complexity, possible concurrency or parallelism, loops, decision points or specific conditions of process realisation). Looking more in detail on process characteristic, for its modelling and optimisation purposes, usage of process-oriented analytic techniques is more relevant. Nowadays, new possibilities in this area are process mining techniques enabling process modelling and analysis, based on event logs from IT systems in organisations.

In the paper we present process-oriented approach for analysis of mining process based on data from longwall monitoring system. We present in detail process of an event log creation enabling usage of process mining techniques for modelling and analysis of a selected process in a longwall face.

2 Process Mining

Process mining is a rapidly growing discipline, described as a bridge between process science and data science [23]. Process mining techniques directly relate event data to end-to-end business processes [22]. Various process mining algorithms enable discovering the real processes from event data, automatically identification of bottlenecks, analysis of deviations and sources of non-compliance [24]. The main source of data for process mining is an event log.

Event log include structured data about process performance. Each event in such log refers to an activity and is related to a particular case. The events in a case are ordered and can be seen as one iteration of the process (process instance). The sequence of activities executed for a case is called a trace and an event log can be viewed as a multiset of traces. Often event logs store additional information about events. For example, resource (i.e. person or device) executing or initiating the activity, data elements recorded with the event (e.g., the size of an order) [24]. Obligatory elements of event log are as follows: case id, event/activity name and timestamp.

There are three main types of process mining tasks [23]:

1. process discovery,
2. conformance checking,
3. enhancement.

Process discovery techniques use an event log and create a model without using any *a-priori* information. Typically the focus of process discovery techniques in on the control-flow aspect of a process. Various process model formalisms could be used, e.g. Petri nets, process trees, BPMN, state charts etc. [24].

During conformance checking an existing process model is compared with an event log of the same process to detect and locate deviations between process model and real process execution (measuring the alignment between model and reality).

The process enhancement enables extension of analysis or improvement of the process by use of the additional information recorded in the event log i.e. involved resources or adding other perspectives to the process model (i.e. organizational, time or case perspective).

Event logs are very often obtained directly from existing IT systems, so events granularity can be different. For any process mining technique ability to identify execution of activities based on events is crucial [11]. Too low level of an event log might result in spaghetti-type of the process model containing uninterpretable mess of nodes and connections with too specific (non-meaningful) names [21]. Moreover, events that do not correspond directly to activities are unsuitable for process analysis if their meaning are not clear to domain experts [11]. In such case, corresponding high-level event log should be used.

Various methods to abstract too low-level event logs into higher level event logs were developed. They can rely on unsupervised learning techniques [2, 5, 8, 25] or supervised abstraction based on available training data with high-level target labels of low-level events [21] or behavioural activity patterns that capture domain knowledge [11].

These considerations are very important in the context of usefulness of industrial sensor data gathered in the monitoring systems for process analysis. Recently, practical problems explored in this area, are mainly related to recognition of human behaviour (activity recognition) based on various sensors for, e.g., self-tracking [20] or improving smart products [25].

In this paper we present approach for creation of high-level event logs based on low-level events from longwall monitoring system in order to model and analyse the mining process in an underground mine.

3 Problem Description

The mining process is a collection of mining, logistics and transport operations. The specificity of this industrial process carried out underground is connected with the changeable conditions of its realisation (geological and mining conditions, natural hazards not occurring on the surface) [17] and the nature of the process performed by machines and devices moving in a workspace and also in relation to each other [18].

Machines and devices within the longwall face (e.g. shearer, conveyor, mechanized roof supports) are equipped with a various number of sensors recording their physical (e.g. temperature, currents, voltages, pressure etc.) and logical parameters (switch on/off) referring to the state of the process at a very low level of abstraction. The data from sensors have different granularity (e.g. some of them are saved every second, some only when the value is changed - 0/1) and different character (most of them are binary variables), that makes their analysis a complex and non-trivial task.

Data in the real form and structure cannot be used directly to analyse the process or improve its effectiveness. It results from the nature of the data but also from the complexity of the analysed process. This process indeed consists of dozen of individual processes of machines and devices, and depending on the dimensions of a mining excavation, it can include up to hundred objects that realise their own processes that finally create the mining process.

Our approach for analysing data from machinery sensors for mining process modelling and analysis purposes consists of stages presented in Fig. 1.

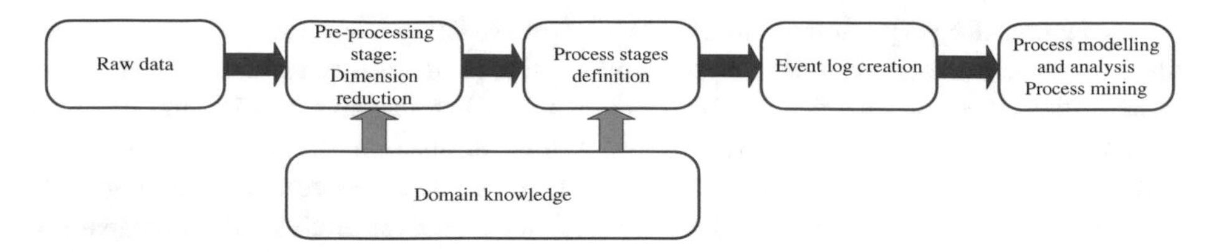

Fig. 1. Stages of the proposed approach

Basic data set include raw data from longwall monitoring system. There are low-level sensor data from various machinery (among other: shearer, conveyors, crusher, transformer, pumps) and devices (mechanized roof support). Characteristic feature of data set is its structure – up to1000 variables mainly binary (approx. 70%).

At the beginning pre-processing stage have to be carried out and data are prepared in a form that allows identification of the stages in the process. It is important to determine the desired data granulation and reduce the set of variables.

After selection of the final data set, unique states in the operation of machines and devices, associated with the stages of the process, have to be distinguished. Due to the presence of quantitative variables in the data set, over-dimensionality problem can be found. Therefore, discretization of quantitative variables should be done (with selected methods and adequate intervals).

After distinguishing unique states of the machinery operation, it is important to group them into more general process stages with use of unsupervised learning techniques (namely cluster analysis). Generalization degree of the defined stages depends on the required detail level of the process model according to process analysis needs and purposes. Clearly defined process stages enable to prepare appropriate event logs, that can be used in process modelling. Considering character of the mining process and longwall machinery operation, to ensure correctness of the defined process stages, domain knowledge should be used, especially at the pre-processing stage (selection of relevant variables) and process stages definition.

Creation of usable event logs (besides timestamps and defined process stages) requires traces indication. It has to be emphasized that in raw data there is no clear process trace definition. In a case of mining process in a longwall face – the shearer operation cycle is the most clear and intuitive choice for trace identification (Fig. 2). For this step it is very important to ensure high quality of data to distinguish traces in data.

Event logs enable further use of various process mining techniques for process model discovery, conformance checking and analysis of process performance.

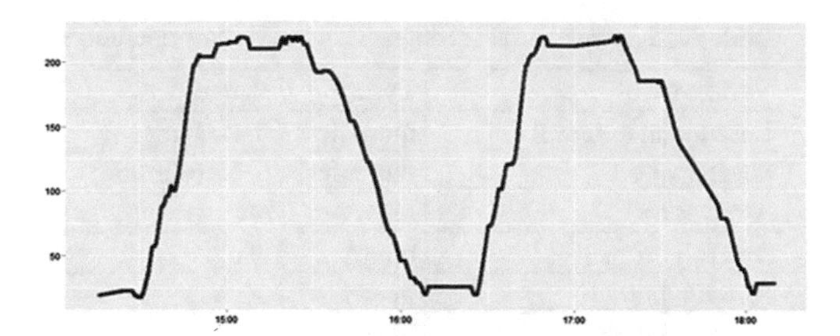

Fig. 2. Example cycles of shearer operation (x axis – time, y axis – location in a longwall face)

In the next section we present an example of event log creation based on low-level sensor data regarding to shearer operation process in an example longwall face.

4 Creation of an Event Log of Shearer Operation Process

Presently data from longwall monitoring system are mainly used for creation of simple daily reports, especially regarding to shearer operation. Report includes simple time statistics related to pre-defined stages of the shearer operation (e.g. shearer stoppage, work with loading, work without loading, work with moving), currents' statistics and other information (e.g. speed under loading, incidents, alarms).

In our approach we assume distinguishing the mining process stages known to experts as well as identifying clearly occurring stages that have not been defined yet, but clearly appear in the process data or exist in the process, but not stored directly in the sensor data (e.g. conveyor drives movements).

Analysed data set was obtained from one of the Polish mining companies. Raw data are related to shearer operation process (include 2,5 million records and 176 variables) from monthly period.

At the pre-processing stage exploratory data analysis was conducted. Firstly, quality of data was investigated. Variables with min. 40% rate of missing values or with only one value (in a case of binary variables) were excluded from further analysis. Afterwards correlation analysis for numerical variables and cross tables for binary variables were prepared. On the basis of obtained results and domain knowledge the most important variables regarding to shearer operation were chosen:

- basic parameters of operation: speed, currents on organs and tractors, position of organs,
- additional parameters related to security systems: overheating, short circuit, overload,
- parameters related to position in a longwall face: location, direction of movement.

In the Table 1 selected variables characterizing shearer operation are presented.

Table 1. Selected variables characterizing shearer operation

Variable	Type	Range
Location in longwall	numerical	0–200 [m]
Shearer speed	numerical	0–20 [obr/min]
Arm left up	binary	0/1
Arm left down	binary	0/1
Arm right up	binary	0/1
Arm right down	binary	0/1
Move in the right	binary	0/1
Move in the left	binary	0/1
Current on the left organ	numerical	0–613 [A]
Current on the right organ	numerical	0–680 [A]
Current on the left tractor	numerical	0–153 [A]
Current on the right tractor	numerical	0–153 [A]
Security DMP left organ	binary	0/1
Security DMP right organ	binary	0/1
Security DMP left tractor	binary	0/1
Security DMP right tractor	binary	0/1

Selected variables were used for distinguishing unique states of the machinery operation and process stages definition.

4.1 Process Stages Definition

Process of stages definition for shearer operation description was conducted in the following steps described below.

Firstly, all continuous variables were discretized. Intervals for currents of organs and tractors were divided into idle running, small load, medium load, high load and overload. The remaining variables, i.e. position in the longwall and speed, were divided into 6 intervals.

Then, the output data set containing both discretized continuous variables and binary variables was reduced to unique, uncommon combinations of shearer operation parameters. This resulted in a matrix, that has been used in the next stage, i.e. in hierarchical clustering.

The hierarchical clustering is based on data in the form of dissimilarity matrix. The challenge in this case was to create dissimilarity matrix for qualitative variables. For this purpose, the Gower distance was used, allowing to determine the difference between the qualitative variables. In this case, the matrix consists of a value of 0 or 1. Gower's distance taking into account distances between pairs of variables over all dataset and then combines those distances to a single value per record-pair. Hierarchical clustering based on created matrix was performed using the Ward's minimum variance method. Results, in a form of dendrogram, are presented in Fig. 3.

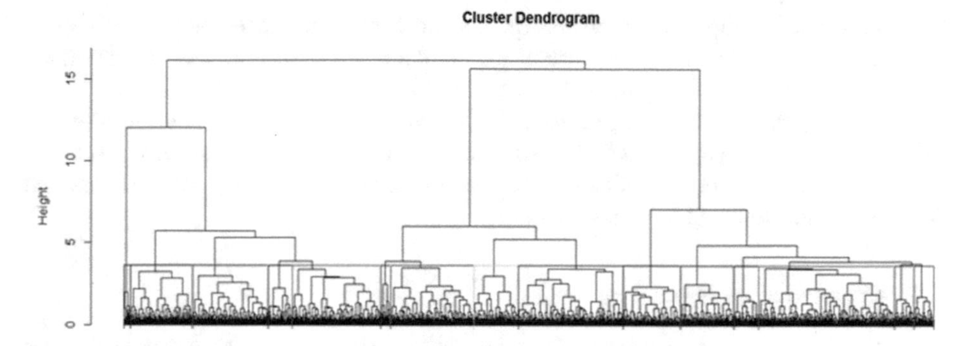

Fig. 3. Cluster dendrogram of unique stages of shearer operation

Hierarchical grouping is unsupervised machine learning method that requires a predetermined number of groups. Determination of the optimal number of groups was based first of all on the computation of silhouette information according to a given clustering in k clusters, and then on the interpretation of the practical use of the obtained groups to determine the characteristic process states of the shearer operation. On the basis of the results of the average silhouette value, 6 groups were pre-selected, however, due to the specificity of the shearer operation and its location in the longwall their number was increased to 14.

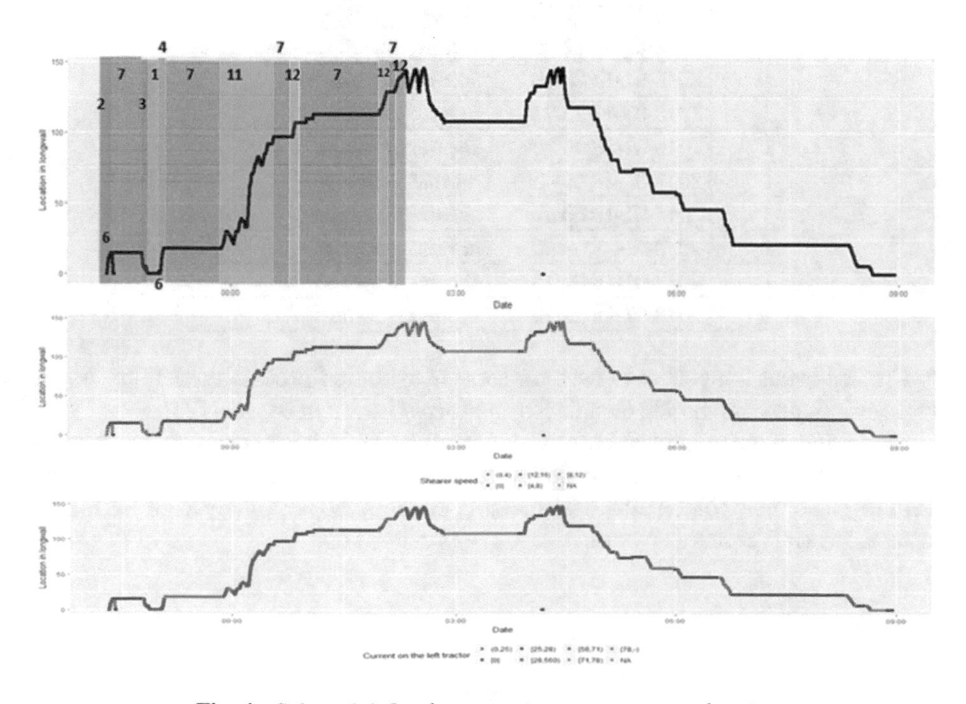

Fig. 4. Selected defined process stages on an example trace

The selected groups have been described as process stages based on the statistics of the analyzed parameters with the use of expert knowledge. In Fig. 4 selected defined stages of shearer operation process are presented.

These stages include: (1) operational activities near drive unit, (2) cutting towards drive unit, (3) slotting by shearer, (4) cutting towards reverse drive, (6) shearer reversion, (7) shearer stoppage (e.g. roof problems, cleaning), (11) intense working (diversified operation conditions), (12) normal working.

4.2 Creation of an Event Log

The characteristic process stages of shearer operation were used to create high-level event log (Table 2). The main variable for event log creation was “location”, enabling identification of traces.

Table 2. Event log for an example trace

Trace	Timestamp	Activity/process stage
1	2018-01-17 21:00	cutting towards drive unit
1	2018-01-17 21:05	shearer reversion
1	2018-01-17 22:05	shearer stoppage
1	2018-01-17 22:15	slotting by shearer
1	2018-01-17 22:30	operational activities near drive unit
1	2018-01-17 22:50	shearer reversion
1	2018-01-17 22:55	cutting towards reverse drive
1	2018-01-17 23:10	shearer stoppage
1	2018-01-17 23:45	intense working
1	2018-01-18 00:30	shearer stoppage
1	2018-01-18 00:45	normal working
1	2018-01-18 00:50	shearer stoppage
1	2018-01-18 02:00	normal working
1	2018-01-18 02:15	shearer stoppage
1	2018-01-18 02:20	normal working

Created event log enable to perform shearer operation process modelling. At this stage various process mining techniques and formalisms can be used [23].

Modelling of the mining process in a longwall face requires the knowledge about all machinery operation processes, so similar efforts have to be taken for creation of other event logs and their combination into one log, enabling holistic analysis of the mining process. Recently research work in this scope is carried out.

5 Conclusions

Process management is based on the knowledge about process real performance. Such knowledge in the case of a mining process in the longwall face is recorded in machinery monitoring system and can be used for management purposes, however it requires process-oriented analysis.

In the paper we present process mining as modern possibility of process analysis based on event logs. In our approach we propose to use low-level sensor data from machinery monitoring system to create high-level event logs in order to model and analyse the mining process in an underground mine. We used combination of unsupervised data mining techniques as well as domain knowledge to discover stages in an example process and create an event logs for further process analysis with process mining techniques. These techniques enable modelling of a process, its diagnostics and advanced analyzes aimed at process enhancement.

Created event logs can be used not only for process modelling but also for detail reporting purposes or predictive maintenance as input in a form of time series of process activities.

Presented approach for creation of useable event logs from monitoring system and its low-level event logs could be applied to analysis of sensor data in other industries. However, such extension requires adequate data quality and domain knowledge about analysed process.

Acknowledgements. This paper presents the results of research conducted at AGH University of Science and Technology – contract no 11.11.100.693.

References

1. Bednarz, T., James, C.A.R., Widzyk-Capehart, E., Caris, C., Alem, L.: Distributed collaborative immersive virtual reality framework for the mining industry. In: Billingsley, J., Brett, P. (eds.) Machine Vision and Mechatronics in Practice, pp. 39–48. Springer, Berlin Heidelberg (2015)
2. Bose, R.P.J.C., van der Aalst, W.M.P.: Abstractions in process mining: a taxonomy of patterns. In: 7th International BPM Conference Proceedings, Germany, pp. 159–175 (2009)
3. Brodny, J., Alszer, S., Krystek, J., Tutak, M.: Availability analysis of selected mining machinery. Arch. Control Sci. **27**(2), 197–209 (2017). https://doi.org/10.1515/acsc-2017-0012
4. Brzychczy, E., Lipiński, P., Zimroz, R., Filipiak, P.: Artificial immune systems for data classification in planetary gearboxes condition monitoring. In: Dalpiaz, G. (ed.) Advances in Condition Monitoring of Machinery in Non-stationary Operations, CMMNO 2013. Lecture Notes in Mechanical Engineering. Springer, Heidelberg (2014)
5. Cook, D.J., Krishnan, N.C., Rashidi, P.: Activity discovery and activity recognition: a new partnership. IEEE Trans. Cybern. **43**(3), 820–828 (2013)
6. Demetgul, M.: Fault diagnosis on production systems with support vector machine and decision trees algorithms. Int. J. Adv. Manuf. Technol. **67**, 2183–2194 (2013). https://doi.org/10.1007/s00170-012-4639-5

7. Erkayaoğlu, M., Dessureault, S.: Using integrated process data of longwall shearers in data warehouses for performance measurement. Int. J. Oil Gas Coal Technol. **16**(3) (2017). https://doi.org/10.1504/ijogct.2017.10007433
8. Guenther, C.W., van der Aalst, W.M.P.: Mining Activity Clusters from Low-Level Event Logs. BETA Working Paper Series, WP 165, Eindhoven University of Technology, Eindhoven (2006)
9. Korbicz, J., Koscielny, J.M., Kowalczuk, Z., Cholewa, W. (eds.): Fault Diagnosis: Models, Artificial Intelligence, Applications. Springer (2004)
10. Le, T., Luo, M., Zhou, J., Chan, H.L.: Predictive maintenance decision using statistical linear regression and kernel methods. In: Proceedings of the 2014 IEEE Emerging Technology and Factory Automation (ETFA), Barcelona, pp. 1–6 (2014). https://doi.org/10.1109/etfa.2014.7005357
11. Mannhardt, F., de Leoni, M., Reijers, H.A., van der Aalst, W.M.P., Toussaint, P.J.: From low-level events to activities - a pattern-based approach. In: International Conference on Business Process Management, pp. 125–141. Springer (2016)
12. Michalak, M., Sikora, B., Sobczyk, J.: Correlation and association analysis in wall conveyor engines diagnosis. Studia Informatica [S.l.] **36**(3), 43–60 (2015). https://doi.org/10.21936/si2015_v36.n3.737
13. Neustupa, Z., Benes, F., Kebo, V., Kodym, O.: New trends of automation and control of opencast mining technology. In: 13th SGEM GeoConference on Science and Technologies in Geology, Exploration and Mining, vol. 1, pp. 555–562 (2013). https://doi.org/10.5593/sgem2013/ba1.v1/s03.044
14. Reid, P.B., Dunn, M.T., Reid, D.C., Ralston, J.C.: Real-world automation: new capabilities for underground longwall mining. In: Proceedings of the Australasian Conference on Robotics and Automation Brisbane, Australia (2010)
15. Sammouri, W.: Data mining of temporal sequences for the prediction of infrequent failure events: application on floating train data for predictive maintenance. Signal and Image processing. Université Paris-Est (2014)
16. Singh, R.D.: Principles and Practices of Modern Coal Mining. New Age International, New Delhi (2005)
17. Snopkowski, R., Napieraj, A., Sukiennik, M.: Method of the assessment of the influence of longwall effective working time onto obtained mining output. Arch. Min. Sci. **61**(4), 967–977 (2016). https://doi.org/10.1515/amsc-2016-0064
18. Stecula, K., Brodny, J.: Application of the OEE Model to analyse the availability of the mining armored face conveyor. In: Conference: 16th International Multidisciplinary Scientific Geoconference (SGEM 2016). Exploration and Mining, Albena, Bulgaria, vol. II, pp. 57–64 (2016)
19. Susto, G.A., Beghi, A.: Dealing with time-series data in predictive maintenance problems. In: IEEE 21st International Conference on Emerging Technologies and Factory Automation (ETFA), Berlin, pp. 1–4 (2016). https://doi.org/10.1109/etfa.2016.7733659
20. Sztyler, T., Carmona, J., Völker, J., Stuckenschmidt, H.: Self-tracking reloaded: applying process mining to personalized health care from labeled sensor data. In: Koutny, M., Desel, J., Kleijn, J. (eds.) Transactions on Petri Nets and Other Models of Concurrency XI. LNCS, vol. 9930, pp. 168–180. Springer, Heidelberg (2016)
21. Tax, N., Sidorova, N., Haakma, R., van der Aalst, W.M.P.: Event abstraction for process mining using supervised learning techniques. In: Bi, Y., Kapoor, S., Bhatia, R. (eds.) Proceedings of SAI Intelligent Systems Conference (IntelliSys) 2016. IntelliSys 2016. Lecture Notes in Networks and Systems, vol 15, pp. 251–269. Springer, Cham (2018)

22. van der Aalst, W.M.P.: Process Mining in the Large: A Tutorial. In: Zimanyi, E. (ed.) Business Intelligence (eBISS 2013). Lecture Notes in Business Information Processing, vol. 172, pp. 33–76. Springer, Berlin (2014)
23. van der Aalst, W.M.P.: Process Mining: Data Science in Action. Springer, Berlin (2016)
24. van der Aalst, W.M.P., Bolt, A., van Zelst, S.J.: RapidProM: mine your processes and not just your data. In: Hofmann, M., Klinkenberg, R. (eds.) RapidMiner: Data Mining Use Cases and Business Analytics Applications. Chapman and Hall/CRC Press (2016)
25. van Eck, M.L., Sidorova, N., van der Aalst, W.M.P.: Enabling process mining on sensor data from smart products. In: IEEE RCIS, pp. 1–12. IEEE Computer Society Press, Brussels (2016) https://doi.org/10.1109/rcis.2016.7549355
26. Widodo, A., Yang, B.S., Han, T.: Combination of independent component analysis and support vector machines for intelligent faults diagnosis of induction motors. Expert Syst. Appl. **32**, 299–312 (2007)
27. Zorychta, A., Burtan, Z.: Conditions and future directions for technological developments in the coal mining sector. Min. Resour. Manage. **24**(1), 53–70 (2008)

Analysis of Moving Averages of BWEs Actual Capacity

Leszek Jurdziak(✉)

Wroclaw University of Science and Technology, Wyb. Wyspianskiego 27, 50–370 Wrocław, Poland
leszek.jurdziak@pwr.edu.pl

Abstract. Analysis of productivity data from different lignite mines leads to the conclusion that theoretical BWE capacity is utilized only in a small percentage. Therefore, it is possible to achieve a substantial reduction of energy consumption by optimal dimensioning of BWEs and belt conveyor parameters to actual loads generated during the cyclical excavation of subsequent slices, terraces, blocks, and benches. Based on the analysis of BWEs efficiency and time series of their actual volume capacities in a Polish opencast lignite mine, it is proposed to apply the Extreme Value Theory (EVT) for dimensioning belt width and for selecting required power of conveyor drives. The asymptotic decrease of high order percentiles of actual capacity averages with the increase of loading time (the length of receiving conveyors) shows that there is a big potential for energy savings.

Keywords: Bucket wheel excavator · BWE capacity · Extreme value theory

1 Utilization of BWEs Theoretical Capacity in Lignite Mines

Output from the analysis of utilization of calendar time of Bucket Wheel Excavators (BWEs) and their efficiency measured by the ratio of averaged volume capacity (volume of material transported per operating time) to the theoretical one (designed capacity of an excavator) in Polish lignite mines is shown in Table 1 (Kasztelewicz 2004). The data were collected for a few dozen years from the construction of mines and start of the operation of particular BWEs. It can be seen that the general calendar time utilization is at low level. As it is shown (Table 1) the average operating time in 4 lignite mines examined during this period varied between 35.87% and 40.76% of the total calendar time. For individual BWEs the range is greater: from 19.0% (Rs540 in Turów) up to 49.9% (SchRs800/1 in Adamów). This means that the non-working times, which include the annual inspection and repair of the machine, holidays, standstill due to belt conveyor shifting near the new mining front position, standstill due to breakdown or repair of the BWE, the conveyor system or the stacker and also time for transporting a machine, strong wind, rocks embedded in the mining face and other reasons for stopping the work took more than 50% of calendar time (even about 60% in average). Similar but slightly better results are reported by Kolosov (2004) for Greek mines. The average operating time of the 5 BWEs examined during the 15-year period at the North Field Mine varied between 49.6% and 60.8% of the total calendar time. Standstill time of BWE machines

© Springer Nature Switzerland AG 2019
A. Burduk et al. (Eds.): ISPEM 2018, AISC 835, pp. 622–632, 2019.
https://doi.org/10.1007/978-3-319-97490-3_59

and the connected conveyors shows that in fact they were working for only half of calendar time and it should be taken into account during dimensioning the conveyors for new mines with BWEs. Increasing this value to 60% is possible but greater value of operating time could be difficult to attain in longer periods.

Table 1. Productivity of BWEs in Polish lignite mines (Kasztelewicz 2004).

Lignite mines	No. of BWEs	Total			No. of BWEs	Overburden			No. of BWEs	Lignite		
		T operating per T calendar				Q average per Q theoretical				Q average per Q theoretical		
		Min.	Weighted average	Max.		Min.	Weighted average	Max.		Min.	Weighted average	Max.
Adamów	9	27,20	40,76	49,90	6	31,50	39,70	53,10	3	27,20	64,40	56,20
Bełchatów	13	30,50	36,96	46,30	13[a]	35,10	44,65	57,60	13[a]	35,10	44,65	56,20
Konin	17	36,00	40,37	43,20	11	31,60	51,35	62,90	6	40,00	48,74	56,20
Turów	14	19,00	35,80	43,80	14[a]	20,60	36,58	56,20	14[a]	20,60	36,58	56,20

[a] - almost all BWEs excavated both overburden and lignite.

The average current volume capacity in 4 lignite mines examined during the whole life of BWEs varied between 36.58% and 51.35% for overburden and for lignite between 36.58% and 63.4% of their total theoretical capacity. Similar results were reported by Kolovos (2004). The average output of the 5 BWEs examined during the 15-year period at the North Field Mine varied between 21.24% and 39.69% of their theoretical capacity, which means very low utilization of theoretical capacity within long periods. On the other hand, maximum average capacities of particular BWEs can be as high as 75.6% (in lignite in the Adamow mine) and 52.3% in the Greek mines for maximum annual capacity.

Detailed analysis of actual volume capacities of lignite for BWE SRs2000 over the period of 1977–2004 in one of Polish opencast lignite mines was 2 146 m^3/h, which corresponds to 35.8% of their theoretical capacity (6 000 m^3/h). During the analysis period (08.03.05–18.04.05 and 23.04.05–01.05.05) maximum current volume capacity was 6 885 m^3/h (114.5%). According to German authors (Scheffzyk and Jahn 2004) maximum actual volume capacity can attain a level as high as 135% of the theoretical capacity. The range of actual volume capacity is therefore great – from 0 to 135% of its theoretical BWE capacity, and attained averaged operational capacity is sometimes very low – 36%. Additionally, if a likewise low level of calendar time utilization (e.g. 39.2% for K37 BWE SRs2000 in period 1977–2004) is taken into account, doubts arise whether the machinery was properly adjusted to mining targets. This is especially true for belt conveyors which were designed to take maximum BWE actual capacity (135% of theoretical) and in fact were utilized in about ¼. It means that by maximizing BWEs capacity the operational time can be reduced to 1/8 of the day – machines could work only 3 h daily (= 1/4 * 50% 24 h) – or belt conveyors capacity and BWEs theoretical capacity can be reduced 4 times if conveyors work 12 h daily at full but reduced capacity. The latter option would require a stable stream of material from BWEs which is difficult to attain due to the character of their work.

Low level of theoretical BWE capacity utilization increases costs enormously due to transporting material with 30% level of belt capacity utilization and is twice as expensive as in the 110% case (Bednarczyk 1976). In consequence, the energy usage in the

"Belchatow" mine was 4.23 kWh/Mg and is 2x higher than it could be. In Laubag Vatenfall Europe AG it was lower, at 1.1 kWh/Mg (Köhler and Lehmann 2002) albeit not only due to better efficiency and lower resistance but also easier mining conditions.

Of course, as the variations of current BWE volume capacity are great, it is impossible to properly adjust conveyor belt capacity to BWE capacity without detailed analysis of the time series of BWE capacity.

2 BWE Actual Volume Capacities as a Time Series

BWE capacity is a consequence of its parameters such as bucket volume, frequency of their discharge, BWE size and manoeuvre capabilities etc., but not only. There are a lot of other parameters, dependent or independent of human will, such as: a shortwall and its blocks cut designs (including the height of excavated slice), technology of BWE excavation, efficiency targets, individual style of BWE operator and mining conditions.

Observations of this process (Fig. 1) show that it is cyclical but not regular. In different periods BWE works with different efficiencies, which could be a consequence of cut design, applied targets (e.g. in connection with coal blending) or working conditions. Expected value of bucket discharge $E\{u_n\} = U$ is constant only locally and even then actual output is oscillating according to different BWE cycles of work and is different for different cycle phases (Fig. 1). The shortest cycle is the discharge of material from buckets within time 60/n (n = the number of buckets discharged within 1 min). Excavation of one slice (called also chips e.g. by Fries 2010) in a terrace is slightly longer (about 3 min, Fig. 1B), then excavation of one terrace in a block (about 1 h), and then excavation of full block (few hours), full shortwall (several weeks) and full level. For the dimensioning of belt conveyors, it is important that cycles are shorter than their loading periods due to cyclical reduction (sometimes even to zero) of actual volume capacities (Fig. 1), which is connected with cyclical movements of BWE.

The measurements of specific digging and breakout forces, as well as of the capacity on the excavator K 2000, were carried out in the 5th overburden section of the Bílina Mine on the 27th of September 2007. The measurements (Fig. 1C) were performed on the 4th bench of the K2000 BWE. Its theoretical capacity after reconstruction was 5500 m^3/h. During measurements, the average output was 2692 m^3/h (48.95% of theoretical capacity). Zhao-Xue and Yan-Long (2014) argue that the theoretical capacity of BWEs is oversized. The use of "1/cosφ" to adjust the speed of a bucket wheel excavator will result in rapid speeds, which may cause non-uniform flow from the distribution of material flow, and decreased capacity utilization.

A closer look at the distribution of actual volume capacities registered at the receiving belt conveyor (Koch 2005) has shown that it is not a normal one. The histogram is 2-modal and is almost uniform on left side (<3 300 m^3/h) with two frequency peaks: close to zero and a bit below 5 000 m^3/h (83.3% of theoretical capacity) (Fig. 2).

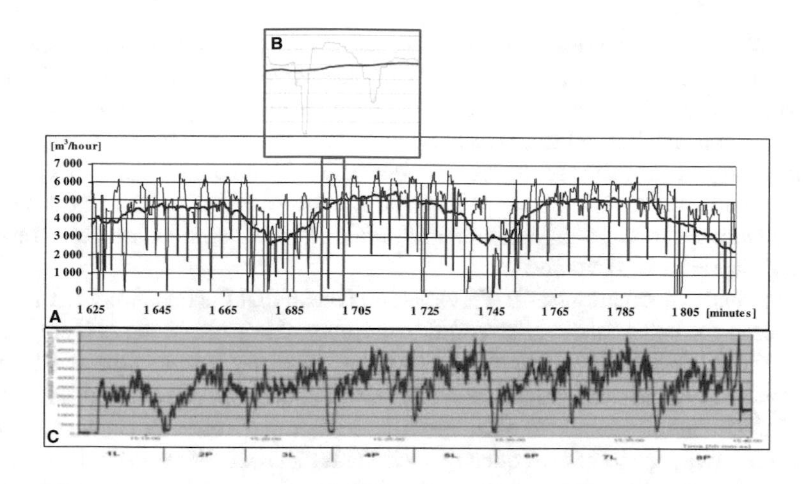

Fig. 1. Different cycles of work of BWE SRs2000 - terraces excavation (A - 1 h, 14-min moving average) and slices (B - 3 min, actual capacity measured every 5 s) (Jurdziak 2008). Capacity value of K 2000 excavator upon 4th bench extraction (C - 8 chips) (Fries et al. 2010)

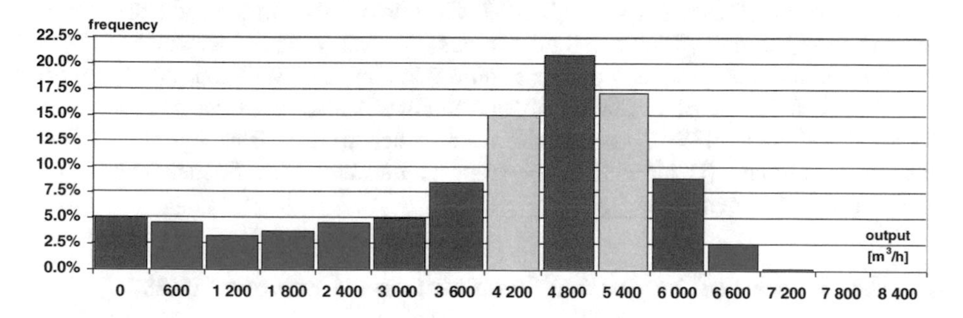

Fig. 2. Histogram of actual volume capacities of BWE SRs2000 (Jurdziak 2006).

This non-standard distribution is a consequence of different BWE working cycles. Local maximum near zero is a result of idle movements of excavator wheel between different cycles. It is estimated that idle work can take 2–36% of BWE time. In order not to influence the outcomes of research, zero capacities longer than 2 min have been removed from the source data. Their removal should not influence the dimensioning of belt conveyors as it should be based rather on maximum than minimal capacities.

3 Belts and Conveyors Dimensioning

High actual volume capacity of BWE, which achieves even 135% of the theoretical capacity level, has to be received by moving belt on the conveyor. It means that the capacity of receiving conveyors ought to be greater than the maximum BWE capacity. Otherwise a spillage of material at loading points can occur. However the biggest capacities are rather very rare events (Fig. 2) and it is worth to consider what is better:

energy losses due to low level of utilization of conveyors capacities (transporting of almost empty belt) or incurring expenditure connected with removing the spilled material and threat of conveyors stoppage and failures from overspills. In order to decide such a dilemma, detailed data must be collected about how much material can be spilled off in a given period of time for a given system of BWE and the cooperating conveyors. It would let tie together investment and operating costs of BWE and conveyor with their capacity parameters and with determined confidence level concerning the takeover of excavated material by the conveyor.

Additionally, fluctuations of BWE capacities mean that the total amount of material on the conveyor belt is also variable and it is impossible to have the belt fully loaded without special dosing devices with some storage capacity between the BWE and the conveyor. Analysis of time series of BWE actual capacities or rather their moving sums should help in finding answer to what maximum amount of material can be supplied by BWE onto conveyor and how frequently such loads can occur. Therefore, it is necessary to determine probability distribution of total amount of material on conveyor and properties of its time series. For the dimensioning of conveyors, only the distribution of maximum data over time and the high threshold excesses are interesting.

Time series of BWEs capacities were analyzed by several authors for dimensioning joint ore streams (Dworczynska et al. 2012; Czaplicki 2014), optimised selection of a belt conveyor loaded by BWEs in lignite mines (Gladysiewicz and Kawalec 2006) and by LHDs in an underground copper ore mine (Chlebus and Stefaniak 2012; Kruczek et al. 2017). Influence of variability of such loads was also investigated in regards to idlers durability (Krol 2017), stress relaxation in belt splices (Blazej et al. 2017) and their fatigue strength (Bajda et al. 2016). Important is also their influence on belt puncture resistance (Komander et al. 2014).

4 Time Series of Amount of Material on a Conveyor Belt

Total amount of material on the conveyor belt is variable and depends on BWE output during the loading period t_L. Knowing belt speed c we can determine the time required to fully load the conveyor of length L:

$$t_L = \frac{L}{c} = Lc^{-1} \tag{1}$$

Total amount of material V_L on the conveyor belt in time t can be calculated as follows:

$$V_L(t) = \int_{t-t_L}^{t} v(t)dt \tag{2}$$

where $v(t)$ is an actual volume capacity of BWE.

Lignite mines store data about capacities. However, these are not available as a continuous function but in a form of discrete series, so Eq. (2) should be as follows:

$$V_L(t) = \sum_{i=0}^{N_L-1} v(t - it_0)t_0 = t_0 \sum_{i=0}^{N_L-1} v(t - it_0) = t_0 S_{N_L}(t) \tag{3}$$

$$N_L = \frac{t_L}{t_0}, S_{N_L}(t) = \sum_{i=0}^{N_L-1} v(t - it_0) \tag{4}$$

$$V_L(t) = t_L \frac{S_{N_L}(t)}{N_L} = t_L MA_{N_L}(t)\,, \quad \text{where} \quad MA_{N_L}(t) = \frac{S_{N_L}(t)}{N_L} \tag{5}$$

Total amount of material $V_L(t)$ loaded on belt conveyor (having length L) till time t is calculated as a sum of N_L preceding actual volume capacities $v(t - it_0)$ (where $i = 0, \ldots, N_L - 1$) multiplied by time of measurement interval t_0 (here 5 s). N_L capacity measurements are done during loading time t_L preceding moment t and we can calculate sum of those capacities $S_{NL}(t)$ and moving average $MA_{NL}(t)$ (see Figs. 1 and 4).

5 Moving Averages of BWE Actual Volume Capacity

The volume of material on a belt conveyor having the length L can be calculated by multiplying the moving average of preceding capacity measurements by loading time (5). It is important therefore to analyze moving averages of actual BWE volume capacities for different loading periods as random variables. In Fig. 3 it can be seen how histograms of moving averages change with the increase of loading period (from bimodal,

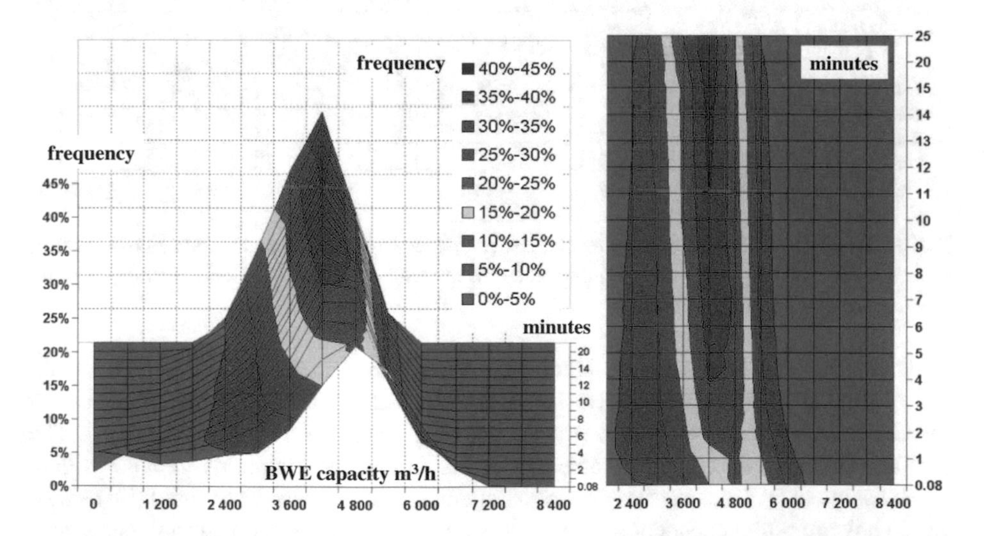

Fig. 3. Histograms of moving averages of actual BWE's capacities shown as a 3D view (left) and 2D map (right) for different loading time (changing length of a conveyor) (Jurdziak 2006).

for $t_L = 0.08$ min, to very narrow unimodal, for $t_L = 20$–25 min). The data have been taken during 2nd period 23.04.05–01.05.05.

The comparison of descriptive statistics of actual capacities and their 14-min moving averages for 2 different periods and different intervals of measurements: 1 min and 5 s is shown in Table 2. As it is seen mean for the 2nd period is greater by over 1 000 m^3/h, which proves that the time series of capacities is not stationary (Fig. 4).

Table 2. Descriptive statistics of moving averages for different loading time and periods.

Statistics	1 min[a]	14-min (1st)	5 s.[ab]	14-min (2nd)
Minimum	0.0	102	0.0	1168
1st Quartile	1440.0	2033	2840.6	3211
Median	2889.4	2673	4200.0	3804
3rd Quartile	3815.6	3253	4918.1	4275
Maximum	6885.0	5361	6787.5	5685.0
Kurtosis	−0.9240	−0.2465	−0.2327	−0.0162
Skewness	−0.2780	−0.0969	−0.8630	−0.5245
Stnd. deviation	1520.8	899.7	1703.9	802.5
Mean deviation	1279.4	723.6	1368.0	640.7
Mean	2637.3	2637.2	3705.0	3702.5

BWE capacities measured every 1 min ([a] 1st period 8.03–18.04.05) and every 5 s. ([ab] 2nd period 23.04–1.05.05).

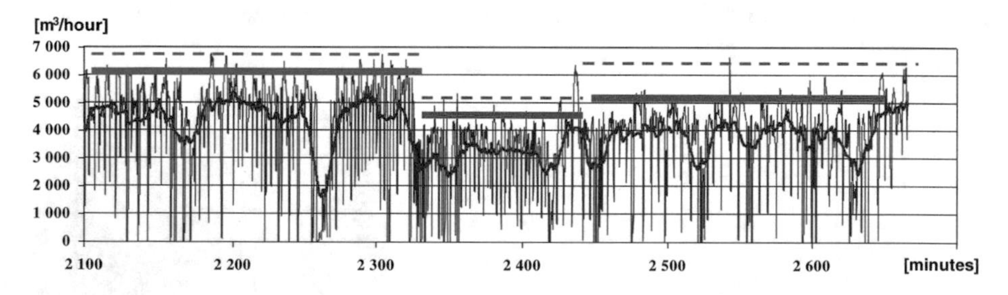

Fig. 4. Diversity of BWE capacity levels. Broken lines – maximum level, solid line – level of the often achieved high capacity, red line –14-min moving average, blue line – current capacity measured every 5 s (Jurdziak 2008)

Histogram of 14-min moving averages is more symmetrical and steeper (see reduction of standard deviation and kurtosis closer to 0), but due to the dependency between subsequent capacities we cannot expect that they have the normal distribution. The central limit theorem is not applicable here as the entrance assumptions are not fulfilled.

Subsequent BWE capacities are not independent, identically distributed (iid) variables. They are both dependent (capacity depends on the excavated material and localization of the BWE's wheel - phase of the excavation cycle, see Fig. 1 and they do not have the same df. At least the parameters of dfs are different as different amounts of material can be taken into the bin at different BWE's positions. Capacity fluctuations are easily seen in the picture (Fig. 3).

In the case of belt conveyor dimensioning and selection of optimal power for long conveyors, the most important factor is the shape of the right side of the histogram, where time for calculating moving averages increases. The changes of 95%, 98% and 100% percentiles of actual capacities moving averages (the last is the maximum) for the 1st period of data analysis (08.03.05–18.04.05) can be seen in Fig. 5.

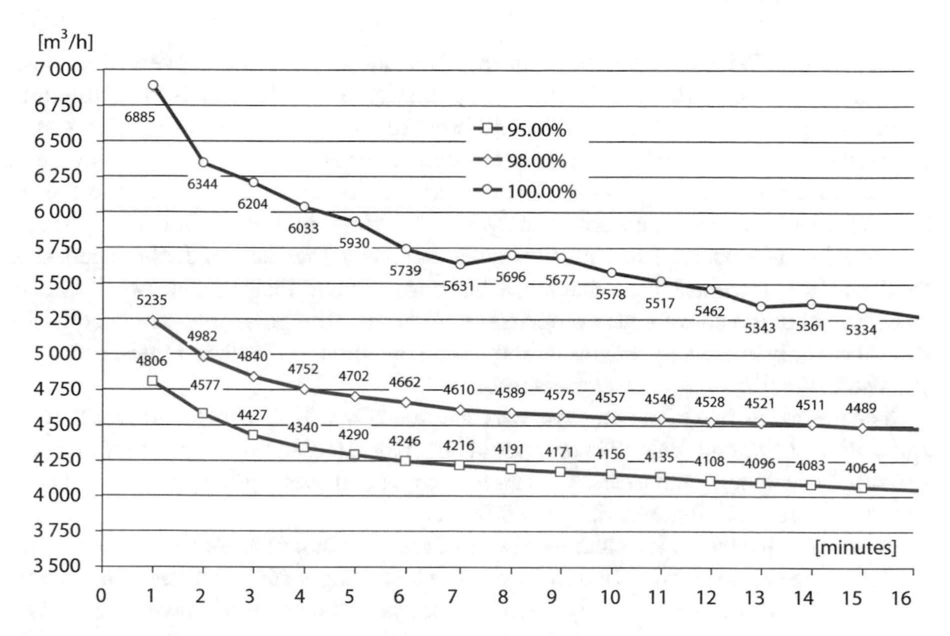

Fig. 5. The influence of conveyor loading time (period of moving averages calculation) on asymptotic behavior of 95%, 98% and 100% percentiles of moving averages of actual volume capacities (Jurdziak 2006).

The hypothetical 4.5 km conveyor receiving material from the BWE SRs2000 with 14 min loading time for 98% of time would have the averaged capacity lower than 4511 m^3/h and the current capacity would never exceed 5361 m^3/h (Fig. 5). It would be enough to have the drives prepared to start up a conveyor with 1251 m^3 (14 × 5361/60) of material instead of 1607 m^3 (14 × 6885/60). Preparation of the drives for maximum utilization of belt capacity (maximum actual volume excavated by BWE during loading time) demonstrates that they are overpowered as maximum volume on the 4.5 km conveyor was 28.4% lower than maximum actual capacity multiplied by loading time.

It should be stressed that historic percentiles are not a very good measure of expected BWE capacity especially when the period of analysis is short. For different reasons, BWE operators can work and receive mining targets, which would be well below potential BWE capacity (e.g. excavation of shortwall nests, selective mining, cleaning of floor, excavation in difficult geotechnical conditions etc.). Longer periods of lowered capacities have a big influence on percentile levels (e.g. compare 3rd quartile in Table 2). Therefore, it is better to concentrate only on maximum capacities. The proper value of expected (within long time, e.g. operating time of a mine) maximal amount of material which can be supplied by BWEs on a conveyor can be found through application of Extreme Value Theory to the analysis of BWE actual volume capacities and their moving averages.

6 Conclusions

Capacities of BWEs were measured in the Belchatow lignite mine every 5 s. at all receiving conveyors. However, lack of stationarity of BWE capacities in different mining and geology conditions as well as different regime of BWE work (Fig. 1) suggests that further research should be carried out in longer periods covering a broader range of different conditions of BWE work. The data should cover a full year of operation in order to take into account all potential cycles including differences in capacity levels among different seasons. The analysis should also cover how the length of the period of local maxima influences the behavior of distributions and their parameters. It is also worth to study the convergence properties to find out all regularities, which could be helpful not only in capacity prognoses both for the maxima and for the r-largest statistics, but also in the distribution of exceedances.

The presented data of moving averages changes (Figs. 3, 5) seem to indicate that the application of Extreme Value Theory should be an ideal tool to find the required answers and supply designers with necessary data for proper and even optimal dimensioning of belt conveyors receiving material from BWEs.

Extreme value theory has emerged as one of the most important statistical disciplines for the applied sciences over the last 50 years. The distinguishing feature of an extreme value analysis is the objective to quantify the stochastic behavior of a process at unusually large — or small levels. In particular, extreme value analyses usually require estimation of the probability of events which are more extreme than any that have already been observed (Coles 2001). Based on these answers, it should be possible to build an overall model of conveyor investment and operational costs taking into account stochastic nature of volatility of BWE current capacities. The proper selection of conveyor and BWE parameters for the given confidence levels can be done through economic optimization, taking into account all distributions estimated by the EVT.

Additionally, EVT can provide data in the form of time series of the given threshold excesses for dimensioning of the chute storage capacity to receive peak loads from BWEs. This should allow loading the conveyor with stable stream of material and reducing the need for wide belts. It is currently popular to use DEM simulations to design the shape and size of chutes and even to optimize them for a given stream of material. Regardless of the elegance of this method and its efficiency, it has to rely on data about fluctuations of the loading material stream in order to properly reflect real conditions so the analysis of BWE capacities is necessary. Additionally, concluding about stream of material leaving the chute should also be based on EVT applied to output data from simulations, as even a high number of simulations is always a finished number, while

EVT is based on asymptotic properties and therefore can predict the unpredictable, which means it can conclude about the future relying on the present, restricted data.

References

Bajda, M., Blazej, R., Hardygóra, M.: Impact of selected parameters on the fatigue strength of splices on multiply textile conveyor belts. In: IOP Conference Series: Earth and Environmental Science, vol. 44, no. 5, p. 052021 (2016)

Bednarczyk, J.: Control and automation in opencast mines, Slask (1976). (in Polish)

Blazej, R., Bajda, M., Hardygóra, M.: Monitoring creep and stress relaxation in splices on multiply textile rubber conveyor belts. Acta Montanist. Slovaca **22**(2), 116–125 (2017)

Chlebus, T., Stefaniak, P.: The concept of intelligent system for horizontal transport in a copper ore mine. In: Hybrid Artificial Intelligent Systems, HAIS 2012. Lecture Notes in Computer Science, vol. 7209, pp. 267–273. Springer, Heidelberg (2012)

Coles, S.: An Introduction to Statistical Modelling of Extreme Values. Springer Series in Statistics. Springer, London (2001)

Czaplicki, J.: Statistical analysis of a stream of mined rock generated by bucket wheel excavator. Gornictwo Odkrywkowe (Opencast Mining) **55**(1), 22–28 (2014). (in Polish)

Dworczyńska, M., Gladysiewicz, L., Kawalec, W.: Model of a load stream for the purpose of designing joint belt conveyors. School of Opencast Mining, 27–28 March 2012. (in Polish)

Fries, J., Helebrant, F., Klouda, P.: Operational Measuring on Bucket Wheel Excavators. Paper presented during the Mechanical Structures and Foundation Engineering 2010, Ostrava, 13 September 2010

Gladysiewicz, L., Kawalec, W.: Optimised selection of a belt conveyor loaded by a BWE. In: Proceedings of the 8th International Symposium of Continuous Surface Mining ISCSM, Achen, pp. 353–357 (2006)

Jurdziak, L.: Methodology of BWE efficiency analysis for power reduction of conveyor drives. In: Proceedings of the 8th International Symposium Continuous Surface Mining, ISCSM 2006, Aachen, Mainz, 24th–27th September 2006, pp. 125–131 (2006)

Jurdziak, L.: Application of extreme value theory for joint dimensioning of BWEs and long distance belt conveyors in lignite mines. In: International Conference on Bulk Europe 2008, Prague, Czech Republic, 11–12 September 2008. Vogel Industrie Medien, Wurzburg (2008)

Kasztelewicz, Z.: Comparison of Polish lignite mines. Gornictwo Odkrywkowe (Opencast Mining) **46**(2), 21–32 (2004). (in Polish)

Koch, R.: Utilization of conveyors cooperating with BWEs in KWB “Bełchatów” S.A., diploma thesis (not published), Faculty of Geoengineering, Mining and Geology, Wroclaw University of Technology (2005). (in Polish)

Köhler, U., Lehmann, L.B.: Belt conveyors with variable speed of transportation. Transport Przemyslowy (Industrial Transport) **3**(9), 48–51 (2002)

Kolovos, Ch.: Efficiency of a bucket wheel excavator lignite mining system. Int. J. Surf. Min. Reclam. Environ. **18**(1), 21–29 (2004)

Komander, H., Hardygóra, M., Bajda, M., Komander, G., Lewandowicz, P.: Assessment methods of conveyor belts impact resistance to the dynamic action of a concentrated load. Maint. Reliab. **16**(4), 579–584 (2014)

Krol, R.: Studies of the durability of belt conveyor idlers with working loads taken into account. In: IOP Conference Series: Earth and Environmental Science, vol. 95, no. 4 (2017)

Kruczek, P., Polak, M., Wyłomańska, A., Kawalec, W., Zimroz, R.: Application of compound Poisson process for modelling of ore flow in a belt conveyor system with cyclic loading. Int. J. Min. Reclam. Environ, 1–16 (2017). https://doi.org/10.1080/17480930.2017.1388335

Scheffzyk, P., Jahn, W.: Modern technology in belt conveyor drives. Transport Przemyslowy (Industrial Transport) **3**(17), 18–20 (2004). (in Polish)

Zhao-Xue, C., Yan-Long, C.: Determination and analysis of the theoretical production of a bucket wheel excavator. Arch. Min. Sci. **59**(1), 283–291 (2014)

Application of the Discrete Element Method (DEM) for Simulation of the Ore Flow Inside the Shaft Ore Bunker in the Underground Copper Ore Mine

Piotr Walker, Witold Kawalec, and Robert Król(✉)

Faculty of Geoengineering, Mining and Geology,
Wroclaw University of Science and Technology, Wroclaw, Poland
{Piotr.Walker,Witold.Kawalec,Robert.Krol}@pwr.edu.pl

Abstract. Identification of the stream of transported mine output has a great importance for the effectiveness of ore processing and for the control of mining operations. The above fact entails the growing importance of various types of scanners and on-line sensors which serve to determine the measurable parameters of a random sample and to use this information as a basis for the classification of the whole stream of mine output. An alternative method consists in effective processing of large data sets (the "big data" technology), i.e. in combining the previously generated information on the deposit (geological data), information on the locations of currently operated mine fields (mining data), and information on the flow of the stream of mined ore in the underground transportation system. The flow of mine ore is modeled with the use of simulation methods, whose reliability largely depends on how accurately a particular transportation system and its operation are reproduced in the model. Particular attention must be paid to accurate modeling of ore behavior in the nodes of the transportation system, i.e. in the bunkers. The bunkers, apart from serving a retaining function, also effect the averaging of ore parameters, as the portions of mined material subsequently added to the bunker change their position and mix. Modeling the behavior of mined ore inside the bunker is the object of tests performed with the use of Discrete Element Method (DEM).

This paper includes the description of DEM and the method used for parameterizing the mined ore data (granulation, ore stream efficiency) and the bunker data (geometry, discharge technology), which are required to build a DEM model of an ore bunker following the design of an actual large-capacity shaft station bunker located in an underground copper ore mine.

Keywords: DEM · DISIRE · Ore bunker · Ore flow · Grain size distribution

1 Introduction

Increased profit in mining business may be generated by optimizing the mining processes and by increasing the yield of valuable elements from the extracted ore. Increasing cost of electric energy effects the efforts to reduce its consumption. In the case of the KGHM Polska Miedź S.A. company, the largest electricity consumption is observed in the ore

© Springer Nature Switzerland AG 2019
A. Burduk et al. (Eds.): ISPEM 2018, AISC 835, pp. 633–644, 2019.
https://doi.org/10.1007/978-3-319-97490-3_60

concentration plant, in which most of the electric energy is consumed when ore is prepared for processing, i.e. crushed and milled. As Polish copper ore deposits are composed of three lithology types and as the amount of electric energy used in ore processing largely depends on the type of mined material subjected to processing (Krzeminska and Malewski 2011), the proposed solution consisted in improving the efficiency of ore processing operations by providing the ore concentration plant with information on the quality of the supplied ore (Jurdziak et al. 2017), in order to enable the plant to adjust the parameters of mechanical processing to a particular rock type. This concept is researched within the DISIRE project.

The DISIRE project, being part of the Horizon 2020 program, consists in simulating the transportation of mined ore using the FlexSim Software, which enables the monitoring of ore composition and quality along the whole length of the transportation line. Knowing the composition and quality of ore in the rock mined in a particular deposit and also knowing the flow of ore between the mining plant and the concentration plant, it is possible to determine the quality of ore fed to the concentration plant during the subsequent production shifts (Jurdziak et al. 2017; Kawalec et al. 2016). Owing to drill-hole sampling, mining companies are able to predict the properties of the part of the deposit which is to be exploited, but during transportation the mined material is frequently mixed. Various types of retention bunkers, here including shaft station bunkers, are the locations where most of the information on the quality of the transported ore is lost.

This paper focuses on the representation of one of the elements of the transportation line in the Lubin Mining Plant: the P1 south shaft station bunker. It is the only bunker in the transportation line to include two chutes, thus causing significant distortions to the flow of mined ore. In addition, the bunker must allow the ore to be constantly fed to the skip, which requires at least one of the chutes to remain in operation. The modeling of the retention point will allow gathering information on ore flow at various bunker fill levels and identifying the behavior of individual fractions during the process of both feeding ore into the bunker and discharging it from the bunker.

In order to research the processes occurring in the investigated environment, ore flow was represented using Discrete Element Method (DEM). Its advantage over the previously used methods lies in the fact that it allows ore flow to be modeled with respect to its granulation. The choice of this method was also dictated by its other advantages. It enjoys a growing popularity in both research and business circles as it involves aspects of physics and engineering. Potential DEM Software have been recently described in a growing number of publications. According to Google Scholar, 2500 research works on DEM were published in 2015 alone (Katterfeld and Wensrich 2017). The method has been mostly applied in the analysis of the behavior of bulk materials and fine fractions, especially in relation to their transportation and to their interaction with technological machines. (Mendyka 2017; Minkin 2012; Dewicki and Mustoe 2002; Kessler and Prenner 2009).

The advantages of DEM will ensure that tests of ore flow through the skip station bunker will provide complete information on the process. Additionally, by including the grain size distribution of mined ore in the simulations, it will be possible to examine the behavior of individual fractions during the processes of filling and emptying the bunker.

2 Discrete Element Method - Theoretical Grounds

In order to describe the mechanics and flow of bulk materials, either analytical or numerical approaches are used. In the analytical approach, bulk material is treated on the basis of empirically determined coefficients. Such an approach allows basic calculations, but does not reflect the actual mechanics involved in systems composed of a number of elements (Coetzee 2009).

As the analytical methods entail excessive simplifications, in order to model the behavior of bulk materials with particular focus on the interaction between the particles, numerical methods were employed. Numerical methods can be divided into methods based on continuous mechanics (FEM) and discrete mechanics (DEM). In the case of Finite Element Method, the simulated ore is treated as a continuous medium having elastic and plastic properties. This method of describing the material offers an improved representation of fine media, such as sands or dusts. Finite Element Methods, however, do not allow the identification of either rock fragmentation or the mixing of the investigated fractions, because the material is treated as uniform mass. Due to the above limitations, the modeling and analysis of the behavior of fine particle materials started to be performed with the use of Discrete Element Method (Coetzee 2009).

DEM is based on the principle stating that over a short period of time disturbances only occur between particles which remain in direct interaction. As the disturbance propagation speed in bulk material depends on its parameters, it is possible to select a time step which will satisfy the above principle (Mendyka 2017). The algorithm used in DEM calculations is cyclical, and the calculations are performed at minimal (discrete) time intervals. At each time step, the calculations should be repeated for all particles and should include the identification of interactions, the force of interactions and particle displacement. After the interactions between the simulated elements have been identified (this stage consumes approximately 70% to 80% of the computing power), the DEM algorithm solves two types of equations: motion equations, which are based on Newton's second law and which allow finding the acceleration, speed and position of the investigated elements, and interaction equations, which are calculated on the basis of a selected rheological model to find forces acting on the elements in contact. Each subsequent calculation is performed on the basis of the previous calculation. In order to enable and facilitate calculating the contact events, most of the available computer applications represent particles as spheres or combinations of several spheres. These spheres are considered as solid bodies, but may overlap each other, causing contact elastic forces to act on them (Czuba et al. 2010).

As the calculations are performed for very short time intervals, building a simulation which lasts approximately one minute requires several million computation cycles. Great demand for computing power causes simulation generation process to be time consuming. The simulation presented below, although simplified, was generated for almost one hundred hours.

3 Preparation of the Simulation Environment

The shape and size of the bunker have a significant impact on ore flow, and therefore the simulation environment was prepared using data obtained by performing three-dimensional scans of the underground bunkers. Based on the available vertical and horizontal sections (Fig. 1), simulation environment was prepared in AutoCAD by combining horizontal sections with the LOFT function. The generated model was simplified, because the visible partial separation of the two bunker legs was removed from the 9-9 section. Instead, a new section was added (below the 9-9 section), which shows full separation of the bunker legs. Thus, it was possible to build to solid bodies which will represent different types of material. The intake of the bunker was not modeled, as it is not relevant for the simulation and the bunker should not be filled above 85% of its capacity. Figure 2 shows the simulation environment and its vertical section.

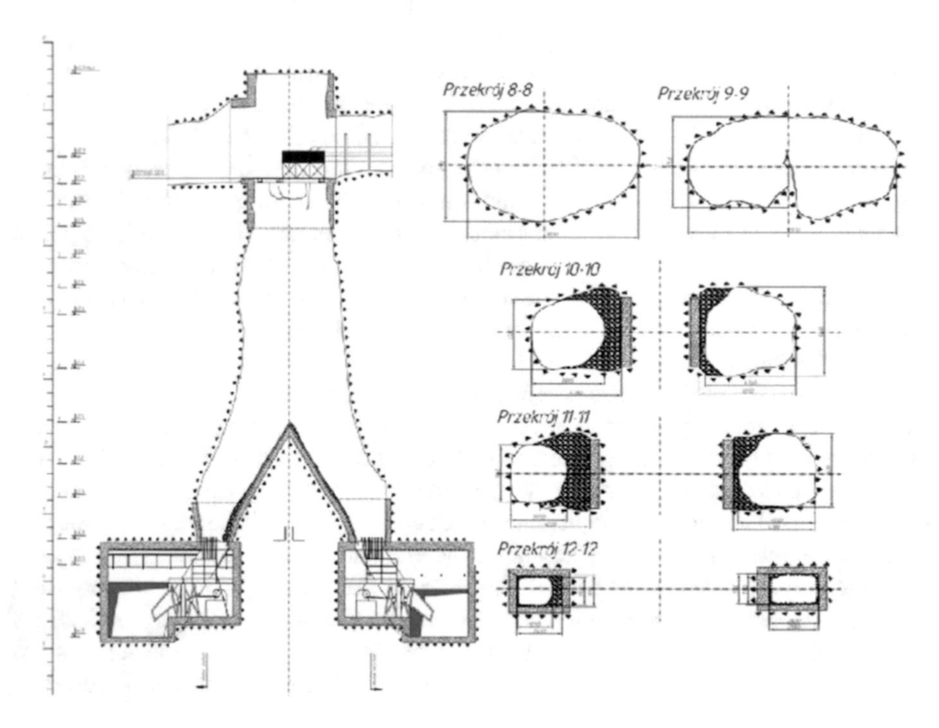

Fig. 1. Data used in the construction of the model: vertical section and selected horizontal sections

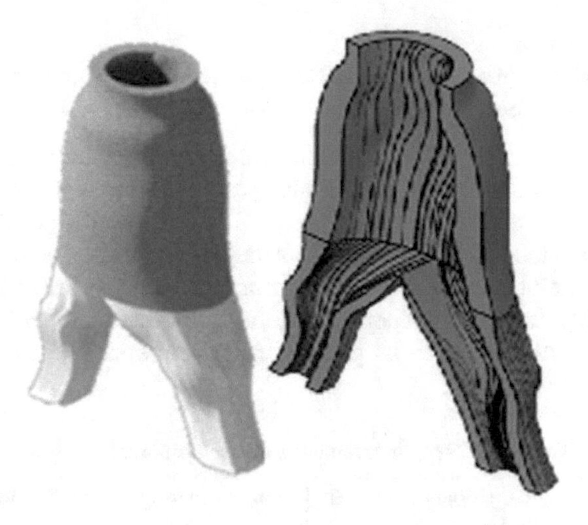

Fig. 2. Modeled bunker and its vertical section

In order to investigate the bunker filling process, the simulation included representing a fragment of the feeding belt conveyor with actual dimensions and locations preserved. The feeding belt conveyor was modeled with actual operating conditions, and thus an actual discharge trajectory was obtained (Fig. 3).

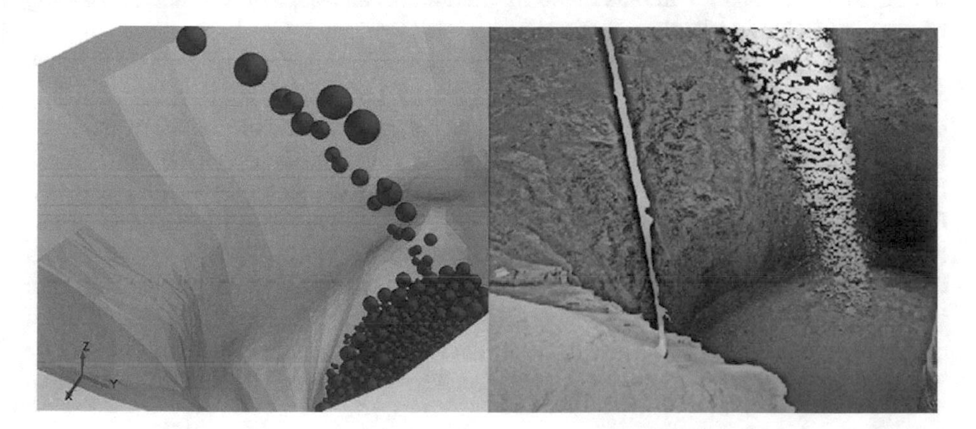

Fig. 3. DEM-simulated discharge trajectory vs actual discharge trajectory

In order to perform a DEM-based simulation, input parameters should include each type of materials used and the interactions between the materials. In the case of the EDEM Software which was used to perform the simulations, the required parameters include Poisson ratio, Young's modulus and volume density of the material. In the simulation here presented, five materials were modeled:

- the conveyor belt,
- the housing of the main part of the bunker, built of ore-bearing rock,

- chutes made of steel,
- the transported ore (material flowing through the bunker) – sulfuric copper ore,
- pellets carrying information on the quality of the deposit, made of PET.

Table 1 shows a comparison of strength parameters describing the above materials. As determining the values shown in the table would be time-consuming in laboratory conditions, and as it would not significantly affect the simulation result, data on the materials were provided on the basis of the available literature (Kozlowski and Kudelko 2014; Polyethylene terephthalate material information 2017; Prawoto 2013; Fedorko and Ivancob 2012). In the case of copper ore, the data were obtained from a publication which describes ore-bearing rocks present in the mine of interest (Kozlowski and Kudelko 2014).

Table 1. Strength parameters of the simulated materials.

	Copper ore	Pellet	Bunker housing	(Steel) chute	Conveyor belt
Poisson ratio [-]	0.2	0.35	0.22	0.285	0.5
Young's modulus [GPa]	22.54	2.28	36.32	190	0.01
Volume density [kg/m^3]	2565	1500	2565	7860	2000

Each type of the simulated particles should be provided with values describing their interactions with other materials in the simulation, as: static friction coefficient, rolling friction coefficient, reflection coefficient.

These values, after preliminary tests in the laboratory, should be calibrated in the Software so as to ensure that the simulated material shows identical behavior to the actual material (Schott and Katterfeld 2011). In this simulation, the values of the coefficients describing the interactions between the elements were based on the values taken from tables (Friction and Friction Coefficients 2017; Rocscience Coefficient of Restitution Table 2017; Coefficient of Friction 2017), included in the available publications regarding a particular phenomenon (Fedorko and Ivancob 2012) and on the values offered by the producer of the Software. As no data was found on the investigated ore type, it was decided to use data available for geologically similar rock types. The selected coefficients were adjusted to ensure the behavior of the simulated material would be identical to the behavior of the actual material. These values are shown in Tables 2 and 3.

Table 2. Interactions between copper ore and other simulated materials

Ore interactions			
Material	Coefficient of restitution	Static friction coefficient	Rolling friction coefficient
Conveyor belt	0.1	0.65	0.01
Bunker housing	0.1	0.75	0.01
Ore	0.1	0.75	0.01
Steel	0.1	0.3	0.01
Pellet	0.2	0.2	0.01

Table 3. Interactions between pellet and other simulated materials

Pellet interactions			
Material	Reflection coefficient	Static friction coefficient	Rolling friction coefficient
Conveyor belt	0.1	0.7	0.01
Bunker housing	0.2	0.2	0.01
Ore	0.2	0.2	0.01
Steel	0.2	0.3	0.01
Pellet	0.3	0.2	0.01

In order to fully simulate the ore flow process, photogrammetric method was used to reproduce grain size distribution in the ore fed to the bunker. Grain size distribution slope was produced by taking a series of photographs over the belt conveyor and subsequently by processing the photographs in the Split Desktop 4.0 Software. However, as the value of the time interval is directly related to the radius of the smallest particle used in the simulation, it was decided to model only four fractions having the greatest size. Finer particles were substituted by increasing the share of the finest fraction. Table 4 shows the percentage share of individual fractions in the transported ore, while Fig. 4 is a grain size distribution slope for rocks fed to the bunker.

Table 4. Simulated grain size composition

Rock fragment diameter [mm]	Rock fragment radius [m]	Percentage share of rock fragments in the ore [%]
635	0.32	1
381	0.19	4
254	0.13	10
203	0.1	85

As the simulated material is a dry rock, adhesion forces are not of concern. Therefore, it was decided to use the basic Hertz-Mindlin model to simulate both the particle-particle interactions and the interactions between the particles and the material of the environment.

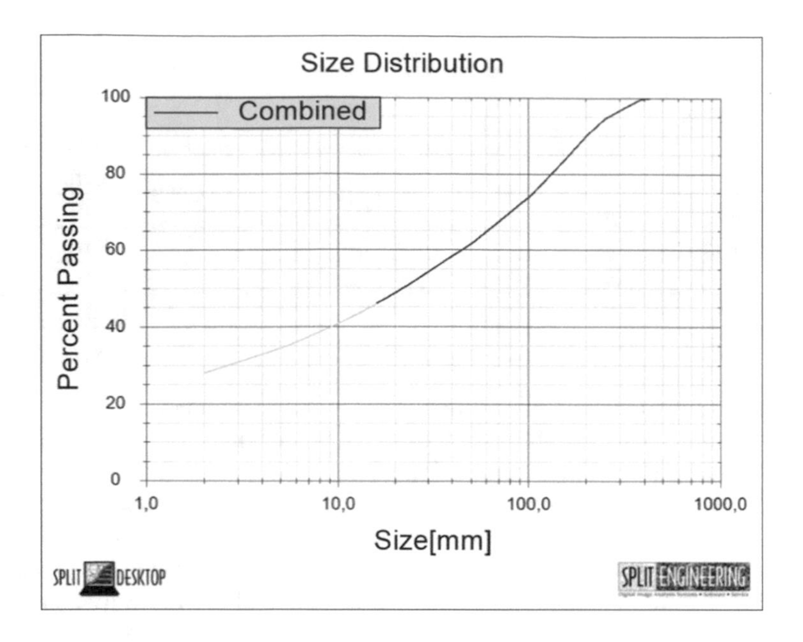

Fig. 4. Grain size distribution for filled bunker

4 Simulation Summary and Discussion

The simulation was performed in the EDEM Software, using the Euler method. Time interval value was $7,33 * 10^{-5}$ s, which corresponds to 40% of its theoretical value. The mesh used to detect particle-particle interactions had a size equal to three times the minimal particle radius length ($3R_{min}$). The simulation shows the bunker filling and emptying process, which lasts 1577 s. The ore was fed via the belt conveyor with efficiency equal to 950 kg/s, to shorten the simulation time (actual belt conveyor capacity is approximately 222 kg/s). Filled of the first leg of the bunker takes 216 s. The second leg was filled after next 259 s. The bunker was filled over the following 1045 s. The filling of the bunker was stopped in 1516.6 s, at 85% of its capacity and with the mass of the stored ore equal to 1406 Mg (Fig. 7a). After the bunker was filled, the chute gates in both legs were opened simultaneously. The ore was completely discharged from the bunker in 57 s (Fig. 7). Importantly, the sections of the two bunker legs have different surface areas, which influenced the discharge efficiencies and the fluency of ore flow.

In the initial phase of bunker filling, all of the ore is fed to the first leg, which accounts for approximately 12% of the bunker's capacity. Subsequently, when the base of the natural pile of material rises above the separation between the two legs, ore starts to drop into the second leg; the process is shown in Fig. 5.

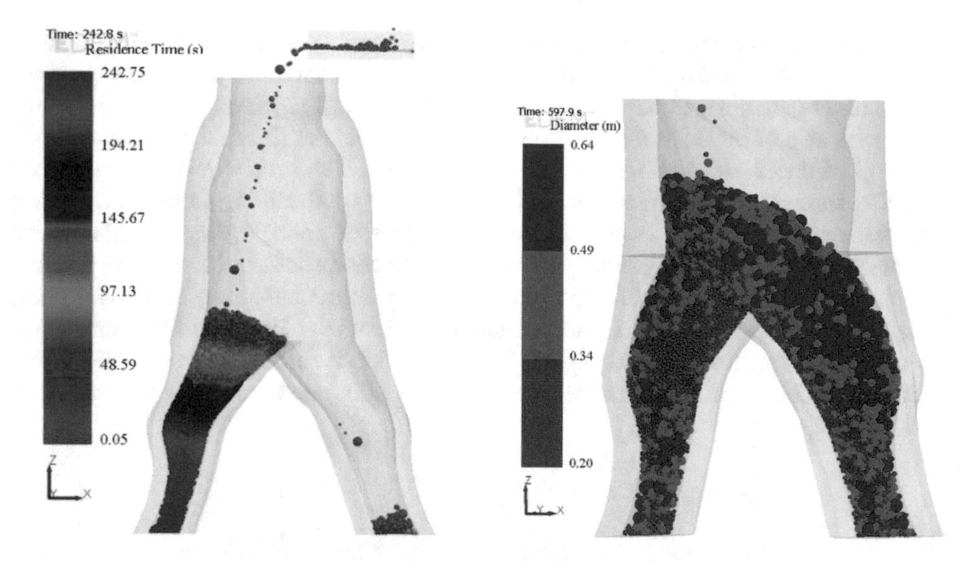

Fig. 5. The process of filling the second leg of the bunker. Particles are assigned a color representing their residence time in the bunker.

Fig. 6. Fraction distribution in the bunker. Particles are assigned a color representing their diameter.

As the legs of the bunker are being filled with ore, material fractions undergo natural separation (Fig. 6). In the first leg, the separation is dictated by the inertia of ore particles at the moment of being discharged from the belt conveyor. As the measure of the body's inertia is its mass (McCulloch 2013), larger particles drop further from the conveyor. In the case of the second leg, coarse ore fragments slide down the conically shaped pile of material and collect at the outer wall of the bunker. In both cases, another phenomenon also occurs, which is similar to grain segregation, the so-called Brazil nut effect (Fan et al. 2017). As was observed, finer particles are collected in the bottom part of the legs, almost completely filling the surface of the inclined plane, while larger ore fragments are located above and collected at the outer walls of the bunker. This phenomenon occurs when the whole system, naturally tending to lower its energy state, moves or vibrates. Additionally, finer particles, due to their size, are able to fit in the gaps and voids which prove too small for the larger particles. In the subsequent stages of the bunker filling process, a naturally inclined conical pile is formed, whose tip overlaps with the curve formed by the falling ore. Due to the fact that, as was mentioned above, larger lumps of rock are subjected to greater inertia, it may be assumed that such lumps will collect in the left part of the bunker.

The process of emptying the bunker in the proposed methodology was distorted by the uneven throughput of the discharge chutes. A significantly larger portion of the mined ore was discharged through the right leg. The bunker emptying process is shown in Fig. 7. At the maximum bunker filling degree, one leg of the bunker can be observed to fill at a greater pace than the other leg. The illustration also demonstrates how important this process is for the subsequent forming of the natural conical pile. In the bottom part of the bunker, the angle of slide is greater and gradually decreases as the bunker is filled. At 70% fill degree it can be clearly seen how greatly the increased throughput of the right leg affects the ore flow. The material on the right-hand side of the bunker moves much faster than the material on the opposite side. Nevertheless, faster displacement of the ore is also observed at the walls on the left-hand side of the bunker (purple color). This phenomenon seems intuitive, but its detailed description and understanding requires further investigations. In the subsequent stages of the filling process, the mixing of ore is observed to be consistent with the effect of grain segregation (the above mentioned Brazil nut effect). Fine particles of the fractions positioned at the top become displaced deeper into the bunker, while the largest fragments move at a lower pace, which eventually leads to their accumulation in the upper layer. At 30% bunker fill degree, dead regions can be observed. The ore which is located in those regions and which was delivered at early stages of the bunker filling process leaves the bunker in

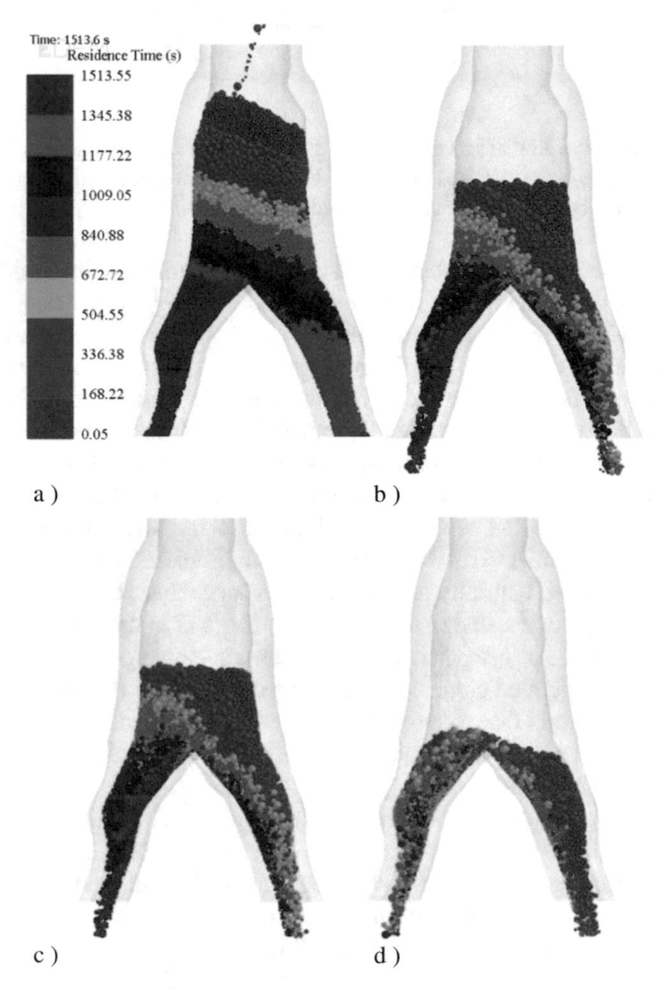

Fig. 7. The process of emptying the bunker at subsequent fill degree: (a) 85%, (b) 70%, (c) 50%, (d) 30% (particles are assigned by a color representing their residence time in the bunker)

the final emptying stages. Dead regions are located on the inclined planes of the legs and in the area where the legs meet. However, as can be observed, in the middle part of the inclined plane the material moves faster than at its edges.

5 Conclusion

An important aspect of the DEM modeling is to visually inspect the results of simulations. In the case of the model here presented, this inspection allowed investigations into the mechanisms involved in the movement of ore particles fed into the bunker and discharged from it.

Acknowledgements. This work was partly supported by the Framework Programme for the Research and Innovation Horizon 2020 under the grant agreement No. 636834 (DISIRE) and by the Polish Ministry of Science and Higher Education as scientific project No.0401/0131/17.

References

Coetzee, C.: The modelling of bulk materials handling using the discrete element method. In: 1st African Conference on Computational Mechanics – An International Conference, 7–11 January 2009, pp. 1–38. Department of Mechanical & Mechatronic Engineering, Sun City, South Africa (2009)

Czuba, W., Gospodarczyk, P., Kulinowski, P.: Zastosowanie Metody Elementów Dyskretnych (DEM) do symulacji odstawy urobku przez ścianowy przenośnik zgrzebłowy. Symulacja Bad. Rozw. **1**(3), 213–221 (2010)

Coefficient of Friction (2017). http://www.roymech.co.uk/Useful_Tables/Tribology/co_of_frict.htm. Accessed 23 Dec 2017

Dewicki, G., Mustoe, G.: Bulk material belt conveyor transfer point simulation of material flow using DEM. In: Third International Conference on DEMs, pp. 1–11. Santa Fe, New Mexico (2002)

Fan, F., Liu, J., Parteli, E.J., Poschel, T.: Vertical motion of particles in vibration-induced granular capillarity. In: Powders and Grains 2017 – 8th International Conference on Micromechanics on Granular Media (2017). https://doi.org/10.1051/epjconf/201714016008

Fedorko, G., Ivancob, V.: Analysis of force ratios in conveyor belt of classic belt conveyor. Procedia Eng. **48**, 123–128 (2012)

Friction and Friction Coefficients (2017). https://www.engineeringtoolbox.com/friction-coefficients-d_778.html. Accessed 23 Dec 2017

Jurdziak, L., Kawalec, W., Król, R.: Application of flexsim in the disire project. Stud. Proc. Pol. Assoc. Knowl. Manag. **84**, 87–96 (2017)

Katterfeld, A., Wensrich, C.: Understanding granular media: from fundamentals and simulations to industrial application. Granular Matter **19**, 83 (2017). https://doi.org/10.1007/s10035-017-0765-y

Kawalec, W., Jurdziak, L., Król, R., Kaszuba, D.: Project DISIRE (H2020) – an idea of identification of copper ore with the use of process analyser technology sensors. In: E3S Web of Conferences, vol. 8, p. 01058 (2016)

Kessler, F., Prenner, M.: DEM – simulation of conveyor transfer chutes. FME Trans. **37**, 185–192 (2009)

Kozłowski, T., Kudełko, J.: Weryfikacja doboru obudowy kotwowej w warunkach zaburzeń tektonicznych w kopalni Lubin. CUPRUM – Czas. Nauk. Tech. Gór. Rud **73**(4), 55–71 (2014)

Krzemińska, M., Malewski, J.: Energochłonność operacji przygotowania rud do wzbogacania w kopalniach LGOM. Prz. Gór. **67**(7–8), 143–147 (2011)

McCulloch, M.E.: Inertia from an Asymmetric Casimir Effect. https://arxiv.org/abs/1302.2775. Accessed 23 Dec 2017

Mendyka, P.: Inżynierskie zastosowania metody elementów dyskretnych. Napędy i sterowanie **11**, 117–124 (2017)

Minkin, A.: Analysis of transfer stations of belt conveyors with help of discrete element method (DEM) in the mining industry. Transp. Logist. **12**(24) (2012). ISSN 1451-107X. http://www.sjf.tuke.sk/transportlogistics/wp-content/uploads/31.Minkin.pdf

Polyethylene terephthalate material information (Polyester, PET, PETP) (2017). http://www.goodfellow.com/E/Polyethylene-terephthalate.html. Accessed 23 Dec 2017

Prawoto, Y.: Integration of Mechanics into Material Science Research: A guide for Material Researchers in Analytical Computational and Experimental Methods. Johor, Malezja (2013)

Rocscience Coefficient of Restitution Table (2017). https://www.rocscience.com/help/rocfall/webhelp/RocFall.htm#baggage/rn_rt_table.htm. Accessed 23 Dec 2017

Schott, D., Katterfeld, A.: Influence of the software on the calibration parameters for DEM simulations. Bulk Solids Handl. **31**(7–8), 396–400 (2011)

Conveyor Belt 4.0

Leszek Jurdziak(✉), Ryszard Blazej, and Miroslaw Bajda

Faculty of Geoengineering, Mining and Geology, Wrocław University of Science and Technology, ul. Na Grobli 13/15, 50–421 Wrocław, Poland
{leszek.jurdziak, ryszard.blazej, miroslaw.bajda}@pwr.edu.pl

Abstract. Industrial revolution known as Industry 4.0 is present in all branches of industry and in all areas of production. It also stimulates changes in the mining industry, known as Mining 4.0. Transportation processes have a significant impact on the technological chain, from the working of minerals to the selling of the final products. Continuous transportation with the use of belt conveyors allows cost reductions and increased range. It also allows fully automatic operation as well as remote monitoring and control from the centralized control room. Presently, the conveyor belt is the least sensor-monitored system in Polish mines. This paper demonstrates that appropriate methods and devices for the monitoring of conveyor belts and splices already exist. Polish mines have been collecting and recording data on the installed belts, splices and repairs. The authors propose to introduce diagnostic systems and integrate data from a number of sources in order to include the idea of conveyor belt 4.0 into the digitally controlled conveyors 4.0, and – in wider perspective – into the Mining 4.0 intelligent solutions.

Keywords: Industry 4.0 · Mining 4.0 · Belt conveyor 4.0

1 The Fourth Industrial Revolution

Industry 4.0 is a notion referring to the fourth industrial revolution, which we currently witness. The current revolution was preceded by three stages of industrial development (Fig. 1).

Their description is provided in a publication under [22]:

Industry 1.0 – Mechanization – invention and deployment of mechanical means of control (cams), and especially of a steam engine, led production processes into the industrial era.

Industry 2.0 – Electrification – electricity rendered steam power obsolete, and enabled first production lines. Large batches of goods could be manufactured and information started to be recorded on perforated cards.

Industry 3.0 – Digitization – increasingly efficient computers and data processing systems allowed control over machines to be exerted with the use of software. Thus, the efficiency, precision and versatility of machines was improved and the digitization process led to increased levels of automation. The first planning and control systems were designed in order to coordinate production processes.

© Springer Nature Switzerland AG 2019
A. Burduk et al. (Eds.): ISPEM 2018, AISC 835, pp. 645–654, 2019.
https://doi.org/10.1007/978-3-319-97490-3_61

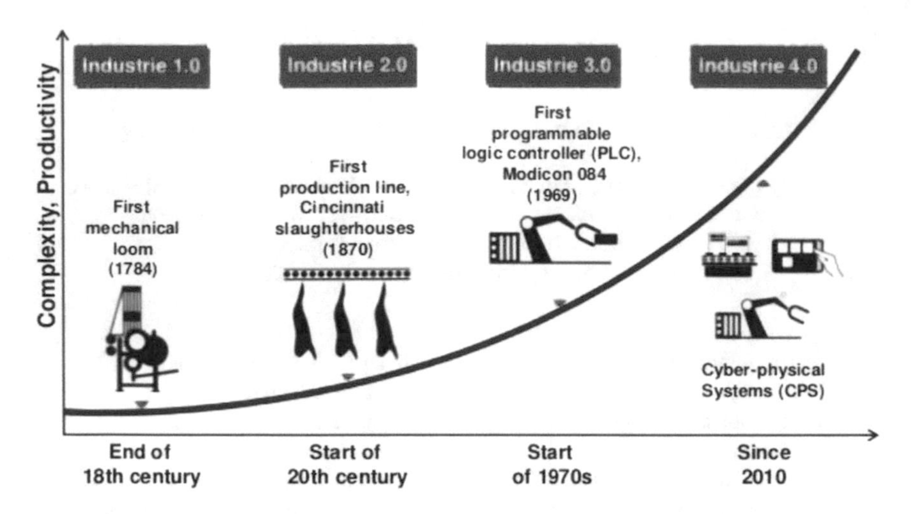

Fig. 1. Subsequent industrial revolutions according to the Cassantec company [16].

Industry 4.0 – Integration of intelligent machines and systems into networks. Integration of people and digitally controlled machines with the Internet and information technologies. The manufactured materials or the materials used in manufacturing processes always allow identification and remain in independent communication with one another. The flow of information is vertical: from individual components to the IT department in a company and from the IT department to the components. Information flow is also horizontal: between machines engaged in manufacturing processes and the manufacturing system in the company.

The notion was introduced in Germany in 2011, when the representatives of businesses, as well as political and academic institutions, started to promote this idea as an approach to increased competitiveness of German manufacturing industry. German federal government supported the idea, pronouncing Industrie 4.0 to be an integral part of "Action Plan High-Tech Strategy 2020" – an initiative aimed for Germany to take a leading role in technological innovation [7]. Literature studies led the authors of the above paper to identify the following 6 rules of designing implementations for Industry 4.0: interoperability, virtualization, decentralization, ability to work in real time, focus on providing services and modularity.

2 Mining 4.0

Mining 4.0 is an idea of Industry 4.0 transplanted to the field of mining industry. Professor Paschedag describes Mining 4.0 as follows [18]:

1. Mining machines, devices, sensors and humans can network and communicate with one another.
2. Sensor data widen the information systems of the digital mine to create a virtual view of the mine.

3. Technical assistance systems support miners using aggregated, visual and understandable information to allow fact-based decisions and to solve problems quicker; furthermore miners are physically supported in tiring, awkward or dangerous work.
4. Cyber-physical systems in mining are capable of making their own decisions and of doing their job as autonomous as possible; only in exceptional situations (i.e. incidents or target conflicts) are tasks handed over to a human.

A lot of the above is technically possible already today. The above ideas have been already implemented in the continuous transportation systems, including in one of the most important replaceable elements of the belt conveyor, i.e. in the conveyor belt [17].

3 Belt Conveying 4.0

Modern technical solutions allow belt conveyor lines to be remotely controlled from technically advanced control centers relying on a great amount of data provided by an increasing number of sensors and related to the operation of machines and devices. Intelligent application of big data analytics becomes a challenge if production targets need to be met with lowest possible costs, and not only from the perspective of the quantity but also the quality of the transported material.

This challenge is already addressed in the **Central Control Room** of the PGE GiEK SA lignite mine in Belchatow, and solutions are implemented for the Belchatow and Szczercow mining fields. The Mine Supervision System developed by Merrid Controls [25] ensures complex monitoring of the technological systems in both open pits. Constant monitoring covers 25 main machines and 180 conveyors which form the LCS (Loader – Conveyor – Stacker) systems in both mines.

Apart from performing real-time diagnostics, the system also triggers warnings and alarms. A large mimic display panel presents the operation of all extraction and transportation systems (Fig. 2). It is possible to provide adequate security and to start and stop their operation remotely.

Fig. 2. Central Control Room at PGE GiEK SA [27].

The *Material Tracking Module* offers significant functionality. Based on the received information about excavator efficiencies, speeds of belt conveyors and configurations of the transfer points, the module defines the amount and quality of coal at any of the plants in the mine [13]. It also allows the operator to monitor in real time the aggregated data on the transported mass (in Mg) as well as temporary parameters such as efficiency (in m^3/h) and the composition of coal fed to the power plant (calorific value in kJ/kg, ash and sulfur content in %) and to control loader efficiency so that the power plant receives mixed fuel having required parameters. Thus, the overall efficiency of electricity production is increased in the technological chain from the mine to the power plant (in a vertically integrated mining and power generation complex).

Mimic panels at the KGHM PM SA copper ore mines display the operating parameters of conveyors and their components. Remote predictive diagnostics is possible owing to data from a number of sensors. The monitoring covers the amount of the transported material and the filling degree of underground bunkers. Similar systems are also developed for hard coal mines [14].

Monitoring ore quality and predicting the composition of ore fed to the processing plant remains a problem, although a 3D lithological model already exists for the deposit, as is the case in KWB Belchatow. The blasted and mixed ore from many faces is fed via numerous division conveyors to main haulage conveyors and ultimately to the shaft station bunker. The number of grizzlies, ore transfer stations and bunkers is so large that the composition of ore fed to the concentration plant cannot be predicted accurately. The Faculty of Geoengineering, Mining and Geology at Wroclaw Univ. of Science and Technology carried out an international DISIRE project (a part of the Horizon 2020 program [11]). The results of both the experiment with the use of e-pellets carrying data on ore quality and the simulations of ore flow and quality distribution are successively published [10].

Costs are another important issue in mining. Transportation accounts for ca. 50% of production costs in mines [19]. They also account for ca. 40% of electric energy consumed in mines. A lot has been done in recent years to lower the energy consumption by introducing new designs of idlers, energy-saving belts and variable speed drives which adjust to load changes. Objective comparison of the effectiveness of such solutions requires introducing some analytical indicators [12].

Due to high cost of conveyor belts and to their influence on the reliability of transportation systems in mines [2] operators in the Central Control Room should also be provided information on the condition of conveyor belts. Mines store large amounts of data on conveyor belts, but they are not integrated and presented in a way that would allow identifying potential disruptions to belt and idlers [15] operation.

4 Conveyor Belt 4.0

Conveyor belts account for up to 60% of transportation costs [17]. Due to increased specialization and outsourcing tendencies, mining companies delegate the responsibility for belt management to subcontractors as part of long-term agreements. Such policy allows the mine to reduce own staff, while belts are maintained by specialists. Service workers must also ensure that transport in the mine is uninterrupted.

A contract between RWE Power and REMA TipTop may serve as an example [23]. The service contract has a term of 10 years. It comprises on-site support for the continuous surface mining operations in the Rhenish lignite mining area (surface mines Inden, Hambach and Garzweiler). Repairs and maintenance of steel cord conveyor belts in the transportation system (including vulcanization of belts, joints and belt reconditioning for all conveyors, BWEs and stackers) will be carried out by 120 locally employed service staff, who will be quickly available 24/7 from their local office. The continuous and trouble-free operation of the conveyor systems is one of the most important prerequisites for reliable power generation and supply of energy to the highly industrialized region. If a conveyor belt loop fails, the entire operation is stopped. It is therefore crucial to identify wear and tear as early as possible before a failure occurs, in order to schedule maintenance periods in a targeted manner. Any damage needs to be repaired as soon as possible to avoid further troubles.

When the operated conveyors have a total length of several hundred kilometers (e.g. at RWE Power it is ca. 500 km and at Belchatow – ca. 160 km), these tasks cannot be effectively performed without monitoring belt condition, planning repairs and replacements with the use of digital diagnostic tools [4].

At the Bergbau 4.0 conference held in Aachen, one of the world's leading manufacturers of conveyor belts presented a complex solution for gathering, processing and cloud storage of data related to belt operation. The data can be accessed at any time by the users and service workers via the Internet to facilitate rational belt management, eliminate downtime and reduce excessive belt reserves [17]. Such a policy, however, ties the mine with one belt supplier, which also services belts, and granting an external contractor access to operating data may not be in line with the mining company's policy on sensitive data protection. The mining company may be also charged with monopoly rents related to eliminating competition mechanisms.

The information system presented at the conference [17] comprised data collection and storage systems for conveyors & belts (Fig. 3a).

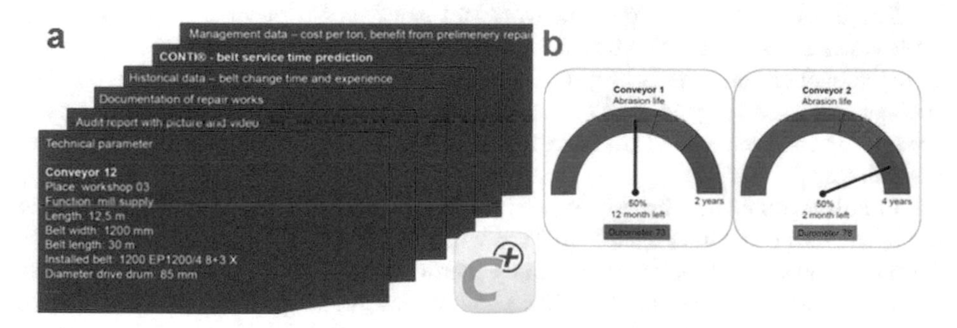

Fig. 3. System Conti + contains full spectrum of information about each conveyor: a – parameters, files, reports; costs etc.; b – belt service time meters/belt abrasion life [17].

The system offers:

- increased information recording speed,
- easy organization and reports on conveyor data,
- estimation of belt lifetime and replacement data (Fig. 3b), and
- accessibility at any time and place

owing to cloud and web processing of all gathered information (Fig. 4).

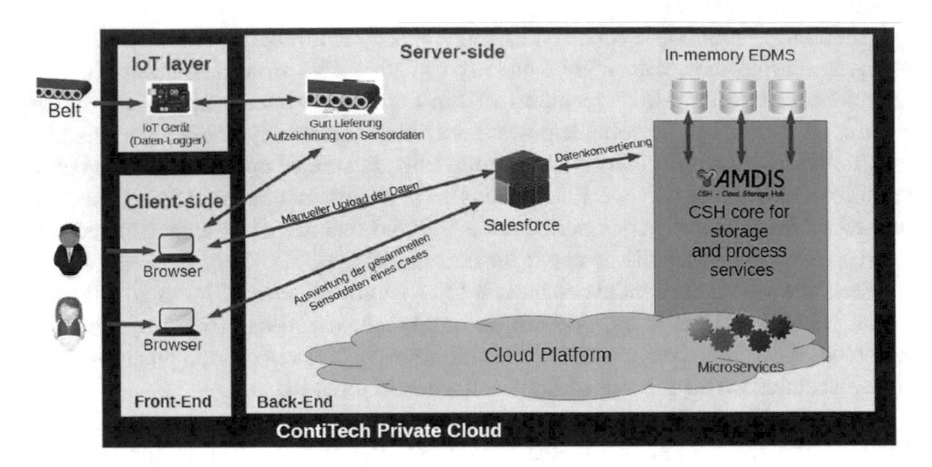

Fig. 4. Automation of belt service process using web and cloud platform [17].

On its website [26], the company informs that the recently started cloud solution aids sales processes. It handles 100,000 offers a year. Conti Cloud Platform (Fig. 4) is based on the SAP Cloud for Sales solution. The core process of tendering was ultimately the driver of the project. With over 100,000 offers per year, a small gain in efficiency and better operations brings a significant amount of time and money. The two most important goals are the optimization of in-house processes as well as faster adaptation of new business areas.

The usage of databases on conveyors and belts, predictive maintenance and belt replacements (Fig. 4) are also beneficial to the clients (belt users). The ordering process is much shorter and thus belts can be shipped „*just in time*". In effect, the costs of maintaining large emergency belt storehouses is minimized. As a result, Total Costs of Ownership and Production Losses are minimized due to reduced number of unexpected downtimes.

In Poland, computers have been used to aid belt management in lignite mines since the mid-1980s (the Sufler system in Turow and the Tasma system in Belchatow [9]). Steel cord belts have been diagnosed in the Turow mine since 2000. Databases on belts and conveyors are used in the KGHM PM SA mines (the Diagmanager system and the system developed by AGH University of Science and Technology in Cracow). These systems were convinced and co-developed by researchers from Wroclaw Univ. of Science and Technology. This is where the High Resolution Diagnostic System (HRDS) was modernized for the KWB Turow mine and where the DiagBelt system was designed [3, 24]. It is hoped that these systems can be used on a larger and more complex scale than at present (e.g. [1, 6]). The advantages of the system for an open pit mine and for a service company were discussed in papers [2, 8].

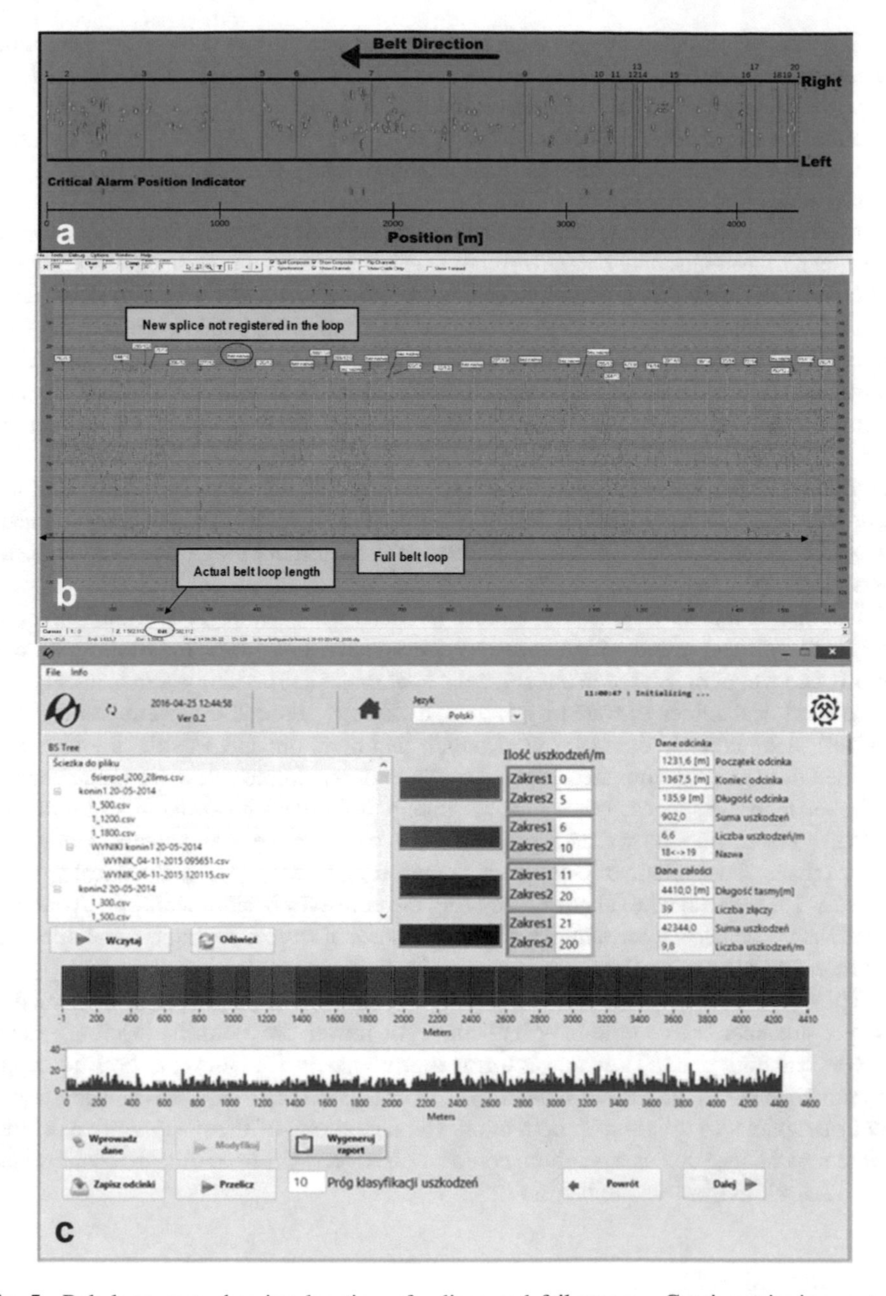

Fig. 5. Belt loop map showing location of splices and failures: **a** - Conti monitoring system [17]; **b** - Source data for the DiagBelt system (image from the BeltGuard system [21]); **c** - Processed data – belt loop image from the DiagBelt system with the histogram of damage density along belt axis [24].

The benefits of using IT systems and diagnostics have been also noticed by a belt manufacturer from Germany. They are used in the Conti + solution. Damage is mapped (Fig. 5) similarly in various high-resolution systems, e.g. Belt Guard, DiagBelt, Veyance and Contitech [5]. Importantly, the system must facilitate identifying the locations, sizes and intensity of defects (e.g. damage density per 1 running meter). Color coding allows the condition of all belt loop sections to be immediately evaluated.

Presently, Polish belt manufacturers offer neither their own diagnostic tools nor belt management systems. Only lignite mines and copper ore mines have their own IT solutions. Diagnostic tools have also been introduced recently (the HRDS at KWB Turow). The cost of conveyor belts also stimulated a more rational belt management in underground hard coal mines. For example, in the Marcel mine, belts on main haulage conveyors (located in the dip-heading which functions as an output shaft) are regularly scanned to prevent downtime and production losses. With very high fixed costs, mines may reduce unit cost of extracting a tone of minerals by avoiding time waste. Each downtime reduces mine output and increases unit production cost. Reliability and lack of downtimes (belt replacements during non-production shifts) are necessary to ensure cost-effective production. Therefore, KGHM PM SA orders diagnostics of steel cord belts (e.g. [3]).

National Robotics Engineering Center (NREC) from Carnegie Mellon University designed, built and tested a high-speed machine vision system for monitoring the condition of conveyor belts used in underground coal mines [28]. This system is in everyday use at hard coal mines operated by CONSOL Energy, Inc.. The system, called “Belt Vision”, incorporates a computer workstation that monitors and records digital images obtained from cameras mounted above the conveyor belt to provide continuous imaging of the belt loop and splices. Splice tears are among the biggest threats to underground belt systems. The device allows managing belts by preventative maintenance rather than costly repairs. One minute of emergency stop to replace destroyed mechanical splice is valued $ 1 000 and it takes about 4 h to bring the conveyor back to work. To avoid such situations the device captures high resolution images of splices during normal operation and monitors belts to perform necessary repairs before a break occurs.

The costs of breakdowns can be even higher. In an open-pit copper mine owned by a major producer, a five mile-long belt conveyor moves ore from the in-pit crusher to the concentrating plant. The mine extracts approximately $320,000 per hour in copper and molybdenum. Mechanical breakdowns of this belt conveyor were causing significant disruptions to the mine’s production. Replacing one of the many conveyor drive train components, such as a gearbox, could result in an eight-hour outage, costing more than $2.5 million in lost production [20].

5 Conclusions

Belt conveyor transport systems currently used in mines are already fully automated and can be controlled from control rooms. Individual mechanical parts, especially drive trains, are already monitored and on-line data are aggregated from many sensors for diagnostic purposes and to predict repairs and replacement of individual components and their parts.

Moreover, the quality of fuel delivered to the power plant can already be controlled in lignite mines by adjusting the efficiency of BWEs on individual benches and appropriately mixing lignite in the combined output stream transported on main haulage conveyors. Thus, conveyors are intelligent, as they "know" not only the quantity of material they transport (owing to data from scales and from devices monitoring the volume of transported material) but also its quality (owing to scanners and mined material quality monitoring at the location of the excavator wheel equipped with GPS).

In Poland, mimic panels and monitors in central control rooms do not yet provide information on conveyor belt condition. Belts account for 60% of the transportation costs and their failures may effect costly downtimes. Production losses due to emergency splice repairs and belt replacements (e.g. after cuts) may be counted in hundreds of thousands of dollars. Therefore, their condition and operation should be closely monitored to prevent damage development. Also, individual defects should be repaired as they occur, and worn belt segments or splices should be replaced before they fail. Diagnostic tools and IT solutions corresponding to industry 4.0 are already available commercially. Some 4.0 ready Polish modular diagnostic tools are also available (compare DiagBelt). They should be deployed and integrated with other conveyor automatic systems in order to ensure fully reliable operation of transportation systems. Eventually, integration should include the whole mining machinery systems.

With high fixed costs, a mine may reduce unit production costs by making full use of its production capacity and this means operating without breaks or down-times. Such efficiency may be obtained as a result of the digital revolution described here as **Conveyor Belt 4.0** being a part of **Smart mining** solutions and **Mining 4.0**.

Acknowledgements. The publication was partially financed from the funds of a project realized under the Applied Research Programme in path A, titled "*Joints of multiply conveyor belts with increased functional durability*" No. PBS3/A2/17/2015.

References

1. Bajda, M., Blazej, R., Jurdziak, L.: A new tool in belts resistance to puncture research. Min. Sci. **23**, 173–182 (2016)
2. Blazej, R., Jurdziak, L.: Condition-based conveyor belt replacement strategy in lignite mines with random belt deterioration. IOP Conf. Ser. Earth Environ. Sci. **95**(4), 042051 (2017)
3. Blazej, R., Jurdziak, L., Kirjanow, A., Kozlowski, T.: Core damages increase assessment in the conveyor belt with steel cords. Diagnostyka (Diagnostics) **18**(3), 93–98 (2017)
4. Blazej, R., Jurdziak, L., Kozlowski, T., Kirjanow, A.: The use of magnetic sensors in monitoring the condition of the core in steel cord conveyor belts - tests of the measuring probe and the design of the DiagBelt system. Measurement (London) **123**, 48–53 (2018)
5. Blazej, R.: Review of the newest NDT equipment for conveyor belt diagnostics. Diagnostyka (Diagnostics) **4**(64), 21–24 (2012)
6. Fedorko, G., Molnar, V., Ferkova, Z., Peterka, P., Kresak, J., Tomaskova, M.: Possibilities of failure analysis for steel cord conveyor belts using knowledge obtained from non-destructive testing of steel ropes. Eng. Fail. Anal. **67**, 33–45 (2016)

7. Hermann, M., Pentek, T., Otto, B.: Design principles for industrie 4.0 scenarios. In: 49th Hawaii International Conference on System Sciences (HICSS), 5–8 January 2016
8. Jurdziak, L., Blazej, R.: Economic analysis of steel cord conveyor belts replacement strategy in order to undertake profitable refurbishment of worn out belts. In: 17th International Conference on Multidisciplinary Scientific GeoConference SGEM 2017, Conference Proceedings, ISBN 978-619-7105-00-1/ISSN 1314-2704, 29 June–5 July 2017, vol. 17, no. 13, pp. 283–290 (2017). https://doi.org/10.5593/sgem2017/13/s03.036
9. Jurdziak, L.: Conveyor belts management in mines - present state and prospective Conveyor belts management in mines - present state and perspectives (in Polish). Gor. Odkryw. (Opencast Min.) **40**(5/6), 63–81 (1998)
10. Jurdziak, L., Kawalec, W., Krol, R.: Study on tracking the mined ore compound with the use of process analytic technology tags. In: Advances in Intelligent Systems in Production Engineering and Maintenance - ISPEM 2017, pp. 418–427. Springer, Cham (2018)
11. Kawalec, W., Krol, R., Zimroz, R., Jurdziak, L., Jach, M., Pilut, R.: Project DISIRE (H2020) – an idea of annotating of ore with sensors in KGHM Polish Copper S.A. underground copper ore mines. In: MEC2016, E3S Web of Conferences, vol. 8, p. 01058 (2016)
12. Kawalec, W., Wozniak, D.: Energy efficiency of the bottom cover of a conveyor belt – the first step to the new classification of belts (in Polish). Min. Sci. **21**(2), 47–60 (2014)
13. Kopertowski, A., Kolodziejczak, M.: Practical aspects of the use of the coal tracking function on ECS systems in the KWB Belchatow mine (in Polish). Wegiel Brunatny **3**(96), 15–20 (2016)
14. Kozielski, M., Sikora, M., Wróbel, L.: DISESOR - decision support system for mining industry. In: Proceedings of the FedCSIS 2015 Conference, Article no. 2015F168, pp. 67–74 (2015)
15. Krol, R., Kisielewski, W.: Research of loading carrying idlers used in belt conveyor - practical applications. Diagnostyka (Diagnostics) **15**(1), 67–74 (2014)
16. Kuznetsov, M.: Combining expert knowledge and advanced data analytics for predictive maintenance. In: Smart Mining Conference Forum Bergbau 4.0, Aachen, 14–15 November 2017
17. Neumann, T.: Conveyor belt group. Mining 4.0 - our digital journey. In: Smart Mining Conference Forum Bergbau 4.0, Aachen, 14–15 November 2017
18. Paschedag, U.: Mining 4.0 –new challenges to international cooperation. In: Smart Mining Conference Forum Bergbau 4.0, Aachen, 14–15 November 2017
19. Roumpos, C., Partsinevelos, P., Agioutantis, Z., Makantasis, K., Vlachou, A.: The optimal location of the distribution point of the belt conveyor system in continuous surface mining operations. Simul. Model. Pract. Theor. **47**, 19–27 (2014)
20. Schools, T.: Condition monitoring of critical mining conveyors. Eng. Min. J. **216**(3), 50–54 (2015)
21. http://www.beltscan.com/
22. http://przemysl-40.pl/index.php/2017/03/22/czym-jest-przemysl-4-0/. Accessed 4 June 2018
23. http://www.bulk-solids-handling.com/?q=topics/conveying-transportation/mechanical-belt-conveyor-systems/rema-tip-top-wins-conveyor. Accessed 25 May 2018
24. http://www.diagbelt.pwr.edu.pl/index.php/pl
25. http://www.merrid.com.pl/pl/oferta/systemy/system-wydobywczy. Accessed 4 June 2018
26. https://news.sap.com/germany/contitech-100-000-angebote-pro-jahr-nun-aus-der-cloud/
27. https://www.merrid.com.pl/pl/portfolio/realizacja-3/. Accessed 4 June 2018
28. https://www.nrec.ri.cmu.edu/solutions/mining/other-mining-projects/belt-inspection.html. Accessed 4 June 2018

Probabilistic Modeling of Mining Production in an Underground Coal Mine

Edyta Brzychczy(✉)

Faculty of Mining and Geoengineering, AGH University of Science and Technology, al. Mickiewicza 30, 30-065 Cracow, Poland
brzych3@agh.edu.pl

Abstract. The article presents probabilistic modeling of mining production in hard coal mines with the application of stochastic networks. The paper includes basic definitions and assumptions of the evolved method and the systematic enabling the description of network activities. As a results of calculations according to the mathematical model the paper presents probability distributions of output for each production flow and for a whole mine in each of the calculated variants. To estimate the risk of output results the standard deviation of the distribution was assumed. Probability distributions form the basis of the optimization procedure. An application of the developed method in a coal mine is presented.

Keywords: Modeling · Mining production · Stochastic networks GERTS

1 Introduction

Mining exploitation is a peculiar process, first of all, due to its internal conditions. Depending on the mine, the geological and mining conditions, the organizational and technical specifications as well as the applied technologies determine the achieved output levels to a considerable degree, being the major reason for the peculiar division of mines into better and worse ones. In the times of free market economy a mining enterprise, in order to survive on the free market, has to prepare production plans well beforehand and optimize its production process so as to be able to guarantee output with a strong focus on meeting the quality standards required by customers. Deep mining of mineral deposits is an extremely complex and expensive process and therefore all the activities aiming at the future production levels estimation should be carefully and meticulously elaborated on the basis of reliable data pertaining to other fields of a particular mine's activity, which are stored on numerous computer data systems. Designed activities should also take into account the hitherto experience in the existing mining exploitation in a given mine. It should be emphasized that the mining process has a random nature, which has to be considered in the course of planning both the mining operations and the output levels as well.

© Springer Nature Switzerland AG 2019
A. Burduk et al. (Eds.): ISPEM 2018, AISC 835, pp. 655–667, 2019.
https://doi.org/10.1007/978-3-319-97490-3_62

Analyses of different types of risk and uncertainty sources in mining project are presented in [1, 6, 14]. The most often specified sources are [7]:

- Internal sources (dictated by the deposit itself i.e. grade distribution, ground conditions as well as workforce, equipment, infrastructure)
- External sources determined by outside considerations (i.e. market prices, environmental conditions, political risk, country risk, community relations, industrial issues legislation).

The presence of risk determined by internal sources is frequently disregarded in carried out analyses. Therefore, on the basis of the above-mentioned considerations, the concept of creating a method facilitating future production planning in a mine under the free market economy conditions, taking the random nature of the mining process into account, has been originated. Developed method makes use of stochastic networks for the modeling and optimization of mining operations in an underground hard coal mine [3].

2 Stochastic Networks, GERTS Method

One of the essential elements of production planning is an adequate mapping of the analyzed process. The spatial structure of a mine (mining excavations network) can facilitate the choice of the appropriate representation of the planned activities in the form of activity networks. These techniques are often used for managing projects at the planning and realization stages [4, 8–11].

Activity networks have three different levels of complexity and they can be represented in three different forms. The complexity levels range from deterministic (for example CPM, in which the activity's duration and the cost are constant) through probabilistic (for example PERT in which the activity's duration and the cost are represented by probability distribution functions) and generalized activity networks (GAN) [2]. Detailed information on network model classification is presented in [15].

For the purpose of analyzing the GAN type networks the method called GERT (Graphical Evaluation and Review Technique) has been developed [12, 13] as well as its simulation version GERTS, which allows for handling extremely complex networks of this kind.

An arc of the activity network in the GERTS method is described as:

$$arc_i = [p_i, t_i]$$

where:

p_i – the probability of i arc realization,
t_i – the duration of activity or other characteristic of i arc (distribution function $f_i(t)$).

In practice, the following distribution functions are used: deterministic distribution, normal distribution, rectangular distribution, Erlang distribution, log-normal distribution, Poisson distribution, Beta distribution, gamma distribution and Beta (as in PERT method).

The GERTS method is extremely interesting for a designer due to the probabilistic nature of its network structure (activities are realized with a certain probability and their characteristics are represented as random variables).

Such a structure enables an analysis of the project according to the probabilistic structure of activities and its probable sequences. This possibility has encouraged the author of this paper to create a method with the application of stochastic networks to mining operations modeling which could include some aspects of internal uncertainty sources in this part of hard coal mine activity. Available techniques, for example PERT or CPM models, enclosed in different software packages (presented in [5]) are limited to analyze such a complex and specific problem as underground mining operations (e.g. the possibility of different sequences of longwall extraction). Even more advanced VERT technique without some modifications related to a problem specification is useless, but its possibilities to analyze the risk are powerful.

In the next sections of this paper the main definitions and assumptions of the evolved method shall be presented.

3 Modeling and Optimization Method of Mining Operations with an Application of Stochastic Networks

The main stages of the evolved method, presented in detail in [5], are as follows:

1. Input data introduction (i.e. mining and geological, technical, organizational and economical) related to the past and future mining operations.
2. Data analysis:
   (a) For the activities carried out in the past, the data shall be divided into the determined data and the random variables for which a probability distribution shall be found after the statistical analysis.
   (b) For the planned activities, possible realization variants shall be determined, and the probabilistic and the deterministic nodes of the network shall be defined.
3. Creation of a stochastic network for planned mining operations.
4. Calculations based on mathematical model of the stochastic network with use of Monte Carlo simulation (activity duration, other determined characteristics, e.g. output)
5. Selection of the optimal variant.

Having made the simulation calculations in accordance with a mathematical model, we obtain the probability distributions of the examined characteristics pertaining to, among other things, the output for the particular production flows as well as for the whole mine in each of the calculation variants. The standard deviation is used to assess the risk of carried out exploitation operations and to obtain the desired values of the examined characteristics. The probability distributions of the given characteristics form the basis for the selection of the best possible method of carrying out mining exploitation in accordance with the selected optimization criteria.

3.1 Main Definitions and Assumptions

The basic spatial elements include: maingate (CHP), tailgate (CHN), starter entry (P), longwall face (S).

Each spatial object can be characterized by a set of characteristics which are relevant because of its specific nature and which take into account possible technological variants related to carrying out mining operations.

The basic spatial elements are grouped (with reference to their territorial belonging) to form the so-called longwall object (LO), which can be expressed as follows:

$$LO = [CHN, CHP, P, S]$$

At a given point in time the longwall object is represented by the basic object according to the stage of the production process.

Each longwall object assigned to a production flow (PF) can be expressed as:

$$PF_i = \left[LO_{i1}, LO_{i2}, \ldots, LO_{ij}\right]$$

where:

i – the successive number of the production flow, for $i = 1, 2, \ldots, n$,
j – the successive number of the longwall object in the i-th production flow, for $j = 1, 2, \ldots, m$.

Each longwall object can be assigned to only one production flow (Fig. 1).

With regard to the equipment, the diversity of machines involved in the mining production process has to be considered.

The following technical objects have been distinguished on the basis of technological conditions:

- objects designed for dog headings (MCH): combined miners,
- objects designed for longwall working (MS): shearers (MSK), longwall chain conveyors (MSPS), main entry chain conveyors (MSPP), mechanized support set (ZOb), crushers (MSKr).

Each technical object is described by a set of characteristics referring to its peculiar character. In case of basic objects of MS type, complex objects, the so-called longwall sets (Z), which shall be allocated to particular longwall faces in longwall objects should be created. Each longwall set Z_i consists of the following elements: MSK_i, $MSPS_i$, $MSPP_i$, ZOb_i, $MSKr_i$.

Technical objects are assigned to spatial objects, in which they can be utilized with regard to particular mining and geological conditions as well as technical and organizational specifications, creating so-called allocation matrixes of a certain type of equipment.

The following kinds of mining operations have been distinguished:

- driving the maingate and tailgate entries (RPp, RPn),
- driving the starter entry (RO),
- installation of longwall face equipment (ZB),

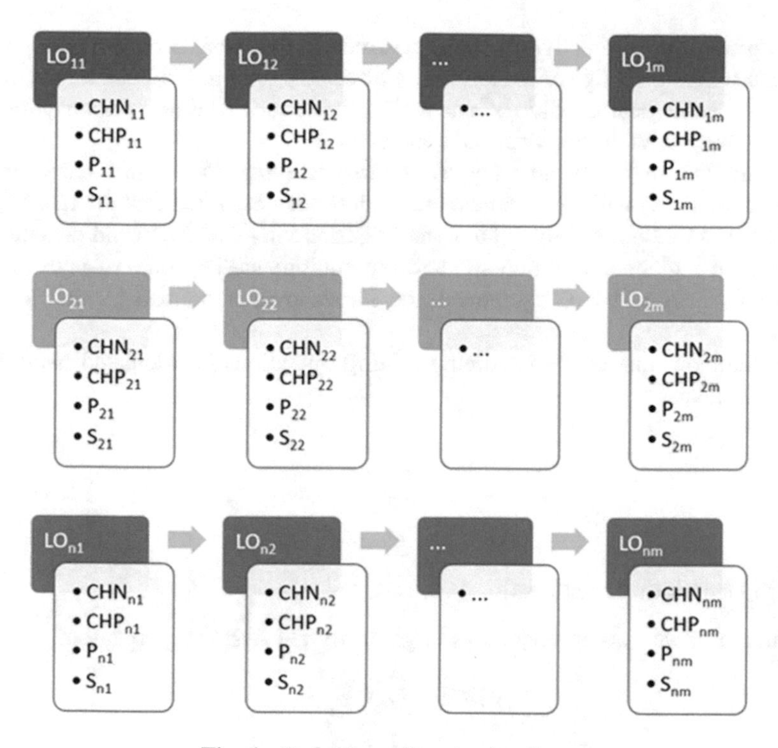

Fig. 1. Definition of production flows

- coal extraction (E),
- removal of longwall face equipment (LIK).

Having determined both the precise location for carrying out a given kind of mining operations and the suitable equipment (on the basis of the allocation matrix), we can proceed to define the characteristics of the planned mining operations.

3.2 Characteristics of Mining Operations (Network Activities)

Each arc of the network is described by the following parameters:

$$a_i = [p_i, Pos_i, t_i, Wdeb_i, Wden_i]$$

where:

i – the arc number,
p – the probability of the arc i realisation,
Pos – the rate of mining operations [m/d],
t – the duration of mining operations described by the arc i [d],
$Wdeb$ – the gross output from mining operations [Mg/d],
$Wden$ – the net output from mining operations [Mg/d],

The probabilities of activities (arcs) assumed by the designer are related to his/her experience and knowledge of the feasibility of carrying out mining operations in certain mining and geological conditions and with specific technical and organizational constraints as well as with the given equipment.

Driving the maingate and tailgate entries, driving the starter entry and coal extraction are given with an adequate rate. It has been assumed that the rate of mining operations is a random variable of a normal distribution - *Pos* (μ,σ) and depends on the conditions in a given excavation site and the equipment. The time of both the installation of the longwall face equipment and its removal is defined by the determined value T.

The duration time of the remaining mining operations is calculated using the following formula:

$$t = \frac{l}{Pos}$$

where:

l – the length of the excavation [m].

Gross output from mining operations is given by the following equation:

$$Wdeb = Pos \cdot d \cdot f \cdot g$$

where:

d – the vertical dimension of the excavation [m],
f – the horizontal dimension of the excavation [m],
g – the coal bulk density [Mg/m^3].

Net output from mining operations is given by the followed equation:

$$Wden = Wdeb \cdot z$$

where:

z – the coefficient of exploitation losses.

After calculation according to the mathematical model as a result the following characteristics, among others are given:

- the total net output in the analyzed period,
- the monthly net output in the analyzed period in production flows and a whole mine.

In the next part of this paper a practical application of the evolved method in the selected coal mine shall be presented.

4 An Example of the Production Modelling in a Coal Mine

In the selected coal mine output is realized in four exploitation panels. The longwall panels constituting an individual exploitation panel are combined to form a production flow. Seven longwall sets: Z_1–Z_7 can be allocated to planned longwall faces and eight combined miners MCH_1–MCH_8 can be used in the driving type of mining operations (RPp, RPn, RO). The time of the installation of the longwall face equipment and its removal is equal to 40 working days.

In order to illustrate possible variants of the planned mining operations in the selected coal mine a stochastic network was drawn. The created network consists of 96 arcs and 69 nodes (Fig. 2).

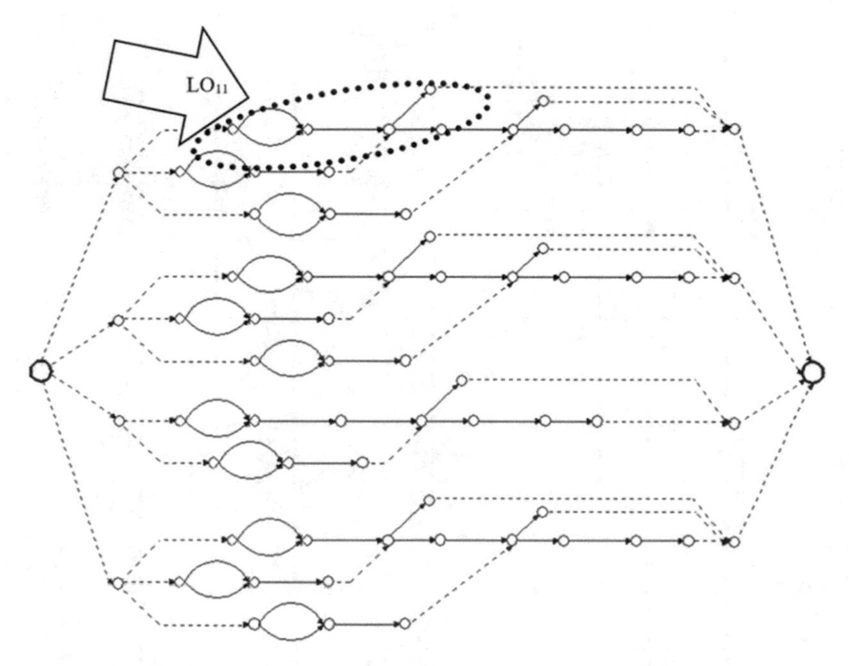

Fig. 2. Stochastic network for 4 production flows. (Note: the numbers of the nodes and arcs have been neglected due to clarity of the illustration)

Created network was used for analysis of two variants: A and B. Characteristic parameters of network's arcs and planned mining operations are presented in Table 1.

Variants are differ in the allocation of machinery to planned mining operations with given probability of usage in a particular conditions of an excavation.

In Table 2 probability distributions' parameters of the rate of mining operations used in simulation are presented. These values were assumed on basis of multidimensional comparative analysis between conditions of the planned excavations and conditions of the mining operations carried out in the past (according to mathematical model given in [3]).

Table 1. Characteristic of network's arcs

Arc no	Mining operation	p_i – Variant A	p_i – Variant B	Spatial element	Technical object	Arc no	Mining operation	p_i – Variant A	p_i – Variant B	Spatial element	Technical object
1–4	0	1	1	–	–	48	0	1	1	–	–
5	ZB_{11}(a)	0,3	1	S_{11}	Z_1	49	E_{23}	1	1	S_{23}	*
6	ZB_{11}(b)	0,7	0	S_{11}	Z_2	50	L_{23}	1	1	S_{23}	*
7	RPp_{12}(a)	0,8	0	CHP_{12}	MCH_5	51–55	0	1	1	–	–
8	RPp_{12}(b)	0,2	1	CHP_{12}	MCH_1	56	RO_{31}(a)	0,2	1	P_{31}	MCH_3
9	RPn_{13}	1	1	CHN_{13}	MCH_5	57	RO_{31}(b)	0,8	0	P_{31}	MCH_7
10	RPp_{13}	1	1	CHP_{13}	MCH_1	58	RPn_{32}(a)	0,8	0	CHN_{32}	MCH_7
11	E_{11}	1	1	S_{11}	*	59	RPn_{32}(b)	0,2	1	CHN_{32}	MCH_3
12	RO_{12}	1	1	P_{12}	*	60	ZB_{31}	1	1	S_{31}	Z_5
13	0	1	1	–	–	61	RO_{32}	1	1	P_{32}	*
14	RO_{13}	1	1	P_{13}	*	62	E_{31}	1	1	S_{31}	Z_5
15	L_{11}	1	1	S_{11}	*	63	0	1	1	–	–
16	ZB_{12}	1	1	S_{12}	*	64	L_{31}	1	1	S_{31}	Z_5
17	0	1	1	–	–	65	ZB_{32}	1	1	S_{32}	Z_5
18	E_{12}	1	1	S_{12}	*	66	0	1	1	–	–
19	0	1	1	–	–	67	E_{32}	1	1	S_{32}	Z_5
20	L_{12}	1	1	S_{12}	*	68	L_{32}	1	1	S_{32}	Z_5
21	ZB_{13}	1	1	S_{13}	*	69-74	0	1	1	–	–
22	0	1	1	–	–	75	ZB_{41}(a)	0,7	0,5	S_{41}	Z_7
23	E_{13}	1	1	S_{13}	*	76	ZB_{41}(b)	0,3	0,5	S_{41}	Z_6
24	L_{13}	1	1	S_{13}	*	77	RPn_{42}(a)	0,8	0	CHN_{42}	MCH_4
25-30	0	1	1	–	–	78	RPn_{42}(b)	0,2	1	CHN_{42}	MCH_8
31	ZB_{21}(a)	0,3	0,5	S_{21}	Z_3	79	RPn_{43}	1	1	CHN_{43}	MCH_8
32	ZB_{21}(b)	0,7	0,5	S_{21}	Z_4	80	RPp_{43}	1	1	CHP_{43}	MCH_4
33	RPp_{22}(a)	0,8	1	CHP_{22}	MCH_6	81	E_{41}	1	1	S_{41}	*
34	RPp_{22}(b)	0,2	0	CHP_{22}	MCH_2	82	RO_{42}	1	1	P_{42}	*
35	RPp_{23}(a)	0,2	0	CHP_{23}	MCH_6	83	RO_{43}	1	1	P_{43}	*
36	RPp_{23}(b)	0,8	1	CHP_{23}	MCH_2	84	0	1	1	–	–
37	E_{21}	1	1	S_{21}	*	85	L_{41}	1	1	S_{41}	*
38	RO_{22}	1	1	P_{22}	*	86	ZB_{42}	1	1	S_{42}	–
39	RO_{23}	1	1	P_{23}	*	87	0	1	1	–	–
40	0	1	1	–	–	88	E_{42}	1	1	S_{42}	*
41	L_{21}	1	1	S_{21}	*	89	0	1	1	–	–
42	ZB_{22}	1	1	S_{22}	*	90	L_{42}	1	1	S_{42}	*
43	0	1	1	–	–	91	ZB_{43}	1	1	S_{43}	*
44	E_{22}	1	1	S_{22}	*	92	0	1	1	–	–
45	0	1	1	–	–	93	E_{43}	1	1	S_{43}	*
46	L_{22}	1	1	S_{22}	*	94	L_{43}	1	1	S_{43}	*
47	ZB_{23}	1	1	S_{23}	*	95–96	0	1	1	–	–

* - depends on the choice (a) or (b)

Table 2. Parameters of probability distributions of the mining operations' rate

Mining operation	Pos μ [m/d]	Pos σ [m/d]	Mining operation	Pos μ [m/d]	Pos σ [m/d]
RPp_{12}(a)	8,77	2,57	E_{11}(a)	1,90	0,52
RPp_{12}(b)	7,92	2,11	E_{11}(b)	2,62	0,77
RPp_{13}	9,90	1,86	E_{12}(a)	2,67	0,63
RPn_{13}	7,86	1,78	E_{12}(b)	2,89	0,54
RPp_{22}(a)	9,11	2,96	E_{13}(a)	2,38	0,47
RPp_{22}(b)	7,86	1,78	E_{13}(b)	2,89	0,54
RPp_{23}(a)	6,90	1,30	E_{21}(a)	3,96	0,96
RPp_{23}(b)	9,26	3,15	E_{21}(b)	4,81	1,96
RPn_{32}(a)	9,11	2,96	E_{22}(a)	2,67	0,63
RPn_{32}(b)	8,83	2,34	E_{22}(b)	2,89	0,54
RPn_{42}(a)	7,27	2,99	E_{23}(a)	3,96	0,96
RPn_{42}(b)	6,27	2,89	E_{23}(b)	4,81	1,96
RPn_{43}	8,77	2,57	E_{31}	4,57	1,54
RPp_{43}	6,23	2,45	E_{32}	4,57	1,54
RO_{12}(a)	7,37	2,15	E_{41}(a)	1,90	0,52
RO_{12}(b)	6,74	2,51	E_{41}(b)	1,91	0,67
RO_{13}(a)	6,74	2,51	E_{42}(a)	1,90	0,52
RO_{13}(b)	7,42	2,68	E_{42}(b)	2,89	0,54
RO_{22}(a)	7,42	2,68	E_{43}(a)	1,90	0,52
RO_{22}(b)	6,71	2,95	E_{43}(b)	2,89	0,54
RO_{23}(a)	6,89	2,71			
RO_{23}(b)	7,37	2,15			
RO_{31}(a)	6,17	1,65			
RO_{31}(b)	8,20	2,15			
RO_{32}(a)	8,62	3,75			
RO_{32}(b)	6,17	1,65			
RO_{42}(a)	7,37	2,15			
RO_{42}(b)	6,74	2,51			
RO_{43}(a)	6,74	2,51			
RO_{43}(b)	7,42	2,68			

Results of output calculations for analyzed variants are presented in Figs. 3 and 4 (number of simulations = 100). The interval $< \mu - \sigma, \mu + \sigma >$ around the expected value is drawn by the broken line.

After making the calculations in accordance with a mathematical model an optimization procedure is carried out. The choice of the best possible solution can be made with reference to adjusting the volume of the output for particular months of the analyzed period to the requirements of the technical-economic plan. Analyzed period equals to 24 months. Planned output and expected values of output in each variant are presented in Fig. 5.

Comparison between variants A and B can be performed with use of Euclidean distance given by the following formula:

$$d = \sqrt{\sum_{i=1}^{24} (EWden_i - PWden_i)^2}$$

where:

i – the month number,
$EWden_i$ – the expected value of output in i-th month,
$PWden_i$ – the planned value of output in i-th month.

Values of proposed metric for variants are as follows: d_A = 92024 [Mg], d_B = 98 622 [Mg]. The difference between variants equals to 6 600 [Mg] in favor of variant A.

The choice of variant A means that execution of the mining operations should be performed using machinery presented in Table 3.

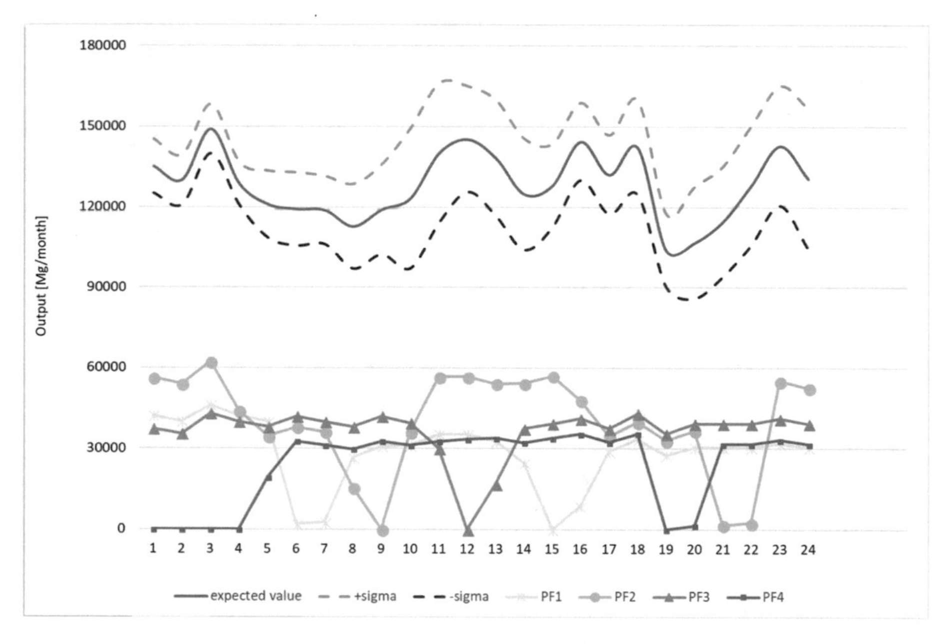

Fig. 3. The monthly coal output in variant A

Choice of the better variant can take also into consideration risk measure related to mining operations. Thus presented comparison can be extended with analysis of monthly output variability of each variant. At this stage it is possible for the designer to decide taking also into account the risk of non-compliance with or exceeding the planned values of an output.

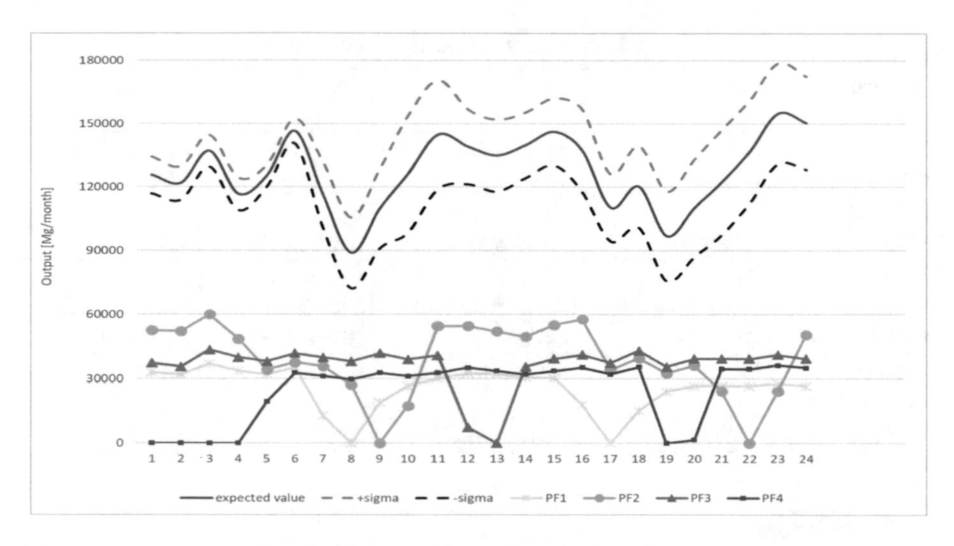

Fig. 4. The monthly coal output in variant B

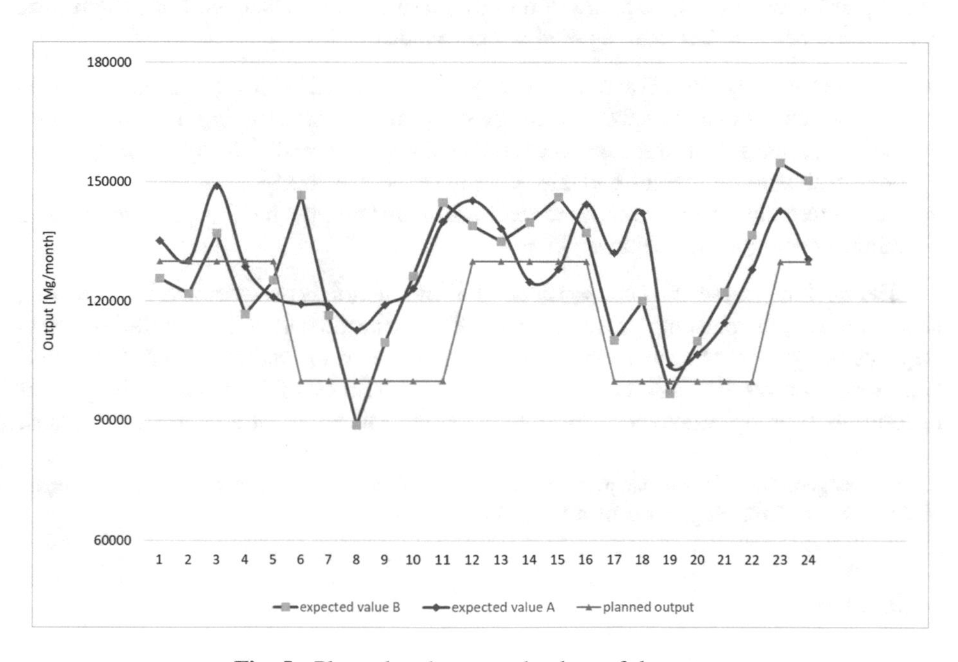

Fig. 5. Planned and expected values of the output

Table 3. Machinery allocation in variant A

Production flow	PF_1	PF_2	PF_3	PF_4
Spatial objects	LO_{11}, LO_{12}, LO_{13}	LO_{21}, LO_{22}, LO_{23}	LO_{31}, LO_{32}	LO_{41}, LO_{42}, LO_{43}
Longwall set:	Z_2	Z_4	Z_5	Z_7
MSK	KGS 600	KSW 500	EDW 300LN	KSW475
MSPS	PSZ 750/3 * 65/200	PSZ 750/3 * 65/200	Glinik 260/724/E	PSZ 750/2 * 55/160
MSPP	GSW 750	GSW 750	Glinik 1024	GSW 750
Combined miner	MCH_5	MCH_6, MCH_2	MCH_7	MCH_4
MCH	AM75	AM75, AM50	AM75	AM50

5 Conclusions

In the paper probabilistic modeling of mining production is presented. For this purpose a stochastic network has been applied which enabled:

- the representation of the activity uncertainty connected with mining operations in underground coal mines (the probability distribution of the mining operations rate),
- the representation on one network model of several possibilities of mining operations realization (different variants of equipment selection),
- the representation of the variant realization uncertainty by the probability of a mining operations realization.

The random character of underground coal production is necessary to be taken into account during the planning process. Usage of a stochastic network gives the chance to improve the quality of mining operations schedules and the estimation of economical and production results. This technique could be a useful tool for designing the process of future mining operations in coal mine, according to technical and economic plans.

Acknowledgements. This paper presents the results of research conducted at AGH University of Science and Technology – contract no 11.11.100.693

References

1. Baynes, F.J.: Sources of geotechnical risk. Q. J. Eng. Geol. Hydrogeol. **43**(3), 321–331 (2010). https://doi.org/10.1144/1470-9236/08-003
2. Brazil, M., Thomas, D.A.: Network optimization for the design of underground mines. Networks **49**(1), 40–50 (2007)
3. Brzychczy, E.: Modeling and optimization method of mining works in hard coal mine with an application of stochastic networks. Ph.D. thesis, AGH University of Science and Technology (2005). (in Polish)

4. Dawson, R.J., Dawson, C.W.: Generalized activity-on-the-node networks for managing uncertainty in projects. Int. J. Proj. Manage. **13**(6), 353–362 (1995). https://doi.org/10.1016/0263-7863(95)00027-5. Elsevier Science Ltd and IPMA
5. Dawson, R.J., Dawson, C.W.: Practical proposals for managing uncertainty and risk in project planning. Int. J. Proj. Manage. **16**(5), 299–310 (1998). https://doi.org/10.1016/s0263-7863(97)00059-8. Elsevier Science Ltd and IPMA
6. Haque, M.A., Topal, E., Lilford, E.: A numerical study for a mining project using real options valuation under commodity price uncertainty. Resour. Policy **39**, 115–123 (2014). https://doi.org/10.1016/j.resourpol.2013.12.004
7. Kazakidis, V.N., Scoble, M.: Planning for flexibility in underground mine production systems. Min. Eng. **55**(8), 33–38 (2003)
8. Kurihara, K., Nishiuchi, N.: Efficient Monte Carlo simulation method of GERT-type network for project management. Comput. Ind. Eng. **42**(2–4), 521–531 (2002). https://doi.org/10.1016/s0360-8352(02)00050-5
9. Laslo, Z.: Activity time-cost tradeoffs under time and cost chance constraints. Comput. Ind. Eng. **44**(3), 365–384 (2003). https://doi.org/10.1016/s0360-8352(02)00214-0
10. Lockyer, K.G., Gordon, J.: Project Management and Project Network Techniques. Financial Times/Prentice Hall, London (2005)
11. Nahmias, S., Olsen, T.L.: Production and Operations Analysis, 7th edn. Waveland Press (2015)
12. Pritsker, A.A.B., Happ, W.W.: GERT – graphical evaluation and review technique. Part 1, fundamentals. J. Ind. Eng. **17**(5), 267–274 (1966)
13. Pritsker, A.A.B., Whitehouse, C.E.: GERT – graphical evaluation and review technique. Part 2, probabilistic and industrial engineering applications. J. Ind. Eng. **17**(6), 293–301 (1966)
14. Summers J.: Analysis and management of mining risk. In: MassMin 2000, Brisbane (2000)
15. Voropajev, V.I., Ljubkin, S.M., Titarenko, B.P., Golenko-Ginzburg, D.: Structural classification of network models. Int. J. Project Manage. **18**(5), 361–368 (2000). https://doi.org/10.1016/s0263-7863(99)00032-0

Random Loading of Blasted Ore with Regard to Spatial Variations of Its Actual Lithological Compound

Piotr J. Bardziński(✉), Robert Król, Leszek Jurdziak, and Witold Kawalec

Faculty of Geoengineering, Mining and Geology, Wrocław University of Science and Technology, ul. Na Grobli 13/15, 50-421 Wrocław, Poland
{piotr.bardzinski, robert.krol, leszek.jurdziak, witold.kawalec}@pwr.edu.pl

Abstract. The simulation model of the mine transportation system with regard to actual parameters of haul truck inter-arrival times was built in FlexSim. Haul trucks classified by their payload were sampling the rock material from corresponding mining faces. Two main simulation variants (V1 and V2) with a constant overall or mining face specific copper content and lithology were analyzed. The new algorithm was developed, based on the random sampling of lithologic factions from the muck pile by haul trucks. The variant V2b illustrates the interesting feature of this algorithm, when first more material from the bottom layer of the mined seam was drawn, causing the depletion of its deposit in the subsequent haul truck courses. The study demonstrated that both approaches can be used interchangeably to simulate the economic outcome of the metal production on the hourly basis. The difference between V1 and V2 variants in the total amount of copper produced after full simulation cycle changes linearly with the amount initially blasted rock mass declared in the V2 variant. The algorithm allows to choose the maximum amount of rocks to be blasted that can be delivered to the OEP in a given time.

Keywords: Empirical model · Rock blasting works · Ore transport FlexSim

1 Introduction

The improvement of the whole mining & processing chain with the help of an on-line recognition of the actual run-of-mine material has been recently addressed by the real-time-process reconciliation and optimization (RTRO) framework (Benndorf et al. 2015). Not always, however, any on-line ore recognition is available. The lithological composition of the copper ore delivered to the mills, when recognised in advance, is considered as the most important factor for the proper settings of grinding/milling equipment that can both decrease the specific energy consumption of ore processing and increase metal recovery (Kawalec et al. 2016). The ore quality control and prediction became a significant problem since it is difficult to predict it based on composite samples taken from a transfer conveyor linking two mines (Jurdziak et al. 2016a, 2016b).

© Springer Nature Switzerland AG 2019
A. Burduk et al. (Eds.): ISPEM 2018, AISC 835, pp. 668–677, 2019.
https://doi.org/10.1007/978-3-319-97490-3_63

The highest variability of Cu content is observed on the transfer conveyor (coefficient of variation *cov* = 11.85%) and the smallest for the feed to the processing plant (*cov* = 5.33%, Jurdziak et al. 2017a). These results are a consequence of almost perfect ore mixing from different mining blocks and bunkers in the final feed to the mills and only partial mixing of ore on the transfer and section conveyors due to smaller number of ore sources and time separation between subsequent loads from mining faces (cyclical loading). It can also be observed also as the decreasing range of significant autocorrelations of Cu content in the ACF (autocorrelation function) and partial ACF in different places along ore route from mining faces to ore enrichment plant (Jurdziak et al. 2018a).

In order to address problems with control of Cu content variability and prediction of ore lithological composition (which now is only determined in-situ - in mining faces by channel samples) the idea of tracking the mined ore compound with the use of Process Analytic Technology (PAT) tags carrying information about the original location of the ore (Jurdziak et al. 2018b). In order to avoid costly experiments (DISIRE Deliverable D5.3) the application of simulation techniques has been proposed (Jurdziak et al. 2017c).

Ore tracking using tags bearing information about the place of ore extraction (that provides parameters of the mined block of the balance ore seam) allows to follow the process of mixing different streams. It should be remembered, however, that values of ore parameters for the place of loading are averaged. The variability of the feed stream parameters, both observed during the experiment (DISIRE) and simulations (Bardziński and Król 2018c) will be smaller than the actual variability, as the conveyor is fed with ore portions taken from the heap of blasted ore and the ore mixing is not perfect. The copper ore deposit in Poland has layered structure, so the distribution of the quality of the blasted ore in the heap as well as the degree of its mixing may not be random, as it is a consequence of the selection of a blasting technique. In order to better determine the variability of the ore composition, a model of random loading of a single loading point served by various machines (loaders and haulage trucks) with variable capacity and cyclic work was developed. The exploitation data from ZG Lubin were used to define the parameter of the model. It is the first step to estimate variability of ore compound and Cu content in feed to ore enrichment plant on continuous basis instead of shift basis as it was done in all described hereinafter analyses.

2 Computational Details

The model of the ZG Lubin copper mine was built in the FlexSim environment. Ore portions were discretized according to haul truck payload. Total simulation time was an equivalent of one working day - 4 shifts of 4.5 h each (64800 s or 18 h). As shown in Fig. 1, the farthest from the OEP) loading point LP 71 located at the end of A183 conveyor was chosen for simulations. It assures that the ore flow path from the mining front (represented in the model by the corresponding haul truck) to the OEP is the longest and adequately simulates conditions of the actual mine. All conveyors in the mine have scheduled operational time marked with vertical bars in the Fig. 2.

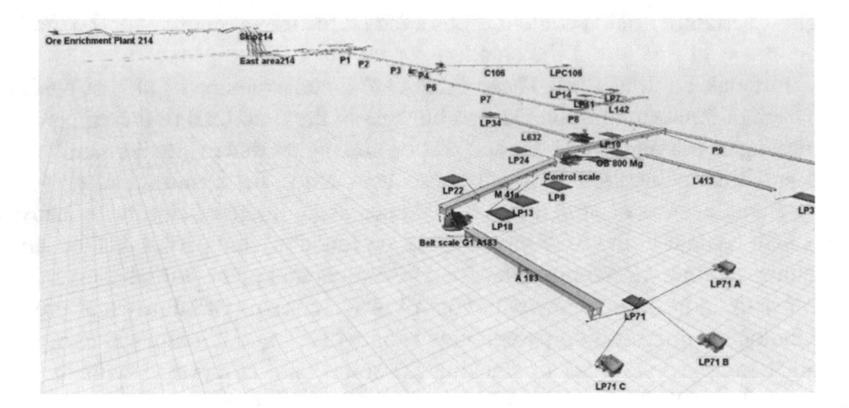

Fig. 1. Three-dimensional representation of the Lubin mine model built in FlexSim

Operational time of the skips was set to 18 h, from 0:00 to 17:00. Each of two skips in the system has loading-unloading cycle of 176 s with a maximum speed of 16 ms^{-1}. Belt speed of main haulage conveyors (P1 – P9 in the Fig. 1) is set to 2.5 ms^{-1}, while all division haulage conveyors to 2 ms^{-1}.

	:00	:05	:10	:15	:20	:25	:30	:35	:40	:45	:50	:55
12 AM	IIIIIIIIII	IIIIIIIIII	IIIIIIIIII	IIIIIIIIII	IIIIIIIIII	IIIIIIIIII	IIIIIIIIII	IIIIIIIIII	IIIIIIIIII	IIIIIIIIII	IIIIIIIIII	IIIIIIIIII
1 AM	IIIIIIIIII	IIIIIIIIII	IIIIIIIIII	IIIIIIIIII	IIIIIIIIII	IIIIIIIIII	IIIIIIIIII	IIIIIIIIII	IIIIIIIIII	IIIIIIIIII	IIIIIIIIII	IIIIIIIIII
2 AM	IIIIIIIIII	IIIIIIIIII	IIIIIIIIII	IIIIIIIIII	IIIIIIIIII	IIIIIIIIII	IIIIIIIIII	IIIIIIIIII	IIIIIIIIII	IIIIIIIIII	IIIIIIIIII	IIIIIIIIII
3 AM	IIIIIIIIII	IIIIIIIIII	IIIIIIIIII	IIIIIIIIII	IIIIIIIIII	IIIIIIIIII	IIIIIIIIII	IIIIIIIIII	IIIIIIIIII	IIIIIIIIII	IIIIIIIIII	IIIIIIIIII
4 AM	IIIIIIIIII	IIIIIIIIII	IIIIIIIIII	IIIIIIIIII	IIIIIIIIII	IIIIIIIIII	IIIIIIIIII	IIIIIIIIII	IIIIIIIIII	IIIIIIIIII	IIIIIIIIII	IIIIIIIIII
5 AM												
6 AM												
7 AM												
8 AM	IIIIIIIIII	IIIIIIIIII	IIIIIIIIII	IIIIIIIIII	IIIIIIIIII	IIIIIIIIII	IIIIIIIIII	IIIIIIIIII	IIIIIIIIII	IIIIIIIIII	IIIIIIIIII	IIIIIIIIII
9 AM	IIIIIIIIII	IIIIIIIIII	IIIIIIIIII	IIIIIIIIII	IIIIIIIIII	IIIIIIIIII	IIIIIIIIII	IIIIIIIIII	IIIIIIIIII	IIIIIIIIII	IIIIIIIIII	IIIIIIIIII
10 AM	IIIIIIIIII	IIIIIIIIII	IIIIIIIIII	IIIIIIIIII	IIIIIIIIII	IIIIIIIIII	IIIIIIIIII	IIIIIIIIII	IIIIIIIIII	IIIIIIIIII	IIIIIIIIII	IIIIIIIIII
11 AM	IIIIIIIIII	IIIIIIIIII	IIIIIIIIII	IIIIIIIIII	IIIIIIIIII	IIIIIIIIII	IIIIIIIIII	IIIIIIIIII	IIIIIIIIII	IIIIIIIIII	IIIIIIIIII	IIIIIIIIII
12 PM	IIIIIIIIII	IIIIIIIIII	IIIIIIIIII	IIIIIIIIII	IIIIIIIIII	IIIIIIIIII	IIIIIIIIII	IIIIIIIIII	IIIIIIIIII	IIIIIIIIII	IIIIIIIIII	IIIIIIIIII
1 PM	IIIIIIIIII	IIIIIIIIII	IIIIIIIIII	IIIIIIIIII	IIIIIIIIII	IIIIIIIIII	IIIIIIIIII	IIIIIIIIII	IIIIIIIIII	IIIIIIIIII	IIIIIIIIII	IIIIIIIIII
2 PM	IIIIIIIIII	IIIIIIIIII	IIIIIIIIII	IIIIIIIIII	IIIIIIIIII	IIIIIIIIII	IIIIIIIIII	IIIIIIIIII	IIIIIIIIII	IIIIIIIIII	IIIIIIIIII	IIIIIIIIII
3 PM	IIIIIIIIII	IIIIIIIIII	IIIIIIIIII	IIIIIIIIII	IIIIIIIIII	IIIIIIIIII	IIIIIIIIII	IIIIIIIIII	IIIIIIIIII	IIIIIIIIII	IIIIIIIIII	IIIIIIIIII
4 PM	IIIIIIIIII	IIIIIIIIII	IIIIIIIIII	IIIIIIIIII	IIIIIIIIII	IIIIIIIIII	IIIIIIIIII	IIIIIIIIII	IIIIIIIIII	IIIIIIIIII	IIIIIIIIII	IIIIIIIIII
5 PM	IIIIIIIIII	IIIIIIIIII	IIIIIIIIII	IIIIIIIIII	IIIIIIIIII	IIIIIIIIII	IIIIIIIIII	IIIIIIIIII	IIIIIIIIII	IIIIIIIIII	IIIIIIIIII	IIIIIIIIII
6 PM												
7 PM	IIIIIIIIII	IIIIIIIIII	IIIIIIIIII	IIIIIIIIII	IIIIIIIIII	IIIIIIIIII	IIIIIIIIII	IIIIIIIIII	IIIIIIIIII	IIIIIIIIII	IIIIIIIIII	IIIIIIIIII
8 PM	IIIIIIIIII	IIIIIIIIII	IIIIIIIIII	IIIIIIIIII	IIIIIIIIII	IIIIIIIIII	IIIIIIIIII	IIIIIIIIII	IIIIIIIIII	IIIIIIIIII	IIIIIIIIII	IIIIIIIIII
9 PM	IIIIIIIIII	IIIIIIIIII	IIIIIIIIII	IIIIIIIIII	IIIIIIIIII	IIIIIIIIII	IIIIIIIIII	IIIIIIIIII	IIIIIIIIII	IIIIIIIIII	IIIIIIIIII	IIIIIIIIII
10 PM	IIIIIIIIII	IIIIIIIIII	IIIIIIIIII	IIIIIIIIII	IIIIIIIIII	IIIIIIIIII	IIIIIIIIII	IIIIIIIIII	IIIIIIIIII	IIIIIIIIII	IIIIIIIIII	IIIIIIIIII
11 PM	IIIIIIIIII	IIIIIIIIII	IIIIIIIIII	IIIIIIIIII	IIIIIIIIII	IIIIIIIIII	IIIIIIIIII	IIIIIIIIII	IIIIIIIIII	IIIIIIIIII	IIIIIIIIII	IIIIIIIIII

Fig. 2. Schedule of the belt conveyor system

Haul trucks of 3 different payload: A – 15 Mg, B – 12 Mg and C – 8 Mg were considered in the simulations. Loading capacity or class–dependent empirical histograms of haul truck inter–arrival times were presented in the Fig. 3. They are based on block–crushing device operators' reports, who count the courses of each truck type in three mining departments of ZG Lubin mine in a time window of 15 min by five consecutive work days. The obtained distributions were applied to the model built in FlexSim in order to simulate the frequency of haul truck courses. In the current study, haul trucks with each capacity were sampling the rock material from its own corresponding mining front. In the case of V1 simulation variant, the every single mining face could be represented by the same copper content and lithology, so only the haul truck loading capacity will vary. In the case of variants V2a–c, the copper content remains constant while blasted rock mass is mining face specific, as it might occur in the real conditions.

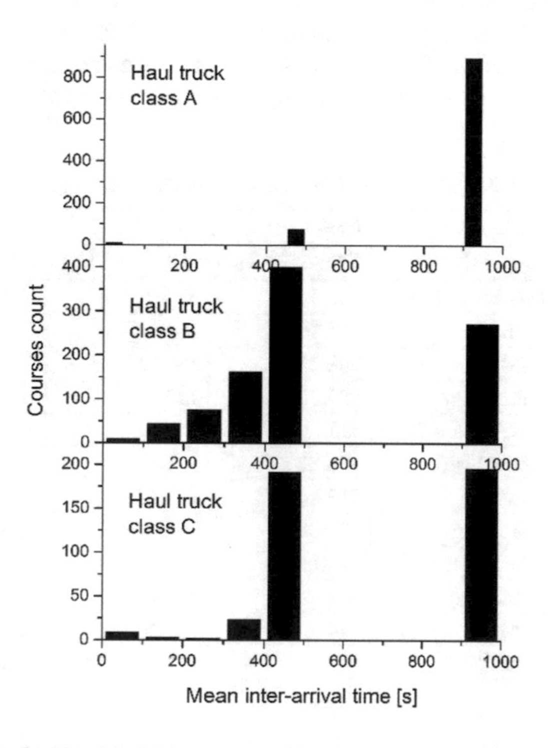

Fig. 3. Empirical histograms of haul truck inter–arrival times

In the V1 simulation variant, values corresponding to ore lithology were fixed and their mean values after single hypothetical rock blasting works were collected in Table 1. Copper content in each lithological faction was based on the MOPRONA database in order to represent the real conditions occurring in the mine (Kulbacki 2007). Lithologic contributions were altered in relation to their real data counterparts to better visualize the variations in ore lithology. In real conditions, the studied mining faces were consisting primarily of sandstone (72.8%) and shale (25.4%).

Table 1. Ore lithology in the V1 variant of the simulation

Lithological fraction [%]	Sandstone	42
	Shale	34
	Dolomite	24
Copper content in rock types [%]	Sandstone	0.9
	Shale	1.7
	Dolomite	1.3

In case of simulation variants V2a–c initial rock mass after one rock blasting event was characteristic for each of three mining fronts and larger for each consecutive variant. It was collected in Table 2. Total rock mass was arbitrarily chosen, and the lithological fractions were calculated according to their mean values given in Table 1.

Table 2. Ore lithology in V2a–c variants of the simulation

Simulation variant		Haul truck class		
		HT A (15 Mg)	HT B (12 Mg)	HT C (8 Mg)
V2a	Total rock mass [Mg]	789	532	671
	Sandstone [Mg]	331.38	223.44	281.82
	Shale [Mg]	268.26	180.88	228.14
	Dolomite [Mg]	189.36	127.68	161.04
V2b	Total rock mass [Mg]	921	770	820
	Sandstone [Mg]	386.82	323.4	344.4
	Shale [Mg]	313.14	261.8	278.8
	Dolomite [Mg]	221.04	184.8	196.8
V2c	Total rock mass [Mg]	1100	890	940
	Sandstone [Mg]	462	373.8	394.8
	Shale [Mg]	374	302.6	319.6
	Dolomite [Mg]	264	213.6	225.6

In the variant V1, the rock sources were infinite and the amount of excavated ore delivered to the loading point was controlled only by the haul truck inter-arrival times. Instead, in case of variants V2a–c the amount of blasted rocks were fixed after each blasting works. The algorithm was presented in the scheme shown in the Fig. 4. The amount of rocks of each lithological member were termed here as a "container", by an analogy to containers from which the material is drawn in a portions defined by haul truck shovel capacity up to the emptying of each container. The similar situation occurs in the actual conditions where mining face, containing of three layers of different rock types (each with certain width) was blown away. The rock debris from each lithological layer can fall in various ways. If direction-oriented blasting was not applied, the rocks will fall in a random manner, the same as in the model chosen in the current study. A random integer from a discrete uniform distribution is drawn using a *duniform* Flexsim function.

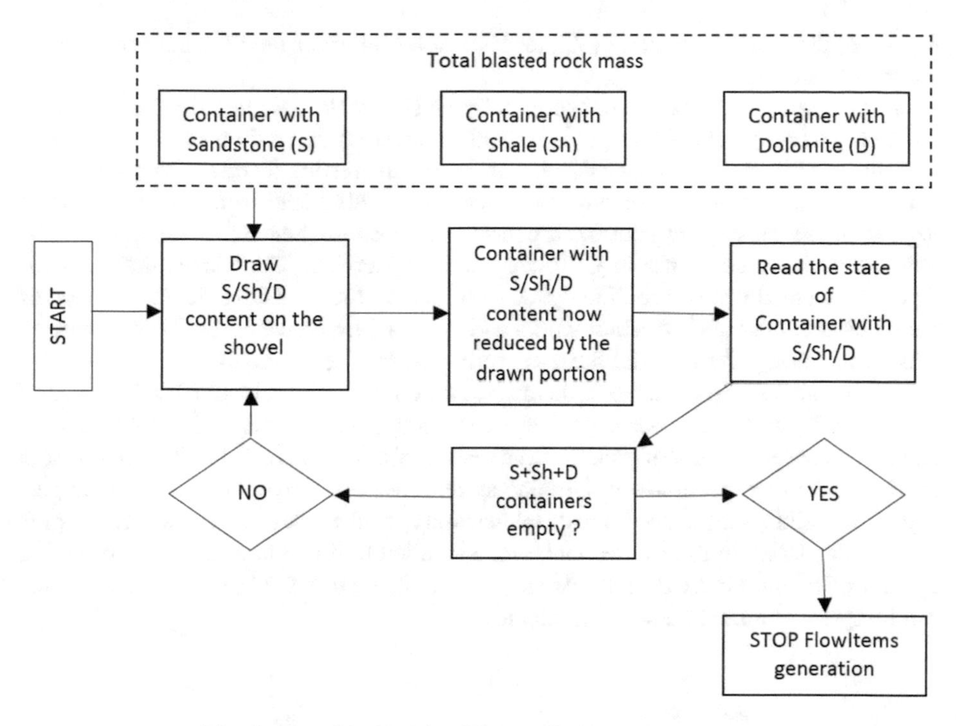

Fig. 4. Ore drawing algorithm applied in variants V2a–c

Container content levels were corresponding to lithology as examined by in-situ channel sampling and stored in MOPRONA system as well as to total blasted rock mass from single rock blasting works.

3 Results and Discussion

Number of FlowItems or discretized ore portions, that were delivered to the final station in the mine, i.e., to the OEP, recorded for different simulation variants was compared in the Fig. 5. Black symbols represent FlowItem count from each mining front (or delivered by haul truck with different class), while open symbols stands for total mass, defined as mass of sum of lithological factions within an hour from each FlowItem source. The solid and dashed lines are guide to the eye. In every case, the ore production was stable, then between 10 and 16 h of the simulation, no ore was delivered to OEP, then the ore production rocketed. This was caused by the down time of the skips, that resulted in FlowItems buildup in the ore bunkers located under the skips. When the latter were online, the FlowItems stored in the bunkers have shorter path to the OEP, resulting in the outburst of the FlowItem count on the figures. Close examination of the graphs revealed that most of the ore mass was delivered by the haul truck class B, owed to its inter-arrival times distribution shown in the Fig. 3. As we can see from the Fig. 1, the use of the second algorithm (V2) leads to faster return of the FlowItem count to its medium values, of c.a. 5–6.

In the variant V2, the amount of initially blasted rock mass has significant impact on the ore lithology at the OEP, especially after the skips down time, as can be seen by studying the Fig. 6. The variant V2b illustrates the interesting feature of this algorithm, when more sandstone was drawn at the beginning, what lead to the depletion of its deposit in the subsequent haul truck courses. As a consequence, the lithology of the run-of-mine ore that comes to OEP changed considerably. This shows the real condition, when the layer of one lithological faction, here the sandstone, falls on the top of the muck pile after rock blasting works and is thus delivered first by the haul trucks, which are hauling the material from top to the bottom of the muck pile.

As can be seen from Table 3, there was some blasted rock material left in case of variants V2b and V2c after complete simulation cycle. It might indicate that the arbitrarily chosen amount of rock material (exact values given in Table 2) was too large for the transport capabilities of the mine system, as it occurred for one of the largest values of initial rock masses blasted. Interestingly, haul truck class A was able to pick most of the debris in the V2c variant, while it failed to do so in such extent when the amount of initially blasted rock material was smaller by 179 Mg (V2b). Such results can be owed to a random schemes applied.

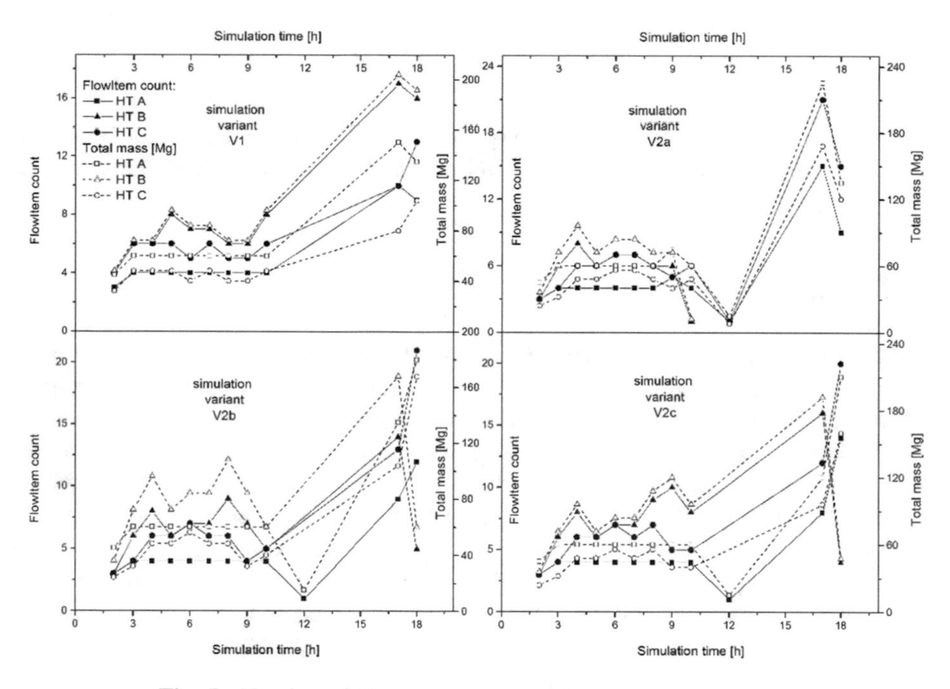

Fig. 5. Number of FlowItems generated by different ore sources

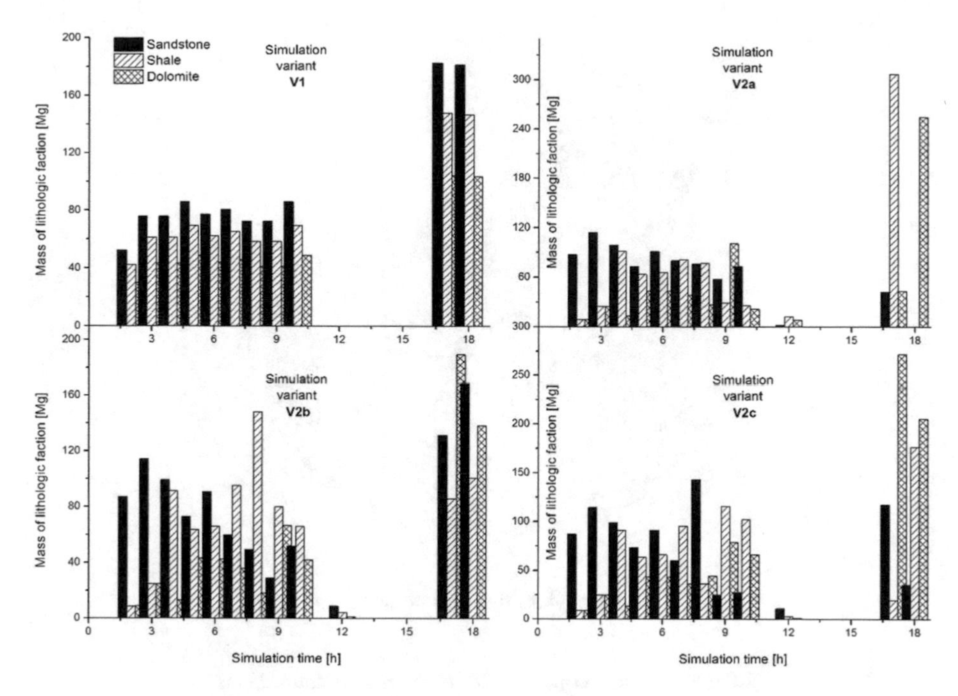

Fig. 6. Ore lithology in different simulation variants

Table 3. Residual rock material after completed simulations

Simulation variant		Haul truck class		
		HT A (15 Mg)	HT B (12 Mg)	HT C (8 Mg)
V2b	Sandstone [Mg]	314	–	–
	Shale [Mg]	74.82	–	–
	Dolomite [Mg]	193.14	–	–
V2c	Sandstone [Mg]	–	–	–
	Shale [Mg]	–	–	44.6
	Dolomite [Mg]	88	–	59.6

Total copper yield in a hourly basis was compared in the Fig. 7. Despite the differences in lithology that two algorithms (V1 and V2) produced, what are important from the scope of choice of the correct ore processing parameters at the OEP, the amount of total produced copper was comparable among all studied variants. This demonstrated that both approaches can be used interchangeably to simulate the economic outcome of the metal production in the mine.

Subtle differences between the amount of copper produced after 18 h of the simulations were compared in the Table 4. The latter shows that the difference between V1 and V2 variants changes linearly with the amount initially blasted rock mass declared in the V2 variant. This allows to tailor the model to obtain comparable result with different approach giving opportunity to describe various conditions.

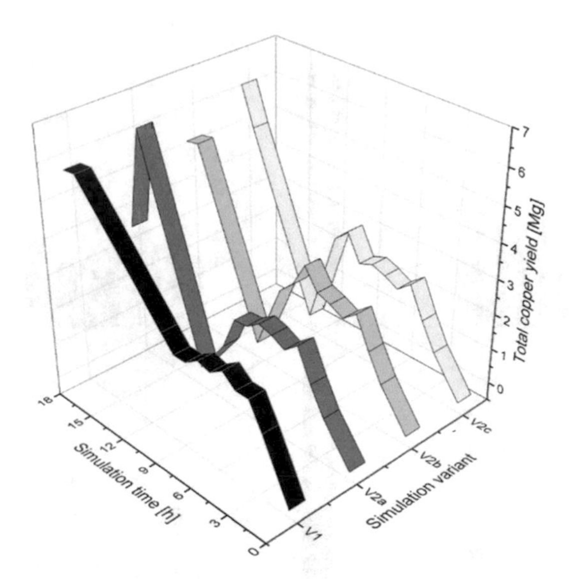

Fig. 7. Copper yield from all sources at the OEP

Table 4. Total copper yield after 18 h of simulations

Vi	Simulation variant			
	V1	V2a	V2b	V2c
Total Cu, M_{Cu} [Mg]	32.11	28.71	31.03	32.51
ΔM_{Cu} (V1-Vi) [%]	–	−10.6	−3.4	1.3

4 Conclusions

- The use of the second algorithm (V2) leads to faster return of the FlowItem count to its medium values, after the FlowItem buildup in the bunkers located under the skips, caused by the skip down time.
- In the variant V2, the amount of initially blasted rock mass alters the ore lithology at the OEP, especially after the skips down time. The variant V2b illustrates the interesting feature of this algorithm, when more sandstone was drawn at the beginning, what lead to the depletion of its deposit in the subsequent haul truck courses. The result demonstrated the resulting changes in the lithology of run-of-mine ore that comes to OEP after such random condition.
  The study demonstrated that both approaches can be used interchangeably to simulate the economic outcome of the metal production in the mine in a hourly basis. The difference between V1 and V2 variants in the total amount of copper produced after full simulation cycle changes linearly with the amount initially blasted rock mass declared in the V2 variant.

- In case of V2b and V2c variants there was a residue of the rock material in 2 out of 3 mining faces, which situation cannot occur in the properly functioning mine. Although some randomness of the simulation cannot be figured out, this result could demonstrate, that the model could be also applied to predict the maximum amount of blasted rock material that could be transported to OEP after 4 shift work day.

Acknowledgements. This work was partly supported by the Horizon 2020 Framework Programme under the grant agreement No. 636834 (DISIRE) and by the Polish Ministry of Science and Higher Education as scientific project No. 0401/0131/17.

References

Benndorf, J., et al.: RTRO–Coal: real-time resource-reconciliation and optimization for exploitation of coal deposits. Minerals **5**, 546–569 (2015)

Jurdziak, L., Kawalec, W., Król, R.: Variation of ore grade transported by belt conveyors to processing plant. Physicochem. Probl. Miner. Proc. **53**(1), 656–669 (2017a)

Jurdziak, L., Kawalec, W., Król, R.: Current methods and possibilities to determine the variability of Cu content in the copper ore on a conveyor belt in one of KGHM Polska Miedz SA mines. In: Mineral Engineering Conference (MEC 2016). In: E3S Web of Conferences, vol. 8, UNSP 01055 (2016a)

Jurdziak, L., Kawalec, W., Król, R.: Autocorrelation analysis of Cu content in ore streams in one of KGHM Polska Miedź S.A. mines. Adv. Intell. Syst. Comput. **637**, 418–427 (2018a)

Kawalec, W., Krol, R., Zimroz, R., Jurdziak L., Jach, M., Pilut, R.: Project DISIRE (H2020) – an idea of annotating of ore with sensors in KGHM Polska Miedz S.A. underground copper ore mines. In: E3S Web of Conferences, vol. 8, p. 01058 (2016)

Jurdziak, L., Kawalec, W., Krol, R.: Study on tracking the mined ore compound with the use of process analytic technology tags. Adv. Intell. Syst. Comput. **637**, 418–427 (2018b)

DISIRE Deliverable D5.3, PAT based IPC demonstration on conveyor network control, Wroclaw University of Science and Technology, January 2018 (not published)

Jurdziak, L., Kaszuba, D., Kawalec, W., Król, R.: Identification of loading parameters for belt conveyors operating in KGHM's mines. SGEM **17**(13), 971–978 (2017c)

Jurdziak, L., Kaszuba, D., Kawalec, W., Król, R.: Idea of identification of copper ore with the use of process analyser technology sensors. IOP Conf. Ser. Earth Environ. Sci. **44**(4), 042037 (2016b)

Kulbacki, A.: MOPRONA system – Ore Production Monitoring – an Important Element of the Strategy of Production in the Copper Ore Mines, School of Underground Mining Conference Proceedings, Szczyrk (2007)

Bardziński, P.J., Król, R.: Ilościowa ocena urobku na potrzeby symulacji procesu jego rozpływu kopalnianą siecią transportową. Transp. Przem. Masz. Robocze **1**, 15–18 (2018c)

Multi-faceted Modelling of Networks and Processes

BPMN Update Proposal for Non-expert Users

Marek Szelągowski(✉)

Vistula University, 3 Stokłosy St., 02-787 Warsaw, Poland
marek.szelagowski@dbpm.pl

Abstract. The aim of the article is to demonstrate that changes to business operations and the emergence of new ICT solutions necessitate the introduction of corresponding changes in the mode of describing processes, including the update of the most commonly used BPMN notation.

The first part of the article analyses the proposal to use multiple process description notations to model dynamically managed processes, as well as presents the direction of the development of process-driven systems (BPMS/ACMS). The second part of the article consists of an analysis of the requirements tied to updating the BPMN notation and presents an overview of the relevant literature in the scope of the proposed changes to the BPMN notation. The third part of the article consists of a proposal of updating the BPMN notation, allowing it to respond to the observed changes to business operations.

The article analyzes the requirements set before process description notations by dynamic business process management. It also contains a proposal of updating the BPMN notation. As a result of the implementation of the changes proposed in the article with respect to the principles of describing processes and updating the BPMN notation standard, the means of communicating knowledge with the use of process models will be improved. Accommodating the form of their presentation to the user interfaces of applications supporting management (ERP/CRM/EHR systems) will facilitate the intuitive understanding of process models by a broad range of knowledge workers.

Keywords: Process model · Dynamic BPM · BPMN · CMMN

1 Introduction

Process management is at present the most commonly adopted practical concept of process management. Organizations are increasingly more often resorting to process management on the level of strategic design, operational management, and the day-to-day management of specific processes. In all of these diverse fields, one broadly used fundamental vehicle for knowledge increasingly rests in business process descriptions. The efficiency of their creation and updating, and – first and foremost – their presentation and use by a broad range of employees plays a decisive role in the actual efficiency of using the knowledge contained therein, and, in a more practical aspect, often has a direct influence on the success or failure of process performance, which translates to attaining the goals of the organization. For this reason, organizations are

© Springer Nature Switzerland AG 2019
A. Burduk et al. (Eds.): ISPEM 2018, AISC 835, pp. 681–691, 2019.
https://doi.org/10.1007/978-3-319-97490-3_64

faced with the extremely crucial decision of selecting the method of describing business processes, which from the point of the user should primarily:

- allow for the cohesive entry of diverse pieces of information
- be intuitive, simple, and readable
- be adaptable to the individual context of process performance
- allow for the collection of information with a view to verifying the status of current information and discovering new information [1, 2].

In other words, knowledge contained in process descriptions cannot be described in an esoteric manner, which is dedicated to specialists alone. One of the fundamental requirements pertaining to business process description is the possibility of communicating the knowledge contained in process descriptions to a broad range of recipients. Because both the organization's management and the process performers themselves, and increasingly more often the recipients of process-based products and services as well, have a direct influence on the scope and the quality of the processed information, the method of presenting process descriptions should remain as close as possible to the way information is presented in devices, IT systems, and websites supporting process performers. For instance, diagnostic-therapeutic processes should possibly be described in a way which closely reflects the interfaces of medical systems (Electronic Healthcare Record – EHR), that is, with the use of analogous nomenclature, references to actual process roles (or even specific positions), etc.

2 Methodology

The article takes the perspective of the non-expert user and presents changes in the approach to modeling processes and using systems supporting BPM on the basis of participant observations and non-participant observations held in healthcare units and Polish institutions of higher education in the course of projects pertaining to modeling clinical pathways and implementing education quality management in the years 2016–2018. In the course of preparations for this article, the author conducted a study of literature with respect to defining the causes and directions of the unification of BPMS and ACMS systems, differences in the scope of using BPMN, CMMN, and DMN notations, and the barriers limiting the use of a single selected notation as an additional driver for systems unification. The resulting proposals of updating BPMN were subjected to still ongoing practical verification in the scope of user readability and the possibility of implementation within BPMN and ACMS systems.

3 Related Works

In the knowledge economy, about 70% of business processes require dynamic management [3–5]. Each performance of an R&D process, decision on the introduction of new products or the cannibalization of existing products, each performance of a therapeutic-diagnostic process, or management of a client complaint requires organizations to account for the individual context of performance, which is usually unpredictable in nature [6]. The method of description requires organizations as early as in the modeling stage to account for the necessity of empowering process performers to introduce changes in the course of performance itself.

3.1 Proposals and Limitations of Selecting the Dynamic Process Modeling Notation

Modeling dynamic processes must a priori account for the possibility of process individualization or improvement in the course of performance itself, including process performers making decisions dependent on the context of performance and their own knowledge. This naturally points us toward Adaptive Case Management (ACM) systems, as well as to the Case Management Model and Notation (CMMN) [7, 8]. However, one should be mindful of the following two practical problems in this regard:

1. The necessity of selecting a notation which would enable the description of both dynamic and static processes. When implementing business processes in organizations in the knowledge economy, one should remain mindful of the 30% of static processes, as well as static processes with ad-hoc exceptions, which still require description and support in the course of performance [9].
2. The small popularity of the CMMN notation, which may result in the prepared process models being misunderstood by representatives of business, administration, or e.g. the medical personnel who participate in standard development programs, in which the BPMN notation is at least mentioned in passing, and in which CMMN is practically nonexistent.

One of the solutions enabling organizations to tailor the method of describing processes to the requirements of dynamic process management is, on the recommendation of The Business Process Management in Health Workgroup (BPMHW) of the Object Management Group (OMG), the use of the following related models in appropriate contexts:

1. An Architectural Scope Diagram
2. A Glossary/5WH Model
3. BPMN Process Model(s)
4. CMMN Case Model(s)
5. DMN Decision Model(s) [1]

Unfortunately, this is an idea which is absolutely impracticable. As the authors themselves have realized, it would require the mapping of model elements between particular notations, which would further complicate and inhibit maintaining the cohesiveness of the overall model created in three (or even five) different notations. First and foremost, it would require:

- users (including management and medical personnel) to learn as many as three different business process modeling notations: BPMN, CMMN, and DMN.
- execution systems (workflow/document management/BPMS or domain-specific systems, e.g. EHR or ERP) to implement an execution engine encompassing as many as 3 process description notations

which is obviously virtually impracticable.

In effect, is a good solution in this respect to use a single specialized notation, such as The User Requirements Notation (URN)? This notation combines two complementary notations: the Goal-oriented Requirement Language (GRL) and Use Case Maps (UCM) which are used for modeling goals and processes respectively [10]. Notwithstanding the complex character of the URN notation itself, it would seem that today the decisive argument against the URN notation, or, more generally speaking, against dedicated notations, is their minuscule, niche popularity. In the case of administration or business, diagrams prepared with its use would be hard to understand. Furthermore, it would be hard to find BPMS systems with execution engines allowing for the creation of applications using models prepared e.g. in the URN notation.

3.2 Changes in the Functionality of Systems Supporting Process Management

An entirely different direction has been adopted by vendors of systems supporting process management. Responding to client pressure, they started creating solutions combining the functionality of BPMS and CMS systems (including Appian, BMP'online, Bizflow, IBM, K2, Kofax, Pegasystems, PNMsoft). This has been noticed by market analysts. On March 12, 2015, Gartner published the first special report Magic Quadrant for BPM-Platform-Based Case Management Frameworks [11], which defined the requirements and the cope of using systems enabling the management of both traditional, static processes, as well as processes requiring the use of dynamic management. Another such report was published on October 24, 2016 [12], and from 2017 [13] onward the requirement of case management functionality has been included by Gartner in the requirements of Intelligent Business Process Management Systems (iBPMS). Among the critical possibilities of iBPMS systems, the report included the following:

- process discovery and optimization
- context and behavior history

which means that in 2017 process mining techniques and machine learning – as well as simulation research – have become an inseparable part of BPMS systems. Requirements for iBPMS systems have also been supplemented with:

- citizen developer application composition
- case management.

Case management and the possibility of introducing changes to applications, including process models, by the users themselves, are according to Gartner an obligatory element, the lack of which in the year 2017 disqualifies a BPMS system on the market. The evolution of Gartner's (and Forrester's) reports in the last 3–4 years demonstrates that vendors, analysts, and, first and foremost, practitioners using systems supporting process management, have realized the necessity of unifying (or rather, reunifying) BPMS and ACMS systems. It would seem natural, therefore, that the notation in which processes will be described following unification cannot be an amalgamate of pieces of three (or even as many as five) different notations. Instead, the solution rests in a single, unified notation responding to the needs of modeling both static and dynamic processes.

3.3 Proposals of Expanding the BPMN Notation in Accordance with Its Standard

Given the limitations presented in Sect. 1 and the direction of the evolution of systems supporting process management presented in Sect. 2, it would seem that the inevitable solution of the posed problem rests in the development of dedicated extensions of the unquestionably currently most popular BPMN notation. This solution is in accordance with the principles of the BPMN notation.

As far back as in 2011 relevant literature in the matter described in detail the methodology of preparing extensions [14], in compliance with the wordage of the newest 2.02 version of the BPMN notation [15]. Among other uses, it was used to propose extensions dedicated to diagnostic-therapeutic processes (clinical pathways, such as):

- Modeling tasks divided between different roles (shared tasks) [16].
- BPMN4CP enabling for a more detailed description of models by adding additional data and resource models and simplifying process views by enabling the creation of different process views for different groups of users [17, 18].

4 Proposal of a BPMN Notation Update

Notwithstanding the unquestionable significance of dedicated BPMN extensions, it would nevertheless seem that according to previous argumentation, from the perspective of the user of BPMS/ACMS systems the sought-after proposals of expansion should:

- primarily focus on expanding the possibility of modeling dynamically managed business processes
- accommodate the method of describing dynamic processes to the interfaces of IT systems supporting the planning, performance, and ex-post analysis thereof, in which all participants of implementations of process management deal with in practice on a daily basis
- account for time relations between performed tasks or events [19, 20]. This factor would significantly raise the quality of process management and enable a broader use of the possibilities offered by increasingly more adopted ICT technologies, such as process mining, robotic process automation (RPA), or machine learning (ML).

The sought-after proposals of expanding the principles of describing processes, including expansions of the BPMN notation, should not lead to lowering the readability or the intuitiveness of the interpretation of process descriptions from the perspective of the recipients and users of the knowledge contained within. As the proposal designed by BPMHW OMG [1] demonstrates, it is impractical to focus on the habits an requirements of a narrow group of specialists in the selection of the notation for process descriptions. Such an approach could result in the repository of knowledge on processes becoming inefficient due to the confusing nature of the information contained therein for the broader personnel of the organization. In order to avoid this threat, organizations should work in two seemingly mutually exclusive directions:

I. they should limit rather than extend the scope of the possibilities offered by the BPMN standard
II. they should supplement the BPMN notation with standard constructs allowing for a more intuitive description (modeling) of dynamic processes.

4.1 Dedicated Process Views

In order to fulfill requirement I without limiting the current possibilities of the BPMN notation, it is proposed to limit the scope of notation use for particular groups of users, and in particular – recipients and users of the knowledge contained in process descriptions (models) throughout the entire process lifecycle in the organization. The most simple method, which is well known from solutions from the field of architecture (including enterprise architecture – EA), is tailoring the presented diagrams to the level and scope of competence of their recipients, by enabling the creation of process views dedicated to particular groups of users [21]. In architecture, different parts of a building will be of interest to the construction team, the electrical installation crew, the gas crew, or the hydraulic crew. For this reason, architectural systems allow for the selection of displayed layers of the project. Similar process descriptions and models are not prepared with the modelers themselves in mind, who by definition may be interested in all of the details, but for a much broader range of recipients, who in the scope of their competences will use the descriptions and models in their day-to-day work and submit their improvements. For this reason, tools supporting process management should offer the possibility of:

I.1 showing preconfigured process views and the option of configuring dedicated views by systems users
I.2 changing the form of presentation (and perhaps also the form of description) of processes in different stages of the process lifecycle in the organization [22].

4.2 The Subprocess Checklist – an Additional Standard Form of an Ad Hoc Process

According to Forrester [23], key vendors of BPMS and CMS systems devote 50% of their effort and resources to accommodating the user interfaces of BPMS/CMS systems to the habits of the users of transactional ICT systems who are not IT professionals or specialists in the field of management.

At present, one of the form which is closest to the practice of data entry, and at the same time to managing the process workflow in transactional systems, is the checklist, and, increasingly more often, a prioritized checklist [24, 25]. It allows for the simple and intuitive:

- presentation of available and upcoming tasks
- selection of performed tasks
- analysis of the sequence of task performance, and, when supplemented with an additional column, the duration or deadlines of tasks

At present, checklists are one of the basic elements of the user interface in ACMS/DCMS systems supporting case management. Furthermore users of traditional transactional systems such as MRPII, ERP, CRM, or EHR are acquainted with this form of communication with their software.

One proposed extension of the BPMN notation has the form of an additional ad hoc subprocess form based on mechanisms used in the course of implementing a checklist in the form of a subprocess checklist. Below, the use of such a solution is presented for subsequent stages of the process lifecycle in the organization on the example of a DMEMO cycle (anagram from the first letters of the names of subsequent stages: Design, Model, Execute, Monitor i Optimize) (BPM Resource Center, 2014) (Fig. 1).

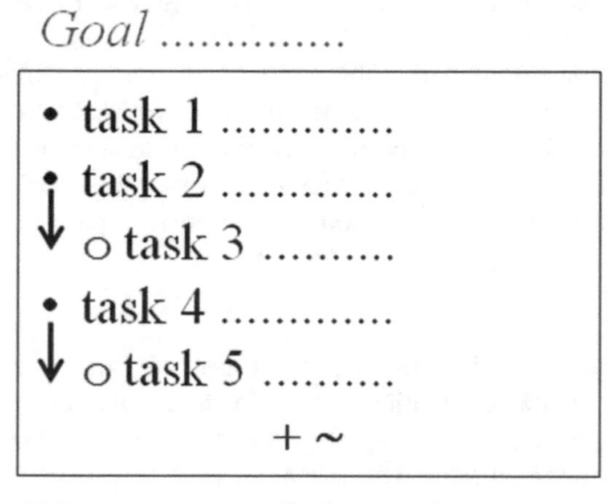

Fig. 1. Proposal of the "Subprocess checklist" object, own elaboration

The Design Stage

In this stage, the basis elements of the subprocess checklist description are:

1.1. The goal of the subprocess – the goal which should be pursued in the course of performing the subprocess and the attainment of which de facto ends the subprocess regardless of whether the checklist is exhausted.
The proposed location is above the standard subprocess symbol, and for the sake of readability it is proposed for the text to be presented in a non-standard form, e.g. in cursive and/or in another color.

1.2. The name of the subprocess – a name identifying the subprocess. In the case of reusable subprocesses the name of the process definition.
The proposed location is below the standard subprocess symbol.
1.3. (optional) link to an ontology describing all possible tasks performed within the subprocess checklist. The ontology should account for limitations with respect to broadly understood knowledge in a given field, which may result from e.g. The availability of resources, the skills and the privileges of the personnel, available authorizations, concessions, and limitations resulting from local law. For example, despite compliance with medical knowledge it is pointless to include "magnetic resonance" in an ontology encompassing diagnostic research when the particular healthcare unit does not have access to the corresponding diagnostic equipment.

The Model Stage
In this stage, beside the abovementioned elements, the basic proposed elements of the subprocess checklist description are:

1.4. Elements of the checklist – names of tasks available to be performed by the process performer.
The proposed location is within the standard subprocess symbol.
1.5. (optional) prioritization of process performance, which blocks the possibility of performing the selected task prior to completing another (or several) task. For example, it is impossible to confirm the loading of goods and the departure of the transport prior to the loading of goods or to provide patients with prescriptions and recommendations prior to their preparation.

Suggestions:

- symbol on a checklist ("↓") pointing to the necessity of performing a task from which the arrow symbol is pointing prior to the task which the arrow is pointing to.
- In the Performance and monitoring stage elements of the checklist which are not available for performance should be written in gray or be set on a gray background.

1.6. (optional) limitation of the possibility of selecting tasks to:
1.6.1. tasks included in the subprocess checklist diagram
1.6.2. tasks included in the ontology indicated in the subprocess checklist description
1.6.3. no limitation (e.g. the process performer may introduce and perform any task in the executive system)

The Execute Stage
In this stage, the following are of crucial importance for the process performers dealing with subprocess checklists:

- readability for different groups of users, particularly those who are not specialists with respect to modeling business processes
- ease and pace of adaptation in accordance with actual process performance by a broad range of process performers

It is proposed that tasks which due to prioritization cannot be performed, as preceding tasks have not been completed, be presented in the user interface of BPMS/ACMS systems in gray or on a gray background, and tasks added by the user (unmodeled to date) in another defined color.

The Monitor and Optimize Stages

In these stages, process performers accessing subprocess checklists are primarily interested in possibilities pertaining to presentation, analysis, and machine learning, used with a view to collecting, verifying, and distributing knowledge on process performance. From the perspective of a user participating in process performance, BPMS/ACMS systems should e.g. for the subprocess checklist enable:

1.8. detailed analysis of performed and unperformed tasks, tasks added by the user and their influence on the efficiency of processes
1.9. analysis of process performance, including the broad context of performance with the use of data derived from BigData analyses (part of the data essential for the evaluation of process performance may be collected in e.g. e-mail or social media)
1.10. changes to the analyzed form of process descriptions (e.g. changing a BPMN diagram to a Gantt diagram presenting the time relations between the performed tasks)
1.11. the possibility of collecting knowledge in the form of proposals for the optimization of process descriptions and/or predefined on-line advice for their performers.

5 Conclusion and Future Works

This article consists of an attempt at approaching process modeling from the perspective of the users themselves, who are not specialists with respect to modeling, but who in the course of their work should nonetheless use process descriptions as a source of knowledge and a repository of knowledge from process performance. Or this reason, the discussed proposals at updating the BPMN notation first and foremost pertain to the expectations of the process performers themselves, which have been noticed as far back as 3–4 years ago by vendors of BPMS and ACMS systems, followed by consulting companies. According to the author, the time is ripe for reflections on the results of research, which show that **users expect simple forms of presenting processes, which are tailored to the particular stage in the lifecycle, or even the specific stage of process performance** [22]. As the article demonstrated, in order to maintain the cohesive nature of data describing the process, one natural solution known from other fields is the possibility of configuring process views in accordance with the requirements of particular groups of user. Due to the rapid emergence of new ICT technologies supporting process performance, which are being equally rapidly adapted by business, more significance is given to the Process performance and monitoring stage. Updating the BPMN notation proposed in the article uses use of the years-long experiences of case management, but does not duplicate its inconveniences and errors [22]. It should also prepare an interface allowing for the intuitive cooperation with ICT technologies supporting process performance, such as artificial intelligence, machine learning, or BigData.

References

1. OMG: Field Guide to Shareable Clinical Pathways. BPMN, CMMN & DMN in Healthcare. Version: 1.0, 10 April 2018
2. Szelągowski, M.: Zarządzanie procesowe w legislacji. Zarządzanie publiczne **3**(35), 169–179 (2016). https://doi.org/10.4467/20843968ZP.17.014.5516
3. Hidden costs of unstructured processes #GartnerBPM. http://column2.com/2009/10/hidden-costs-of-unstructured-processes-gartnerbpm/. Accessed 03 Apr 2016
4. Gartner Group 2020: The De-routinization of Work. http://isismjpucher.wordpress.com/2010/11/12/the-future-of-work/. Accessed 05 Apr 2016
5. Adaptive Case Management over Business Process Management. http://it.toolbox.com/blogs/lessons-process-management/adaptive-case-management-over-business-process-management-40002. Accessed 07 April 2016
6. vom Brocke, J., Zelt, S., Schmiedel, T.: On the role of context in business process management. Int. J. Inf. Manage. (IJIM) **36**(3), 486–495 (2016)
7. Swenson, K.: Mastering the Unpredictable: How Adaptive Case Management Will Revolutionize the Way That Knowledge Workers Get Things Done. Meghan-Kiffer Press, Tampa (2010)
8. OMG: Case Management Model and Notation (CMMN). http://www.omg.org/spec/CMMN/1.1/. Accessed 03 Feb 2017
9. Di Ciccio, C., Marrella, A., Russo, A.: Knowledge-intensive processes: an overview of contemporary approaches?. In: 1st International Workshop on Knowledge-intensive Business Processes (KiBP 2012), 15 June 2012, Rome, Italy. http://ceur-ws.org/Vol-861/KiBP2012_paper_2.pdf. Accessed 02 Apr 2017
10. Pourshahid, A., Amyot, D., Peyton, L., Ghanavati, S., Chen, P., Weiss, M., Forster, A.: Business process management with the user requirements notation. Springer (2009)
11. Gartner: Magic Quadrant for BPM-Platform-Based Case Management Frameworks, 12 March 2015
12. Gartner: Magic Quadrant for BPM-Platform-Based Case Management Frameworks, 24 October 2016
13. Gartner: Magic Quadrant for Intelligent Business Process Management Suites, 24 October 2017
14. Stroppi, L.J., Chiotti, O., Villarreal, P.D.: Extending BPMN 2.0: Method and Tool Support. Business Process Model and Notation, pp. 59–73 (2011)
15. OMG: Business Process Model and Notation (BPMN). http://www.omg.org/spec/BPMN/2.0.2/. Accessed 03 Feb 2017
16. Müller, R., Rogge-Solti, A.: BPMN for Healthcare Processes (2011)
17. Zerbato, F., Oliboni, B., Combi, C., Campos, M., Juarez, J.: BPMN-based representation and comparison of clinical pathways for catheter-related bloodstream infections. In: International Conference on Healthcare Informatics, pp. 346–355 (2015). https://doi.org/10.1109/ichi.2015.49
18. Braun, R., Schlieter, H., Burwitz, M., Esswein, W.: BPMN4CP Revised – Extending BPMN for Multi-Perspective Modeling of Clinical Pathways (2016). https://doi.org/10.1109/hicss.2016.407

19. BPLOGIX BPMN vs Process Timeline. https://www.bplogix.com/blog/bpmn-vs-process-timeline. Accessed 18 Apr 2018
20. Gartner: Magic Quadrant for Intelligent Business Process Management Suites, 18 March 2015
21. Sobczak, A.: Architektura korporacyjna. Aspekty teoretyczne i wybrane zastosowania praktyczne. Ośrodek Studiów nad Państwem Cyfrowym, Warszawa (2013)
22. Szelągowski, M.: Zarządzanie procesowe w gospodarce wiedzy: tworzenie wartości z kapitału intelektualnego. LINIA, Warszawa (2018)
23. Forrester Wave™: Dynamic Case Management, Q1 '16 (2016). http://reprints.forrester.com/#/assets/2/85/%27RES121382%27/reports. Accessed 13 Sept 2016
24. Gawande, A.: The Checklist Manifesto How to Get Things Right. Metropolitan Books, New York (2009)
25. Making the case for BPM. ftp://ftp.software.ibm.com/software/uk/itsolutions/dynamic-bpm/wp_making_the_case.pdf. Accessed 07 May 2018

Envisioning Spread Page Applications: Network-Based Computing Documents for Decision Support in Operations Management

Tomasz Tarnawski(✉)

Kozminski University, ul. Jagiellońska 57/59, Warsaw, Poland
ttarnawski@kozminski.edu.pl

Abstract. Operations management (OM) professionals are faced with a multitude of decisions on different levels and horizons of planning. In their decision making they are supported by various methods and tools, from visualization through computable heuristics, computer simulation models and optimization algorithms. What seems to be scarce, though, is a common framework for connecting all the above "bits and pieces". All the various fragmentary and thematic indicators leading to particular decisions could (and should) be gathered and structured in a single 'document' to provide aggregate insight into the whole story.

The article proposes a vision of applying a computable-document framework, developed under the name of Spread Page, to set up such comprehensive working environment to support decision making in operations management – especially at the high, strategic level.

The 'new kind of document' postulated in the Spread Page project is to present content as multi-dimensional, multi-layered, scalable, interactive, user aware and more. The Spread Page view of the problem at hand, as proposed in the article, could be based on the well-known Ishikawa diagrams supplemented with computational elements drawn from Bayesian belief networks (BBN). This mode of representation would provide computational support in arriving at the final decision, leverage involvement of persons from different departments, allow for presenting the case interactively – at different scales and through numerous aspects. This way one could come up with a single, network based model of the decision problem at hand, aggregating all the identified relevant factors.

Keywords: Spread Page · Decision support · Operations management

1 Introduction

Spread Page is a proposed name denoting concepts, technologies and solutions aiming at representing content in a new way: replacing text-based documents and leveraging existing technological capabilities. It was initially proposed in "Spread Page Initiative Manifest" [1] and then elaborated on in e.g. [2] and [3]. The bottom line of our assertions presented there is that in light of recent technological advancements in

© Springer Nature Switzerland AG 2019
A. Burduk et al. (Eds.): ISPEM 2018, AISC 835, pp. 692–700, 2019.
https://doi.org/10.1007/978-3-319-97490-3_65

electronic devices (computational power, rich multimedia capabilities, touchscreen, prospective 3D displays, tactile feedback and more) it is becoming increasingly inefficient to keep imparting scientific or technical knowledge by means of static, text based documents (books, articles). It is important to note that our focus lies on the areas and applications, where the form is merely a tool for transmitting information and knowledge, such as organizational directives, technical documentation, scientific descriptions, management memos etc. In certain areas, most notably connected with art (e.g. poems) and fiction (novels) such mode of communication presents a value in its own right and is not to be forsaken.

The general idea behind the Spread Page approach is to provide content using model-based representations of knowledge: three-dimensional (or, in fact, multi-dimensional), multi-layered, interactively navigable, scalable, animated (or otherwise time-dependent), aspect-oriented, user-aware, and much more. Depending on the current user's needs, general or detailed aspects would be presented (zooming in and out), particular aspects ('layers') of the topic turned on and off, dynamic relationships emphasized through animation etc. On top of that, built-in calculation would allow to see results computed on the fly for acquired real-life data or assumed hypothetical scenarios.

Various fields have already been pointed out as potential areas of Spread Page applications – from computer science through anatomy and medicine all the way to law and legislation. Management presents itself as another natural arena of applying the above concepts and features. A 'body of information' for an organizations, describing its current (and projected) state is a complex, structured entity and needs to be presented differently to different auditoria and adequately to the purpose at hand. Top level managers look at aggregated performance indicators with strategic planning horizons in mind while frontline workers operate on their daily, detailed data. Role-based differentiation of access to, and presentation of, information is naturally among the key principles in modern ERP-class systems, where the most advanced solutions include interactive "management dashboard" capabilities displaying aggregated KPIs but at the same time allowing for seamless drill-down to more granular data (see e.g. [4] for examples and design details). It is an instance of Spread Page principles already implemented – at least to a degree – in a concrete field.

In this article we concentrate on another prospective application of Spread Page in Operations Management that aims at providing structured support in decision-making. Although managers are already supported by various methods and tools (from visualization through computable heuristics, computer simulation models and optimization algorithms) it seems that a common documentation and exploration framework for connecting all the "bits and pieces" of decision making is still lacking. All fragmentary and thematic indicators leading to particular decisions could (and should) be gathered and structured in a single 'document' to provide aggregate insight into the whole story while the decision is being made, and documenting reasonings and arguments after the call has been made and implemented.

The situation can be compared to medical diagnostic process where a doctor is presented with multiple results of detailed medical tests and has to make up his mind about the likely causes and most prospective treatment. In support of diagnosticians, mathematical tools – specifically based on Bayesian belief networks (BBN) – are being developed, which enable gathering all the information and prompting probability-based and/or rule-based decisions. The methods can be seen as relatives to the well-known in OM, Ishikawa cause-and-effect analysis, with the reservation that they are actually full-blown quantitative computational tools geared with formal probabilistic reasoning. Hence, it seems that similar methods and tools could be just as applicable in OM.

The article proposes a vision of applying a Spread Page-based framework, to support decision making in operations management – especially at the high, executive level. As the backbone of the problem's (/decision's) model we propose a derivative of the Ishikawa cause-and-effect diagram supplemented with computational features of Bayesian belief networks and incarnated into Spread Page flesh. The following section provides a brief overview of the Ishikawa diagram with its relationship to BBN and proposes that a merger of the two could serve as an underlying construct. Section 3 goes on to discuss further features within such framework, aligned with the Spread Page paradigm and desirable from the point of view of efficient conveyance of knowledge, facilitating teamwork or documenting responsibility for decisions. Section 4 provides a short illustrative example related to operations management practice, dealing with medium term production capacity planning.

In the end, we conclude that such Spread Page incarnation – a network-based representation based on a merger of Ishikawa and Bayesian network, fitted with interactivity, computability, multi-scale/multi-aspect presentability, consolidation/drill-down capacity, etc. would be a remarkable platform for efficient and accurate team-based decision making with wide applications in operations management, e.g. in areas such as capacity and production planning, managing logistics capabilities or product development.

2 Ishikawa Diagrams and Bayesian Belief Networks

The use of fishbone diagrams was pioneered by Karou Ishikawa in the 1960's as a tool in investigating quality problems with complex causes. The diagram provides a systematic way of breaking down the multiple (potential) causes in a hierarchical manner and facilitates an ordered, team-based investigation of their contribution to the observed result [7]. Its key advantages play out when a sizeable problem can be broken down into manageable parts to be analyzed by a team of specialists with complementary expertise. Due to its main application, it is customarily referred to as the cause-and-effect diagram. An example is shown below (see Fig. 1).

Although Ishikawa diagram was initially designed for analyzing observed quality problems, its applicability can be much more general – as a help in tracing causes to any observed (or projected) effect. As a further development of the method, it was also proposed to supplement it with computational features and hence move from qualitative analysis into more strict, and quantitative. The authors in [8] provide a case for adding numerical probability values in the defined links between causes and effects, so that the chances of a given result happening could be automatically and (more) objectively computed. Such transformation moves the Ishikawa diagram much closer to Bayesian belief networks – a tool in statistical cause-and-effect analysis that can be

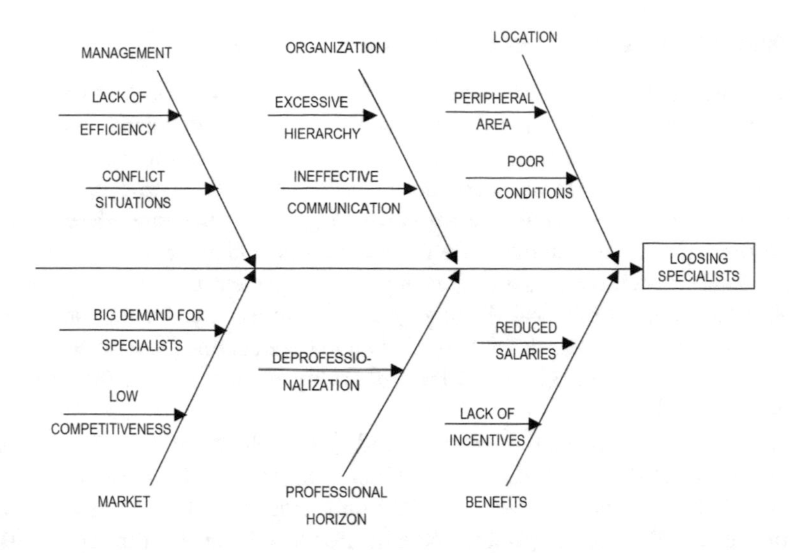

Fig. 1. An example of Ishikawa diagram constructed for identifying causes of an organization's problem of losing specialists; adopted from [8].

fitted with efficient computing algorithms capable of inferencing probabilities of analyzed outcomes or state of affairs.

Bayesian belief networks have become a widely used tool for modeling uncertain cause-and-effect relationships. Next to applications in areas like medical diagnosis or speech recognition they are also used in diagnosis of malfunctioning systems [6] which deepens their analogy with Ishikawa diagrams. However, there are also important distinctions:

- in Ishikawa diagrams, the final result has its high level causes which in turn become results of lower level causes – and so on; in effect, there is a strict hierarchy where a given cause can contribute to only one outcome (the one directly above it in the cause-and-effect hierarchy);
- in BBNs, the nodes (variables) are not (necessarily) distinguished within a higher/lower-level hierarchy and hence all variables are equal-in-rights model elements. Also, one causing factor may influence more than one result.

In effect, Bayesian networks are more general, as they are freed from the enforcement of the fishbone (or tree-like) structure. It seems worthwhile to merge both approaches to join their advantages: facilitated understanding and teamwork from Ishikawa and probability-based computability of Bayesian network. Adding formal restrictions to BBN by enforcing hierarchical structure for causes on different levels of abstraction while relaxing the requirement in Ishikawa diagram (that any factor may influence only one outcome) could create a balanced and useful framework for cause-and-effect analysis – acceptable to both managers accustomed to fishbone diagrams and statisticians fluent in BBN.

2.1 Ishikawa-BBN Diagrams as Interactive Documents

If implemented within a Spread-Page-principles enabled tool, the internal representation of the cause-and-effect probabilistic model would be abstracted away from the presentation layer and, in fact, different modes of presentation could be possible depending on the user's role and preferences. For the purpose of probabilistic modeling and analysis, the network could be visualized e.g. along the lines described in [5], where various graphical methods for enhancing BBN understandability are proposed. The very same model, for presenting to upper management executives, could be projected into the form of classical fishbone graph – through spatial rearrangement and, possibly, dropping some less crucial variables and connections (as defined by specially designed projection rules, e.g. based on probability values, role of the user or other contextual factors).

What is more, within an interactive and zoomable document, the hierarchical structure of causal dependencies could be made arbitrarily deep (as opposed to the typical three-level structure in static Ishikawa diagrams) without the loss of readability. As with digital maps, closing onto a certain area would reveal further details about related causal factors. Such 'drill-down to first causes' would not need to stop anywhere short of accessing individual records of the organization's real-life transactional data stored in its DBMS.

The computable aspect within such Spread Page Ishikawa-BBN representation, calculating the chances of particular outcomes, would take into account both elements: the specified probabilities assigned to particular dependencies and the genuine values queried from business-related, life data. Such probabilistic, hierarchical aggregation-disaggregation mechanism could be seen as a special case for the abovementioned Business Intelligence drill-down functionality within management dashboards.

The final point to note is that uses of such hierarchical/probabilistic cause-and-effect framework do not need to be limited to cases of investigating already observed (present) results. Equally well, an analogical causal structure can be defined to model enabling factors for achieving desired (future) effects. Instead of viewing Ishikawa-BBN diagrams as a (special-purpose) device for investigating quality problems, let us see it as a more general decision making tool.

3 Feature-Rich Framework for Teamwork Decision Making

The envisioned implementation of decision making framework, featured with key aspects of the proposed Spread Page paradigm, would to our mind constitute a key qualitative change from the current uses and applications of Ishikawa diagrams and Bayesian belief networks.

3.1 Multi-role Divide and Conquer

The process of constructing an instance of such framework – for the purpose of making decisions within a given area – requires judgement and expertise in various levels and fields. The strategic goal(s) and a coarse set of the main success enablers are to be defined by the high level executives – together with high-scale indicators for measuring performance. Subsequent tiers of lower level management focuses, in turn, on the enabling factors (defined above) as their goals and adds more detailed branches to the hierarchical structure – all the way down to the nodes connected to company's life data.

Multi-scale, Multi-aspect Content. Among the key features of Spread Page documents is the scalable structure allowing to zoom in from the most general view into the arbitrary depth of increasing levels of detail. What is more, on each level the same elements can be presented in different ways, by showing selected aspects (layers) and filtering out others. For instance, a relationship between a goal (effect) and its enablers (causes) can be analyzed in terms of estimating probabilities, but also in terms of computing aggregated performance (/impact) measures or financial connections.

This way, the whole reasoning structure laying behind the elaborated strategic decision can be divided into fair-sized pieces, so that each one can be assigned to appropriate (by level and expertise) individuals within the organization – some will be made responsible for splitting the problem further, some for defining ways for measuring indicators, some for probabilistic aspects of the reasoning.

User-Awareness. In line with the above, the principle of user-awareness in presenting Spread Page content, furtherly enhances the efficiency of conceptual work. Next to acting appropriately to the user's role (i.e. displaying the correct default aspect and level of detail), the environment must also keep track of the user's actions and preferences so that the presented view is automatically adjusted to the tasks at hand, most likely further needs, etc.

3.2 Scenario-Based What-if Analyses

Another key aspect of the Spread Page paradigm is **multi-dimensionality** allowing the analyzed entity to spread into an arbitrary number of independent (orthogonal) logical dimensions. In the case of corporate decision making, one application of multi-dimensionality is for embedding the Ishikawa-BBN model within the realm of different possible scenarios. If we choose to see each scenario as a different path that the future can take, then our decision model becomes an entity that lives across the dimension(s) of different possible futures.

The range of scenarios under consideration would include not only different possible unfoldings of external factors but also a set of feasible, alternative internal courses of action and, finally, different assessment of probability values defining causal relationships within the cause-and-effect network. Within such-defined sub-space of scenarios, some dimension will be continuous – e.g. for a spectrum of smoothly changing probability vales within an assumed range – while others: discreet, e.g. containing distinct external events that may occur or representing implementation of alternative strategies.

Presentation techniques advocated by the Spread Page Initiative are in particular aimed at **interactivity**, **animation** and **3D graphics** to enhance visibility and understandability of multi-dimensional entities. Animated view seems particularly fitted for showing the changes across a continuously-valued dimension as, in the discussed case, 'travelling' through the dimension of changing probability ('morphing through' values of model's parameters). On the other hand, three-dimensional rendering can be put to use for comparing side-by-side distinct alternatives (from discreet-valued logical dimension). Both techniques would be further enhanced through interactivity (changing time lapse, rotating point of view, etc.).

3.3 Fully Documented Reasoning and Judgment

The elements proposed so far compose a multi-dimensional, multi-faceted, computationally enabled, interactive model representing a case for making key decisions in the context of an organization. As already discussed, disaggregation of the problem allows for facilitating teamwork through distributing modeling decisions across numerous members of the organization. This way, development of particular elements of the model will rely on expertise and value judgement of designated individuals, as far as:

- deriving values of performance measures (KPI) based on lower level figures and (eventually) detailed real-life data;
- assessment of probabilistic implications in cause-and-effect relationships;
- constructing feasible scenarios and expected courses of actions (together with probabilities of them playing out).

A multi-user role-based Spread Page environment not only provides a natural platform to carry out such process but also opens the way to achieve an additional quality. All the calls, decisions and judgements made throughout the development can easily be documented with clear attribution of responsibility to the persons that made them. For increased security, including digital signatures or timestamping is also a viable option and – in special cases, when that might be required – cryptographic mechanisms for ensuring consistency and irreversibility (e.g. with blockchain-related technologies) can also be implemented.

4 An Illustrative Example: Medium Term Capacity Planning

For a simple, illustrative example of applying the proposed concepts, let us consider medium term planning of production with the aim to best matching supply with demand and under the organization's ultimate goal of assuring long-term profitability. On the top level of analysis, the management defines a number of key factors contributing to that goal.

Minding, that the aim of the example is to illustrate the proposed approach and not to provide a rigorous business analysis, let us assume that for the organization the key enabling factors for reaching long-term profitability are (among others):

- maximizing profit/minimizing loss within an accounting period (a year) – to be achieved through proper balancing of the company's cash flow;
- minimizing employee turnover – to avoid disruption in the production and minimize training costs;
- ensuring client satisfaction – through timely deliveries, adequate quality, etc.;

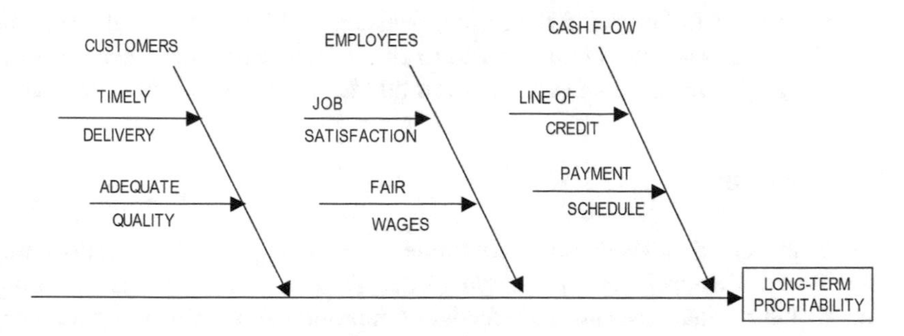

Fig. 2. A hypothetical case of attributing long-term profitability to (among others): customer satisfaction, low employee turnover, and balanced cash flow – presented on a partial Ishikawa diagram.

as presented on a partial Ishikawa diagram, shown on Fig. 2. The decision making objective is to come up with a medium-term production capacity plan that would be best aligned with the overall objective – and hence with its enabling factors.

Following the divide-and-conquer approach, a dedicated group of Sales Department members is to deal with the area of customer satisfaction. Ideally, both timely deliveries and adequate quality objectives should be met 100% of the time, but with limited capacity and variable demand if might be far from feasible. At times, cutting corners might be unavoidable, resulting in delays and/or quality issues – which raises more detailed questions about acceptable tradeoffs.

Investigating in more detail could then proceed, e.g. in the directions of:

- customer stratification, by their importance and contribution to long term profit – to identify the crucial clients but also the ones that the company can afford to upset (through higher prices, quality issues, delayed deliveries etc.);
- managing demand through seasonal pricing, special offers, etc.

and for each, a number of alternative scenarios would be prepared, possible impact quantified and probabilities of fulfilment assessed. Each scenario would assume certain demand pattern, courses of action and probability distribution for customers' responses.

At the same time, parallel departments would carry out their detailed analyses, filling out other parts of the fishbone diagram and the scenario space – e.g.:

- Manufacturing Department would asses feasibilities of various approaches to production capacity: level (produce to stock) vs chasing demand (work overtime, defer maintenance, lay off workers during low season) – see e.g. [9];
- Accounting would attempt to predict cash-flow impacts of the investigated courses of action;
- HR would deal with prospected influence on the employees.

In result, a whole network would be built, connecting the general goal on top with all detailed factors derived from the raw data – on the bottom. A single 'document' would represent the entire, multi-dimensional decision-making problem and allow for displaying its selected aspects and dimensions at any level of detail, interactively and with animations enhancing comprehensibility.

Such Ishikawa-BBN model of the current decision problem, once verified and field-tested, could serve as a pattern to be applied to similar future cases and hence become a part of the organizational knowledge, or even add to the industry's best practices.

5 Conclusions

The article presented a vision of a corporate decision-support environment implementing the core Spread Page paradigm ideas. To our mind, high level, complex decision making situations can be effectively represented within a network-based, computational environment. One mechanism to achieve that could be a specific merger of Ishikawa diagram and Bayesian belief network. Next to (probabilistic) computability derived from its Bayesian relative, it can be furtherly enhanced with the Spread-Page-postulated features of multidimensionality, scalability (consolidation/drill-down), interactivity, animation, etc. – to the point, where a single 'document' aggregates all relevant aspects (on all levels of detail) of a given decision-making problem and presents its content in an optimally-comprehensive manner, most appropriate to the needs of the concrete person viewing it.

Such Spread Page incarnation of an Ishikawa-BBN network would provide a remarkable platform for efficient and accurate team-based decision making with wide applications in operations management, e.g. in areas such as capacity and production planning, managing logistics capabilities, product development, etc.

References

1. Spread Page Initiative: "Spread Page Initiative" Manifest. http://spreadpage.org/2017/06/07/spread-initiative-manifest/. Accessed 03 May 2018
2. Tarnawski, T., Kasprzyk, R., Waszkowski, R.: Foundations for Spread Page: review of existing concepts, solutions, technologies capable of improving effectiveness of conveying knowledge. Comput. Sci. Math. Model. **6**, 33–44 (2017)
3. Waszkowski, R.: Spread Page approach to document management. In: Goossens, R. (eds) Advances in Social and Occupational Ergonomics, AHFE 2017, Advances in Intelligent Systems and Computing, vol. 605, pp. 52–61. Springer, Cham (2017)
4. Hacking, X., Lai, D.: SAP BusinessObjects Dashboards 4.0 Cookbook. Packt Publishing, Birmingham (2011)
5. Zapata-Rivera, D., Neufeld, E., Greer, J.: Visualization of Bayesian belief networks. In: IEEE Visualization 1999 Late Breaking Hot Topics Proceedings (1999)
6. Pearl, J.: Graphical models for probabilistic and causal reasoning. In: Tucker, A. (eds) The Computer Science and Engineering Handbook, pp. 699–711. CRC Press, Boca Raton (1997)
7. Watson, G.: The legacy Of Ishikawa. Qual. Prog. **37**(4), 47–54 (2004)
8. Ilie, G., Ciocoiu, C.N.: Application of fishbone diagram to determine the risk of an event with multiple causes. Manag. Res. Pract. **2**(1), 1–20 (2010)
9. Slack, N., Brandon-Jones, A., Johnston, R.: Operations Management, 7th edn. Pearson, Harlow (2013)

Security and Risk as a Primary Feature of the Production Process

Jerzy Stanik, Maciej Kiedrowicz, and Robert Waszkowski(✉)

Faculty of Cybernetics, Military University of Technology,
Urbanowicza St., 01-908 Warsaw, Poland
{jerzy.stanik,maciej.kiedrowicz,
robert.waszkowski}@wat.edu.pl

Abstract. The paper discusses selected problems and issues in the scope of safety and risk of the production process. Various categories of risk factors have been classified and characterized, affecting the level of functional and information security of the production process. A model of quality and safety of the production process was proposed, which indicates the areas determining the risk level of the production process, as well as ensuring the completeness of the risk assessment process of the production process. The advantages and disadvantages of currently used risk models of the production process to assess the risk of the production process are also discussed.

Keywords: Risk factor · Risk · Safety · Production process

1 Introduction

Issues regarding the safety and risk of production processes and operating systems are constantly and will be developed. Research on production processes covers a variety of areas, of which the most important are:

- quality testing,
- reliability test,
- business continuity test,
- functional[1] and information[2] safety
- testing risk assessment in the scope of work of both the entire production process and individual elements.

The production process (production process) is the whole of phenomena and purposefully undertaken actions, which cause that the desired changes gradually take

[1] Functional safety - part of total safety related to the controlled object and the object control system, which depends on the correct operation of electrical/electronic/programmable electronic systems related to safety and external risk mitigation measures; the concept of functional safety is closely related to the process of reducing risk from the current level of risk of the production process to the level of tolerated risk.

[2] Information security is a set of actions, methods, procedures undertaken by authorized entities, aimed at ensuring the integrity of collected, stored and processed information resources, by protecting them against undesired, unauthorized disclosure, modification, destruction…

© Springer Nature Switzerland AG 2019
A. Burduk et al. (Eds.): ISPEM 2018, AISC 835, pp. 701–709, 2019.
https://doi.org/10.1007/978-3-319-97490-3_66

place in the object subjected to their influence. Accumulating, they cause the object to gradually acquire features that approximate it and make it similar to the intended product. The end of the production process occurs when all necessary elements have been achieved. For the purposes of this article, we assume that:

(1) the production process is an organized team of activities coordinating the flows of materials, information and energy during the implementation of the technological process,
(2) is of the type - continuous flow - type of production line from which the product can not be removed before the process is completed,
(3) acts as a primary production process aiming at the direct implementation of the product to which the enterprise was established and its sale is the main source of revenue. The stages between the warehouse of materials and the warehouse of finished products along with technological processes, what the subject of work is considered as the primary production process, when the result is a product that is the main activity of the enterprise,
(4) from the point of view of the production system is the transformation process, i.e. the transformation of the input vector into the output vector.

Diagram of such a defined and conditioned production process, reflected in Fig. 1.

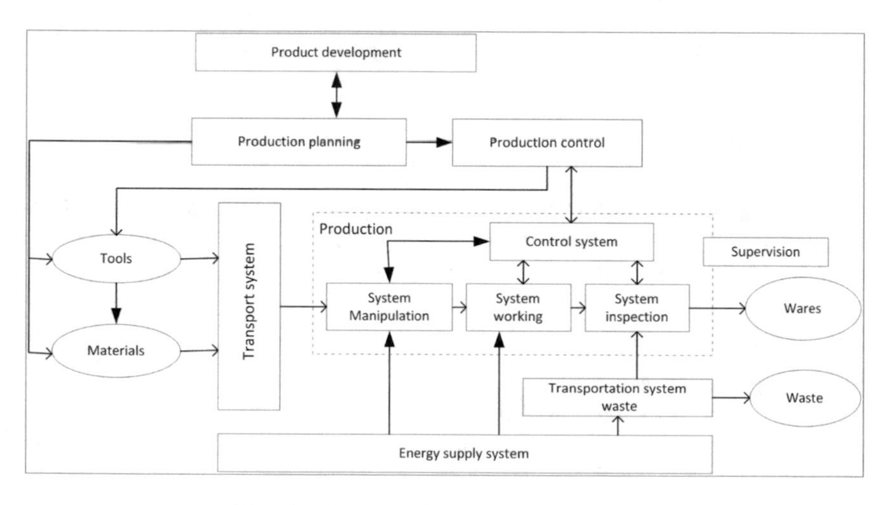

Fig. 1. Diagram of the production process

For the purpose of this article, the risk of a production process is defined as a threat, vulnerability or gap (e.g. information, technological, functional) that various types of technologies, e.g. design, production or information used in a given organization (regardless of its type and scale of activity)) do not meet business requirements or rules, do not ensure adequate quality, security, integrity and availability or continuity of processes, have not been properly implemented and do not operate in accordance with the assumptions or requirements of adopted business policies. In connection with the above, the risk of processes is considered in the division into various categories

(ISO 31000, 2012), e.g.: safety, quality, efficiency, business continuity and architecture of the production process.

The production process is characterized by relatively high production risk, due to a number of factors determining it. The presence of these factors means that the production cycle can be destabilized. The decisions made regarding this process are therefore closely related to the conditions of uncertainty and the risks that can be understood as the effects of this uncertainty. Factors affecting the production process can be divided into the following categories, groups or areas (Fig. 2):

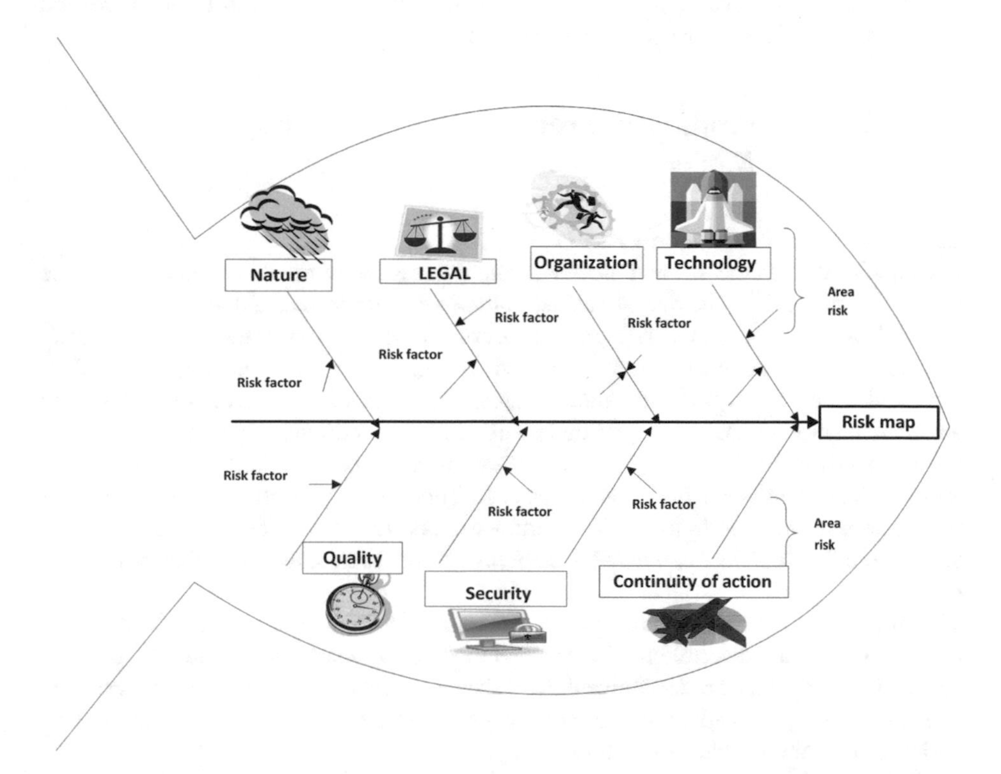

Fig. 2. Basic groups of risk factors for the production process

The approach presented in the work takes into account the following categories of risk factors:

- security (B) - information security factors in accordance with the ISO 27000: 2015 standard and functional safety in accordance with the PN-EN 61508 and PN-EN 61511 standards,
- quality (J) - quality factors in accordance with ISO 9001: 2015,
- business continuity (C) - business continuity factors in accordance with ISO 22301 regarding business continuity,

- technology (T) - a risk factor in accordance with ISO 31000 in the field of risk management
- complexity (S) - factors resulting from the architecture of the production process itself,
- reliability (N) - reliability factors in accordance with PN-EN 61078 and PN-IEC 60300-3-4.

Adequate selection of factors is a difficult task, but its correct implementation will allow the development of a multi-criteria risk assessment model of the production process, which will enable a thorough assessment of this process in terms of several aspects and within each aspect several criteria.

2 Risk of the Production Process and Classic Risk Assessment Models

The Risk of the Production Process

A characteristic feature of each process is the occurrence of the so-called risk. For the purpose of this article, the risk of business processes is defined as a threat, vulnerability or gap (e.g. information, technological, functional) that different types of technologies, e.g. design, production or information used in a given organization (regardless of its type and scale activity) do not meet business requirements or rules, do not ensure adequate quality, security, integrity and availability or continuity of processes, have not been properly implemented and do not operate in accordance with the assumptions or requirements of adopted business policies. In connection with the above, the risk of processes is considered in the division into various categories (ISO 31000, 2012), e.g.: safety, quality, efficiency, business continuity and architecture of the production process.

The production process is characterized by relatively high production risk, due to a number of factors determining it. The occurrence of these factors means that the production cycle can be destabilized. The decisions made regarding this process are therefore closely related to the conditions of uncertainty and the risks that can be understood as the effects of this uncertainty.

Deviating from any attempts to define the concept of risk, its perception in a big simplification means:

- perspective of consequences resulting from the decisions made,
- the threat that technologies used in a given production process (regardless of its type and scale of operations):

  (a) do not meet business requirements,
  (b) have not been adequately implemented and do not operate in accordance with the assumptions,
  (c) they do not ensure adequate efficiency or integrity, security and availability of the production process,
  (d) do not meet the principles or requirements contained in policies such as: security policy, quality policy, business continuity policy, etc.

and at the same time it can be a measure of the danger:

- occurrences,
- the occurrence of the situation,
- maintaining assigned attributes, attributes or properties.

Well targeted and systematic prevention is the best action to avoid the effects of potential threats and increase the chances of achieving higher efficiency or effective implementation of the objectives of the production process or the entire organization. The adoption of an appropriate definition of the concept of production process risk and its model is a starting point to indicate any methods of risk assessment and management of its level.

Review of Risk Models of the Production Process

Numerous risk assessment models for the production process are described in the professional literature, based on standards, experience or expert knowledge. Their specification is based on the number of components included in the risk assessment process. The first and at the same time the basic component is a wide list of risks or resources of risks that will affect the defined modeling objectives - regardless of whether their sources are under the control of the organization or not. The second composition is specified risk assessment criteria that will be used at the stage of risk analysis and assessment, and later its evaluation. Most often the set of these criteria includes, among others:

- definitions of consequences (nature and type and method of measurement),
- probability definitions (type and method of measurement),
- categories of risk level (method of determination),
- evaluation categories (determining the level of acceptable or tolerated risk).

The next group are domain models, e.g. information security, functional security or business continuity or reliability. These models represent a "good" starting point for the construction of multi-component risk assessment models.

By analyzing different models of reliability or risk assessment of the production process, it is easy to notice significant differences in approaches adopted in them.

This leads to the division of models into groups of applications. In the most general case, models perfectly suited to "simple" production processes can be distinguished, where an alternative to the lack of risk assessment of the process is a simple and easy to use method of its evaluation, based on the so-called. risk matrices (Fig. 3).

There are of course models suitable for use both for simple and complex production processes. An example of such a model may be a model [8] or the model proposed in this article. However, one cannot speak of a universal model allowing to precisely determining the level of risk of production processes, because each of the risk assessment models refers only to a certain slice of reality that models and takes into account only selected factors affecting the safety and risk of production processes.

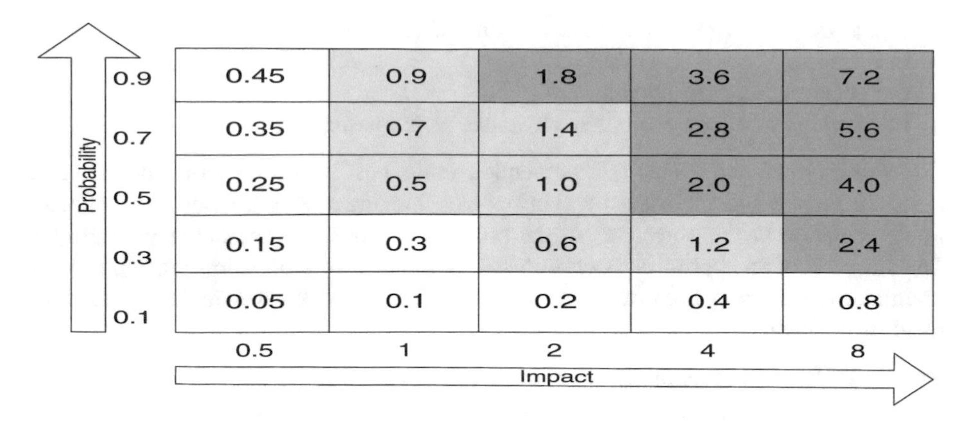

Fig. 3. An example of risk matrices

3 The Safety Model of the Production Process in the Aspect of Risk Analysis

The safety of production processes should be considered in several aspects. The starting point for constructing the risk model of the production process, proposed in this article, reflecting the systematic character of the presented approach, is the quality and safety model of the production process, shown in Fig. 4. The model presented in Fig. 4 is based on the following four components:

- Policies - policies and procedures defining the rules of conduct in terms of safety, quality and business continuity and reliability,
- Principles - current regularities and interdependencies occurring in actual production processes and experience, allowing to determine methods, methods of efficient and economical legitimate course of the production process; the so-called. "Principles of rational organization of the production process" [2],
- Actions - solutions guaranteeing the correct operation of production processes and ensuring safety, quality [7], reliability, business requirements regarding business continuity as well as those responsible for responding to violations of policies and procedures,
- Control - ongoing monitoring of production processes and verification of the level of compliance with safety/quality rules and their coherence and adequacy, as well as solutions guaranteeing reduction of risk in relation to assigned attributes of information security (availability, integrity, accountability), functional safety, quality, efficiency and business continuity of the production process.

The following relationships occur between the individual pillars of the above model:

- Policies or principles included in strategies, different types of plans or procedures determine the operation of security solutions along with the methods of monitoring their work,

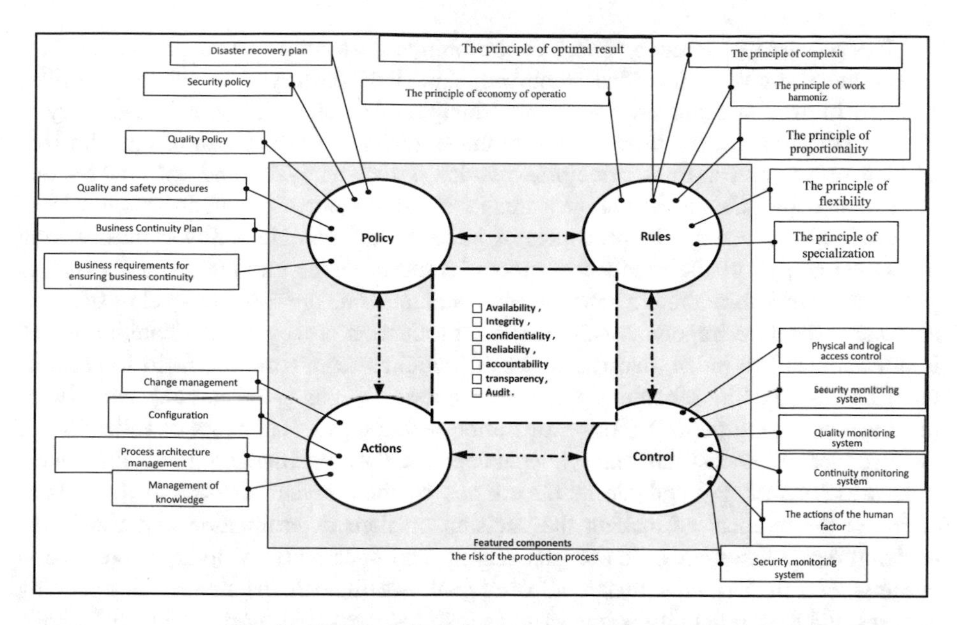

Fig. 4. Model of the safety of a production process.

- Implementation of the aforementioned policies and rules is conditioned by many factors such as: the type of production and manufactured products, manufacturing processes implemented and the technology used.
- The functioning of solutions ensuring safety, quality, efficiency and business continuity determines the emergence of new security, quality and business continuity principles and influences the scope of monitoring,
- Events detected through monitoring affect the modification of security solutions and provide recommendations for modifying existing and creating new policies and procedures.

The model defined in this way indicates the areas determining the risk level of production processes as well as ensuring the completeness of the approach.

The central element of the production process safety model presented above is the security attributes of the production process: Availability, Integrity, Confidentiality, Reliability, Accountability, Transparency, and Audit.

The impact of confidentiality and availability on the risk of production processes depends on the expected level of these attributes for a given process. If the expectations of the production process as to its availability or confidentiality are high, then the attribute will have a significant impact on the process of risk assessment for this process. If, however, availability or confidentiality for a certain production process is not a critical factor, then the impact of the given attribute on the risk of this process will be marginal. In the case of integrity, the authors do not believe that it is possible to accept the fact that the integrity of the production process or information processed in a given process is deliberately accepted, therefore it can be assumed that expectations in relation to each production process in terms of integrity are comparable.

The principle of integrity refers to the obligation to ensure that production processes have not been modified, removed, added or destroyed in an unauthorized manner. In turn, in accordance with the principle of confidentiality, it is necessary to prevent situations in which processes are made available or disclosed to unauthorized entities or processes. Both principles require a risk analysis, and then adapt and implement appropriate technical measures to ensure the integrity and confidentiality of the production process. The principles of integrity and confidentiality of information processed as part of the production process impose on the entities the obligation to process data or information in a manner that guarantees an appropriate level of security.

One of the most important features of the production process is its reliability, which is quite difficult to judge, because it usually requires teamwork, the need to separate other factors affecting reliability (human, organization, management, etc.) and failures in longer operating periods (during operation technology). Indicators of reliability of the production process are partly dependent on the specification of technological processes (technology) and can be determined by the relevant services of the organization at the moment of making the decision on their determination and use in the decision-making system of the organization. The specificity of today's production systems, and in particular their complexity, allows them to be treated as operating systems and then reliability is one of their features measured by the degree of implementation of the determined indicators, parameters and characteristics. Therefore, the reliability of production systems can be determined in a very flexible way, assuming, depending on the needs of the analysis, for "acting in accordance with the intentions of the user" the size of any indicator that the process according to the "user" should characterize. The most frequently analyzed indicators of the production process are: process duration (t), efficiency (W), productivity (P).

Accountability - a property that ensures that an entity's activities can be uniquely assigned only to that entity. This principle applies to the development of mechanisms and procedures for the accountability of stakeholders or entities involved in the production process from the implementation of specific tasks or activities included in the production process. Adherence to the accountability principle requires the implementation of appropriate procedures and reliable documentation of the production process.

Process transparency means that all stakeholders of production processes have access to process documentation and results obtained in processes.

The audit is a modern tool that supports the effective management of the company through the assessment of production processes and risk management processes, control and organizational order. It is a tool enabling more efficient management of business processes and the entire enterprise. In order to ensure high quality, safety, business continuity and reliability, production processes are verified. These activities are carried out by means of an audit.

The safety model of production processes (from Fig. 5) is aimed at protecting the production process against the hazards to which it is exposed. One of such threats is the actions of the human factor, both from the inside and from the inside of the organization. The significance of this factor, as well as the awareness of its existence, is confirmed by the latest global research into the level of information security.

4 Conclusions

Each enterprise operates based on the implementation of a series of production processes. Some of these processes are critical from the point of view of the basic activity of the company, and some are of an auxiliary nature. Smaller processes can be part of larger, much more complex ones, which naturally forces, for example, their re-usability. Some of them are local in nature, while others may occur on the scale of the whole enterprise. Undoubtedly, however, the element that connects them all is the constant need to improve them and measure the level of risk. Knowledge of the risk level of individual production processes allows for effective risk management through the use of dedicated models and methods. It is not possible to completely eliminate the risk from the production process but, according to the authors, you can construct appropriate models, methods or methodologies for risk assessment and risk management, in particular production process risks, the use of which will contribute to its reduction, which is the subject of this article. The issues and considerations presented in this article may be the starting point for the development of multifaceted or multi-dimensional business risk analysis models, which in turn may be the input to the development of the risk management methodology of the production process or operating system.

References

1. Adamska, A.: Risk in the activity of the company - essential issues. In: Firela, A. (ed.) Risk in the Activity of Companies. Selected aspects. Publ. House of Main School Commercial, Warsaw (2009)
2. Brzeziński, M.: Industrial Engineering in the Enterprise. Difin Publishing Company Warsaw (2013)
3. Kaczmarek, T.: The Risk and Remedying with Risk. Difin Publishing Company Warsaw (2009)
4. Korombel, A.: Apettite for risk in business management. Publishing Company of the Częstochowa Technical University, Częstochowa (2013)
5. SIEMENS: Overview of Safety-Related Parameters for Siemens Components in Accordance with ISO 13849-1 and IEC 62061 (2014)
6. Stanik, J., Hoffmann, R., Kiedrowicz, M.: Risk management system as the basic paradigm of the information security management system in an organization. In: MATEC Web of Conferences, CSCC 2016, vol. 76, p. 04011 (2016). https://doi.org/10.1051/matecconf/20167604010
7. Stanik, J., Protasowicki, T.: Methodology of the forming of the risk in the life cycle of a computer system, KKIO "from processes to the software: examinations and the practice" (2015)
8. Stanik, J., Napiórkowski, J., Hoffmann, R.: Risk management in the safety management system of the organization. Scientific booklets of the Szczecin University, Economic Problems of Services (2016)

Spread Page Challenges for Accessibility in Business Modeling

Grzegorz Złotowicz[1] and Robert Waszkowski[2(✉)]

[1] Spread Page Initiative, 103a Kasprowicza St., 01-823 Warsaw, Poland
grzegorz@zlotowicz.pl
[2] Faculty of Cybernetics, Military University of Technology, 2 Urbanowicza St., 01-908 Warsaw, Poland
robert.waszkowski@wat.edu.pl

Abstract. Nowadays, almost every modeling notation uses graphical representation to explain the scope of the model. However, graphical symbols are not suitable in every situation. In computer to computer exchange, for instance, there is no need to use graphical symbols. What more, they are not welcome, because it produces only difficulties with writing, exchanging and understanding models by machines. The ideal way would be to construct models as multi-layer objects that contain both the graphical and the descriptive representations. The main challenge in such an approach is model's ability to follow, on every layer, transitions between elements in a way it is done in graphical layer.

The paper describes the Spread Page approach to construct models that are understandable not only in their graphical layers. Authors take advantage of the experience of blind people working with IT modeling tools.

Keywords: Spread Page · Modeling · Notations · Knowledge representations · Blindness

1 Introduction

Accessibility of information for blind people is a way of non-visual content presentation. Text read aloud, braille writing, or tactile graphic are few simple examples of such approach.

In nowadays world, where information is electronically stored and processed, accessibility became a matter of automated data conversion to alternative, spoken or tactile forms. To make it possible, straight forward and accurate, there is a need of storing data in a way, where its semantic is machine-readable.

The main goal of Spread Page Initiative is to achieve freedom from conventional, two-dimensional data presentation, and to make it readable for all people and even machines.

To achieve it we must understand how to read models that have no visual layer. People with blindness are facing such a problem every day. Working with different models used in science and industry, they are able to see all the disadvantages of currently used notations.

© Springer Nature Switzerland AG 2019
A. Burduk et al. (Eds.): ISPEM 2018, AISC 835, pp. 710–719, 2019.
https://doi.org/10.1007/978-3-319-97490-3_67

Talking about the possibility to read models by people and machines, in case of blind people, is talking about the interaction, or even cooperation, between human and computer. All models that have their visual layer, what means almost all of them, have to be converted into spoken language. This kind of audio description should be done automatically by programs that can translate pictures into words.

But it is not only translation of pictures to words that can make models readable to blind people. There are also dynamic aspects of models. That means, it is not enough to "see" model, but it is also necessary to move through it. The main components of almost every science notation, especially in IT science, are elements and relations. And that's why the main point is to have possibility to move from one element to another that remains in relationship with this first. So, to fully read and understand a model one has to understand what element he or she analyzes at the moment, what are its properties and what relationships to other elements it has. Then, it should be possible to move to another element that is connected to the one being analyzed.

Subsequent chapters describe the state of art in different content accessibility for blind people. It is to understand how important is to prepare models that are not limited to their visual layers.

2 Operating System and Standard Applications Accessibility

Screen Reader Application

An application called screen reader, is a software which works in a background of operating system and monitors activity of opened programs in aspect of user interface creation. Each opened window's content is reflected by the screen reader and can be presented to the blind user in form of text spoken by the TTS engine [1], or send to braille display [2].

The content of a window is not only a text, but also semantic information about user interface, so in order to work properly, the screen reader must know as much as possible about each control placed in the window which means every button, edit field, list box and so on. It is necessary to know what type of control it is, what is its name or label, what is its value or content, and in what state it is (pressed button, checked radio button, or list element, etc.). After collecting this information, screen reader can present it to user in his or her preferred form.

Some information is announced automatically (for example, new title window after switching to another application), some can be announced on-demand (e.g. status bar, screen position of current element), using special commands, specific for used screen-reading tool. Using these commands user can read on-demand different interesting parts of window, track objects hierarchy, hear information about current text formatting, and switch the way of presentation – e.g. punctuation level, speed and volume of speech.

Screen reader software can be used on computers, but also on mobile devices – tablets or smartphones operated by Android and IOS systems.

Using screen reading software and a standard keyboard, a blind person can effectively use a computer. Touch screens presented an interesting challenge for developers which was solved in similar way on each operating system: screen reader captures all gestures performed by user, and executes a command if it was a single tap, or reads aloud the content under finger if the user slides by the screen. Double tapping activates the last announced element. Some additional gestures activate special screen reader commands.

Standard and Custom Programs Accessibility

Accessibility to information for blind users is a matter of accessible software, which presents this information. In some cases, where information being presented is highly graphical, such as maps, there is a need of using some specialized tool, which would present this information in different way.

But in most cases, there is not such need, because standard tools can be perfectly usable if they are accessible. Web browsers, mail clients, text editors are only few, simple examples.

To achieve interface accessibility, developers should know and follow guidelines specific for a given operating system.

In case of standard interface controls, everything accessibility-related is maintained by the operating system.

When creating a custom-drawn control, there is a need of accompany it with proper accessibility interface, e.g. Iaccessible2 [3] under Windows and Unix OS.

For mobile operating systems, there are accessibility guidelines created by their vendors as a part of SDK.

Software Interaction Abilities

Ability to accurately read interesting content presented by used application, is one of crucial accessibility aspects. Another one is the possibility of interacting with applications using keyboard. It's important to have keyboard access to all software elements such as menus, toolbars, and other controls.

Every control should be operable via keyboard, and for better effectiveness – important operations should have quick shortcut keys letting user invoke them from the main window without a need to navigate through the system menu.

In case of standard controls, keyboard operability is maintained by an operating system, but it should be ensured when using custom controls.

Even when developer uses custom controls, there is still possibility of mistakes, when order of controls during navigation using Tab key e.g. in dialog box is improper, or if some control captures the Tab key, not letting user go to next or previous control.

Another problematic control is the shortcut key entering control, which captures the Tab key instead of navigating to the next dialog element.

Fortunately, accessibility guidelines for software developers, are well-documented by operating system vendors, but unfortunately – the consciousness of programmers and their trainers concerning the accessibility still needs improvement to became more propagated.

Word Processing Software Accessibility

In case of editing documents, simple knowledge about textual content is not enough. User needs to know as much as possible about text formatting, font size and other attributes, typing errors, page numbers, footnotes and endnotes presence, structure elements such as table row or column, text selection and currently selected text and so on.

To make it possible, word processing software must expose this information to the accessibility framework, used by screen reader.

All commands of a word processing software need to be accessible with keyboard. It gives a great advantage not only to blind but also to standard users, giving them the effective way of changing the formatting attributes, styles, and performing other editing operations with the use of keyboard shortcuts only, which is extremely effective when a user does not take his hands out of keyboard during typing.

3 Web Content Accessibility

The fact that web browsers are able to cooperate with the screen reader application, is not enough to say that web browsing is accessible for blind people, and people with disabilities in general. Even more important is that web services should be built in an accessible way. The World Wide Web consortium [4] constituted the Web Accessibility Initiative (WAI) [5], which developed the Web Content Accessibility Guidelines (WCAG) [6] - a document that standardize the way of making web pages accessible.

Some key aspects of web accessibility are conformance with standards (HTML/CSS), separating content and presentation layers, providing alternative textual descriptions for images which are not decorative.

A very bad, but unfortunately common, practice is to mark required or mistaken form fields with a color only (with no additional caption), navigation controls made of images without textual "alt" attribute, unlabeled form fields, inaccessible Flash animations used for page navigation or content presentation purposes, graphical Captcha [7] challenges without alternative (textual or audio) hints.

Web Pages

Cooperating with web browser, screen reader acquires information about currently viewed document, and presents it in a special way called virtual buffer. Document view is flattened to one dimension, so there is a simple order of elements, where each element in a document is preceded and followed by exactly one other element. Generally speaking, elements' order is taken from HTML stream, and CSS positioning rules are ignored.

Letters and digits keys are captured by the screen reader and used for quick navigation inside the virtual buffer, so e.g. letter H moves to the next heading, shift+H moves to the previous one, T - moves to the next table, F - to the next form field, and so on. Some special keys are also used to navigate inside tables and easily move to next or previous row/column.

Alternative text is presented in virtual buffer instead of corresponding images, form fields labels are read when tabbing through the form.

Dynamic content updates are reflected in the virtual buffer, and – if Accessible Rich Internet Applications (ARIA) standard is used – some important elements can be automatically announced after change.

Element visibility is also reflected in virtual buffer – the hidden part of a page disappears from the buffer.

Virtualization of document doesn't affect real on-screen view of it, so on touch screen, also the spatial structure of a web page can be explored by blind users.

Web Applications

In order to assure web application accessibility there are some standards that have to be taken into account when designing and building web-based systems. One of them is the Accessible Rich Internet Applications (ARIA) [8], developed by the WAI. It is a set of extended attributes and mechanisms, allowing to achieve better accessibility of dynamic web pages and web applications using JavaScript.

For example, the ARIA role attribute can be used to inform screen reader, that some DOM [9] element, being e.g. a DIV [10], semantically is a menu, tree view or edit control.

Other attributes can inform about element's value, number of child elements, checked state, and even instruct screen reader if text change inside given element should be automatically announced to user.

4 Math Notation Accessibility

Storing math formulas as images is probably most popular, and most inaccessible option. Such graphical objects cannot be converted to textual representation, because even the Optical Character Recognition (OCR) software [11] doesn't handle math formulas properly.

Math can be stored in textual form, readable both by humans and machines. Good example in that field is the Latex [12] markup.

Interesting solution is used e.g. by the Wikipedia project – each math formula is internally stored in Latex format in article source code. MathML and graphical formula in image format for older browsers, is automatically generated from the Latex source. Finally, in the HTML page, formula image has accompanying "alt" attribute, containing original Latex expression, which is presented for screen reader users.

Such technique is both good-looking and accessible.

The MathML [13] language is also machine and human readable, but syntax nesting and complexity makes reading by human difficult.

The Math player [14] tool is an in-place reader of MathML expressions for blind users allowing to browse expression on a different level of detail. At the beginning, whole expression is read; after going deeper – only a part of it, e.g. numerator. Then – the only one bracket content, finally – a single symbol.

Unfortunately, there is no software that allows editing and processing complex expressions in an accessible way.

5 UML Diagrams Accessibility

A project called TeDUB (Technical Drawings Understanding for the Blind) [15] funded by the Information Society Technologies Programme of the European Commission, promised to deliver tools for exploring UML diagrams, electronic circuits and architectural floor plans.

Unfortunately, after 6 years, project failed, and the only usable result is the "Accessible UML" software giving limited possibilities of reading UML diagrams stored in XMI format [16].

Another tool, independent from the above-mentioned project, is the PlantUML [17] – a software for drawing UML diagrams based on textual description language.

The plantUML language allows describing many types of diagrams (class, sequence, use case, activity, and others), but unfortunately does not support data modeling diagrams.

6 Electronic Circuit Scheme Accessibility

Schemes of electronic circuits, similarly to math formulas, are graphical in its nature.

There are also, like in case of Math, some ways of describing circuits in textual form.

The SPICE simulator [18] has a textual language describing analog circuits. Individual element of the circuit is stored in a single line of file, starting from its type and symbol, next a list of enumerated nodes to which this element is connected, and ending with elements's additional parameters.

Similarly to Latex or MathML, this language is readable by human and machine, and although it can be hard to understand for complex circuits, it is far more accessible than a graphical drawing.

Verilog [19] and VHDL [20], used to describe digital circuits, have a programming language syntax, and as such, are accessible.

Similarly to the situation with math representing, the biggest problem faced by blind computer science students and hobbyists, is lack of accessible tools for building, viewing, editing and simulating electronic circuits.

7 Geographical Maps Accessibility

Geographical maps, similarly to above examples, are also of graphical nature. Some way of presenting it to blind users is a tactile graphic. Due to a low resolution of such drawings, determined by technical possibilities of embossing on paper or plastic, and also due to human finger sensitivity, it is not possible to place too much details on such map.

An interesting approach was presented in a project called TEAM (Tutor for Editable Audio Maps) [21]. An open-source software was developed, trying to present geographic data using spatial sound and texts spoken by the screen reader. Due to some inconveniences in use, map data import errors and failed tries of contacting project leader, author of this article decided to build in 2012 his own software from scratch, targeting similar goals, but much more sophisticated.

Grmapa – the talking map [22] utilizes map data available in two open-source projects: UMP (Universal Map of Poland [23]) and OpenStreetMap [24].

The UMP data is converted to Grmapa map format on a daily basis and is available for users to download.

OpenStreetMap data of interesting area can be downloaded on-demand and automatically imported to the program by user from the one of available OSM XAPI [25] servers.

After importing and opening Grmapa map, user can explore it. Virtual space presented by program is split into square boxes. Initially, the square size is 10 m. User faces north, and with cursor arrow keys can move one square in a chosen direction.

Square size (zoom level) and user direction, can be changed at any moment, using quick shortcut keys.

A name of the current square is automatically announced by the screen reader during navigation. The nearest squares, around the current user position, are illustrated by spatially placed sounds. There are about 30 different sound symbols, one for each most frequently used map object types, such as roads, footways, tram and train rails, forest areas, industry areas, food courts, shops, rivers, sea, and so on.

There are also generic "other point", "other line" and "other area" sounds, used when there are objects in surroundings that have not their individual sound defined.

Apart from described exploration mode, there is also a possibility of displaying all points, lines, or areas nearby, optionally filtered by some keywords. User can also search for any object in the whole map scope.

For easier navigation, user can put unlimited number of bookmarks on map, and quickly jump to selected bookmark. User bookmarks can be exported to file in Loadstone-GPS [26] format, and further used for a real outdoor GPS navigation. It is possible to use bookmarks to prepare a route, which user wants to follow.

Loadstone-GPS file format is used by other, modern software that enables blind users to navigate. Good examples are Ariadne GPS [27] and Seeing Assistant – Move [28].

Experimental feature is a possibility of connecting Grmapa with a GPS device, or playing NMEA 0183 [29] log file.

Grmapa is ready for user interface internationalization. Thank to that more blind people from around the world would be able to use it.

8 Production Engineering Models' Accessibility

There are different models in production engineering. They utilize different technics and methods to describe the matter. For ensuring the stability of systems artificial neural networks are used [30]. They are in use also as tools for controlling production systems and ensuring their stability [31]. Process coordination mechanisms can be modeled with business process notation [32, 33]. Maintenance processes can be modeled using lean thinking methods [34] such as Value Stream Mapping (VSM), 5S Practices, Standardization, Total Productive Maintenance. Technological processes are also described with neural models [37].

Every of above mentioned technics utilizes graphical notations. These notations are also subjects to optimization in terms of their accessibility. Screen readers would be able to properly read these models and provide them to blind people if the structure of their graphical representation is supported with appropriate description.

9 Spread Page in Business Modeling Accessibility Assurance

Above described possibilities in reading different types of notations for modelling tools have to be adopted in the Spread Page knowledge representations. There are different ways for presenting content, that include Three-dimensional Representation, Time-varying Representation, Layered Representation, Multi-resolution Representation, and Aspect-oriented Representation.

The Table 1 below shows how different technics for knowledge representation in Spread Page can be useful to address problems with delivering knowledge in model-to-blind-person and model-to-computer modes.

Table 1. Graphical modeling accessibility aspects that can be addressed by Spread Page

Graphical modeling accessibility aspects	Spread Page knowledge representations				
	Three-dimensional	Time-varying	Layered	Scaled Detail	Aspect-oriented
Reading static parameters		X	X	X	X
Announcing possible connections	X			X	X
Describing color marking			X		X
Summarizing and annotating			X	X	X
Element arrangement	X			X	X

10 Conclusions

The article outlines the use of particular methods of knowledge representation in accordance with the Spread Page Initiative assumptions for technical modeling in a way that assures accessibility for bind people. Such approach can be also useful in providing standards for machine-to-machine model exchange.

The research conducted shows that the ideal way for modeling would be to construct models as multi-layer objects that contain both graphical and descriptive representations. The main challenge in such an approach is model's ability to follow, on every layer, transitions between elements in a way it is done in graphical layer.

The paper describes the Spread Page approach to construct models that are understandable not only in their graphical layers. Authors take advantage of the experience of blind people working with IT modeling tools.

References

1. Lemmetty, S.: Review of Speech Synthesis Technology. Helsinki University of Technology (1999). http://research.spa.aalto.fi/publications/theses/lemmetty_mst/thesis.pdf
2. Afb.org.: Refreshable Braille Display - American Foundation for the Blind (2018). http://www.afb.org/info/refreshable-braille-display-3652/5. Accessed 10 May 2018
3. Linuxfoundation.org.: accessibility:iaccessible2:overview [Linux Foundation Wiki] (2018). http://www.linuxfoundation.org/collaborate/workgroups/accessibility/iaccessible2/overview. Accessed 13 May 2018
4. W3.org.: World Wide Web Consortium (W3C) (2018). http://www.w3.org/. Accessed 13 May 2018
5. Web Accessibility Initiative (WAI) (2018). http://www.w3.org/WAI/. Accessed 13 May 2018
6. Web Accessibility Initiative (WAI).: Web Content Accessibility Guidelines (WCAG) Overview (2018). http://www.w3.org/WAI/intro/wcag. Accessed 13 May 2018
7. Web.stanford.edu.: Captcha (2018). https://web.stanford.edu/~jurafsky/burszstein_2010_captcha.pdf. Accessed 13 May 2018
8. Web Accessibility Initiative (WAI).: WAI-ARIA Overview (2018). https://www.w3.org/WAI/intro/aria. Accessed 13 May 2018
9. W3.org.: W3C Document Object Model (2018). http://www.w3.org/DOM/. Accessed 13 May 2018
10. W3schools.com.: HTML div tag (2018). http://www.w3schools.com/tags/tag_div.asp. Accessed 13 May 2018
11. Nicomsoft.com.: Optical Character Recognition (OCR) - How it works (2018). https://www.nicomsoft.com/optical-character-recognition-ocr-how-it-works/. Accessed 13 May 2018
12. Latex-project.org.: LaTeX - A document preparation system (2018). http://latex-project.org/. Accessed 13 May 2018
13. W3.org.: W3C Math Home (2018). https://www.w3.org/Math/. Accessed 13 June 2018
14. Dessci.com.: MathPlayer (2018). http://www.dessci.com/en/products/mathplayer/. Accessed 13 June 2018
15. forte.fh-hagenberg.at: Hortsmann (2018). http://forte.fh-hagenberg.at/Project-Homepages/Blindenhund/conferences/granada/papers/HORSTMANN/horstmann.html. Accessed 13 June 2018
16. alasdairking.me.uk: TEDUB (2018). http://www.alasdairking.me.uk/tedub/index.htm. Accessed 13 June 2018
17. Plantuml.com: PlantUML (2018). http://plantuml.com/. Accessed 13 June 2018
18. Berkeley.edu: Spice (2018). https://ptolemy.berkeley.edu/projects/embedded/pubs/downloads/spice/index.htm. Accessed 13 June 2018
19. IEEE.org: Standard 1800-2017 (2018). http://standards.ieee.org/findstds/standard/1800-2017.html. Accessed 13 June 2018
20. IEEE.org: IEEE Explore (2018). https://ieeexplore.ieee.org/document/26487/. Accessed 13 June 2018
21. Sourceforge.net: TEAM Project (2018). http://sourceforge.net/projects/team/. Accessed 13 June 2018
22. Zlotowicz.pl: GRMapa (2018). http://grmapa.zlotowicz.pl/. Accessed 13 June 2018
23. UMP.waw.pl: UMP (2018). http://www.ump.waw.pl/. Accessed 13 June 2018
24. Openstreetmap.org: Open Street Map (2018). http://www.openstreetmap.org/. Accessed 13 June 2018

25. Openstreetmap.org: XAPI (2018). http://wiki.openstreetmap.org/wiki/Xapi. Accessed 13 June 2018
26. Loadstone-gps.com: LoadStone GPS (2018). http://loadstone-gps.com/. Accessed 13 June 2018
27. Ariadnegps.eu: Ariadne GPS (2018). http://www.ariadnegps.eu/. Accessed 13 June 2018
28. Seeingassistant.tt.com.pl: Seeing Assistant (2018). http://seeingassistant.tt.com.pl/en/move/. Accessed 13 June 2018
29. NMEA.org: NMEA Standard 0183 (2018). http://www.nmea.org/content/nmea_standards/nmea_0183_v_410.asp. Accessed 13 June 2018
30. Burduk, A.: The role of artificial neural network models in ensuring the stability of systems. In: Advances in Intelligent Systems and Computing, vol. 368. Springer (2015)
31. Burduk, A.: Artificial neural networks as tools for controlling production systems and ensuring their stability. In: Springer, Lecture Notes in Computer Science, vol. 8104. Springer (2013)
32. Grzybowska, K., Kovács, G.: The modelling and design process of coordination mechanisms in the supply chain. J. Appl. Log. **24**, 25–38 (2017)
33. Grzybowska, K.: Selected activity coordination mechanisms in complex systems. In: Highlights of Practical Applications of Agents, Multi-Agent Systems, and Sustainability - The PAAMS Collection Communications in Computer and Information Science, vol. 524, pp. 69–79. Springer (2015)
34. Jasiulewicz-Kaczmarek, M.: Practical aspects of the application of RCM to select optimal maintenance policy of the production line. In: Nowakowski, T., Mlynczak, M., JodejkoPietruczuk, A., et al. (eds.) Safety and Reliability: Methodology and Applications, Proceedings of the European Safety and Reliability Conference, ESREL Wroclaw, POLAND, 14–18 September 2014, pp. 1187–1195 (2015)
35. Wieczorek, T.: Neural Models of Technological Processes, Monograph. Publishing House of the Silesian University of Technology, Gliwice (2008)

The Essence of Reflexive Control and Diffusion of Information in the Context of Information Environment Security

Rafał Kasprzyk(✉)

Faculty of Cybernetics, Military University of Technology,
2 Urbanowicza St., 00-908 Warsaw, Poland
rafal.kasprzyk@wat.edu.pl

Abstract. In this paper, a particular emphasis is put on discussing the theory of reflexive control as a model for forecasting the decision-making process and influencing this process. Universal access to the Internet and the growing role of social media in shaping opinions, and more broadly the perception of the world, requires work to develop methods for identifying and recognizing reflexive control in social media.

The main task that is necessary to solve in order to identify and recognize reflective control processes and to prevent them is to understand the spread of information. The paper also presented the essence of models for the spread of information in network systems, exemplified by contemporary social media.

Keywords: Operation information · Reflexive control
Diffusion of information

1 Introduction

Dynamics of changes in every aspect of life of a modern man is associated with the role played by ICT systems, and widely Cyberspace. It is rightly noticed that cyber-space has a significant impact on potential threats to national security, existing both on the technical as well as information level. While awareness of threats at the technical level is systematically enhancing, and competences in this area were consolidated and managed within CERT teams, the awareness of threats at the information level is still scarce.

The main goal of any information operation is to interfere with the decision-making processes of an opponent (offensive perspective) and per contra do not permit to interfere with the self-decision-making processes (defensive perspective). One of the most advanced approach for modeling information operation is the theory of reflexive control introduced by Lefebvre in 60s [13]. Nowadays, this theory seems to be more applicable than ever before due to the way in which we obtain information and validate their sources of information. Universal access to the Internet and the growing role of social media in shaping opinions, and more broadly the perception of the world, lead to the creation of an unprecedented huge and decentralized information environment.

© Springer Nature Switzerland AG 2019
A. Burduk et al. (Eds.): ISPEM 2018, AISC 835, pp. 720–728, 2019.
https://doi.org/10.1007/978-3-319-97490-3_68

This observation will be even more prominent as modern, state-of-the-art ways of communication and collaboration platforms overtake legacy solutions. This information environment being in fact a part of cyberspace constitutes a novel and unbelievable fertile soil for reflexive control over a single person, a group of people or even whole society.

2 The Essence of Reflexive Control

2.1 Reflexion

The idea behind the reflexive control is the concept of reflexion [14, 19]. Beyond the objective reality, if it exists at all, there are only images in human minds, and these images vary. So, there might not be a universal and objective description of reality but many subjective descriptions. This phenomenon relates to the reflextion term, which is traditionally defined as the ability to take the position of an observer in regard to one's own beliefs (thoughts but also feelings) and principles of one's own actions. The core of reflexion is its recursive character. Having certain beliefs about reality, one constructs new beliefs by performing reflexion. As a matter of fact, the reflexion is an infinite process leading to one's own reflexive reality. Reflexion defined in this way is known as a self-reflection or reflexion of the first kind. The sense of the reflexion have been extended and generalized by Lefebvre toward the ability to take the position of an observer in regard to other person's beliefs (thoughts but also feelings) and principles of other person's actions (as the person beliefs the principles are). Reflexion defined in this way is known as reflexion of the second kind. Actually, this process of generalization of the reflexion term and recursive character of reflexion itself, lead to a so called hierarchy of realities (presented as a tree of images). In this way there is a possibility to define any finite or even infinite hierarchy of realities (see Fig. 1).

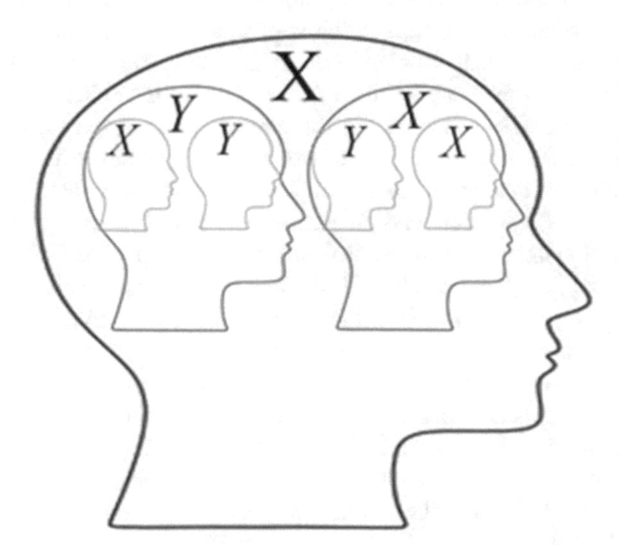

Fig. 1. Hierarchy of realities as outcome of reflexion process (based on [14]).

The largest outline of a face depicts a so called reflexive agent viz. an agent being able to perform reflexion. Let's assume that this agent X is in a relation (interact, play) with other agent Y (partner, opponent). The agent X performs reflexion in order to achieve an "image of the self" and an "image of the agent Y". It is depicted by two medium faces within the largest one. Obviously, the output of the agent X reflexion is also "image of the relations between agents". As Lefebvre indicates, their status of images exists only for the external observer (being also a reflexive agent), while for the agent X, they enact de facto the role of the "real me" and the "real other". For agent X, the images of the reality were depicted by four small faces. Right pair of faces are the agent X an "image of the self" and an "image of the agent Y", while left pair of faces are the agent Y an "image of the self" and an "image of the agent X", from the agent X point of view. Presented hierarchy of reality is a model of a situation awareness achieved via reflexion.

From formal point of you, reflexion is equivalent to modelling e.g. constructing images called models. Central for further consideration is that the result of reflexion as well as reflexion itself can be control in many cases and in many ways. This process is known as reflexive control.

2.2 Reflexive Control

Lefebvre consider reflexive control as a process by which one party (object, agent) transmits the reasons or bases for making decision to another [13, 26]. The reasons or bases are transmitted as a special prepared piece of (dis)information. Thus, reflexive control could be defined as a deliberate impact exerted on a partner or an adversary, so-called controlled object (any party performing decision-making processes), via (dis) information to incline a partner or an adversary to make the decision and in consequence required behavior predetermine by the control subject.

The Reflexive control, as defined, occurs when control subject is able to guide controlled object to make a decision which is profitable for that subject while the controlled object is convinced that decision itself is made autonomously and in its favor [26].

The most important presumption of reflexive control is to change the way how the decision-making processes of agents "in the game" are analyzed [12]. The traditional approach is to predict the agents' decisions in different situations, while reflexive control approach is to determine agents' decisions by affecting their perception of the situation viz. agents' reflexion and in consequence situation awareness of agents. Reflexive control demands from control subject the capability of interfering with reflexion of controlled object and it will be describe in more detail in next section.

It is worth paying attention to the fact that reflexive control can be considered on two distinct but strictly connected stages [19]. The basic stage is the information control in which control subject tries to control information reflexion and in the result control beliefs (thoughts and feelings) of controlled object. In this stage the controlled object yet make no decisions but in fact create their own situation awareness which is in certain sense projected by the control subject. The complementary stage is decision-making control. In this stage, the control subject tries to control strategic reflexion performed by the controlled object. The Strategic reflexion of a party (object, agent) is a reflexion in relation to establishing which decision-making principles parties (objects, agents) employ under the situation awareness achieved via information reflexion.

In the simplest case, there are only two parties i.e. control subject and controlled object. In general, there are many parties (objects, agents) performing simultaneously reflexive control, "playing" at the same time the role of control subject and controlled object, in the "peer-to-peer game".

2.3 The Framework of Reflexive Control Processes

At a high level of generality, the process of reflexive control is presented on Fig. 2. There is a control subject aiming to control or at least influence the decision-making processes of controlled object.

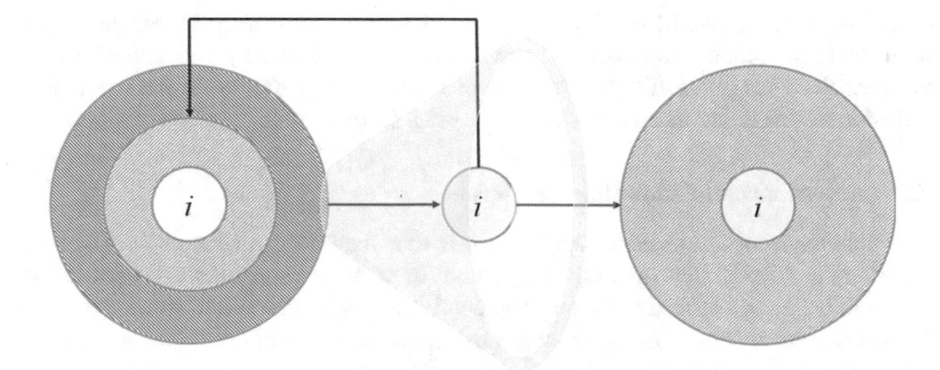

Fig. 2. The concept of reflexive control process (based on [12]).

First of all, to achieve this goal, control subject must have (got or built) a model of controlled object. Then, based on the model of controlled object, a special piece of (dis) information is produced by control subject. Finally, the piece of dis(information) is transmitted to control object. The last step is almost equivalent to receiving, by the control subject, a novel piece of information about the controlled object. Using the reflexive control process in this way, the control subject is able to influence the decision-making processes of controlled object, much better than if control subject did not send the piece of (dis)information.

2.4 Reflexive Control in Modern Strongly Connected World

Universal access to the Internet and the growing role of social media in shaping opinions, and more broadly the perception of the world, require an effort to develop methods for identifying and recognizing reflexive control in social media. The main task that needs to be solved in order to identify and recognize reflexive control processes and to prevent them, is to obtain the understanding of a way the information spreads. The models for the spread of information in network systems (exemplified by contemporary social media), should take into account at least two important determinants, i.e. underling network topology (who is connected to whom) as well as the behavior of agents on networks itself (how social media account manager adopts and disseminates (dis)information).

3 Modelling Social Media and Diffusion of (Dis)Information

3.1 Diffusion and Standard Approach for Modelling This Process

Diffusion is a process by which information, viruses, gossips and any other phenomena spread over networks in particular social networks.

Unfortunately, the standard approach is a simplified assumption that phenomena (information, viruses, gossips) spread in the environment which is modelled using a very simple construction of regular graph like GRID-based graph or similar, very rarely random graphs [4, 5]. In this way, standard approaches do not explain the real dynamic of diffusion in real-world networks, in particular: Why even slightly infectious phenomena (e.g. rumors, conspiracy theories, contagious diseases) can spread over a network for a long time; What is the role of particular agent/node for the dynamic of diffusion; What is the mechanism behind arising secondary phenomena centers.

3.2 Importance of Connections Patterns

The drawbacks of the standard approaches is that they do not take into account the underling real-world networks topology. Who (or what) is connected to whom (what), seems to be the fundament question. Apparently, networks derived from real life cases (most often: networks growing spontaneously), are neither regular graphs nor random ones. As it turned out, real networks, which have been intensively studied recently, have some interesting features [10, 18, 23, 27]. These features, whose origins are nowadays discovered, modelled and examined, significantly affect dynamics of the diffusion processes within real-world networks [8, 9, 11, 16, 17, 20] (Fig. 3).

3.3 Modelling of Social Media

In the late 90s very interesting and what is more important adequate model of real-world networks was introduced, a so called Scale Free network [1–3]. The Scale Free network model takes into account the fact that real networks are not static, but constantly evolving structures. The real networks "grow" by adding the following nodes, whereas new nodes are attached, with higher probability, to the nodes with higher degree. This type of behavior is known as preferential attachment, which states that the nodes are attached to the existing network according to the predefined hierarchy. There are numerous modifications of the basic algorithm for generating the Scale Free network and new modifications are made all the time. This reveals the growing interest in the complex networks. The modifications mainly refer to the change of the linear rules of preferential attachments into other (sometimes very complex) non-linear rules. Another idea is to include in the linear rule of preferential attachments the so-called initial "attractiveness" of the nodes or the "aging" effect of the nodes as well as a possibility of their deactivation. Additionally, the same network evolution algorithm is

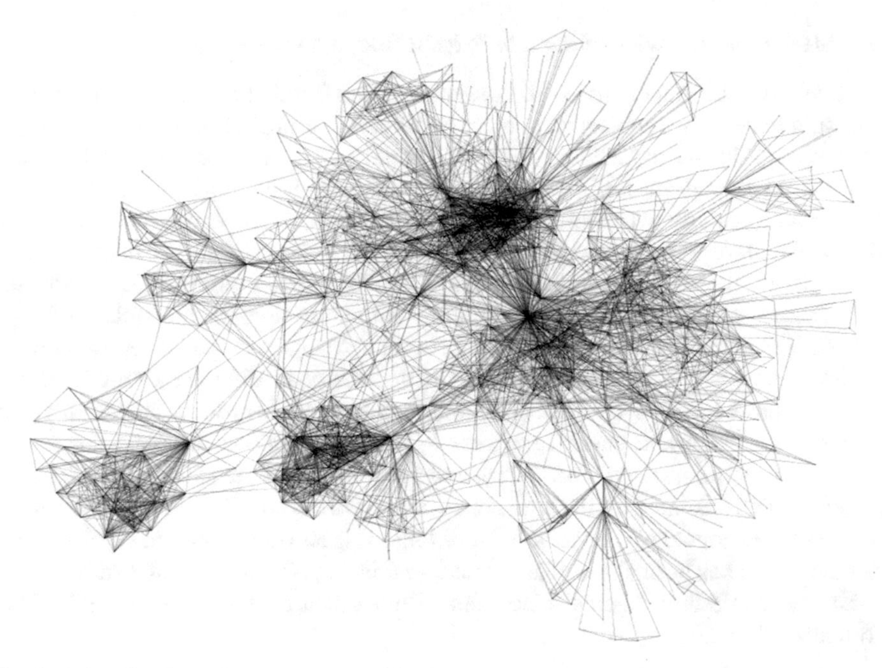

Fig. 3. Visualization of an extract of social interaction on popular social media platform [9].

modified in different ways. Thus, for example, at the following stages of evolution, we may have to deal not only with adding new nodes with new edges, but also with adding only new edges to the existing nodes or re-attaching some of the selected edges. These modifications quite well describe what is happening on social media (like Twitter, Facebook, Instagram).

In case of adaptation of the algorithm for generating the Scale Free network to the process of social media formation, a number of parameters could be included (e.g. geographical location, average "lifetime" of trolls or bots within the network before they are detected, temporary deactivation, etc.).

3.4 Identifying the Roles of Agents in Social Media

Beyond the underlying network topology of social media, the next basic question is: How to identify the most important agents/nodes. The answer can help maximize or on the other hand minimize diffusion dynamic of any phenomena within networks. No single measure of importance is suited for all applications [7, 9, 10, 24, 25]. Several noteworthy centrality measures of nodes are used to search for the key individual in social networks (e.g. degree, radius, closeness, betweenness) and to create rankings of websites (e.g. PageRank, HITS). Therefore, the centrality measures are often called significance measures. Thanks to these measures we can show for example how to disintegrate the network with minimum number of steps and in consequence minimize diffusion area and thus coverage of (dis)information being the product of an information operation.

3.5 Modelling Behavior of Agents Within Social Media

On a final note, it is important to draw attention to the fact that behavior of agents itself, being simultaneously a subject and object of diffusion, is crucial. In the context of (dis) information diffusion modelling on social media, the question one can ask is: How social media account manager adopts and disseminates (dis)information. All kinds of phenomena spreading over the network have their unique properties, and these properties if vital have to be taken into consideration. The notion of a state machine seems to be useful in this modelling situation [21, 22]. Using probabilistic finite-state machines, it becomes possible to model an agent's behaviors in the context of a vast variety of phenomena spreading over the networks. Many cases described in literature, concern contagious diseases, spreading in the population or different kind of malwares spreading across Internet. The idea behind probabilistic finite-state machines entails the identification of relevant states of agents (e.g. for contagious diseases: susceptible, infected, carrier, immunized, dead, etc.), and probabilities of transitions from a state to another resulting from social interactions (contacts). Again, the underlying contacts (social network topology) seems to have a huge impact on the dynamics of diffusion processes, what has been already mentioned. In this way, the concept of a state machine can be used to describe agents' behaviors within social media in the context of (dis) information diffusion.

4 Summary

In the paper, particular emphasis is put on discussing the theory of reflexive control as a model for forecasting the decision-making process and influencing this process. The theory of reflexive control is important for assessing the potential effects of information operations and their optimization. In this context, it is worth to realize a very close ties between reflexive control and advertising [6, 15].

Universal access to the Internet and the growing role of social media in shaping opinions, and more broadly the perception of the world, requires work to develop methods for identifying and recognizing reflexive control in social media. In the next step, it is also reasonable to develop models and methods to counter this type of phenomenon, and their implementation. However, this issue is debatable in democratic countries for obvious reasons. The problem in fact is how to distinguish an adversary information operation from a typical and legitimate public debate in democratic states.

The main task necessary to solve in order to identify and recognize reflective control processes and to prevent them is to understand the spread of information. The paper also presents the essence of models for the spread of information in huge network systems, exemplified by modern social media. Modelling social media and diffusion of (dis)information make possible to assess the role of individual "agents/nodes" in the social network and to forecast the dynamics of diffusion and coverage of (dis)information being the product of an information operation. Naturally, it is the very first step in analyzing reflexive control processes over the social media, yet required and essential.

References

1. Barabási, A.L., Albert, R.: Emergency of scaling in random networks. Science **286**, 509–512 (1999)
2. Barabási, A.L., Albert, R.: Topology of evolving networks: local events and universality. PRL **85**(24), 5234–5237 (2000)
3. Barabási, A.L., Albert, R.: Statistical mechanics of complex networks. Rev. Mod. Phys. **74**, 47–97 (2002)
4. Erdös, P., Rényi, A.: On random graphs. Publ. Math. **6**, 290–297 (1959)
5. Erdös, P., Rényi, A.: On the evolution of random graphs. Publ. Math. Inst. Hung. Acad. Sci. **5**, 17–61 (1959)
6. Godin, S.: Unleashing the Ideavirus. Hyperion, New York (2001)
7. Kasprzyk, R.: The vaccination against epidemic spreeding in complex networks. no. 3 (1/2009), pp. 39–43. Biuletyn ISI, Warszawa (2009). ISSN 1508-4183
8. Kasprzyk, R., et al.: CARE - creative application to remedy epidemics, no. 3(1/2009), pp. 45–52. Biuletyn ISI, Warsaw (2009). ISSN 1508-4183
9. Kasprzyk, R.: Diffusion in networks. J. Telecommun. Inf. Technol. **2**, 99–106 (2012)
10. Kasprzyk, R.: Complex systems evolution models and methods to investigate their characteristics for computer identification of potential crises (Polish title: Modele ewolucji systemów złożonych i metody badania ich charakterystyk dla potrzeb komputerowej identyfikacji potencjalnych sytuacji kryzysowych), Ph. D. thesis, Military University of Technology, Warsaw (2012)
11. Kasprzyk, R., Paź, M., Tarapata, Z.: Modeling and simulation of botnet based cyber-threats. In: MATEC Web Conferences, vol. 125, p. 03013 (2017)
12. Kramer, X.H., Kaiser, T.B., Schmidt, S.E., Davidson, J.E., Lefebvre, V.A.: From prediction to reflexive control. Int. Interdiscip. Sci. Pract. J. Reflex. Process. Control **2**(1), 86–102 (2003)
13. Lefebvre, V.A.: Basic Ideas of the Reflexive Games Logic, Problems of Systems and Structures Researches. USSR Academy of Science, Moscow (1965). (in Russian)
14. Lefebvre, V.A.: Sketch of Reflexive Game Theory, School of Social Sciences. University of California (1998)
15. Leskovec, J., Adamic, L., Huberman, B.A.: The dynamics of viral marketing. ACM Trans. Web **1**(1), 39 (2007). Article 5
16. Lloyd, A.L., May, R.M.: How viruses spread among computers and people. Science **292** (5520), 1316–1317 (2001)
17. López-Pintado, D.: Diffusion in complex social networks. Games Econ. Behav. **62**(2), 573–590 (2008)
18. Newman, M.E.J.: The structure and function of complex networks. SIMA Rev. **45**(2), 167–256 (2003)
19. Novikov, D.A., Chkhartishvili, A.G.: Mathematical models of informational and strategic reflexion: a survey. Adv. Syst. Sci. Appl. **14**(3), 254–278 (2014). ISSN 1078-6236
20. Pastor-Satorras, R., Vespignani, A.: Epidemic spreading in scale-free networks. PRL **86**(14), 3200–3203 (2001)
21. Sokolova, A., de Vink, E.P.: Probabilistic automata: system types, parallel composition and comparison. In: LNCS, vol. 2925, pp. 1–43. Springer, Heidelberg (2004)
22. Vidal, E., Thollard, F., de la Higuera, C., Casacuberta, F., Carrasco, R.C.: Probabilistic finite-state machines – part I. IEEE Trans. Pattern Anal. Mach. Intell. **27**(7), 1013–1025 (2005)
23. Strogatz, S.H.: Exploring complex networks. Nature **410**, 268–276 (2001)

24. Şen, F., Wigand, R., Agarwal, N., Tokdemir, S., Kasprzyk, R.: Focal structures analysis: identifying influential sets of individuals in a social network. Soc. Netw. Anal. Min. **6**, 17 (2016). ISSN 1869-5450 (Print), 1869-5469 (Online). Springer https://doi.org/10.1007/s13278-016-0319-z
25. Tarapata, Z., Kasprzyk, R.: Graph-based optimization method for information diffusion and attack durability in networks. In: LNCS, vol. 6086, pp. 698–709. Springer, Heidelberg (2010)
26. Timothy, L.T.: Russia's reflexive control theory and the military. J. Slavic Military Stud. **17**(2), 237–256 (2004). ISSN 1351-8046
27. Wang, X., Chen, G.: Complex networks: small-world, scale-free and beyond. IEEE Circuits Syst. Mag. **3**(1), 6–20 (2003)

The Supply Chain of the Future – Intelligent and Sustainable Supply Chain Management

Flexibility Strategy in Delayed Differentiation Model of Steel Products

Marzena Kramarz(✉)

Faculty of Organization and Management, Silesian University of Technology, Akademicka 2A, Gliwice, Poland
mkramarz@polsl.pl

Abstract. Distribution enterprises that implement delayed product differentiation establish network relations, invest in flexible resources, accumulate inventory of base products, ensure flexible transport. The aim of this paper is to identify the sensitivity of flexible distribution system with deferred production to the demand fluctuations. The flexibility of distribution system was analyzed in three forms: production infrastructure flexibility, sourcing dedicated resources from a cooperator based on formal relations and sourcing production capacity from the cooperator based on informal cooperation. The research was carried out in the steel products distribution sector. The flagship enterprise of distributions channels were identified, that organize the flows of steel pipes. A flagship enterprise in steel products distribution sector carries out deferred production tasks, selecting machinery flexibility levels and shaping network relations with sub-contractors. Structured interviews were carried out, resources strategies were indicated on the basis of collected primary and secondary data, a reliability indicator was defined and simulation experiments in systems dynamics technique were carried out. The results of the research carried out allowed for the conceptualization of flexible resources strategy of the main link of distribution channels and to direct future research on variables not included in the model, especially disruptions in different models of flexible resources strategy and atmosphere of network cooperation.

Keywords: Flexible resources · Delayed differentiation · Reliability of orders Simulation distribution network

1 Introduction

Identification and measurement of key factors influencing the flexibility of enterprises cooperating within distribution networks is important due to the progressing products customization, changing demand of customers that more and more often concern not only the form of a product, but also its place and time of availability. Delayed product differentiation is a strategic concept identified by scientists both as a chance to strengthen the supply chain resilience and to improve the flexibility of production and distribution system that carries out customized orders in terms of product form adjustment. This concept requires the deferred production strategy to be carried out in selected distribution channel links. Distribution enterprises that implement delayed product differentiation

© Springer Nature Switzerland AG 2019
A. Burduk et al. (Eds.): ISPEM 2018, AISC 835, pp. 731–741, 2019.
https://doi.org/10.1007/978-3-319-97490-3_69

establish network relations, invest in flexible resources, accumulate inventory of base products, ensure flexible transport. The aim of this paper is to identify the sensitivity of flexible distribution system with deferred production to the demand fluctuations.

The research was carried out in the steel products distribution sector. A focus has been put on a selected product – steel pipes and operations related with cutting and chamfering of pipes. Both of these activities are carried out in service center (commercial company delayed with deferred production). In the decisive variants, four types of machinery of varied levels of flexibility were included. The operation duration time of each machine was measured, as well as measurement of operation implementation time in the variant with sub-contracting. Simulation experiments in management system dynamics technique were carried out (Vensim DSS). The sensitivity of three variants of delayed product differentiation to the demand fluctuations was analyzed out.

2 Flexibility of Distribution System in Delayed Product Differentiation Model

Flexibility, both of enterprises and of whole supply chains is considered a crucial element in the turbulent environment. When approached intuitively, flexibility means the ability to react to various abnormal behaviors. Krupski [6] views flexibility as a category encompassing two dimensions of spacetime: the reaction (or creation) speed and the level of match for every organization element separately and for all of them. He also pointed out that flexibility is understood as reaction to impulses either from the environment or from the inside of the company and is used when confronted with resources potential of an organization. Basic tools for ensuring flexibility are redundancies of selected resources (keeping them in case an opportunity emerges) on the one hand, and on the other hand diversification of activities and resources that allows to maintain the ability to continue the operations in the discontinuous turbulent environment. Flexibility of an enterprise in an attribute approach refers to an ability of a system that is the result of its key competences and resources to react in a given time to customized customers' needs. In literature on logistics, flexibility is defined as and ability to react to customized customer's order concerning the form of product, implementation time or requested production batch [2]. Modern research in the scope of manufacturing and logistics flexible systems shaping underline the importance of network relations as being a key resource influencing the configuration of flexibility components. Therefore, flexibility should be considered in a wider scope, in reference to supply chain or distribution networks, sourcing networks or production networks. Flexibility of a supply chain is the ability of an enterprise to manage its processes, deliveries and customers in such a way that will allow to effectively response to sudden changes in deliveries, product and demand [8, 11].

Gattorna [4] indicate four basic (covering 80% of customers) configurations of supply chains: continuous replenishment supply chain, lean supply chain, agile supply chain and fully flexible supply chain. The selection of the appropriate configuration is subject to the type of demand and product. The greater the demand uncertainty, meaning its volatility and unpredictability, the stronger the supply chain should be shaped towards agile supply chain. In turn, in case of product what matters are – among others – the level of its standardization, value, as well as the duration and stage of its life cycle [10]. Agile supply chains are created for products of rich versatility and value as well as of short life cycle and for those introduced to or recalled from the market. Rich versatility is understood as availability of multiple variance obtained thanks to customization of products on various stages of value adding.

Product customization may take place during early products differentiation and delayed product differentiation variants that are carried out in distribution channels [1]. Delayed product differentiation can be carried out in deferred production variant and in module products variant [7]. Postponement production is understood as postponement in time of the last stage of production process until an order is placed [3]. This variant requires inventory of base products (not differentiated yet) in the central chain (differentiation point). Module products are yet another answer to the need to decrease the costs of products customization. In this variant, production enterprise manufactures different variants of modules of a given product. The central chain of distribution channels is responsible for maintain appropriate levels of modules inventories and for configuration of a product in line with the incoming customer's order. In this article, the emphasis is put on the first variant, analyzing the flexibility of a distribution system created by central chain of distribution channel. In terms of flexibility, the resources strategy was analyzed that takes into account both flexible infrastructure and network relations. Thus, the examined distribution system was analyzed at the microsystem level (central organization's resources in distribution channel) as well as at the level of meta-system (relations established vertically within distribution channel and horizontally in distribution network). By referring to model configurations of supply chain by Wieteska [10] and Gattorna [4], in the research carried out the focus was not fully on flexible supply chains that are addressed for humanitarian logistics, so crucial in crisis management, but on agile supply chains, the flexibility of which is obtained by combination of resources strategy of the central link of distribution channel carrying out deferred production and creating network relations.

3 The Concept of Distribution System Flexibility Research

The problem discussed in the article is a complex one and covers the analysis of cause and effect relations between flexibility, delayed differentiation, relational resources, central link infrastructure flexibility and reliability. The concept of research combines the usage of internal potential of an organization to create flexibility with the use of relational resources for that purpose. The research was carried out in the steel products distribution sector. The process of pipes flow in distribution channel was analyzed, taking into account their customization in service center. Own resources of the service center (being the central link in steel products distribution channels) are often insufficient for the purpose of comprehensive implementation of more and more complex customers' orders [9]. Therefore, an important element that strengthens the competitive advantage in terms of flexibility are both the relations within the chain and in the supply network. The core of cooperation within networks and supply chains is the sharing of risk and costs between partners or a co-operative covering complementary competences, technologies or resources. In this context, both at the level of network structures and at the level of chain structures, it is worth differentiating between informal (more flexible) relations and formalized (increasing resilience) ones. The type of distribution channels central link indicated in the research – service center – is a flagship example that allows for an analysis of all identified components of research problem.

Product customization in distribution channels requires flexible systems that are resilient to disruptions and consistent in reliable delivery of product indicated by the customer to the right place at the right time [5]. Reliability is here understood as the ability to implement complete and dependable orders on time.

$$reliability = \frac{punctuality + completeness + dependability}{3} \quad (1)$$

where:

Punctuality = (orders implemented on time/total orders) * 100
Completeness = (completely implemented orders/total orders) * 100
Dependability = (orders implemented without errors/total orders) * 100

The first stage of research was to indicate the resource strategy of a service center implementing the postponed production tasks. Research in this scope was carried out in 12 distribution enterprises dealing in deferred production of steel pipes. The total population comprised 17 organizations being members of Polish Union of Steel Distributors that met the assumptions of steel pipes differentiation point in distribution channel. In organizations that agreed for the research, the formalized direct interviews were carried out. Organizations participating in this stage of research, provided data that allowed to calculate the reliability of implemented orders. On this basis, flagship strategies of central organizations in distribution channels responsible for differentiation of steel pipes were indicated.

Next, an analysis of changes in the resource strategy of the central distributor due to the demand fluctuations was carried out. The flagship strategies were complemented with the data obtained by way of simulation experiments, indicating limit values of parameters that allows to obtain results that increase the reliability level. The problem of central enterprise of distribution network of deferred production adjustment to changes in the environment, taking into account the flexibility of resources and cooperation within a network, is within the character of system dynamics. The application of system dynamics allows to indicate the changeability of studied parameters, depending on demand fluctuations at the same time by indicating the feedback loops. The Vensim DSS used for the purpose of modeling is a product that allows to create simulators for complex systems of varied interactions between components and to include significant number of variables. This tool also allows for using constant and secret delays templates and for taking into account the smoothing of demand fluctuations, so important for this research, by indicating the achievement of system balance by adjusting variables defined by the researchers.

4 Assumptions of Flexible Implementation Strategy Modeling of Steel Pipes Orders in Delayed Product Differentiation

In the designed strategy variant, emphasis was put on flexibility of infrastructure necessary to implement deferred production tasks in differentiation point and on relational resources that allow to decrease the resource flexibility at differentiation point by way of implementation of selected operations on dedicated sub-contractor's resources.

During the research, emphasis was put on a selected product: steel pipes and on operations associated with cutting and chamfering of pipes. In the decisive concepts, four types of machinery were included:

- machinery for simultaneous cutting and chamfering (these devices are designed for simultaneous cutting and chamfering of pipes of varied diameters and wall thickness as well as of material the pipes are made of. It is the highest level of flexibility of a device that allows for maximum product differentiation, flexibility level = 4)
- machinery for pipes chamfering (beveling machines make it possible to prepare various types of pipes for welding. They carry out the following operations: chamfering, planning, rolling. These machines are characterized by average level of flexibility, level 3)
- machinery for pipes planning (these machines are mostly intended for planning pipes of thin walls (made of stainless steel). It is also possible to plan, chamfer and roll pipes made of carbon steel. These machines are characterized by average level of flexibility, level 2)
- high performance machines for pipes chamfering (intended for work at high production loads. Thanks to the use of tools made of carbide, the processing time was reduced to minimum. It is a dedicated machine of the lowest level of flexibility and the shortest time of operation implementation, flexibility level = 1)

Basic enterprise (differentiation point – service center) implements the orders of key recipients from four different industries. Each industry places order for different product variants. The recipients indicated the following elements as key elements of logistics customer service [6]: punctuality, completeness and dependability. Within the same model, the reliability parameter was included meaning the percentage of complete orders implemented without any errors within the deadlines stipulated in the agreement as compared with all orders placed. Lost sales concern orders that were not implemented. The cost of lost sales was assessed according to unit margin of a product. Moreover, the results of measurements of processes implemented in the flag enterprise and industry analyses of statistical data allow to assume the following theories that were initially taken into account in simulative model:

- Along with the increase of machinery flexibility, the time of operation implementation as well as costs associated with its acquisition increase.
- Lowering the level of machinery flexibility results in extending of cooperative relations and not in resignation of part of the orders.

- Presented results of experiments concern average demand fluctuations up to 50%. Resources and assortment production plan takes into account daily machinery load taking into account the accumulated production plan for various product variants.

In a model that from the very beginning simplifies the decision problems to criteria analyzed by the researches, the logistics costs associated with the transport of product from cooperator to the warehouse of finished products of a material decoupling point, finished products storage costs in material decoupling point as well as downtime costs of a resource in material decoupling point were taken into account.

Stage 2 is implemented with the use of a tool to develop simulative model in management system dynamics technique. The division into assessment of formal cooperation (cooperation agreements) and informal cooperation, indicated in the previous stage, was taken into account in the research. In the analyzed resources strategy, additional characteristics of resources were taken into account such as substitutability of resources in which the central link invests, the scarcity of resources, number of different types of cooperators, number of various motives for cooperation were taken into account. The research carried out [6] narrowed down the analysis of resources strategy to machineries flexibility and form of relation establishing with a cooperator, and allowed to make detailed assumptions concerning the flexibility research.

The research carried out in the scope of resources strategy of central enterprises of steel products distribution networks indicate that distributors own flexible machinery that allow to adjust the product to the needs of many recipients' industries as well as rare devices that distinguish flagship enterprises among other steel products distributors and that allow them to create competitive advantage (average ratio of devices rarity at the level of wsprzm = 0.2[1]). Moreover, flagship distributors focus on selected, narrowed down aims of cooperation (average number of different motives behind cooperation at the level of Licztyp mot = 2.56[2]) thus building network relations with a small group of partners (average number of various types of cooperators: licztyp koop = 2.9[3]) as a result increasing, by way of formal cooperation, the potential in a more complimentary way but not to such extent as in the case of informal cooperation (coefficient of substitutability of resources sourced by cooperation under cooperation agreement at the level of Submasz F1 = 0.4[4]). Complimentary resources are sourced mainly by cooperation on the basis of informal cooperation (coefficient of substitutability of potential increased by informal cooperation at the level of Submasz F2 = 0.3).

[1] Coefficient of machinery rarity amounts to (0; 1> where wsprzm = 0 – rare machines owned by only one enterprise, 1 – widely available machines owned by all enterprises in the database.

[2] Number of different motives for cooperation can take values in the range <1; 6>.

[3] Number of different cooperators' types can take values in the range <1; 6>.

[4] Coefficient of substitutability of increased potential can take values in the range (0; 1> where: SubsZPP = (0 – potential increased solely complementarily; 1 – potential increased solely by way of substitution).

The results of these research translated into the subsequent assumptions made in this paper:

- Features of resources held by central distributor strongly influence the assessment of elements of logistics customer service (and also the reliability parameters) in case of informal cooperation (R = 0.98) in case of high level of relevance p = 0.02.

The increase of machinery flexibility coefficient influences positively the reliability level of orders implemented with the use of informal cooperation (canonical weight 0.73). On the other hand, increase of machinery flexibility coefficient strongly decreases the assessment of order implementation time (canonical weight –0.76). Central enterprises dealing with products differentiation in distribution channels of steel products owning rare resources (decreasing coefficient of resources rarity), selecting partners on the basis of informal relations are not able to ensure short time of orders implementation (availability of partners' resources is not guaranteed by the agreement) but for the defined standards on the basis of own production capacity extended with the use of informal cooperation, they offer reliable and complete order implementation.

- Flexibility of machinery is positively correlated with the level of reliability of implemented orders within networks based on formal cooperation. A stronger influence was observed in case of coefficient of substitutability of machines in which the central link invests (R: 0.82; p = 0.82).

In general, the features of resources held by central enterprise, such as increase of coefficient of substitutability of resources as well as increase of flexibility coefficients and rarity of machinery strengthen the positive influence of formal relations on the decrease of orders implemented past the deadline.

- Features of network created by central enterprise on the basis of informal relations influence positively the reliability of implemented orders (R: 0.85; p = 0.88).

The correlation between these two canonical variables is influenced the most by the number of various types of cooperators included in the network, as well by substitutability of resources sourced by way of informal cooperation. Central enterprises assess that increase of variety of cooperators in distribution channels with simultaneous domination of informal relations in the network weakens such elements of service as time and reliability, increasing at the same time the market penetration.

- Assessing the influence of features of network shaped by the central distributor on the assessment of logistics customer service elements for the cooperation on the basis of cooperation agreement a very high correlation coefficient R was obtained (R: 0.96; p = 0.10).

The highest canonical weight was assigned to coefficient of substitutability of resources obtained as a result of informal cooperation (negative canonical weight). Interpretation of this result allows to state that if the informal cooperation generates substitutionary resources for flagship distributors and cooperation based on cooperation agreements generates complimentary resources, this weakness slightly the positive influence of cooperative cooperation, especially on such elements as order implementation time and width of distribution channels, and strengthens the assessment of reliability and implemented orders completeness improvement being the result of the cooperation based on cooperation agreements.

5 Strategy of Flexible Reaction to Demand Changes

To assess the effectiveness of the adopted variant of resources strategy implementation by the central link of distribution network implementing the tasks of deferred production, the reliability indicator was used. It means that in the model assumptions not the shortest order implementation time was taken into account but meeting the terms and conditions of the agreement signed with the customer concerning punctuality, completeness and dependability. An implemented order is about complete product delivery to the customer in a time indicated by them. In the simulation model:

- The cycle of order implementation is defined on the basis of negotiations between the central enterprise and the customer; it takes into account the production cycle in the central chain, transport cycle to the cooperator, production cycle at the cooperator, transport cycle from the cooperator to the warehouse of finished products of the central distributor, delivery cycle to the customer.
- Transport cycle is the transport time from central distributor's warehouses to the cooperator and from cooperator's warehouse to the central distributor's finished products warehouse.
- The model takes into account one product (steel pipes) and different levels of its differentiation, therefore, delays in order implementation are accumulated, causing a "queue of waiting customers".
- Due to the implementation of orders based on pull system, a variable concerning customers' orders plan was developed, where: Customers orders plan takes into account both quantitative data concerning orders, as well as implementation time set in the agreement.
- Order implementation time in each analyzed case requires the use of complimentary resource. The pull system requires to take into account sales plan in the demand corrected in the next step by the available production capacity that allows to prepare an actual sales plan that takes into account schedule of orders implementation.

Simulation experiments were aimed to indicate a decisive variant concerning the level of flexibility of a given resource that is most rational in terms of reliability indicator and logistics costs at demand fluctuations up to 50%. A rational selection of flexibility level assumed that having high level of reliability indicator (above 0.95), a key criterion deciding on the variant selection are logistic costs. Efforts were made to choose decisive variant that ensures the lowest costs and reducing the level of unimplemented orders. Three decisive variants were analyzed. The first variant concerned the purchase made by central enterprise of distribution network of a resource of the highest level of flexibility (flexibility = 4, device for cutting and chamfering pipes). Flexibility of this resource makes it possible to implement all orders in the central link of the network. In the second variant, it was assumed that the enterprise implements 70% of orders it receives and 30% is subcontracted at a subcontractor who has a dedicated resource at their disposal (strong specialization, flexibility at level 1, device 4). In this variant, the central enterprise has a machinery of level 2 or 3 that allows for complete implementation of orders of some of the customers and creates with the sub-contractor, having a dedicated resource, formal relations. This cooperation is based on cooperation agreement where the availability of cooperator's resources is defined for the needs of tasks assigned by the central enterprise (variant 2) or for the needs of informal cooperation (variant 3). Table 1 presents summary of experiments results carried out for resources flexibility strategy of the central enterprise.

Table 1. Summary of simulation analysis results

Criterion	Demand fluctuations	Cooperative agreement	Informal relations	Flexible resources
Lost sales	Stable (up to 20%)	Low	Low	Low
	Unstable (more than 20%)	Low	Average	High
Warehousing costs at the cooperator	Stable (up to 20%)	Low	Average	–
	Unstable (more than 20%)	Low	Low	–
Warehousing costs at the central enterprise	Stable (up to 20%)	Low	Average	Low
	Unstable (more than 20%)	Average	Average	High
Transport costs	Stable (up to 20%)	Average	Average	Low
	Unstable (more than 20%)	Average	High	High

Experiments confirm the results of the research carried out in the previous stages. In stable conditions of insignificant turbulences in the environment, the distributor carrying out flexibility strategy the implementation of which requires complimentary assets will try to source these assets to own them, or – if the level of orders is lower or if the complimentary resource is rare – it will try to establish strong relations with an enterprise owning such an asset. Therefore, for stable orders where the fluctuations do not exceed 20% it is always more profitable to cooperate on the basis of cooperative agreement which limits the delays of orders implementation.

The experiments carried out also indicated that with the increase of demand fluctuations, maintaining high reliability (reliability indicator above 0.95) the logistics costs of investment variant into fully flexible resource increase. It is associated with the prolonged cycle of order implementation and resources securing various product variants, thanks to which the continuity of sales is maintained. Demand fluctuations above 30% with maintaining a high level of customer service indicate lower logistics costs for 2 variant including cooperator having at its disposal a dedicated resource, no matter the cooperation form.

6 Conclusions

The increase of flexibility importance is proportional to the increase of uncertainty and unpredictability of business environment. In this context, one must understand that the approved flexibility dimensions are not fixed and do not concern only the current factors. Flexibility must be analyzed in dynamic approach, complementing the flexibility with new so far omitted factors (unpredictability). The obtained results indicate that a flagship distributor, by way of the manner of resources configuration influences the results that are a proof of its effectiveness. At the same time, it can be noticed that there is a simultaneous influence of market conditions (measured by demand fluctuation) and material and relational resources configuration on the effectiveness of central distributor measured by the reliability of implemented orders. The simulation experiments carried out do not take into account the full limitations concerning the shaping of cross organizational relations resulting from the cooperation or disruptions present in the system and influencing the resilience of the whole supply chain. The research will be extended by these aspects. The developed simulative models should be treated as initial conceptualization of systems supporting central distributor's decision making within the material and relational resources configuration to satisfy the needs of different recipients' segments.

References

1. Anand, K., Girota, K.: The strategic perils of delayed differentiation. Operations, Information and Decisions Papers, No. 5, University of Pennsylvania Scholarly Commons, pp. 1–32. http://ssrn.com/abstract=1574130 (2007)
2. Bozarth, C., Hanfield, R.: Introduction to Operations and Supply Chain Management. One Press, Warszawa (2007)
3. Cheng, T., Li, J., Wang, S.: Postponement Strategies in Supply Chain Management. International Series in Operation Research & Management Science. Springer, New York (2010)
4. Gattorna, J.: Dynamic Supply Chain Alignment. Gower Publishing Limited, MPG Books Group, Farnham (2009)
5. Kramarz, M.: Adaptation Strategy of Flag Enterprises of Distribution Network with Postponement Production. Steel Products Distribution. Publishing of Silesian University of Technology, Gliwice (2012)
6. Krupski, R.: Company's flexibility. Publishing of Wrocław University of Economics, Wrocław (2008)
7. Mutha, A., Pokharel, S.: Strategic networks design for reverse logistics and remanufacturing using new and old product modules. Comput. Ind. Eng. **56**(1), 334–346 (2009)
8. Sajad, F., Zutshi, A., O'Loughlin, A.: Developing an analytical framework to assess the uncertainty and flexibility mismatches across the supply chain. Bus. Process Manag. J. **20**(3), 362–391 (2014)

9. Saniuk, S., Samolejova, A., Saniuk, A., Lenort, R.: Benefits and barriers of participation in production networks in a metallurgical cluster – research results. Metalurgija **54**(3), 567–570 (2015)
10. Wieteska, G.: Effective responses to disruptions – flexibility supply chain. Research Papers of Wrocław University of Economics, vol. 382, pp. 143–153 (2015)
11. Yu, K., Cadeaux, J.: Flexibility and quality in logistics and relationships. Ind. Mark. Manag. **62**(1), 211–225 (2017)

Key Competencies of Supply Chain Managers – Comparison of the Expectations of Practitioners and Theoreticians' Vision

Katarzyna Grzybowska[1(✉)] and Anna Łupicka[2]

[1] Faculty of Engineering Management, Chair of Production Engineering and Logistics, Poznan University of Technology, Strzelecka 11, 60-965 Poznan, Poland
katarzyna.grzybowska@put.poznan.pl
[2] Poznan University of Economics and Business, Al. Niepodległości 10, 61-875 Poznan, Poland

Abstract. This paper attempts to answer the following question: what competencies seem essential for future managers of supply chain? The pharmaceutical and automotive sector were selected for the purpose of study. Both sectors are oriented toward ongoing improvement of competencies. In the article a comparative analysis of the expectations of practitioners and visions of scientists and theoreticians was carried out.

Keywords: Future competencies · Smart supply chain
Sustainable supply chain · Multimodal networks

1 Introduction

The supply chains are in a state of transformation. Sustainability will be of paramount importance in supply chain. A new wave of factory automation (Industry 4.0 and Logistics 4.0) will be supported by the next generation of low-cost robotics. The transformation of the supply chain of today to the supply chain of the future is an enormous task. The better we understand the future needs, the better smart and sustainable supply chain will function.

This paper is to answer the following question: what competencies seem necessary for future managers of supply chain? The answer is crucial for research and teaching centers, whose aim is to educate future managers at the highest level of the specific competencies and skills.

The paper is divided into two main parts. The first main part is a theoretical introduction to the identification of future supply chain.

The second main part is a theoretical introduction to identification of core managerial competencies. It presents three basic categories of competence: technical, managerial and social. In next part presents the methodology of the research conducted and shows and interprets the results.

© Springer Nature Switzerland AG 2019
A. Burduk et al. (Eds.): ISPEM 2018, AISC 835, pp. 742–751, 2019.
https://doi.org/10.1007/978-3-319-97490-3_70

2 Supply Chains of the Future – Intelligent and Sustainable

Supply chains are inherently complex and dynamic systems [1]. In web-based global business arena witnessing Industry 4.0, collaboration across SC partners has to be smart, innovative and socially responsible to form value-creating networks. So, we see a different kind of supply chain emerging. We envision a supply chain of the future: (1) The supply chain will be an incredibly complex and dynamic. Integration of sustainability principles will increase complexity. (2) Information will increasingly be machine-generated (instrumented). (3) The whole supply chain will be connected. The supply chains take interaction (with customers, suppliers and IT systems in general). We see a more holistic view of the supply chain, this extensive interconnectivity will also facilitate collaboration on a massive scale (interconnected). Enterprises have to react very quickly to challenges and opportunities of the business world [2].

Recent research emphasizes the role of advanced ICT. Ivanov and Sokolov [3] consider the next generation supply chain as a cyber-physical system (CPS). Thoben, Wiesner, and Wuest (2017) discussed the main characteristics of cyber-physical systems [4]. Gaynor et al. [5] proposes a smart supply chain using wireless sensors. Bendavid and Cassivi [6] suggest that the higher-level integration of inter-organizations, and e-commerce within a self-managed smart supply chain.

Butner [7] claims that a smarter supply chain would have the analytic capability to evaluate myriad alternatives in terms of supply, manufacturing, and distribution – and the flexibility to reconfigure flows as conditions change. Butner suggest that smarter supply chain has three properties: (1) instrumented – more data in a supply chain would be generated by various devices such as sensors, RFID, and actuators; (2) interconnected – more objects in a supply chain would be extensively connected and lead to massive collation among them; and (3) intelligent – more intelligent systems would help people by making real-time decisions and predicting future events. Recently, Wu et al. [8] added three additional characteristics, including (4) automated, (5) integrated and (5) innovative.

Sustainability will be of paramount importance in supply chain. The goals of sustainability efforts will be materials, manufacturing processes, energy and pollutants attributed to manufacturing and logistics. Sustainable Supply Chain Management (SSCM) is defined as the strategic, transparent integration and achievement of an organization's social, environmental, and economic goals in the systemic coordination of key inter-organizational business processes for improving the long-term economic performance of the individual company and its supply chains [9].

Recent research emphasizes the role of Replacing the sustainable demand chain management (SDCM). Ganji et al. [10] claims that SDCM with supply perspective, sustainability will be integrated into activities beyond the main supply chain concepts such as involving customers to management of by-products generated after home or industrial use and product life extension [11].

There are a number of enablers affecting the implementation of sustainability in the Supply Chain. The analysis ISM-based model [12] reveals that one of six enablers 'Commitment from top management' is ranked as Independent enabler as they possess the maximum driver power. The most important among others them is 'Commitment from top management'. This means that the competences of supply chain managers are an important element of management.

3 Competencies of Future Managers

The strategic revolutions in today's rapidly changing business environment clearly mandate a new set of future competences. The term "competency" has been defined variously, though a common theme has emerged that includes knowledge, skills, attitudes, and behaviors necessary to meet complex demands of a task in the particular context [13, 14]. Personal competence and being capable of performing a job in a way that generates added value are development factors which make the individual an attractive job candidate. Global connectivity, smart machines and industry 4.0 are some of the drivers reshaping how organizations think about work, what constitutes work, and what competences do they need to be efficient contributors in the future.

Individuals need a wide range of competencies in order to face the complex challenges of today's world [13]. The study is focused on exploring the technical, managerial and social competencies of future managers. Based on existing studies and analyses, a total of eight competencies were identified [15–27]. Competences, discussed and selected for analysis, are presented below (Tables 1, 2 and 3).

Table 1. Technical competencies [14]

No.	Competencies	Description
1	IT knowledge and abilities	Information technology is a growing field. Commonly referred to as IT, there are many job titles in the field
2	Knowledge Management	Knowledge management does not just happen. To ensure that organizations can acquire, create, organize, share, use and build on the knowledge
3	Computer programming/coding abilities	One of the basic skill sets an employer will look is the ability to write code; takes more than just proficiency with the coding language, integrating different technology, and having a broad understanding of information systems
4	Data and information processing and analytics	Data science and analytics are among the most in demand and fast growing disciplines. However, due to the fact that the field straddles boundaries of many disciplines and its skillset continues to evolve, it is often difficult to delineate its specific skillset and competencies
5	Specialized knowledge of manufacturing activities	Specialized knowledge must be expressed in terms of 'common knowledge'

(continued)

Table 1. (*continued*)

No.	Competencies	Description
6	Organizational and processual understanding	Important competencies include a strong command of business workflow, knowledge management, feasibility assessment, data-driven change leadership, and business impact assessment
7	Interdisciplinary/generic knowledge about technologies	Expert performance draws on knowledge that is hard to express and structure explicitly. Information and knowledge have dimensions which are never recorded, but which are powerful determinants of organizational behavior
8	Statistical knowledge	Refers to the core capability of analyzing data for insights and solutions that address business challenges. This is done using advanced knowledge of statistics, data visualization.

Table 2. Managerial competencies [14]

No.	Competencies	Description
1	Creativity	Creativity is characterized by the ability to perceive the world in new ways, to find hidden patterns, to make connections between seemingly unrelated phenomena, and to generate solutions. We are naturally creative and as we grow up we learn to be uncreative
2	Entrepreneurial thinking	Entrepreneurial thinking skills refer to the ability to identify marketplace opportunities and discover the most appropriate ways and time to capitalize on them
3	Problem solving	Involves both analytical and creative skills. Analytical thinking includes skills such as comparing, evaluating and selecting. Creative thinking is a divergent process, using the imagination to create a large range of ideas for solutions
4	Conflict solving	Resolving conflict is a key part of a manager's role. Managing and resolving conflict requires emotional maturity, self -control, and empathy
5	Decision making	Decision making is the process of making choices by identifying a decision, gathering information, and assessing alternative resolutions. Essentially, rational or sound decision making is taken as primary function of management
6	Analytical skills	Analytical skills are the thought processes required to evaluate information effectively; are the ability to visualize, gather information, articulate, analyze, solve complex problems
7	Research skills	Can be from need to be able to use reliable sources for continuous learning in changing environments. Doing research in the world of work is all about stepping back from your day-to-day work and looking at ways you can improve. The most successful people tend to develop research skills early and use them consistently
8	Efficiency orientation	An 'efficiency' approach is one that stresses the efficient use of resources as the main determinant decision and action. Efficiency orientation is inevitable

Table 3. Social competencies [14]

No.	Competencies	Description
1	Intercultural skills	Intercultural competences are response to the existence of cultural diversity; can be understood as resources put to work during intercultural dialogue (to use various forms of communication)
2	Language skills	Language skills allow for establishing cooperation with foreign partners, which increases the possibilities for company growth through internationalization
3	Communication skills	It is the ability to listen without prejudice and send convincing messages. Especially important thing here is the so-called emotional intelligence, which every manager should develop
4	Leadership skills	Leadership skills are focused on inspiring employees, directing them, as well as mastering the methods of effective persuasion by managers
5	Ability to be compromising and cooperative	Ability to be compromising and cooperative is experiencing feelings from the point of view of other employees and trade partners. It is also an active interest in their concerns, anxieties and worries
6	Ability to work in a team	Ability to work in a team is primarily the ability to secure cooperation between all members of the group to achieve a collective goal
7	Ability to transfer knowledge	The ability to transfer knowledge is one of the most difficult ones to achieve. It is also the ability to sense the need for growth in other people and to develop abilities in employees
8	Accepting change	Accepting change is related to the ability to initiate changes or to manage them. It is also mediating in implementation of the changes, so as not to cause disputes

4 The Research Results

In 2017 a questionnaire survey was conducted amongst selected experts in pharmaceutical sector and automotive industry. These experts are high qualified managers employed in transnational companies. Respondents were asked to indicate of selected competencies. There were 20 experts who filled in questionnaire.

The pharmaceutical and automotive sector were selected for the purpose of study. The choice was motivated primarily by the specific characteristics. Pharmaceutical industry is one of the most rapidly growing industrial sectors both in Poland and across the world. It is characterized by a growing degree of automation, robotization and computerization. The automotive sector relies on novel technologies, and product and process innovations. Also, both sectors search for and employ qualified personnel. Furthermore, they are oriented toward ongoing improvement of competencies.

In 2018 a questionnaire survey was conducted amongst selected experts in scientists of supply chain and Industry 4.0. These scientists are high qualified specialists. There were 20 experts who filled in questionnaire. Respondents were asked to indicate of selected competencies.

Technical competencies are underestimated. These competencies are assessed by experts at a very low level (Fig. 1). The greatest diversity was noted in the assessment of IT competencies. Researchers and pedagogues look at the question of IT competencies in different ways. However, each time they emphasize the necessity of their acquisition or consolidation. But it turns out that in practice, these skills are not the most important. Computer programming/coding abilities assessment are below the average. There is one conclusion. Managing a modern supply chain in pharmaceutical or automotive company requires good and professional managers who do not necessarily have to be familiar with technical knowledge.

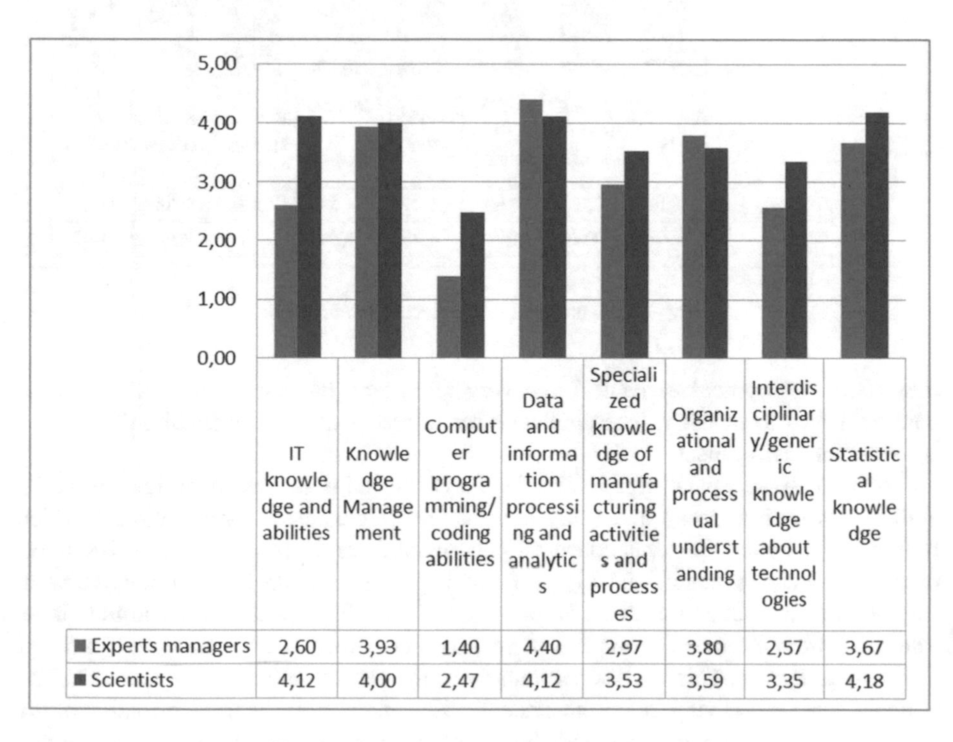

	IT knowledge and abilities	Knowledge Management	Computer programming/coding abilities	Data and information processing and analytics	Specialized knowledge of manufacturing activities and processes	Organizational and processual understanding	Interdisciplinary/generic knowledge about technologies	Statistical knowledge
Experts managers	2,60	3,93	1,40	4,40	2,97	3,80	2,57	3,67
Scientists	4,12	4,00	2,47	4,12	3,53	3,59	3,35	4,18

Fig. 1. Technical competencies; own work

In the group of management competencies, both groups of respondents are characterized similar evaluations of selected managerial competencies (Fig. 2). This applies in particular to competencies related to entrepreneurial thinking, analytical skills, conflict solving, decision making and efficiency orientation. The greatest diversity was noted in the assessment of research skills and creativity. It demonstrates that business views are in sharp contrast to those of the scientists. Universities generally consider

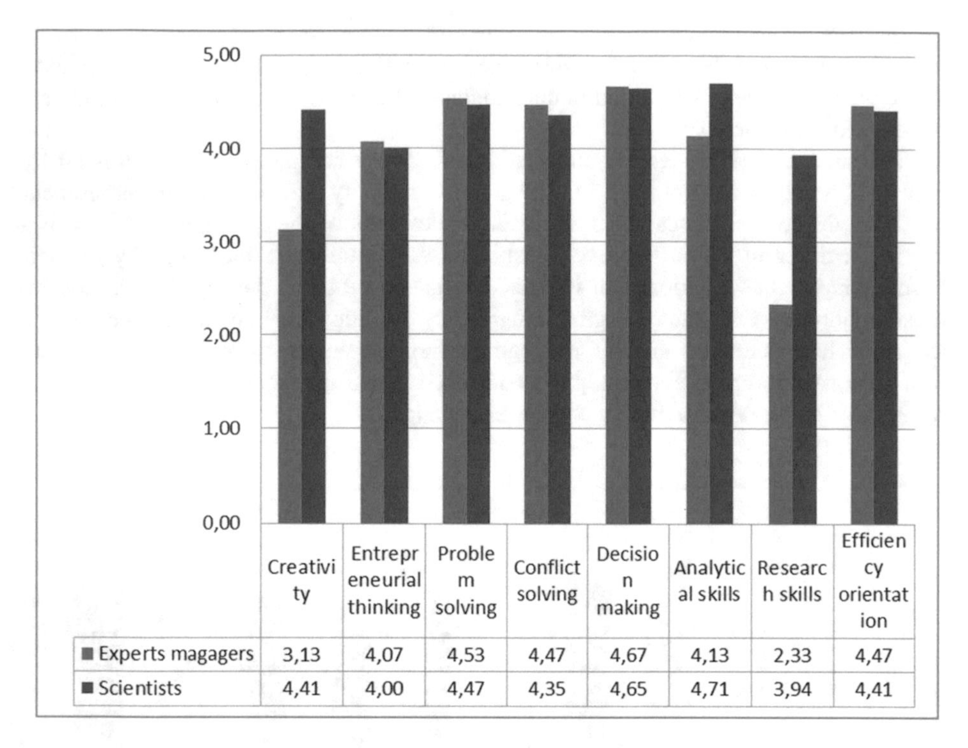

	Creativity	Entrepreneurial thinking	Problem solving	Conflict solving	Decision making	Analytical skills	Research skills	Efficiency orientation
Experts magagers	3,13	4,07	4,53	4,47	4,67	4,13	2,33	4,47
Scientists	4,41	4,00	4,47	4,35	4,65	4,71	3,94	4,41

Fig. 2. Managerial competencies; own work

these two competences as natural and very important, but maybe it could be better communicated to industry. The challenge for universities is to better align them to a commercial environment.

In the case of social skills (Fig. 3), language skills and communication skills, ability to work in a group and accepting change were assessed as the highest, which fully reflects the requirements concerning the candidates for managers in both groups, practitioners and scientists. Ability skills to be compromising and cooperative were assessed low in both groups. To a large extent this may be related to the cultural circle. Managers from Eastern Europe are more withdrawn than their colleagues from e.g. Germany or France. This applies particularly to the generation of 50–60 year olds. The younger generation living in Poland since the 90s, that is after the transformation of the Polish economy, are more open and show higher self-esteem. The diversity was noted in the assessment of intercultural and leadership skills. Scientists focus on intercultural skills, because in their opinion workers, especially leaders need to be competent at working with foreign business partners. Since their employees will be working with people from different cultures, they'll need to leverage the unique skills of all. On the other hand for scientists not important enough are leadership skills as for experts managers. Practitioners know that the winners of tomorrow will be those organizations with strong leaders who demonstrate agility and authenticity.

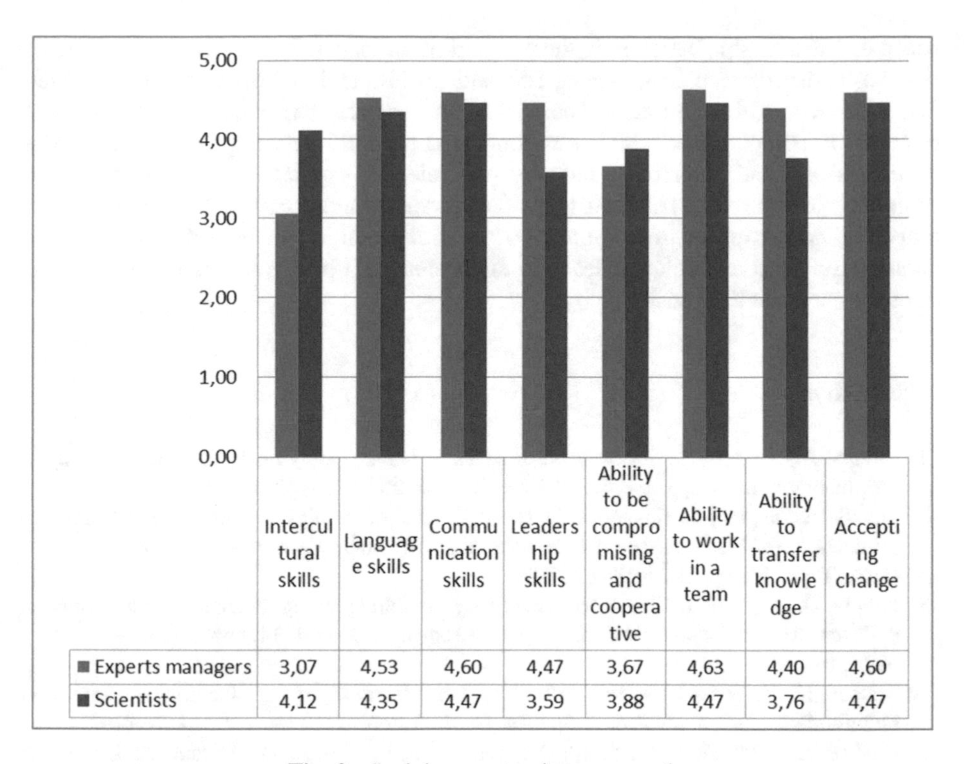

	Intercultural skills	Language skills	Communication skills	Leadership skills	Ability to be compromising and cooperative	Ability to work in a team	Ability to transfer knowledge	Accepting change
Experts managers	3,07	4,53	4,60	4,47	3,67	4,63	4,40	4,60
Scientists	4,12	4,35	4,47	3,59	3,88	4,47	3,76	4,47

Fig. 3. Social competencies; own work

5 Conclusion

At this moment a large change in industry takes place, the so called fourth industrial revolution. Fast technology development, combined with the globalization and fast changes in customer demand, implies that a competitive advantage of a company can be only temporary [28]. This revolution involves advances in underlying technologies, e.g. production and ICT, changes in the business models of firms and is likely to have a deep impact on our society. New manufacturing technologies, extensive digitization, interweaving of machines and organizations (Internet of Things) have a huge impact on industry. In Industry 4.0, dynamic business and engineering processes enable last-minute changes to production and deliver the ability to respond flexibly to disruptions [29, 30].

The business world and supply chain are changing rapidly. The complexity of the problem is connected with the necessity of taking into account some factors that might increase the level of risk, example: supply chain globalization [31, 32]. The environment of business is no more predictive and stable. Knowledge and learning plays a crucial role in that process. Intellectual property becomes the company's most valuable asset. In modern business economy knowledge becomes one of the most important competitive advantages. This paper was to identify what competencies seem necessary for future managers and engineers to be efficient contributors in the future. It was possible to achieve through the development of an extensive list of key competencies based on the available literature in the first part of the paper. Three areas of competence were established: technical, managerial and social, and 8 key competencies were identified within these groups. The second part of the paper shows the results of research on competencies which was conducted amongst selected experts in pharmaceutical sector and automotive industry and selected experts in scientists of supply chain and Industry 4.0. This part helps the reader to understand the requirements for individual competencies from the point of view of practitioners and scientists. There is demand for more research that leads to understand and bridge the gap between practitioners and scientists' vision.

References

1. Surana, A., Kumara, S., Greaves, M., Raghavan, U.N.: Supply chain networks: a complex adaptive systems perspective. Int. J. Prod. Res. **43**(20), 4235–4265 (2005)
2. Saniuk, A., Saniuk, S., Caganova, D. et al.: Control of strategy realization in metallurgical production, In: 23rd International Conference on Metallurgy and Materials Location, Brno, Czech Republic, pp. 1876–1881 (2014)
3. Ivanov, D., Sokolov, B.: The inter-disciplinary modelling of supply chains in the context of collaborative multi-structural cyber-physical network. J. Manuf. Technol. Manage. **23**, 976–997 (2012)
4. Thoben, K.-D., Wiesner, S., Wuest, T.: Industrie 4.0 and smart manufacturing – a review of research issues and application examples. Int. J. Autom. Technol. **11**(1), 4–16 (2017)
5. Gaynor, M., Moulton, S.L., Welsh, M., LaCombe, E., Rowan, A., Wynne, J.: Integrating wireless sensor networks with the grid. IEEE Internet Comput. **8**, 32–39 (2004)
6. Bendavid, Y., Cassivi, L.: Bridging the gap between RFID/EPC concepts, technological requirements and supply chain e-business processes. J. Theor. Appl. Electron. Commer. Res. **5**, 1–16 (2010)
7. Butner, K.: The smarter supply chain of the future. Strateg. Leadersh. **38**, 22–31 (2010)
8. Wu, L., Yue, X., Jin, A., Yen, D.C.: Smart supply chain management: a review and implications for future research. Int. J. Logist. Manage. **27**, 395–417 (2016)
9. Carter, C.R., Rogers, D.S.: A framework of sustainable supply chain management: moving toward new theory. Int. J. Phys. Distrib. Logist. Manage. **38**(5), 360–387 (2008)
10. Ganji, E.N., Shah, S., Coutroubis, S.: Sustainable supply and demand chain integration within global manufacturing industries. In: Proceedings of the 2017 IEEE International Conference on Industrial Engineering and Engineering Management (IEEM), pp. 1807–1811. IEEE (2018)
11. Linton, J.D., Klassen, R., Jayaraman, V.: Sustainable supply chains: an introduction. J. Oper. Manage. **25**(6), 1075–1082 (2007)
12. Grzybowska, K.: Supply chain sustainability – analysing the enablers. In: Golinska, P., Romano, C.A. (eds.) Environmental Issues in Supply Chain Management – New Trends and Applications, pp. 25–40. Springer (2012). https://doi.org/10.1007/978-3-642-23562-7_2
13. Organization for Economic Cooperation and Development (OECD): The definition and selection of key competencies (2005)
14. Grzybowska, K., Łupicka, A.: Future competencies in the automotive industry – practitioners' opinions. In: Soliman, K.S. (ed.) 30th International Business Information Management Association Conference (IBIMA); Vision 2020: Sustainable Economic development, Innovation Management and Global Growth, XLVI, pp. 5199–5208 (2017)

15. Köffer, S.: Designing the digital workplace of the future – what scholars recommend to practitioners. In: International Conference on Information Systems, Fort Worth (2015)
16. Nelson, G.S., Horvath, M.: The Elusive Data Scientist: Real-World Analytic Competencies. ThotWave Technologies, Chapel Hill (2017)
17. Relich, M.: Identifying relationships between eco-innovation and product success. In: Technology Management for Sustainable Production and Logistics, pp. 173–192. Springer, Heidelberg (2015)
18. Joerres, J., McAuliffe, J., et al.: The Future of Jobs – Employment, Skills and Workforce Strategy for the Fourth Industrial Revolution (2016)
19. Ten Hompel, M., Anderl, R., Gausemeier, J., Meinel, C., et al.: Kompetenzentwicklungsstudie Industrie 4.0 - ErsteErgebnisse und Schlussfolgerungen, München (2016)
20. Davies, A., Fidler, D., Gorbis, M.: Future Work Skills 2020, Palo Alto (2011)
21. Bauer, H., Baur, C., Camplone, G., George, K., et al.: Industry 4.0 – How to navigate digitization of the manufacturing sector (2015)
22. Störmer, E., Patscha, C., Prendergast, J., Daheim, C., Rhisiart, M., Glover, P., Beck, H.: The Future of Work: Jobs and skills in 2030 (2014)
23. Pompa, C.: Jobs for the Future, London (2015)
24. Morgan, J.: The Future of Work – Attract New Talent, Build Better Leaders, and Create a Competitive Organization. Wiley, Hoboken (2014)
25. Sitek, P., Wikarek, J.: A hybrid programming framework for modeling and solving constraint satisfaction and optimization problems. Sci. Program. **2016**, Article ID 5102616 (2016). https://doi.org/10.1155/2016/5102616
26. Grzybowska, K., Gajdzik, B.: SECI model and facilitation on change management in metallurgical enterprise. Metalurgija **52**(2), 275–278 (2013). DOI: 65.01:669.013.003: 658.5:658.8=111
27. Shirani, A.: Identifying data science and analytics competencies based on industry demand. Issues Inf. Syst. **17**(IV), 137–144 (2016)
28. Oleśków-Szłapka, J., Stachowiak, A., Batz, A., Fertsch, M.: The level of innovation in SMEs, the determinants of innovation and their contribution to development of value chains. Procedia Manuf. **11**, 2203–2210 (2017)
29. Jasiulewicz-Kaczmarek, M., Saniuk, A., Nowicki, T.: The maintenance management in the macro-ergonomics context. In: Goossens, R.H.M. (eds.) Advances in Social and Occupational Ergonomics Proceedings of the AHFE 2016, Florida, USA. Advances in Intelligent Systems and Computing, vol. 487, pp. 35–46 (2016). https://doi.org/10.1007/978-3-319-41688-5
30. Burduk, A.: The role of artificial neural network models in ensuring the stability of systems. In: 10th International Conference on Soft Computing Models in Industrial and Environmental Applications. Advances in Intelligent Systems and Computing, pp. 427–437. Springer (2015)
31. Wicher, P., Lenort, R., Krausova, E.: Possible applications of resilience concept in metallurgical supply chains. In: Proceedings of the METAL Conference, Brno, Czech Republic (2012)
32. Tubis, A., Nowakowski, T., Werbińska-Wojciechowska, S.: Supply chain vulnerability and resilience-case study of footwear retail distribution network. Logist. Transp. **33**, 15–24 (2017)

Logistical Aspects of Transition from Traditional to Additive Manufacturing

Patrycja Szymczyk, Irina Smolina(✉), Małgorzata Rusińska, Anna Woźna, Andrea Tomassetti, and Edward Chlebus

Wroclaw University of Science and Technology, Wroclaw, Poland
iryna.smolina@pwr.edu.pl

Abstract. The intensive growth and development of additive manufacturing affects all operations across the production chain. This article focuses on aspects of changes in logistics and supply chains during the transition from or replacement of traditional manufacturing with additive manufacturing, in new or existing enterprise. Today, hardware being shipped from far away and stored in a warehouse, that is still far from the end-user destination, generates high costs, but the digital file that just could be sent to one of the operational units closest to the customer could change that. Additive fabrication of an urgently required component located at place could overcome the lead time, shipping cost, inventory requirements and transport vulnerability.

Keywords: Additive manufacturing · Logistics · Supply chains
Production management

1 Introduction

In the last two decades, the manufacturing world has gone through various technological, methodological and process innovations that are now at the roots of the way the goods are produced and assembled and delivered to the market. Those changes had an immense effect on lowering production costs, better recourse utilization, quality improvement, production flexibility, value chain integration that led to development of new business models. The innovation process didn't stop and new technological solutions are emerging continuously. Along with them the Additive Manufacturing (AM) technologies have been arising, gaining interest from production industry. The idea of manufacturing products by addition of material in "layer by layer" manner instead of material subtraction opened the possibilities that traditional technologies couldn't support, especially when it came to complex external and internal geometry. Moreover no specific tools were needed - the self-sustaining technologies were designed to manufacture without extra equipment, avoiding the need to shift between many devices. Instead of controlled removal of undesired material through cutting, drilling or milling to achieve the desired forms one process was used, lowering the costs, time, materials and resource usage. Additive manufacturing (AM) have a key role in contingency plans, since its fastness when production line problems occur or yet, they give flexibility to the company, which will be able to easily switch in

© Springer Nature Switzerland AG 2019
A. Burduk et al. (Eds.): ISPEM 2018, AISC 835, pp. 752–760, 2019.
https://doi.org/10.1007/978-3-319-97490-3_71

production designs. AM represent a great opportunity to create an inventory of occasionally needed parts and components.

Besides all the achievements, AM technologies in most cases are still perceived as aid for traditional manufacturing and the possibility to run a company based exclusively on it is yet to be found. The potential is clear and it is opinion of many that it could represent, thanks to further development on its concept and probably a lower cost of the machines, the future way to make products. But there are also many problems to be solved and more detailed research is needed, as besides all the advantages AM technologies also have drawbacks that limit their implementation in large scale production. Many challenges have to be faced and solved and the basic question from industry is: "When it is beneficial to switch from traditional to AM manufacturing?". AM is still addressed with many challenges, where some of them are: large scale production, parts topology optimization, energetic costs, material heterogeneity and structural reliability and AM standardization.

In this article the general procedure usable, under specific limitations, for replacement of traditional technologies or to start over their business with AM is presented.

1.1 AM Technologies

Nowadays several additive manufacturing processes are available. AM processes can be categorized by the type of material used, the deposition technique or by the way the material is fused or solidified [1]. In accordance with ASTM F42 they are divided on the 7 main groups: vat photopolymerisation (VT), material jetting (MJ), binder jetting (BJ), material extrusion (common known as FDM or FFF), powder bed fusion (PBF, where the most known technologies are SLM/DLMS and EBM), sheet lamination (SL) and direct energy deposition (DED or LENS). They differ from some technical aspect and working principles, but the idea behind them is always the same: proceeding layer by layer by the deposition or fusion of material until the desired shape is obtained [2]. Each method has anyway drawbacks and advantages and the choice on whether using a process rather than another one should be detected in the individuation of functional goals [3].

The variety of materials for AM is continuously expanding, but it (the variety of available materials) is still one of the challenges as well as opportunities to be developed. The problems behind question about switching to AM manufacturing is not lack of materials, but their price and the obtained material properties after the production process. Therefore research race is now directed on delivery of proper materials with guaranty of quality and cost effectiveness challenging the demands of standardization processes.

1.2 Areas of Application

With the high potential and numerous applications the AM technologies cut across many branches of industry and that is what makes its potential so compelling. Research and development in AM are already driving aerospace, automotive, medical and consumer products into the future. Vast investments in those industries propel AM technology to new heights. Other industries, including the military, jewelry, maritime, energetics, construction, furniture, home accessories and toys, will also play an important role in the future development of AM technology and its application. The Wohlers Report 2017 [4] presents the share of AM applications between branches (Fig. 1).

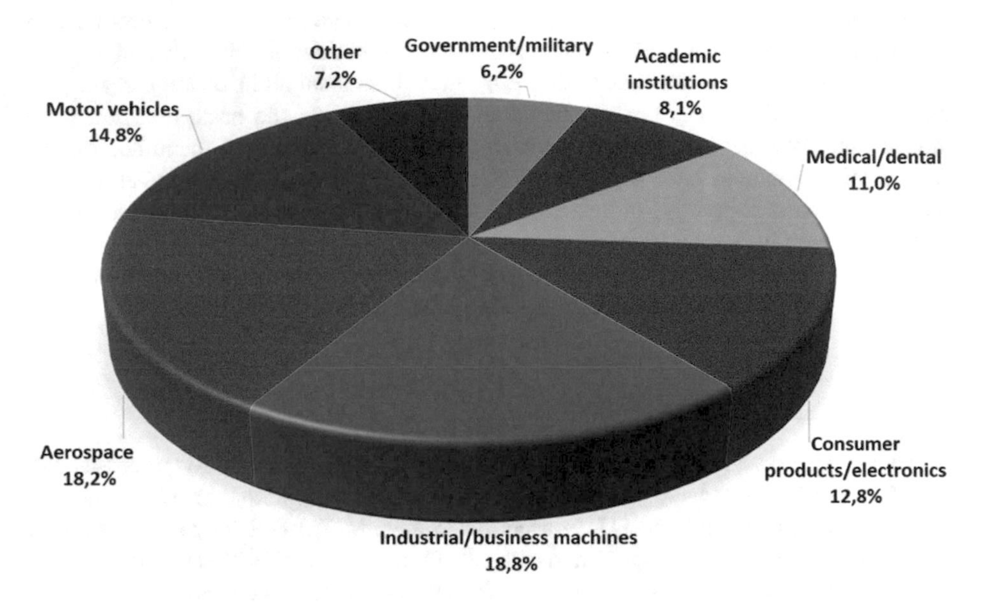

Fig. 1. Market share of AM application in industry by branches [4].

Manufacturing of custom, semicustom and limited product editions is desirable trend nowadays. Other ways of development is production or replacement parts for military, marine, land and air vehicles and other products. AM introduction into industry brings new types of businesses that unleash new types of products that were previously impractical due to high cost and risk, or lack of manufacturability. Companies slowly discover that AM is a strategic bridge in shortening delivery to market time. In developing industry the case for the use of AM instead of conventional manufacturing methods, numerous factors must be taken into account. Starting with production costs estimation: fixed cost, the cost of process qualification and component certification, logistical costs, and the cost of time have to be analyzed. Moreover, the costs connected directly to product lifecycle costs have to be taken into account, as the, for example, better mechanical properties will assure the extended potential value of manufactured goods. Other key factors to be compared are: range and availability of materials, manufacturing process performance especially its throughput compared with product demand, the existence of closed-loop control systems to reduce process variance, time and cost for AM designers, the existence of industry standards for AM qualification in selected technology.

2 Effects of AM Implementation in Logistics

The introduction of AM technologies into manufacturing company will cause many changes in its operational and strategical structure, which will highly depend on the type of production and the product itself. Nevertheless, it is possible to present more universal perspective in selected areas. The logistic aspects can be standardized and applicable more widely. The key benefits for logistics due to implementation of additive manufacturing in production process are as follow [5, 6]:

1. Shortening delivery time:
   a. Due to the possibility of parts consolidation/merging,
   b. Decreasing the number of manufacturing operations for the item/part;
2. Production on location:
   a. Manufacturing facilities back to the domestic countries of inventory (for example, USA or Germany, etc.),
   b. The new opportunities for production spare parts straightaway. Especially it is important for the marine and military industries,
   c. The decreasing of transport costs;
3. The easiness of creation and production of customer-specific, complex items;
4. The changes of relations between manufacturers and retailers. The role of stocks and retails will decrease, due to implementation of conception of on-demand manufacturing (also, named "agile manufacturing")
5. Possibility of launching low volume production at costs comparing to the mass production;
6. Decreasing the supplier risks;
7. The reduction of weight and volume parameters. In aviation – buy-to-fly ratio;
8. Improvement of decision making process (agility).

Figure 2 represents the quantity of components for which SLM and one of the methods of traditional manufacturing (high-pressure die casting/HPDC) has a similar total cost. After that point, due to the economy of scale, the HPDC (traditional manufacturing) cost per piece become lower than for the SLM (additive manufacturing).

AM technologies reduce times and costs of product development in the designing and production phases, because investments on tooling are not required. However, this is not the only advantage shown by the study [7]. According to the results, it is seen how AM technology is competitive for small and medium-sized production in metallic components.

It is worth to underline how the greater costs' percentage in SLM process is attributed to the machine, while the other factors do not have an important effect. Nevertheless, AM technologies will spread even more within the common production processes, it is possible to imagine a reduction on the machines' cost and, as a consequence, a modification of the break-even point towards bigger production's volumes [7].

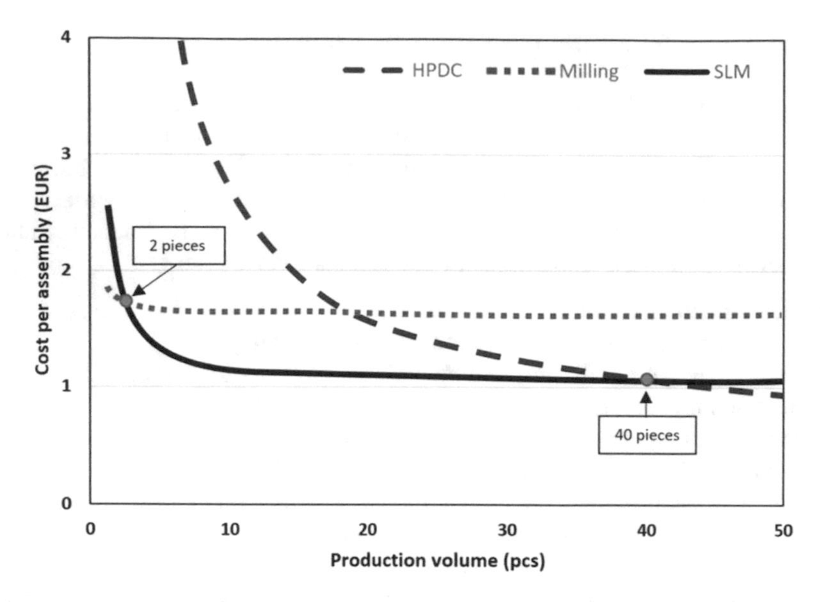

Fig. 2. The break-even point analysis [7].

2.1 Analyzed Areas

Logistics is much more complicated and takes a huge area of application involved in several aspects of company's management. That is why it is important to divide the logistical domain into sub-areas at first. The sub-areas chosen to ease the comprehension of the procedure are divided on three processes: (1) supply chain network/management; (2) storage management; (3) packaging and delivery management.

Supply Chain Network. It represents the interconnections between the characters that concur to the manufacture of the product, such as goods' and part's suppliers, carrier and producers. The supply chain network is not simply the movement of the product and its components among them, though; it is a huge quantity of information needed to make all the operation possible and to realize a complete satisfaction of customers. It also concerns the potential of logistical platforms (if needed) around the market geographical zone and it establishes where to place them.

As previously stated, additive manufacturing might represent a breakthrough in the way supply chain networks are conceived. One of the main advantages of AM is possibility to be located close to the customer, since this technology does not require any associate manufacturers to collect parts or any other auxiliaries: this lead to more favorable cost-related factors.

Another positive side of AM influence of changes in supply chain network is the absence of costly set up of the machines, which allows the production of small batches particularly affordable. A huge advantage is also represented by the reduction, on the first stage of the supply chain (from suppliers to producers), of the resources' volume transported. This is caused by a better buy-to-fly ratio, the weight ratio between the raw material used to produce the part and the final weight of the part after the process.

Finally, it can be highlighted the enrichment obtained by the flow of information and communication processes using AM technologies. The digital file can be sent from any part of the world through the service platforms to the high-qualified supplier and printed within a few hours [3, 8–13].

Storage Management. It refers to all the issues related with the coordination and organization of the warehouse, from the optimization of the layout to the determination of the stock level, going through the positioning of the goods in the platform.

The storage management aspect can be deeply influenced by implementing AM to the production process. First of all, the materials, often thermoplastic filaments or powders are reduced in size. Secondly, demand, in case of comparison with traditional technologies, is much lower for producing the same part. The result is a more lean manufacturing process.

Another important aspect is represented by the high efficiency of AM machines: thanks to their chambers, which can be fully utilized, there is a possibility to make more than one part (even different between them in design) at the same time, in parallel manufacturing. As a clear consequence, the reduction of inventory's costs is relevant [3, 8–13].

Packaging and Delivery. It designs the possible way of the final product packaging, which should be done following the standards set by the ISO and which changes according to the materials or the shape of the product. It shows solutions of shipping as well as takes in to account costs and timing.

The biggest advantage concerning the packaging is that, since undercuts are allowed in AM and even the most complicated components can be done as one consolidated part (without any assembly). Therefore, instruction manuals, screws and nails are not needed. Thus, the packaging will result easier to design, lighter and less bulky [3, 8–13].

According to all aspects mentioned above, it will be possible to analyze how can the transition from traditional do additive manufacturing affect all those sub-areas and define the critical factor that will be important in the development of a general procedure [3, 8, 10–14].

2.2 Processes and Procedures

Limitations

To come up with general and standardized procedures, the certain limitations should be done. This is due by the fact that many experts in the AM technologies' world claim that this method cannot be universal and it is not possible to implement it in to a mass production.

The limitations, which were set, based on literature research, are following:

- **Spare parts companies, most likely working with niche product**. There is awareness of the inconvenience of AM over traditional technologies for mass production (huge number of products and batches to be sold give important advantages to those methods which facilitate economies of scale instead of economies of one in case of AM);

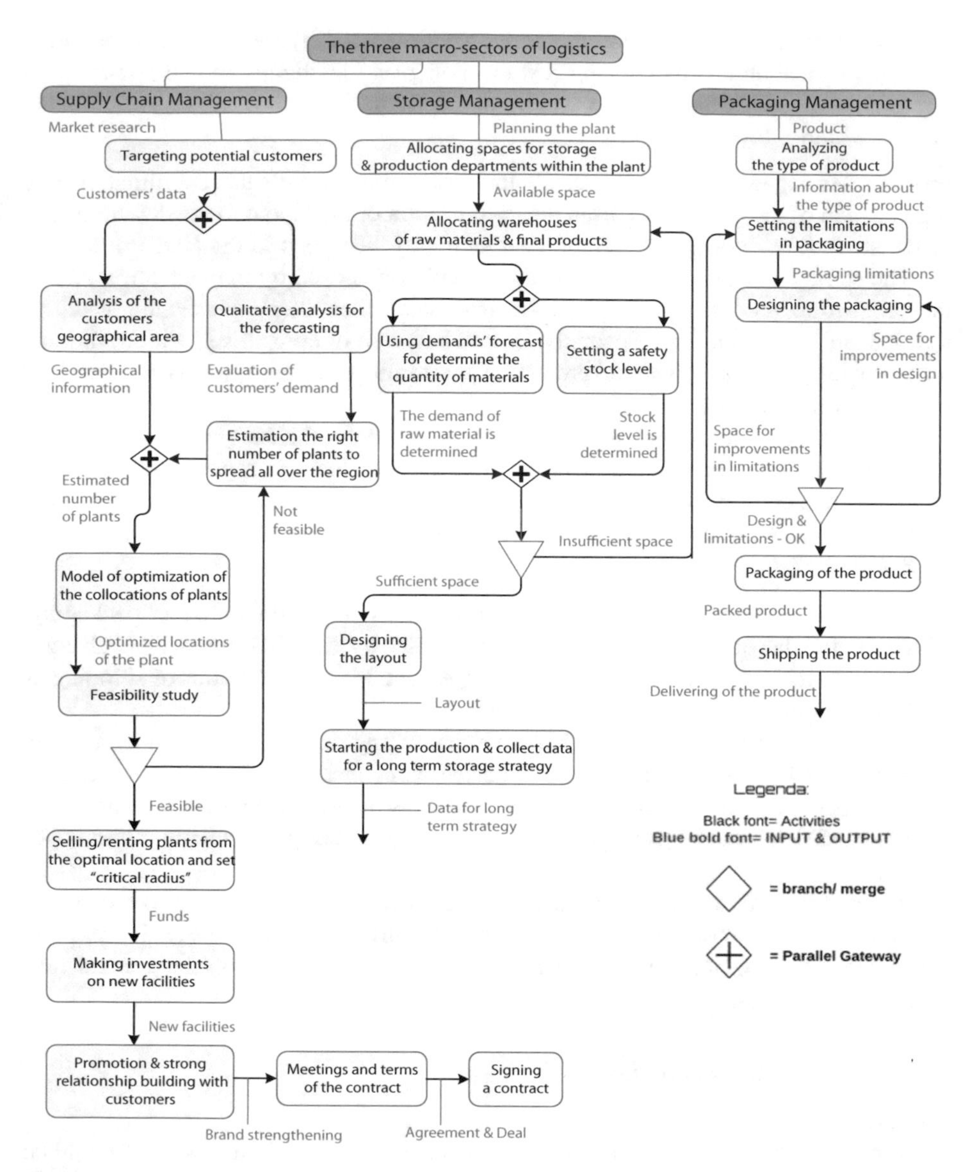

Fig. 3. Processes, divided according to the three macro-sectors of logistics (supply chain management, storage management, packaging), to be carried out in case of manufacturing technology replacement.

- **Companies willing to change their business model**. AM is particularly effective with customer-centric business models, where the satisfaction of the client is even more important than revenues;
- **Companies with the possibility to invest in facilities and plants**. Strictly connected to the previous two points, since the procedure rely on decentralized production.

Processes

Figure 3 refers to the processes, divided according to the three macro-sectors of logistics (supply chain management, storage management and packaging), to be carried out in case of replacement of manufacturing technology.

Supply Chain Management. This process can divided into 11 sections, which represent the activities needed to set a proper supply chain for the company. Each section is displayed in a chart, where is possible to identify four parts:

a. Input/Output chart shows the activity being taken into consideration;

b. Evaluation parameters set the requirements for a positive outcome from the activity;

c. Procedures list all the actions needed to run the activity;

d. Added value explains the obtained benefits after completing the activity;

Storage Management. This process consists of 6 sections, which give all the needed information to perform the storage management of products and raw materials in case of additive technologies in production.

Packaging. The packaging management explores the packaging process and analyses of its activities through the tools already used in the previous ones (Input/Output chart, evaluation parameters, procedures, added value).

The above figure presents the process of implementation of additive manufacturing to the enterprise under some specific limitations, but the format can be replicated indiscriminately: in fact, every section can properly work for different kind of processes. The working procedure and its logical system have been standardized but are also open for modifications in special cases. The universal value of this methodology, as a consequence, gives even more credit to the procedures previously presented.

3 Conclusions

The gathered information about the possibilities of AM technologies, costs of implementation and production, new business models but also existing challenges to be solved and barriers to be overtaken gave the fundamental knowledge for further development of presented procedure. The analyzed logistic area of production companies show that it is possible to introduce standards for AM based production, nevertheless the presented method do not solve many technical and economic issues of manufacturing itself based on new technologies and therefore more detailed studies have to be conducted. From supply chain network, storage management and packaging and delivery processes point of view introducing the AM based production will minimize costs and space for logistic procedures. The supply chain grants from its integration, there is a strong need of customer-producer cooperation that is the key factor for specific production that contributes to make the flow of information and materials/products easier. The big competitive advantage of AM technologies is plant location with optimized "buy-to-fly ratio". In storage management area there were also strong benefits identified, such as: lower material stock levels, optimized location of all assets in the layout, which means time and cost saving. The packaging processes would stay similar to traditional production.

The factories of the future and, respectfully, the supply chains of the future will synergize the available resources and benefits of the additive manufacturing incorporating them in fast-emerging markets and businesses.

References

1. AM Platform: Additive Manufacturing: Strategic Research Agenda (2014)
2. Designation: F2792 − 12a StandardTerminology for Additive Manufacturing Technologies 1,2. https://doi.org/10.1520/f2792-12a
3. Marchese, K., Crane, J., Haley, C.: Transforming the supply chain with additive manufacturing | Deloitte Insights. https://www2.deloitte.com/insights/us/en/focus/3d-opportunity/additive-manufacturing-3d-printing-supply-chain-transformation.html
4. Wohlers Associates, I.: Wohlers report 2017 : 3D printing and additive manufacturing state of the industry : annual worldwide progress report. Wohlers Associates, Inc. (2017)
5. Knulst, R.: 3D printing of marine spares A case study on the acceptance in the maritime industry (2016). https://dspace.ou.nl/bitstream/1820/7942/1/Knulst_R%scriptie_v8%dspace.pdf
6. Pilot Project 3D printing of Marine spares | Port of Rotterdam, Rotterdam (2016)
7. Iuliano Eleonora Atzeni, L., Minetola, P., Salmi, A.: Confronto tra tecnologie additive laser e processi tradizionali nella produzione di componenti aerospaziali. In: ExpoLaser, Piacenza (2013)
8. Bogers, M., Hadar, R., Bilberg, A.: Additive manufacturing for consumer-centric business models: Implications for supply chains in consumer goods manufacturing. Technol. Forecast. Soc. Change. **102**, 225–239 (2016). https://doi.org/10.1016/j.techfore.2015.07.024
9. Barz, A., Buer, T., Haasis, H.-D.: A study on the effects of additive manufacturing on the structure of supply networks. IFAC-PapersOnLine. **49**, 72–77 (2016). https://doi.org/10.1016/J.IFACOL.2016.03.013
10. Pour, M.A., Zanardini, M., Bacchetti, A., Zanoni, S.: Additive manufacturing impacts on productions and logistics systems. IFAC-PapersOnLine. **49**, 1679–1684 (2016). https://doi.org/10.1016/J.IFACOL.2016.07.822
11. Scott, A., Harrison, T.P.: Additive manufacturing in an end-to-end supply chain setting. 3D Print. Addit. Manuf., 2, 65–77 (2015). https://doi.org/10.1089/3dp.2015.0005
12. Emelogu, A., Marufuzzaman, M., Thompson, S.M., Shamsaei, N., Bian, L.: Additive manufacturing of biomedical implants: A feasibility assessment via supply-chain cost analysis. Addit. Manuf. **11**, 97–113 (2016). https://doi.org/10.1016/J.ADDMA.2016.04.006
13. Khajavi, S.H., Partanen, J., Holmström, J.: Additive manufacturing in the spare parts supply chain. Comput. Ind. **65**, 50–63 (2014). https://doi.org/10.1016/J.COMPIND.2013.07.008
14. Barz, A., Buer, T., Haasis, H.-D.: Quantifying the effects of additive manufacturing on supply networks by means of a facility location-allocation model. Logist. Res. **9**, 13 (2016). https://doi.org/10.1007/s12159-016-0140-0

Reconfiguration to Renovation in a Sustainable Supply Chain

Katarzyna Grzybowska(✉)

Faculty of Engineering Management, Chair of Production Engineering and Logistics, Poznan University of Technology, Strzelecka 11, 60-965 Poznan, Poland
katarzyna.grzybowska@put.poznan.pl

Abstract. The article presents the issue of reverse logistics in the configuration of a sustainable supply chain. The focus was on optimizing the supply chain as a complex and dynamically changing system. In order to analyze the problem of reuse of the component/finished product, a simulation model was built based on the actual realities of the industry. This model takes into account aspects of both coordination and interdependence of links in the supply chain and material flow. By conducting experiments on the supply chain system, recommendations were developed, which radically changed the nature of reverse logistics in the analyzed complex system.

Keywords: Sustainable supply chain · Reconfiguration · Model

1 Introduction

An increasing number of manufacturers are gradually realizing the potential economic, social and environmental benefits from the research and development of sustainable products [1], processes [2] and logistics systems as a supply chain [3]. As a consequence, the significance of a sustainable supply chain has been increasingly recognized by both industry practitioners and academic researchers [1]. The interaction between products and supply chains is critical. The research [4–9] shows an emphasis on sustainable management.

The aims of the study are to present the results of a simulation experiment conducted and to perform an empirical analysis and optimizing of simulated supply chain. The publication has the following structure. Part two outlines the background information related to the sustainable supply chain. Part three describes the brief description of the industry and product harmfulness. Next part presents supply chain characteristics and simulation model used. Part fifth focuses on the optimizing the supply chain as a complex and dynamically changing system. The final section contains a summary and conclusions.

© Springer Nature Switzerland AG 2019
A. Burduk et al. (Eds.): ISPEM 2018, AISC 835, pp. 761–770, 2019.
https://doi.org/10.1007/978-3-319-97490-3_72

2 Sustainable Supply Chain

The concept of sustainable development is the result of the growing awareness of the global links between mounting environmental problems, socio-economic issues to do with poverty and inequality and concerns about a healthy future for humanity [10]. Sustainability in the Supply Chain (or the Sustainable Supply Chain (SSC), the Environmentally Responsible Supply Chain, Green Supply Chain (GSC), green logistics and reverse logistics) is a key component of corporate responsibility [3]. Sustainability in the supply chain is the management of environmental, social and economic impacts, and the encouragement of good governance practices, throughout the lifecycles of goods and services [11]. Now, many organizations have considered environmental, social, and economic concerns and have measured their suppliers' sustainability performance resulting from the adoption of sustainable supply chain initiatives [12–14].

The sustainable supply chain is a popular area for researchers (Fig. 1). After 2010 year number of publication definitely increases. It is both complex and dynamic. It is both subjected to random events and stochastic processes. They can both become unstable if not managed well.

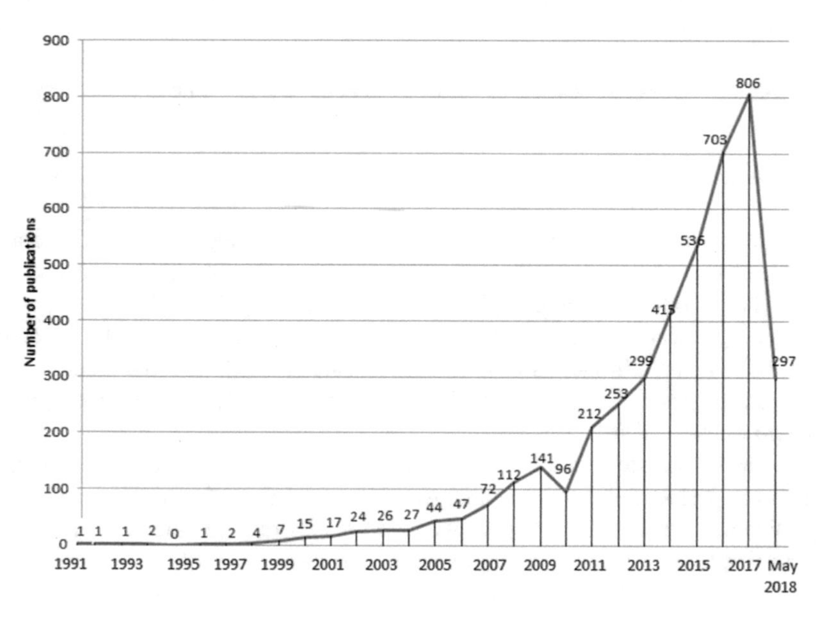

Fig. 1. Number of publications on the sustainable supply chain in Web of Science (ISI); own work

To give an overview on the current research on sustainable supply chain, conducted an analysis, including papers of international journals, conference papers and book chapters.

In the first stage of the analysis database Web of Science were scanned for papers, which dealt in any way with sustainable supply chain. To set limits, I used keywords such as "supply chain" and "sustainable". The dataset selected for this study was the "ISI Web of Science" that is one of the most consistent repositories of business and management papers [15].

This process resulted in an output of 4,261 papers, all of them associated with the topic of sustainable supply chain. Then analyzed this sample to learn about the content and the quintessence of the various papers. Due to research goal, I limited the research references to journal articles and conference papers. Other sources like unpublished material, newspaper articles and monographs were not considered in our sample.

Table 1 presents the list of the first ten journals where sustainable supply chain research has been published. Journal of Cleaner Production, Sustainability and International Journal of Production Economics lead the ranking with 368, 163 and 116 publications, respectively.

Table 1. Ranking of journals by number of publications; own work

No.	Journal title	Record count	% of 4.261
1	Journal of Cleaner Production	368	8.636%
2	Sustainability	163	3.825%
3	International Journal of Production Economics	116	2.722%
4	Acta Horticulturae	56	1.314%
5	International Journal of Production Research	56	1.314%
6	Supply Chain Management an International Journal	52	1.220%
7	Business Strategy and the Environment	50	1.173%
8	Computer Aided Chemical Engineering	50	1.173%
9	IFIP Advances in Information and Communication Technology	45	1.056%
10	Resources Conservation and Recycling	44	1.033%

The geographic diversity of scholars presents the leadership of North American (811 publications), China (596 publications) and England (520 publications) academic institutions that contribute substantially equally to the research field development.

With an increasing awareness, stakeholders especially consumers are more concerned about the environmental and social issues associated with the development and the use of products. Over the past few decades, sustainability within the operations of organizations as well as within the supply chain has become an important area of research. The implementation of sustainability initiatives not only improves the environmental and social performance of organizations but also provides them a competitive advantage by acquiring a new set of competencies [16].

3 A Brief Description of the Industry and Product Harmfulness

In the period from 2005 to 2016, employment in the household appliances sector increased more than twice, while the production of large household appliances increased by three times. The market has grown significantly over the last 10 years. No frost fridges as well as induction hobs and built-in appliances gained high popularity during the discussed period. Poland is one of the main exporters of household appliances to the European Union. Domestic products are most often exported to Germany (24%), France and the United Kingdom (12% each) as well as Italy (9%) [17].

Among the main challenges, the following aspects that are already crucial for the industry today are often mentioned: (1) improvement of production and implementation of innovations – activities aimed at reasonably reconciling the automation of production with the level of employment of personnel with appropriate qualifications and getting involved in research and development activities, (2) improving operations related to electro-waste management, (3) improving supply chain operations.

For the purposes of the work, product harmfulness was assessed in terms of the product's whole life cycle, both on the environment and on people. Both impacts were determined as well as assessed on a five-point scale.

It can be concluded that the greatest direct impact is recorded at the stage of manufacturing of the finished product as well as its packing and distribution (Table 2).

Table 2. Harmfulness of the product during its life cycle; student project under the direction of K. Grzybowska

Product life cycle phase	Negative impact on the environment	Impact assessment
Production	CO_2 emission in production (e.g. glassworks); impact of machines and devices (noise); high energy consumption; drainage of cooling liquids and other liquid waste (including harmful waste) to sewage or to rivers; production waste generation at the stage of production of semi-finished products and finished products; waste disposal in places not intended for this purpose; formerly, the use of harmful freons.	Very high – 5
Packing, distribution	CO_2 emission in production; waste generation in the form of cartons, foil, tape, styrofoam; impact of packaging machines on the employee (noise).	Very high – 5
Use of the product	possibility of leaking harmful (cooling) substances; high energy consumption; overheating of the equipment; correlation between operation and emission of gases contributing to the enlargement of the ozone hole.	Average – 3
End of life	the problem of separation of components and their utilisation; the real impact on the environment of harmful coolants; a long time of decomposition of production materials used, which are thrown away together with the used product.	Average – 3

The links of component and finished product manufacturers are relatively sensitive points characterised by a high level of use of raw materials and energy for production. This results in a large amount of output emissions, substances adversely affecting the environment and people, as well as a large scale of noise and exhaust. All this causes the manufacturers of components and finished products to become links in the supply chain, which have a considerable negative impact on the natural environment.

4 Supply Chain Characteristics

The supply chain consists of 12 links, the joint actions of which carry out the necessary tasks to meet the demand for refrigerators and the disposal of products. The main purpose of the supply chain is to deliver the goods to the client in the appropriate quantity, quality and in accordance with the specification, at a specified time and at a competitive price. Four companies supplying raw materials and semi-finished products, a manufacturing company, an independent service company, a dismantling station, a household appliances collection point, a transport company and a logistics operator work and co-operate with one another in the supply chain (Fig. 2).

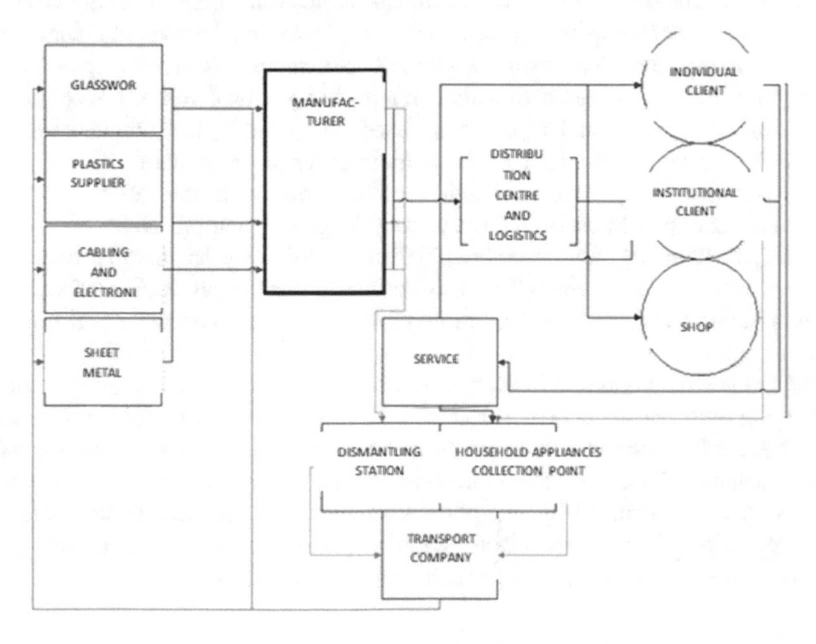

Fig. 2. Supply chain configuration; student project under the direction of K. Grzybowska

A supply chain configured in such a way supports three types of clients (individual clients, institutional clients, and shops). The client has a dominant role in the supply chain, and the main goal of the entire network of cooperating companies is striving to achieve 100% compliance with the client's expectations.

The glassworks manufacture tempered glass and glass doors for refrigerators and refrigerated display cases. Glass tempering is a thermal process that comes down to processing flat glass at a temperature of approximately 640–700 °C. The heated glass sheets are rapidly cooled with a stream of cold air. This creates an internal stress system that changes the properties and technical parameters of the material. Tempered glass is characterised by above-average strength, which makes it exceptionally safe. The plastics supplier manufactures and supplies plastics to a production company, i.e. a set of wheels for a restaurant fridge, frames for shelves and sheets made of ABS plastic. Plastics are also used in the interior of refrigerators. The cabling and electronics supplier is responsible for manufacturing and delivery of high quality components, including evaporators, thermostats, heat sinks, compressors, LED panels, control panels and fans for the refrigerator manufacturer. The sheet metal supplier is responsible for smelting of stainless and acid-resistant steel as well as the production of 2 m, 4 m and 6 m wide sheet metal rolls and the supply of semi-finished products to the manufacturing company. The refrigerator manufacturer is responsible for manufacturing a defect-free finished product. The logistics operator supports the manufacturer. They are responsible for handling orders, contacting clients, collecting used appliances from clients and handing them over to the household appliances collection point or the dismantling and service station. The service deals with the repair of electronic parts. The service receives fridges submitted for complaint by the client. The service obtains materials for the repair from the cabling and electronics supplier. The dismantling station deals with the recovery of components from complained or returned products. It is also responsible for dismantling refrigerators and regenerating used components. Regenerated assemblies are handed over to refrigerator manufacturers in order to reuse them.

The household appliances collection point deals with the collection of used refrigerators and processes of recovery, recycling and neutralisation of toxic substances. The secondary raw materials obtained in this way are handed over to processing plants, thanks to which they can be reused for the production of new items. A largely variable and unstable system may become unbalanced making it impossible to execute a task [18].

One of the most important indicators for a sustainable and green supply chain is the component/product reuse indicator. For the analysed supply chain, this indicator is the highest for the Service link and amounts to 100%. Among suppliers, only the cabling and electrical supplier did not reuse the components. The largest reuse of products from suppliers was demonstrated by: the plastics supplier and the manufacturer (Table 3). The weak point of the entire chain is the low reuse of products. Therefore, new solutions should be considered in order to improve this indicator.

Table 3. The selected Key Performance Indicator (KPI) – reverse logistics of the analysed supply chain; student project under the direction of K. Grzybowska.

Key Performance Indicator	Glassworks	Plastics supplier	Cabling and electronics supplier	Sheet metal supplier	Manufacturer	Logistics operator	Service	Dismantling station	Household appliances collection point	Transport company
Product reuse rate [%]	32.62	57.07	0.00	36.15	56.56	20.00	100.00	0.00	0.00	–

5 Renovation

As a result of the KPI analysis, the problematic links were identified. In order to improve the supply chain, all the complications encountered were analysed. The following key difficulties were listed: (1) problem with communication between the links, (2) extended time of raw materials delivery from the dismantling station to the raw material suppliers, (3) untimely execution of client's orders, (4) a large number of returns received from the client, (5) locally inefficient use of resources owned, (6) unsatisfactory rate of product/semi-finished product reuse.

Actions were taken to eliminate returns from the client on the scale of the entire supply chain. This means that the defective product should not be allowed to leave the manufacturing company. It is necessary to expand the service's activity. Most complaints did not include service repairs that could be carried out by this link. As a result, the products were returned and handed over for recycling. Another weakness of the supply chain, which should be improved, is the completion of orders on time. Only half of the orders were completed in accordance with the client's expectations. This may indicate a weak allocation of resources at the manufacturer of the finished product.

Specific investments were decided upon in order to improve the entire supply chain. Matrices for assessing investment impacts (Leopold matrix) and interaction matrices were carried out for the investments. The system shows a clear tendency to solve the problem of handing over the used ready product to a specific point, which in turn determines the further handling of the used product, the proper disposal of which guarantees a lower impact on the environment.

As a result of identifying the problems, selecting the most favourable scenario and developing improvement measures, the reconfiguration of the supply chain was decided upon. An independent service link was given up. The dismantling station was also transformed into a link dealing with the dismantling of products and service.

The logistics operator is placed on the distribution side in the supply chain, being the last link that has contact with final clients. In the supply chain, the disassembly station and the household appliances collection point are responsible for reverse logistics. It was decided to close and tighten the reverse logistics processes in the analysed supply chain. Coordination, control of physical receipt and delivery of components and ready products for processing and recycling or other dispositions, as well as return to the area of application, were improved. In the reverse logistics of the supply chain, to date, it was decided to include material flows related to both warranty and post-warranty repairs. As a result of the reconfiguration of the supply chain and reorganisation of work, an integrated supply chain link of the so-called dismantling and service station was created. The service and temporary withdrawal of valuable goods from the logistics system takes place there, e.g. in connection with their moral aging or seasonality of demand. It is a link in the supply chain in which it is possible to recover values from the withdrawn products and situations where the output constitutes a supply for the new supply chain.

As a result of introducing the improvements of individual links in the supply chain, reconfiguration of the entire supply chain as well as flow of finished products and reverse logistics, the use of an external dismantling station for new refrigerators was given up in order to create a structure inside the company. This allowed to reduce the costs associated with the purchase of raw materials and subassemblies from ready-made refrigerators. Up to now, the external dismantling station bought fridges at a lower price than production costs, and then sold the recovered materials and sub-assemblies to the manufacturer. In this case, the manufacturer was making significant losses (incurring twice the costs of recovering materials). The separation of a new unit inside the production company allowed to reduce the number of complaints as the quality control department could reject the products, which were then repaired by the dismantling station functioning as the service at the same time.

The nature of reverse logistics changed dramatically as well. The share of reuse of the product increased from 0% to 63.63% (Table 4). The use of materials in the recovery process alone increased by 19.23%, which is highly beneficial, especially considering the fact that the recovered materials did not require involvement of the company's capital. In the case of operational management indicators, the time of completing client orders was radically reduced by more than half. The recovered materials could quickly return to production and supply work stations, while repaired products could quickly meet the demand of clients within the set deadline.

Table 4. The selected Key Performance Indicator (KPI) - reverse logistics of the reconfigured supply chain.

Key Performance Indicator	Glassworks	Plastics supplier	Cabling and electronics supplier	Sheet metal supplier	Manufacturer	Logistics operator	Dismantling station and service	Household appliances collection point	Transport company
Product reuse rate [%]	40.06	84.59	14.00	39.5	58.79	20.00	63.63	0.00	–

The expansion of the service activity proved to be a positive change. This action relieved the Logistics Operator from decision-making and reduced transport. Thanks to this, the product could be returned to the client after repair. Orders executed on time amounted to 76.92% of all orders. After the reconfiguration of the supply chain, this rate increased by 12% mainly due to the introduction of production in advance.

It is worth mentioning that the average delivery time decreased by almost a half, which was influenced by a better integration of the supply chain. The logistics opera-tor participated not only in distribution processes, but also directly transferred information concerning the demand to raw material suppliers, which shortened the time of material flow in the supply chain. Shortening the time of order execution also resulted from the increase of the product reuse rate.

Designing as well as subsequent reorganisation and configuration of the supply chain constituted the basis for carrying out extensive analyses. The analysis of reverse logistics, i.e. the development of the company and process balance for each of the links in the supply chain and the assessment of product harmfulness to the environment and people, allowed to assess the harmfulness of the product. It was concluded that the greatest negative impact on the environment is exerted by the product in the first phases of its life cycle, i.e. at the stage of manufacturing processes of both the components and the finished product. Packing and distribution processes were assessed negatively as well. In these phases, environmental toxicity was assessed as a very high negative impact. The process of producing a refrigerator has a negative impact on people, as significant amounts of gases and wastewater are emitted. The creation of a process and plant balance showed that the most important factors affecting the environment include waste, sewage and noise as well as huge amounts of energy used. Carrying out the reverse analysis constituted a broad issue; therefore, it was presented in a synthetic way in order to illustrate the scale of the phenomenon. The subsequent analysis was related to the issue of sustainable development. A higher emphasis is put on the sustainable aspect of company activities, while the related expenses are treated not as costs, but as investments.

6 Conclusion

The work has shown that reverse logistics play a key role in the operation of companies, especially those that have a significant negative impact on the natural environment. Closing the flow of materials in reverse logistics as well as reconfiguring the warranty and post-warranty service system should be considered as key activities positively affecting the supply chain, the environment and the consumer. The application of the principles of sustainable development in everyday business operations is problematic, as there is a lack of the clearly defined sustainability indicators, which might be used in the assessment of remanufacturing activities [19].

References

1. Guo, Z., Liu, H., Zhang, D., Yang, J.: Green supplier evaluation and selection in apparel manufacturing using a fuzzy multi-criteria decision-making approach. Sustainability **9**(650), 1–13 (2017)
2. Jasiulewicz-Kaczmarek, M., Saniuk, A., Nowicki, T.: The maintenance management in the macro-ergonomics context, In: Goossens, R.H.M. (eds.) Advances in Social & Occupational Ergonomics Proceedings of the AHFE 2016, Florida, USA. Advances in Intelligent Systems and Computing, vol. 487, pp. 35–46 (2016). https://doi.org/10.1007/978-3-319-41688-5
3. Grzybowska, K.: Supply Chain Sustainability – analysing the enablers. In: Golinska, P., Romano, C.A. (eds.) Environmental Issues in Supply Chain Management – New Trends and Applications, pp. 25–40. Springer (2012). https://doi.org/10.1007/978-3-642-23562-7_2
4. Drake, D.F., Spinler, S.: OM forum-sustainable operations management: an enduring stream or a passing fancy? Manuf. Serv. Oper. Manag. **15**, 689–700 (2013)
5. Plambeck, E.L., Toktay, L.B.: Introduction to the special issue on the environment. Manuf. Serv. Oper. Manag. **15**, 523–526 (2013)
6. Relich, M.: Case-based reasoning for selecting a new product portfolio. In: Knowledge for Market Use, pp. 410–421 (2016)
7. Li, Z., Xu, Y., Deng, F., Liang, X.: Impacts of power structure on sustainable supply chain management. Sustainability **10**(55), 1–10 (2018). https://doi.org/10.3390/su10010055
8. Linton, J.D., Klassen, R., Jayaraman, V.: Sustainable supply chains: an introduction. J. Oper. Manag. **25**, 1075–1082 (2007)

9. Kot, S.: Sustainable supply chain management in small and medium enterprises. Sustainability **10**(4), 1143 (2018)
10. Hopwood, B., Mellor, M., O'Brien, G.: Sustainable development: mapping different approaches. Sustain. Dev. **13**, 38–52 (2005)
11. Supply Chain Sustainability a practical Guide for Continuous improvement. UN Global Compact Office and Business for Social Responsibility (2010)
12. Bai, C., Sarkis, J.: Integrating sustainability into supplier selection with grey system and rough set methodologies. Int. J. Prod. Econ. **124**, 252–264 (2010)
13. Buyukozkan, G., Çifçi, G.: A novel fuzzy multi-criteria decision framework for sustainable supplier selection with incomplete information. Comput. Ind. **62**, 164–174 (2011)
14. Seuring, S., Müller, M.: From a literature review to a conceptual framework for sustainable supply chain management. J. Clean. Prod. **16**, 1699–1710 (2008)
15. Shepherd, C., Gunter, H.: Measuring supply chain performance: current research and future directions. Int. J. Prod. Perform. Manag. **55**(3–4), 242–258 (2005)
16. Adebanjo, D., Teh, P.-L., Ahmed, P.K.: The impact of external pressure and sustainable management practices on manufacturing performance and environmental outcomes. Int. J. Oper. Prod. Manag. **36**(9), 995–1013 (2016)
17. www.cecedpolska.pl
18. Burduk, A.: Artificial neural networks as tools for controlling production system and ensuring their stability. In: 12th IFIP TC8 International Conference on Computer Information Systems and Industrial Management, CISIM, pp. 487–498 (2013)
19. Golinska-Dawson, P., Kosacka, M., Werner-Lewandowska, K.: Sustainability Indicators System for Remanufacturing. In: Golinska-Dawson, P., Kübler, F. (eds.) Sustainability in Remanufacturing Operations. EcoProduction (Environmental Issues in Logistics and Manufacturing). Springer, Cham (2018)

The Framework of Logistics 4.0 Maturity Model

Joanna Oleśków-Szłapka(✉) and Agnieszka Stachowiak

Poznan University of Technology, Strzelecka 11, 60-965 Poznan, Poland
{joanna.oleskow-szlapka, agnieszka.stachowiak}@put.poznan.pl

Abstract. The term Industry 4.0 is widely recognized, not only among academics, but also among business. The term that has not gained that popularity and publicity yet is Logistics 4.0. As industry needs material flows, and in global economy they are of high complexity, authors believe that Logistics 4.0 will be the field of research and solutions within it will be sought for and implemented by companies. The paper presents the framework of Logistics 4.0 Maturity Model, developed to provide companies with opportunity to assess current status with respect to Logistics 4.0 and develop a road map for improvement process. The model is developed basing on the literature research on Logistics 4.0 and maturity models and design to offer measures that can be translated into set of guidelines or recommended solutions towards Logistics 4.0. The research is at the modeling stage, however the pilot research among business to recognize their awareness of Logistics 4.0 as a source of competitive advantage and tools and methods within Logistics 4.0 was already conducted and is to be the basis for further research stages.

Keywords: Logistics 4.0 · Maturity model · Maturity assessment

1 Motivation

Today, companies striving to survive in an increasingly ambitious environment need to undergo substantial transformations. The implement organizational and technical solutions to improve their flexibility, respond to customer requirements quickly and efficiently, be competitive. The range of the solutions available is wide and the most advanced in terms of technology are within Industry 4.0 concept [1]. Industry 4.0 offers a broad variety of options and selecting the best/most suitable requires from a company defining its goals at the strategic level. However, it is believed, that nowadays, business models rather than strategy, should be implemented as a navigation instrument towards sustainable success in the market.

Dynamic development of manufacturing Industry 4.0 is a result of some processes, for example: internationalization, information technology development and also hyper competition [2]. The example of such models are maturity models developed to measure the degree of progress and advancement in the given field. Their goal is to provide insight into continuous process improvement and status quo analysis. Hence, a

© Springer Nature Switzerland AG 2019
A. Burduk et al. (Eds.): ISPEM 2018, AISC 835, pp. 771–781, 2019.
https://doi.org/10.1007/978-3-319-97490-3_73

maturity model can guarantee confirmation of business model management's performance and measure its capability.

In the literature authors have found numerous examples of maturity models for business processes, as well as Industry 4.0 [3–6], nevertheless, there is a gap in the field of Logistics 4.0 [7] - the approach complementary to Industry 4.0, covering solutions for designing and managing material flows in contemporary economic systems (namely global supply chains) – which gives opportunities for developing new business models. To fill the gap in, authors present a framework of the Logistics 4.0 Maturity Model. The model is developed to provide companies with a methodology that allows both to assess their Logistics 4.0 current status and to outline a road map for improvement, considering three key elements: Management, Material Flows, and Information Flows.

2 Research Design and Methodology

Industry 4.0 and Logistics 4.0 are inherent elements of contemporary business environment, even if companies are not aware that they benefit from solutions within their scope. The research strives to develop an evaluation methodology to assess whether the current status of a company (processes it performs, information systems it benefits from and management approach it uses) corresponds in any way with Logistics 4.0 solutions. Such a feedback can make managers aware that they already are benefiting from Logistics 4.0 and be an incentive to implement more solutions from its scope.

The methodology is to focus on assessing level of maturity in the context of Logistics 4.0 – Logistics 4.0 Maturity Model. The model is based on authors knowledge and experience in the field. Theoretical framework is to be implemented into IT environment to make the methodology available for wide spectrum of potential users. The main stages of the research are presented in the Fig. 1.

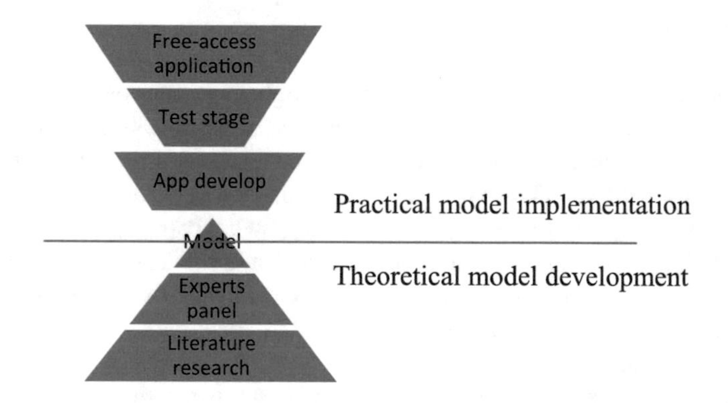

Fig. 1. Brief outline of methodology of research implemented

The paper covers mainly the basis, which theoretical part. Literature research was carried among world literature in Logistics 4.0 and maturity models. Experts panel brainstormed the most important elements of the model and produced the framework of

Logistics 4.0 Maturity Model. The model is to be transformed into an online application, tested and published as free-access tool.

3 Literature Review

3.1 Logistics 4.0 and Smart Supply Chain Management

According to World Economic Forum [8] by 2025, the global supply chain will mature as a vast network of interconnected companies, processes, and data flows that will support new business designs and models. End-to-end visibility is a necessity, in fact, complete visibility of the entire supply chain could achieve true demand-driven planning, allowing efficient response to changes in sourcing, supply, capacity and demand [9]. Demand for individualized products is continually increasing hence, supply chain processes (inbound and outbound logistics) have to adapt to this changing environment, since due to the increasing complexity, it cannot be handled with ordinary planning and control practices [10].

The Logistics Trend Radar report, developed by the DHL company in 2014, drew attention to the possibility of using modern technologies within the field of Industry 4.0 as part of the provision of logistic services. The paradigm shift caused by Industry 4.0 is bringing about a fundamental change in the associated information flow and this affects the entire logistics delivery process. As a result, logistics becomes closely integrated into the overall value chain and shifts towards service-oriented logistics on demand [11]. DHL decided that the Internet of Things will support the introduction of intelligent machinery and devices that enable the control of modern logistics processes to the logistics market [12, 13], contributing to development of Logistics 4.0 idea.

Logistics 4.0 definitions are vague as the concept is not homogenous. They focus on large amounts of data flow management and integration of decentralised complex systems [13–19] From operational perspective, Logistics 4.0 status is presented in the reports by research centers and Logistics services providers [20–22].

The definition of Logistics 4.0 combines two aspects: processual (supply chain processes are a subject of the Logistics 4.0 actions) and technical (tools and technologies that support internal processes in the supply chains) [23]. The tools and technologies are mostly within IT range as digitization creates plenty of advantages for the supply chain. They comprise among other things following issues: reduced complexity, increased reliability, predictability and thus minimized risks, reduced errors, reduced transport cost, creating new business areas and thus turnover potential, increased innovation capability, increased agility and flexibility, e.g. in case of new market requirements [24]. As a result Logistics 4.0 plays the same role in managing supply chains as Industry 4.0 for modern manufacturing factories, and in Industry 4.0 solutions spectrum is often called Smart Logistics, Logistics Management or Supply Chain Management [25], whereas in the literature on the subject Smart Logistic is defined as the logistics system, which can improve the flexibility, the adaptation to the market changes and bring the company closer to customer needs [26] however the definition of Smart Logistics assumes that a specific level of technological development is temporary [27, 28], so it is not an accurate reflection of Logistics 4.0, which requires an additional defined level of technological development, introduction of regular technological changes with respect to currently binding standards and techniques.

Disregarding small terminological differences between Smart Logistics and Logistics 4.0, the latter is also related to such concepts as Smart Services and Smart Products, which are autonomously implemented, without human participation, to increase the availability of employee time. The concept of Smart Products refers to products that have the ability to collect, store and share relevant data, while the idea of Smart Services is based on providing modern methods of measuring and analyzing information [23, 26]. In the logistics industry, where companies manage millions of daily shipments, large amounts of data is obtained via orders transcripts, smart low battery consuming sensors, GPS, RFID tags, weather forecasts or even social media. But transport is not the only area that benefits from advanced technical solutions. Warehousing processes can be improved by combining automation, the Internet of things, drones, 3D printing and innovative applications, becoming smarter, more connected, automated and robotized.

The effort of implementation of Logistics 4.0 brings numerous benefits to companies, nevertheless, there are some disadvantages of the solutions as well. Advantages and disadvantages of Logistics 4.0 application are presented in Table 1.

Table 1. Advantages and disadvantages of Logistics 4.0

The advantages of Logistics 4.0	The disadvantages of Logistics 4.0
Full integration of reality and virtual world	High implementation cost
Opportunity of real-time communication between system users, machines and other systems	Strict requirements concerning advanced IT hardware implementation
Improvement of all the processes performed in supply chain	Strict requirements concerning implementation of process-oriented management methods (i.e. Just-in-Time or Lean Management)
Opportunity of lead times decreasing for products and services directly responding to customers'needs	Requirements concerning implementation of Industry 4.0 technologies
Decrease in cost of product design thanks to implementation of Digital Twins	Problems with availability of data with no methods to process them with
Decrease in risk of structural or organizational mistakes in processes performed	Novelty of the approach and low level of awareness among companies
Availability of advanced technologies for analysis of unlimited amount of data	Strict requirements concerning integration of company's sub-systems or supply chain elements
Increased performance and availability of machines and operators	
Opportunity to make autonomous decisions by all the system users	
Increased visibility and flexibility of supply chains	

Source: Own work based on [23, 26, 27, 29]

The advantages listed in the table refer mostly to potential Logistics 4.0 offers while the disadvantages are the consequence of Logistics 4.0 requirements concerning technical and organizational solutions that need to be implemented prior to benefiting from the potential identified. Hence, implementation of the solutions is crucial to minimize the disadvantages and benefit from the concept, however it is not simple as Logistics 4.0 is based on advanced knowledge and high-end solutions. This is why the ability to identify, assimilate and use the available knowledge is one of the factors creating companies' sustainable competitive advantage [29–31].

3.2 Overview on Maturity Models and Tools

Maturity can be defined as "the state of being complete, perfect or ready" [32–34]. Maturity is referred to growth, as in Maier et al. [35], where the process of bringing something to maturity means bringing it to a state of full growth, and to improvement and excellence.

According to Cookie-Davies [36] the term "maturity" has a number of usages; but when used in conjunction with organization or industry development it signifies full development or perfection. Maturity models give companies indicators as well as guidance to analyze and subsequently improve their processes.

The concept of maturity is not new in the industrial engineering and management field [37]. Crosby was among the first to propose, in 1979, a quality management model with fives levels of maturity [39, 40].

Maturity models are now widely spread in Project Management (PM), Knowledge Management, Information Systems (IS) and Supply Chain Management (SCM) industries [4]. Maturity models in literature have different characteristics: they can be of moderate or high complexity, maturity levels can be described in simple or complex terms, and so on. Fraser et al. [3] presented a first clear classification per typology of maturity models. In particular, they distinguish three types of maturity models: (1) maturity grids, (2) Likert-like questionnaire, (3) CMM-like models.

Probably the most disseminated maturity model is the Capability Maturity Model (CMM) developed by researchers at the Software Engineering Institute of the Carnegie Mellon University. This model supports the management of the software development process [38] Bowersox et al. [41] emphasize integration and collaboration in mature supply chain.

In the field of Industry 4.0 many researchers have developed their maturity models (f.ex. A. Schumacher et al. [5].) whereas authors of the proposal have not found any maturity models referring specifically to Logistics 4.0.

Hence authors believe that developing Logistics 4.0 Maturity Model could increase recognition of the Logistics 4.0 concept, structure knowledge acquired so far and produce original interpretation of the phenomena.

The Logistics 4.0 Maturity Model will be based on authors' of the proposal interpretation of existing literature. The literature search on Logistics 4.0 publications in Scopus and Web of Science databases, gave the feedback presented in the diagram below (see Fig. 2).

The number of references found is not large (max 27), but growing, which shows on one hand the cognitive gap and on the other proves the potential of the field. The

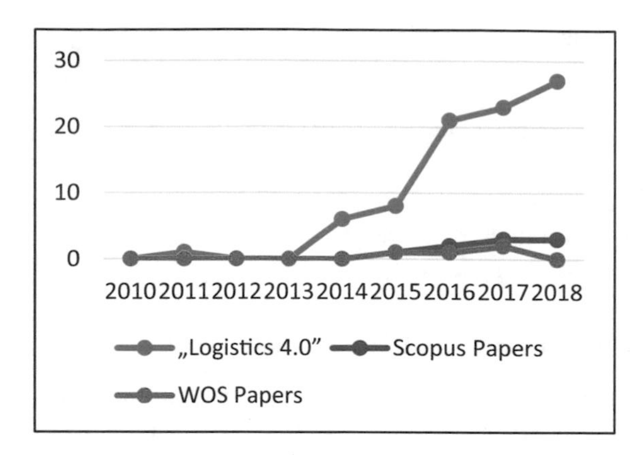

Fig. 2. The number of publications referring to Logistics 4.0

total number of references include press material and professional trade publications. The list was filtered to identify scientific papers and the diagram presents those indexed by Scopus and WoS.

4 The Framework of Logistics 4.0 Maturity Model

4.1 Dimensions

Following the introduced research procedure, authors carried out the literature research to characterize the terms "Logistics 4.0", "maturity" and "maturity models". Due to limited size of the paper only brief presentation of the most distinctive features of the terms were presented in the previous sections. Based on literature review of existing maturity models, and Logistics 4.0 enablers, challenges and trends, the draft model was developed, including maturity indicators and elements. Authors conformed their ideas with academics specializing in the field of logistics during the experts panel to foster introduction of new items and indicators for the maturity framework. The feedback from experts panel and the literature review enabled definition of the final structure of Logistics 4.0 Maturity Model.

The Logistics 4.0 Model was used to define criteria with which companies are classified into five types. This classification is based on the following three aspect of logistics[1]: (1) management (2) flow of material, (3) flow of information, which become naturally three dimensions for Logistics 4.0 solutions, as presented in the Table 2.

[1] According to Council of Supply Chain Management Professionals (previously the Council of Logistics Management) logistics is the process of planning, implementing and controlling (altogether referred to as **management**) procedures for the efficient and effective transportation and storage of goods (**material flows**) including services and related information (**information flows**) from the point of origin to the point of consumption for the purpose of conforming to customer requirements and includes inbound, outbound, internal and external movements.

Table 2. Logistics 4.0 dimensions and areas of evaluation

Logistics 4.0 dimensions	Areas of evaluation
Management	Investments, innovations management, integration of value chains
Flow of material	Degree of automation and robotization in warehouse and transportation, Internet of things, 3D printing, 3D scanning, advanced materials, augmented reality, smart products
Flow of information	Data driven services, Big data (data capturing and usage), RFID, RTLS (real time locating systems), IT systems (ERP, WMS, cloud systems)

Source: Own work

The three dimensions constituting the Model Area can be used to assess the maturity and awareness of managers in terms of solutions within Logistics 4.0. Depending on the number and scope of solutions implemented, the conclusion on present Logistics 4.0 status can be drawn. Moreover, recommendation concerning the status improvement, and maturity increase, can be defined basing on gaps identification and analysis, making the model useful not only in terms of diagnosis, but also in terms of management.

4.2 Maturity Levels

According to authors, the term 'Logistics 4.0 maturity' reflects the level to which a company or a supply chain has implemented Logistics 4.0 concepts. Authors distinguish five maturity levels: Ignoring, Defining, Adopting, Managing and Integrated. In the Fig. 3 maturity levels are confronted with Logistics 4.0 dimensions.

The assessment of maturity level is based on analysis of Logistics 4.0 dimensions. The aspects to be assessed include the need and the level of integration of internal processes, and when applicable, the supply chain; number and scope of advanced solutions improving materials and information flows. The assessment criteria should reflect the range of solutions implemented, whether they cover all the needs and requirements of the company. The presented theoretical approach assumes some homegenity in companies performance, namely the same status of integration, material and information flows, nevertheless it is possible, that company is advanced in the area of information technology but material flows are realized with traditional equipment (or opposite: material flows are automized and information flows are based on traditional documents flow, however such situation seems less likely). Hence, authors decided that the most important determinant of maturity is management, and if integration level is coherent with at least one form of the flow (either material or information) the maturity level the two represent is assigned to the company, assuming that the latter dimension is soon to be upgraded. Such recommendation should be presented to the company, according with gaps analysis and general guielines for reaching the next level of maturity.

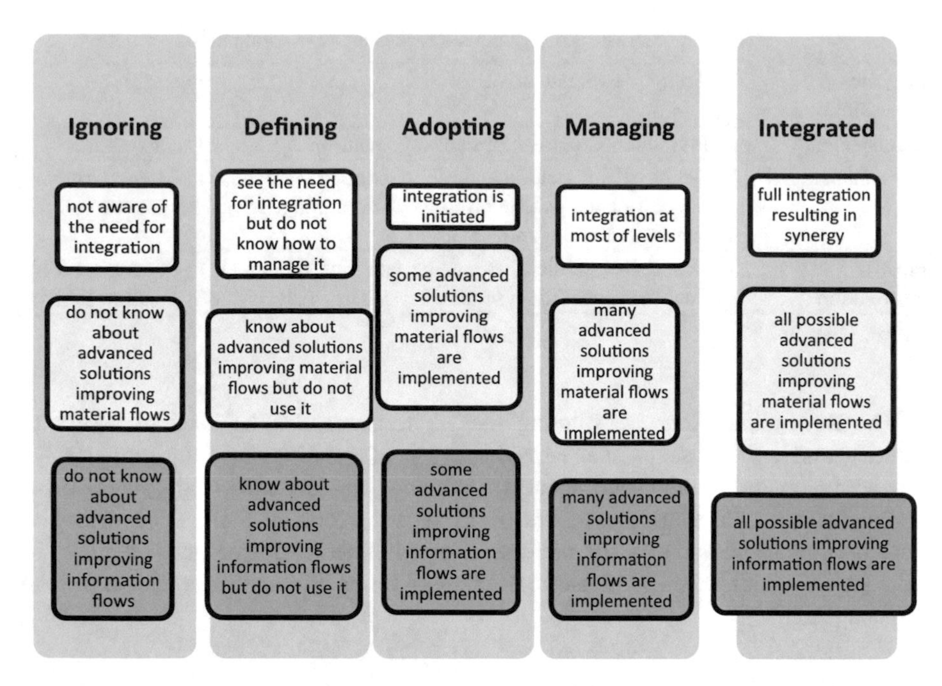

Fig. 3. Logistics 4.0 maturity levels Source: Authors' own elaboration

5 Conclusions and Outlook

By that year, according to the World Economic Forum, the implementation of digital platforms that enable cross-border trade and crowdsourcing of logistics routes could together will contribute to substantial development of logistics. On top of integrating Industry 4.0, supply chain and procurement leaders will also need to redistribute roles and responsibilities between humans and machines, taking the right cues and best practices from across industries as they too evolve alongside the fourth industrial revolution. Most companies associate digitization with the goal of greater delivery reliability. Digitization provides various levers for this purpose, such as more transparency, better predictability and planning, risk reduction or customer-specific products. At the same time, modern IT enables companies to integrate greater flexibility and agility in their processes and therefore to react more promptly to customers and market requirements. Increasing competition causes that data is progressively becoming an asset. The goal of connected digitization and Industry 4.0 and Logistics 4.0 are therefore the learning, agile company that is able to adapt to changing requirements. Unfortunately, many enterprises have little knowledge about the characteristics of both concepts, the solutions they propose, as well as the benefits of using the latest methods of process digitization. The aforementioned thesis is confirmed by research carried out by the Ministry of Development and Siemens in the years 2016–2017 collected as part of the Smart Industry Polska Report [42].

By knowing where companies are today, they can easily find their future goal – and how they will get there. Assessing Logistics 4.0 maturity will help visualize companies' path forward and set priorities for process improvement. First stage of planned research is to carry out survey between Polish companies – the practical, implementation part of the research. Based on preliminary results from survey conducted by authors in logistics and manufacturing companies from among the surveyed companies 33% know the term Logistics 4.0, 50% of companies know the concept of big data, 83% companies want to apply automated data exchange systems and are willing to automate their processes as well as introduce partial robotization of the processes.

Identification of logistics maturity of companies will enable assessment of the industry condition in Poland and will provide data for analysing correlation between the maturity level of a company, and its competitive position, size, development dynamics, number of services offered, structure of capital and level of internationalization of operations. The analysis is supposed to lead to general conclusion and development of system dynamics model presenting system, that a company providing logistics services undoubtedly is, behaviour in the form of causal loop. The model will add dynamic layer to the static concept of logistics maturity levels.

References

1. Kagermann, H., Wahlster, W., Helbi, J.: Recommendations for implementing the strategic initiative Industrie 4.0. Final report of the Industrie 4.0 Working Group (2013)
2. Grzybowska, K., Łupicka, A.: Key managerial competencies for Industry 4.0. Econ. Manag. Innov. **1**(1), 250–253 (2017)
3. Fraser, P., Moultrie, J., Gregory, M.: The use of maturity models/grids as a tool in assessing product development capability. In: IEEE International Engineering Management Conference (2002)
4. Banyani, M.A., Then, D.S.S.: Development of the industry maturity framework facilities management. In: Proceedings of the 5th Built Environment Conference (2010)
5. Schumacher, A., Erol, S., Sihn, W.: A maturity model for assessing industry 4.0 readiness and maturity of manufacturing enterprises. Procedia CIRP **52**, 161–166 (2016)
6. Plan digitalization precisely with the Industry 4.0 CheckUp, Fraunhoffer IFF, p. 3, October 11 2016
7. Strandhagen, J.O., Vallandingham, L.R., Fragapane, G., et al.: Logistics 4.0 and emerging sustainable business models. Adv. Manuf. **5**, 359 (2017)
8. Lanng, Ch.: Global supply chains in 2025: Industrial Internet, Next-generation logistics and space tech (2017). www.insightsucess.com
9. Oleśków-Szłapka, J., Lubiński, P.: New technology trends and solutions in logistics and their impact on processes. In: 3rd International Conference on Social Science, pp. 408–413 (2016). ISBN 978-1-60595-410-3
10. Premm, M., Kirn S.: A multiagent system perspective on Industry 4.0 supply networks. In: Muller, J.P., Ketter, W., Kaminka, G., Wargner, G., Bulling, N. (eds.) Multiagent System Technologies (2015)
11. Schuh, G., Anderl, R., Gausemeier, J., ten Hompel, M., Wahlster W. (eds.): Industry 4.0 maturity index. Managing the digital transformation of companies, Acatech – National Academy of Science and Engineering (2017)

12. Yilmaz, L., Chan, W.K.V., Moon, I., Roeder, T.M.K., Macal C., Rossetti D. (Eds.): Logistics 4.0 – a challenge for simulation. In: Proceedings of the 2015 Winter Simulation Conference, IEEE Press Piscataway, NJ, USA, pp. 3118–3119 (2015)
13. Hompel, M., Kerner, S.: Logistics 4.0: The vision of the Internet of Autonomous hings [Logistik 4.0: Die Vision vom Internet der autonomen Dinge]. Informatik-Spektrum **38**(3), 176–182 (2015)
14. Wang K.: Logistics 4.0- New challenges and opportunities. In: 6th International Workshop of Advanced Manufacturing and Automation (2016)
15. Barreto, L., Amaral, A., Pereira, T.: Industry 4.0 implications in logistics: an overview. Proceedia Manuf. **13**, 1245–1252 (2017)
16. Strandhagen, J.O., Vallandingham, L.R., Fragapane, G., Strandhagen, J.W., Stangeland, A. B.H., Sharma, N.: Logistics 4.0 and emerging sustainable business models. Adv. Manuf. **5**(4), 359–369 (2017)
17. Wrobel-Lachowska, M., Wisniewski, Z., Polak-Sopinska, A.: The role of the lifelong learning in logistics 4.0. Adv. Intell. Syst. Comput. **596**, 402–409 (2018)
18. Glistau, E., Machado, N.I.C.: Industry 4.0, logistics 4.0 and materials - Chances and solutions. Mater. Sci. Forum **919**, 307–314 (2018)
19. Dussmann Group (2016). Logistics 4.0, https://news.Dussmanngroup.com/en/multimedia/news/logistics-40/. Accessed 22 May 2018
20. DHL: Internet of Things in Logistics. www.dhl.com/content/dam/Local_Images/g0/New_aboutus/innovation/DHLTrendReport_Intemet_of_things.pdf. Accessed 22 May 2018
21. Fraunhoffer: Logistics 4.0 and challenges for the supply chain planning and IT (2014). https://www.iis.fraunhofer.de/content/dam/iis/tr/Session%203_5_Logistics_Fraunhofer%20IML_Akinlar.pdf. Accessed 21 Apr 2018
22. Bubner, N., Helbig, R., Jeske, M.: Logistics trend radar, Delivering insight today. Creating value tomorrow!, Pub. DHL Customer Solutions & Innovation, Troisdorf (2014)
23. Szymańska, O., Cyplik, P., Adamczak, M.: Logistics 4.0 A new paradigm or a set of solutions, Research in logistics and production, vol. 7, No. 4, pp. 299–310, Poznan University of Technology Publishing House, Poznan (2017)
24. https://www.i-scoop.eu/industry-4-0/supply-chain-management-scm-logistics/. Accessed 21 Mar 2018
25. Digitalization in logistics, A practical guide on the way into the digital world, Axit connecting logistics, October 2016, https://axit.de/images/Whitepaper_Download/AXIT-Expert-Paper_Digitalization-in-Logistics_EN.pdf. Accessed 23 May 2018
26. Galindo, L.D.: The challenges of Logistics 4.0 for the supply chain management and the information technology. Norvegian University of Science and Technology, master thesis, May 2016
27. Heistermann, F., ten Hompel, M., Mallée, T.: Digitisation in logistics, BVL International (2018). www.bvl.de/en
28. Lichtenthaler, U., Ernst, H.: The role of champions in the external commercialization of knowledge. J. Prod. Innov. Manag. **26**(4), 371–387 (2009)
29. Lev, S., Fiegenbaum, A., Shoham, A.: Managing absorptive capacity stocks to improve performance: Empirical evidence from the turbulent environment of Israeli hospitals. J. Bus. Res. 61 (2008)
30. Jelonek, D.: Zdolność absorpcji wiedzy a innowacyjność małych i średnich przedsiębiorstw. Studia Ekonomiczne. Zeszyty naukowe Uniwersytetu Ekonomicznego w Katowicach, nr 281 (2016). ISSN 2083-8611
31. Gubán, M., Kovács, G.: Industry 4.0 Conception. In: Acta Technica Corviniensis – Bulletin of Engineering Tome X, Timisoara (2017)

32. Simpson, J.A., Weiner, E.S.C.: The Oxford English Dictionnary. Oxford University Press, Oxford (1989)
33. Mettler, T.: A design science research perspective on maturity models in information systems (2009)
34. Karkkainen, H., Myllarniemi, J., Okkonen, J., Silventoinen, A.: Assesing maturity requirements for implementing and using product lifecycle management. Int. J. Electr. Bus. **11**(2), 176–198 (2014)
35. Cooke-Davies, T.J.: Measurement of organisation maturity: what are the relevant questions about maturity and metrics for a project-based organisation to ask and what do these imply for project management research?, J. Innov.-Proj. Manag. Res., 1–19, (2004)
36. Green, A., Price, I.: Whither FM? a delphi study of the profession and the industry. J. Facil. **18**, 281–292 (2000)
37. Jasiulewicz-Kaczmarek, M., Drożyner, P.: Preventive and Pro-Active Ergonomics Influence on Maintenance Excellence Level, In: M.M. Robertson (Ed.): Ergonomics and Health Aspects, HCII 2011, LNCS 6779, pp. 49–58 (2011)
38. Maier, A.M., Moultrie, J., Clarkson, P.J.: Assesing organizational capabilities: Reviewing and guiding the development of maturity grids. IEEE Trans. Eng. Manag. **59**, 138–159 (2012)
39. Vaidyanathan, K., Howell, G.: Construction supply chain maturity model –Conceptual framework. In: Proceedings IGLC-15, July 2007, pp 170–180 (2007)
40. Kwak, Y.H., Ibbs, W.C.: Project management process maturity model. J. Manag, Eng. **18**(3), 150–155 (2002)
41. Bowersox, D.J., Closs, D., Stank, T.: Ten mega-trends that will revolutionize supply chain logistics. J. Bus. Logist. **21**, 1–16 (2000)
42. Ministerstwo Rozwoju (Ministry of Regional Development), Siemens Sp. z. o. o., Smart Industry Polska 2017, Warszawa (2017)

Author Index

© Springer Nature Switzerland AG 2019

A. Burduk et al. (Eds.): ISPEM 2018, AISC 835, pp. 783–785, 2019.
https://doi.org/10.1007/978-3-319-97490-3